Multigenic and Induced Systemic Resistance in Plants

T0189273

Sadik Tuzun
Elizabeth Bent
(Editors)

Multigenic and Induced Systemic Resistance in Plants

 Springer

Sadik Tuzun
Department of Entomology
  and Plant Pathology
Auburn University
Auburn, AL
USA

Elizabeth Bent
Department of Plant Pathology
University of California
Riverside, CA 92521
USA

e-ISBN: 0-387-23266-4

ISBN: 978-1-4419-3593-9   e-ISBN: 978-0-387-23266-9

9 8 7 6 5 4 3 2 1

springeronline.com

To the memories of
Dr. Rebii Tuzun, Mrs. Muveddet Tuzun, and Mr. Fred Bent, Jr.
who played a major role in our lives
and left us recently

Preface

Plants have developed very sophisticated mechanisms to combat pathogens and pests using the least amount of reserved or generated energy possible. They do this by activating major defense mechanisms after recognition of the organisms that are considered to be detrimental to their survival; therefore they have been able to exist on Earth longer than any other higher organisms. It has been known for the past century that plants carry genetic information for inherited resistance against many pathogenic organisms including fungi, bacteria, and viruses, and that the relationship between pathogenic organisms and hosts plants are rather complex and in some cases time dependent. This genetic information has been the basis for breeding for resistance that has been employed by plant breeders to develop better-yielding disease resistant varieties, some of which are still being cultivated.

Single gene resistance is one type of resistance which has been extensively studied by many research groups all around the world using biotechnological methodologies that have been the subject of many books and journal articles; therefore, it is beyond the scope of this book. This type of resistance is very effective, although it can be overcome by the pressure of pathogenic organisms since it depends on interaction of a single elicitor molecule from the pathogen with a single receptor site in the host. It is race specific and under any favorable conditions for development of new races of the pathogenic organisms, similar to development of resistance to single site systemic pesticides, resistance achieved by a single gene in plant can be diminished leading to development of major epidemics. It is, therefore, the constant effort of pyramiding of the resistance genes that is needed to overcome pathogenicity achieved by the new races of the pathogen.

Breeders have been crossing between disease resistant wild ecotypes of plants with the cultivated ones to achieve disease resistance for centuries. This type of resistance is governed by multiple genes and many years of breeding effort is needed to eliminate bad genes coming from the wild ecotypes to increase the quality of cultivated crops. Although it is very stable and not race specific, unfortunately, this type of resistance has not gathered attention from the scientific community and we are just beginning to understand how this type of resistance works. In fact, our group has been one of the first to clearly demonstrate the involvement of elicitors released by the activity of pre-existing hydrolytic enzymes to activate further defense reactions. This is very energy efficient and stable since the pathogenic organisms have to modify the whole cell structure to overcome the activity of a battery of constitutively-expressed hydrolytic enzymes.

In a similar way, plants carry inducible mechanisms that protect plants in a time-dependent manner. Induced systemic resistance [ISR, *syn* systemic acquired resistance (SAR)], therefore, has probably been involved in survival of plants for millennia and indeed is the reason why susceptibility is an exception rather than the rule, considering that plants are infected by a minority of possibly pathogenic microorganisms in their environment. ISR has been known since the beginning of the 20th century but really became an area of scientific interest around thirty years ago. When I started to work on ISR during the mid-to-late 1970s the general attitude among the scientists was that this phenomenon is nothing more than an exception and the concept was rejected by most of them. In contrast, today it is well established and has become one of the major areas of study by many laboratories, including the ones which originally concentrated on biological control. It is not unlikely to hear most biological control organisms indeed induce some sort of disease resistance in plants at least as a part of their mechanism of action. For a scientist who dedicated his whole professional life to this very subject area it is very pleasing to live through this change. However, we must never forget that we are just at the beginning to discover survival mechanisms of plants in a very complex environment.

We have decided that it is very timely to publish a book on the subject area of durable resistance, multigenic, and induced systemic resistance, since similarities mechanistically among them are becoming more understood. It is important to compile the information into one book as a text or reference to researchers. The greatest encouragement came from our colleagues who actually read our first book on "Induced Plant Defenses Against Pathogens and Herbivores" and how well it was received by the scientific community. It was important to develop a very comprehensive book so we have carefully chosen the subjects for the chapters and contributors and emphasized the importance of extensive peer review of each chapter.

The importance of consistent terminology and its usage was foremost among the comments of the reviewers. It is carefully covered in the first chapter, which considers the meaning of various terms and the involvement of various groups who pioneered the area of induced systemic resistance—we sincerely urge our colleagues to read this chapter and consider the suggestions very seriously, since it is time to use common terms to describe the same phenomenon, whatever the mechanism of action is. As I mentioned earlier, we are at the beginning stage of this very important subject area and we will find out that many mechanisms are involved in this fascinating phenomenon that leads susceptible plants to become resistant to pathogenic organisms. The terminology has to be flexible to accommodate new mechanisms and yet should not create confusion with the old publications.

Finally, I would like to thank many colleagues who served as reviewers to make this book scientifically as accurate as possible. I hope that this book will serve as a comprehensive reference to many professionals, students, and scientists in plant pathology, entomology, plant physiology, or biochemistry to enhance their knowledge, research, and teaching efforts.

SADIK TUZUN
Auburn, Alabama, USA

Contents

Contributors

ANNE J. ANDERSON
Department of Biology
Utah State University
Logan, UT 84322
USA

ERIN BAKKER
Laboratory of Nematology
Department of Plant Sciences
Wageningen University
Binnehaven 5, 6709 PD Wageningen
The Netherlands

JAAP BAKKER
Laboratory of Nematology
Department of Plant Sciences
Wageningen University
Binnehaven 5, 6709 PD Wageningen
The Netherlands

NICOLE BENHAMOU
Recherches en sciences de la vie et de la santé
Pavillon C.E. Marchand
Université Laval
Sainte-Foy Québec G1K 7P4
Canada

ELIZABETH BENT
Department of Plant Pathology
University of California
Riverside, CA 92521
USA

KRIS A. BLEE   Department of Biological Sciences
                               California State University
                               Chico, CA 95929
                               USA

JÖRG BOHLMANN   Department of Forest Sciences
                               Biotechnology Laboratory
                               Wesbrook Building
                               University of British Columbia
                               #237-6174 University Boulevard
                               Vancouver, British Columbia V6T 1Z3
                               Canada

JOHN P. CARR   Department of Plant Sciences
                               University of Cambridge
                               Downing Street, Cambridge CB2 3EA
                               United Kingdom

RACHAEL A.J. CARSON   Department of Plant Sciences
                               University of Cambridge
                               Downing Street, Cambridge CB2 3EA
                               United Kingdom

UWE CONRATH   Department of Plant Molecular Biology
                               Institute of Botany
                               University of Bonn
                               1 Kirschallee
                               D-53115 Bonn
                               Germany

ROBERT DEES   Laboratory of Molecular Recognition
                                 and Antibody
                               Technology
                               Department of Plant Sciences
                               Wageningen University
                               Binnehaven 5, 6709 PD Wageningen
                               The Netherlands

NEIL EMANS   Institute for Molecular Biotechnology
                               RWTH Aachen
                               Worringerweg 1
                               D-52074 Aachen
                               Germany

RAINER FISCHER

Institute for Molecular Biotechnology
RWTH Aachen
Worringerweg 1
D-52074 Aachen
Germany

THIERRY GENOUD

Department of Biology
Unit of Plant Biology
University of Fribourg
10 ch. du musée, 1700 Fribourg
Switzerland

JEANNIE GILBERT

Agriculture and Agri-Food Canada
Cereal Research Centre
195 Dafoe Rd.
Winnipeg, Manitoba R3T 2M9
Canada

ANDROULLA GILLILAND

Department of Plant Sciences
University of Cambridge
Downing Street, Cambridge CB2 3EA
United Kingdom

ASKA GOVERSE

Laboratory of Nematology
Department of Plant Sciences
Wageningen University
Binnehaven 5, 6709 PD Wageningen
The Netherlands

ERKKI HAUKIOJA

Section of Ecology
Department of Biology
University of Turku
FIN-20014 Turku
Finland

DEZENE P.W. HUBER

Department of Botany
Biotechnology Laboratory
University of British Columbia,
Vancouver, British Columbia V6T 1Z3
Canada

M. JORDAN

Agriculture and Agri-Food Canada
Cereal Research Centre
195 Dafoe Rd.
Winnipeg, Manitoba R3T 2M9
Canada

JAMES KELLY

Crop and Soil Sciences
Michigan State University
East Lansing, MI 48824
USA

KENNETH L. KORTH

Department of Plant Pathology
217 Plant Science Building
University of Arkansas
Fayetteville, AR 72701
USA

JOSEPH KUĆ

Professor Emeritrus, University of Kentucky
5502 Lorna St.
Torrance, CA 90503
USA

ERIC LAM

Biotech Center
Foran Hall
Rutgers – State University of New Jersey
New Brunswick, NJ 08903
USA

BRIGITTE MAUCH-MANI

Institute of Botany
University of Neuchâtel
Rue Emile-Argand 9
Case postale 2
2007 Neuchâtel
Switzerland

JEAN-PIERRE MÉTRAUX

Department of Biology
Unit of Plant Biology
University of Fribourg
10 ch. du musée, 1700 Fribourg
Switzerland

ALEX M. MURPHY

Department of Plant Sciences
University of Cambridge
Downing Street, Cambridge CB2 3EA
United Kingdom

CHRISTIANE NAWRATH

Department of Biology
Unit of Plant Biology
University of Fribourg
10 ch. du musée, 1700 Fribourg
Switzerland

MASARU OHTA

Institution of Agricultural and Forest
   Engineering
University of Tsukuba
Tsukuba, Ibaraki 305-8572
Japan

CORNÉ M.J. PIETERSE

Section Phytopathology
Faculty of Biology
Utrecht University
PO Box 800.84
3508 TB Utrecht
The Netherlands

ZAMIR K. PUNJA

Department of Biological Sciences
Simon Fraser University
Burnaby, British Columbia V5A 1S6
Canada

ANDREAS SCHALLER

Institute of Plant Physiology
   and Biotechnology
University of Hohenheim
25 Emil-Wolff St.
70593 Stuttgart
Germany

STEFAN SCHILLBERG

Fraunhofer Institute for Molecular Biology
   and Applied Ecology IME
Worringerweg 1
52074 Aachen
Germany

KAREN S. SCHUMAKER

Department of Plant Sciences
University of Arizona
Tucson, AZ 85721
USA

ARAVIND SOMANCHI

Department of Biological Sciences
Cellular and Molecular Biology Program
101 Life Sciences Bldg
Auburn University
Auburn, AL 36849
USA

D.J. SOMERS

Agriculture and Agri-Food Canada
Cereal Research Centre
195 Dafoe Rd.
Winnipeg, Manitoba R3T 2M9
Canada

JURIAAN TON

Institute of Botany
Department of Biochemistry
University of Neuchâtel
Rue Emile-Argand 9,
CH-2007, Neuchâtel
Switzerland

GARY A. THOMPSON

Department of Applied Science
575 ETAS Building
University of Arkansas at Little Rock
2801 S. University Ave.
Little Rock, AR 72204-1099
USA

SADIK TUZUN

Department of Entomology and Plant Pathology
Cellular and Molecular Biosciences Program
209 Life Sciences Building
Auburn University
Auburn, AL 36849
USA

VERONICA VALLEJO

Crop and Soil Sciences
Michigan State University
East Lansing, MI 48824
USA

L.C. VAN LOON

Faculty of Biology,
Section Phytopathology,
Utrecht University,
P.O. Box 800.84,
3508 TB Utrecht,
The Netherlands

NAOHIDE WANTANABE

Biotech Center
Foran Hall
Rutgers – State University of New Jersey
New Brunswick, NJ 08903
USA

CHUI ENG WONG   Department of Plant Sciences
                           University of Cambridge
                           Downing Street, Cambridge CB2 3EA
                           United Kingdom

T. XING   Assistant Professor
                           Department of Biology
                           Carleton University
                           Ottawa, Ontario K1S 5B6
                           Canada

KWANG-YEOL YANG   Department of Biology
                           Utah State University
                           Logan, UT 84322
                           USA

JIAN-KANG ZHU   Department of Plant Sciences
                           University of Arizona
                           Tucson, AZ 85721
                           USA

SABINE ZIMMERMANN   Institute for Molecular Biotechnology
                           RWTH Aachen
                           Worringerweg 1
                           D-52074 Aachen
                           Germany

1

Terminology Related to Induced Systemic Resistance: Incorrect Use of Synonyms may Lead to a Scientific Dilemma by Misleading Interpretation of Results

SADIK TUZUN

1.1 Introduction

During the review process of the book, several reviewers suggested that the same terminology should be used throughout the book to describe the same phenomenon. Since I am a firm believer of academic freedom and freedom of expression, no changes to the chapter will be made. Instead, to comply with the second suggestion of the reviewers and to eliminate misunderstanding due to multiple ways of using the terms induced systemic resistance (ISR) and systemic acquired resistance (SAR) which has been accepted as synonyms, a comprehensive chapter dealing with the terminology is included. As described by Kuć in Chapter 2, inducible defense responses in plants have been observed since early 1900s (Beauverie, 1901; Ray, 1901) and reviewed by Chester as early as 1930s (Chester, 1933). Chester called the phenomenon "acquired physiological immunity", since his review was based on "observations" rather than "scientific experiments", and indeed this term was correct since he was describing disease resistance clearly "acquired" by plants. Later on, studies conducted by Kuć and his colleagues (Kuć et al., 1959; Maclennan et al., 1963) on apple and by Ross (1961, 1966) on tobacco, which lead to the induction of local and systemic resistance gave first evidences that indeed otherwise susceptible plants have inducible defense responses if they are previously treated with some chemicals or pathogens which are unspecific in nature, although both phenomenon involves salicylic acid as mediator (Ryals et al., 1996).

During the past 40 or more years nearly a thousand journal articles have been published calling the phenomenon "induced" or "acquired" systemic resistance. The mechanisms of resistance against viruses are still not understood well. Nevertheless, the elegant work of Kuć and his coworkers and several other research groups using cucurbits and many other plant species explained the broad nature of resistance, and the term induced systemic resistance (ISR) was used in these pioneering publications using pathogens or chemicals as inducers which clearly involve salicylic aid as mediator, whereas systemic acquired resistance (SAR) was mainly used in recent publications by scientists using the *Arabidopsis* model system. ISR was proposed to be the correct term to describe the active nature of "inducible defense mechanisms in plants" regardless of the inducing agent or the pathway which they use to achieve the resistant state by Kloepper et al. (1992), and in the introduction to the book "Biology and Mechanisms of Induced Resistance to Pathogens and Insects" by Agrawal et al. (1999), considering the pioneering work of Kuć and many scientists trained in his lab who led the area for many years mainly used ISR as the term to describe the phenomenon. Induced resistance is still the most widely accepted terminology in meetings and workshops related to "inducible" defense mechanisms against pathogens and insects in plants.

1.2 Differentiation of ISR and SAR

As mentioned above, the terms "induced" and in some cases "acquired" systemic resistance were used interchangeably by the different research groups until Ryals et al. (1996) defined the type of resistance induced by pathogenic organisms and/or chemicals involving salicylic acid as mediator as systemic acquired resistance (SAR) as a tribute to Ross, disregarding many earlier publications describing entirely the same phenomenon using ISR as a synonym. Furthermore, a series of about 25 journal articles mainly published by Van Loon's research group used ISR as the term solely to describe resistance mediated by plant growth-promoting rhizobacteria (PGPR) (Pieterse et al., 1996, 1998, 2000, 2002; Van Loon et al., 1998) while at least as much published by others indicating PGPR-mediated ISR used it as a synonym to SAR. This use of terminology by disregarding at least 10 times more publications using ISR to describe the phenomenon that is induced by many pathogenic organisms and chemicals actually created a dilemma leading to a misunderstanding of earlier literature and confusion among scientists.

1.2.1 ISR and SAR are Decided to be Used as Synonyms

PGPR is a generic term which includes many plant associated organisms some of which may be partially pathogenic to plants to be recognized by them (Tuzun and Bent, 1999) using the salicylate pathway for induction of resistance (see Chapter 10). Therefore, the results obtained on *Arabidopsis* plants using a few PGPR strains to use ISR as the term to describe only "jasmonate-mediated resistance" creates a major problem in scientific literature as mentioned above. This subject was extensively discussed amongst the attendees in detail during the "1st International Symposium of Induced Resistance" which was held in Greece in 1999. During this symposium, it was unanimously agreed by the participants that these terms that are describing essentially the same phenomenon should be used as "synonyms". Indeed, the paper came as a result of these meetings and authored by Ray Hammerschmid, Jean-Pierre Metraux, and Kees Van Loon (Hammerschmidt et al., 2001) clearly stated that induced systemic resistance and systemic acquired resistance are "synonyms" and should be used in the scientific literature as synonyms and treated as the same.

Scientists hold a big responsibility when they introduce "new uses" for the "old terms" and, although we are not linguists, it is essential that we *must* understand and adhere to the meaning of the words before using it. In this chapter, the meaning of various words used in the literature are described using the 2003 Electronic Edition of Merriam Webster's Collegiate Dictionary containing over 250,000 words. The meaning of "synonym" is "one of two words or expressions of the same language that have the same or nearly the same meaning in some or all senses". The scientific problem becomes apparent if one or more groups of scientists decide to use synonyms to describe essentially two different phenomenon which involves different pathways that may even crosstalk amongst themselves (Kunkel and Brook, 2002). Even though this is used just for "convenience" (see Chapter 9), the use of existing terms (ISR and SAR) to differentiate two independent phenomenon activated by different pathways (see Figure 9.1 and 9.2 in Chapter 9) will cause even more confusion in the future since recent research also indicates that there are more than two biochemical pathways by which induced resistance can be activated (e.g., Bostock et al., 2001; Dong and Beer, 2000; Mayda et al., 2000a,b; Zimmerli et al., 2000).

1.2.2 Contradicting Results in the Literature with the Use of "Synonym"

In scientific literature, synonyms should be describing exactly the same phenomenon resulting from activation of the same pathways. So, there was no reason to call PGPR-mediated induction of systemic resistance as ISR and all others as SAR (Van Loon et al., 1998). Since ISR and SAR are accepted as synonyms, by no means should they interfere with each other, neither should they work synergistically nor should they inhibit each other's expression. However, there is ample evidence that pathways leading to ISR and SAR actually work synergistically (Van Wees et al., 2002) by enhancing disease resistance or in a contradicting fashion by inhibiting each other (Doares et al., 1995; Ryan, 2000). Therefore, the use of synonyms "ISR and SAR" in the same publication to describe entirely different phenomenon is not scientifically correct and contradicts the meaning of synonym (see Chapters 8 and 9). The use of ISR and SAR to describe separate biochemical pathways relating to induced resistance is misleading for a variety of reasons. First, it contradicts the decision of the scientific community to use these terms as synonyms. Furthermore, it is the fact that researchers will tend to make the (erroneous) assumption that ISR must be distinct from every phenomenon referred to as SAR, regardless of whether any work has ever been done to actually characterize the biochemical pathway(s) involved in each system. Therefore, it is hoped that

ISR will no longer be used solely as the term to describe induced resistant state mediated by plant growth-promoting rhizobacteria (PGPR).

1.3 Definitions Used in the Literature to Describe Inducible Defense Responses in Plants

1.3.1 Acquired Immunity

This term was first used by Chester to describe achievement of resistant state in otherwise susceptible plants during 1930s (Chester, 1933). Although Chester did not perform experiments, he clearly described the phenomenon with various examples. Immunization of plants was used in several articles that experimentally described Chester's original observations. The term "immunity" was not widely accepted by the scientific community since it creates confusion with the immunization of animals. Nevertheless the term immunity that is being used since 14th century means: "a condition of being able to resist a particular disease especially through preventing development of a pathogenic organism or by counteracting the effect of its products" and immunization simply means "to make immune". Chester called the phenomenon "acquired immunity" since he thought that these plants acquired a state of being immune where acquired means: "to come into possession or control of often by unspecific means" or "to come to have as a new or added characteristic, trait or ability (as sustained effort or natural selection as in bacteria acquire resistance to antibiotics)". To Chester this definition was correct since he thought that plants were acquiring a state of resistance by an unspecified means of natural phenomenon.

1.3.2 Systemic Acquired Resistance

This term was first used by Ross (1961) to describe a phenomenon where he observed protection against TMV both local and systemically upon treatment of either the same leaf or leaves below the protected leaf upon treatment with live TMV. He described systemic nature of the phenomenon; however, plants did not passively acquire the resistant state as indicated in the definition above. Neither plants obtained resistance in a genetically inherited fashion as in bacterial resistance to antibiotics nor the phenomenon occurred naturally as in the observations of Chester and others. Ross actually induced a state of resistance in tobacco against TMV by using TMV which was not inherited by the offsprings of the tobacco plant. Although the term is widely used by scientists working in the area as attribute to Ross, it is not correct by any means to describe an active phenomenon which involves activation of many genes leading to the development of resistant state in otherwise susceptible plants. Indeed, experiments conducted by Kuć and his colleagues described the phenomenon of chemically induced resistance against scab disease in apple much earlier than Ross (Kuć et al. 1959, see Chapter 2), which clearly involves salicylic acid as mediator.

1.3.3 Induced Systemic Resistance

The term first used by Kuć and his coworkers in numerous publications (see Chapter 2) is actually the correct way of describing the phenomenon. The meaning of Induced is: "to call forth or bring about by influence or stimulation" or "to cause the formation of", in this particular case ISR indicates an active phenomenon which causes the formation of systemic resistance in otherwise susceptible plants. According to Van Loon and his colleagues, only a few PGPR strains which induce the systemic state of resistance via salicylate-independent pathway (see Chapters 8 and 9) are justified to be called initiators of ISR whereas others as inducers of SAR as suggested by Ryals et al. (1996). This use of terminology is neither correct nor the common use of term SAR is fair to the overall contributions of Joseph Kuć who actually "for the first time" experimentally demonstrated the induction of systemic resistance using various derivatives of amino acids. If anyone "fathered this area" it must be him, not only through his contributions but also through numerous scientists, students, post-docs, collaborators etc. who published hundreds of papers, using the term ISR to describe "induced state of resistance in plants by biological or chemical inducers" which definitely uses salicylic acid as mediator. These are the pioneering scientists who led the field of ISR to become a "common phenomenon" found to be a part of the overall protection achieved by many biological and chemical agents including the organisms known for a long time as biological control organisms as described throughout this book. If we must be honest, no student or co-worker has actually followed the initial experiments of Ross against viruses until mid to late 1980s when the scientists working in then Ciba-Geigy started to work on ISR. Needless to say, most of these scientists also performed their initial experiments on induced systemic resistance in Kuć's lab while he was collaborating with Ciba on this project. It is interesting that we still do not know the mechanism of resistance against viruses in plants.

1.4 Proposed Use of Terminology

Considering that ISR and SAR are well accepted by the scientific community as terms to describe inducible defense responses in plants, the use of all other terms such as systemic induced resistance (SIR) or acquired systemic resistance (ASR) should be avoided. ISR indicates actively-inducible defense mechanisms which may involve one or more metabolic pathways, as indicated above. Therefore, ISR is the correct term to describe "activated defense mechanisms" whether the inducers are pathogenic or nonpathogenic organisms or chemicals. SAR, however, should be indicated as synonym in each case when ISR is used for the first time in any article. If an author prefers to use SAR as the term, it is expected that ISR is indicated as synonym in the same fashion.

Certainly, the phenomenon can be differentiated by stating the inducer, i.e., PGPR-induced systemic resistance or PGPR-mediated systemic acquired resistance; or chemically induced systemic resistance or chemically mediated systemic acquired resistance (actually, using the term "induced" in "induced systemic resistance" will eliminate the use of "mediated" while describing different inducers), however, one type of inducer may induce different pathways. It is the most correct way, therefore, we should follow the terminology where the phenomenon was described according to which pathway the induction of resistance is activated through, either jasmonate or salicylate, as ISR appears to involve these two major pathways (Spoel et al., 2003).

It is proposed that the different variants of induced systemic resistance should be distinguished according to the pathway they activates, i.e., "salicylate-dependent" ISR (or SA-ISR) and "jasmonate-dependent" ISR (or JA-ISR), as our knowledge increases new terms could be added in the same fashion.

1.5 Conclusion

As scientists we have to stick to the scientific guidelines when creating definitions, whether they are scientifically correct or not and the definitions must adhere to linguistic meanings, otherwise once mistakes are made it becomes very difficult to rectify them. It is unfortunate that the terminology used in publications may become part of textbooks misleading young minds and future scientists, whom we have the responsibility to educate with an open mind, without leading to any assumption. This requires respect of the previous use of terms to describe the same phenomenon yet the terms, which are introduced must be flexible enough to accommodate definitions as our knowledge base broadens by the development of new technologies that may not be currently available.

It is certainly hoped that this attempt to correct the terminology will be recognized by colleagues as a friendly suggestion and will be used in coming publications to further avoid any confusion that may arise by using synonyms to describe different phenomenon and every attempt to correct this error should be made.

References

Agrawal, A., Tuzun, S., and Bent, E., eds. 1999. *Induced Plant Defenses Against Pathogens and Herbivores*. St. Paul, MN: American Phytopathological Society.

Beauverie, J. 1901. Essais d'immunization des végétaux contre de maladies cryptogramiques. *CR Acad. Sci. III* 133:107–110.

Bostock, R.M., Karban, R., Thaler, J.S., Weyman, P.D., and Gilchrist, D. 2001. Signal interactions in induced resistance to pathogens and insect herbivores. *Eur. J. Plant Pathol.* 107:103–111.

Chester, K. 1933. The problem of acquired physiological immunity in plants. *Quar. Rev. Biol* 8:129–154, 275–324.

Doares, S.H., Narvaez-Vasquez, J., Conconi, A., and Ryan C.A. 1995. Salicylic acid inhibits synthesis of proteinase inhibitors in tomato leaves induced by systemin and jasmonic acid. *Plant Physiol.* 108:1741–1746.

Dong, H., and Beer, S.V. 2000. Riboflavin induces disease resistance in plants by activating a novel signal transduction pathway. *Phytopathology* 90:801–811.

Hammerschmidt, R., Métraux, J.P., and Van Loon, L.C. 2001. Inducing resistance: a summary of papers presented at the First International Symposium on Induced Resistance to Plant Diseases, Corfu, May 2000. *Eur. J. Plant Pathol.* 107:1–6.

Kloepper, J.W., Tuzun S., and Kuć J. 1992. Proposed definitions related to induced disease resistance. *Biocontrol Sc. Technol.* 2:347–349.

Kuć, J., Barnes, E., Daftsios, A., and Williams, E. 1959. The effect of amino acids on susceptibility of apple varieties to scab. *Phytopathology.* 49:313–315.

Kunkel, B., and Brook, D.M. 2002. Cross talk between signaling pathways in pathogen defense. *Curr. Opin. Plant Biol.* 5:325–331.

Maclennan, D., Kuć, J., and Williams, E. 1963. Chemotherapy of the apple scab disease with butyric acid derivatives. *Phytopathology.* 53:1261–1266.

Mayda, E., Marqués, C., Conejero, V., and Vera, P. 2000a. Expression of a pathogen-induced gene can be mimicked by auxin insensitivity. *Mol. Plant Microb. Interact.* 13:23–31.

Mayda, E., Mauch-Mani, B., and Vera, P. 2000b. Arabidopsis *dth9* mutation identifies a gene involved in regulating disease susceptibility without affecting salicylic acid-dependent responses. *Plant Cell* 12:2119–2128.

Pieterse, C.M.J., Van Wees, S.C.M., Hoffland, E., Van Pelt, J.A., and Van Loon, L.C. 1996. Systemic resistance in *Arabidopsis* induced by biocontrol bacteria is independent of salicylic acid accumulation and pathogenesis-related gene expression. *Plant Cell* 8:1225–1237.

Pieterse, C.M.J., Van Wees, S.C.M., Van Pelt, J.A., Knoester, M., Laan, R., Gerrits, H., Weisbeek, P.J., and Van Loon, L.C. 1998. A novel signaling pathway controlling induced systemic resistance in *Arabidopsis*. *Plant Cell* 10:1571–1580.

Pieterse, C.M.J., Van Pelt, J.A., Ton, J., Parchmann, S., Mueller, M.J., Buchala, A.J., Métraux, J.-P., and Van Loon, L.C. 2000. Rhizobacteria-mediated induced systemic resistance (ISR) in *Arabidopsis* requires sensitivity to jasmonate and ethylene but is not accompanied by an increase in their production. *Physiol. Mol. Plant Pathol.* 57:123–134.

Pieterse, C.M.J., Van Wees, S.C.M., Ton, J., Van Pelt, J.A., and Van Loon, L.C. 2002. Signalling in rhizobacteria-induced systemic resistance in *Arabidopsis thaliana*. *Plant Biol.* 4: 535–544.

Ray, J. 1901. Les maladies cryptogramiques des végétaux. *Rev. Gen. Bot.* 13:145–151.

Ross, A.F. 1961. Systemic acquired resistance induced by localized virus infections in plants. *Virology* 14:340–358.

Ross, A. 1966. Systemic effects of local lesion formation. In *Viruses of Plants*, eds. A. Belmster, and S. Dykstra, pp. 127–150. Amsterdam: North Holland.

Ryals, J.A., Neuenschwander, U.H., Willits, M.G., Molina, A., Steiner, H.-Y., and Hunt, M.D. 1996. Systemic acquired resistance. *Plant Cell* 8:1808–1819.

Ryan, C.A. 2000. The systeminsignaling pathway: differential activation of plant defensive genes. *Biochim. Biophys. Acta* 1477:112–121.

Spoel, S.H., Kornneef, A., Claessens, S.M.C., Korzellius, J.P., Aan Pelt, J.A., Mueller, M.J., Buchala, A.J., Metrausx, J.-P., Brown, R., Kazzan, K., Van Loon, L.C., Dong, X., and Pieterse, C.M.J. 2003. NPR1 modulates crosstalk between salicylate- and jasmonate-dependent defense pathways through a novel function in the cytosol. *Plant Cell* 15:760–770.

Tuzun, S., and Bent, E. 1999. The role of hydrolytic enzymes in multigenic and microbially-induced resistance in plants. In *Induced Plant Defenses Against Pathogens and Herbivores: Biochemistry, Ecology and Agriculture,* eds. A.A. Agrawal, S. Tuzun, S., and E. Bent, pp. 95–115. St. Paul, MN: American Phytopathological Society.

Van Loon, L.C., Bakker, P.A.H.M., and Pieterse, C.M.J. 1998. Systemic resistance induced by rhizosphere bacteria. *Annu. Rev. Phytopathol.* 36:453–483.

Van Wees, S.C.M., De Swart, E.A.M., Van Pelt, J.A., Van Loon, L.C., and Pieterse, C.M.J. 2000. Enhancement of induced disease resistance by simultaneous activation of salicylate- and jasmonate-dependent defense pathways in *Arabidopsis thaliana. Proc. Natl. Acad. Sci. USA* 97:8711–8716.

Zhang, S., Reddy, M.S., Kokalis-Burelle, N., Wells, L.W., Nightengale, S.P., and Kloepper, J.W. 2001. Lack of induced resistance in peanut to late blight spot disease by plant growth-promoting rhizobacteria and chemical elicitors. *Plant Dis.* 85: 879–884.

Zimmerli, L., Jakab, G., Métraux, J.-P., and Mauch-Mani, B. 2000. Potentiation of pathogen-specific defense mechanisms in Arabidopsis by β-aminobutyric acid. *Proc. Natl. Acad. Sci. USA* 97:12920–12925.

2

What's Old and What's New in Concepts of Induced Systemic Resistance in Plants, and its Application

JOSEPH KUĆ

2.1 Historical Perspective

Disease and induced resistance to disease in plants and animals has been with us as long as plants, animals, and their pathogens have coevolved. Observations of induced resistance in plants were reported as early as the late 1800s and early 1900s (Beauverie, 1901; Ray, 1901; Chester, 1933). Muller and Borger (1940) described carefully conducted experiments which established the phenomenon of induced local resistance (ILR) in potatoes to late blight (*Phytophthora infestans*). Inoculation of potato tubers with cultivar-nonpathogenic races of the fungus induced local resistance to cultivar-pathogenic races. This work, and subsequent studies by Muller and coworkers also established the concept of active defense for resistance, a response after infection, and this proved to be the foundation for work with phytoalexins.

Induced systemic resistance (ISR) was analytically established by Kuć et al. (1959) and Ross (1966). Kuć et al. (1959) and Maclennan et al. (1963) demonstrated that apple plants were made systemically resistant to apple scab (*Venturia inaequalis*) by infiltrating lower leaves with D-phenylalanine, D-alanine and aminoisobutyric acid (AIB). The amino acids did not inhibit the growth of *V. inaequalis in vitro* at concentrations used for infusion. Ross (1966) and coworkers demonstrated that inoculation of lower leaves of tobacco with a local lesion strain of tobacco mosaic virus (TMV) systemically enhanced resistance to the same strain of the virus. They also established the time required between induction and inoculation for ISR and its persistence. The continued research by Kuć and coworkers verified the reports by Ross and expanded and defined our understanding of ISR and its application for disease control in the greenhouse and field. They demonstrated that ISR was not specific with respect to the nature of the inducer or the biological spectrum of the diseases it protects against. Thus, unrelated fungi, bacteria, viruses, or chemicals induced resistance systemically

9

against all three classes of pathogens (bacteria, fungi, and viruses), and in some experiments, even protected plants against damage caused by herbicides and oxidants (Kuć, 1982, 1995a, 1995b, 1997, 1999; Dalisay and Kuć, 1995a, 1995b; Fought and Kuć, 1996; Gottstein and Kuć, 1989; Karban and Kuć, 1999; Lusso and Kuć, 1999; Mucharromah and Kuć, 1991; Strobel and Kuć, 1995). ISR was demonstrated with different plants, including cucumber, watermelon, muskmelon, tobacco, tomato, green bean, apple and pear, and was found to be effective in these plants against bacterial, fungal, and viral pathogens. An important aspect of ISR established by this body of work is that it sensitizes (or primes) plants to respond rapidly to a pathogen after infection. The molecular basis for sensitization is still unclear, but it appears that the phenomenon is even more important for defense against disease than the initial accumulation of defensive compounds, observed upon induction of systemic resistance (Kuć, 1984, 2001; Conrath et al., 2001).

Research with ISR has expanded rapidly, with contributions from many laboratories worldwide. ISR has now been reported in plants as diverse as *Arabidopsis thaliana* to coffee, and ISR is also effective against insects and nematodes (Agrawal et al., 1999; Schmidt and Huber, 2002; Hammerschmidt and Kuć, 1995).

A key to the evolution of ISR was the early research with phytoalexins, pioneered by Cruickshank, Kuć, Uritani, Tomiyama, and Metilitskii and their coworkers (reviewed in Kuć, 1995a; Hammerschmidt, 1999). The research with phytoalexins assigned chemical structures to the putative defense compounds and established a close relationship between the localized early accumulation of phytoalexins and inhibition of pathogen development and disease. The research also established that phytoalexin accumulation was elicited by simple inorganic and organic chemicals, as well as by microorganisms and their products. Phytoalexins accumulated in resistant as well as susceptible interactions. The difference between resistant and susceptible plants was evident in the timing of phytoalexin accumulation: in resistant plants accumulation was rapid and in susceptible plants, accumulation was delayed. The early experiments conducted with phytoalexins established a foundation for ISR research, and the similarities between phytoalexin accumulation and ISR in plants are evident. Whether phytoalexins are major factors for resistance has been reviewed (Kuć, 1995a; Hammerschmidt, 1999). Most of the research with phytoalexins has indicated that their accumulation is most often associated with resistance to fungal diseases, is less so for bacterial diseases, and is unlikely to be associated with resistance to viruses, though ISR is effective against some viral diseases.

The discovery of the central role of salicylic acid (SA) in some mechanisms for ISR opened the door to investigations of the regulation of, and mechanisms involved in, ISR on a molecular and genetic level (Metraux, 2001).

2.2 The Phenomenon of Induced Resistance

Pertinent to an understanding of the phenomenon if ISR is a consideration of the question about why plants and animals are susceptible to infectious diseases. Disease resistance in plants and animals requires multiple components (see

Section 2.3). The antibody-based, or humoral, immune system in animals is highly specific, both in terms of the elicitors (specific antigens) that generate a humoral response, and in the nature of the response (the production of antibodies that recognize and bind to the antigen). The first time an animal is exposed to an antigen, the humoral response is sluggish. Upon subsequent exposure to the antigen, the response is much more rapid and results in the production of greater quantities of antigen-specific antibodies. These antibodies work in concert with cell-mediated defense responses in animals to limit pathogen attacks.

ISR in plants lacks the specificity of the humoral immune system: ISR can be generated by a wide variety of structurally unrelated elicitors, and once activated, it is effective against a wide variety of organisms. Some plant–pathogen interactions are, however, highly specific, as is observed in gene-for-gene interactions and host specificity.

Excluding genetic faults, animals and plants express genes for resistance mechanisms, and both have demonstrated resistance to the bacteria, fungi, and viruses in their environment throughout the ages of evolution. The mechanisms by which plant and animal defense, or immune, response systems function are clearly very different, but in one principle they are similar: unless activated sufficiently in a timely manner, the responses will fail to contain a given pathogen, even when all the required components needed to contain a pathogen are present. In animals, and seemingly also in plants, immune or defense responses may fail when (1) there has been no prior exposure to the pathogen, or another elicitor, which can prime the immune system to produce a more rapid and effective response, (2) the plant or animal is subjected to stresses (e.g., poor nutrition, developmental or environmental stress) which decrease its ability to mount an immune or defense reaction, or (3) the pathogen dose is too high and defenses, while activated, are simply inadequate to deal with the number of infectious agents. To use our species as an example, human disease epidemics have occurred in the past when groups of people were exposed to novel pathogens they had never encountered before (e.g., smallpox, new strains of influenza), or when changes in human living conditions or the environment brought people into contact with greater numbers of pathogens (e.g., bubonic plague), and it is commonly observed that the malnourished, the elderly, and the very young tend to be more susceptible to diseases than healthy adults. Genetic variation between individuals also exists, and some human immune systems are simply more effective at dealing with pathogens than others.

In plants, particularly in natural communities of plants, their defense responses are extremely effective at combating pathogens. To my knowledge, a plant species has not disappeared from the earth as a result of disease, unless human activity can be considered a disease. However, plants that survive diseases in the wild are not necessarily perfectly fit, lush, and healthy. A disease-tolerant plant may be able to fulfill its evolutionary prerogative and reproduce, and is in terms of evolution a success; but unless the quality and yield of produce from the plant is high, this plant is not useful to current agricultural production. A distinction should be made between disease resistance needed for the survival of a species, and disease resistance necessary to minimize economic losses when growing the plants commercially. When we speak of the need to increase plant resistance to disease, we are actually referring to the latter, since plants in natural communities already have the defenses they need to survive.

2.3 Single and Multigenic Resistance, ISR and Defense Compounds

The literature contains references to many defense compounds and their alleged importance in plant disease resistance. However, nounequivocal case has been made for the necessity of any one defense compound for resistance, and many compounds accumulate after infection. More information is necessary concerning the contribution of defense compounds to resistance, individually and collectively, as well as the timing, magnitude, and localization of their accumulation relative to pathogen development. More research is also necessary to determine the mode of action of defense compounds, whether they inhibit development of a pathogen and/or reduce damage caused by a pathogen, and whether there is an interdependence or synergy in their activity. Until this information is available, the reported defense compounds are at best associated with resistance and are putative defense compounds/mechanisms (PDCM).

The PDCM include those that are preformed, as well as those that are produced in response to wounding, and those that accumulate locally or systemically after infection, ISR, or infection after ISR. PDCM include simple inorganic and organic compounds, peptides, proteins, enzymes, and phenolic and carbohydrate polymers (Table 2.1). It is evident, therefore, that many different pathways, loci, and compartments are involved in their synthesis and different mechanisms are required for the regulation of their accumulation and mode of action. As important as activation of resistance mechanisms is to disease resistance, it is equally vital to the plant's survival that the regulated, though apparently chaotic, metabolic processes that were put into motion can be redirected to normal.

From the above it seems reasonable to conclude that the mechanisms for ISR and disease resistance/susceptibility are multicomponent and, therefore, their regulation will be multicomponent. Since the genes for PDCM are present in susceptible and resistant plants, what is it that regulates single gene resistance, and its frequent loss, as well as multigenic resistance and ISR?

TABLE 2.1. Putative defense compounds/systems for disease resistance in plants

Passive and/or wound responses
    Waxes, cutin, phenolic glycosides, phenols, quinones, steroid glycoalkaloids, suberin, terpenoids and proteins

Increases after infection
    Phytoalexins, reactive oxygen species/free radicals, calcium, silicon/silicates, polyphenoloxidases, peroxidases, phenolic cross-linked cell wall polymers, hydroxyproline and glycine-rich glycoproteins, thionins, antimicrobial proteins and peptides, chitinases, β-1,3-glucanases, ribonucleases, proteases, callose, lignin, lipoxygenases and phospholipases

Evidence is not available, and it is highly unlikely, that single gene resistance is due to the production of a single PDCM. The response of a plant with single gene resistance to a pathogen is multicomponent, and differs from the susceptible plant lacking the gene for resistance only in the timing of the response. The magnitude of response is often greater in the susceptible plant lacking the gene, but the response is delayed until after the pathogen has been established. Regardless of the presence of single gene or multigenic resistance, many unrelated organisms and chemicals can elicit the same metabolic responses in a plant and elicit ISR to a broad spectrum of pathogens and environmental stresses.

One interpretation of the above observations is that the resistance gene, via its product, regulates the timing of the expression of multiple mechanisms, either directly or indirectly, via a master switch(es), which eventually leads to the multistep mechanisms for the synthesis and accumulation of PDCM. It is likely that a master switch(es) would regulate many other switches, or cascades, which activate or de-activate signals for individual pathways and their interaction. Thus, it is important to differentiate resistance genes which regulate expression of a master switch(es) from the genes for steps within the pathways for the synthesis of PDCM.

When resistance is "lost" in a plant with single gene resistance, it is not the gene itself which is lost. What is lost is the gene's effectiveness. The genes for the PDCM are still present, as is the potential for their activation. The pathogen overcoming single gene resistance may do so by a number of mechanisms: (1) avoid activation of the resistance gene product (or receptor), and thereby a factor is not produced to activate the master switch(es) and trigger a defense response. Pathogen avirulence gene products which do not bind to plant receptors, or which bind but do not activate or fully activate the receptor, would accomplish this, (2) a product that modifies and thereby inactivates the plant receptor, (3) a product that inactivates the master switch(es), (4) a product(s) that inactivates all or many of the pathways producing PDCM. This latter possibility is highly unlikely, given the diversity of PDCM.

With multigenic resistance, the PDCM are likely to be identical to those utilized in single gene resistance. The difference between the two types of resistance would be the presence of multiple host genes, which may encode receptors, capable of binding and detecting nonspecific pathogen products (i.e., fragments of cell wall polymers such as chitin and peptidoglycan, or other conserved structural components, such as lipopolysaccharides or flagellin). To avoid activating resistance, the pathogen would have to produce structural components that do not bind to any plant receptor (which is unlikely), or find a way to inactivate all the plant receptors, or the master switch(es). It is possible that binding of a nonspecific elicitor to a receptor results in less efficient activation of these receptors, but there are also a greater diversity of receptors. Upon encountering initial plant defense responses, cells of an invading pathogen may be damaged or lyse and release a great quantity of nonspecific elicitors (i.e., cell wall fragments), amplifying the original signal. Multigenic resistance is therefore much more difficult to overcome than single-gene resistance. If there are multiple and redundant master switches governing plant defense responses, it is possible that they do not regulate PDCM equally, resulting in qualitative and quantitative differences in PDCM and the timing of their appearance.

Since ISR has the same PDCM as those associated with single gene and, probably, multigenic resistance, and as ISR lacks specificity with respect to the nature of the inducers and spectrum of its biological activity, it is possible that inducers of ISR, directly or indirectly, regulate a master switch(es) governing the timing of PDCM production. The factors activating a master switch(es) have yet to be fully elucidated, but could include those produced by single and multigenic resistance, i.e., reactive oxygen species (ROS). The difference between gene-based resistance and ISR would therefore be the site of action. In gene-based resistance, the expressed host receptors (resistance gene products) govern resistance. In ISR, resistance may be governed via the priming of master switch(es).

The agents causing ISR, whether microorganisms or chemicals, could affect a master switch directly by causing metabolic perturbations that generate a signal affecting that switch, i.e., ROS. During the induction of ISR in Plant A, the plant's master switch(es) are activated and PDCM are produced. A susceptible, noninduced plant (Plant B) that is infected by a pathogen could also generate ROS, activating the master switch(es) and the production of PDCM, but this response would be delayed, allowing the pathogen to spread and cause further damage. Upon subsequent infection by a pathogen in each of Plants A and B, the master switch(es) react more quickly. The difference is that Plant B has suffered greater damage and may not even have survived its first infection.

There may be many paths leading to PDCM, plant disease resistance and ISR. The key may be the levels of incompatibility/compatibility between a microorganism or chemical and the plant during the early stages of their interaction, and this may be determined by the ability to generate, tolerate, or inactivate ROS.

2.4 Induction of ISR

The inducers of ISR vary greatly and include fungi, bacteria, viruses, nematodes, insects, components, and products of pathogens and nonpathogens, organic and inorganic polymers and simple organic and inorganic compounds (Table 2.2). It is not possible to assign a unique chemical structure as being necessary for the induction of ISR (Fought and Kuć, 1996). Compounds as simple as phosphate salts and ferric chloride have been reported to induce ISR (Gottstein and Kuć, 1989; Mucharromah and Kuć, 1991; Reuveni et al., 1996, Reuveni and Reuveni, 1998; Manandhar et al., 1998). Therefore, inducers are active not because of what they are, but rather for what they do, and they are likely to have common features in how they affect plants. Not all inducers have been reported active in all plants against all diseases, but it is clear that biologically-induced ISR is active with the same microorganism as inducer in unrelated plants against unrelated diseases (Kuć, 1982; Kuć, 2001). The commercially available compound Bion (benzo(1,2,3)thiadiazole-7-carbothioic acid S-methyl ester) is active in many unrelated plants against many unrelated pathogens and some nematodes and insects (Oostendorp et al., 2001).

TABLE 2.2. Agents reported to elicit induced systemic resistance in plants

- fungi, bacteria, viruses, nematodes, insects
- fungal, bacterial and plant cell wall fractions, intercellular plant fluids and extracts of plants, fungi, yeasts, bacteria and insects
- potassium and sodium phosphates, ferric chloride, silica
- glycine, glutamic acid, α-aminobutyric acid, β-aminobutyric acid, γ-aminobutyric acid, α-aminoisobutyric acid, D-phenylalanine, D-alanine and DL tryptophan
- salicylic acid, m-hydroxybenzoic acid, p-hydroxybenzoic acid, phloroglucinol, gallic acid, isovanillic acid, vanillic acid, protocatecheic acid, syringinc acid, 1,3,5 benzene tricarboxylic acid
- D-galacturonic acid, D-glucuroinic acid, glycollate, oxalic acid and polyacrylic acid
- Oleic acid, linoleic acid, linolenic acid, arachdonic acid, eicosapentaenoic acid
- Paraquat, acifluorfen, sodium chlorate, nitric oxide, reactive oxygen species
- 2,6-dichloroisonictonic acid, benzo (1,2,3) thiadiazole-7-carbothioic acid s-methyl ester
- jasmonic acid, methyl jasmonate, ethylene
- ethylene diamine tetraacetic acid (EDTA), riboflavin
- probenazole and 2,2-dichloro-3,3-di-methyl cyclopropane carboxylic acid
- -dodecyl DL-alanine and dodecyl-L-valine
- phenanthroline and pththalocyanine metal complexes (cobalt, iron and copper)

The acceptance of the non-specificity of inducers of ISR is a key to an understanding of the mechanisms responsible for ISR and its induction and regulation. Metabolic perturbation resulting in the generation of ROS may be one feature in common amongst the great diversity of ISR inducers. Many current reports support an important role for ROS in resistance and ISR (Averyanov et al., 2000; Dempsey et al., 1999; Lamb and Dixon, 1997; McDowell and Dangl, 2000; Murphy et al., 2001; Kim et al., 2001; Kiraly, 1998).

2.5 Application of ISR

Microorganisms and chemicals that induce ISR are commercially successful and available for the control of plant diseases (Oostendorp et al., 2001; Kim et al., 2001; Zhender et al., 2001; Reuveni et al., 1996; Bednarz et al., 2002). These include such diverse agents as rhizobacteria, Bion, Messenger, inorganic phosphates, ROS, and Probenazole. The development of new commercial agents for ISR depends upon several factors, some of which are favorable for development, and some unfavorable.

Favorable factors include:

(1) Problems with the resistance of pathogens to classical pesticides.

(2) The necessity to remove some pesticides from the market, the increased testing and cost of testing to meet requirements of regulatory agencies and the lack of substitutes for removed compounds.

(3) Health and environmental problems, real and perceived, associated with pesticides and the increased popularity of "organic" crops and "sustainable agriculture."

(4) The inability of pesticides to effectively control some pathogens, e.g., virus and soilborne pathogens.

(5) Classical pesticides may not be economically feasible for farmers in developing countries. In these countries, the level of awareness for the safe and effective application of classical pesticides is low, thus creating dangers to human health and the environment.

(6) Resistance of the public to genetically modified plants. In ISR, foreign genes are not introduced. The innate genes for resistance in the plant are those that are expressed.

(7) ISR has a broad spectrum of activity and its effectiveness persists for an extended period.

(8) Since many defenses are activated, pathogens are less likely to develop resistance to ISR.

Unfavorable factors include:

(1) Some plant pathologists still scoff at the applicability of ISR.

(2) Only high profit, patented and complex inducers make the major markets. Who champions the simple, nonpatented yet equally effective compounds?

(3) Lack of sufficient information exchange and financial support for non mega-agribusiness-oriented scientists, and a lack of adequate information flow to farmers and the public.

(4) Unlike classical pesticides which directly kill or inhibit the development of a pathogen, ISR depends upon the expression of genes for resistance in the plant. Therefore ISR is more subject to physiological and environmental influences that may alter its effectiveness.

(5) Public and farmer's apprehension of new technologies.

2.6 Directions for Future Research

Priorities for research include investigations that should have and could have been completed years ago as well as those that require new information and technologies for their initiation.

Which of the putative defense compounds contribute to resistance? Is the timing of their appearance important? Is the synthesis of the compounds and the timing of their appearance regulated differently? More attention should be given to individual plant–pathogen interactions to determine which inducers and their doses, as well as which putative defense compounds and the timing of their appearance, are important.

Do plants respond to the pathogen per se or to the stress (metabolic perturbation) caused by the pathogen, or a combination of both? What is the translocated signal(s) in ISR? What causes the synthesis or release of the signal(s)?

Is it possible to develop plants with enhanced ISR through plant breeding? When breeding for resistance, are we also often breeding for enhanced ISR? What are the genetic and metabolic bases for the cascade of events associated with defense compounds, ISR, and sensitization (priming)?

What are the molecular and practical significances of the nonspecificity of the agents which elicit ISR?

Are the mechanisms for the different types of resistance (nonhost, agerelated, organ specific) the same or different, and do they have components in common with ISR? Can the genes for the different types of resistance be selectively expressed without detrimentally influencing plant development, e.g., express genes for age-related resistance without prematurely aging the plant?

What are the roles of oxidative stress, ROS, and nitric oxide as defenses against disease and initiators of defense mechanisms? In mammals, hydrogen peroxide and superoxide anion are the major microbiocides produced by circulating phagocytic leukocytes. However, hydrogen peroxide and ROS may function alone or together with NO to enhance death of pathogens, as well as triggering transcriptional activation of plant defense genes and the hypersensitive response (Delledone et al., 1998). Elevated levels of Ca^{2+} can enhance NO synthase activity, and perhaps this partially explains the frequent association of calcium with resistance. Averyanov and colleagues (2000) reported that phenanthroline and phthalocyanine metal complexes induced ISR to rice blast when applied to foliage or the soil. Both compounds produced ROS, and the authors suggest that increased ROS resulted in ISR, sensitization, and the hypersensitive response. In addition, metal complexes of phthalocyanine stimulated ISR when applied to rice seeds before sowing, and the protection lasted for at least one month in seedlings. More emphasis should be placed on effective seed treatments for ISR.

Can defensins and protegrins be utilized effectively for ISR? Defensins and protegrins are antimicrobial peptides found in plants and animals ranging from insects to humans. They are part of an innate immune system which evolved before antibodies and lymphocytes. Since antimicrobial peptides are reported in plants, ISR may provide a mechanism to enhance production of the peptides in plants without the introduction of foreign genes.

Do DNA-binding proteins (zinc fingers) and cell-permeable polyamides have a role as agents for the selective expression of genes for ISR? Synthetic transcription factors have been developed which are designed proteins containing DNA-binding elements, or zinc fingers (Borman, 2000). Similar structures are found in some natural transcription factors. Zinc fingers are independently folding domains of about 30 amino acid residues centered on a zinc ion. These proteins and synthetic polyamides can turn endogenous genes on and off in living cells in a very specific manner.

Does the progress made with bacterial harpin indicate the presence of many similar proteins for ISR? Harpin produced by the pathogenic bacterium responsible for fire blight (*Erwinia amylovora*), induces systemic resistance in plants against many diseases caused by fungi, bacteria, and viruses, as well as some insects (Brasher, 2000; Bednarz et al., 2002). It also promotes root growth, reducing the need for water. The protein can be sprayed on plants before they are attacked by pathogens and it degrades so quickly that it cannot be detected within two hours of application. Other pathogens and even some nonpathogens are reported to produce harpin-like proteins and it is likely that proteins other than harpins have a capability for ISR.

2.7 Conclusions

Though resistance and susceptibility to pathogens are often specific and biochemicals determining this specificity have specific structures and receptors, nonspecific agents and multiple signals and pathways for their transduction can also induce resistance to unrelated pathogens and toxicants. This makes the possibility of finding additional effective agents for ISR and disease control highly promising. The agents need not be patented, expensive, or complex. Much more research is needed on the use of ISR agents to reduce dependence on chemical pesticides and enhance utilization of high-yielding plants that presently have a level of resistance that is inadequate for disease control under high pathogen pressure. ISR does not depend upon introducing genes into the plants, and it would not meet the resistance from the public engendered by genetically modified plants. ISR should be increasingly incorporated into integrated pest management practices. Increased funding and information exchange is needed to better utilize and direct the rapidly emerging information concerning signals, receptors, signal transduction, and gene expression for the practical control of plant disease.

Acknowledgments

I acknowledge the input of Dr. Elizabeth Bent and her valuable assistance in the preparation of this chapter. Without her help the chapter would not have been possible.

References

Agrawal, A., Tuzun, S., and Bent, E. eds. 1999. *Induced Plant Defenses Against Pathogens and Herbivores*. St. Paul, MN: American Phytopathological Society.

Averyanov, A.A., Lapikova, V.P., Gaivornsky, L.M., and Lebrun, M.H. 2000. Two step oxidative burst associated with induced resistance to rice blast. *First International Symposium on Induced Resistance to Plant Diseases*, Corfu, Greece, May 22–27, 2000, pp. 125–126 (Abstract).

Beauverie, J. 1901. Essais d'immunization des végétaux contre de maladies cryptogramiques. *CR Acad. Sci.* 133:107–110.

Bednarz, C.W., Brown, S.N., Flanders, J.T., Tankersley, T.B., and Brown, S.M. 2002. Effects of foliar applied harpin protein on cotton lint yield, fiber quality, and crop maturity. *Comm. Soil Sci. Plant Anal.* 33:933–945.

Borman, S. 2000. DNA-binding proteins turn genes on and off. *Chem. Eng. News*, 34–35.

Brasher, P. 2000. Protein for replacing pesticide approved. *San Diego Tribune*, April 29, p. A7.

Chester, K. 1933. The problem of acquired physiological immunity in plants. *Q. Rev. Biol.* 8:129–154, 275–324.

Conrath, U., Thulke, O., Katz, V., Schwindling, S., and Kohler, A. 2001. Priming as a mechanism in induced systemic resistance in plants. *Eur. J. Plant Pathol.* 107:113–119.

Dalisay, R., and Kuć, J. 1995a. Persistence of induced resistance and enhanced peroxidase and chitinase activities in cucumber plants. *Physiol. Mol. Plant Pathol.* 47:315–327.

Dalisay, R., and Kuć, J. 1995b. Persistence of reduced penetration by Colletotrichum lagenarium into cucumber leaves with induced systemic resistance and its relation to enhanced peroxidase and chitinase activities. *Physiol. Mol. Plant Pathol.* 47:329–338.

Delledonne, M., Xia, Y., and Lamb, C. 1998. Nitric oxide functions as a signal in plant disease resistance. *Nature* 394:585–588.

Dempsey, D., Shah, J., and Klessig, D. 1999. Salicylic acid and disease resistance in plants. *Crit. Rev. Plant Sci.* 18:547–575.

Fought, L., and Kuć, J. 1996. Lack of specificity in plant extracts and chemicals as inducers of systemic resistance in cucumber plants to anthracnose. *J. Phytopathol.* 144:1–6.

Gottstein, H., and Kuć, J. 1989. The induction of systemic resistance to anthracnose in cucumber plants by anthracnose. *Phytopathology.* 79:271–275.

Hammerschmidt, R. 1999. Phytoalexins: what have we learned after 60 years? *Annu. Rev. Phytopathol.* 37:285–306.

Hammerschmidt, R. and Kuć, J. eds. 1995. *Induced Systemic Resistance to Disease in Plants.* Dordrecht, The Netherlands: Kluwer.

Karban, R., and Kuć, J. 1999. Induced resistance against herbivores and pathogens: an overview. In *Induced Plant Defenses Against Pathogens and Herbivores*, eds. A. Agrawal, S. Tuzun, and E. Bent, pp. 1–16. St. Paul, MN: American Phytopathological Society.

Kim, Y., Blee, K., Robins, J., and Anderson, A. 2001. Oxycom under field and laboratory conditions increases resistance responses in plants. *Eur. J. Plant Pathol.* 107:129–136.

Kiraly, Z. 1998. Plant infection-biotic stress. *Ann. New York Acad. Sci.* 851:233–240.

Kuć, J. 1982. Induced immunity to plant disease. *Bioscience* 32:854–860.

Kuć, J. 1984. Phytoalexins and disease resistance mechanisms from a perspective of evolution and adaptation. In *Origin and Development of Adaptation*, pp. 100–118. London: Pitman.

Kuć, J. 1995a. Phytoalexins, stress metabolism and disease resistance in plants. *Annu. Rev. Phytopathol.* 33:275–297.

Kuć, J. 1995b. Induced systemic resistance: an overview. In *Induced Systemic Resistance to Disease in Plants*, eds. R. Hammerschmidt, and J. Kuć, pp. 169–175, Dordrecht, The Netherlands: Kluwer.

Kuć, J. 1997. Molecular aspects of plant responses to pathogens. *Acta Physiol. Plantarum.* 19:551–559.

Kuć, J. 1999. Specificity and lack of specificity as they relate to plant defense compounds and disease control. In *Modern Fungicides and Antifungal Compounds, 12th Internat. Symposium, Rheinhardsbrunn,* eds. P. Russell, and H. Dehne, pp. 31–37. UK: Intercept Ltd.

Kuć, J. 2001. Concepts and direction of induced systemic resistance in plants and its application. *Eur. J. Plant Pathol.* 107:7–12.

Kuć, J., Barnes, E., Daftsios, A., and Williams, E. 1959. The effect of amino acids on susceptibility of apple varieties to scab. *Phytopathology.* 49:313–315.

Lamb, C., and Dixon, R. 1997. The oxidative burst in plant disease resistance. *Annu. Rev. Plant Physiol. Plant Mol. Biol.* 48:251–275.

Lusso, M., and Kuć, J. 1999. Plant responses to pathogens. In *Plant Responses to Environmental Stresses from Phytohormonse to Genome Reorganization*, ed. H. Lerner, pp. 683–706. New York: Marcel Dekker.

Maclennan, D., Kuć, J., and Williams, E. 1963. Chemotherapy of the apple scabe disease with butyric acid derivatives. *Phytopathology.* 53:1261–1266.

Manandhar, H., Lyngs-Jorgensen, H., Mathur, S., and Smedgaard-Peterson, V. 1998. Resistance to rice blast induced by ferric chloride, dipotassium hydrogen phosphate and salicylic acid. *Crop Prot.* 17:323–329.

McDowell, J., and Dangl, J. 2000. Signal transduction in the plant immune response. *Trends Biochem. Sci.* 25:79–82.

Metraux, J.-P. 2001. Systemic acquired resistance and salicylic acid: current state of knowledge. *Eur. J. Plant Pathol.* 107:13–18.

Mucharromah, E., and Kuć, J. 1991. Oxalate and phosphate induce systemic resistance against diseases caused by fungi, bacteria and viruses. *Crop Prot.* 10:256–270.

Muller, K., and Borger, H. 1940. Experimentelle untersuchungen uber die *Phytophthora*-resistance der kartoffel. *Arbeiten Biologischen Anst.* Reichsanstalt, Berlin 23:189–231.

Murphy, A., Gilliland, C., Wong, J., West, D., Singh, D., and Carr, J. 2001. Signal transduction in resistance to plant viruses. *Eur. J. Plant Pathol.* 107:121–128.

Oostendorp, M., Kuz, W., Dietrich, B., and Staub, T. 2001. Induced resistance in plants by chemicals. *Eur. J. Plant Pathol.* 107:19–28.

Ray, J. 1901. Les maladies cryptogramiques des végétaux. *Rev. Gen. Bot.* 13:145–151.

Reuveni, M., Agapropov, V., and Reuveni, R. 1996. Controlling powdery mildew caused by *Sphaerotheca fuliginea* in cucumber by foliar sprays of phosphate and potassium salts. *Crop Prot.* 15:49–53.

Reuveni, R., and Reuveni, M. 1998. Foliar fertilizer therapy: a concept in integrated pest management. *Crop Prot.* 17:111–118.

Ross, A. 1966. Systemic effects of local lesion formation. In *Viruses of Plants*, eds. A. Belmster, and S. Dykstra, pp. 127–150. Amsterdam: North Holland.

Schmidt, A., and Huber, J. 2002. Bulletin IOBC/WPRS 25: 6.

Strobel, N., and Kuć, J. 1995. Chemical and biological inducers of systemic resistance to pathogens protect cucumber and tobacco plants from damage caused by Paraquat and cupric chloride. *Phytopathology.* 85:1306–1310.

Zhender, G., Murphy, J., Sikora, E., and Kloepper, J. 2001. Application of rhizobacateria for induced resistance. *Eur. J. Plant Pathol.* 107:39–50.

3

QTL Analysis of Multigenic Disease Resistance in Plant Breeding

James D. Kelly and Veronica Vallejo

3.1 Introduction

Multigenic or quantitative disease resistance has challenged plant breeders working to develop disease resistant crop cultivars. The challenge to incorporate into new cultivars equivalent levels of resistance that existed in the original genetic resistance stock(s) is formidable, given the apparent complexity of quantitative resistance. Environmental factors, complex multigenic inheritance, plant avoidance, and escape mechanisms combine to hamper the efforts of breeders working to incorporate multigenic resistance into future cultivars. Breeding for quantitative resistance is more formidable than for qualitative resistance traits as more complex and lengthy breeding procedures are needed to effectively incorporate adequate levels of quantitative resistance into new crop cultivars. The expression of quantitative resistance in many instances is partial, not absolute, and the control of resistance appears to be governed by many genes acting cumulatively. The rating of genotypes for disease development in field or greenhouse becomes more subjective due to interactions with environmental and plant morphological factors, requiring additional testing and replications to validate their accuracy. In the literature, many nonspecific and complex resistance mechanisms associated with quantitative resistance have been grouped under the broad general headings of horizontal resistance, polygenic resistance, partial resistance, or durable resistance, which suggests complexity but contributes little to resistance breeding. Current analytical molecular tools, however, are making the breeding of quantitative resistance more effective and new insights on the magnitude and location of such resistance loci may assist plant breeders in better exploiting this type of resistance in future crop cultivars. Quantitative Trait Loci (QTL) analysis is a valuable tool for genome exploration and the investigation of multigenic traits. The focus of this chapter is to review the body of work devoted to the identification of QTL controlling quantitative disease resistance in crops and the exciting implications of the implementation of QTL analysis to dramatically enhance disease resistance breeding. QTL analysis is rapidly changing the way scientists view disease resistance and the time-held concepts and importance of major and minor gene resistance. In order to discuss the implications of QTL analyses in resistance breeding, we first attempt to bring some clarity to the terminology and controversial theories that have historically competed for recognition in the breeding literature.

3.2 Terminology

3.2.1 Complex Multigenic/Quantitative Traits and Durable Resistance

The terms multigenic and quantitative are somewhat interchangeable but multigenic implies knowledge of gene action, hence genotype, whereas quantitative implies characterization based on observation, hence phenotype. Not all quantitative traits are multigenic in terms of gene action as environmental factors combine to influence phenotypic expression of complex traits. As authors, we favor the use of the term quantitative; in most instances, breeders base decisions on phenotype, since gene action of complex traits is not always known. Most, but not all, complex resistance traits are controlled by multiple loci. A complex trait is one that does not fit simple Mendelian ratios (Young, 1996). Resistance phenotypes that do not fit discrete categories and are measured quantitatively are assumed to be controlled by multiple loci referred to as QTL. QTL for resistance refer to locations on the genome that are involved in quantitative resistance, but are not informative of the function (Lindhout, 2002). Quantitative resistance has been assumed to be more durable than resistance conferred by a single dominant gene (Parlevliet, 2002). Durable resistance is resistance that remains effective during prolonged and widespread use in environments favorable for the spread of the pathogen (Johnson, 1984). This definition does not imply gene action or resistance mechanism. It is generally assumed, however, that for resistance to be durable it must be under polygenic control. This term implies the role of many genes with the implication that each "gene" has a small but cumulative effect on the expression of resistance in the host. The explanation is based on the inability of scientists to identify clearly major gene effects controlling resistance. Causes for the inability to identify major gene effects are based on (1) absence of major genes and role of minor genes in resistance expression, (2) large environmental effects on major or minor genes which result in non discrete resistance categories, (3) mixtures of pathogenic races that obscure major gene effects, (4) pathogen interactions, (5) interaction with plant morphological avoidance mechanisms or disease escape due to difference in phenology between genotypes, and (6) possible confusion with tolerance mechanisms where specific genotypes tolerate higher levels of disease infection without a corresponding reduction in productivity.

Durability of resistance is viewed as a quantitative trait as it can range from ephemeral to highly durable (Parlevliet, 2002). Despite the clear recognition that ephemeral resistance is characterized by major gene resistance to those pathogens known as specialists (Lamb et al., 1989), the nature of durable resistance is less clear. Durable resistance can be oligogenic particularly against viral pathogens

(Harrison, 2002), but more commonly resistance is quantitative and durable to those pathogens known as generalists that pathogenize a wide host range. The long held theory that polygenic resistance is more durable (van der Plank, 1968) is now being refuted due to the ability of certain pathogens (*Mycosphaerella graminicola*) to overcome both qualitative and partial resistance in wheat (*Triticum aestivum*; Mundt et al., 2002) and reports that monogenic resistance can be durable (Eenink, 1976). For example, the genetic control of bean common mosaic virus (BCMV) conditioned by the dominant *I* gene (Ali, 1950) has been effective in common bean (*Phaseolus vulgaris*) for over 40 years. No reports exist of breakdown of the *I* gene resistance to new evolving strains of BCMV, despite the extensive deployment of the *I* gene in bean cultivars worldwide.

3.2.2 Horizontal and Vertical Resistance

Since its introduction by van der Plank (1968), the concept of vertical and horizontal disease resistance has been an invaluable hypothesis for plant breeders needing to conceptualize the nature of the disease resistance in a specific crop/pathogen interaction. The need to understand the interaction is essential to formulate a strategy for resistance breeding based on the type of resistance (qualitative or quantitative) present in the host, and the nature and type of variability in the pathogen. Breeders rarely choose the type of resistance with which they work, as factors outside their control influence that decision. Such factors include: the nature of the pathogen (specialist or generalist), host range (wide or species specific), type and availability of resistance mechanisms present in the host (gene-for-gene vertical resistance, nonspecific avoidance), the level of resistance (complete or partial) needed in the crop, and the difficulty of distinguishing partial resistance in the presence of major resistance genes.

van der Plank (1968) defined vertical resistance as race-specific and horizontal resistance as race-nonspecific. The terminology used to describe these types of resistance can be confusing as it includes both the genetic control of resistance and observations on the performance of resistance in the field. The term "quantitative resistance" has often been synonymous with "horizontal resistance", implying, by van der Plank's definition, that quantitative resistance is race-nonspecific. QTL for resistance can be identified using specific races of the pathogen that behave as specialists, but the more common instance is the association of QTL with resistance to a pathogen that is a generalist in its mode of action. Certain QTL are related to strain-specific resistance whereas others are strain-nonspecific (Young, 1996). Qi et al. (1998) mapped QTL for resistance to leaf rust (*Puccinia hordei*) in barley (*Hordeum vulgare*) and identified several QTL (*Rphq1-6*) linked to resistance using a single isolate of *P. hordei*. A subsequent study using another isolate (24) (Qi et al., 1999) found that four other QTL (*Rphq7-10*) were specific for isolate 24 and two QTL (*Rphq5* and *6*) were specific to a different isolate of *P. hordei*. Isolate-specific QTL for resistance have also been found for bacterial wilt (*Pseudomonas solanacearum*) in tomato (*Lycopersicon esculentum*; Danesh and Young, 1994) and late blight (*Phytophthora infestans*) in potato (*Solanum tuberosum*;

Leonards-Schippers et al., 1994). These studies lend support to the "minor-gene-for-gene" hypothesis that there exist small but significant cultivar/isolate interactions (Parlevliet and Zadoks, 1977) that appear qualitative on an individual basis but behave cumulatively in a quantitative manner.

The discovery that QTL for resistance can be race-specific opens the possibility that these QTL are involved in similar resistance mechanisms as major race-specific R-genes. In the concept of race-specificity of major R-genes, elicitor molecules encoded by an *Avr* gene in the pathogen are perceived by the plant cell by binding of this ligand to a receptor encoded by the R-gene. Binding of this ligand by the receptor triggers a signal transduction pathway leading to the hypersensitive response (HR), which is characterized as accelerated, localized, plant cell death, and an incompatible reaction (Hammond-Kosack and Jones, 1997). Vleeshouwers et al. (2000) studied the interactions between *P. infestans* and *Solanum* spp. by examining the differential reactions of a diverse series of wild species. They found that in partially resistant species, HR was induced between 16 and 46 hours, and had variable lesions of five or more dead cells from which, in some cases, hyphae were able to escape and establish disease. These results, and other studies discussed in Sections 3.8 and 3.9 of this chapter, indicate that the HR of the partially resistant *Solanum* species used was quantitative in nature. Partial resistance refers to quantitative resistance not based on HR (Parlevliet, 1975); thus, partial resistance should not be used synonymously with QTL unless the type of gene action is known.

The types of mechanisms functional in horizontal resistance are commonly referred to in the literature as multigenic or polygenic. A more appropriate terminology that would benefit breeders in distinguishing the types of host resistance is based on the classification of the trait as either qualitative or quantitative resistance. Breeders are familiar with these types of traits and can formulate effective breeding schemes to incorporate such traits into new cultivars. When treated as quantitative resistance, the breeder has a body of information on the expression of these types of traits and methodologies to effectively manipulate such traits (Hallauer and Miranda, 1981). The basis for quantitative inheritance is as complex as the traits being studied since the range of traits under quantitative control in most crops plants dramatically out-number those under qualitative control. Progress in the improvement of quantitative traits has lagged behind similar efforts to improve simply inherited traits due to their complexity, lack of complete expression, inconsistent screening methods, and the need for widespread multilocation testing. The lack of progress is best understood when differences in inheritance patterns between qualitative and quantitative resistance are compared.

The relative contribution and stability of the QTL to disease resistance is another important criterion of QTL analysis. Quantitative genetic theory implies that many minor genes control quantitative traits, but what is not known, is the differential effect of different minor genes. In the case of disease resistance, QTL analysis reveals that resistance may be controlled by a few QTL with major effect (high coefficient of determination, $R^2 > 35\%$), and a number of QTL with relatively minor effects ($R^2 < 15\%$) (Kolb et al., 2001; Young, 1996). For example, one QTL conditioning resistance to downy mildew (*Sclerospora graminicola*) in pearl millet (*Pennisetum glaucum*) accounted for 60% of the phenotypic variation whereas another accounted for only 16% of the variation associated with resistance (Jones et al., 2002). Clearly such information provides breeders with a clear choice on which QTL to emphasize in breeding for resistance, along with the tools to achieve that objective.Other factors that influence the effectiveness of QTL analysis are the potential interaction between QTL, and their stability across environments and populations, and possible linkages with other traits. Generally breeders shy away from population and environmentally sensitive QTL as they are too restrictive to the overall goals of most breeding programs (Asins, 2002).

3.3 Historical Perspective

Due to the complex nature of inheritance, classical Mendelian techniques were not applicable to quantitative traits, and in the early part of the 20th century, quantitative genetics emerged as a specialized branch of genetics to address issues related to traits under quantitative genetic control. Until recently, quantitative genetics relied on biometrical approaches that deal mainly with the characterization of multiple factors affecting a quantitative trait and partition the phenotypic variance into its genotypic and environment components (Hallauer and Miranda, 1981; Sprague, 1966). From these statistical procedures, several parameters could be estimated including the approximate number of loci influencing the trait, gene action, gain from selection estimates, and the degree to which the loci interacted with other loci and the environment to produce the observed phenotype. These approaches, however, were limited in the sense that they were not able to characterize any *one* specific locus that contributed to the trait, either its location or size of effect. The biometrical information did provide breeders with information on type of gene action that suggested the most appropriate breeding methods to use to optimize or fix favorable gene action controlling the quantitative trait. Many of the mating procedures, however, were limited to specific crops such as maize (*Zea mays*) due to the pollination mechanisms and reproductive biology of the crop.

Sax (1923), accredited with being the first to describe the theory of mapping quantitative traits, showed that loci involved in a quantitative trait (seed size in common bean) were associated with a qualitative trait (seed-coat pigmentation). Another pioneer in the characterization of quantitative traits, Thoday (1961), suggested the need to exploit the association with qualitative traits as a means to locate the polygenes involved in the control of a complex trait. He astutely noted, however, that the limiting factor in using this strategy was the availability of suitable markers. With the advent of molecular markers that are sufficiently numerous to provide adequate genome coverage, this is no longer a limitation and therefore, QTL mapping, at least in theory, can resolve any additive gene of small effect as Mendelian through associations with a marker locus. The era of molecular markers commenced with the discovery of isozyme techniques (Hunter, 1957; Smithies, 1955) and quickly progressed to DNA-based marker systems, first of which were

RFLP (Botstein et al., 1980) followed by PCR-based molecular markers (RAPD, SCAR, SSR, AFLP; Michelmore et al., 1991; Vos et al., 1995; Weber and May, 1989; Williams et al., 1990). For a more complete review of the different marker systems available for mapping, see Staub et al. (1996).

The basic concept of QTL mapping is very simple: to find significant associations between marker genotypes and quantitative phenotypes in a large controlled, experimental cross between two parental genotypes. A conceptual diagram of QTL mapping is provided by Young (1996). In practice, however, there are many issues: (1) population size, (2) parental selection, (3) population type, (4) marker efficiency, (5) phenotypic data that breeders need to consider, (6) map density, and (7) data analysis that influence QTL analysis.

3.4 Mapping Considerations

3.4.1 Population Size

The purpose of a mapping population, in essence, is to simplify partitioning of genetic variance components to provide a clear genetic interpretation and genomic data analysis. The mating design of a mapping population is important for making the relationships among the polymorphic markers and traits of interest detectable and tractable. The effective population size for QTL analysis is a very important consideration that has a direct impact on the resolution of the map and the accuracy of the QTL location. Population size also affects the genetic gains breeders achieve using marker-assisted selection (MAS). If the population is not large enough (<100 individuals) in a QTL analysis, certain putative QTL will not be detected and therefore gains using these candidate QTL in MAS will be reduced. Large population sizes (>200 individuals) are not always feasible due to the space and time constraints on the researcher, therefore, some strategies have been implemented to maximize information from smaller populations, including selective genotyping (Lander and Botstein, 1989) and DNA pooling of similar phenotypes (Michelmore et al., 1991).

In QTL analyses, selective genotyping and bulked segregant analysis (BSA) (Michelmore et al., 1991) have been utilized to efficiently screen large numbers of polymorphic markers, without having to genotype entire segregating populations. Selective genotyping involves the identification of a subset, usually 10–14% of the genotypes that possess extreme phenotypes of the population. By this method, breeders can obtain equal or greater information about QTL than from mapping of randomly chosen individuals. A small percentage of the total genotypes that exhibit extreme phenotypic values for the trait of interest can be grouped (bulked) together, and either analyzed as individuals, or through BSA, where the DNA of the similar phenotypes are pooled. BSA is most often used when mapping genes with major effect. BSA may have limited application to QTL analysis due to factors such as dominance and non-Mendelian segregation that decrease the effectiveness. Selective genotyping and BSA has been used successfully in the identification of QTL for quantitatively-inherited traits related to disease resistance (Chen et al., 1994; Miklas et al., 1996; Schneider et al., 2001). Another application of BSA in

QTL analysis is in fine mapping of a QTL position. To find additional markers linked to a particular genomic region, pools are created based on alternate alleles at a marker locus, providing a very efficient method for screening large numbers of markers to saturate a QTL region (Giovannoni et al., 1991). Paterson (1998) states that rare QTL with large effects can be fixed in the phenotypically extreme individuals, and therefore may be detected as a chromosome segment polymorphic between contrasting DNA pools. Most QTL with smaller effects, however, will remain heterogeneous in the DNA pools and will not be detected. To detect many QTL with smaller effects, Paterson suggests a comprehensive mapping approach. Despite the view that DNA pooling might be useful in the identification of QTL of very large effect but unlikely to permit the comprehensive identification of the majority of QTL affecting a complex trait (Wang and Paterson, 1994), breeders have successfully used BSA in the identification of QTL for disease resistance (Miklas et al., 1996; Young, 1996).

3.4.2 Selection of Parents

When the objective of the research is to search for genes controlling a particular disease resistance trait, adequate genetic variation for resistance must exist between the parents. There must be sufficient variation between the parents at the DNA sequence level and at the phenotypic level for the trait of interest. The choice of parents may be restricted by the availability of resistance but breeders usually make a decision as to the level of diversity of the parents of the mapping population. Wider diversity between parents may be desirable to allow the mapping of traits in addition to the targeted resistance source, or breeders may need to work with genetically similar parents to avoid the interaction of other traits such as plant morphology and phenology on the expression of resistance in the field (Lindhout, 2002).

3.4.3 Population Type

The most commonly used mating types in QTL analyses are F_2 and backcross (BC) populations. The disadvantage of these types of populations is that they are unique and progeny cannot be propagated, so breeders are unable to recreate the same population for further testing. Recombinant inbred lines (RIL) and double haploid (DH) homozygous lines can be used to avoid this problem because the lines are maintained by selfing, allowing marker-trait associations to be scored across multiple environments in a completely homozygous background. RIL are developed initially by self-pollinating the F_2 generation for up to 10 generations using the single-seed descent method (Burr and Burr, 1991). DH lines are produced by the induction of diploid gametes by tissue culture. In this case, haploid gametes from F_1 parents are chemically treated to induce the doubling of the chromosome number (Jensen, 1989; Knapp, 1991; Knapp et al., 1990). The technology to generate DH lines, however, is not available in all crops. Although RIL populations take longer to generate, they have become the cornerstone of many QTL analyses as they can be easily duplicated for widespread testing. The utility of phenotypic-based DNA pools on the isolation of QTL in different genetic populations was assessed by Wang and Paterson (1994). The effects of population size, portion of population selected, magnitude of phenotypic effects of individual QTL alleles (QTL allele effects) and effects of both dominance and deviations from Mendelian segregation ratios were considered. Backcross populations were better than F_2 populations, but were less efficient than RIL or DH populations in detecting QTL. To detect QTL using phenotypic-based DNA pools, Wang and Paterson (1994) suggested using wide crosses, large homozygous populations such as RIL and DH populations where the replication of phenotypic data is easily facilitated by the use of homozygous populations.

3.4.4 Marker Efficiencies

The choice of markers is dependent on those available in each crop, but PCR markers are the clear choice over RFLP markers because of cost and convenience. Many of the major crops such as soybean (*Glycine max*) have numerous microsatellite or SSR markers (Cregan et al., 1999a) and/or SNP and CAPS markers available for mapping. In minor crops where sequence-based markers are not yet available, breeders may utilize AFLP markers or even RAPD markers. Different marker systems have varying levels of resolution to detect genome variations. Codominant markers are generally preferred over dominant markers in certain populations. Dominant marker types are not recommended for F_2 populations because in repulsion linkage phase the dominant markers provide low information content on linkage (Paterson, 1998). This disadvantage is less acute when mapping more homozygous RIL populations. In a BC population, if the recurrent parent is recessive for the dominance loci, dominant and codominant markers are equivalent in terms of genomic analysis.

3.4.5 Phenotypic Data

Limitations of QTL analyses rarely lie in the lack or inability to find useful polymorphic markers associated with disease resistance, but reside in the accuracy of trait analysis. The collection of the phenotypic data used to conduct the analysis is challenging in terms of the establishment of rating scales for disease evaluation, actual evaluations and data collection, seasonal and location effects of the environment and the structure and size of the genetic population being evaluated. All of these factors can contribute unexplained variability to the data set and need to be considered by the researcher conducting the analysis. In the vast majority of cases the weakness of the QTL analysis resides in the phenotypic data used to conduct the analysis and less in the density of markers available for mapping. The most common rationale in mapping disease resistance traits is to generate a segregating population where individuals exhibiting the extreme expression(s) of the resistance trait can be identified for mapping purposes. In the case of the oligogenic traits, such contrasting individuals can easily be identified in early generations such as the F_2, whereas in mapping of quantitative resistance, individuals can only be identified on a progeny basis in later, more homozygous, generations. Since the expression of quantitative resistance can be effected by environmental conditions, the resistance trait needs to be measured over locations and/or years. The need to create, replicate and evaluate more homozygous lines results in significant delays in all QTL analyses of quantitative resistance. All sound QTL analyses must be based on clear reproducible quantitative phenotypic data generated from the genetic population segregating for the resistance trait. Breeders need to be aware that many QTL analyses fail to identify true or significant effects simply due to weak or questionable phenotypic data collected on the disease resistance trait. Marker-assisted selection must be based on a data set that is uncompromised in quality and reproducibility.

3.4.6 Map Density

QTL discovery may be conducted with or without using an existing genetic linkage map. Not all crop species, such as the octoploid strawberry (*Fragaria* x *ananassa*), have a well-saturated linkage map with even distribution of markers across the genome. In such instances, QTL discovery is accomplished by simply finding a statistically significant association between a phenotype and a marker. The marker is often detected by screening random primers against a population segregating for the quantitative trait. Although this approach may appear inefficient, valuable QTL for resistance to root rot (*Fusarium solani* f. sp. *phaseoli*) have been discovered in common bean by this method (Schneider et al., 2001). In crop species such as soybean that do have linkage maps with even distribution of markers across the genome (Cregan et al., 1999b), marker density can have an impact on the accuracy or resolution of the QTL location. In general, markers should be evenly distributed with at least one marker every 5 cm. Genome coverage and map density can be influenced by a number of different factors: size of the genome, population size and type, mapping strategy used, distribution of crossovers in the genome, and number of markers (Liu, 1998).

3.4.7 Data Analysis

Three main methods of data analysis are generally used in evaluating linkage between markers and a phenotype. These methods include: single-marker analysis, interval mapping, and composite interval mapping.

Single-Marker Analysis

In single-marker analysis, the trait value distribution is examined separately for each marker locus. This can be done using a simple t-test, analysis of variance, linear regression, or likelihood ratio test and maximum likelihood estimation. Due to the simplicity of this analysis, SAS (Statistical Analysis Software, SAS Institute, Cary, NC) can be used. There are a few disadvantages of this type of analysis. One disadvantage is that the QTL location and the putative QTL genotypic means are confounded, which reduces the statistical power of this analysis. This is a particularly important consideration when working with a low-density map. Another disadvantage to single-marker analysis is that the QTL cannot be precisely mapped due to the non-independence among the hypothesis tests for linked markers that confound QTL effect and position (Liu, 1998). This method is therefore more suited to a study where the goal is to simply detect QTL linked to a marker rather than to accurately map and estimate their effects.

Interval Mapping

The limitations of single-marker analysis prompted Lander and Botstein (1986a; 1986b; 1989) to propose an interval mapping (IM) method to position QTL. In IM, a separate analysis is performed for each pair of marker loci using one of the three approaches: likelihood (Lander and Botstein, 1989), regression (Knapp et al., 1990), and/or a combination of both methods. The IM method provides increased power of detection of QTL and more accurate QTL positioning when compared to single-marker analysis (Liu, 1998). The disadvantages of this method are that the number of QTL cannot be resolved, the exact position of the QTL cannot be determined, and the statistical power, although higher than single-marker analysis, is still relatively low. These problems can result from linkage or interactions between QTL, and limited information in the model (Liu, 1998). The outcome of this method is highly influenced by background QTL that result in low wide peaks which mask the appearance and positioning of multiple linked QTL.

Composite Interval Mapping

Composite interval mapping (CIM) is a combination of interval mapping and multiple linear regression (Zeng, 1993, 1994). This method considers a marker interval and a few chosen markers in each analysis. These chosen markers are used to reduce background effects of other linked QTL in the analysis of a marker interval. The result of CIM is to define the most likely position of the QTL with more precision and greatly increase the resolution of the analysis, which is the most important advantage of CIM over single-marker analysis and IM. Since there are more variables in the model, CIM is more informative and efficient, and results can be presented using the log likelihood ratio test statistic plot and the LOD score plot for all possible genome positions.

3.5 Applications of QTL Analysis to Disease Resistance Breeding

QTL analysis has enhanced our understanding of quantitative resistance in a number of key areas by revealing the location and size of loci controlling disease resistance. Locating resistance loci has confirmed the interaction between resistance traits that control physiological processes and those traits influencing plant morphology and phenology that control disease avoidance and/or escape in a field setting. By locating loci for quantitative disease resistance on different linkage groups, QTL analyses provide unique opportunities to pyramid resistance loci in order to restore higher levels of resistance lost in many cases after crossing with a highly resistant source (Vertifolia effect; van der Plank, 1968). While the practical application of MAS for quantitative traits has yet to be fully realized in breeding, many studies recognize its potential to facilitate improved disease resistance controlled by quantitative traits (Asins, 2002; Faris et al., 1999; Jones et al., 2002; Kolb et al., 2001; Lindhout, 2002; Lubberstedt et al., 1998; Mangin et al., 1999; Miklas et al., 1998; Pilet et al., 1998; Schechert et al., 1999). QTL-marker associations may also provide a basis for a greater understanding of quantitative disease resistance through the identification of loci that influence resistance to more than one disease (Ariyarathne et al., 1999). The application of MAS in breeding for quantitative resistance should have the most impact in breeding for resistance to soilborne pathogens such as *Fusarium* and *Sclerotinia*. Screening for resistance in the field is both destructive and complicated by the interaction of other soil borne pathogens (root rot complex), seasonal environmental factors, and plant morphological traits that contribute to disease avoidance or escape which hinders the normal selection procedures (Tanksley et al., 1989). Replacing laborious screening of quantitatively inherited traits with MAS has several advantages in a breeding program. Breeding for quantitative resistance can be enhanced with the discovery of QTL for resistance that would allow for the indirect selection of resistance without confounding effects of environmental factors. In the absence of candidate QTL, breeders were often forced to cross "blindly" in the hope that they were combining resistance sources (loci) but with the discovery of QTL, breeders can target specific loci on different linkage groups and combine these in future resistant cultivars. The breeding literature has many examples of attempts to transfer quantitative resistance to potential new cultivars that have resulted in the transfer of partial levels of resistance. When breeders lack the tools to identify putative QTL involved in resistance they are equally ineffective in transferring all the resistance QTL to future cultivars.

3.6 Use of Multitrait Bulking Methods in QTL Analysis

Disease development can be influenced by plant morphological and phenological factors that must be considered by breeders working with quantitative disease resistance. For example, a number of agronomic traits, including growth habit, canopy height and width, branching pattern, lodging, days to flower and maturity have been shown to be significantly associated with white mold (*Sclerotinia sclerotiorum*) development in common bean (Kolkman and Kelly, 2002). The interaction of such traits on the expression of the disease resistance trait complicates breeding for resistance. Morphological traits such as plant architecture afford disease avoidance, whereas phenological traits such as early flowering afford disease escapes in many instances (Coyne, 1980; Kolkman and Kelly, 2002). Since both types of traits influence disease reactions in the field, both need to be considered in a QTL analysis of specific disease resistance traits. In the selective genotyping of quantitative resistance, the identification of individuals with extreme expression of disease resistance may result in the selection of individuals that exhibit undesirable morphological and/or phenological traits due to interaction of these traits on the expression of disease resistance. Highly resistant individuals may result not from the expression of true physiological resistance but from a combination of such agronomically undesirable traits as short plant stature, or extremely early flowering or maturity that result in individuals with no agronomic or yield potential for commercial production. Such individuals serve no potential as parents, or cultivars as their agronomic weaknesses outweigh their low disease resistance ratings. This problem becomes particularly acute in QTL analyses where selective genotyping is used to assist the breeder in identifying the extreme expression(s) of disease resistance, but results in an analysis of the extreme expression of agronomic traits that escape or avoid the disease, resulting in the mapping of traits associated with agronomically inferior individuals.

In the mapping of QTL associated with white mold resistance in common bean, DNA bulks comprised solely of a small number of lines in the extreme phenotypes may not adequately represent useful resistant genotypes in the population. Since the use of selective genotyping for a trait as complex as resistance to white mold may be hindered by the limitation of a set of DNA bulks based on disease reaction alone, Kolkman and Kelly (2003) compared the efficiency of single and multitrait bulking strategy for the identification of QTL associated with white mold resistance. The multitrait bulking strategy utilized multiple traits (MT) to develop contrasting DNA bulks for use in genotyping as opposed to traditional single trait bulks. The traits selected in the MT bulks included disease reaction and also flowering range and yield to avoid the indirect selection of resistant, low-yielding genotypes with inferior agronomic traits such as very early or late flowering that would effect local adaptation (Kolkman and Kelly, 2003). The results of the study indicated that both single- and multi- trait bulking strategies identified QTL for resistance to white mold on one linkage group. However, eight molecular markers on a second linkage group B7 were identified using the MT bulks, whereas the single-trait bulk for disease incidence alone would not have identified the most closely linked markers to the QTL conferring resistance to white mold on B7. The disease ratings in the selected individuals within the resistant MT bulks were higher than those of the single-trait disease resistant bulk, suggesting that the disease resistant bulk may have included genotypes with greater avoidance mechanisms that significantly reduced yield but were potentially commercially unproductive. The authors concluded that genotyping a chosen set of individuals with specific phenotypes, based on a priori knowledge of the traits that are segregating in the population that may affect the desired phenotype, was an efficient method to detect markers linked to the resistance phenotype which would not have been detected in the single-trait disease resistant bulks alone (Kolkman and Kelly, 2003).

In soybean, two of the three QTL associated with disease resistance to *S. sclerotiorum*, were also associated with plant avoidance mechanisms, such as plant height, lodging, and date of flowering (Kim and Diers, 2000). The authors speculate that the third QTL, which was not significantly associated with escape mechanisms, may be involved in physiological resistance to *S. sclerotiorum*. Plant avoidance mechanisms may also play an important role in resistance to *S. sclerotiorum* in sunflower (*Helianthus annuus*). QTL accounted for up to 60% of the leaf resistance and up to 38% of the capitulum resistance in sunflower. Apical branching pattern was suggested as exhibiting the best resistance to infection of the capitulum (Mestries et al., 1998), whereas the association between days to flowering and resistance to *S. sclerotiorum* in sunflower was dependent upon the population (Castano et al., 1993). Clearly, MAS allows for the identification and selection of superior genotypes without having to employ undue effort in phenotyping large number of individuals. The difficulty in detection of desirable phenotypes, due to factors such as environmental variation, hinders normal selection procedures for important quantitative traits, and increases the importance of MAS. DNA bulks comprised solely of a small number of lines in the extreme phenotypes may not adequately represent resistant genotypes in the population. DNA pooling strategies based on a priori knowledge about the population should help resolve useful markers linked to QTL, and discern the location of QTL regions (Wang and Paterson, 1994). Genotyping multiple traits that are related to the trait of interest have been effective in identifying QTL that may not be detected through screening extreme phenotypes (Kolkman and Kelly, 2003; Ronin et al., 1998).

3.7 Identification of Novel Disease Resistance Sources Using QTL Analysis

Interspecific hybridization has been used to improve disease resistance in many crop species (Hadley and Openshaw, 1980). The inheritance of the resistance is not always known as breeders rarely conduct genetic studies in the alien species but focus on the successful transfer of the resistance to the cultivated species. As a result the assumption is often made that resistance in the alien species is novel and worth the substantial efforts needed to transfer resistance. Mapping of QTL has provided new information on resistance sources integrated from other species. Lack of adequate levels of resistance to common bacterial blight (CBB; *Xanthomonas axonopodis* pv. *phaseoli*) in common bean has forced bean breeders to find resistance in the related tepary bean (*P. acutifolius*) species. Impetus to use interspecific crosses came from early work by Honma (1956) who reported a successful interspecific hybrid between common and tepary bean that has become the focus of CBB resistance breeding for the last 40 years. Progress in breeding for resistance to CBB in common bean has been modest as resistance is quantitative, largely influenced by environment and pathotype, and functional in different organs, leaf, seed, or pod depending on resistance source(s). The complexity of resistance to CBB where different QTL conditioned resistance in young and adult tissues to different strains of the pathogen, or where one genomic region possessed a factor(s) which influenced resistance in all three tissues, seeds, leaves, and pods, while another QTL only influenced resistance within a single plant organ has been demonstrated (Jung et al., 1999; Kelly et al., 2003). Such complexity in disease expression has limited progress in breeding for resistance to CBB.

Despite these difficulties, QTL analyses of resistance to CBB in common bean has resulted in the identification of four major QTL associated with resistance on four different linkage groups that provides breeders with the possibility of combining QTL to enhance resistance. One of the most revealing findings provided by QTL analyses concerns the resistance source originally believed to have been derived from the tepary bean (Honma, 1956). This source has proven to be of common bean origin, not tepary as previously thought (Miklas et al., 2003). The QTL for resistance on linkage group B10 is found only in common bean germplasm and is absent from all tepary bean resistance sources tested (Miklas et al., 2003). The resistance QTL on linkage group B10 co-segregated with resistance in common bean progeny tested for reaction to CBB confirming that resistance was not derived from tepary bean in the original cross. QTL mapping, therefore, provides an opportunity to verify the uniqueness of resistance sources prior to using them directly in breeding programs.

Another advantage of QTL analysis is the identification of previously unknown resistance sources. QTL have revealed that different genetic sources present in related species may not always represent new or novel resistance loci. These exotic sources may be assumed to be unique, as the resistance sources are not characterized if present in a related species. Genetic studies are not routinely conducted on alien or exotic species to determine their relationship, so a savings in time and resources results from knowing if an exotic resistance source does or does not carry a unique QTL. Given the lack of adequate resistance sources to CBB in common bean, resistance has been successfully introgressed from different tepary accessions into common bean (McElroy, 1985; Scott and Michaels, 1992). QTL analyses of these resistance sources for CBB derived from different interspecific tepary bean sources mapped to linkage groups B6 and B8 on the common bean map (Miklas et al., 2000). These resistance sources are clearly derived from tepary bean, as the QTL are absent in susceptible common bean genotypes and present in resistant tepary bean germplasm. One of these sources with QTL on B6 and B8 is XAN 159 (McElroy, 1985), the most widely deployed source of resistance currently used by bean breeders. A second tepary derived resistance source OAC 88-1 was developed independently (Scott and Michaels, 1992). Since no genetic studies were conducted on the tepary bean sources, the assumption that different resistance sources had been successfully introgressed into common bean persisted. This assumption has proved false as the QTL from XAN 159 known as SU91 mapped to the same location on B8 as the R7313, the QTL from OAC 88-1 (Miklas et al., 2000). This represents a duplication of effort and resources, given the difficulty of making interspecific crosses between tepary and common bean and the need to employ embryo rescue in the procedure. Apparently, the same source of resistance was independently introgressed into common bean without prior knowledge of the genetic similarities of the tepary bean accessions. QTL analyses can serve a vital role in distinguishing resistance sources based on their location in the genome.

Those QTL that map to the same location most likely condition similar resistance; with that knowledge, intelligent decisions can be made on the choice of sources to introgress when time consuming interspecific crosses are required. Finally, the potential to pyramid four QTL, two from common bean and two from tepary bean, into a single bean genotype opens up the exciting possibility of developing common bean cultivars with CBB resistance levels (Singh and Munoz, 1999) equivalent to those in the original tepary bean sources.

3.8 Colocalization of QTL with Resistance Genes

The focus of QTL analysis has changed recently from simply discovery of QTL associated with quantitative disease resistance to determining the biological function underlying the QTL. Knowledge about the biological functions of QTL will help breeders develop cultivars with more durable resistance as well as elucidate the mechanisms behind quantitative resistance. Understanding the function of genes that confer quantitative resistance will provide breeders with mechanistic information that can be used to make more informed and prudent decisions as to why QTL for resistance may be more durable. The term QTL is not very descriptive, only referring to a specific genomic location involved in the quantitative disease resistance, and does not provide information about the function of those genes. By studying the function of other genes that map to the same genomic regions as QTL, information on the mechanisms influencing resistance conferred by QTL may be elucidated. The role that some QTL play in resistance through their association with the HR (Vleeshouwers et al., 2000) may provide information on the biological function of QTL in the resistance reaction.

3.8.1 QTL that Colocalize with Major Genes for Resistance

QTL may be Allelic Variants of Qualitative Resistance Genes

There are two broad categories of genes involved in plant defense response: *R*-genes (those involved in the recognition of the pathogen), and defense response (DR) genes (those involved in the general defense response of the plant). One instance of colocalization is the mapping of a QTL to the same genomic regions where previously mapped *R*-genes reside. The existence of quantitative and qualitative resistance genes in the same genomic regions favors the consideration that QTL, which confer intermediate resistance, may correspond to allelic versions of qualitative resistance genes. This is consistent with the hypothesis that mutant alleles of qualitative genes that affect quantitative traits are one extreme in the spectrum of alleles (Robertson, 1989). A possible explanation at the molecular level is that qualitative mutants may result from loss of function mutations whereas quantitative alleles may result from mutations that produce a less efficient gene product resulting in differences in phenotypes. Support for this theory comes from a study on rice (*Oryza sativa*) using 20 RFLP marker loci associated with quantitative resistance to rice blast *Pyricularia oryzae* (Wang et al., 1994). Among the markers, *RG16*, located on chromosome 11, was also associated with complete resistance to the same rice blast isolate. In addition, three other marker loci associated with partial resistance, *RG64*, *RG869B*, and *RG333*, were found to be linked to the previously mapped *R*-genes *Pi-2(t)*, *Pi-4(t)*, and *Pi-zh*, respectively (Yu et al., 1991). The results of this study indicate that more than one resistance gene for rice blast may reside in this region of chromosome 11. In other crops, QTL for *P. infestans* resistance and the dominant race specific allele *R1* have been identified on chromosome V of potato, in the interval formed by markers *GP21* and *GP179* (Leonards-Schippers et al., 1994). The possibility that these QTL are alleles of *R*-genes suggests that the QTL may have a similar function in the resistance mechanism. The fact that some QTL have been discovered to be race-specific (Qi et al., 1999) and involved in HR (Vleeshouwers et al., 2000) also lends support to this theory.

Numerous genes involved in resistance have been isolated and cloned. Sequence analysis has revealed that there exist four major classes of *R*-genes and that their functional domains are highly conserved (Bent, 1996). Degenerate primers, designed from these consensus sequences, are used to amplify *R*-gene analogues (RGAs) in the candidate gene approach. Some of these RGAs have also mapped to regions containing quantitative and/or qualitative resistance loci. A candidate gene approach was used to identify and map QTL for resistance to anthracnose (*Colletotrichum lindemuthianum*) in common bean (Geffroy et al., 2000). Using a RIL population, and candidate genes, 10 QTL for resistance were identified. The candidate genes that were used included pathogen recognition genes, such as *R*-genes and RGAs, and general DR genes. Three of the QTL, linked to marker loci *D1020*, *D1861*, and *D1512*, on linkage groups B3, B7, and B11 respectively, were also associated with previously mapped QTL for resistance to CBB (Nodari et al., 1993). *D1512*, on linkage group B11, is also located in the same genomic region as the qualitative *Co-2* gene for resistance to anthracnose, and a family of leucine-rich repeat (LRR) sequences (Geffroy et al., 1998).

QTL have also been mapped to resistance gene clusters in different crops. Comparative mapping is a strategy that has been increasingly more feasible as more maps are generated across diverse taxa. The fundamental concept is based on the finding that diverse taxa with common taxonomic families often share similar gene order over large chromosomal segments. Therefore, QTL mapped in one species may be located at the same chromosomal region in another evolutionarily related species (Grube et al., 2000). Using comparative mapping, four QTL for resistance to *Erwinia carotovora* ssp. *atroseptica*, mapped to genomic segments in potato containing RGAs, qualitative and quantitative factors conditioning resistance to different pathogens that attack potato, tomato, and tobacco (*Nicotiana tabacum*; Zimnoch-Guzowska et al., 2000). Shared markers between the potato linkage map generated by this study and other potato and tomato maps, were used as anchors to align the maps and allow positional comparisons. *Eca1A*, a major QTL for resistance to *E. carotovora* ssp. *atroseptica*, is located in a similar genomic region in potato as the *Cf* gene family in tomato. The *Cf* genes in tomato are *R*-genes that confer resistance to the fungal pathogen *Cladosporium fulvum*. The QTL, *Eca6A*, is situated in a genomic segment, which in tomato contains the qualitative genes for nematode (*Meloidogyne* spp.) resistance, *Mi*, and the *Cf-2* gene. In addition, *Eca11A* maps to the same region in potato as another QTL for resistance to *P. infestans*, the virus resistance gene *Ry*, the *Synchytrium endobioticumm* resistance gene, and to the virus resistance *N* gene in tobacco. Several factors may contribute to the clustering effect that has been observed between resistance-related genes (resistance-related includes quantitative, and qualitative genes and RGAs). Clustering could be an anomaly resulting from small population sizes or insufficient number of markers used to precisely map the loci. If the genes were located in an area of reduced recombination rate this would also result in a cluster at that region. Another important factor is that not all resistance-related genes have been identified and mapped, therefore, it is not possible to make absolute conclusions about the clustering of resistance-related genes. The clustering of resistance-related loci in species of the *Solanaceae* family was supported by another comparative mapping study in tomato, potato, and pepper (*Capsicum annuum*) where several cross-generic clusters were observed (Grube et al., 2000).

Most pathologists would agree that the plant developmental stage used to evaluate resistance influences the type of resistance that is detected. Adult plant resistance is generally considered to be quantitative and distinct from the qualitative resistance detected in a seedling assay. Two QTL that confer seedling resistance to three isolates of stripe rust (*Puccinia striiformis* f. sp. *hordei*) in barley mapped to the same region as two of the four QTL that conferred adult plant resistance (Castro et al., 2002). Coincident QTL detected in distinct assays in different plant stages suggest different gene action, yet QTL analysis illustrated colocalization. Linkage mapping of quantitative resistance has revealed other examples of colocalization of major R-genes and QTL for resistance in a wide array of host/pathogen interactions: powdery mildew (*Erysiphe graminis*) in barley (Backes et al., 1996), potyvirus in pepper (Caranta et al., 1997), northern corn leaf blight (*Exserohilum turcicum*) in maize (Freymark et al., 1993), powdery mildew (*Blumeria graminis*) in wheat (Keller et al., 1999), and cyst nematode (*Globodera* spp.) in potato (van der Voort et al., 1998).

QTL may be Defeated Qualitative Resistance Genes

The term defeated, or ghost genes, was first introduced by Riley (1973) to explain the minor contribution to resistance of major genes that were defeated by virulent strains of a pathogen. Defeated genes were visualized as contributing to quantitative resistance controlled by polygenes. Martin and Ellingboe (1976) proposed that defeated major genes may conserve residual resistance effects. They showed that *Pm* genes that had been overcome by virulent isolates of powdery mildew still contributed to the partial resistance in wheat. Nass et al. (1981) also found residual resistance effects for two *Pm* resistance loci, *Pm3c* and *Pm4b*, but not for *Pm2* and *Pm5* loci in wheat. Keller et al. (1999), however, reported that the *Pm5* gene showed a large effect, despite the use of virulent races of powdery mildew being present in the mixture of isolates used. They concluded that the detected effect of the *Pm5* gene could be explained by the reduced growth of isolates that contain *avrPm5* virulence gene due to reduced fitness and/or the delayed spread of the *Pm5* virulent isolates due to the residual effect of the defeated gene. This residual effect could result from the limited expression of the overcome gene (Nelson, 1978). Li et al. (1999) used a population of 315 RILs and a linkage map that consisted of 182 RFLP markers to map the major gene (*Xa4*) and 10 QTL linked to resistance to bacterial blight, caused by *Xanthomonas oryzae* pv. *oryzae* (*Xoo*) in rice. They found that most QTL mapped to genomic regions where major genes or other QTL for *Xoo* resistance were located. In addition, they discovered that the $Xa4^T$ locus, an allele of *Xa4* from the cultivar, "Tequing", behaved as a dominant major resistance gene against strains CR4 and CX08 and as a recessive QTL against strain CR6 of *Xoo*. The resistance conferred by the $Xa4^T$ allele, however, was overcome by the mutation at the *avrXa4* locus in the virulent strain CR6. The *Xa4* gene is considered to be a defeated major resistance gene that plays a key role in rice–*Xoo* interaction (Narayanan et al., 2002). These data suggest that the major genes and QTL for resistance, in this instance, may be the same gene and supports the hypothesis that defeated major resistance genes have residual effects against different races of the same pathogen.

QTL may be Members of Multigene Families

An alternative hypothesis is that QTL and major genes that map to the same genomic region are different members of a cluster of resistance gene families. Many studies have demonstrated that major genes are often members of clustered multigene families (Michelmore and Meyers, 1998; Parniske et al., 1997; Pryor and Ellis, 1993; Ronald, 1998; Song et al., 1997; Yu et al., 1996). These clusters are composed of linked, and evolutionarily related, resistance specificities, therefore, a potential for a structural and functional similarity between qualitative genes and the resistance QTL that map to the same region exists. Grube et al. (2000) showed that major resistance genes can occur in transgeneric clusters with QTL in the *Solanaceae*, suggesting that sequence similarity and probably function similarities exist between qualitative and quantitative genes for resistance. Therefore, QTL that map to the same genomic regions as specific resistance genes may be involved in pathogen recognition. In rice, the major resistance gene *Xa21* is a member of a multigene family located on chromosome 11 that confers race specific resistance to *Xoo*. *Xa21* encodes an extracellular LRR domain and a serine/threonine kinase that is believed to determine the race-specific resistance response (Ronald, 1997). Another member *Xa21D* of the same gene family displays the same resistance spectrum as *Xa21* but confers only partial resistance. *Xa21D* only encodes an extracellular LRR domain due to a retrotransposon insertion. The LRR domain was shown to control race-specific pathogen recognition. This study lends support to the theory that changes in major genes could produce a gene, which confers partial resistance that breeders recognize as quantitative in function.

3.9 Colocalization of QTL with Defense Response Genes

Plant defense response is a complex mechanism that is triggered by pathogen attack. The DR is highly conserved between mono- and dicotyledonous plants and is responsive to different types of pathogens. Numerous DR genes have been cloned (Lamb et al., 1989) and several colocalizations with QTL have been observed, lending support to the hypothesis that colocalization may reflect a functional relationship between the QTL and the DR genes. In common bean, Geffroy et al. (2000) mapped a QTL for stem resistance against *C. lindemuthianum* strain A7 of bean anthracnose near a locus for phenylalanine ammonia-lyase (*Pal-2*). This enzyme is a critical branch point control for the biosynthetic pathways of certain antimicrobial phenolic compounds. Another QTL was mapped on linkage group B7 near the hydroxyproline-rich glycoprotein locus, *Hrgp36*. These types of proteins are believed to contribute to the formation of a structural barrier to block pathogen invasion. The researchers conclude that allelic variants of *Pal-2* and *Hgp36* may be responsible for the differences in quantitative resistance to *C. lindemuthianum*. This evidence supports the theory that molecular polymorphisms within the DR genes result in allelic diversity and may relate to differences in resistance levels (Pflieger et al., 2001). Pflieger et al. (2001) mapped several DR genes to genomic regions corresponding with QTL for resistance to different pathogens in pepper. A class-III chitinase gene colocalized with a QTL conferring resistance to *Phytophthora capsici* in pepper. In addition, three pathogenesis-related protein (PR) loci mapped within the region containing QTL to *P. capsici*, Potato virus Y, and potyvirus E in pepper.

QTL located on linkage group B2 of the common bean map (Freyre et al., 1998; Kelly et al., 2003) spanned a region that encompasses the *Pv*PR2 locus, and suggested a role for this PR protein in resistance to *Fusarium* root rot and white mold in common bean (Kolkman and Kelly, 2003; Schneider et al., 2001). Defense response genes, such as the *P. vulgaris* pathogenesis-related gene, *PvPR-2* (Walter et al., 1990), a polygalacturonase-inhibiting protein, *Pgip* (Toubart et al., 1992), and the chalcone synthase gene, *ChS* (Ryder et al., 1987) located on B2 invites speculation that fungal defense-related genes are triggered as a general resistance response to *Fusarium* and *Sclerotinia* infection, suggesting that physiological resistance is associated with a generalized host defense response. *Pv*PR2 and its counterpart *Pv*PR1 are low molecular weight acidic proteins induced during fungal elicitation (Walter et al., 1990). These bean PR proteins share similarities with PR proteins in crops such as potato, parsley (*Petroselinum crispum*), and pea (*Pisum sativum*). Linkage was also reported between QTL conferring partial resistance to Ascochyta blight (*Mycosphaerella pinodes*) of pea and candidate genes including DR and RGA located on the pea linkage map (Timmerman-Vaughan et al., 2002). The role of *Pv*PR proteins in *Fusarium* resistance in common bean is further confirmed by the significant association observed between QTL that map to B3 in the region of *Pv*PR1 gene. Differences in *Pv*PR gene arrangements were detected between anthracnose (*C. lindemuthianum*) resistant and susceptible bean genotypes indicating that polymorphism between *Pv*PR as well as other defense response-related genes may contribute to our understanding of quantitative resistance (Walter et al., 1990). QTL associated with resistance to the late blight fungus of potato have also been reported to colocalize with DR genes for specific PR proteins in potato (Gebhardt et al., 1991; Leonards-Schippers et al., 1994). To capitalize on the assumption that defense proteins may be associated with quantitative resistance, a method of candidate gene analysis where genes known to be involved in host defense responses are used as markers to identify potential QTL associated with disease resistance, has been evaluated in maize (Byrne et al., 1996; Causse et al., 1995; Goldman et al., 1993) and wheat (Faris et al., 1999), and could be employed to improve root rot and white mold resistance in common bean.

In summary, there appears to be two kinds of coincident QTL: those that map with major genes and those that map with DR genes. The QTL that map with major genes could be allelic versions of those genes. Some of those alleles may confer partial resistance rather than complete resistance as a result of a mutation in the pathogen that now overcomes the original resistance gene. In the other instance, QTL that map with major genes may also be members of a multigene family that is involved with recognition of the pathogen. QTL that map with DR genes may be involved in a general defense mechanism. To differentiate between the various hypotheses, fine mapping of the region containing the QTL is needed to determine if the relationship between the QTL and the colocalized gene is allelic or not.

3.10 Conclusions

In a computer simulation study, Bernardo (2001) concluded that genomics is of limited value in the selection for quantitative traits in hybrid (and self-pollinated) crops. Despite this dire prognosis, Bernando (2001) stated that gene information (we equate "gene information" with "QTL analysis") is most useful in selection when fewer than 10 loci control the trait and becomes imprecise when the number exceeds 50 loci. Although the actual number of loci controlling multigenic disease resistance is not generally known, the number is most likely to be under 10 than exceed 50 loci, and therefore, be responsive to selection using QTL analysis. Performance based traits are more likely to exceed 50 loci than those conditioning quantitative disease resistance (Young, 1996). QTL analysis provides plant breeders with the tools to reassemble, into future cultivars, the multigenes influencing quantitative resistance, previously not possible through routine disease screening. Unlike qualitative resistance, where genotype and phenotype are one and the same, breeders struggle with reassembling, in new cultivars, all the genes that controlled quantitative resistance after they were "disassembled" in crossing. The breeding literature is fraught with examples of the partial recovery of quantitative resistance from unique genetic stocks. With the knowledge of the location and size of the QTL controlling quantitative resistance breeders can use MAS to reassemble that resistance in new genetic backgrounds and restore it to levels present in the original sources. Due to the complexity of quantitative resistance, breeders often failed to adequately compare resistance sources for uniqueness. QTL analysis provides breeders with a tool to compare the location and size of individual effects to determine if new resistance sources are unique prior to undertaking the long and arduous process of utilizing quantitative resistance in breeding. One area where QTL analysis offers exciting opportunities is in the utilization of wild germplasm in resistance breeding. Prebreeding using markers linked to resistance traits to introgress genomic regions from the wild to cultivated species is being investigated in many crops using breeding methods such as the advanced-backcross QTL analysis (Tanksley and McCouch, 1997). Combining backcrossing with MAS allows breeders to evaluate the potential of specific genomic regions from the wild species in the genetic background of the cultivated species, as the expression of quantitative resistance in the field can only be tested in adapted lines.

Aside from the practical utility and efficiency that QTL analysis brings to breeding of quantitative resistance, the identification and location of QTL is providing new insights to many of the age-old theories and controversies that have competed for importance in the literature. Many of these theories have impeded the process of resistance breeding as they classified resistance into two clear camps, implying that selection for one form of resistance would impede the use of the other. New information is being provided by QTL analyses on such questions as: (1) the actual durability of quantitative resistance sources, (2) the possible distinction between the resistance detected by seedling assays and adult plant resistance, (3) the potential of defeated major genes in resistance breeding, (4) the actual similarities between qualitative and quantitative resistance sources, (5) the nature of the differences in resistance may reside in expression due to interaction with genetic backgrounds or pathotypes, and (6) the opportunity to clone underlying genetic factors that confer quantitative resistance as was demonstrated for fruit size in tomato (Frary et al., 2000).

Genetic mapping, in general, has provided breeders with new insights into old problems. The evidence that resistance gene clusters exist in plants is widely reported (Michelmore and Meyers, 1998), so why should qualitative and quantitative resistance mechanisms be different? Resistance gene clusters imply that in the plant the DR genes are localized and can be shared in response to attack by different pathogens and/or stress factors. QTL analyses are adding to the body of evidence that in many instances qualitative and quantitative resistance reside in the same regions and are differential responses to different pathotypes, and the methods used by scientists to detect and measure their effect. For example, the literature is fraught with implications that seedling resistance is qualitative and adult plant resistance is more complex, hence quantitative and distinct. QTL for seedling resistance to stripe rust in barley has been shown to share common QTL for adult plant resistance (Castro et al., 2002). Defeated major resistance genes are receiving renewed attention as QTL analyses position partial or quantitative resistance in those regions of the genome where major genes reside (Li et al., 1999). The importance of defeated genes in resistance breeding is obvious as the underlying suggestion that breeders are not always finding unique or new sources of resistance but differential expression of existing sources. Breeders may need to consider how best to utilize existing resistance sources rather than search for new sources that may prove to be elusive. If quantitative resistance is distinct from major R-genes functional in a specific crop/pathogen system, there is an increasing body of evidence that supports the role of DR genes in quantitative resistance. DR genes have been shown to play a key role in resistance (Lamb et al., 1989) but their effect is only partial, not unlike the effect(s) that defines quantitative resistance (Parlevliet, 1975). QTL analyses are placing partial resistance sources in genomic regions where DR genes are located (Pflieger et al., 2001). The partial resistance detected in QTL analysis may not be due to actual resistance genes similar to R-genes but may be due to an enhanced expression of DR genes (Schneider et al., 2001). The role that DR genes can enhance resistance is known, so links between quantitative resistance and DR genes is interesting and could benefit breeders as DR genes have similar functionality across plant species.

Finally, mapping studies have shown that cereal genomes exhibit a high degree of synteny (Gale and Devos, 1998). Based on this information, QTL for resistance to Fusarium head blight (*Fusarium graminearum*) in wheat and barley appear to reside in syntenous locations on chromosome 3 in both crops (Kolb et al., 2001). Breeders can use the conservation of gene order and position among related species to assist in the identification of resistance sources that may be absent from their crop. QTL markers could be used to probe for resistance in one species based on the presence of QTL for resistance in a related species and provide breeders with the opportunity to use alternative resistance sources in the development of future cultivars with adequate levels of disease resistance.

References

Ali, M.A. 1950. Genetics of resistance to the common bean mosaic virus bean virus 1) in the bean (*Phaseolus vulgaris* L.). *Phytopathology* 40:69–79.

Ariyarathne, H.M., Coyne, D.P., Jung, G., Skroch P.W., Vidaver, A.K., Steadman, J.R., Miklas, P.N., and Bassett,. M.J. 1999. Molecular mapping of disease resistance genes for halo blight, common bacterial blight, and bean common mosaic virus in a segregating population of common bean. *J. Am. Soc. Horticult. Sci.* 124:654–662.

Asins, M.J. 2002. Present and future of quantitative trait locus analysis in plant breeding. *Plant Breed.* 121:281–291.

Backes, G., G. Schwarz, Wenzel, G., and Jahoor, A. 1996. Comparison between QTL analysis of powdery mildew resistance in barley based on detached primary leaves and on field data. *Plant Breed.* 115:419–421.

Bent, A.F. 1996. Plant disease resistance genes: function meets structure. *Plant Cell* 8:1757–1771.

Bernardo, R. 2001. What if we knew all the genes for a quantitative trait in hybrid crops? *Crop Sci.* 41:1–4.

Botstein, D., White, R.L., Skolnick, M., and Davis, R.W. 1980. Construction of a genetic-linkage map in man using restriction fragment length polymorphisms. *Am. J. Hum. Genet.* 32:314–331.

Burr, B., and Burr, F.A. 1991. Recombinant inbreds for molecular mapping in maize – theoretical and practical considerations. *Trends Genet.* 7:55–60.

Byrne, P.F., McMullen, M.D., Snook, M.E., Musket, T.A., Theuri, J.M., Widstrom, N.W., Wiseman, B.R., and Coe, E.H. 1996. Quantitative trait loci and metabolic pathways: Genetic control of the concentration of maysin, a corn earworm resistance factor, in maize silks. *Proc. Natl. Acad. Sci. USA* 93:8820–8825.

Caranta, C., Lefebvre, V., and Palloix, A. 1997. Polygenic resistance of pepper to potyviruses consists of a combination of isolate-specific and broad-spectrum quantitative trait loci. *Mol. Plant Microbe Interact.* 10:872–878.

Castano, F., Vear, F., and Delabrouhe, D.T. 1993. Resistance of sunflower inbred lines to various forms of attack by *Sclerotinia sclerotiorum* and relations with some morphological characters. *Euphytica* 68:85–98.

Castro, A.J., Chen, X.M., Hayes, P.M., Knapp, S.J., Line, R.F., Toojinda, T., and Vivar, H. 2002. Coincident QTL which determine seedling and adult plant resistance to stripe rust in barley. *Crop Sci.* 42:1701–1708.

Causse, M., Rocher, J.P., Henry, A.M., Charcosset, A., Prioul, J.L., and Devienne, D. 1995. Genetic dissection of the relationship between carbon metabolism and early growth in maize, with emphasis on key- enzyme loci. *Mol. Breed.* 1:259–272.

Chen, F.Q., Prehn, D., Hayes, P.M., Mulrooney, D., Corey, A., and Vivar, H. 1994. Mapping genes for resistance to barley stripe rust (*Puccinia striiformis* f. sp. *hordei*). *Theor. Appl. Genet.* 88:215–219.

Coyne, D.P. 1980. Modification of plant architecture and crop yield by breeding. *HortScience* 15:244–247.

Cregan, P.B., Mudge, J., Fickus, E.W., Marek, L.F., Danesh, D., Denny, R., Shoemaker, R.C., Matthews, B.F., Jarvik, T., and Young, N.D. 1999a. Targeted isolation of simple sequence repeat markers through the use of bacterial artificial chromosomes. *Theor. Appl. Genet.* 98:919–928.

Cregan, P.B., Jarvik, T., Bush, A.L., Shoemaker, R.C., Lark, K.G., Kahler, A.L., Kaya, N., VanToai, T.T., Lohnes, D.G., Chung, L., and Specht, J.E. 1999b. An integrated genetic linkage map of the soybean genome. *Crop Sci.* 39:1464–1490.

Danesh, D., and Young, N.D. 1994. Partial resistance loci for tomato bacterial wilt show differential race specificity. *Rep. Tomato Genet. Coop.* 44:12–13.

Eenink, A.H. 1976. Genetics of host-parasite relationships and uniform and differential resistance. *Neth. J. Plant Pathol.* 82:133–145.

Faris, J.D., Li, W.L., Liu, D.J., Chen, P.D., and Gill, B.S. 1999. Candidate gene analysis of quantitative disease resistance in wheat. *Theor. Appl. Genet.* 98:219–225.

Frary, A., Nesbitt, T.C., Grandillo, S., van der Knaap, E., Cong, B., Liu, J.P., Meller, J., Elber, R., Alpert, K.B., and Tanksley, S.D. 2000. Fw2.2: a quantitative trait locus key to the evolution of tomato fruit size. *Science* 289:85–88.

Freymark, P.J., Lee, M., Woodman, W.L., and Martinson, C.A. 1993. Quantitative and qualitative trait loci affecting host-plant response to *Exserohilum turcicum* in maize (*Zea mays* L). *Theor. Appl. Genet.* 87:537–544.

Freyre, R., Skroch, P.W., Geffroy, V., Adam-Blondon, A.F., Shirmohamadali, A., Johnson, W.C., Llaca, V., Nodari, R.O., Pereira, P.A., Tsai, S.M., Tohme, J., Dron, M., Nienhuis, J., Vallejos, C.E., and Gepts, P. 1998. Towards an integrated linkage map of common bean. 4. Development of a core linkage map and alignment of RFLP maps. *Theor. Appl. Genet.* 97:847–856.

Gale, M.D., and Devos, K.M. 1998. Plant comparative genetics after 10 years. *Science* 282:656–659.

Gebhardt, C., Ritter, E., Barone, A., Debener, T., Walkemeier, B., Schachtschabel, U., Kaufmann, H., Thompson, R.D., Bonierbale, M.W., Ganal, M.W., Tanksley, S.D., and Salamini, F. 1991. RFLP maps of potato and their alignment with the homoeologous tomato genome. *Theor. Appl. Genet.* 83:49–57.

Geffroy, V., Creusot, F., Falquet, J., Sevignac, M., Adam-Blondon, A.F., Bannerot, H., Gepts, P., and Dron, M. 1998. A family of LRR sequences in the vicinity of the *Co-2* locus for anthracnose resistance in *Phaseolus vulgaris* and its potential use in marker-assisted selection. *Theor. Appl. Genet.* 96:494–502.

Geffroy, V., Sevignac, M., De Oliveira, J.C.F., Fouilloux, G., Skroch, P., Thoquet, P., Gepts, P., Langin, T., and Dron, M. 2000. Inheritance of partial resistance against *Colletotrichum lindemuthianum* in *Phaseolus vulgaris* and co-localization of quantitative trait loci with genes involved in specific resistance. *Mol. Plant Microbe Interact.* 13:287–296.

Giovannoni, J.J., Wing, R.A., Ganal, M.W., and Tanksley, S.D. 1991. Isolation of molecular markers from specific chromosomal intervals using DNA pools from existing mapping populations. *Nucleic Acids Res.* 19:6553–6558.

Goldman, I.L., Rocheford, T.R., and Dudley, J.W. 1993. Quantitative trait loci influencing protein and starch concentration in the Illinois long-term selection maize strains. *Theor. Appl. Genet.* 87:217–224.

Grube, R.C., Radwanski, E.R., and Jahn, M. 2000. Comparative genetics of disease resistance within the *Solanaceae. Genetics* 155:873–887.

Hadley, H.H., and Openshaw, S.J. 1980. Interspecific and intergeneric hybridization. In *Hybridization of Crop Plants*, eds. W.R. Fehr, and H.H. Hadley, pp. 133–159. Amdison, WI: American Society Agronomy.

Hallauer, A.R., and Miranda, J.B. 1981. *Quantitative Genetics in Maize Breeding.* Ames IA: Iowa State University Press.

Hammond-Kosack, K.E., and Jones, J.D.G. 1997. Plant disease resistance genes. *Ann. Rev. Plant Phys. Plant Mol. Biol.* 48:575–607.

Harrison, B.D. 2002. Virus variation in relation to resistance-breaking in plants. *Euphytica* 124:181–192.

Honma, S. 1956. A bean interspecific hybrid. *J. Hered.* 47:217–220.

Hunter, R.L., and Markert, C.L. 1957. Histochemical demonstration of enzymes separated by zone electrophoreses in starch gels. *Science* 125:1294–1295.

Jensen, J. 1989. Estimation of recombination parameters between a quantitative trait locus (QTL) and two marker gene loci. *Theor. Appl. Genet.* 78:613–618.

Johnson, R. 1984. A critical analysis of durable resistance. *Annu. Rev. Phytopathol.* 22:309–330.

Jones, E.S., Breese, W.A., Liu, C.J., Singh, S.D., Shaw, D.S., and Witcombe, J.R. 2002. Mapping quantitative trait loci for resistance to downy mildew in pearl millet: Field and glasshouse screens detect the same QTL. *Crop Sci.* 42:1316–1323.

Jung, G., Skroch, P.W., Nienhuis, J., Coyne, D.P., Arnaud-Santana, E., Ariyarathne, H.M., and Marita, J.M. 1999. Confirmation of QTL associated with common bacterial blight resistance in four different genetic backgrounds in common bean. *Crop Sci.* 39:1448–1455.

Keller, M., Keller, B., Schachermayr, G., Winzeler, M., Schmid, J.E., Stamp, P., and Messmer, M.M. 1999. Quantitative trait loci for resistance against powdery mildew in a segregating wheat x spelt population. *Theor. Appl. Genet.* 98:903–912.

Kelly, J.D., Gepts, P., Miklas, P.N., and Coyne, D.P. 2003. Tagging and mapping of genes and QTL and molecular marker-assisted selection for traits of economic importance in bean and cowpea. *Field Crops Res.* 82:135–154.

Kelly, J.D., and V.A. Vallejo. 2005. QTL analysis of multigenic disease resistance in plant breeding. *In* S. Tuzun and E. Bent (ed.) Multigenic and Induced Systemic Resistance in Plants (in press).

Kim, H.S., and Diers, B.W. 2000. Inheritance of partial resistance to Sclerotinia stem rot in soybean. *Crop Sci.* 40:55–61.

Knapp, S.J. 1991. Using molecular markers to map multiple quantitative trait loci - models for backcross, recombinant inbred, and doubled haploid progeny. *Theor. Appl. Genet.* 81:333–338.

Knapp, S.J., Bridges, W.C., and Birkes. D. 1990. Mapping quantitative trait loci using molecular marker linkage maps. *Theor. Appl. Genet.* 79:583–592.

Kolb, F.L., Bai, G.H., Muehlbauer, G.J., Anderson, J.A., Smith, K.P., and Fedak, G. 2001. Host plant resistance genes for fusarium head blight: mapping and manipulation with molecular markers. *Crop Sci.* 41:611–619.

Kolkman, J.M., and Kelly, J.D. 2002. Agronomic traits affecting resistance to white mold in common bean. *Crop Sci.* 42:693–699.

Kolkman, J.M., and Kelly, J.D. 2003. QTL conferring resistance and avoidance to white mold in common bean. *Crop Sci.* 43:539–548.

Lamb, C.J., Lawton, M.A., Dron, M., and Dixon, R.A. 1989. Signals and transduction mechanisms for activation of plant defenses against microbial attack. *Cell* 56:215–224.

Lander, E.S., and Botstein, D. 1986a. Mapping complex genetic-traits in humans: new methods using a complete RFLP linkage map. *Cold Spring Harb. Symp. Quant. Biol.* 51:49–62.

Lander, E.S., and Botstein, D. 1986b. Strategies for studying heterogeneous genetic-traits in humans by using a linkage map of restriction fragment length polymorphisms. *Proc. Natl. Acad. Sci. USA* 83:7353–7357.

Lander, E.S., and Botstein, D. 1989. Mapping mendelian factors underlying quantitative traits using RFLP linkage maps. *Genetics* 121:185–199.

Leonards-Schippers, C., Gieffers, W., Schaferpregl, R., Ritter, E., Knapp, S.J., Salamini, F., and Gebhardt, C. 1994. Quantitative resistance to *Phytophthora infestans* in potato – a case-study for QTL mapping in an allogamous plant-species. *Genetics* 137: 67–77.

Li, Z.K., Luo, L.J., Mei, H.W., Paterson, A.H., Zhao, X.Z., Zhong, D.B., Wang, Y.P., Yu, X.Q., Zhu, L., Tabien, R., Stansel, J.W., and Ying, C.S. 1999. A "defeated" rice resistance gene acts as a QTL against a virulent strain of *Xanthomonas oryzae* pv. *oryzae*. *Mol. Gen. Genet.* 261:58–63.

Lindhout, P. 2002. The perspectives of polygenic resistance in breeding for durable disease resistance. *Euphytica* 124:217–226.

Liu, B.H. 1998. *Statistical genomics: Linkage, Mapping, and QTL Analysis.* Boca Raton, FL: CRC Press.

Lubberstedt, T., Klein, D., and Melchinger, A.E. 1998. Comparative QTL mapping of resistance to *Ustilago maydis* across four populations of European flint-maize. *Theor. Appl. Genet.* 97:1321–1330.

Mangin, B., Thoquet, P., Olivier, J., and Grimsley, N.H. 1999. Temporal and multiple quantitative trait loci analyses of resistance to bacterial wilt in tomato permit the resolution of linked loci. *Genetics* 151:1165–1172.

Martin, T.J., and Ellingboe, A.H. 1976. Differences between compatible parasite/host genotypes involving the *Pm4* locus of wheat and the correspondiing genes in *Erysiphe graminis* f. sp. *tritici*. *Phytopathology* 66:1435–1438.

McElroy, J.B. 1985. Breeding dry beans, *Phaseolus vulgaris* L. for common bacterial blight resistance derived from *Phaseolus acutifolius* A. Gray. Ph.D. dissertation., Cornell University, Ithaca, N.Y.

Mestries, E., Gentzbittel, L., de Labrouhe, D.T., Nicolas, P., and Vear, F. 1998. Analyses of quantitative trait loci associated with resistance to *Sclerotinia sclerotiorum* in sunflowers (*Helianthus annuus* L.), using molecular markers. *Mol. Breed.* 4:215–226.

Michelmore, R.W., and Meyers, B.C. 1998. Clusters of resistance genes in plants evolve by divergent selection and a birth-and-death process. *Genome Res.* 8:1113–1130.

Michelmore, R.W., Paran, I., and Kesseli, R.V. 1991. Identification of markers linked to disease-resistance genes by bulked segregant analysis: a rapid method to detect markers in specific genomic regions by using segregating populations. *Proc. Natl. Acad. Sci. USA* 88:9828–9832.

Miklas, P.N., Stone, V., Urrea, C.A., Johnson, E., and Beaver, J.S. 1998. Inheritance and QTL analysis of field resistance to ashy stem blight in common bean. *Crop Sci.* 38:916–921.

Miklas, P.N., Johnson, E., Stone, V., Beaver, J.S., Montoya, C., and Zapata, M. 1996. Selective mapping of QTL conditioning disease resistance in common bean. *Crop Sci.* 36:1344–1351.

Miklas, P.N., Smith, J.R., Riley, R., Grafton, K.F., Singh, S.P., Jung, G., and Coyne, D.P. 2000. Marker-assisted breeding for pyramided resistance to common bacterial blight in common bean. *Ann. Rep. Bean Improv. Coop.* 43:39–40.

Miklas, P.N., Coyne, D.P., Grafton, K.F., Mutlu, N., Reiser, J., Lindgren, D.T., and Singh, S.P. 2003. A major QTL for common bacterial blight resistance derives from the common bean great northern landrace cultivar Montana No. 5. *Euphytica* 131:137–146.

Mundt, C.C., Cowger, C., and Garrett, K.A. 2002. Relevance of integrated disease management to resistance durability. *Euphytica* 124:245–252.

Narayanan, N.N., Baisakh, N., Vera Cruz, C.M., Gnanamanickam, S.S., Datta, K., and Datta, S.K. 2002. Molecular breeding for the development of blast and bacterial blight resistance in rice cv. IR50. *Crop Sci* 42:2072–2079.

Nass, H.A., Pedersen, W.L., Mackenzie, D.R., and Nelson, R.R. 1981. The residual effects of some "defeated" powdery mildew resistance genes in isolines of winter wheat. *Phytopathology* 71:1315–1318.

Nelson, R.R. 1978. Genetics of horizontal resistance to plant diseases. *Annu. Rev. Phytopathol.* 16:359–378.

Nodari, R.O., Tsai, S.M., Guzman, P., Gilbertson, R.L., and Gepts, P. 1993. Toward an integrated linkage map of common bean. 3. Mapping genetic-factors controlling host-bacteria interactions. *Genetics* 134:341–350.

Parlevliet, J.E. 1975. Disease resistance in plants and its consequences for plant breeding, *In Plant Breeding II*, ed K.J. Frey., pp. 309–364. Ames, IA: Iowa State University Press.

Parlevliet, J.E. 2002. Durability of resistance against fungal, bacterial and viral pathogens; present situation. *Euphytica* 124:147–156.

Parlevliet, J.E., and Zadoks, J.C. 1977. Integrated concept of disease resistance - new view including horizontal and vertical resistance in plants. *Euphytica* 26:5–21.

Parniske, M., Hammond-Kosack, K.E., Golstein, C., Thomas, C.M., Jones, D.A., Harrison, K., Wulff, B.B.H., and Jones, J.D.G. 1997. Novel disease resistance specificities result from sequence exchange between tandemly repeated genes at the *Cf-4/9* locus of tomato. *Cell* 91:821–832.

Paterson, A.H. 1998. Of blending, beans, and bristles: the foundations of QTL mapping, In *Molecular Dissection of Complex Traits*, ed. A.H. Paterson, pp. 1–10. Boca Raton, FL: CRC Press.

Pflieger, S., Palloix, A., Caranta, C., Blattes, A., and Lefebvre, V. 2001. Defense response genes co-localize with quantitative disease resistance loci in pepper. *Theor. Appl. Genet.* 103:920–929.

Pilet, M.L., Delourme, R., Foisset, N., and Renard, M. 1998. Identification of QTL involved in field resistance to light leaf spot (*Pyrenopeziza brassicae*) and blackleg resistance (*Leptosphaeria maculans*) in winter rapeseed (*Brassica napus* L.). *Theor. Appl. Genet.* 97:398–406.

Pryor, T., and Ellis, J. 1993. The genetic complexity of fungal resistance genes in plants. *Adv. Plant Pathol.* 10:281–305.

Qi, X., Niks, R.E., Stam, P., and Lindhout, P. 1998. Identification of QTLs for partial resistance to leaf rust (*Puccinia hordei*) in barley. *Theor. Appl. Genet.* 96:1205–1215.

Qi, X., Jiang, G., Chen, W., Niks, R.E., Stam, P., and Lindhout, P. 1999. Isolate-specific QTLs for partial resistance to *Puccinia hordei* in barley. *Theor. Appl. Genet.* 99: 877–884.

Riley, R. 1973. Genetic changes in hosts and the significance of disease. *Ann. Appl. Biol.* 75:128–132.

Robertson, D.S. 1989. Understanding the relationship between qualitative and quantitative genetics, In *Development and Application of Molecular Markers to Problems in Plant Genetics*, eds. T. Helentjaris, and B. Benjamin, pp. 81–87. Cold Spring, Harbor, NY: Cold Spring Harbor Laboratory.

Ronald, P.C. 1997. The molecular basis of disease resistance in rice. *Plant Mol. Biol.* 35:179–186.

Ronald, P.C. 1998. Resistance gene evolution. *Curr. Opin. Plant Biol.* 1:294–298.

Ronin, Y.I., Korol, A.B., and Weller, J.I. 1998. Selective genotyping to detect quantitative trait loci affecting multiple traits: interval mapping analysis. *Theor. Appl. Genet.* 97:1169–1178.

Ryder, T.B., Hedrick, S.A., Bell, J.N., Liang, X.W., Couse, S.D., and Lamb, C.J. 1987. Organization and differential activation of a gene family encoding the plant defense enzyme chalcone synthase in *Phaseolus vulgaris*. *Mol. Gen. Genet.* 210:219–233.

Sax, K. 1923. The association of size differences with seed-coat pattern and pigmentation in *Phaseolus vulgaris*. *Genetics* 8:552–560.

Schechert, A.W., Welz, H.G., and Geiger, H.H. 1999. QTL for resistance to *Setosphaeria turcica* in tropical African maize. *Crop Sci.* 39:514–523.

Schneider, K.A., Grafton, K.F., and Kelly, J.D. 2001. QTL analysis of resistance to fusarium root rot in bean. *Crop Sci.* 41:535–542.

Scott, M.E., and Michaels, T.E. 1992. *Xanthomonas* resistance of *Phaseolus* interspecific cross selections confirmed by field performance. *HortScience* 27:348–350.

Singh, S.P., and Munoz, C.G. 1999. Resistance to common bacterial blight among *Phaseolus* species and common bean improvement. *Crop Sci.* 39:80–89.

Smithies, O. 1955. Zone electrophoreses in starch gels. *Biochem. J.* 61:629.

Song, W.Y., Pi L.Y, Wang, G.L., Gardner, J., Holsten, T., and Ronald, P.C. 1997. Evolution of the rice *Xa21* disease resistance gene family. *Plant Cell* 9:1279–1287.

Sprague, G.F. 1966. Quantitative genetics in plant improvement. In *Plant Breeding*, ed. K.J. Frey, pp. 315–357. Ames, IA: Iowa State University Press.

Staub, J.E., Serquen, F.C., and Gupta, M. 1996. Genetic markers, map construction, and their application in plant breeding. *HortScience* 31:729–741.

Tanksley, S.D., and McCouch, S.R. 1997. Seed banks and molecular maps: unlocking genetic potential from the wild. *Science* 277:1063–1066.

Tanksley, S.D., Young, N.D., Paterson, A.H., and Bonierbale, M.W. 1989. RFLP mapping in plant-breeding - new tools for an old science. *Bio/Technology* 7:257–264.

Thoday, J.M. 1961. Location of polygenes. *Nature* 191:368–370.

Timmerman-Vaughan, G.M., Frew, T.J., Russell, A.C., Khan, T., Butler, R., Gilpin, M., Murray, S., and Falloon, K. 2002. QTL mapping of partial resistance to field epidemics of ascochyta blight of pea. *Crop Sci* 42:2100–2111.

Toubart, P., Desiderio, A., Salvi, G., Cervone, F., Daroda, L., Delorenzo, G., Bergmann, C., Darvill, A.G., and Albersheim, P. 1992. Cloning and characterization of the gene encoding the endopolygalacturonase-inhibiting protein (PGIP of *Phaseolus vulgaris* L. *Plant J.* 2:367–373.

van der Plank, J.E. 1968. *Disease Resistance in Plants.* Academic Press, New York.

van der Voort, J.R., Lindeman, W., Folkertsma, R., Hutten, R., Overmars, H., van der Vossen, E., Jacobsen, E., and Bakker, J. 1998. A QTL for broad-spectrum resistance to cyst nematode species (*Globodera* spp.) maps to a resistance gene cluster in potato. *Theor. Appl. Genet.* 96:654–661.

Vleeshouwers, V., van Dooijeweert, W., Govers, F., Kamoun, S., and Colon, L.T. 2000. The hypersensitive response is associated with host and nonhost resistance to *Phytophthora infestans. Planta* 210:853–864.

Vos, P., Hogers, R., Bleeker, M., Reijans, M., Vandelee, T., Hornes, M., Frijters, A., Pot, J., Peleman, J., Kuiper, M., and Zabeau, M. 1995. AFLP – a new technique for DNA-fingerprinting. *Nucleic Acids Res.* 23:4407–4414.

Walter, M.H., Liu, J., Grand, C., Lamb, C.J., and Hess, D. 1990. Bean pathogenesis-related (PR) proteins deduced from elicitor-induced transcripts are members of a ubiquitous new class of conserved PR proteins including pollen allergens. *Mol. Gen. Genet.* 222:353–360.

Wang, G.L., and Paterson, A.H. 1994. Assessment of DNA pooling strategies for mapping of QTL. *Theor. Appl. Genet.* 88:355–361.

Wang, G.L., Mackill, D.J., Bonman, J.M., McCouch, S.R., Champoux, M.C., and Nelson, R.J. 1994. RFLP mapping of genes conferring complete and partial resistance to blast in a durably resistant rice cultivar. *Genetics* 136:1421–1434.

Weber, J.L., and May, P.E. 1989. Abundant class of human DNA polymorphisms which can be typed using the polymerase chain-reaction. *Am. J. Hum. Genet.* 44:388–396.

Williams, J.G.K., Kubelik, A.R., Livak, K.J., Rafalski, J.A., and Tingey, S.V. 1990. DNA polymorphisms amplified by arbitrary primers are useful as genetic-markers. *Nucleic Acids Res.* 18:6531–6535.

Young, N.D. 1996. QTL mapping and quantitative disease resistance in plants. *Annu. Rev. Phytopathol.* 34:479–501.

Yu, Y.G., Buss, G.R., and Maroof, M.A.S. 1996. Isolation of a superfamily of candidate disease-resistance genes in soybean based on a conserved nucleotide binding site. *Proc. Natl. Acad. Sci. USA* 93:11751–11756.

Yu, Z.H., Mackill, D.J., Bonman, J.M., and Tanksley, S.D. 1991. Tagging genes for blast resistance in rice via linkage to RFLP markers. *Theor. Appl. Genet.* 81:471–476.

Zeng, Z.B. 1993. Theoretical basis for separation of multiple linked gene effects in mapping quantitative trait loci. *Proc. Natl. Acad. Sci. USA* 90:10972–10976.

Zeng, Z.B. 1994. Precision mapping of quantitative trait loci. *Genetics* 136:1457–1468.

Zimnoch-Guzowska, E., Marczewski, W., Lebecka, R., Flis, B., Schafer-Pregl, R., Salamini, F., and Gebhardt, C. 2000. QTL analysis of new sources of resistance to *Erwinia carotovora* ssp. *atroseptica* in potato done by AFLP, RFLP, and resistance-gene-like markers. *Crop Sci.* 40:1156–1167.

4

Ultrastructural Studies in Plant Disease Resistance

NICOLE BENHAMOU

4.1 Introduction

Fungi which could potentially be phytopathogenic are ubiquitous soil and phyllosphere microorganisms of many crop plants including cereals, vegetables, and fruits. Although their epidemiology and pathogenesis have been studied extensively for several decades, producers are still faced with few options for effectively treating fungus-incited plant diseases. In spite of strong restrictions being imposed to protect food quality and environmental safety, agrochemicals are still the method of choice used worldwide to control major crop diseases. However, the resistance of many fungal strains to currently used chemicals and the increasing demand of consumers for pesticide-free products have reduced the appeal for agrochemicals and have stimulated the exploration for efficient alternatives that are safer for the environment. In that context, research in recent years has witnessed the discovery of several new approaches for enhancing the resistance of plants to disease through biotechnology. Beside the use of potential microbial antagonists (Chet and Inbar, 1994) and beneficial microorganisms such as plant growth promoting rhizobacteria (PGPR) (Kloepper, 1993) to control pathogen populations, the possibility of stimulating the plant's "immune system" to increase the speed of response to subsequent attack by pathogenic organisms has opened novel avenues for plant disease management. Evidence has been provided that physical, chemical, and biological agents could trigger defense-related reactions in plants. UV-C light, gamma radiation, silicon (Cherif et al., 1992), bacterial endophytes (Benhamou et al. 1996a,b,c), biological products such as chitin and chitosan (Benhamou and Theriault, 1992; Benhamou et al., 1994b), and chemical compounds such as salicylic acid (Raskin, 1992), 2-6-dichoro-isonicotinic acid (Metraux et al., 1991) and benzothiadiazole (Benhamnou and Belanger, 1998a b) have been reported to stimulate plant defense genes leading to the production and accumulation of an array of new molecules. Today, plant induced resistance attracts much attention, mainly because it offers the potential for nonchemical control of plant pathogens.

Recently, significant advances have been made toward unraveling the major events governing the plant response to pathogenic attack through molecular cytology approaches (Benhamou, 1996). Taken together, cyto- and immunocy-tochemical methods have opened up exciting and promising angles to study the mechanisms by which plants protect themselves from pathogen invasion. In combination with biochemical and molecular data, such ultrastructural approaches have provided useful and often unique information on various topics, including the structure–function relationship of a particular cell compartment, the spatio-temporal distribution of newly synthesized molecules during the course of plant tissue invasion, the antimicrobial potential of some plant defense molecules (i.e., phenolic compounds, pathogenesis-related proteins), the reinforcement of the plant cell wall by structural compounds such as callose and lignin, and the crucial role played by secondary metabolism in the resistance process. Based on investigations of the cytologically visible consequences of the plant responses to biotic stresses, several studies have contributed to shed more light on the functional activity of the plant cell during the resistance process (Benhamou, 1996; Heath, 2000; Moerschbacher and Mendgen, 2000). Recent advances in the isolation and purification of plant molecules and in the preparation of specific gold-complexed probes have provided opportunities for the development of ultracytological approaches which not only allow an accurate localization of these molecules in their respective cell compartments but also help elucidate their functions (Benhamou and Nicole, 1999). Innovative developments in plant immunocytochemistry appear with increasing frequency and it is expected that improvements in both tissue processing and probe specificity will extend the applicability of this approach to more and more research areas in plant disease resistance.

This review highlights some of the recent findings associated with the spatio-temporal localization of molecules involved in plant induced resistance with special emphasis on how an elicitation stimulus may confer increased plant resistance to pathogen attack. To put this information in context, we will concentrate on two host-pathogen systems that have been the focus of much interest as models in both basic and applied research. For more than two decades, the tomato-*Fusarium oxys-porum* f. sp. *radicis-Lycopersici* and the cucumber-*Pythium ultimum* interactions have received increasing attention mainly because losses from the diseases can be considerable in some greenhouse districts and also because chemical control proved unsuccessful for controlling the pathogen populations (Jarvis, 1988). Since the rhizosphere provides the first line defense for roots against attack by pathogens and because plants have sophisticated defense mechanisms that can be naturally activated by environmental factors and microorganisms, the possibility of enriching the rhizosphere with adapted microorganisms and/or specific eliciting products has become a challenging priority for plant pathologists. In that context, the shift from synthetic chemicals to biological control products, the so-called "green products", has become essential not only because of cost and impact of fungicides but also because commercially acceptable tomato and cucumber cultivars with strong resistance to soil-borne pathogens are not readily available. Among the microbial agents that have shown satisfactory degrees of control against pathogens causing root rot diseases in tomato and cucumber, *Trichoderma* spp. (Chet, 1987) and PGPR (Kloepper, 1993) have been reported to reduce disease incidence by inhibiting pathogen growth and development in the rhizosphere. Recently, two other microorganisms, *Pythium oligandrum* Dreschsler (Benhamou et al., 1997) and *Verticillium Lecanii* (Benhamou and Brodeur, 2001), have received increasing attention as new potential biocontrol agents. Recent investigations have provided evidence that, in addition to exerting an antagonistic activity against a wide range of fungal pathogens (Askary et al., 1998; Benhamou et al., 1999; Benhamou and Brodeur, 2000), both *P. oligandrum* and *V. Lecanii* display the ability to penetrate the plant root system and to trigger an array of structural and chemical defense-related reactions (Benhamou et al., 1997; Benhamou and Brodeur, 2001). Beside the potential use of beneficial microorganisms, a number of biological, chemical, and natural products such as chitosan (Lafontaine and Benhamou, 1995), benzothiadazole (Benhamou and Belanger, 1998b), and silicon (Cherif et al., 1992) have been reported to protect tomato and cucumber plants from root rot diseases. Thus, today, ecological pressures have considerably reduced the use of chemicals and there is no doubt that this process will continue under the present socio-political situation. However, this trend toward introducing new management approaches depends on an array of criteria, one of them being a deep knowledge of the mode of action of the selected biocontrol agents on the plant and its pathogens.

4.2 Diseases Caused by Root Pathogens in Tomato and Cucumber

Fusarium crown and root rot and *Pythium* damping off cause widespread, heavy economic losses in commercially grown greenhouse crop plants. Wilt symptoms of Fusarium crown and root rot appear in infected tomato plants during late winter and are noticeably more severe on sunny days (Jarvis, 1988; Malathrakis and Goumas, 1999). Signs of Fusarium infection are characterized by a marked wilting of the upper leaves followed by a gradual chlorosis and stunting of the lower leaves. Other typical symptoms noted when plants are removed from the soil includes severe rotting or even loss of the primary seminal root and occurrence of numerous brown lesions along the lateral roots. When the outer layers of crown and lower stem are sliced off, a chocolate-brown vascular discoloration, frequently found to extend upward in the stem for 15 to 20 cm, is typically observed (Lafontaine and Benhamou, 1995). In cucumber, *Pythium* damping off is associated with typical root symptoms characterized by the formation of brownish lesions preceding severe root rot and leaf wilting (Benhamou et al., 1996a). Various strategies for controlling *Fusarium* and *Pythium* spp. have been introduced over the years (i.e., soil disinfestation, cultural practices, fungicide treatments, and allelopathy) but serious losses still occur, largely because the effectiveness of these approaches is short-lived (Jarvis, 1988).

In the past decade, much effort has been directed toward elucidating the cytological events underlying the process of plant colonization by either *F.o. radicis-lycopersici* or *P. ultimum* (Brammal and Higgins, 1988). With the refinement of cytological and molecular techniques and the application of these approaches to the pathosystems under study, several unanswered questions regarding the relationships established between the pathogen and its target host have been addressed.

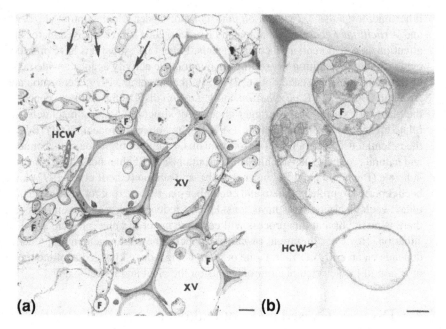

FIGURE 4.1. Transmission electron micrographs of *Fusarium*-infected tomato root tissues. (a) *Fusarium* hyphae (F) colonize rapidly the root tissues causing extensive cell damage and host cell wall (HCW) alterations (arrows). Xylem vessels (XV) are invaded through penetration of the pit membranes. Bar = 2 μm. (b) *Fusarium* (F) ingress in the root tissue coincides with extensive host cell wall (HCW) damage. Bar = 1 μm.

Using gold-complexed probes for the localization of cellulose and pectin, Benhamou et al. (1987, 1990) reported that hydrolysis of the wall-bound pectin and, to a lesser extent, cellulose was one of the key mechanisms involved in fungal ingress toward the vascular stele. Indeed, in all cases, marked cell wall damage involving loosening of the fibrillar layers (Figure 4.1a, arrows), disruption of the primary walls and middle lamella matrices (Figure 4.1b), and, in some cases, complete wall breakdown leading to tissue maceration was observed. Evidence was provided that typical host reactions such as wall appositions, intercellular plugging, and xylem vessel coating seldom occurred in the heavily parasitized root tissues (Benhamou and Lafontaine, 1995). Understandably, the massive fungal colonization and alteration of the root tissues always correlated with the presence of numerous dark brown lesions on the root system and the expression of typical symptoms including leaf chlorosis and wilting.

4.3 Microbially-Mediated Induced Resistance in Tomato and Cucumber

With the development of more and more pesticide-resistant pathogen strains, the replacement of chemicals by the controlled use of alternative agents has become the focus of considerable interest in the context of a sustainable, economically profitable agriculture (Gullino et al., 1999). Much of the development of this biotechnology has benefited from the discovery of biocontrol agents with antagonistic activity (Handelsman and Stabb, 1996; Harman, 2000). However, recent advances in our understanding of the molecular basis of plant disease resistance have led to the concept that plants could protect themselves against the harmful impact of pathogens through the coordinated stimulation of defense genes (Ward et al., 1991). The work on induced resistance over the past decade has led to a remarkable awareness of the pivotal role being played by some microbial agents in stimulating defense gene expression and disease resistance in plants (Kuc, 1987; van Peer et al., 1991; Tuzun and Kloepper, 1995). From these fundamental studies, it has become more and more realistic that sensitizing a plant to respond more rapidly to infection could confer increased protection against virulent pathogens. The stage is now set for this investment in knowledge to generate novel biocontrol approaches for protecting plants against microbial diseases while reducing environmental pollution. Based on the explosive rate of developments in this field, microbially-mediated induced resistance offers good prospects as a long-lasting, safe option for disease control (Paulitz and Matta, 1999).

4.3.1 Induction of Resistance by Beneficial Rhizobacteria

One of the most promising options in the context of microbially-mediated induced resistance concerns the use of plant growth-promoting rhizobacteria (PGPR) as potential elicitors of plant defense mechanisms (Tuzun and Kloepper, 1995). Surprisingly, the role played by these bacterial endophytes was not fully appreciated until recent years, although reduction of disease incidence and severity following soil amendment with such bacteria was often observed (Dimock et al., 1989; Chen et al., 1995). Recent progress in the purification and identification of antifungal metabolites has led to the consideration that plant protection against virulent pathogens relied, at least partly, in the production of bacterial antibiotics (Fiddaman and Rossal, 1993), associated with a possible competition for nutrients and iron in the rhizosphere (Bakker et al., 1990). Despite the extensive research devoted to the antimicrobial activity of bacterial endophytes, our knowledge regarding the involvement of the plant itself in the observed reduction of disease incidence was until recently elusive, although an increasing number of reports indicated that bacterially-mediated induced resistance was likely a crucial event in the complex process of disease protection (Wei et al., 1994).

This view was substantiated further by the biochemical and cytological demonstration that marked host metabolic changes, including production of phytoalexins (van Peer et al., 1991), accumulation of pathogenesis-related (PR) proteins (Zdor and Anderson, 1992), and deposition of structural barriers (Benhamou et al., 1996a,c), occurred at the onset of plant colonization by the rhizobacteria. These findings by themselves highlight the concept that prior inoculation with selected rhizobacteria stimulates a number of plant defense reactions culminating in the creation of a fungitoxic environment and in the elaboration of permeability barriers that prevent fungal spread in the plant tissues (Benhamou et al., 1998).

Recent cytochemical investigations have clearly indicated that PGPR-mediated induced resistance was a multifaceted process requiring not only the synergistic contribution of several mechanisms (i.e., control of the pathogen populations in the rhizosphere through antibiosis and/or competition and induction of plant defense reactions), but also taking into account the intricate relationship established between the plant, the bacteria, and the pathogen species (Benhamou et al., 1996a,b).

Effect of *Pseudomonas fluorescens* and *Bacillus pumilus* on the Induction of Resistance Against *Pythium ultimum* and *Fusarium oxysporum fsp. pisi* in Ri T-DNA Transformed Pea Roots

As a prelude to further investigations on tomato plants, the influence exerted by *Pseudomonas fluorescens* strain 63-28R, in stimulating plant defense reactions was ultrastucturally investigated using an in vitro system in which Ri T -DNA transformed pea roots were subsequently infected with *Pythium ultimum or Fusarium oxysporum* fsp. *pisi* (Benhamou et al., 1996a,b). Transformed roots, obtained by inoculating plant tissues with virulent strains of the soil bacterium *Agrobacterium rhizogenes* and then isolating the adventitious roots arising from the wound sites (Savary and Flores, 1994), offer the advantages of being genetically and biochemically stable and exhibiting fast growth as compared to untransformed root systems. Such transformed roots have been extensively used to study the biosynthetic pathways of phenolic compounds (Flores and Curtis, 1992) and have also proven useful for investigating the influence of endomycorrhizal infection on pathogen-induced resistance (Benhamou et al., 1994a). These ultracytological studies provided the first evidence that not only *P. fluorescens* multiplied abundantly at the root surface but also was able to colonize a small number of epidermal and cortical cells.

Upon inoculation with the pathogens, strong differences in the extent of fungal damage were observed at the root surface. The rapid collapse and loss of turgor of *P. ultimum* hyphae was taken as an indication that *P. fluorescens* produced antifungal metabolites that could play an important role in controlling the pathogen population in the rhizosphere (Benhamou et al., 1996a). However, the finding that *Fusarium* hyphae were not affected to the same extent by the presence of bacteria corroborated the current concept that a large number of PGPR strains, found to enhance plant protection against a broad-range of pathogens, produce metabolites that have very specific effects, and target selected microorganisms only (Benhamou et al., 1996b). Reduction in the rate of host cell colonization by either *P. ultimum* or *F. oxysporum radicis-lycopersici*, restriction of fungal cell growth to the epidermal and outer cortical root tissues, and marked decrease in pathogen viability were typical features observed only in roots previously inoculated with *P. fluorescens* (Figure 4.2). Support for the close association between the presence of bacterial cells and induced resistance came from the observation that host cell wall damage in advance of invading hyphae, a typical feature of invasion monitored in nontreated plants, was absent from bacterized roots. In these pretreated roots, the apparent preservation of the cell wall architecture as well as the

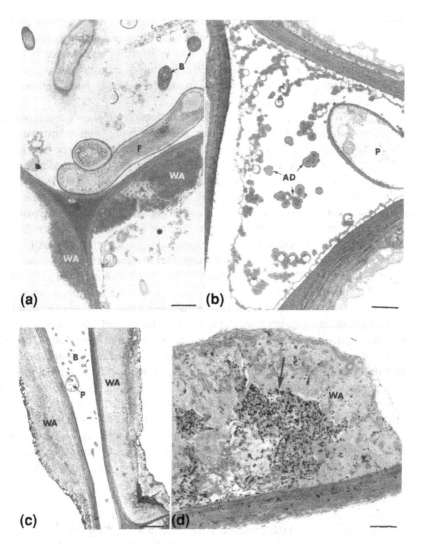

FIGURE 4.2. Transmission electron micrographs of *P. fluorescens*-inoculated pea root tissues, collected 2 days after challenge with *P. ultimum* or *Fusarium oxysporum* f. sp. *radicis lycopersici*. (**a–c**). In bacterized roots, *Fusarium* (F) and *Pythium* (P) cells are essentially found in the epidermis. Pathogen ingress in the root epidermis is associated with the deposition of electron opaque wall appositions (WA) (**a** and **c**) at sites of potential host cell wall penetration and with the accumulation of amorphous deposits (AD) in most intercellular spaces (**b**). Phenolic compounds, labeled with a laccase-gold complex, are detected in a wall apposition (WA) (**d**). **a**, bar = 1 μm. **b**, bar = 0.5 μm; **c**, bar = 1 μm; **d**, bar = 0.5 μm.

massive accumulation of structural barriers at sites of attempted penetration indicated that host cell walls were likely protected against both physical and biochemical contact with the pathogen (Figures 4.2a, b). Such cellular changes, characterized by the deposition onto the inner surface of the cell walls of callose-enriched wall appositions, were apparently efficient in preventing fungal ingress toward the vascular stele and probably also in shielding the inner root tissues from phytotoxic, diffusible products such as hydrolytic enzymes and toxins (Figure 4.2a). Hyphae of the pathogen were markedly altered in intercellular spaces colonized by the rhizobacteria (Figure 4.2b).

Using a laccase-gold complex for the localization of phenolic compounds, Benhamou et al. (1996b) reported that phenolics were widely distributed in *Fusarium*-challenged, bacterized roots. Because phenolic substances are known to confer strong rigidity to cell wall structures through peroxidase-mediated cross-linking with constitutive (i.e., hemicellulose and pectin) and newly formed (i.e., callose) wall carbohydrates (Fry, 1986), it is tempting to speculate that these compounds contribute to the elaboration of physical barriers restricting pathogen spread. Support to this speculation is provided by earlier observations indicating that plant root colonization by PGPR promoted peroxidase activity (Albert and Anderson, 1987) and enhanced lignin accumulation (Anderson and Guerra, 1985) in bean. In addition to their infiltration at strategic sites of potential penetration, phenolic compounds were also detected in the host cells as amorphous aggregates often interacting with the fungal cell surface (Figure 4.2b). This abnormal accumulation of phenolic-enriched deposits in the fungal cell walls was associated with morphological changes and cytological alterations of the invading hyphae, suggesting that these compounds were laid down by the plant to restrict pathogen growth through fungitoxic activity. According to their wide pattern of distribution, phenolic compounds may thus play a key role in PGPR-mediated induced resistance by directly inhibiting fungal growth and indirectly protecting the plant cell walls from the deleterious effect of microbial toxins and enzymes. This confirms and extends earlier results suggesting that the increased accumulation of phytoalexins triggered by *P. fluorescens* in carnation was, at least partly, responsible for the enhanced resistance of the plants to *Fusarium* infection (van Peer et al., 1991).

These results, obtained with transformed pea roots, were the first to provide a detailed picture of the intricate interaction established between the plant, the beneficial bacteria, and the pathogen at the cellular level. They illustrated that pea root bacterization with *P. fluorescens* strain 63-28 induced a set of plant defense reactions that culminated in the elaboration of physical barriers and in the creation of a fungitoxic environment which adversely affected pathogen growth and development.

Effect of *Pseudomonas fluorescens* on the Induction of Resistance in Tomato

The induction of resistance obtained with the in vitro root system was also observed to occur in whole plants (Mpiga et al., 1997). The authors reported that tomato plants treated with *Pseudomonas fluorescens* strain 63-28 gained increased protection against tomato crown and root rot caused by *F. oxysporum* f. sp. *radicis-iycopersici*. Again, the restriction of fungal growth to the epidermis and the outer root cortex coincided with a marked decrease in pathogen viability and striking cellular changes, mainly characterized by the deposition onto the inner cell wall surface of callose-enriched wall appositions. The massive deposition of such structures at sites of attempted fungal entry as well as the accumulation of phenolic substances suggested that epidermal and cortical host cells were signaled to mobilize a number of defense strategies including activation of secondary responses with direct impact on the pathogen (Mpiga et al., 1997).

Effect of *Serratia plymuthica* on the Induction of Resistance in Cucumber

Recently, another endophytic bacterium, *Serratia plymuthica*, proved to be a powerful inducer of resistance in cucumber plants (Benhamou et al., 2000). In a way similar to most endophytic bacteria, S. *plymuthica* displayed the ability to penetrate the cucumber root epidermis and to induce a number of host defense reactions that were amplified after challenge with *P. ultimum* (Benhamou et al., 2000). Reduction in fungal biomass and increase in hyphal structural alterations leading to the frequent occurrence of empty fungal shells were also typical features of the reactions observed in bacterized roots only. The observation that fungal cells, trapped in the osmiophilic material accumulating in most intercellular spaces of reacting host cells, were markedly damaged at a time when the cellulose component of their cell walls was preserved led Benhamou et al. (2000) to raise the hypothesis of a specific plant defense reaction. While cucumber has been frequently used as a model for induced resistance (Siegrist et al., 1994), little information is available on the relative contribution of phenolic compounds in the protection of cucumber against fungal attack, although enhanced activity of some enzymes involved in the phenylpropanoid pathway (i.e. peroxidase and phenylalanine ammonia-lyase) has been monitored in *Pseudomonas*-treated cucumber plants (Chen et al., 1997). In a recent report, Daayf et al. (1997) showed that several phenolic compounds with antifungal activity accumulated in cucumber root tissues at the onset of induced resistance. In line with this work, the cytological results reported by Benhamou et al. (2000) support the idea that secondary metabolites, either as peroxidase-converted preformed phenolics or as newly-formed phytoalexins, are potentially involved in the resistance process expressed in bacterized cucumber plants in response to *Pythium* attack. Beside the accumulation of osmiophilic deposits likely composed of phenolics, another striking feature of host reaction was the formation of wall appositions at sites of potential pathogen penetration (Figure. 4.2c). Incubation with the laccase-gold complex revealed that the core of the wall appositions was enriched with phenolic-like compounds, likely corresponding to lignin (Figure 4.2d, arrow).

4.3.2 Induction of Resistance by Antagonistic Fungi

In the past decade, major advances have been made in understanding the sequential events taking place in the regulation and expression of mycoparasitism (Shirmbock et al., 1994). Progress in characterizing the mechanisms of cell-to-cell signaling and identifying the cascade of biochemical events leading to antagonist establishment in a particular fungal host has led to the consideration that mycoparasitism could provide a conceptual basis to confer enhanced plant protection to microbial attack (Chet, 1987). However, attempts to exploit fungal antagonists as potential biological control agents have recently led to the proposal that beside their recognized antifungal properties (Deacon, 1976), such organisms could also act as elicitors of plant defense reactions, thereby promoting the overall plant protection (Rey et al., 1998; Yedidia et al., 1999). The exact mechanisms underlying plant protection by antagonistic fungi have been little documented and, in many cases, remain controversial. Several hypotheses have been put forward, but very few have been convincingly assessed through biochemical and cytological investigations of plant tissues challenged by these fungal agents. Among the fungal agents that have shown satisfactory degrees of control against root rot pathogens, *Trichoderma* spp. (Chet, 1987), *Pythium oligandrum* Dreschsler (Benhamou et al., 1997), and more recently *Verticillium lecanii* (Zimm.) Vegas (Benhamou and Brodeur, 2001) have been reported to reduce disease incidence by inhibiting pathogen growth and development in the rhizosphere in addition to triggering the plant defense system. Research to identify microbial control agents that may function in both plant colonization and pathogen antagonism thus appears crucial for a more effective selection of agents capable of operating not only through an antimicrobial activity but also by sensitizing the plant to respond more rapidly and efficiently to subsequent pathogen attack. Since the rhizosphere provides the first line of defense for roots against attack by pathogens and because plants have sophisticated defense mechanisms that can be naturally activated by environmental factors and microorganisms, the possibility of enriching the rhizosphere with selected microorganisms has become a challenging priority for plant pathologists (Cook, 1993).

Trichoderma harzianum-Mediated Plant Induced Resistance

In spite of the increasing amount of research devoted to the antimicrobial activity of *T. harzianum* in vitro (Benhamou and Chet, 1997), our knowledge of the exact mechanisms responsible for the observed growth promotion and reduction of disease incidence following soil treatment with *Trichoderma* propagules was, until recently, elusive. If there is no doubt that reduction of the pathogen population densities through a direct antimicrobial activity exerted by the antagonist as well as through indirect effects, such as improved nutrient and mineral uptake (Harman and Bjorkman, 1997), are responsible, at least partly, for the enhanced plant growth and protection described by several authors (Baker, 1989), it appears surprising that attention has only been focused recently on the cytological and physiological changes occurring at the onset of plant colonization by this mycopathogen. The concept of an altered plant physiology following root tissue colonization by *T. harzianum,* together with the earlier demonstration that infection with beneficial fungi such as endomycorrhizal fungi sensitized host plants to respond more rapidly and efficiently to pathogen attack (Morandi et al., 1983; Spanu et al., 1989;

Benhamou et al., 1994a), raised the question as to what extent nonpathogenic fungi such as *Trichoderma* spp. could signal the plant to mobilize its defense strategy.

Ultrastructural investigations of root tissues from cucumber plants inoculated with *T. harzianum* have proved useful in delineating the events underlying the colonization process (Yedidia et al., 1999). Convincing evidence was provided that root colonization by *T. harzianum* involved a sequence of events, including fungal proliferation along the elongating root and local penetration of the epidermis at the junctions of adjacent epidermal walls. Penetration of the epidermis and subsequent ingress in the outer cortex suggested that at least small amounts of cell wall hydrolytic enzymes such as cellulases were produced by the fungus to locally weaken or loosen the epidermal cell wall, thereby facilitating fungal spread into the root tissues. However, the regular pattern of cellulose distribution in the internal root tissues was taken as an indication that cellulases were only very slightly or not at all produced inside the plant. Interestingly, penetration of *Trichoderma* hyphae into the root tissues was found to be associated with cellular changes mainly characterized by the deposition onto the inner cell wall surface of callose-enriched wall appositions (Figure 4.3a). This phenomenon was even amplified by the impregnation of osmiophilic substances in the host cell walls and in the intercellular spaces of reacting host cells (Figure 4.3b). The massive deposition of such structures at sites of attempted fungal entry as well as the accumulation of osmiophilic deposits was taken as an indication that epidermal and cortical host cells were signaled to mobilize a number of defense strategies. Such cellular changes were apparently efficient in preventing fungal ingress toward the vascular stele since *Trichoderma* hyphae were seldom seen in the innermost root tissues. These cytological observations were further confirmed by biochemical analyses of the enzymatic activity occurring at the onset of plant colonization by *T. harzianum* (Yedidia et al., 2000). Increases in chitinase, β-1,3-glucanase and peroxidase activities were detected in *Trichoderma*-colonized root tissues with a peak at 72 h post-inoculation. However, a significant decrease of enzyme activity was observed after establishment of the fungus in the root tissues. In a way similar to what is known to occur with endomycorrhizal fungi (Spanu et al., 1989), one may consider that *T. harzianum* may be capable of evoking the transcriptional activation of plant defense genes, the expression of which may be subsequently suppressed by an unknown mechanism, and restimulated upon perception of signals originating from contact with a potential pathogen.

Pythium oligandrum-Mediated Plant Induced Resistance

Since the initial mention that *P. oligandrum* could be a secondary invader of diseased plant root tissues in addition to acting as a hyperparasite on primary pathogens (Drechsler, 1943), the antagonistic activity of this nonpathogenic *Pythium* strain has been abundantly documented (Foley and Deacon, 1986; Martin and Hancock, 1987; Benhamou et al., 1999). Although mycoparasitism is currently considered to be the primary mechanism involved in biocontrol by *P. oligandrum,*

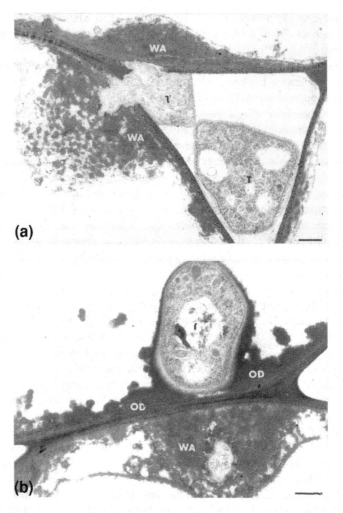

FIGURE 4.3. Transmission electron micrographs of *T. harzianum*-inoculated cucumber root tissues. (**a**). Heterogeneous wall appositions (WA) are formed in the noninfected host cells adjacent to invaded cells. Unsuccessful attempts of the *Trichoderma* hyphae (T) to penetrate wall appositions are observed. bar = 0.5 μm. (**b**). In addition to wall appositions, osmiophilic deposits, lining the host cell wall and surrounding *Trichoderma* hyphae, are frequently seen. Bar = 0.5 μm.

the possibility that antibiosis and competition for nutrients in the rhizosphere as well as plant induced resistance may play key roles in the microbial interactions has recently gained increased popularity (Benhamou et al., 1997).

Cytological investigations of root samples from *P. oligandrum-* inoculated tomato plants revealed that the fungus displayed the ability to colonize the root tissues without inducing extensive cell damage (Benhamou et al., 1997; Rey et al., 1998). However, and unlike *T. harzianum* (Yedidia et al., 1999), the invading

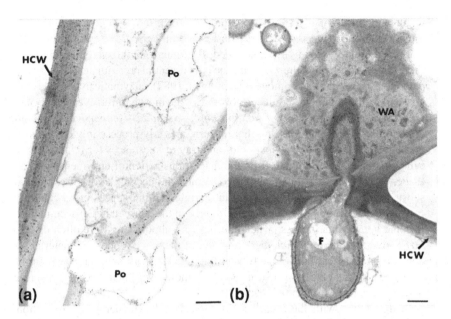

FIGURE 4.4. Transmission electron micrographs of *P. oligandrum*-inoculated tomato root tissues. (a) Hyphae of *P. oligandrum* (Po) colonize the root tissues without inducing extensive host cell wall (HCW) damage. However, these invading hyphae degenerate during the colonization process as evidenced by the frequent occurrence of empty fungal shells in the root tissues. Bar = 0.5 µm. (b) Deposition of heterogeneous wall appositions (WA) beyond the infection sites are the main visible features of the cellular response to pathogen attack in *P. oligandrum*-inoculated tomato plants. Bar = 0.5 µm.

hyphae of *P. oligandrum* were found to degenerate during the colonization process as evidenced by the frequent occurrence of empty fungal shells in the root tissues (Figure 4.4a). Whether such alterations are attributable to the creation of a fungitoxic environment associated with the synthesis and accumulation of antimicrobial compounds by the reacting host cells or simply relate to a specific behavior of the fungus *in planta* has not been elucidated yet.

When tomato plants were challenged with *F. oxysporum* f. sp. *radicis-iycopersici*, *Pythium* ingress in the root tissues was associated with substantial host metabolic changes. Strong differences in the rate and extent of pathogen invasion were observed whether the roots were previously infected or not with *P. oligandrum*. Interestingly, an antagonistic process similar to that observed in vitro (Benhamou et al., 1997) was observed *in planta*. The specific labeling patterns obtained with the exoglucanase-gold complex and the WGA-ovomucoid-gold complex confirmed that *P. oligandrum* successfully penetrated invading cells of the pathogen without causing substantial cell wall alterations as judged by the intense labeling of chitin.

Restriction of *Fusarium* growth to the outermost root tissues, together with deposition of newly-formed barriers beyond the infection sites, were the main visible features of the cellular response to pathogen attack in *P. oligandrum*-inoculated tomato plants (Figure 4.4b). These host reactions appeared to be amplified as compared to those seen in nonchallenged, *P. oligandrum* infected plants. Considering that ingress toward the vascular stele is an essential prerequisite for successful pathogenesis by vascular pathogens (Beckman, 1987), enzymatic hydrolysis of the host cell wall components is conceivably one of the most harmful events associated with the infection process by *F. oxysporum* f. sp. *radicis-lycopersici* (Benhamou et al., 1990). It is not surprising that, in turn, an early process in the expression of plant resistance is the production of an array of substances for reinforcing the cell walls and shielding them from the deleterious action of enzymes and toxins (Ride, 1983). Another host reaction concerned the accumulation of osmophilic substances in the host cell walls and in the intercellular spaces of reacting host cells. The observation that fungal cells trapped in the dense material accumulating in some intercellular spaces or neighboring wall appositions were often disorganized suggested that the biological function of the newly-deposited material was not only mechanical but probably also fungicidal. This assumption was further confirmed by the detection of considerable amounts of phenolic-like substances in these structures (Benhamou et al., 1997).

In direct line with these earlier cytological and cytochemical observations, attempts were recently made to identify the trigger involved in *P. oligandrum*-mediated induced resistance. The isolation of a proteinaceous metabolite bearing the "elicitin signature", as shown by the amino acid composition of its N-terminal end and by its migration profile within the plant tissues (Ponchet et al., 1999) led to include this peptide, termed oligandrin, into the elicitin family (Picard et al., 2000). While treatment of tomato plants with oligandrin failed to elicit the HR-associated necrotic response, a reaction consistently found to occur in tobacco plants treated with true elicitins (Ricci et al., 1993; Ponchet et al., 1999), a substantial level of protection against the oomycete fungus, *Phytophthora parasitica*, was obtained, thus substantiating the concept that oligandrin could be considered as a resistance elicitor (Picard et al., 2000). When applied to decapitated tomato plants, oligandrin displayed the ability to induce plant defense reactions that restricted stem cell invasion by *P. parasitica*. Ultrastructural investigations of the infected tomato stem tissues from nontreated plants showed a rapid colonization of all tissues associated with a marked host cell disorganization. In stems from oligandrin-treated plants, restriction of fungal growth to the outermost tissues and a decrease in pathogen viability were the main features of the host-pathogen interaction. Invading fungal cells were markedly damaged at a time when the cellulose component of their cell walls was quite well preserved, thus favoring the hypothesis of a fungitoxic effect (Figure 4.5a). The observation that a large number of invading hyphae were filled with electron-opaque inclusions led Picard et al. (2000) to suggest that incorporation of phenolics in fungal cells was a mechanism involved in the alteration of essential physiological functions such as respiration.

To determine whether the signaling role of oligandrin as an inducer of resistance against a foliar pathogen (Picard et al., 2000) was operational against a soilborne pathogen, the effectiveness of a stem treatment with oligandrin in inducing systemic resistance in tomato plants against the root pathogen, *F. oxysporum* f. sp.

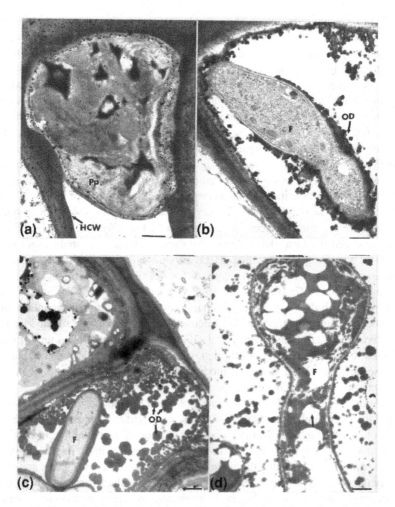

FIGURE 4.5. Transmission electron micrographs of stem tissue from decapitated tomato plants treated with the oligandrin of *P. oligandrum*. (a) Upon stem inoculation with *Phytophthora parasitica*, fungal growth in planta is mainly restricted to the outermost cell layers. Hyphae of the pathogen (Pp) are highly disorganized as evidenced by their cytoplasm which is filled with dense inclusions. The host cell wall (HCW) is apparently well preserved. Bar = 0.5 μm. (a–d). Upon root inoculation with *F. oxysporum radicis-lycopersici*, host cells in the cortical area are filled with osmiophilic deposits (OD) which accumulate in the cell lumen as well as at the cell surface of invading fungal cells (b and c). Fungal cells (F), surrounded by the osmiophilic deposits (OD), are severely damaged and exhibit marked changes including increased vacuolation (d, arrow), and densification of the cytoplasm. b, bar = 0.5 μm; c, bar = 1 μm; d, bar = 0.5 μm.

radicis-lycopersici was investigated (Benhamou et al., 2001). In this study, the authors examined ultrastructurally the outcome of the tomato-*Fusarium* interaction upon oligandrin treatment and compared the cytologically visible consequences of the induced response to those triggered by other biotic elicitors, including chitosan

(Benhamou and Lafontaine, 1995). Evidence was provided that oligandrin had the potential to induce a systemic resistance in *Fusarium*-infected tomato plants mainly associated with a massive accumulation of antifungal compounds in infected cells and intercellular spaces (Figures 4.5b,c). Interestingly, this oligandrin-mediated induced response was found to differ from that observed in chitosan-treated tomato plants where the formation of structural barriers at sites of attempted pathogen penetration was the main feature of reaction. The authors suggested that such a differential host response could be explained by the possibility that different levels of Ca^{2+} influx in oligandrin- or chitosan-treated tomato plants stimulated different signaling pathways resulting in distinct cellular changes. This view was substantiated by the reported correlation between the formation of callose-enriched wall appositions and the enhanced intracellular concentration of free Ca^{2+}, known to control the activity of one of the key structural enzymes, β-1,3-glucan synthase (Kohle et al., 1985). Recent investigations of the effect of fungal elicitins on the host cell responses have shown that these proteinaceous molecules triggered early effects (e.g., Ca^{2+} influx, H_2O_2 production) similar to those induced by oligosaccharides with, however, some differences in terms of Ca^{2+} uptake intensity (Pugin et al., 1997; Binet et al., 1998). The extent of the Ca^{2+} influx did not correlate with the basic or acidic characters of the elicitins but rather could be associated with the activation of distinct or additional signaling pathways leading to differential cell responses (Binet et al., 1998). Although the reasons why oligandrin failed to induce massive callose deposition are still speculative, it is clear that these defense-related structural modifications are not major components of the oligandrin-mediated induced response in tomato.

Active synthesis and accumulation of phenolic compounds in both the cell walls and the intercellular spaces were to be the most striking feature of the plant reaction to oligandrin treatment (Figure 4.5c). Although the phenolic-enriched material may indirectly contribute to disease resistance by reinforcing the mechanical strength of the cell walls, the main role played by these substances appeared to rely on a direct antifungal activity, as evidenced by the strong alteration of most invading hyphae of the pathogen. In an attempt to determine whether fungal wall hydrolysis was associated with the frequent disorganization of fungal hyphae colonizing the outer root tissues in oligandrin-treated plants, chitin was ultrastructurally localized by using the WGA/ovomucoid-gold complex. Analysis of the labeling pattern over invading fungal cells clearly revealed that chitin was altered in severely damaged hyphae (Figure 4.5d). This was taken as an indication that the plant cells were signaled to produce chitinases that accumulated at invaded sites. However, the finding that chitin molecules were still present over cell walls of hyphae showing obvious signs of degradation led us to suggest that production of hydrolytic enzymes such as chitinases was not an early event in the expression of oligandrin-mediated induced resistance. According to these cytochemical observations, a scheme of events could be drawn including an early synthesis of toxic substances (i.e., phenolics) with direct incidence on the pathogen followed by the production of chitinases which probably contributed to a more complete disintegration of the fungal cells. The ultrastructural work performed on the ability of either *P. oligandrum* itself (Benhamou et al., 1997) or its proteinaceous molecule, oligandrin (Picard et al., 2000; Benhamou et al., 2001), to reduce disease incidence caused by foliar and root pathogens in tomato opens new avenues for the development of integrated protection programs. Because of its antimicrobial activity (Benhamou et al., 1999) and its ability to produce large amounts of oligandrin (Picard et al., 2000), the incorporation of selected strains of *P. oligandrum* into the arsenal of strategies currently developed for controlling diseases caused by soil-borne fungi is a promising step toward elaborating integrated pest management programmes which will allow greenhouse producers to reduce yield losses caused by root rot pathogens while participating in the current trend toward reducing the use of chemical pesticides.

Verticillium lecanii-Mediated Plant Induced Resistance

Although increasing expectations are emerging in the area of plant disease management for new strategies mediated by nonpathogenic fungi such as some *Trichoderma* spp. (Chet, 1987; Chet and Inbar, 1994) and *Pythium oligandrum* Dreschler (Benhamou et al., 1997; Picard et al., 2000), another facet that is attracting much attention concerns the potential of the hyphomycete, *Verticillium lecanii* (Zimm.) Viegas (Askary et al., 1997; 1998). Since the first demonstration that this well-known entomopathogenic fungus (Hall, 1981) could also parasitize rusts (Spencer and Atkey, 1981), convincing evidence has been provided that *V. lecanii* could be a promising biocontrol agent of rusts and powdery mildew pathogens (Verhaar et al., 1996; Askary et al., 1997). The rationale for such an interest toward the remarkable properties of *V. lecanii* was that not only could this fungus be a powerful candidate in an integrated arthropod pest management strategy, but it could also become a valuable option for plant disease management. Recent ultrastructural investigations have shown that the beneficial effect of *V. lecanii* at exploiting insects and fungal pathogens involved a series of coordinated events including recognition, antibiosis and production of hydrolytic enzymes such as chitinases (Askary et al., 1997; 1999). More recently, evidence was provided that the antagonistic properties of *V. lecanii* were not restricted to rusts and powdery mildews, but could extend to other fungal pathogens such as *Penicillium digitatum*, highlighting the remarkable and atypical potential of this fungus as a promising biocontrol agent of a wide array of pathogens (Benhamou and Brodeur, 2000).

In a recent study, Benhamou and Brodeur (2001) demonstrated that cucumber roots, grown in the presence of *V. lecanii*, afforded increased protection against *Pythium ultimum* attack. These observations were of particular relevance since they highlighted, for the first time, the dual properties of *V. lecanii*, which, in addition to being a strong antagonist in the rhizosphere, was also capable of evoking biochemical events characteristic of the natural plant disease resistance process.

The beneficial effect of *V. lecanii* in repressing *Pythium* ingress in the root tissues appeared to rely on a strong antifungal activity associated with an induction of structural and biochemical barriers in the host tissues. Under the conditions of the experimental system used, *V. lecanii* was found to proliferate at the root

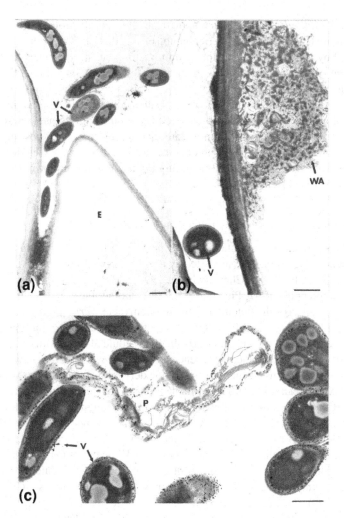

FIGURE 4.6. Transmission electron micrographs of cucumber root tissues inoculated with *Verticillium lecanii*. (**a**) Cells of *V. lecanii* (V) occur at the junction of adjacent epidermal cell walls. Direct epidermis (E) penetration is not observed. Bar = 1μm. (**b**) A heterogeneous wall apposition (WA) is formed in a region proximal to a *V. lecanii*-colonized intercellular space. Bar = 1 μm. (**c**) A highly damaged *Pythium* cell (P) is surrounded by several hyphae of *V. lecanii* (V). Cellulose labeling over the *Pythium* cell wall is markedly altered. Bar = 1 μm.

surface and to interact with the pathogen, causing marked hyphal alterations. The occurrence of numerous cells of *V. lecanii* at the junction of adjacent epidermal walls indicated that these areas were preferential sites of penetration, subsequently leading to colonization of some intercellular spaces (Figure 4.6a). The authors suggested that at least small amounts of cell wall hydrolytic enzymes such as pectinases and cellulases were produced by the fungus to locally weaken or loosen the epidermal cell wall to facilitate entry into the root tissues. The relationship established between the host plant and *V. lecanii* appeared highly specific and the possibility that this fungus may behave as a fungal endophyte has been raised.

The main facets of the altered metabolism induced by *V. lecanii* concerned the abnormal formation of wall appositions beyond the infection sites and the filling of some intercellular spaces with an electron-dense material (Figure 4.6b). It is assumed that these plant reactions are related to a protective mechanism elaborated by the plant to restrict massive spread of *V. lecanii* in the root tissues, thus preventing fungal pathogenesis. When *V. lecanii*-inoculated cucumber roots were challenged with *P. ultimum*, a marked decrease in the rate of host cell colonization by the pathogen was observed. Observations of the root surface showed that *V. lecanii* established close contact with hyphae of the pathogen, leading to a series of cellular disturbances. The ability of this fungus to produce a wide range of metabolites and enzymes led the authors to suggest that antibiosis as well as enzymatic hydrolysis operated synergistically to reduce the pathogen population densities that, in turn, delayed root colonization. In addition to a direct inhibitory effect on hyphal growth in the rhizosphere, the few *Pythium* hyphae that could penetrate the root epidermis were markedly altered, thus indicating that antibiosis likely contributed to the biocontrol activity of *V. Lecanii in planta* (Figures 4.6a–c). Several host defense reactions, including phenolic compounds, were seen in the infected root tissues, providing evidence that antibiosis, mycoparasitism, and host defense reactions could operate synergistically or concomitantly to control pathogen colonization in *V. Lecanii* inoculated plants.

4.4 Elicitor-Mediated Induced Resistance in Tomato and Cucumber

The term biotic elicitor usually refers to macromolecules, originating from either the host plant or the plant pathogen, capable of inducing structural and/or biochemical responses associated with the expression of plant disease resistance (Ward et al., 1991). A wide range of compounds including defined oligosaccharides (Ryan and Farmer, 1991), glycoproteins and peptides, as well as fungal toxins, the so-called elicitins (Ricci et al., 1993) have been suggested to play a key role in mediating the induction of plant defense reactions. Among the oligosaccharides, strong evidence was provided that chitin and chitosan were elicitors of plant defense reactions (Hadwiger et al., 1988; Lafontaine and Benhamou, 1995). In the early 1980s, Pearce and Ride (1982) reported that treatment of wheat plants with fungal chitin, a polymer of β-1,4-N-acetyl-D-glucosamine, resulted in a rapid induction of cell wall lignification, a structural process considered to be a defense reaction designed to prevent pathogen penetration. In a subsequent report, Barber et al. (1989) convincingly showed that only chitin oligomers with a DP between four and six possessed significant lignification-eliciting activity. Similarly, chitosan, the partially deacetylated derivative of chitin, was found to play a major signaling role in plant-fungus interactions (Benhamou, 1996).

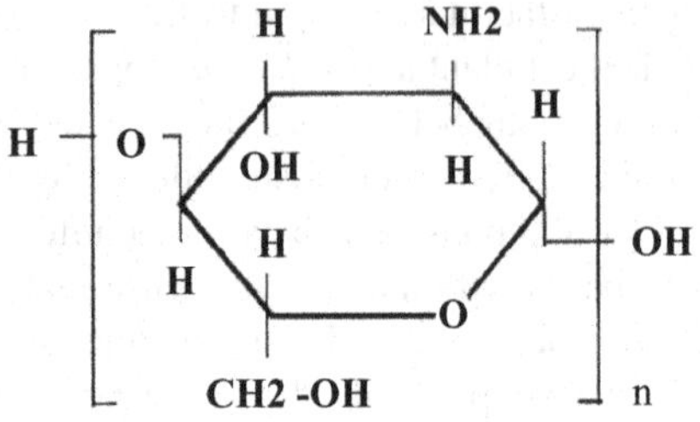

FIGURE 4.7. Schematic representation of an elicitor-active chitosan molecule.

In recent years, a question that has challenged many scientists was how the plant cell could perceive and respond to resistance elicitors. Several lines of evidence have shown that the earliest event in the elicitor-induced transduction pathway was the recognition of a signaling molecule by a specific receptor. Although our knowledge of the molecular structure of plant receptors is rudimentary, the interaction of elicitor molecules with membrane receptors has been associated with complex responses including the production of reactive oxygen intermediates (Boller and Keen, 2000). Beside the oxidative burst, the response of plants to elicitors has been shown to provoke a rapid change in the permeability of the plasma membrane to such ions as Ca^{2+}, H^+, K^+, and Cl^-. Calcium ion fluxes are thought to play a key role in the elicitor signaling pathways. Support of this concept is provided by the observation that treatment of cultured soybean cells with a calcium ionophore induces phytoalexin synthesis (Stab and Ebel, 1987) while calcium channel blockers prevent elicitation of these secondary metabolites. An increase in the intracellular Ca^{2+} concentration is considered to be of prime importance to a number of processes, such as activation of the membrane-bound β-1,3-glucan synthase leading to the formation of callose (Kauss et al., 1989), activation of both protein phosphorylation and cyclic AMP, formation of Ca^{2+}-calmodulin complexes associated with the cytoskeleton, and transduction of the elicitor-mediated signal.

4.4.1 Chitosan: General Properties

Chitosan, a by-product from the seafood industry, is an insoluble high molecular weight cationic polysaccharide (Figure 4.7), considered to be one of the most abundant polymers in nature. Chitosan has several industrial applications and is used in waste water purification, cosmetics and fruit juices, and chelation of transition metals because of its polycationic nature. In addition, the specific properties of this compound, including biodegradability, toxicological safety (Hirano et al., 1990), and bioactivity, have suggested that it could be an ideal pathogen control product in agriculture and horticulture (Hadwiger et al., 1988). In the postharvest industry, the ability of chitosan to form films was explored to coat fruits and vegetables in order to improve resistance against a number of fungal pathogens including *Botrytis cinerea* (Wilson et al., 1994). Reduction of disease incidence by chitosan coating has been reported in tomato, cucumber, strawberry, and bell pepper fruits (El Ghaouth et al., 1992a,b, 1994). Such a reduction of decay has been attributed to the synergistic effect between the antifungal activity of chitosan and its resistance-eliciting properties.

The mechanisms by which chitosan contributes to decrease disease incidence in both plants and post-harvest commodities is still unclear. It is likely that the interaction between the positive charges along the chitosan chain and the negative charges of some molecules at the membrane level is responsible for strong alteration in cell permeability, leading to a series of events involved in the establishment of plant defense reactions. Kauss et al. (1989) reported that the effect of chitosan was not related to specific binding to receptor-like molecules but rather to a more general change in membrane properties. Oligomer size is also known to be an important aspect of chitosan action. Chitosan oligomers with seven or more sugar units have been shown to possess the greatest antifungal and eliciting properties (Kauss et al., 1989).

4.4.2 Chitosan: Antimicrobial and Eliciting Properties

Antimicrobial Properties of Chitosan

Two decades ago, Allan and Hadwiger (1979) were the first to clearly show that chitosan could inhibit the growth of several pathogenic fungi, with the exception of zygomycetes (fungi containing chitosan as a major cell wall component). Later on, Stössel and Leuba (1984) and Leuba and Stössel (1986) confirmed the antifungal potential of chitosan and showed, by using UV absorption analyses, that chitosan caused marked leakage of proteinaceous material from *Pythium paroecandrum* at pH 5.8, thus indicating that fungal cell permeability was altered. More recently, ultrastructural investigations of fungal cultures grown in the presence of chitosan provided new insights into the cytological changes associated with reduction of radial growth. Studies conducted on *F. oxysporum* f. sp. *radicis-lycopersici*, the causal agent of tomato crown and root rot, revealed that not only did chitosan inhibit fungal growth at concentrations ranging from 0 to 2 mg/ml, but also it induced marked morphological changes, structural alterations, and molecular disorganization (Benhamou, 1992). The antifungal properties of chitosan were further demonstrated in a number of pathogens including *Rhizopus stolonifer* and *B. cinerea* (ElGhaouth et al., 1992a; 1997), *Pythium ultimum, Rhizoctonia solani* and *Fusarium graminearum* (Benhamou, unpublished observations). Chitosan was also shown to affect bacterial growth and survival (Sudarshan et al., 1992). A number of pathogenic bacteria, including *Escherichia coli, Enterobacter aerogenes*, and *Staphylococcus aureus* were also found to be strongly affected by chitosan. In all cases, leakage of intracellular components was associated with the observed bacterial alterations.

Eliciting Properties of Chitosan

Since the first demonstration that chitosan could increase membrane permeability of suspension cultured cells, several lines of evidence have indicated that treatment with chitosan induced a number of plant defense reactions including the formation of callose (Kauss et al., 1989), the accumulation of chitinases (Mauch et al., 1984), the production of phytoalexins (Kendra and Hadwiger, 1984), and the synthesis of protease inhibitors (Walker-Simmons and Ryan, 1984).

More recently, the cytologically visible consequences of chitosan treatment were investigated in the tomato-*Fusarium oxysporum* f. sp. *radicis-lycopersii* interaction. In control tomato plants (absence of chitosan treatment), the pathogen ramified rapidly through much of the root tissues causing extensive host cell damage. Application of chitosan oligosaccharides as root coatings, foliar sprays, or seed treatments resulted in enhanced seedling protection against Fusarium attack (Benhamou and Theriault, 1992; Benhamou et al., 1994b). This induced protection was found to correlate with a marked restriction of fungal colonization and with the rapid expression of a number of defense responses, including accumulation of phenolic compounds and formation of structural barriers at sites of attempted fungal penetration (Figure 4.8a). Fungal growth was limited to the epidermis and the outer cortex in which large amounts of phenolic-like compounds accumulated. Hyphae, surrounded by this material, suffered from severe damage and were often reduced to empty shells (Figure 4.8b). Fungal hyphae did not penetrate the innermost cortical cells. However, these host cells exhibited marked changes, mainly characterized by the accumulation of deposits varying in size, shape, and texture (Figure 4.8b). Although indirect evidence of the production of chitinases was provided by the altered pattern of chitin distribution over the cell walls of the invading hyphae, the observation that chitin molecules still occurred over the walls of highly disorganized fungal cells was taken as an indication that production of chitinases was probably preceded by other defense mechanisms. The heavy accumulation of densely-stained, phenolic-like deposits, encircling pathogen hyphae in the colonized cells and also accumulating in the noninfected inner cortex, appeared to be the earliest feature of the host defense response.

Formation of structural barriers at sites of attempted fungal penetration is also an important feature of host reactions to chitosan treatment (Benhamou and Lafontaine, 1995). The most commonly encountered were heterogeneous wall appositions, which formed in the invaded epidermis and outer cortex (Figure 4.9a). Callose and to a lesser extent phenolics were detected in these wall appositions (Figure 4.9b) (Benhamou and Lafontaine, 1995). Conceivably, enrichment of both the host cell walls and the wall appositions with phenolics is likely to contribute to the elaboration of permeability barriers preventing pathogen spread and enzymatic degradation. Several studies have convincingly shown that phenolic structures could confer strong rigidity to the host cell walls through peroxidase-mediated cross-linking with pre-existing wall carbohydrates such as hemicellulose, pectin, and callose (Fry, 1986). In line with these studies, we found that chitosan treatment initiated a marked increase in phenylalanine ammonia-lyase (PAL) and peroxidase activities (unpublished observations). The time-course of PAL and peroxidase levels suggested a coordinated action of these enzymes. Induction of PAL and peroxidase activities was chronologically related to phenolic accumulation and/or polymerization in elicitor-treated plants. Such a relationship was not observed in

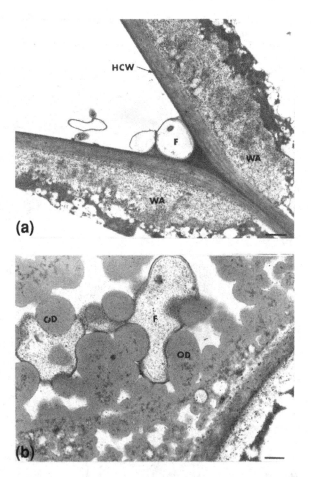

FIGURE 4.8. Transmission electron micrographs of *Fusarium*-infected root tissues from chitosan treated tomato plants. (**a**). Fungal colonization is reduced and restricted to a few cortical cells. Hemispherical and elongated wall appositions (WA) are formed onto the inner surface of the cell wall in some host cells. Bar = 4 μm. (**b**) In the outer root cortex, an osmiophilic material (OD) accumulates along the cell walls and extends toward the inside of the cell, encircling highly damaged hyphae (F). Bar = 0.5 μm.

control plants although some stimulation of the two enzymes could be recorded by one or two days after inoculation.

In light of these observations, Benhamou and Lafontaine (1995) concluded that the wall appositions formed in tomato root tissues upon chitosan treatment and fungal challenge were made of a polysaccharidic matrix mainly composed of callose on which phenolic compounds (likely lignin) were sequentially deposited, probably to build a more impervious composite. Recently published studies agreed on the key role of lignification and phenolic deposition in resistance to disease and speculated on the secondary importance of callose and other polysaccharides in defense. It is obvious that callose, unlike phenolics, does not contribute to the

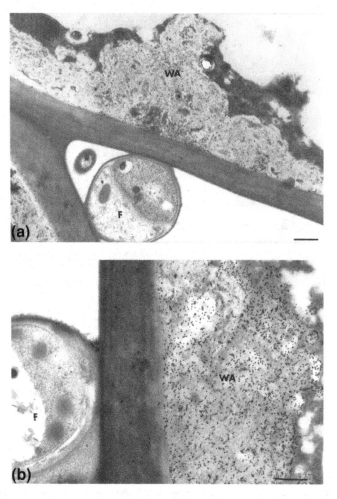

FIGURE 4.9. Transmission electron micrographs of *Fusarium*-infected root tissues from chitosan treated tomato plants, and identification of callose with gold-complexed tobacco β-1,3-glucanase. A strong deposition of gold particles is detected over the matrix of wall appositions. **a** and **b**, bars = 0.5μm.

creation of a fungitoxic environment. However, it may contribute to the establishment of lignin-like compounds by providing potential binding sites.

4.5 Induced Resistance and Integrated Crop Protection Strategy

Although there has been great effort devoted to the identification and testing of microbial antagonists, biological control has relied mainly on the use of single microbial inoculants to suppress diseases. Research over the last few years has led to the concept that application of a single biocontrol agent was not always the most appropriate approach for obtaining sustainable disease management. By contrast, biological mixtures, containing two or more microbial products or agents with complementary modes of action, appear to offer more promise to achieve long-term stable and persistent biological control (Hoitink et al., 1996; Benhamou et al., 1998). During the last decade, much attention has been paid to disease suppressive systems such as naturally suppressive soils (Alabouvette, 1999) and composts (Phae et al., 1990), mainly because of their potential to offer environmental conditions favoring the growth and development of effective microbial antagonists while providing the energy required to support the metabolic activity of these beneficial organisms (De Ceuster and Hoitink, 1999). The modes of action by which suppressive composts operate have been the subject of intensive studies, but it is only recently that plant induced resistance has been suggested to contribute to the biocontrol process mediated by compost-amended substrates (Zhang et al., 1996, 1998). Using a cucumber split root system, Zhang et al. (1996) reported that composted pine bark and composted cow manure induced systemic resistance to *Pythium ultimum* in cucumber whereas a sphagnum peat mix failed to trigger plant disease resistance. This compost mediated induced resistance appeared to correlate with an increase in the activity of some enzymes such as peroxidases and β-1,3-glucanases, known to be important components in the overall plant defense strategy. Recently, Zhang et al. (1998) provided evidence that amendment of composts with a selected biocontrol agent was a valuable option to confer increased resistance against anthracnose in cucumber and bacterial speck in *Arabidopsis*.

In an attempt to investigate the effect of compost obtained from pulp and paper mill residues on the cytologically visible consequences of the response induced in susceptible tomato plants infected by *F. oxysporum* f. sp. *radicis-lycopersici*, Pharand et al. (2002) showed that the beneficial effect of compost in reducing disease symptoms was associated with increased plant resistance to fungal colonization (Figure 4.10). Accumulation of osmiophilic deposits encircling *Fusarium* hyphae (Figure 4.10a) and formation of physical barriers at sites of attempted fungal penetration were important features of reaction in tomato plants grown in compost-amended mix. Understandably, reinforcement of the mechanical properties of the plant cell wall is a prerequisite to successful protection of the internal root tissues. Because it does not require transcriptional activation, callose synthesis, a Ca^{2+}-dependent process, is probably one of the first key events leading to plant cell wall modifications (Moerschbacher and Mendgen, 2000). While it is clear that an increased Ca^{2+} influx is responsible for the observed callose deposition in wall appositions and is involved in the complex network of cell signaling that generate the second messengers and trigger the inducible host response, the exact mechanisms leading to accelerated Ca^{2+} entrance are largely unknown, although a number of possibilities may be raised. First, the enzymatic activity (i.e., chitinases, chitosanases, and β-1,3-glucanases) of the microflora present in the compost may promote the release of fungal and bacterial wall fragments that, in turn, are recognized by specific membrane receptors of the host plant. This would lead to a complex response in which a number of early events including oxidative burst, rapid changes in the permeability of the plasma membrane to such ions as Ca^{2+}, H^+, K^+, and Cl^-, and production of salicylic acid (Boller and Keen, 2000)

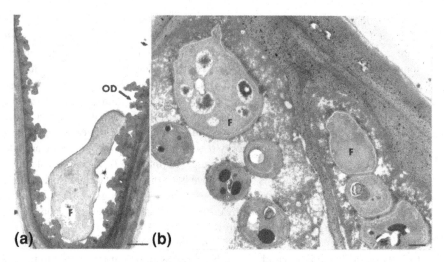

FIGURE 4.10. Transmission electron micrographs of root tissues from tomato plants grown in a mixture of peat moss and pulp and paper mill sludge compost. Samples were collected 7 days after inoculation with *Fusarium oxysporum* f. sp. *radicis-lycopersici*. (**a**) Osmiophilic deposits (OD), accumulating in colonized intercellular spaces, interact with *Fusarium* (F) hyphae causing some alterations. Bar = 2 μm. (**b**) In root tissues from tomato plants grown in a mixture of peat moss and pulp and paper mill sludge compost amended with *Pythium oligandrum*, hyphae of *Fusarium oxysporum* f. sp. *radicis-lysopersici* (F) are embedded in an electron-dense matrix and are surrounded by multitextured wall appositions (WA). Bar = 2 μm.

are stimulated. Second, other enzymes, including pectinases, xylanases, and xyloglucanases, produced by the microorganisms to obtain food sources from paper and mill residues present in the compost, may have generated potential elicitors of plant defense reactions. In recent years, the eliciting properties of various oligosaccharides of plant origin have been abundantly documented in a number of plants and have been shown to play a key role in priming the plant defense system (Cote and Hahn, 1994). Third, fungal proteinaceous molecules, such as the elicitins produced by a number of *Phytophthora* and *Pythium* spp. (Ricci, 1997), may have been the stimulus required for signaling the plant to mobilize its defense strategy.

The most striking and interesting was the amplified response detected in tomato plants grown in *P. oligandrum*-fortified compost. Pharand et al., (2002) showed a substantial increase in the extent and magnitude of the compost-induced cellular changes when *P. oligandrum* was supplied to the potting substrate (Figure 4.10b). This finding corroborates the current concept that amendment of composts with specific antagonists is a valuable option for amplifying their beneficial properties in terms of plant disease suppression (Hoitink et al., 1996). Because *Trichoderma* spp. are known to exhibit antagonism against a large number of soil-borne pathogens (Chet, 1987) and to massively colonize cellulose-containing composts (Thorton and Gilligan, 1999), most attempts to improve the performance of composts have focused on the integration of these fungal microorganisms (Hoitink et al., 1996) and have ignored the potential of other antagonists such as *P. oligandrum* at controlling *Fusarium* wilts (Benhamou et al., 1997). The antimicrobial and elicit-ing properties of *P. oligandrum*, as recently evidenced by the identification of the inducing factor, oligandrin, (Picard et al., 2000) should favor the incorporation of this nonpathogenic organism into compost-amended mixes.

Over the past decades, a number of different approaches have been considered by plant pathologists toward enhancing plant disease resistance. Among these, the use of naturally suppressive composts as part of an integrated disease control strategy offers exciting opportunities, although it is clear that unequivocal answers to key questions are lacking, including the stability and persistence of the induced host response, the efficiency of such substrates under commercial conditions, and their suitability in an integrated crop protection system. In spite of these limi-tations, the recent advances in our fundamental understanding of the nature of compost-mediated induced resistance in plants highlights the great potential of these peat substitutes in greenhouse crop protection. The cytological demonstra-tion that pathogen growth and development were restricted or even halted and that structural and biochemical barriers were elaborated in plant tissues underlying ar-eas of pathogen penetration gives reason to believe that compost-amended potting mixes may play a key role in the control of a large number of root diseases. Com-posts are gaining increased popularity not only from an ecological point of view as a means to recycle pollutants, but also because their potential in suppressing soilborne plant diseases clearly meets with the current needs toward sustainable agriculture at a lower environmental cost.

4.6 Conclusion

Over the past two decades, a number of different approaches have been considered by plant pathologists toward enhancing plant disease resistance. Among these, the use of non-specific resistance elicitors as part of an integrated disease control strategy offers exciting opportunities. However, it is clear that unequivocal an-swers to key questions, including the stability and persistence of the induced host response, the efficiency of such agents, products and/or molecules under commer-cial conditions, and their suitability in an integrated crop protection system, need to be answered before elicitors can be considered as powerful crop protectants. In spite of these limitations, the recent advances in our fundamental understanding of the nature of microbially- and chitosan-mediated induced resistance in plants highlights the great potential of induced resistance in plant protection. The demon-stration that pathogen growth and development were restricted or even halted and that structural and biochemical barriers were elaborated in plant tissues underlying areas of pathogen penetration gives reason to believe that induced resistance may be active against a wide array of pathogens and even insects, thereby increasing the level of resistance. It is clear that exploiting plant induced resistance as an alternative strategy of disease and pest management clearly meets with the cur-rent needs toward sustainable agriculture at a lower environmental cost. However, coordinated research efforts are still needed to develop programmes dealing with molecular genetic analyses, formulation studies, and large-scale experiments.

Acknowledgments

The author acknowledges the financial support of research in her laboratory by the Fonds Québécois pour la Formation de Chercheurs et l'Aide à la Recherche (FCAR) and by the Natural Sciences and Engineering Council of Canada (NSERC). The author wishes also to thank Alain Goulet and Chantal Garand for excellent technical assistance.

References

Alabouvette, C. 1999. *Fusarium* wilt suppressive soils: an example of disease-suppressive soils. *Aust. Plant Pathol.* 28:57–64.

Albert, F., and Anderson, A.J. 1987. The effect of *Pseudomonas putida* colonization on root surface peroxidase. *Plant Physiol.* 85:537–541.

Allan, C.R, and Hadwiger, L.A. 1979. The fungicidal effect of chitosan on fungi of varying cell wall composition. *Exp. Mycol.* 3:285–287.

Anderson, A.J., and Guerra, D. 1985. Responses of bean to root colonization with *Pseudomonas putida* in a hydroponic system. *Phytopathology* 75:992–995.

Askary, H., Benhamou, N., and Brodeur, J. 1997. Ultrastructural and cytochemical investigations of the antagonistic effect of *Verticillium lecanii* on cucumber powdery mildew. *Phytopathology* 87:359–368.

Askary, H., Carriere, Y., Belanger, R.R., and Brodeur, J. 1998. Pathogenicity of the fungus *Verticillium lecanii* to aphids and powdery mildew. *Biocontrol Sci. Technol.* 8:23–32.

Askary, H., Benhamou, N., and Brodeur, J. 1999. Ultrastructural and cytochemical characterization of aphid invasion by the hyphomycete *Verticillium lecanii*. *J. Invertebr. Pathol.* 74:1–13.

Barber, M.S., Bertram, R.E., and Ride, J.P. 1989. Chitin oligosaccharides elicit lignification in wounded leaves. *Physiol. Mol. Plant Pathol.* 34:3–12.

Baker, R. 1989. Improved *Trichoderma* spp. for promoting crop productivity. *Trends Biotech.* 7:34–38.

Bakker, P.A. H.M., van Peer, R., and Schippers, B. 1990. Specificity of siderophores and siderophore receptors in biocontrol by *Pseudomonas* spp. In *Biological Control of Soilborne Pathogens*, ed. M.K.Hornby, pp.131–142. Wallingford: CAB International.

Beckman, C.H. 1987. *The Nature of Wilt Diseases of Plants*, 175. American Phytopathological Society.

Benhamou, N. 1992. Ultrastructural and cytochemical aspects of chitosan on *Fusarium oxysporum* f. sp. *radicis-lycopersici*, agent of tomato crown and root rot. *Phytopathology* 82:1185–1193.

Benhamou, N. 1996. Elicitor-induced plant defense pathways. *Trends Plant Sci.* 1:233–240.

Benhamou, N., and Belanger, R.R. 1998a. Induction of systemic resistance to *Pythium* damping-off in cucumber plants by benzothiadiazole: ultrastructure and cytochemistry of the host response. *Plant J.* 14:13–21.

Benhamou, N., and Belanger, R.R. 1998b. Benzothiadiazole-mediated induced resistance to *Fusarium oxysporum* f. sp. *radicis-lycopersici* in tomato. *Plant Physiol.* 118:1203–1212.

Benhamou, N., and Brodeur, J. 2000. Evidence for antibiosis and induced host defense reactions in the interaction between *Verticillium lecanii* and *Penicillium digitatum*, the causal agent of green mold. *Phytopathology* 90:932–943.

Benhamou, N., and Brodeur, J. 2001. Pre-inoculation of Ri T-DNA transformed cucumber roots with the mycoparasite, *Verticillium lecanii,* induces host defense reactions against *Pythium ultimum* infection. *Physiol. Mol. Plant Pathol.* 58:133–146.

Benhamou, N., and Chet, I. 1997. Cellular and molecular mechanisms involved in the interaction between *Trichoderma harzianum* and *Pythium ultimum. Appl. Environ. Microbiol.* 63:2095–2099.

Benhamou, N., and Lafontaine, P.J. 1995. Ultrastructural and cytochemical characterization of elicitor-induced responses in tomato root tissues infected by *Fusarium oxysporum* f. sp. *radicis-Lycopersici. Planta* 197:89–102.

Benhamou, N., and Nicole, M. 1999. Cell biology of plant immunization against microbial infection: the potential of induced resistance in controlling plant diseases. *Plant Physiol. Biochem.* 37:703–719.

Benhamou, N., and Theriault, G. 1992. Treatment with chitosan enhances resistance of tomato plants to the crown and root rot pathogen *Fusarium oxysporum* fsp. *radicis-lycopersici. Physiol. Mol. Plant Pathol.* 41:33–52.

Benhamou, N., Chamberland, H., Ouellette, G.B., and Pauze, F.I. 1987. Ultrastructural localization of β-1,4-D-glucans in two pathogenic fungi and in their host tissues by means of an exoglucanase-gold complex. *Can. J. Microbiol.* 33:405–417.

Benhamou, N., Chamberland, H., and Pauze, F. I. 1990. Implication of pectic components in cell surface interactions between tomato root cells and *Fusarium oxysporum* f. sp. *radicis lycopersici.* A cytochemical study by means of a lectin with polygalacturonic-acid binding specificity. *Plant Physiol.* 92:995–1003.

Benhamou, N., Fortin, I.A., Hamel, E., St Arnaud, M., and Shatill A.A. 1994a. Resistance responses of mycorrhizal Ri T-DNA transformed carrot roots to infection by *Fusarium oxysporum* f. sp. *chrysanthemi. Phytopathology* 84:958–968.

Benhamou, N., Lafontaine, P.IL., and Nicole, M. 1994b. Seed treatment with chitosan induces systemic resistance to *Fusarium* crown and root rot in tomato plants. *Phytopathology* 84:1432–1444.

Benhamou, N., Belanger, R.R., and Paulitz, T.E. 1996a. Ultrastructural and cytochemical aspects of the interaction between *Pseudomonas fluorescens* and Ri T-DNA transformed pea roots: host response to colonization by *Pythium ultimum* Trow. *Planta* 199: 105–117.

Benhamou, N., Belanger, R.R., and Paulitz, T.E. 1996b. Induction of differential host responses by *Pseudomonas fluorescens* in Ri-T DNA transformed pea roots upon challenge with *Fusarium oxysporum* f. sp. *pisi* and *Pythium ultimum. Phytopathology* 86:1174–1185.

Benhamou, N., Kloepper, J.W., Quadt-Hallmann, A., and Tuzun, S. 1996c. Induction of defense-related ultrastructural modifications in pea root tissues inoculated with endophytic bacteria. *Plant Physiol.* 112:919–929.

Benhamou, N., Rey, P., Cherif, M., Hockenhull, J., and Tirilly, Y. 1997. Treatment with the mycoparasite, *Pythium oligandrum,* triggers the induction of defense-related reactions in tomato roots upon challenge with *Fusarium oxysporum* f. sp. *radicis-lycopersici. Phytopathology* 87:108–122.

Benhamou, N., Kloepper, J.W., and Tuzun, S. 1998. Induction of resistance against Fusarium wilt of tomato by combination of chitosan with an endophytic bacterial strain: Ultrastructure and cytochemistry of the host response. *Planta* 204:153–168.

Benhamou, N., Rey, P., Picard, K., and Tirilly, Y. 1999. Ultrastructural and cytochemical aspects of the interaction between the mycoparasite, *Pythium oligandrum,* and soilborne pathogens, *Phytopathology* 89:506–517.

Benhamou, N., Gagne, S., LeQuere D., and Dehbi, L. 2000. Bacterial-mediated induced resistance in cucumber: beneficial effect of the endophytic bacterium, *Serratia plymuthica*, on the protection against infection by *Pythium ultimum*. *Phytopathology* 90:45–56.

Benhamou, N., Belanger, R.R., Rey, P., and Tirilly, Y. 2001. Oligandrin, the elicitin-like protein produced by the mycoparasite *Pythium oligandrum*, induces systemic resistance to Fusarium crown and root rot in tomato. *Plant Physiol. Biochem.* 39:681–696.

Binet, M.N., Bourque, S., Lebrun-Garcia, A., Chiltz, A., and Pugin, A. 1998. Comparison of the effects of cryptogein and oligogalacturonides on tobacco cells and evidence of different forms of desensitization induced by these elicitors. *Plant Sci.* 137:33–41.

Boller, T., and Keen, N. 2000. Resistance genes and the perception and transduction of elicitor signals in host-pathogen interactions. In *Mechanisms of Resistance to Plant Diseases*, eds. A.J. Sluzarenko, R.S.S. Fraser, and L.E. van Loon, pp. 189–229. Dordrecht, The Netherlands: Kluwer Academic Press.

Brammall, R.A., and Higgins, V.J. 1988. A histological comparison of fungal colonization in tomato seedlings susceptible and resistant to Fusarium crown and root rot disease. *Can. J. Bot.* 66:915–925.

Chen, C., Bauske, E.M., Musson, G., Rodriguez-Cabana, R., and Kloepper, L.W. 1995. Biological control of *Fusarium* wilt on cotton by use of endophytic bacteria. *Biol. Control* 5:83–91.

Chen, E., Paulitz, T.E., Belanger, R.R., and Benhamou, N. 1997. Inhibition of growth of *Pythium aphanidermatum* and stimulation of plant defense enzymes in split roots of cucumber systemically induced with *Pseudomonas* spp. *Phytopathology* 87:S 18 (Abstract)

Cherif, M., Benhamou, N., Menzies, J.G., and Belanger, R.R. 1992. Silicon induced resistance in cucumber plants against *Pythium ultimum*. *Physiol. Mol. Plant Pathol.* 41:411–425.

Chet, I. 1987. *Trichoderma*-Applications, mode of action and potential as a biocontrol agent of soilborne plant pathogenic fungi. In *Innovative Approaches to Plant Diseases*, ed. I. Chet, pp. 137–160. New York: John Wiley & Sons.

Chet, I., and Inbar, J. 1994. Biological control of fungal pathogens. *Appl. Biochem. Biotechnol.* 48:37–43.

Cook, R.J. 1993. Making greater use of introduced microorganisms for biological control of plant pathogens. *Annu. Rev. Phytopathol.* 31:53–80.

Cote, F., and Hahn, M. 1994. Oligosaccharins: structures and signal transduction. *Plant Mol. Biol.* 26:1379–1411.

Daayf, F., Schmitt, A., and Belanger, R.R. 1997. Evidence of phytoalexins in cucumber leaves infected with powdery mildew following treatment with leaf extracts of *Reynoutria sachalinensis*. *Plant Physiol.* 113:719–727.

Deacon, J.W. 1976. Studies on *Pythium oligandrum*, an aggressive parasite of other fungi. *Trans. Br. Mycol. Soc.* 66:383–391.

De Ceuster, T.J.J., and Hoitink, H.A.J. 1999. Prospects for composts and biocontrol agents as substitutes for methyl bromide in biological control of plant diseases. *Compost Sci. Uti!.* 7:6–15.

Dimock, M.B., R.M. Beach, and Carlson, P.S. 1989. Endophytic bacteria for the delivery of crop protection agents. In *Biotechnology, Biological Pesticides and Novel Plant-Pest Resistance for Insect Pest Management*, eds. D.W. Roberts, and R.R. Granados, pp. 88–92. Boyce Thompson Institute for Plant Research.

Drechsler, C. 1943. Antagonism and parasitism among some oomycetes associated with root rot. *J. Washington Sci. Acad.* 33:21–28.

El Ghaouth, A., Ponnampalam, R, Castaigne, F., and Arul, J. 1992a. Chitosan coating to extend the storage life of tomatoes. *HortScience* 27:1016–1018.

El Ghaouth, A., Arul, J., Grenier, J., and Asselin, A. 1992b. Antifungal activity of chitosan on two post-harvest pathogens of strawberry fruits. *Phytopathology* 82:398–402.

El Ghaouth, A., Arul, J., Benhamou, N., Asselin, A., and Belanger, R.R. 1994. Effect of chitosan on cucumber plants: suppression of *Pythium aphanidermatum* and induction of defense reactions. *Phytopathology* 84:313–320.

El Ghaouth, A., Arul, J., Wilson, C. and Benhamou, N. 1997. Biochemical and cytochemical aspects of the interactions of chitosan and *Botrytis cinerea* in bell pepper fruit. *Postharvest Biol. Technol.* 12:183–194.

Fiddaman P.l., and Rossall, S. 1993. The production of antifungal volatiles by *Bacillus subtilis*. *J. Appl. Bacteriol.* 74:119–126.

Flores, H.E., and Curtis, W.R. 1992. Approaches to understanding and manipulating the biosynthetic potential of plant roots. *Ann. NY. Acad. Sci.* 665:188–209.

Foley, M.F., and Deacon, J.W. 1986. Susceptibility of *Pythium* species and other fungi to antagonism by the mycoparasite *Pythium oligandrum*. *Soil Biol. Biochem.* 18: 91–95.

Fry, S.C. 1986. Polymer-bound phenols as natural substrates of peroxidases. In *Molecular and Physiological Aspects of Plant Peroxidase*, eds. H.Greppin, C. Penel, and T.H. Gaspar, pp. 169–182. Switzerland: Universite de Geneve.

Gullino, M.L, Alabajes, R., and van Lenteren, J.C. 1999. Setting the stage: chraracteristics of protected cultivation and tools for sustainable crop protection. In *Integrated Pest and Disease Management in Greenhouse Crops*, eds. R. Alabajes, M.L. Gullino, J.C. van Lenteren, and Y. Elad, pp. 1–15. New York: Kluwer Academic Press.

Hadwiger, L.A., Chiang, C., Victory, S., and Horovitz, D. 1988. The molecular biology of chitosan in plant-pathogen interactions and its application to agriculture. In *Chitin and Chitosan: Sources, Chemistry, Biochemistry, Physical Properties and Applications*, eds. G. Skjiik, B.T. Anthonsen, and P. Sandford, pp. 119–138. Amsterdam: Elsevier Applied Sciences.

Hall, R.A. 1981. The fungus *Verticillium lecanii* as a microbial insecticide against aphids and scales. In *Microbial Control of Pests and Plant Diseases*, ed. H.D. Burges, pp. 483–498. New York: Academic Press.

Handelsman, J., and Stabb, E.V. 1996. Biocontrol of soilborne plant pathogens. *Plant Cell* 8:1855–1869.

Harman, G.B. 2000. Myths and dogma of biocontrol: changes in perceptions derived from research on *Trichoderma harzianum* T-22. *Plant Dis.* 84:377–393.

Harman, G.E., and Bjorkman, T. 1997. Potential and existing uses of *Trichoderma* and *Gliocladium* for plant disease control and plant growth enhancement. In *Trichoderma and Gliocladium*, eds. C.K. Kubicek, and G.E. Harman, pp. 229–295. London: Taylor and Francis.

Heath, M.C. 2000. Advances in imaging the cell biology of plant-microbe interactions. *Annu. Rev. Phytopathol.* 38:443–459.

Hirano, S., Itakura, C., Seino, H., Akiyama, Y., Nonaka, I., Kanbara, N., and Kawakami, T. 1990. Chitosan as an ingredient for domestic animal feeds. *J. Agr. Food Chern.* 38:1214–1217.

Hoitink, H.A.J., Stone, A.G., and Han, D.Y. 1996. Suppression of plant diseases by composts. *HortScience* 32:184–187.

Jarvis, W.R. 1988. Fusarium crown and root rot of tomatoes. *Phytoprotection* 69:49–64.

Kauss, H., Jellick, W., and Domard, A. 1989. The degrees of polymerization and N- acetylation of chitosan determine its ability to elicit callose formation in suspension cells and protoplasts of *Catharanthus roseus*. *Planta* 178:385–392.

Kendra, F.D., and Hadwiger, L.A. 1984. Characterization of the smallest chitosan oligomer that is maximally antifungal to *Fusarium solani* and elicits pisatin formation in *Pisum sativum Exp. Mycol.* 8:276–281.

Kloepper, J.W. 1993. Plant growth-promoting rhizobacteria as biological control agents. In *Soil Microbial Technologies*, ed. B. Metting, pp. 255–274. New York: Marcel Dekker Inc.

Kohle, H., Jeblick, W., Poten, F., Blaschek, W., and Kauss, H. 1985. Chitosan-elicited callose synthesis in soybean cells as a Ca^{2+} dependent process. *Plant Physiol.* 77:544–551.

Kuc, J. 1987. Plant immunization and its applicability for disease control. In *Innovative Approaches to Plant Disease Control*, ed. I.Chet, pp. 255–274. New York: John Wiley & Sons.

Lafontaine, P.J., and Benhamou, N. 1995. Chitosan treatment: an emerging strategy for enhancing resistance of greenhouse tomato plants to infection by *Fusarium oxysporum* f. sp. *radicis-lycopersici. Biocontrol Sci. Technol.* 6:111–124.

Leuba, J.L, and Sössel P. 1986. Chitosan and other polyamines: antifungal activity and interaction with biological membranes. In *Chitin in Nature and Technology*, eds. R. Muzzarelli, C. Jeuniaux, and G.W. Gooday, pp. 215–222. New York: Plenum Press.

Malathrakis, N.E., and Goumas, D.E. 1999. Fungal and bacterial diseases. In *Integrated Pest and Disease Management in Greenhouse Crops*, eds. R. Albajes, M.L. Gullino, J.C. van Lenteren, and Y. Elad, pp. 34–47. Kluwer Academic Press.

Mauch, F., Hadwiger, L.A., and Boller, T. 1984. Ethylene: Symptom, not signal for the induction of chitinase and β-1,3-glucanase in pea pods by pathogens and elicitors. *Plant Physiol.* 76:607–611.

Martin, F.N., and Hancock, J.G. 1987. The use of *Pythium oligandrum* for biological control of preemergence damping-off caused by *P. ultimum. Phytopathology* 77:1013–1020.

Metraux, J.P., AhlGoy, P., Stub, T., Speich, J., Wyss-Benz, M. et al. 1991. Induced resistance in cucumber in response to 2-6-dichloroisonicotinic acid and pathogens. In *Advances in Molecular Genetics of Plant Microbe Interactions*, eds. H. Hennecke, and D.P.S.Verma, pp. 432–439. Dordrecht, The Netherlands: Kluwer Academic Press.

Moerschbacher, B., and Mendgen, K. 2000. Structural aspects of defense. In *Mechanisms of Resistance to Plant Diseases*, eds. A.J. Sluzarenko, R.S.S. Fraser, and L.C. van Loon, pp. 2310–277. Dordrecht, The Netherlands: Kluwer Academic Press.

Morandi, D., Bailey, J.A., and Gianinazzi-Pearson, V. 1983. Isoflavonoid accumulation in soybean roots infected with vesicular-arbuscular mycorrhizal fungi. *Physiol. Plant Pathol.* 24:357–364.

Mpiga, P., Belanger, R.R., Paulitz, T.C., and Benhamou, N. 1997. Increased resistance to *Fusarium oxysporum* f. sp. *radicis-lycopersici* in tomato plants treated with the endophytic bacterium, *Pseudomonas fluorescens,* strain 63-28. *Physiol. Mol. Plant Pathol.* 50:301–320.

Paulitz, T.C., and Matta, A. 1999. The role of the host in biological control of diseases. In *Integrated Pest and Disease Management in Greenhouse Crops*, eds. R. Alabajes, M.L. Gullino, J.C. van Lenteren, and Y. Elad, pp. 394–410. New York: Kluwer Academic Press.

Pearce, R.B., and Ride, J.P. 1982. Chitin and related compounds as elicitors of the lignification response in wounded wheat leaves. *Physiol. Plant Pathol.* 20:119–123.

Phae, C.G., Saski, M., Shoda, M., and Kubota, H. 1990. Characteristics of *Bacillus subtilis* isolated from composts suppressing phytopathogenic microorganisms. *Soil Sci. Plant Nutr.* 36:575–586.

Pharand, B., Carisse, O., and Benhamou, N. 2002. Cytological aspects of compost-mediated induced resistance against Fusarium crown and root rot in tomato. *Phytopathology.*

Picard, K., Ponchet, M., Blein J.P., Tirilly, Y., and Benhamou, N. 2000. Oligandrin: A proteinaceous molecule produced by the mycoparasite, *Pythium oligandrum*, induces resistance to *Phytophthora parasitica* infection in tomato plants. *Plant Physiol.* 124:379–395.

Ponchet, M., Panabieres, F., Milat, M.L., Mike, S., Montillet, J.L., Suty, L., Triantaphylides, C., Tirilly, Y., and Blein, J.P. 1999. Are elicitins cryptograms in plant –oomycete communications? *Cell Mol. Life Sci.* 56:1020–1047.

Pugin, A., Frachisse, J.M., Tavernier, F., Bligny, R., Gout, E., Douce, R. et al. 1997. Early events induced by the elicitor cryptogein in tobacco cells: involvement of a plasma membrane NADPH oxidase and activation of glycolysis and the pentose phosphate pathway. *Plant Cell* 9:2077–2091.

Raskin, I. 1992. Role of salicylic acid in plants. *Annu. Rev. Plant Physiol. Plant Mol. Biol.* 43:439–463.

Rey, P., Benhamou, N., Wulff, J., and Tirilly, Y. 1998. Interactions between tomato (*Lycopersicon esculentum*) root tissues and the mycoparasite *Pythium oligandrum*. *Physiol. Mol. Plant Pathol.* 53:105–122.

Ricci, P. 1997. Induction of the hypersensitive response and systemic acquired resistance by fungal proteins: the case of elicitins. 1997. In *Plant-Microbe Interactions*, eds. G. Stacey, and N.T. Keen, pp. 53–75. New York: Chapman & Hall.

Ricci, P., Panabieres, F., Bonnet, P., Maia, N., Ponchet, M., Devergne, J.C., Marais, A., Cardin, L., Milat, M.L., and Blein, J.P. 1993. Proteinaceous elicitors of plant defense responses. In *Mechanisms of Plant Defense Responses*, eds. B. Fritig, and M. Legrand, pp. 121–135. Dordrecht, The Netherlands: Kluwer Academic Publishing.

Ride, J.P. 1983. Cell wall and other structural barriers in defense. In *Biochemical Plant Pathology*, ed. J.A. Callow, pp. 215–235. New York: John Wiley & Sons.

Ryan, C.A., and Farmer, E.F. 1991. Oligosaccharide signals in plants: a current assessment. *Annu. Rev. Physiol. Mol. Biol.* 42:651–674.

Savary, B.J., and Flores, H.E. 1994. Biosynthesis of defense-related proteins in transformed root cultures of *Trichosanthes kirilowii* Maxim. var *japonicum* (Kitam.). *Plant Physiol.* 106:1195–1204.

Siegrist, J., Jeblick, W., and Kauss, H. 1994. Defense responses in infected and elicited cucumber (*Cucumis sativus* L) hypocotyl segments exhibiting acquired resistance. *Plant Physiol.* 105:1365–1374.

Shirmbock, M., Lorito, M., Wang, Y.L, Hayes, C.K., Arisan-Atac, I., Scala, F., Harman, G.E., and Kubicek, C.P. 1994. Parallel formation and synergism of hydrolytic enzymes and peptaibol antibiotics, molecular mechanisms involved in the antagonistic action of *Trichoderma harzianum* against phytopathogenic fungi. *Appl. Environ. Microbiol.* 60:4364–4370.

Spanu, P., Boller, T., Ludwig, A., Wiemken, A., Faccio, A., and Bonfante-Fasolo, P. 1989. chitinase in roots of mycorrhizal *Allium porum*: regulation and localization. Planta 177:447–455.

Spencer, D.M., and Atkey, P.T. 1981. Parasitic effects of *Verticillium lecanii* on two rust fungi. *Trans. Br. Mycol. Soc.* 77:535–542.

Stab, M.R. and Ebel, J. 1987. Effects of Ca^{2+} on phytoalexin induced by fungal elicitors in soybean. *Arch. Biochem. Biophys.* 257:416–423.

Stössel, P., and Leuba, J.L 1984. Effect of chitosan, chitin and some amino sugars on growth of various soilborne pathogenic fungi. *Phytopathol. Z.* 111:82–90.

Sudarshan, N.R., Hoover, D.G., and Knorr, D. 1992. Antibacterial action of chitosan. *Food Biotechnol.* 6:257–272.

Thornton, C.R., and Gilligan, C.A. 1999. Quantification of the effect of the hyperparasite *Trichoderma harzianum* on the saprophytic growth dynamics of *Rhizoctonia solani* in compost using a monoclonal antibody-based ELISA. *Mycol. Res.* 103:443–445.

Tuzun, S., and Kloepper, J.W. 1995. Practical application and implementation of induced resistance. In *Induced Resistance to Disease in Plants*, eds. R. Hammerschmidt, and J. Kuc, pp. 152–168. Dordrecht, Boston, London: Kluwer Academic Press.

van Peer, R., Niemann, G.J., and Schippers, B. 1991. Induced resistance and phytoalexin accumulation in biological control of Fusarium wilt of carnation by *Pseudomonas* sp. strain WCS417r. *Phytopathology* 81:728–734.

Verhaar, M.A., Hijwegen, T., and Zadoks, J.E. 1996. Glasshouse experiments on biocontrol of cucumber powdery mildew (*Sphaerotheca fuliginea*) by the mycoparasites *Verticillium lecanii* and *Sporothrix rugulosa*. *Biol. Control.* 6:353–360.

Walker-Simmons, M., and Ryan, C.A. 1984. Proteinase inhibitor synthesis in tomato leaves. Induction by chitosan oligomers and chemically modified chitosan and chitin. *Plant Pathol.* 76:787–790.

Ward, E.R., Uknes, S.J., Williams, S.C., Dincher, S.S., Wiederhold, D.L., Alexander, D.C., Ahl-Goy, P., Metraux, J.P., and Ryals, J.A. 1991. Coordinate gene activity in response to agents that induce systemic acquired resistance. *Plant Cell* 3:1085–1094.

Wei, G., Kloepper, J.W., and Tuzun, S. 1994. Induced systemic resistance to cucumber diseases and increase plant growth by plant growth-promoting rhizobacteria under field conditions. In *Improving Plant Productivity with Rhizosphere Bacteria*, eds. M.H. Ryder, P.M. Stephens and G.D. Bowen, pp. 70–71. Glen Osmond, Australia: CSIRO Division of soils.

Wilson, C.L., ElGhaouth, A., Chalutz, E., Droby, S., Stevens, C, Lu, J.Y., Khan, V., and Arul, J. 1994. Potential of induced resistance to control postharvest diseases of fruit and vegetables. *Plant Dis.* 9:837–843.

Yedidia, I., Benhamou, N., Chet, I. 1999. Induction of defense responses in cucumber by the biocontrol agent *Trichoderma harzianum*. *Appl. Environ. Microbiol.* 65:1061–1070.

Yedidia, I., Benhamou, N., Kapulnik, Y., and Chet, I. 2000. Activation of plant defense responses following the colonization and penetration of cucumber roots by *Trichoderma harzianum*. *Plant Physiol. Biochem.* 38:863–873.

Zdor, R.E., and Anderson, A.J. 1992. Influence of root colonizing bacteria on the defense responses of bean. *Plant Soil* 140:99–107.

Zhang, W., Dick, W.A., and Hoitink, H.A.J. 1996. Compost induced systemic acquired resistance in cucumber to *Pythium* root rot and anthracnose. *Phytopathology* 86:1066–1070.

Zhang, W., Han, D.Y., Dick, W.A., and Hoitink, H.A.J. 1998. Compost and compost water extract-induced systemic acquired resistance in cucumber and *Arabidopsis*. *Phytopathology* 88:450–455.

5

The Hypersensitive Response in Plant Disease Resistance

Naohide Watanabe and Eric Lam

5.1 Hypersensitive Response: The Phenomenon

5.1.1 Physical Properties of the Hypersensitive Response

Plants can recognize certain pathogens and activate defenses (called the resistance response) that result in the limitation of pathogen growth at the site of infection. One dramatic hallmark of the resistance response is the induction of rapid and localized cell death, a reaction known as the hypersensitive response (HR), when plants are challenged with an incompatible pathogen. HR cell death is also manifested as a collapse of the infected tissue (see Fig. 5.1) and is considered to be involved in pathogen resistance by creating a physical barrier that may impede proliferation and spread of some pathogens (Goodman and Novacky, 1994; Alfano and Collmer, 1996). Furthermore, the HR is important for limiting the nutrient supply of some pathogens, since the dying tissue rapidly becomes dehydrated. Thus, the antimicrobial defense of plant cells is thought to involve the activation of a suicide pathway in infected cells.

Programmed cell death (PCD) is one of the key mechanisms controlling cell proliferation, generation of developmental patterns, and defense of animals against pathogens and environmental insults (Schwartzman and Cidlowski, 1993). One of the most widely studied forms of PCD is apoptosis, a type of PCD that displays a distinct set of physiological and morphological features (Martin et al., 1994). Morphological hallmarks of apoptosis include the condensation of chromatin at the nuclear periphery, the condensation and vacuolization of the cytoplasm and blebbing of the plasma membrane. Despite these cellular changes, the mitochondria remain relatively stable. These changes are followed by breakdown of the nucleus and fragmentation of the cell to form apoptotic bodies (Schwartzman and Cidlowski, 1993). Among the many biochemical changes commonly found in cells undergoing apoptosis is the systematic fragmentation and degradation of nuclear DNA (Bortner et al., 1995). Large fragments of 300 kb and/or 50 kb are first produced by endonucleolytic degradation of nuclear DNA (Oberhammer et al., 1993). These are further degraded by cleavage at linker DNA sites between nucleosomes resulting in DNA fragments that are multimers of about 180 bp

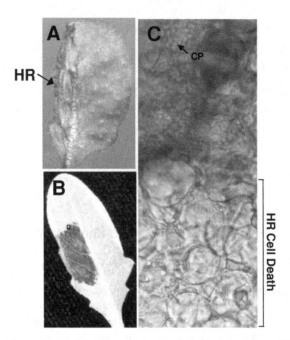

FIGURE 5.1. Morphological observation of collapsed cells of *Arabidopsis* leaf after induction of HR cell death. (**a**) Fully expanded leaf of 4- to 5-week-old *Arabidopsis* plants (ecotype Colombia; Col-0) was infiltrated with *Pseudomonas syringae* pv *maculicola* strain that contains the *avrRpt2* gene (*P.s.m*. ES4326/*avrRpt2*; 10^5 CFU/cm^2) and induced an HR. The leaf was photographed at 20 hour-post-infiltration. (**b**) The leaf was sampled 24 hour-post-infiltration and was fixed in 10% formaldehyde-5% acetic acid-50% ethanol for 3 hours, dehydrated through a graded ethanol series (50, 75, and 100% for 20 min at each step), and incubated in 100% ethanol at 4°C for 3 hours. After rehydrating in water, the collapsed cells were observed by light microscopy. (**c**) shows the higher magnification view of the area which is shown as a red frame in (**b**). CP, chloroplast.

(Wyllie et al., 1984). Degradation of nuclear DNA during apoptosis is coordinated with activation of specific endonucleases that are thought to mediate chromatin cleavage (Peitsch et al., 1993). Cells undergoing HR cell death have some of the features that characterize apoptosis, including condensation and vacuolization of the cytoplasm, blebbing of the plasma membrane, stable mitochondria, and cell shrinkage (Roebuck et al., 1978; Levine et al., 1996; Mittler et al., 1997b; Che et al., 1999; Kawasaki et al., 1999). Moreover, the biochemical events involved in apoptosis, such as activation of specific endonucleases and DNA fragmentation, are also found in plant PCD (Mittler and Lam, 1995; Ryerson and Heath, 1996; Wang et al., 1996b; Mittler and Lam, 1997; Mittler et al., 1997a,b; Sugiyama et al., 2000). These observations indicate that some mechanisms of cell death activation may be conserved between animals and plants.

It is believed that the HR constitutes one of the mechanisms of resistance to plant pathogens. Induction of HR is often associated with elevated levels of salicylic acid (SA), a key regulator of defense responses and pathogen resistance, synthesis of pathogenesis-related (PR) proteins that exhibit antimicrobial activity such as glucanases and chitinases, thickening and hardening of cell walls, and production of antimicrobial compounds called phytoalexins (Hammond-Kosack and Jones, 1996; Ryals et al., 1996). Furthermore, recognition of an incompatible pathogen triggers the rapid production of reactive oxygen intermediates (ROI) superoxide (O_2^-) and hydrogen peroxide (H_2O_2) in an oxidative burst (Lamb and Dixion, 1997). ROI, in turn, drive crosslinking of the cell wall (Bradley et al., 1992), induce several plant genes involved in cellular protection and defense (Chen et al., 1993; Levine et al., 1994; Jabs et al., 1996), and are necessary for the initiation of host cell death in HR (Lamb and Dixion, 1997). However, these signaling intermediates may not be sufficient to activate cell death on their own. Evidence for involvement of nitric oxide (NO) in the activation of HR cell death has recently been reported (Delledonne et al., 1998; Durner et al., 1998; Clark et al., 2000). It was also suggested that SA, which accumulates during the HR, is involved in the production of ROI (Chen et al., 1993; Chamnongpol et al., 1996; Durner and Klessig, 1996; Takahashi et al., 1997). Thus, multiple secondary signals, such as ROIs, SA, and NO, appear to be essential second messengers for the activation and execution of HR cell death. As a result of the multiple biochemical events during the induction of the HR, growth of the pathogen is restricted.

5.1.2 Genetics of Host-Microbe Signaling

The resistance of some plants to infection by certain pathogens reflects the presence of disease resistance (*R*) genes, which are predicted to encode receptors for pathogen-derived molecules (see in recent review: Shirasu and Schulze-Lefert, 2000; Dangl and Jones, 2001). A single gene in the host (the *R* gene) confers resistance only to those pathogen isolates containing a corresponding *Avr* gene (Flor, 1971). This "gene-for-gene" type of resistance is generally interpreted by an elicitor-receptor model: the plant R proteins recognize directly or indirectly particular Avr proteins produced by different pathogen strains. Most *R* gene-triggered resistance appears to be associated with HR cell death. In other cases, the plant, although infected, may outgrow the pathogen long enough to complete its life cycle. Figure 5.2 shows a typical example of the gene-for-gene system in *Arabidopsis*: the compatible/incompatible interactions between *Pseudomonas syringae* and *Arabidopsis thaliana*. *Arabidopsis* plants of the ecotype Columbia (Col-0) contain the resistance (*R*) gene *Rps2* (Bent et al., 1994; Mindrinos et al., 1994). These plants, but not *rps2* plants, can recognize *Pseudomonas* strains that contain the *Avr* gene *avrRpt2* and mount an HR. However, they are unable to recognize a *Pseudomonas* strain that does not contain the *avrRpt2* gene and are therefore unable to mount a defense response in the form of an HR.

Activation of the HR triggers a systemic resistance response known as systemic acquired resistance (SAR). This response includes the accumulation of the signal molecule salicylic acid (SA) throughout the plant and the expression of a characteristic set of defense gene, including PR genes (Malamy et al., 1990; Métraux

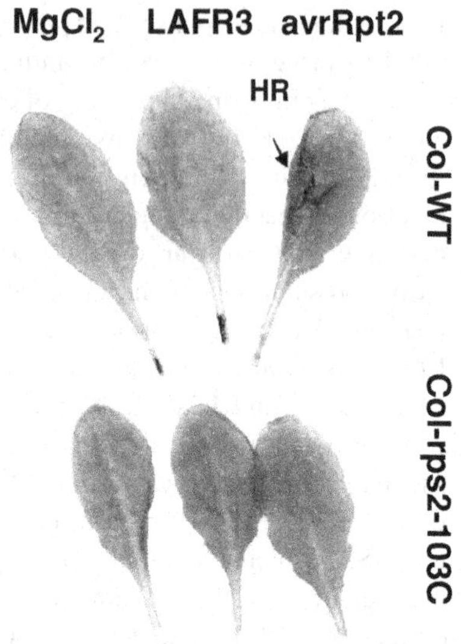

MgCl$_2$ LAFR3   avrRpt2

FIGURE 5.2. Illustration of "gene-for-gene" interaction: activation of HR cell death via the *Rps2/AvrRpt2* pathway. Fully expanded leaves of 4- to 5-week-old *Arabidopsis* plants (Col-0) were infiltrated with *Pseudomonas syringae* pv *maculicola* strain ES4326 (*P.s.m.*; 10^5 CFU/cm^2) that does not induce an HR, *P.s.m.* ES4326/*avrRpt2* (10^5 CFU/cm^2) which induces an HR, or mock infected with 10 mM MgCl$_2$ as described by Greenburg et al. (1994) or Mittler et al. (1997a). Leaves of 4- to 5-week-old *Arabidopsis rps2-103C* plants were infiltrated with *P.s.m.* and *P.s.m.* ES4326/*avrRpt2*, or mock infected with 10 mM MgCl$_2$. These leaves were sampled and photographed at 20 hour-post-infiltration.

et al., 1990; Gaffney et al., 1993) [salicylate-mediated induced systemic resistance is also called ISR in some literature]. Plants expressing SAR are more resistant to subsequent attacks by a variety of otherwise unrelated virulent pathogens (Ryals et al., 1996). Many defense responses that are characteristic of SAR also contribute to local resistance that is mediated by *R* genes, and to the local growth limitation of moderately virulent pathogens. In addition to SA, the *NPR1/NIM1* gene product is a key mediator of SAR as well as gene-for-gene disease resistance (Dong, 2001). The SA signal is transduced through NPR1/NIM1, a nuclear-localized protein that interacts with TGA transcription factors, which may be involved in SA-mediated gene expression (Zhang et al., 1999; Després et al., 2000; Zhou et al., 2000b).

The past decade has seen the isolation of various *R* genes from different plant species that can specify resistance to viruses, bacteria, fungi, nematodes and insects. An important observation from sequence alignment of the encoded proteins is a modular R protein structure. Despite a wide range of pathogen taxa and their presumed pathogenecity effector molecules, *R* genes encode only several classes of structurally related proteins (see reviews: Shirasu and Schulze-Lefert, 2000; Dangl and Jones, 2001). These include the *Cf-X* class of tomato R proteins (Cf-2, Cf-4, Cf-5, Cf-9), containing extracellular leucine-rich repeats (LRRs), a single-span transmembrane domain, and a short cytoplasmic region with no known homologies; Xa21 and FLS2, transmembrane proteins containing extracellular LRRs and a cytoplasmic serine-threonine kinase domain; and Pto, a cytoplasmic soluble serine-threonine kinase. More recently, two genes containing novel structures for a disease resistance gene were cloned. One gene is tomato *Ve*, which encodes a putative cell surface-like receptor that has N-terminal LRR domain containing 28 or 35 potential glycosylation sites, a hydrophobic membrane-spanning domain, and a C-terminal endocytosis-like signal sequence (Kuwchuk et al., 2001). The other gene is barley *Rpg1*, which encodes a receptor-kinase that contains an N-terminal domain that does not resemble any previously described receptor and two tandem protein kinase domains (Brueggeman et al., 2002). Aside from these noted exceptions, the majority of cloned R genes contain nucleotide-binding site (NBS) and LRR motifs. The *Arabidopsis* genome sequence annotation predicted that ~150 genes with homology to the NBS-LRR class of *R* genes exist in this species alone (The *Arabidopsis* Genome Initiative, 2000). Proteins containing LRR motif are thought to be involved in protein–protein interactions, and the specificity of these interactions is likely to be determined by the composition of the variable amino acids in the consensus core of the LRRs (Kobe and Deisenhofer, 1995). In addition, the NBS motif is thought to be critical for ATP or GTP binding, although to date there is no direct biochemical evidence for the postulated nucleotide binding activity via this domain of R proteins. The NBS-LRR class of R proteins can be divided into two subclasses based on the conserved N-terminal motif. One subclass has a coiled-coil domain (CC) that consists of a putative leucine zipper motif: this CC-NBS-LRR subclass includes *Arabidopsis RPM1, RPS2, RPP8, RPS5*, tomato *Prf* and *Mi*, and potato *Rx1*. The other subclass contains an N-terminal domain that has significant homology with the Toll/interleukin receptor domain (TIR): this TIR-NBS-LRR subclass includes the tobacco *N*, flax L_6 and *M*, and *Arabidopsis RPS4, RPP5*, and *Rpp1*. Surprisingly, the NBS region of the *R* genes shares sequence homology with the NBS region of cell death genes such as *CED4* from *Caenorhabditis elegans* and *Apaf-1, FLASH, CARD4*, and *Nod1* from human (Aravind et al., 1999). The presence of conserved TIR, NBS, and LRR structural motifs in different R proteins may imply their involvement in protein complexes that recognize pathogen-derived ligands (Avr products) and trigger signal transduction leading to defense response. Moreover, identification of the TIR domains in the N, L_6, M, RPP5, and RPS4 proteins suggests that plants and animals might use proteins with similar domains to resist infection.

5.1.3 Relationship to Disease Resistance

In most studied cases, HR appears to correlate with activation of resistance to a broad range of pathogens. However, HR cell death does not protect plants against infection by necrosis-causing pathogens (necrotrophic pathogens) such as the fungi *Botrytis cinerea* and *Sclerotina sclerotiorum*, although HR is thought to deprive the pathogens of the supply for food and confine them to initial infection site. The disease is manifested by appearance of necrotic lesions. Necrotrophic pathogens usually kill the host cells before deriving food from them, often through secretion of toxin (Weymann et al., 1995; Hunt et al., 1997). Recent study has provided interesting evidence showing that infection of plants by necrotrophic pathogens can induce an oxidative burst and HR cell death with a marker of apoptosis, such as nuclear condensation, and with induction of HR-specific gene *HSR203J* (Govrin and Levine, 2000). The degree of *B. cinerea* and *S. sclerotiorum* pathogenicity was directly dependent on the level of generation and accumulation of superoxide or hydrogen peroxide. Interestingly, growth of *B. cinerea* can be suppressed in the HR-deficient mutant *dnd1*, and enhanced by HR caused by simultaneous infection with an avirulent strain of *P. syringae* (Govrin and Levine, 2000). Thus, HR induced by incompatible strain of bacterial pathogen (biotrophic pathogen) or elicited by nectrotrophic pathogen can restrict the spread of a biotrophic pathogen, but has an opposite effect against necrotrophic pathogens.

Furthermore, previous studies have provided some evidence that the HR is not always required for gene-for-gene resistance and SA synthesis. Examples have been reported with *Avr*-specific resistance genes that do not provoke macroscopic HR during the restriction of pathogen growth (Goulden and Baulcombe, 1993). Recent evidence that HR cell death and defense gene activation can be uncoupled comes from their apparent separation by the *dnd1* mutation (Yu et al., 1998; Clough et al., 2000) and protease inhibitor studies (del Pozo and Lam, 1998). Although these do not rule out the possibility that HR plays an important role in resistance, it suggests that disease resistance may be activated by a number of mechanisms, and in some cases, a subset of defense mechanisms would be sufficient to stop the growth of particular pathogens in the infected tissue.

HR cell death appears more tightly correlated with viral resistance as compared to resistance against bacterial pathogens. For example, the interaction between tobacco mosaic virus (TMV) and tobacco harbouring the *N* gene is a classic model system for studying gene-for-gene interaction and disease resistance (Holmes, 1938). Recently, Baker and coworkers systemically investigated the precise role of the *N*-encoded TIR, NBS, and LRR domains in conferring TMV resistance by the construction and analysis of a series of deletion and amino acid substitution mutant alleles of *N* (Dinesh-Kumar et al., 2000). Their deletion analysis suggests that TIR, NBS, and LRR domains each play an important role in the induction of resistance response against TMV. Moreover, they found that amino acid residues conserved among the TIR domain and NBS-containing proteins play critical roles in *N*-mediated TMV resistance. Some loss-of-function *N* alleles, such as the TIR deletion mutant and others with point mutations in the NBS region, apparently can interfere with the wild-type *N* function and behave like dominant negative mutations. Interestingly, many amino acid substitutions in the TIR, NBS, and LRR domains of *N* lead to a partial loss-of-function phenotype in which transgenic tobacco plants can mount a delayed HR compared with the wild-type plants but fail to contain the virus to the infection sites.

In animal cells, the ability of many viruses to replicate and spread is dependent on the production of inhibitors of apoptosis such as the p35 protein and Inhibitor of Apoptosis Protein (IAP) of baculovirus that act as inhibitors to caspases, a family of cysteine proteases that serve as the crucial switch for many forms of PCD in animal cells (Green, 2000). Recent evidence from inhibitor studies and biochemical approaches suggests that caspase-like proteases may also be involved in PCD control in plants (Lam and del Pozo, 2000) (also see Section 5.3.4). Evidence for the functional significance of PCD as a plant defense response against viruses is apparent from studies in which baculovirus *p35* was expressed in transgenic tobacco plants and then challenged with TMV (del Pozo and Lam, 2003). Infection of *p35*–expressing transgenic tobacco plants with TMV also can result in systemic spreading of the virus within a resistant background. Transgenic tobacco plants expressing mutant versions of the p35 protein that are defective in caspase inhibition did not show this phenotype. A striking characteristic of these plants is that TMV is able to escape from the primary inoculated leaves and systemically infect the plant in spite of the presence of the *N* resistance gene. Thus, in this particular plant–virus interaction, timely induction of HR cell death is necessary for restricting the pathogen to the primary infection site.

5.2 Approaches for the Characterizaton of the Response

5.2.1 Differential Gene Expression

The processes that determine the outcome of an interaction between plants and pathogens appear to be complex. Identification of genes differentially expressed in the compatible and incompatible interaction would allow a greater understanding of the molecular mechanism of HR cell death. To address this problem, differential library screening has been frequently used in earlier work. For example, Marco et al. (1990) reported the identification of two classes of genes (*str* and *hsr*) that are activated during the HR of tobacco in response to an incompatible isolate of *Pseudomonas solanacearum*, but not in response to an *hrp* mutant of the same bacterial isolate. Among these genes, activation of the tobacco gene *hsr203J* is rapid, highly localized, and specific for incompatible plant–pathogen interactions (Pontier et al., 1994). Its expression is also strongly correlated with PCD occurring in response not only to diverse pathogens but also to various cell-death-triggering extracellular agents (Pontier et al., 1998). On the other hand, using a synchronous HR-inducing system with TMV and resistant tobacco cultivars, Seo et al., (2000) isolated the cDNA of tobacco DS9 the transcript level of which specifically decreased three hours after TMV infection. The DS9 gene encodes a chloroplast-targeted homolog of bacterial FtsH protein, which serves to maintain quality control of some cytoplasmic and membrane proteins. The authors clearly demonstrated that reduced levels of DS9 protein in TMV-infected tobacco leaves accelerate the HR, suggesting that accumulation of damaged protein in the plastids may act as a signal for HR induction (Seo et al., 2000).

Early attempts to document global changes in defense-associated gene expression were limited by the difficulty of identifying the significant genes and their products using differential screening or differential display methods. Although the above studies introduced here provide the identification of some interesting factors that may be involved in HR cell death, many aspects of the response to infection remain uncharacterized. Improvements in technology such as the generation of expressed sequence tag (EST) collections for various plant species and the complete sequencing of the *Arabidopsis* genome offer the potential for a global understanding of the transcriptional response during HR activation. DNA microarrays are powerful tools for a wide range of areas in plant molecular biology and can provide information on the expression patterns for thousands of genes in parallel (Zhu and Wang, 2000; Ahanoni and Vorst, 2001; Kazan et al., 2001). DNA microarrays are currently fabricated and assayed by two main approaches, involving either *in situ* synthesis of oligonucleotides (oligonucleotide microarray) or deposition of presynthesized DNA fragments (cDNA microarray) on solid surfaces (see recent review by Aharoni and Vorst, 2002). The application of this technology is being used to comprehensively profile gene expression networks during the plant defense response that is triggered when a plant encounters a pathogen or an elicitor molecule (Maleck et al., 2000; Schenk et al., 2000; Chen et al., 2002; Scheideler et al., 2002). In addition to identifying new genes induced during various defense responses in a global scale, these studies are providing new insights into the complex pathways governing defense gene regulation.

5.2.2 Biochemical, Pharmacological and Physical Approaches

One of the earliest responses activated after host plant recognition of an Avr protein or nonhost specific elicitor is the oxidative burst, in which levels of ROI rapidly increase (Lamb and Dixon, 1997). Earlier pharmacological and physiological evidence using an inhibitor of the neutrophil NADPH oxidase, diphenylene iodonium (DPI), indicated that DPI can block the oxidative burst in plant cells (Doke, 1983; Doke and Ohashi, 1988; Levine et al., 1994; Auh and Murphy, 1995; Levine et al., 1996). Activation of the oxidative burst is governed by a phosphorylation/dephosphorylation poise because the protein phosphatase 2A inhibitor cantharidin can enhance ROI production in soybean cells in response to avirulent bacteria or elicitor (Levine et al., 1994; Tenhaken et al., 1995). In contrast, the serine/threonine protein phosphatase inhibitor okadaic acid inhibits the oxidative burst and HR cell death induced by TMV (Dunigan and Madlener, 1995). The inhibitor of eukaryotic ribosomes, cycloheximide, can inhibit the oxidative burst of soybean cells in response to avirulent pathogen (Shirasu et al., 1997) and suggests that *de novo* protein synthesis is required for this process. Lastly, mastoparan, a specific activator of G-proteins in mammal, induces H_2O_2 accumulation in soybean cells in the absence of elicitor (Legendre et al., 1992; Chandra and Low, 1995).

Another early signaling event induced in plants during recognition of an invading pathogen is thought to be the enhanced flow of ions across the plasma membrane. This response involves an inward flux of calcium and protons, combined with outward fluxes of potassium and chloride (Atkinson and Baker, 1989). The involvement of ion fluxes in the induction of HR signal transduction pathway was suggested by direct physiological measurement of the particular ion concentrations (Nürnberger et al., 1994; Jabs et al., 1997), as well as by different pharmacological studies (Jabs et al., 1997; Zhou et al., 2000a). In parsley, inhibition of elicitor-stimulated ion fluxes by ion channel blockers prevented ROI production, defense gene activation, and phytoalexin biosynthesis, while artificial induction of ion fluxes, in the absence of the elicitor, stimulated these responses (Jabs et al., 1997). In tomato, treatment with fusicoccin, an activator of the plasma membrane H^+-ATPase pump, was found to cause the acidification of the apoplast and the induction of SA biosynthesis and PR-gene expression (Schaller and Oecking, 1999). Fusicoccin, as well as treatment with a low-pH buffer, was also found to enhance HR cell death in barley (Zhou et al., 2000b). Ca^{2+} influxes also play a crucial role in the execution of the HR. Blocking Ca^{2+} ion channels using calcium channel blocker La^{3+} was shown to inhibit HR in tobacco, *Arabidopsis* and soybean systems (Atkinson et al., 1990; He et al., 1993; Levine et al., 1996; Mittler et al., 1997b). Treatment of plant cells with a Ca^{2+} ionophore can also induce HR-like cell death (Levine et al., 1996). Calcium signals appear to be at least in part mediated through protein phosphorylation and such activity has been implicated in cell culture response to the bacterial nonspecific HR elicitor protein harpin (Pike et al., 1998). However, the genes encoding the channels that mediate these fluxes *in vivo* have not been identified, and no direct genetic evidence currently exists for the involvement of ion fluxes in the induction of the HR.

Salicylic acid (SA) plays a key role in the activation of SAR and gene-for-gene resistance. SA levels increase after pathogen infection, which, in turn, leads to the induction of a number of PR genes (Malamy et al., 1990; Métraux et al., 1990). SAR can also be modulated by treatment with SA and chemical inducers such as 2,6-dichloroisonicotinic acid (INA; Métraux et al., 1990) and benzo(1,2,3)-thiadiazole-7-carbothioic acid *S*-methyl ester (BTH; Friedrich et al., 1996). Depletion of endogenous SA levels in *Arabidopsis* and tobacco by overexpression of the bacterial gene *nahG*, encoding the enzyme salicylate hydroxylase, results in a breakdown of SAR and gene-for-gene resistance (Ryals et al., 1996). INA and BTH do not increase SA concentration in the plant and can activate SAR in both wild-type and NahG plants, suggesting these synthetic analogues of SA act independently or downstream of SA in the SAR signaling pathway (Friedrich et al., 1996; Ryals et al., 1996).

Host cell death can also be caused by pathogen-produced phytotoxic compounds that function as key virulence determinants. Necrotrophic phytopathogenic fungi synthesize a wide range of phytotoxic compounds, including the sphinganine analog mycotoxins, which are produced by at least two unrelated groups of fungi, *Alternaria* and *Fusarium* spp (Gilchrist, 1998). Fumonisin B1 (FB1) is one of several related sphinganine analog mycotoxins produced by *F. moniliforme* and elicits an apoptotic form of PCD in both plants and animal cell cultures (Wang et al., 1996a,b). Ausubel and coworkers have recently established a relatively simple pathogen-free system in *Arabidopsis* involving FB1 that can be used to study the signal transduction events involved in pathogen-elicited cell death (Stone et al., 2000). FB1-induced lesions in *Arabidopsis* are similar to pathogen-induced lesions in many aspects, including deposition of phenolic compounds and callose, production of ROIs, accumulation of the phytoalexin camalexin, and induction of defense-related gene expression. The authors also showed that FB1 can be used to select directly for FB1-resistant mutants, some of which display enhanced resistance to a virulent strain of *P. syringae*, suggesting that pathogen-elicited PCD of host cells may be an important feature for certain compatible plant-pathogen interactions.

In the past several years, indirect evidence from biochemical and physiological studies has pointed to the involvement of proteases as a key player in the activation of HR cell death. For example, in cultured soybean cells, synthetic protease inhibitors effectively suppressed PCD triggered by oxidative stress or by infection with avirulent pathogens (Levine et al., 1996). It is noteworthy that only a subset of the tested protease inhibitors (PMSF, AEBSF, and leupeptin) partially block PCD. No inhibition and in some cases even increased cell death were observed with the serine protease inhibitors TLCK and TPCK, suggesting the stabilization of certain positive factors for HR activation (Levine et al., 1996). On the other hand, it is widely known that caspases are conserved cysteine proteases that regulate animal PCD (White, 1996). The possible involvement of caspase-like protease activities during HR cell death was also implicated by using specific inhibitors and substrates (del Pozo and Lam, 1998; D'Silva et al., 1998) (see also Section 5.3.4). In mammalian systems, cysteine proteases including caspases are major executors of PCD, but other classes of proteases, such as cathepsin D, aspartate proteases, metalloproteases, calcium-dependent proteases (calpain) and the ubiquitin/proteasome system have also been found to be involved in PCD (Beers et al., 2000). It is currently unclear whether these classes of proteases may also be involved in the activation of HR cell death (Heath, 2000). Recent studies with the tomato *Cf-2* resistance system has identified the locus *Rcr3*, which encodes a papain-like cysteine protease, as a specific and critical mediator for elicitation of the HR by Avr2 expressing races of the fungus *Cladosporium fulvum* (Kruger et al., 2002). Although its mode of action remains to be determined, Rcr3 serves as the first clear genetic evidence that a dedicated protease is involved in the HR induction process.

The importance of the mitochondrion in the expression of HR-associated PCD in plants comes from studies of the alternative oxidase (AOX), the mitochondrial enzyme localized in the inner membrane (reviewed in Lam et al., 1999a). AOX can control the generation of reactive oxygen species (ROS) from the mitochondrial electron transport chain when oxidative phosphorylation is inhibited. For example, when activity of the mitochondrial electron transport chain is inhibited by antimycin A treatment, AOX expression is induced and ROS generation is kept to a minimum so that little cell death is activated (Maxwell et al., 1999). Overexpression of AOX has the reverse effects, suggesting that plant mitochondria have an important role as a signal generator for HR-induced cell death, perhaps by generation of ROS derived from electron-transfer intermediates in the inner mitochondrial membrane (Maxwell et al., 1999). Furthermore, Chivasa and Carr (1998) has shown by using an inhibitor of AOX salicylhydroxamic acid, that inhibition of AOX activity causes the inhibition of SA-induced resistance to TMV in tobacco, and antimycin A and KCN also induced AOX transcript accumulation and resistance to TMV. Induction of AOX has also been observed under several stress conditions and a recent study in *Arabidopsis* showed that rapid localized AOX induction by avirulent bacterial pathogens requires SA (Simons et al., 1999). Thus, these features strongly suggest that AOX may act as a safety valve for the control of HR activation and are consistent with its enhanced expression during the latter phase of the HR. AOX is not found in animal cells, thus it may be a specialized regulator to control cell death activation in plants (Lam et al., 1999a). In addition to the above observation, the importance of mitochondria in the expression of HR cell death comes from studies of the mode of action for host-selective toxin, victorin, which is required for pathogenesis and induces rapid cell death in susceptible, toxin-sensitive oat genotype (reviewed in Wolpert et al., 2002).

5.2.3 Genetic Dissection of the Key Factors Involved in HR

It has been difficult to assess experimentally the utility of cell death in gene-for-gene disease resistance because cell death is usually a central feature of this response. A mutational approach was used to shed further light on the relationships between HR cell death and pathogen growth arrest. The *dnd* (defense, no death) class of mutants, including *dnd1*, *dnd2*, and *Y*15, were identified by their reduced ability to produce the HR in response to avirulent *P. syringae* that express avrRpt2, and were isolated in a screen designed to discover additional components of the *avrRpt2-RPS2* disease resistance pathway in *Arabidopsis* (Yu et al., 1998). Among these mutants, the *dnd1* are defective in HR cell death but retain characteristic responses to avirulent bacteria, such as induction of PR gene expression and strong restriction of pathogen growth. Interestingly, progeny lines derived from the *dnd1* mutant also failed to produce an HR in response to *P. syringae* strains expressing avirulence genes *avrRpm1* or *avrB* (Kunkel et al., 1993). Since two separate resistance genes (*RPS2* and *RPM1*) control responsiveness to these three separate avirulence genes (*avrRpt2*, *avrRpm1*, *avrB*), it appears that DND1 is a common component of the plant defense response shared by distinct signal initiators. Recent identification of the *Dnd1* gene by positional cloning revealed that DND1 shows homologies to cyclic nucleotide-gated ion channels and confirmed to have ion channeling activity when expressed in yeast and animal cells (Clough et al., 2000). However, its mode of action *in planta* remains to be defined.

The highly localized nature of the HR suggests that mechanisms must exist to keep cell death contained. A large class of mutation exists in maize that is characterized by the spontaneous formation of discrete or expanding lesions of varying size, shape, and color in leaves (Johal et al., 1995). Because lesions associated with some of these mutants resemble symptoms of certain diseases of maize, they have been collectively called disease lesion mimics. To date, more than 40 independent lesion mimics, both recessive (designed *les*) and dominant (designed *Les*), have been identified in maize (Johal et al., 1995). More recently, systematic screening of similar mutants in *Arabidopsis* yielded a number of mutants called *acd* (accelerated cell death), *lsd* (lesion stimulating disease resistance), and *cpr* (constitutive expresser of PR genes) (Greenberg and Ausubel, 1993; Dietrich et al., 1994; Greenberg et al., 1994; Weymann et al., 1995; Bowling et al., 1997). The expression of lesions in these mutants, generally designated as disease lesion mimics, can be developmentally programmed and is often affected by the environment or genetic background of the plant (Johal et al., 1995; Dangl et al., 1996). In some cases, lesion mimic mutants exhibit SAR and show high, constitutive levels of PR gene expression. Recently, some recessive lesion mimic genes have been cloned from three plant species: *Arabidopsis*, barley, and maize (Büschges et al., 1997; Dietrich et al., 1997; Gray et al., 1997; Hu et al., 1998). For example, the *LSD1* gene of *Arabidopsis*, which encodes a zinc finger protein, may negatively regulate cell death (Dietrich et al., 1997). Likewise, the *Mlo* gene of barley appears to encode a membrane protein whose function may be to negatively regulate both cell death and the disease resistance response (Büschges et al., 1997). The maize *Lls1* gene inhibits cell death, although apparently by degrading a phenolic mediator of cell death (Gray et al., 1997). Also, several recent studies have revealed that genetic disruption of biosynthesis pathway of tetrapyrroles (chlorophylls and heme) causes lesion formation that can lead to the induction of a set of defense response including activation of SAR (Hu et al., 1998; Molina et al., 1999; Ishikawa et al., 2001). Tetrapyrrole biosynthesis is highly regulated, in part to avoid the accumulation of intermediates that can be photoactively oxidized, leading to the generation of ROI and subsequent photosensitized damages. In this case, a light-sensitive ROI cascade mediated by the accumulated tetrapyrrole intermediate can apparently mimic the oxidative burst seen in plant defense response.

5.2.4 Lessons Learnt from Transgenes Which can Activate HR-Like Symptoms

Spontaneous formation of HR-like lesions in the absence of a pathogen has also been reported in a number of transgenic plants that express foreign or modified transgenes. Several transgenes that can activate or affect different components of a signal transduction pathway involved in pathogen recognition or defense response activation have been reported (reviewed in Dangl et al., 1996; Mittler and Rizhsky, 2000). Moreover, activation of HR-like cell death by transgene expression is viewed as important evidence for the existence of PCD pathways in plants.

The induction of proton and ion flux across the plasma membrane during plant-pathogen interaction was found to be one of the primary events that occurs during activation of the HR and other defense mechanisms. However, the genes encoding the channels that mediate ion fluxes *in vivo* have not been identified, and no direct molecular evidence currently exists for the involvement of ion fluxes in the induction of the HR. We have previously shown that expression of the gene encoding the bacterial proton pump bacterio-opsin (*bO*) in transgenic tobacco and potato plants resulted in the induction of multiple defense mechanisms with a heightened state of resistance against pathogen attack (Mittler et al., 1995; Abad et al., 1997). In the absence of a pathogen, bO-expressing plants developed lesions similar to HR lesion, accumulated PR protein, and synthesized SA. Furthermore, our recent study using different mutant forms of bO provided direct molecular evidence that passive leakage of protons through the bO proton channel is likely the cause of the lesion mimic phenotype in transgenic tobacco plants (Pontier et al., 2002). The activation of defense mechanisms by bO expression supports a working hypothesis that enhancing the proton flux across the plasma membrane may mimic the presence of a pathogen, similar to the situation that occurs in a number of disease lesion mutants (Dangl et al., 1996, Mittler and Lam, 1996).

Ca^{2+} signal is also essential for the activation of plant defense responses, but downstream components of the signaling pathway are still poorly understood. Calmodulin (CaM) is known to be a universal Ca^{2+}-binding signal mediator in eukaryotes. Specific CaM isoforms of soybean, SCaM-4 and SCaM-5, are activated by infection or pathogen-derived elicitors, whereas other SCaM genes encoding highly conserved CaM isoforms did not show such response (Heo et al., 1999). Constitutive expression of either isoforms in tobacco resulted in spontaneous lesion formation, constitutive PR gene expression and enhanced resistance against virulent oomycete, bacterial and viral pathogens. Surprisingly, in contrast to SA-dependent activation of these pathogen-induced markers in wild-type tobacco, their lesion formation and PR gene activation in these transgenic tobacco plants did not require SA, suggesting that specific CaM isoforms are components of an SA-independent signal transduction pathway leading to disease resistance (Heo et al., 1999) (see also Section 5.3.1).

GTP-binding proteins (G-proteins) act as molecular signal transducers whose active or inactive states depend on the binding of GTP or GDP, respectively, in the regulation of a range of cellular processes-including growth, differentiation, and intracellular trafficking. In animals and fungi, cholera toxin (CTX) can activate signaling pathways dependent on heterotrimeric G-proteins. Transgenic tobacco expressing a gene encoding the A1 subunit of cholera toxin (CTX) showed greatly reduced susceptibility to the bacterial pathogen *Pseudomonas syringae* pv *tabaci*, accumulated high levels of salicylic acid (SA) and constitutively expressed PR genes, suggesting that CTX-sensitive G-proteins are important in inducing the SAR (Beffa et al., 1995). Furthermore, expression of *rgp1*, a gene encoding a Ras-related small G-protein, in transgenic tobacco was shown to increase resistance to TMV infection through SAR activation pathway (Sano and Ohashi, 1995). Shimamoto and coworkers recently reported that expression of a constitutively active derivative of monometric G-protein Rac of rice (OsRac1) activated ROS production and phytoalexin levels, developed symptoms of HR-like lesion, increased resistance against virulent fungal and bacterial pathogens, and activated cell death with biochemical and morphological features similar to apoptosis in mammalian cells (Kawasaki et al., 1999; Ono et al., 2001). Conversely, a dominant-negative OsRac1 was shown to suppress elicitor stimulated ROI production and pathogen-induced cell death in transgenic rice.

Expression of some metabolism-perturbing transgenes in plants is thought to result in the alteration of cellular homeostasis and the generation of a signal that activates the PCD signaling pathways (Dangl et al., 1996; Mittler and Rizhsky, 2000). For example, tobacco plants expressing yeast-derived vacuolar and apoplastic invertases develop spontaneous necrotic lesions similar to the HR caused by avirulent pathogens; uncontrolled expression of these genes can drastically alter the metabolic balance of cells due to changes in hexose transport or metabolism (Herbers et al., 1996). In animal systems, many perturbations in cellular metabolism were shown to activate an apoptosis-signaling pathway (Bratton and Cohen, 2001). Since infection of plants with avirulent pathogens such as bacteria and viruses is likely to cause general alterations in the metabolic balance of cells (Dangl et al., 1996; Mittler and Lam, 1996), mutation of general housekeeping genes involved in plant cell metabolism can result in PCD in some cases, but not in others.

5.3 Current Mechanistic Understanding of the Response

5.3.1 Recent Studies Related to Calcium and its Homeostasis as Signal for HR

Transient influx of Ca^{2+} constitutes an early event in the signaling cascades that trigger plant defense responses. Since Ca^{2+} signaling is usually mediated by Ca^{2+}-binding proteins such as calmodulin (CaM), identification and characterization of CaM-binding proteins elicited by pathogens could provide insights into the mechanism through which Ca^{2+} regulates defense responses including the HR. Very recently, an interaction between CaM and Mlo proteins was found by screening a rice cDNA expression library in *Escherichia coli* with the use of soybean CaM1 conjugated to horseradish peroxidase as a probe (Kim et al, 2002a). Rice Mlo homologue (OsMlo) has a molecular mass of 62 kDa and shares 65% sequence identity and predicted topology with barley Mlo, a seven-transmembrane-helix protein known to function as a negative regulator of broad spectrum disease resistance and plant cell death (Büschges et al., 1997). These research groups also showed that barley Mlo can bind CaM (HvCaM3) using the above *in vitro* assays, and *in vivo* expression assays using both the yeast split-ubiquitin technique and transient expression system in barley epidermal cells by biolistic methods (Kim et al., 2002b). The significance of barley Mlo-CaM interaction *in vivo* in pathogen defense was also shown by transient expression assays in which Mlo activity is shown to depend on its specific binding to CaM. Likewise, gene suppression of HvCaM3 by RNA interferance (RNAi) in an *Mlo* background quantitatively lowered the susceptibility seen in Mlo wild-type leaves, which is consistent with an enhancing function for CaM in Mlo-mediated defense suppression. Resistance suppression by CaM3 required the presence of wild-type MLO because its expression in the mutant *mlo* background did not influence the resistant phenotype (Kim et al., 2002b). Taken together, these results provide strong evidence that CaM has an activator role for Mlo-mediated defense suppression and places CaM activity upstream of, or coincident with, the action of Mlo. However, the precise connection between a change in cellular Ca^{2+} concentrations and this novel interaction between CaM and Mlo for HR regulation remains to be defined.

5.3.2 Transcriptional Mediators and Lipid Metabolism that are Involved in HR Activation

In animal cells, PCD is controlled through the expression of a number of conserved genes. Some gene products activate PCD, such as caspases, whereas others are inhibitors, such as some members of the Bcl-2 family. In addition to their role in cell-cycle regulation, recent studies have suggested a new role for MYB proteins as regulators of cell survival and/or cell death through the regulation of a new MYB target gene, Bcl-2 (Frampton et al., 1996; Solomoni et al., 1997). A *MYB* gene from tobacco is induced in response to TMV activated HR and it can bind to a consensus MYB recognition sequence found in the promoter for the *PR-1a* gene (Yang and Klessig, 1996). Furthermore, Daniel et al., (1999) have shown that expression of *Arabidopsis MYB30* is closely associated with the initiation of cell death. This gene is thought to be a strong candidate for a component of a regulatory network controlling the establishment of cell death.

Genetic approaches in *Arabidopsis* have been used to identify signaling components involved in HR control. Recent findings have strongly suggested that specific regulation of lipid metabolism may closely associate with HR activation (Falk et al., 1999; Jirage et al., 1999; Brodersen et al., 2002). The *EDS1* gene was cloned by transposon tagging and found to encode a protein that has similarity in its animo-terminal portion to the catalytic site of eukaryotic lipases (Falk et al., 1999). The PAD4 gene was cloned by map-based positional cloning and found to encode another member of the L-lipase class of plant proteins that include EDS1 (Jirage et al., 1999). EDS1 and PAD4 were shown to be required for SA accumulation upon avirulent pathogen and their mRNA levels are upregulated by applications of SA, although EDS1 and PAD4 function upstream of SA accumulation. It should be noted that EDS1 appears to be involved in signaling pathway for specific types of TIR-NBS-LRR resistance genes (Liu et al., 2002; Peart et al., 2002). The recessive *acd11 Arabidopsis* mutant exhibits characteristics of animal apoptosis and defense-related responses that accompany the HR (Brodersen et al., 2002). The *acd11* phenotype is SA dependent, as *acd11* is rescued by NahG gene, and application of BTH to *acd11/nahG* restores cell death. This SA-mediated death pathway requires both functional PAD4 and EDS1, as the *acd11* phenotype is suppressed by the *pad4-2* and *eds1-2* mutations. Molecular cloning, complementation, and biochemical analyses revealed that ACD11 encodes a homolog of mammalian glycolipid transfer protein and has sphingosine transfer activity (Brodersen et al., 2002). Furthermore, it was shown that a putative lipid transfer protein (DIR1) is involved in SAR signaling in *Arabidopsis* (Maldonado et al., 2002). Lipid molecules such as jasmonic acid, phosphatidic acid and N-acylethanolamines are synthesized or released from membranes upon pathogen or insect attack. Some act as second messengers in plant defence signaling (Wasternack and Parthier, 1997; Chapman, 2000; Munnik, 2001). A role for DIR1 and ACD11 in disease resistance signaling would be consistent with the observation that some mammalian lipid transfer proteins act as lipid sensors or are involved in phospholipase-C-linked signal transduction (Wirtz, 1997). Therefore, these molecular genetic studies using *Arabidopsis* mutants implicate the involvement of a lipid derived signal component for HR and SAR signaling and, ACD11 and DIR1 could act as a translocator for release of the mobile signal into the vascular system and/or chaperone the signal through the plant.

5.3.3 Reactive Oxygen Species Generation and Cellular Energy Status as a Rheostat

Many studies document the detection of O_2^- and/or its dismutation product, H_2O_2, during the HR (Lamb and Dixon, 1997). *Arabidopsis lsd1* mutants exhibit impaired control of cell death in the absence of a pathogen and could not control the spread of cell death once it was initiated (Dietrich et al., 1994). Jabs et al. (1996) showed that treatment with superoxide, but not H_2O_2, triggers cell death in *lsd1* mutants. DPI, an inhibitor of neutrophil NADPH oxidase, reduced cell death in the *lsd1* genetic background. This suggests that superoxide is necessary and sufficient to propagate lesion formation in an *lsd1* background, accumulating before the onset of cell death and subsequently in live cells adjacent to spreading *lsd1* lesions. LSD1 encodes a zinc finger protein with homology to mammalian GATA-type transcription factors and it may function either to suppress a pro-death pathway component or to activate a repressor of plant cell death (Dietrich et al., 1997).

One source generating ROI in plants is thought to be produced by enzymatic machinery similar to the mammalian respiratory burst NADPH oxidase complex (Doke, 1985; Lamb and Dixon, 1997). Recently, homologues of gp91[phox] (respiratory burst oxidase homologue [*rboh*]), which is a plasma membrane localized component of neutrophil NADPH oxidase, was isolated from rice (Groom et al., 1996), *Arabidopsis* (Keller et al., 1998; Torres et al., 1998), and tomato (Amicucci et al., 1999). They can encode a protein of about 105 kDa in size, with a C-terminal region that shows pronounced similarity to the 69 kDa apoprotein of the gp91[phox] and a large hydrophobic N-terminal domain that is not present in mammalian gp91[phox]. This domain contains two Ca^{2+}-binding EF hand motifs and has extended similarity to the human Ran GTPase-activating proteins. A recent mutant study provides strong genetic evidence that *Arabidopsis rbohD* and *rbohF* are required for accumulation of ROIs in the plant defense response (Torres et al., 2002). The *AtrbohD* gene is required for most of the ROI observed after inoculation with avirulent bacteria, whereas *AtrbohF* apparently has a limited contribution. In contrast, the *atrboh* mutants exhibit enhanced HR and less sporangiophore formation in response to the weakly avirulent fungi, *Peronospora parasitica*. Interestingly, although *atrbohF* exhibits minor suppression of ROI production, it exhibits strongly enhanced cell death phenotype. A double mutant combination of the two *Atrboh* genes dramatically suppresses the oxidative burst triggered by bacterial and fungal pathogens (Torres et al., 2002). Using a novel activity gel assay, Sagi and Fluhr (2001) also confirmed that a putative plant plasma membrane NADPH oxidase can produce O_2^-. These studies show that pathways for ROI generation and their involvement in the HR can be quite complex.

5.3.4 Mechanism for Cell Death Activation

Identities of the key executioners in HR cell death remain elusive, whereas in animal systems a large number of caspases and their regulators have been defined in the past decade (Aravind et al., 1999). Caspase-like protease activity has been observed to be transiently activated in plants synchronized to undergo the HR (del Pozo and Lam, 1998). Peptide inhibitors of caspases can abolish HR cell death of tobacco induced by avirulent bacteria without affecting the induction of defense-related genes significantly. On the other hand, induction of proteolytic activity that may be relevant to cell death during the HR has also been studied in the cowpea rust fungus/cowpea pathosystem using a bovine poly(ADP-ribose) polymerase (PARP) as substrate (D'Silva et al., 1998). PARP is a well-characterized substrate for caspase-3 and was found to be endoproteolytically cleaved when added to extracts prepared from fungus-infected cowpea plants that were developing a HR, while no PARP cleaving activity could be detected in the presence of extracts from cowpea plants that were undergoing a susceptible interaction. The cleavage of PARP observed in this study could be partially suppressed using caspase inhibitors. Moreover, it was clearly shown that tetrapeptide caspase inhibitors can block or significantly diminish plant cell death associated with compatible plant-bacteria interactions which are activating an HR as part of pathogenesis and, in all cases when death is limited, bacterial multiplication is concomitantly reduced (Richael et al., 2001). These studies provide support for a caspase-like protease(s) in plants, the activity of which is correlated with the induction of HR cell death.

Using iterative database searches, Uren et al. (2000) identified potential relatives of caspases in the *Arabidopsis* genome, which they termed metacaspases. Their homology to mammalian caspases is not restricted to the primary sequence, including the catalytic diad of histidine and cysteine, but extends to the secondary structure as well. The plant metacaspases can be divided into two subclasses based on the sequence similarity within their caspase-like regions and their overall predicted domain structure. Type I plant metacaspases contain a predicted N-terminal prodomain which consists of a proline-rich region and a zinc finger motif that is also found in LSD1, a negative regulator of HR cell death (Dietrich et al., 1997). Type II plant metacaspases possess no obvious prodomain but have a conserved insertion of approximately 180 amino acids between the regions corresponding to the p20 and p10 subunits of activated caspases (reviewed in Lam and del Pozo, 2000). However, it remains unclear whether plant metacaspases are functionally equivalent to classical caspases in terms of their target specificities as well as their involvement in controlling the activation of cell death. Recently, a caspase-like protein has been identified in the budding yeast *Saccharomyces cerevisiae* (Yor197W) and is implicated in cell death induced by H_2O_2, acetic acid and ageing (Madeo et al., 2002). Yeast caspase-1 (YCA1) is a member of the metacaspase family and like Type I plant metacaspases, it also has a proline-rich domain at its N-terminus. Overexpression of YCA1 enhances apoptosis-like death of yeast upon addition of H_2O_2 or acetic acid, whereas targeted ablation of YCA-1 dramatically improves survival. YCA1 protein also seems to undergo proteolytic processing in a manner that is dependent on its active-site cysteine, which is similar to mammalian caspases (Madeo et al., 2002). However, many features of the yeast metacaspase YCA1 remain to be clarified. These include direct demonstration of its protease activity, identification of its substrate specificity, and elucidation of its endogenous targets and regulators. It would be interesting to investigate whether plant metacaspases are functionally equivalent to YCA1 using the yeast *yca1* mutant (Madeo et al., 2002) as well as to define their possible roles in HR cell death using reverse genetic approaches.

In animal cells, mitochondria-mediated PCD acts through the proapoptotic Bax and its related proteins that associate with the outer mitochondrial membrane and can oligomerize to form an ion-conducting channel through which macromolecules and other metabolites can pass. This activity can be blocked by the anti-apoptotic proteins Bcl-2 and Bcl-X_L, which play crucial functions to control PCD activation or suppression (Lam et al., 1999b; Martinou and Green, 2001). Recent comparative genomics revealed that no obvious homologue of mammalian Bcl-2 related proteins exists in *Arabidopsis*. Nonetheless, overexpression of human Bcl-2, nematode CED-9, or baculovirus Op-IAP transgenes can confer resistance to several necrotrophic fungal pathogens and a necrogenic virus in tobacco plants (Dickman et al., 2001). Likewise, in transgenic tobacco plants overexpressing Bcl-X_L, delay of HR cell death as well as UV-induced PCD, has been reported (Mitsuhara et al., 1999). Expression of Bax using a TMV vector triggers cell death in tobacco leaf cells in an *N* gene-independent manner (Lacomme and Santa Cruz, 1999). Bax also confers a lethal phenotype when expressed in yeast with typical hallmarks of Bax-induced PCD in animal cells despite the apparent absence of classical caspases or Bcl-2-related proteins in yeast. Thus, the expression of pro- and anti-apoptotic proteins in these heterologous systems has similar effects to those observed in animal cells. This is consistent with the speculation that a conserved cellular pathway for cell death control may exist in eukaryotes. It is possible that the metacaspase YCA-1 may mediate Bax-induced cell death in yeast. This can now be tested with the *yca1* strain (Madeo et al., 2002).

Two groups have used yeast to isolate suppressors of Bax-induced cell death (Greenhalf et al., 1999; Xu and Reed, 1998). One of these suppressors, Bax inhibitor-1 (BI-1), prevents cell death in yeast and animal cells, suggesting that BI-1 could be a distinct class of PCD regulator for pathways activated by Bax expression. Homologues of BI-1 isolated from *Arabidopsis* and from rice have also been shown to suppress Bax-induced PCD in yeast (Kawai et al., 1999; Sanchez et al., 2000). These BI-1 proteins contain six potential transmembrane helices and it has been proposed that they may form ion-conducting channels or modify the activity of existing channels formed by Bax. Expression of *AtBI-1* was rapidly up-regulated in plants during wounding or pathogen challenge. *AtBI-1* up-regulation appears to be R-gene independent and is not remarkably affected by mutations required for specific classes of R gene, suggesting a ubiquitous role in responses for biotic and abiotic stresses (Sanchez et al., 2000). On the other hand, Kawai-Yamada et al. (2001) demonstrated that AtBI-1 overexpression could rescue transgenic plants expressing Bax gene from lethality, while Bax caused potent PCD symptoms, including leaf chlorosis, cytoplasmic shrinkage, and DNA laddering. Although this finding provides direct genetic evidence that Bax-induced cell death can be down-regulated by overexpression of AtBI-1 protein *in planta*, it remains unclear how AtBI-1 suppresses the activity of Bax, given that no obvious Bcl-2 family members have been found in yeast and plants. Surprisingly, it was recently reported by the same research group that AtBI-1 did not block Bax-induced cell death, but instead triggered apoptotic cell death in certain mammalian cultured cells (Yu et al., 2002). The unexpected apoptotic effect of AtBI-1 was shown to be blocked by the caspase inhibitor XIAP and antiapoptic protein Bcl-X_L, suggesting that the cell death caused by AtBI-1 is similar to that caused by Bax and that AtBI-1 caused apoptosis in this case through a caspase-dependent pathway. Yu et al. (2002) speculated that plant BI-1 may competitively interact with endogenous mammalian BI-1 or with a BI-1 target protein in certain cell types, thus interfering with its function and thereby triggering cell death.

5.4 Future Perspectives

The past few years have seen a steady increase in our knowledge of HR cell death in plants. In spite of all the information described above, research in the past years have added very little to our sparse knowledge of the actual control mechanism for HR cell death in clear molecular terms. In particular, no relative of any classical metazoan regulator of apoptosis (for example, Ced-3/caspases, Ced-4/ Apaf-1, Ced-9/Bcl-2) has been defined structurally and genetically so far. The apparent absence of caspases, which are considered to be the major executors of cell death in metazoans, has been a strong argument against a mechanistic and functional conservation of PCD between plants and animals (Lam et al., 2001). Plant genomic studies have produced large quantities of sequence information that await functional analysis. In particular, metacaspases and BI-1 related proteins could be likely candidates for plant cell death regulators at the present time. *Arabidopsis* contains at least nine possible metacaspase-encoding genes, one BI-1 homologue and two other BI-1-related homologues: AtBI-2 and AtBI-3. Furthermore, *Arabidopsis* contains a new gene family discovered by homology searches that we designated as ABRs (for AtBI-2 related proteins) (Lam et al., 2001). This gene family contains twelve putative genes that encode proteins with five or six predicted membrane-spanning helices, although most of the predicted amino acid sequences are unique for this family. Deployment of reverse genetic approaches such as PTGS/RNAi strategies (Wang and Waterhouse, 2001) and knockout screens using T-DNA or transposon insertion collections (Bouche and Bouchez, 2001), coupled with informatic approaches should help to speed up the first essential step of identifying the important players involved in plant cell death activation. This approach would be complementary to forward genetic approaches that are revealing new regulators which may not have counterparts in other organisms. As a second step for studying the physiological function of these regulators, development of other functional genomic tools, such as global transcriptome profiling by DNA microarrays and proteome analyses, would be of importance if we are to take full benefit of resources generated from the rapidly developing model plant systems such as *Arabidopsis* and rice.

Acknowledgments

We would like to thank the USDA for its support of plant cell death research in the Lam laboratory at Rutgers University (grant #99-35303-8636). Partial support by the New Jersey Commission of Science and Technology is also gratefully acknowledged. N.W. is supported in part by a postdoctoral research fellowship from the Japanese Society for Promotion of Science (JSPS grant #06525).

References

Abad, M.S., Hakimi, S.M., Kaniewski, W.K., Rommens, C.M.T., Shulaev, V., Lam, E., and Shah, D. 1997. Characterization of acquired resistance in lesion mimic transgenic potato expressing bacterio-opsin. *Mol. Plant Microbe Interact.* 10:653–645.

Aharoni, A., and Vorst, O. 2002. DNA microarrays for functional plant genomics. *Plant Mol. Biol.* 48:99–118.

Auh, C.-K., and Murphy. T.M. 1995. Plasma membrane redox enzyme is involved in the synthesis of O_2^- and H_2O_2 by Phytothora elicitor-stimulated rose cells. *Plant Physiol.* 107:1241–1247.

Alfano, J.R., and Collmer, A. 1996. Bacterial pathogens in plants: life up against the wall. *Plant Cell* 8:1683–1698.

Amicucci, E., Gaschler, K., and Ward, J. 1999. NADPH oxidase genes from tomato (*Lycopersicon esculentum* and curly-leaf pondweed (*Potamogeton crispus*). *Plant Biol.* 1:524–528.

Arabidopsis Genome Initiative, The 2000. Analysis of the genome of the flowering plant *Arabidopsis thaliana. Nature* 408: 796–815.

Aravind, L., Dixit, V.M., and Koonin, E.V. 1999. The domains of death: evolution of the apoptosis machinery. *Trends Biochem. Sci.* 24:47–53.

Atkinson, M.M., and Baker, C.J. 1989. Role of the plasmalemma H^+-ATPase in *Pseudomonas syringae*-induced K^+/H^+ exchange in suspension-cultured tobacco cells. *Plant Physiol.* 91:298–303.

Atkinson, M.M., Keppler, L.D., Orlandi, E.W., Baker, C.J., and Mischke, C.F. 1990. Involvement of plasma membrane calcium influx in bacterial induction of the K^+/H^+ exchange and hypersensitive responses in tobacco. *Plant Physiol.* 92:1241–1247.

Beers, E.P., Woffenden, B.J., and Zhao, C. 2000. Plant proteolytic enzymes: possible roles during programmed cell death. *Plant Mol. Biol.* 44:399–415.

Beffa, R., Szell, M., Meuwly, P., Pay, A., Vogeli-Lange, R., Metraux, J.P., Neuhaus, G., Meins, F. Jr., and Nagy, F. 1995. Cholera toxin elevates pathogen resistance and induces pathogenesis-related gene expression in tobacco. *EMBO J.* 14:5753–5761.

Bent, A.F., Kunkel, B.N., Dahlbeck, D., Brown, K.L., Schmidt, R., Giraudat, J., Leung, J., and Staskawicz, B.J. 1994. RPS2 of Arabidopsis thaliana: A leucine-rich repeat class of plant disease resistance gene. *Science* 265:1856–1860.

Bortner, C.D., Oldenburg, N.B.E., and Cidlowski, J.A. 1995. The role of DNA fragmentation in apoptosis. *Trends Cell Biol.* 208:8–16.

Bouche, N., and Bouchez, D. 2001. *Arabidopsis* gene knockout: phenotype wanted. *Curr. Opin. Plant Biol.* 4:111–117.

Bowling, S.A., Clark, J.D., Liu, Y., Klessig, D.F., and Dong, X. 1997. The *cpr5* mutant of *Arabidopsis* expresses both NPR1-dependent and NPR1-independent resistance. *Plant Cell* 9:1573–1584.

Bradley, D.J., Kjellbom, P., and Lamb, C. 1992. Elicitor-induced and wound-induced oxidative cross-linking of a proline-rich plant-cell wall protein: a novel, rapid defense response. *Cell* 70: 21–30.

Bratton, S.B., and Cohen, G.M. 2001. Apoptotic death sensor: an organelle's alter ego? *Trends Pharmacol Sci.* 22:306–315.

Brodersen, P., Petersen, M., Pike, H.M., Olszak, B., Ødum, N., Jørgensen, L.B., Brown, R.E., and Mundy, J. 2002. Knockout of *Arabidopsis ACCELERATED-CELL-DEATH* encoding a shingosine transfer protein causes activation of programmed cell death and defense. *Genes Dev.* 16:490–502.

Brueggeman, R., Rostoks, N., Kilian, A., Han, F., Chen, J., Druka, A., Steffenson, B., and Kleinhof, A. 2002. The barley stem rust-resistance gene *Rpg1* is a novel disease-resistance gene with homology to receptor kinases. *Proc. Natl. Acad. Sci. USA* 99:9328–9333.

Büschges, R., Hollricher, K., Panstruga, R., Simons, G., Wolter, M., Frijters, A., van Daelen, R., van der Lee, T., Diergaarde, P., Groenendijk, J., Topsch, S., Vos, P., Salamini, F., and Schulze-Lefert, P. 1997. The barley *Mlo* gene: a novel control element of plant pathogen resistance. *Cell* 88:695–705.

Chamnongpol, S., Willekens, H., Moeder, W., Langebartels, C., Sandermann, H., van Montagu, M., Inze, D., and van Camp, W. 1996. Transgenic tobacco with a reduced catalase activity develop necrotic lesions and induces pathogenesis-related expression under high light. *Plant J.* 10:491–503.

Chandra, S., and Low, P.S. 1995. Role of phosphorylation in elicination of the oxidative burst in cultured soybean cells. *Proc. Natl. Acad. Sci. USA* 92: 4120–4123.

Chapman, K.D. 2000. Emerging physiological roles for N-acylphosphatidylethanolamine metabolism in plants: signal transduction and membrane protection. *Chem. Phys. Lipids* 108:221–230.

Che, F.-S., Iwano, M., Tanaka, N., Takayama, S., Minami, E., Shibuya, N., Kadota, I., and Isogai A. 1999. Biochemical and morphological features of rice cell death induced by *Pseudomonas avenae*. *Plant Cell Physiol.* 40:1036–1045.

Chen, W., Provart, N.J., Glazebrook, J., Katagiri, F., Chang, H.S., Eulgem, T., Mauch, F., Luan, S., Zou, G., Whitham, S.A., Budworth, P.R., Tao, Y., Xie, Z., Chen, X., Lam, S., Kreps, J.A., Harper, J.F., Si-Ammour, A., Mauch-Mani, B., Heinlein, M., Kobayashi, K., Hohn, T., Dangl, J.L., Wang, X., and Zhu, T. 2002. Expression profile matrix of Arabidopsis transcription factor genes suggests their putative functions in response to environmental stresses. *Plant Cell* 14:559–574.

Chen, Z., Silva, H., and Klessig, D.F. 1993. Active oxygen species in the induction of plant systemic acquired resistance by salicylic acid. *Science* 262:1883–1886.

Chivasa, S., and Carr, J.P. 1998. Cyanide restores *N* gene-mediated resistance to tobacco mosaic virus in transgenic tobacco expressing sacylic acid hydroxylase. *Plant Cell* 10: 1489–1498.

Clark, D., Durner, J., Navarre, D.A., and Klessig, D.F. 2000. Nitric oxide inhibition of tobacco catalase and ascorbate peroxidase. *Mol. Plant Microbe Interact.*13:1380–1384.

Clough, S.J., Fengler, K.A., Yu, I-C., Lippok, B., Smith, R.K.Jr., and Bent, A.F. 2000. The Arabidopsis *dnd1* "defense, no death" gene encodes a mutated cyclic nucleotide-gated ion channel. *Proc. Natl. Acad. Sci. USA* 97:9323–9328.

Dangl, J.L., Dietrich, R.A., Richberg, M.H. 1996. Death don't have no mercy: Cell death programs in plant-microbe interactions. *Plant Cell* 8:1793–1807.

Dangl J.L., and Jones, J.T.G. 2001. Plant pathogens and integrated defense responses to infection. *Nature* 411:826–833.

Daniel, X., Lacomme, C., Morel, J.B., and Roby, D. 1999. A novel myb oncogene homologue in *Arabidopsis thaliana* related to hypersensitive cell death. *Plant J.* 20:57–66.

Delledonne, M., Xia, Y.J., Dixon, R.A., and Lamb, C. 1998. Nitric oxide functions as a signal in plant disease resistance. *Nature* 394:585–588.

del Pozo, O., and Lam, E. 1998. Caspases and programmed cell death in the hypersensitive response of plants to pathogens. *Curr. Biol.* 8:1129–1132.

del Pozo, O., and Lam, E. 2003. Expression of the baculovirus p35 protein in tobacco delays cell death progression and enhanced systemic movement of tobacco mosaic virus during the hypersensitive responses. *Mol. Plant Microbe Interact.* 16:485–494.

Després C., DeLong, C., Glaze, S., Liu, E., and Fobert, P.R. 2000. The *Arabidopsis* NPR1/NIM1 protein enhances the DNA binding activity of a subgroup of the TGA family of bZIP transcription factors. *Plant Cell* 12:279–290.

Dickman, M.B., Park, Y.K., Oltersdorf, T., Clemente, T., and French, R. 2001. Abrogation of disease development in plants expressing animal antiapoptotic genes. *Proc. Natl. Acad. Sci. USA* 98:6957–6962.

Dietrich, R.A., Delaney, T.P., Uknes, S.J., Ward, E.R., Ryals, J.A., and Dangl, J.L. 1994. Arabidopsis mutants simulating disease resistance response. *Cell* 77: 565–577.

Dietrich, R.A., Richberg, M.H., Schmidt, R., Dean, C., and Dangl, J.L. 1997. A novel zinc finger proteins is encoded by the Arabidopsis *LSD1* gene and functions as a negative regulator of plant cell death. *Cell* 88:685–694.

Dinesh-Kumar, S.P., Wai-HongTham, and Baker, B.J. 2000. Structure-function analysis of the tobacco mosaic virus resistance gene *N*. *Proc. Natl. Acad. Sci. USA* 97:14789–14794.

Doke, N. 1983. Generation of superoxide anion by potato tuber protoplasts during the hypersensitive response to hyphal cell wall components of *Phytophthora infestans* and specific inhibition of the reaction by suppressors of hypersensitivity. *Physiol. Plant Pathol.* 23:359–367.

Doke, N. 1985. NADPH-dependent O_2^- generation in membrane fraction isolated from wounded potato tubers inoculated with Phytophthora infestans. *Physiol. Plant Pathol.* 27:311–322.

Doke, N., and Ohashi, Y. 1988. Involvement of an O_2^- generating system in the induction of necrotic lesions on tobacco leaves infected with tobacco mosaic virus. *Physiol. Mol. Plant Pathol.* 32:163–175.

Dong, X. 2001. Genetic dissection of systemic acquired resistance. *Curr. Opin. Plant Biol.* 4: 309–314.

D'Silva, I., Pirier, G.G., and Heath, M.C. 1998. Activation of cysteine proteases in cowpea plants during the hypersensitive response, a form of programmed cell death. *Exp. Cell Res.* 245:389–399.

Durner, J., and Klessig, D.F. 1996. Salicylic acid is a modulator of tobacco and mammalian catalases. *J Biol. Chem.* 271:28492–28501.

Durner, J., Wedndehenne, D., and Klessig, D.F. 1998. Defense gene induction in tobacco by nitric oxide, cyclic GMP, and cyclic ADP-ribose. *Proc. Natl. Acad. Sci. USA* 95:10328–10333.

Dunigan, D.D., and Madlener, J.C. 1995. Serine/threonine protein phosphatase is required for tobacco mosaic virus-mediated programmed cell death. *Virology* 207:460–466.

Flor, H.H. 1971. Current status of the gene-for-gene concept. *Annu. Rev. Phytopathol.* 9:275–296.

Falk, A., Fey, B.J., Frost, L.N., Jones, J.D.G., Daniels, M.J., and Parker, J.E. 1999. *EDS1*, an essential component of *R*-gene-mediated disease resistance in *Arabidopsis* has homology to eukaryotic lipases. *Proc. Natl. Acad. Sci. USA* 96:3292–3297.

Friedrich, L., Lowton, K., Dincher, S., Winter, A., Staub, T., Uknes, S., Kessmann, H., and Ryals, J. 1996. Benxothiadiazole induces systemic acquired resistance in tobacco. *Plant J.* 10:61–70.

Frampton, J., Ramqvist, T., and Graf, T. 1996. v-Myb of Eleukemia virus up-Regulates bcl-2 and suppresses apoptosis in myeloid cells. *Genes Dev.* 10:2720–2731.

Gaffney, T., Friedrich, L., Vernooji, B., Negrotto, D., Nye, G., Uknes, S, Ward, E., Kessmann, H., and Ryals, J. 1993. Requirement for salicylic acid for the induction of systemic acquired resistance. *Science* 261:754–756.

Gilchrist, D.G. 1998 Programmed cell death in plant disease: the purpose and promise of cellular suicide. *Annu. Rev. Phytopathol.* 36:393–414.

Goodman, R.N., and Novacky, A.J. 1994. *The Hypersensitive Reaction in Plants to Pathogens*. St. Paul, MN: APS Press.

Goulden, M.G., and Baulcombe, D.C. 1993. Functionally homologous host components recognize potato virus X in *Gompherena globosa* and potato. *Plant Cell* 5:921–930.

Govrin, E.M., and Levine, A. 2000. The hypersensitive response facilitates plant infection by the necrotrophic pathogen *Boytrytis cinerea*. *Curr. Biol.* 10:751–757.

Gray, J., Close, P.S., Briggs, S.P., and Johal, G.S. 1997 A novel suppressor of cell death in plants encoded by the *Lls1* gene of maize. *Cell* 89:25–31.

Green, D.R. 2000. Apoptotic pathways: paper wraps stone blunts scissors. *Cell* 102:1–4.

Greenberg, J.T., and Ausubel, F.M. 1993. Arabidopsis mutants compromised for the control of cellular damage during pathogenesis and aging. *Plant J.* 4:327–341.

Greenberg, J.T., Guo, A., Klessig, D.F., and Ausubel, F.M. 1994. Programmed cell death in plants: A pathogen triggered response activated coordinately with multiple defense functions. *Cell* 77: 551–563.

Greenhalf, W., Lee, J., Chaudhuri, B. 1999. A selection system for human apoptosis inhibitors using yeast. *Yeast* 15:1307–1321.

Groom, Q.J., Torres, M.A., Forrdam-Skelton, A.P., Hammond-Kosack, K.E., Robinson, N.J., and Jones, J.D.G. 1996. RbohA, a rice homologue of the mammalian *gp91 phox* respiratory burst oxidase gene. *Plant J.* 10:515–522.

Hammond-Kosack, K.N., and Jones, J.D.G. 1996. Resistance gene-dependent plant defense responses. *Plant Cell* 8:1773–1791.

He, S.Y., Huang, H.-C., and Collmer, A. 1993. *Pseudomonas syringae* pv. *syringae* Harpin$_{PSS}$: a protein that is secreted by the *Hrp* pathway and elicits the hypersensitive response in plants. *Cell* 73: 1255–1266.

Heath, M.C. 2000. Hypersensitive response-related death. *Plant Mol. Biol.* 44:321–334.

Heo, W.D,, Lee, S.H., Kim, M.C., Kim, J.C., Chung, W.S., Chun, H.J., Lee, K.J., Park, C.Y., Park, H.C., Choi, J.Y., Cho, M.J. 1999. Involvement of specific calmodulin isoforms in salicylic acid-independent activation of plant disease resistance responses. *Proc. Natl. Acad. Sci USA* 96:766–771.

Herbers, K., Meuwly, P., Frommer, W.B., Metraux, J.P., and Sonnewald, U. 1996. Systematic acquired resistance mediated by the ectopic expression of invertase: possible hexose sensing in the secretory pathway. *Plant Cell* 8: 793–803.

Holmes, F.O. 1938. Inheritance of resistance to tobacco-mosaic disease in tobacco. *Phytopathology* 28:553–561.

Hu, G., Yalpani, N., Briggs, S.P., and Johal, G.S. 1998. A porphyrin pathywa inpairment is responsible for the phenotype of a dominant disease lesion mimic mutant of maize. *Plant Cell* 10:1095–1105.

Hunt, M.D., Delaney, T.P., Dietrich, R.A., Weymann, K.B., Dangl, J.L., and Ryals, J.A. 1997. Salicylate-independent lesion formation in *Arabidopsis lsd* mutant. *Mol. Plant Microbe Interact.* 10:531–536.

Ishikawa, A., Okamoto, H., Iwasaki, Y., and Asahi, T. 2001. A deficiency of coprotoporphyrinogen III oxidese causes lesion formation in Arabidopsis. *Plant J.* 27: 89–99.

Jabs, T., Dietrich, R.A., and Dangl, J.L. 1996. Initiation of runaway cell death in an *Arabidopsis* mutant by extracellular superoxide. *Science* 273:1853–1856.

Jabs, T., Tschope, M., Colling, C., Hahlbrock, K., and Scheel, D. 1997. Elicitor-stimulated ion fluxes and O_2^- from the oxidative burst are essential components in triggering defense gene activation and phytoalexin synthesis in parsley. *Proc. Natl. Acad. Sci. USA* 94:4800–4805.

Jirage, D., Tootle, T.L., Reuber, L., Frost, L.N., Fey, B.J., Parker, J.E., Ausbel, F.M., and Glazebrook, J. 1999. *Arabidopsis thaliana PAD4* encodes a lipase-like gene that is important for salicylic acid signaling. *Proc. Natl. Acad. Sci. USA* 96:13583–13588.

Johal, G.S., Hulbert, S., and Briggs, S.P. 1995. Disease lesion mimic mutations of maize: A model for cell death in plants. *Bioessays* 17:685–692.

Kawai, M., Pan, L., Reed, J.C., and Uchimiya, H. 1999. Evolutionally conserved plant homologue of the Bax inhibitor-1 (BI-1) gene capable of suppressing Bax-induced cell death in yeast. *FEBS Lett.* 464:143–147.

Kawai-Yamada, M., Jin, U., Yoshinaga, K., Hirata, A., and Uchimiya, H. 2001. Mammalian Bax-induced plant cell death can be down-regulated by overexpression of *Arabidopsis* Bax inhibitor-1 (*AtBI-1*). *Proc. Natl. Acad. Sci. USA* 98:12295–12300.

Kawasaki, T., Henmi, K., Ono, E., Hatakeyama, S., Iwano, M., Satoh, H., and Shimamoto K. 1999. The small GTP-binding protein Rac is a regulator of cell death in plants. *Proc. Natl. Acad. Sci. USA* 96:10922–10926.

Kazan, K., Schenk, P.M., Wilson, I., and Manners, J.M. 2001. DNA microarrays: new tools in the analysis of plant defense responses. *Mol. Plant Pathol.* 2:177–185.

Keller, T., Damude, H.G., Werner, D., Doener, P., Dixion, R.A., and Lamb, C. 1998. A plant homolog of the neutrophil NADPH oxidase *gp91phox* subunit gene encodes a plasma membrane protein with Ca^{2+} binding motif. *Plant Cell* 10:255–266.

Kim, M.C., Lee, S.H., Kim, J.K., Chun, H.J., Choi, M.S., Chung, W.S., Moon, B.C., Kang, C.H., Park, C.Y., Yoo, J.H., Kang, Y.H., Koo, S.C., Koo, Y.D., Jung, J.C., Kim, S.T., Schulze-Lefert, P., Lee, S.Y., and Cho, M.J. 2002a. Mlo, a modulator of plant defense and cell death, is a novel calmodulin-binding protein. Isolation and characterization of a rice Mlo homologue. *J. Biol. Chem.* 277:19304–19314.

Kim, M.C., Panstruga, R., Elliott, C., Muller, J., Devoto, A., Yoon, H.W., Park, H.C., Cho, M.J., Schulze-Lefert, P. 2002b. Calmodulin interacts with MLO protein to regulate defence against mildew in barley. *Nature* 416:447–451.

Kobe, B., and Deisenhofer, J. 1995. Proteins with leucine-rich repeats. *Curr. Opin. Struct. Biol.* 15: 409–416.

Kruger, J., Thomas, C.M., Golstein, C., Dixon, M.S., Smoker, M., Tang, S., Mulder, L., and Jones, J.D.G. 2002. A tomato cysteine protease required for *Cf-2*-dependent disease resistance and suppression of autonecrosis. *Science* 296:744–747.

Kunkel, B.N., Bent, A.F., Dahlbeck, D., Innes, R.W., and Staskawicz, B.J. 1993. *RPS2*, an *Arabidopsis* disease resistance locus specifying recognition of *Pseudomonas syringae* strains expressing the avirulence gene *avrRpt2*. *Plant Cell* 5:865–875.

Kuwchuk L.M., Hachey, J., Lynch, D.R., Kulcsar, F., van Rooijen, G., Waterer, D.R., Robertson, A., Kokko, E., Byers, R., Howard, R.J., Fischer, R., and Prüfer, D. 2001. Tomato *Ve* disease resistance genes encode cell surface-like receptors. *Proc. Natl. Acad. Sci. USA* 98:6511–6515.

Lacomme, C., and Santa Cruz, S. 1999. Bax-induced cell death in tobacco is similar to the hypersensitive response. *Proc. Natl. Acad. Sci. USA* 96:7956–7961.

Lam, E., Pontier, D., and del Pozo, O. 1999a. Die and let live—programmed cell death in plants. *Curr. Opin. Plant Biol.* 2:502–507.

Lam, E., del Pozo, O., and Pontier, D. 1999b. BAXing in the hypersensitive response. *Trends Plant Sci.* 4:419–421.

Lam E., and del Pozo, O. 2000. Caspase-like protease involvement in the control of plant cell death. *Plant Mol. Biol.* 44:417–428.

Lam, E., Kato, N., and Lawton, M. 2001. Programmed cell death, mitochondria and the plant hypersensitive response. *Nature* 411 848–853.

Lamb, C., and Dixon, R.A. 1997. The oxidative burst in plant disease resistance. *Anuu. Rev. Plant Physiol. Plant Mol. Biol.* 48:251–275.

Legendre, L., Heinstein, P.F., and Low, P.S. 1992. Evidence for participation of GTP-binding proteins in elicination of the rapid oxidative burst in cultured soybean cells. *J. Biol. Chem.* 267:20140–20147.

Levine, A., Pennell, R.I., Alvarez, M.E., Palmer, R., and Lamb, C. 1996. Calcium-mediated apoptosis in a plant hypersensitive disease resistance response. *Curr. Biol.* 6:427–437.

Levine, A., Tenhaken, R., Dixon, R., and Lamb, C. 1994. H_2O_2 from the oxidative burst orchestrate the plant hypersensitive disease resistance response. *Cell* 79:583–593.

Liu, Y., Schiff, M., Marathe, R., and Dinesh-Kumar S.P. 2002 Tobacco Rar1, EDS1 and NPR1/NIM1 like genes are required for N-mediated resistance to tobacco mosaic virus. *Plant J.* 30:415–429.

Madeo, F., Herker, E., Maldener, C., Wissing, S., Lächelt, S., Herlan, M., Fehr, M., Lauber, K., Sigrist, S.J., Wesselborg, S., and Fröhlich, K.-U. 2002. A Caspase-related protease regulates apoptosis in yeast. *Mol. Cell* 9: 911–917.

Malamy, J., Carr, J.P., Klessig, D.F., and Raskin, I. 1990. Salicylic acid: a likely endogenous signal in the resistance response of tobacco to viral infection. *Science* 250:1002–1004.

Maldonado, A.M., Doerner, P., Dixon, R.A., Lamb, C.J., and Cameron, R.K. 2002. A putative lipid transfer protein involved in systemic resistance signalling in *Arabidopsis*. *Nature*. 419:399–403.

Maleck, K., Levine, A., Eulgem, T., Morgan, A., Schmid, J., Lawton, K.A., Dangl, J.L., and Dietrich, R.A. 2000. The transcriptome of Arabidopsis thaliana during systemic acquired resistance. *Nature Genet.* 26:403–410.

Marco, Y., Regueh, F., Goldlard, L., and Froissard, D. 1990. Transcriptional activation of 2 classes of genes during the hypersensitive reaction of tobacco leaves infiltrated with an incompatible isolate of the phytopathogenic bacterium *Pseudomonas solanacearum*. *Plant Mol. Biol.* 15:145–154.

Martin, S.J., Green, D.R., and Cotter, T.G. 1994. Dicing with death: dissecting the components of the apoptosis machinery. *Trends Biochem. Sci.* 19:26–30.

Martinou, J.-C., and Green, D.R. 2001. Breaking the mitochondrial barrier. *Nature Cell Biol.* 2: 63–67.

Maxwell, D.P., Wang, Y., and McIntosh, L. 1999. The alternative oxidase lowers mitochondria reactive oxygen production in plant cells. *Proc. Natl. Acad. Sci. USA* 96:8271–8276.

Métraux, J.P., Singer, H., Ryals, J., Ward, E., Wyss-Benz, M., Gaudin, J., Raschdorf, K., Schmid, E., Blum, W., and Inverardi, B. 1990. Increase in salicylic acid at the onset of systemic acquired resistance in cucumber. *Science* 250:1004–1006.

Mindrinos, M., Katagiri, F., Yu, G., and Ausubel, F.M. 1994. The *A. thaliana* disease resistance gene *RPS2* encodes a protein containing a nucleotide binding site and a leucine-rich repeats. *Cell* 78:1089–1099.

Mitsuhara, I., Malik, K.A., Miura, M., and Ohashi Y. 1999. Animal cell-death suppressors Bcl-X$_L$ and Ced-9 inhibit cell death in tobacco cells. *Curr. Biol.* 9:775–778.

Mittler, R., and Lam, E. 1995. Identification, characterization, and purification of a tobacco endonuclease activity induced upon hypersensitive response cell death. *Plant Cell* 7:1951–1962.

Mittler, R., Shulaev, V., and Lam, E. 1995. Coordiated activation of programmed cell death and defense mechanisms in transgenic tobacco plants expressing a bacterial proton pump. *Plant Cell* 7:29–42.

Mittler, R., and Lam, E. 1996. Sacrifice in the face of foes: pathogen-induced programmed cell death in higher plants. *Trends Microbiol.* 4:10–15.

Mittler, R., and Lam, E. 1997 Characterization of nuclease activities and DNA fragmentation induced upon hypersensitive response cell death and mechanical stress. *Plant Mol. Biol.* 34:209–221.

Mittler, R., del Pozo, O., Meisel, L., and Lam, E. 1997a. Pathogen-induced programmed cell death in plants, a possible defense mechanism. *Dev. Genet.* 21:279–289.

Mittler, R., Simon, L., and Lam, E. 1997b. Pathogen-induced programmed cell death in tobacco. *J. Cell Sci.* 110: 333–1344.

Mittler, R., and Rizhsky, L. 2000. Transgene-induced lesion mimic. *Plant Mol. Biol.* 44:335–344.

Molina, A., Volrath, S., Guyer, D., Maleck, K., Ryals, J., and Ward, E. 1999. Inhibition of protoporphyrinogen oxidase expression in *Arabidopsis* causes a lesion-mimic phenotype that induces systemic acquired resistance. *Plant J.* 17:667–678.

Munnik, K. 2001. Phosphatidic acid: an emerging plant lipid second messenger. *Trends Plant Sci.* 6:227–233.

Nürnberger, T., Nennstiel, D., Jab, T., Sacks, W.R., Hahlbrock, K., and Scheel, D. 1994. High affinity binding of a fungal oligopeptide elicitor to parsley plasma membranes triggers multiple defense response. *Cell* 78:229–460.

Oberhammer, F., Wilson, J.W., Dive, C., Morris, I.D., Hickman, J.A., Wakeling, A.E., Walker, P.R., and Sikoeska, M. 1993. Apoptotic death in epithelial cells: cleavage of DNA to 300 and/or 50 kb fragments prior to or in the absence of internucleosomal fragmentation. *EMBO J.* 12:367–3684.

Ono, E., Wong, H.L., Kawasaki, T., Hasegawa, M., Kodama, O., and Shimamoto K. 2001. Essential role of the small GTPase rac in disease resistance of rice. *Proc. Natl. Acad. Sci. USA* 98:759–764.

Peart, J.R., Cook, G., Feys, B.J., Parker, J.E., Baulcombe, D.C. 2002. An EDS1 orthologue is required for N-mediated resistance against tobacco mosaic virus. *Plant J.* 29:569–579.

Peitsch, M.C., Polzar, B., Stephan, H., Cromptom, T., MacDonald, H.R. Mannherz, H.G., and Tschopp, J. 1993. Characterization of the endogenous deoxyribonuclease involved in nuclear DNA degradation during apoptosis (programmed cell death). *EMBO J.* 12:371–377.

Pike, S.M., Adam, A.L., Pu, X.-A., Hoyos, M.E., Laby, R., Beer, S.V., and Novacky, A. 1998. Effects of *Erwinia amylovora* harpin on tobacco leaf cell membranes are related to leaf necrosis and electrolyte leakage and distinct from perturbation caused by inoculated *E. amylovora*. *Physiol. Mol. Plant Pathol.* 53: 39–60.

Pontier, D., Godiard, L., Marco, Y., and Roby, D. 1994. *hsr203J*, a tobacco gene whose activation is rapid, highly localized and specific for incompatible plant/pathogen interactions. *Plant J.* 5:507–521.

Pontier, D., Tronchet, M., Rogowsky, P., Lam, E., and Roby, D. 1998. Activation of hsr203, a plant gene expressed during incompatible plant-pathogen interactions, is correlated with programmed cell death. *Mol. Plant Microbe Interact.* 11:544–554.

Pontier, D., Mittler, R., and Lam, E. 2002. Mechanism of cell death and disease resistance induction by transgenic expression of bacterio-opsin. *Plant J.* 30:499–510.

Richael, C., Lincoln, J E., Bostock, R.M., and Gilchrist, D.G. 2001. Caspase inhibitors reduce symptom development and limit bacterial proliferation in susceptible plant tissues. *Physiol. Mol. Plant Pathol.* 59:213–221.

Roebuck, P., Sexton, R., and Mansfield, J.W. 1978. Ultrastructual observations on the development of the hypersensitive reaction in leaves of *Phaseolus vulgaris* cv. Red Mexican inoculated with *Pseudomonas phaseolicola* (race 1). *Physiol. Plant Pathol.* 12: 151–157.

Ryals, J.A., Neuenschwander, U.H., Willits, M.G., Molina, A., Steiner, H.-Y., and Hunt, M.D. 1996. Systemic acquired resistance. *Plant Cell* 8:1809–1819.

Ryerson, D.E., and Heath, M.C. 1996. Cleavage of nuclear DNA into oligonucleosomal fragments during cell death induced by fungal infection or by abiotic treatment. *Plant Cell* 8: 393–402.

Sagi, M., and Fluhr, R. 2001. Superoxide production by plant homologues of the gp91[phox] NADPH oxidase. Modulation of activity by calcium and by tobacco mosaic virus infection. *Plant Physiol.* 126:1281–1290.

Sanchez, P., de Torres Zebala, M., and Grant, M. 2000. AtBI-1, a plant homologue of Bax inhibitor-1, supperss Bax-induced cell death in yeast is rapidly upregulated during wounding and pathogen challenge. *Plant J.* 21:393–399.

Sano, H., and Ohashi, Y. 1995. Involvement of small GTP-binding proteins in defense signal-transduction pathways of higher plants. *Proc. Natl. Acad. Sci. USA* 94:4138–144.

Schaller, A., and Oecking, C. 1999. Modulation of plasma membrane H^+-ATPase activity differentially activates wound and pathogen defense responses in tomato plants. *Plant Cell* 11:263–272.

Scheideler, M., Scjlaich, N.L., Fellenberg, K., Beissbarth, T., Hauser, N.C., Vingron, M., Slusarenko, A.J., and Hoheosel, J.D. 2002. Monitoring the switch from housekeeping to pathogen defense metabolism in *Arabidopsis thaliana* using cDNA microarray. *J. Biol. Chem.* 277:10555–10561.

Schenk, P.M., Kazan, K., Wilson, I., Anderson, J.P., Richmond, T., Somerville, S.C., and Manners, J.M. 2000. *Proc. Natl. Acad. Sci. USA* 97:11655–11660.

Schwartzman, R.A., Cidlowski, J.A. 1993. Apoptosis: the biochemistry and molecular biology of programmed cell death. *Endocr. Rev.* 14:133–151.

Seo, S., Okamoto, M., Iwai, T., Iwano, M., Fukui, K., Isogai, A., Nakajima, N., and Ohashi, Y. 2000. Reduced levels of chloroplast FtsH protein in tobacco mosaic virus-infected tobacco leaves accelerate the hypersensitive reaction. *Plant Cell* 12:917–932.

Shirasu, K., Nakajima, H., Rajasekhar, V.K., Dixon, R.A., and Lamb, C. 1997. Salicylic acid potentiates an agonist-dependent gain control that amplifies pathogen signals in the activation of defense mechanisms. *Plant Cell* 9:1–10.

Shirasu, K., and Schulze-Lefert, P. 2000. Regulator of cell death in disease resistance. *Plant Mol. Biol.* 44:371–385.

Simons, B.H., Millenaar, F.F., Mulder, L., van Loon, L.C., and Lambers, H. 1999. Enhanced expression and activation of the alternative oxidase during infection of Arabidopsis with *Pseudomonas syringae* pv. tomato. *Plant Physiol.* 120:529–538.

Solomoni, P., Perrotti, D., Martinez, R., Franceschi, C., and Calabretta, B. 1997. Resistance to apoptosis in CTLL-2 cells constitutively expressing *c-myb* is associated with induction of BCL-2 expression and Myb-dependent regulation of bcl-2 promoter activity. *Proc. Natl. Acad. Sci. USA* 94:3296–3301.

Stone, J.M., Heard, J.E., Asai, T., and Ausubel, F.M. 2000. Simulation of fungal-mediated cell death by fumonisin B1 and selection of fumonisin B1-resistant (*fbr*) Arabidopsis mutants. *Plant Cell* 12:1811–1822.

Sugiyama, M., Ito, J., Aoyagi, S., and Fukuda, H. 2000. Endonnuclease. *Plant Mol. Biol.* 44:387–397.

Takahashi, H., Chen, Z., Du, H., Liu, Y., Klessig, D.F. 1997. Development of necrosis and activation of disease resistance in transgenic tobacco plants with severely reduced catalase levels. *Plant J.* 11:993–1005.

Tenhaken, R., Levine, A., Brisson, L.F., Dixon, R.A., and Lamb, C. 1995. Function of the oxidative burst in hypersensitive disease resistance. *Proc. Natl. Acad. Sci. USA* 92:4158–4163.

Torres, M. A., Dangl, J.L., and Jones, J.D.G. 2002. Arabidopsis gp91[phox] homologues *AtrbohD* and *AtrbohF* are required for accumulation of reactive oxygen intermediates in the plant defense response. *Proc. Natl. Acad. Sci. USA* 99:523–528.

Torres, M.A., Onouchi, H., Hamada, S., Machida, C., Hammond-Kosack, K.E., and Jones, J.D.G. 1998. Six *Arabidopsis thaliana* homologues of the human respiratory burst oxidase (*gp91[phox]*). *Plant J.* 14:365–370.

Uren, A.G., O'Rourke, K., Aravind, L., Pisabarro, M.T., Seshagiri, S., Koonin, E.V., and Dixit, V.M. 2000. Identification of paracaspase and metacaspases: two ancient families of caspase-like proteins, one of which plays a key role in MALT lymphoma. *Mol. Cell* 6:961–967.

Wang, H., Jones, C., Ciacci-Zannella, J., Holt, T., Gilchrist, D.G., and Dickman, M. 1996a. Sphinganine analog mycotoxins induce apoptosis in monkey cells. *Proc. Natl. Acad. Sci. USA* 93:3461–3465.

Wang, H., Li, J., Bostock, R.M., and Gilchrist, D.G. 1996b. Apoptosis: a functional paradigm for programmed cell death induced by a host-selective phytotoxin and invoked during development. *Plant Cell* 8:375–391.

Wang, M.-B., and Waterhouse, P.M. 2001. Application of gene silencing in plants. *Curr. Opin. Plant Biol.* 5:146–150.

Wasternack, C., and Parthier, B. 1997. Jasmonate-signalled plant gene expression. *Trends Plant Sci.* 2:302–307.

Weymann, K., Hunt, M., Uknes, S., Neuenschwandler, U., Lawton, K., Steiner, H.Y., and Ryals, J. 1995. Suppression and restoration of lesion formation in *Arabidopsis lsd* mutant. *Plant Cell* 7:2013–2022.

White, E. 1996. Life, death, and pursuit of apoptosis. *Genes Dev.* 10:1–15.

Wirtz, K.W. 1997. Phospholipid transfer proteins revisited. *Biochem. J.* 324:353–360.

Wolpert, T.J., Dunkle, L.D., and Ciuffetti, L.M. 2002. Host-selective toxins and avirulence determinants: what's in a name? *Annu. Rev. Phytopathol.* 40:251–285.

Wyllie, A.H., Morris, R.G., Smith, A.L., and Dunlop, D. 1984. Chromatin cleavage in apoptosis: association with condensed chromatin morphology and dependence on macromolecular system. *Mol. Gen. Genet.* 239:122–128.

Xu, Q., and Reed, J.C. 1998. Bax inhibitor-1, a mammalian apoptosis supperssor identified by functional screening in yeast. *Mol. Cell* 1:337–346.

Yang, Y., and Klessig, D.F. 1996. Isolation and characterization of a tobacco mosaic virus-inducible myb oncogene homolog from tobacco. *Proc. Natl. Acad. Sci. USA* 93:14972–14977.

Yu, I-C., Parker, J., and Bent, A.F. 1998. Gene-for-gene resistance without the hypersensitive response in *Arabidopsis dnd1* mutant. *Proc. Natl. Acad. Sci.* 95:7819–7824.

Yu, L.H., Kawai-Yamada, M., Naito, M., Watanabe, K., Reed, J.C., Uchimiya, H. 2002. Induction of mammalian cell death by a plant Bax inhibitor. *FEBS Lett.* 512:308–312.

Zhang, Y.L., Fan, W.H., Kinkema, M., Li, X., and Dong, X. 1999. Interaction of NPR1 with basic leucine zipper protein transcription factors that bind sequences required for salicylic acid induction of the PR-1 gene. *Proc. Natl. Acad. Sci. USA* 96:6523–6528.

Zhou, F., Andersen, C.H., Burhenne, K., Fischer, P.H., Collinge, D.B., and Thordal-Christensen, H. 2000a. Proton extrusion is an essential signaling component in the HR of epidermal single cells in the barley-powdery mildew interaction. *Plant J.* 23: 245–254.

Zhou, J.M., Trifa, Y., Silva, H., Pontier, D., Lam, E., Shah, J., Klessig, D.F. 2000b. NPR1 differentially interacts with member of the TGA/OBF family of transcription factors that bind an element of the PR-1 gene required for induction by salicylic acid. *Mol. Plant Microbe Interact.* 13: 191–202.

Zhu, T., and Wang, X. 2000. Large scale profiling of the *Arabidopsis* transcriptome. *Plant Physiol.* 124:1472–1476.

6

The Possible Role of PR Proteins in Multigenic and Induced Systemic Resistance

SADIK TUZUN AND ARAVIND SOMANCHI

6.1 Introduction

There have been many studies dealing with PR proteins since their first discovery during analyses of the protein composition of tobacco mosaic virus (TMV)-induced hypersensitivity response in tobacco (Van Loon and Van Kammen, 1970) over 30 years ago. The amino acid composition of these proteins was quite variable, showing some of the proteins to be acidic and others to be basic, it was suggested that they may play an important role in the fate of pathogenesis (Kassanis et al., 1974; Van Loon, 1975). Fungi and bacteria were also discovered to induce similar new protein components in various plant species, particularly during incompatible combinations resulting in hypersensitive necrosis (Redolfi, 1983). Analyses of several of these showed a pattern of host responses in that they apparently consisted of one or more families of host-coded proteins, which were induced by different types of pathogens and abiotic stresses, and were most often of relatively low molecular weight, preferentially extracted at low pH, highly resistant to proteolytic degradation, and localized predominantly in the intercellular space of the leaf. These proteins coded for by the host plant but induced only in pathological or related situations have been termed pathogenesis-related (PR) proteins. While constitutively expressed proteins that show increases upon pathogen infection, such as oxidative and enzymes of aromatic biosynthesis are generally excluded, specific isoforms of such enzymes that are induced only as a result of infection have been grouped with PR proteins.

In this chapter, we will present evidences that timely accumulation of PR proteins during pathogenesis is a part of defense mechanisms in plants against pathogens and pests. Some of these proteins may have different roles in plant metabolism and/or may just be produced as a part of more generalized "housekeeping" regulatory systems during a plant–pathogen interaction. Specific isozymes of hydrolytic enzymes, on the other hand, demonstrate differential activity toward the substrate during the release of elicitor molecules from the pathogens. These isozymes may have evolved as a part of a suite of defense mechanisms in "naturally resistant" plants, or plants considered to have higher basal resistance. Isozyme-based higher basal resistance may come about via the sensitization of plants to a particular pathogen, via the isozyme-mediated production of pathogen-derived nonspecific or specific elicitors that initiate the whole battery of defense mechanisms. It is important to recognize that plant defense mechanisms are complex and that more than one factor is involved in the successful existence of a plant species over the centuries in the face of abundant pathogen pressures. Pathogenesis can be considered the result of the failure of the plant's many and redundant defense-related mechanisms to activate in a timely manner to prevent or sharply contain pathogen infection.

6.2 Classification of PR Proteins

On the basis of their properties, the tobacco PR proteins were initially grouped into five families, and this classification is used in other plant species in which PR proteins are identified. The families are numbered and the different members within each family are assigned letters according to the order in which they are described. Thus the same designation for a PR protein in different plant species indicates that they belong to the same PR-family, but only reflects how many proteins of that family had been identified within those plant species. The genes encoding these proteins are designated as *ypr* followed by the suffix that corresponds to the protein. Because PR proteins are generally defined by their occurrence as protein bands on gels, gene and cDNA sequences cannot be fitted into the adopted nomenclature. Conversely, homologies at the nucleotide sequence level may be encountered without information on the expression or characteristics of the encoded protein. This leads to a complexity in comparative analysis of PR proteins from different species. Also, when new genes induced by pathogens or specific elicitors are identified they may be added to the existing families. Thionins (Bohlmann, 1994) and defensins (Broekaert et al., 1995), both families of small, basic, cysteine rich polypeptides were subsequently added to the families of PR proteins, based on this criterion. The identification of several such proteins with disparate properties necessitated the expansion of the classification, and addition of families. The nomenclature currently in use was proposed in 1994, and groups PR proteins into the 17 plant-wide families depicted in Table 6.1, on the basis of sequence homology and similarities in enzymatic and biological activities (Van Loon et al., 2002).

Localization of majority of the PR proteins in the intercellular spaces of leaves seems to guarantee contact with the invading pathogen before penetration. However, *in vitro* and *in vivo* analyses failed to show anti-pathogenic activity in any of the PR proteins associated with systemically induced resistance by a few *Pseudomonas* species (Van Loon, 1997), suggesting that they may not play a major role in defense in some systems (Pieterse et al., 1996). Accumulation of PR proteins to similar amounts in compatible as well as incompatible host interactions (Hoffland et al., 1995) suggests that PR proteins do not determine the resistance response in this particular host-pathogen interaction. However, constitutive accumulation of

TABLE 6.1. Recognized families of pathogenesis-related proteins.

Family	Type member	Properties	Reference
PR-1	Tobacco PR-1a	Antifungal	Antoniw et al. (1980)
PR-2	Tobacco PR-2	β-1,3-glucanase	Antoniw et al. (1980)
PR-3	Tobacco P, Q	chitinase type I,II, IV,V,VI,VII	Van Loon (1982)
PR-4	Tobacco 'R'	chitinase type I,II	Van Loon (1982)
PR-5	Tobacco S	thaumatin-like	Van Loon (1982)
PR-6	Tomato Inhibitor I	proteinase-inhibitor	Green and Ryan (1972)
PR-7	Tomato P69	Endoproteinase	Vera and Conejero (1988)
PR-8	Cucumber chitinase	chitinase type III	Métraux et al. (1988)
PR-9	Tobacco 'lignin-forming peroxidase'	Peroxidase	Lagrimini et al. (1987)
PR-10	Parsley 'PR1'	'ribonuclease-like'	Somssich et al. (1986)
PR-11	Tobacco 'class V' chitinase	chitinase, type I	Melchers et al. (1994)
PR-12	Radish Rs-AFP3	Defensin	Terras et al. (1992)
PR-13	Arabidopsis THI2.1	Thionin	Epple et al. (1995)
PR-14	Barley LTP4	lipid-transfer protein	García-Olmedo et al. (1995)
PR-15	Barley OxOa (germin)	oxalate oxidase	Zhang et al. (1995)
PR-16	Barley OxOLP	oxalate oxidase-like	Wei et al. (1998)
PR-17	Tobacco PRp27	Unknown	Okushima et al. (2000)

low levels of hydrolytic enzymes in disease resistant varieties, as discussed below, may indicate a major role in defense against specific pathogens at least for this group of PR proteins (Lawrence et al., 2000).

Elucidation of the biochemical properties of the major pathogen-inducible PR proteins of tobacco and subsequent cloning of their genes revealed that many proteins with similarities to the classical PR proteins are present even in healthy plants (van Loon and van Strien, 1999). The term "PR-like proteins" was proposed to accommodate such protein homologues of PR proteins induced principally in a developmentally controlled, tissue specific manner. Sequence analyses and development of easily accessible database search tools in recent years have resulted in the identification of several proteins with sequence homology to established PR proteins and PR-like proteins (van Loon and van Strien, 1999). Though their inducibility and stress responses have not yet been established, they have been classified as PR and PR-like proteins based on their similarity. In contrast to the classical PR proteins, which are both intracellular and extracellular proteins, the PR-like proteins are mostly intracellular and localized to the vacuole (Linthorst et al., 1991), possessing enzymatic activities similar to the homologous PR proteins, but with different substrate specificities. The similarities in the activities of PR and PR-like proteins discovered in recent years makes it difficult to maintain their distinction. Some of these proteins have been shown to respond differently to different stimuli, and other proteins have shown organ specific regulation (Lotan et al., 1989; Memelink et al., 1990). The varied locations of the PR and PR-like proteins, and their differential induction by endogenous and exogenous signaling compounds (Memelink et al., 1990) suggests that these proteins may have important functions extending beyond their apparently limited role in plant defense.

6.3 PR Proteins in Multigenic and Induced Systemic Resistance

Multigenic resistance, also known as "horizontal", "quantitative", or "polygenic" resistance, refers to plant disease resistance generated via interactions between the products of multiple plant genes, not a single R gene (Nelson, 1978; Simmonds, 1991). Multigenic resistance is considered to be nonspecific in that the plant and pathogen do not require matching R and *avr* genes for a timely plant defense response to occur. Multigenic-resistant plants which have been bred to resist a specific pathogen tend to resist a greater variety of pathogens and pathogen races than those bred or engineered to express particular R genes (Simmonds, 1991) and physical interaction of molecules present in the pathogens and their host receptors (Tang et al., 1996) lacks in this particular type of broad resistance.

Another category of disease resistance depends upon the induction of defenses following exposure to organisms or compounds. A variety of organisms, including virulent and avirulent pathogens (Tuzun et al., 1986, 1992, Tuzun and Kuć, 1991), mycorrhizal fungi (Borowicz, 1997) and nonpathogenic rhizobacteria (Tuzun and Kloepper, 1995; Benhamou et al., 1998) have all been observed to activate plant defense responses. Abiotic inducing agents include compounds isolated from plant pathogens (Wei and Beer, 1996; Norman et al., 1999) and a variety of chemicals (Fought and Kuć, 1996, Benhamou and Belanger, 1998). This general phenomenon is known as induced systemic resistance (ISR) (a term originally synonymous with systemic acquired resistance, or SAR) and generally results in a nonspecific resistance against a broad spectrum of pathogens and pests (Karban and Kuć, 1999). The extent of protection has sometimes been observed to vary (e.g., Manhandhar et al., 1999; Ton et al., 1999), and may depend upon the genotype and physiological condition of the plant, as well as the nature of the inducing agent used.

PR proteins may be induced in various tissues in response to a variety of stresses or stress-related plant hormones, including ethylene, osmotic stress, wounding, drought, high salt, and absicic acid (Horvath et al., 1998; Ponstein et al., 1994; Xu et al., 1994). The various conditions under which PR proteins occur are reminiscent of the conditions under which general stress response factors such as heat shock proteins are induced. However, PR proteins are not expressed or induced to any detectable levels in response to heat shock, suggesting that these proteins do not act as generic stress response factors. This suggests that the PR proteins may play roles that are more specific than those of general stress response factors.

In the remainder of this article, we intend to review evidence concerning the nature of multigenic and induced plant defense responses in terms of PR protein induction. The induction patterns and possible functions of specific genes, those encoding hydrolase isozymes in particular, related to these forms of resistance in several plant–pathogen systems will be discussed. A wide variety of enzymes have been associated with disease resistance, only a few of which will be discussed in this article. For a more comprehensive review, see Van Loon and Van Strien (1999). For a review of PR proteins identified in cereals, see Muthukrishnan et al. (2001).

6.3.1 Hydrolytic Enzymes: Chitinases and β-1,3-glucanases

Families of PR proteins including hydrolytic enzymes include PR-2 (β-1,3-endoglucanases, Kauffman et al., 1987), PR-7 (endoproteinase, Vera and Conejoero, 1988) and the PR-3,-4,-8 and -11 families (chitinases, Legrand et al., 1987; Ponstein et al., 1994; Metraux et al., 1989; Melchers et al., 1994). The production of hydrolytic enzymes alone may not be sufficient for the protection of all plants from disease (e.g., Dalisay and Kuć, 1995 a,b). However, this does not mean that hydrolase isozymes are not involved in disease resistance, or that they do not play an important role in resistance to some pathogens. Hydrolytic enzymes may have a dual function in disease resistance: some isozymes will have direct antimicrobial effects against an invading pathogen. These isozymes, and/or others, may also accelerate and amplify the disease resistance process by generating hypersensitive response elicitors upon encountering a pathogen. Unfortunately, a great deal of the work regarding the role of specific enzymes in disease resistance fails to distinguish between the different isozymes that are present. Significant changes in the expression of a particular isozyme may go undetected.

Chitinases catalyze the hydrolysis of chitin, a linear polymer of β-1,4-linked N-acetylglucosamine residues that is the predominant constituent of fungal cell walls, nematode eggs, and mid gut layers of insects. Some plant chitinases also exhibit lysozymal activity (Boller, 1985; Dodson et al., 1993). Three classes of plant chitinases have been proposed based upon protein primary structure (Shinshi et al., 1990). The highly variable nature of chitinases, and the multiplicity of chitinase isozymes in plants, suggest that plant chitinase isozymes may carry out specific and differing roles. For example, chitinases, glucanases, and other PR proteins have been found to be induced as a consequence of cold stress and might be involved in resistance of winter wheat to snow mould infections (Gaudet et al., 2000). A *Lubinus albus* chitinase accumulates in response to salicylic acid, wounding, infection, and UV-C light (Regalado et al., 2000); and tobacco chitinases, glucanases and thaumatin-like proteins increased in response to UV-B light (Fujibe et al., 2004). Pre-treatment of tomato fruit with methyl jasmonate or methyl salicylate induces the synthesis of a variety of stress proteins, including chitinase and β-1,3-glucanase PR proteins, and subsequently increases the resistance of the fruit to chilling injury and infection by pathogens (Ding et al., 2002). Expression studies of various *Pinus* chitinase homologues showed the induction of multiple chitinase homologues after challenge by a necrotrophic pathogen (Davis et al., 2002), suggesting that different homologues may serve different functions in the plant. Some chitinase isozymes have antifungal activity while others do not, and the activity of antifungal chitinase isozymes isolated from tobacco (Sela-Buurlage et al., 1993) and tomato (Lawrence et al., 1996) has been found to be specific for certain pathogens.

Many plant pathogenic fungi contain β-1,3-glucans in their cell walls in addition to chitin. Chitinases and β-1,3-glucanases purified from tomato (Lawrence et al., 1996), tobacco (Sela-Buurlage et al., 1993), pea (Mauch et al., 1988) and the tropical forage plant *Stylosanthes guianensis* (Brown and Davis, 1992) have been found to have synergistic antifungal effects *in vitro*. The *in planta* antifungal effects of tomato and tobacco chitinases and β-1,3-glucanases have also been recorded (Benhamou et al., 1990; Benhamou, 1992), and chitinases and glucanases coexpressed in transgenic wheat were found to protect plants from infection by *Fusarium graminearum* under greenhouse conditions, although this resistance did not hold under field conditions (Anand et al., 2003). It has been suggested that the synergistic effects of these enzymes, and the specificity of their effects, may be attributed to the structure of a particular fungal cell wall. For example, the chitin layers of some fungal cell walls appear to be buried in β-glucans, rendering the chitin inaccessible to chitinases unless there is prior hydrolysis with β-1,3-glucanases (Benhamou et al., 1990).

Oligosaccharide elicitors of plant defense responses can be generated by chitinases and β-1,3-glucanases. Soybean β-1,3-glucanases (Keen and Yoshikawa, 1983; Ham et al., 1991) and specific isozymes of tomato chitinase and β-1,3-glucanase (Lawrence et al., 1996, 2000) have been demonstrated to generate elicitors from fungal pathogens. Tomato chitinases have also been shown to generate elicitors from germinating spores *of Alternaria solani*, but not the mature cell walls of this pathogen (Lawrence et al., 2000).

The tobacco PR-2 glucanase isozymes vary up to 250-fold in specific activity on various substrates, suggesting that their normal functions *in planta* may be quite diverse (Cote et al., 1991; Hennig et al., 1993). Interestingly, a β-1,3-glucanase found to accumulate in cultivars of resistant wheat could be involved in resistance to the Russian wheat aphid (Lintle et al., 2002). Glucanase and chitinase isozymes may also govern plant developmental processes not directly related to pathogenesis or stress resistance. For example, the expression studies of PR-2d in transgenic tobacco suggest that this protein functions developmentally in seed germination by weakening the endosperm (Vogeli-Lange et al., 1994). In yeast, a specific chitinase is secreted into the growth medium that is required for cell separation after division has taken place (Kuranda and Robbins, 1991), and has homology to a cucumber PR8 type III chitinase, suggesting that the yeast chitinase has functions in cell separation as well as defense. Specific chitinase homologues of PR3 and PR4 were found to be necessary for somatic embryogenesis to proceed beyond the globular stage (De Jong et al., 1992).

6.3.2 Antioxidant Enzymes

Plant cells are protected against damage from active oxygen species generated during the hypersensitive response by a complex antioxidant system, including enzymatic antioxidants such as superoxide dismutase (SOD), peroxidase, and catalase (Zhang and Kirkham, 1994). Several species of active oxygen (O_2^-, H_2O_2, and OH^-) result from the reduction of molecular oxygen, and there are numerous possible reactions which allow these species to interconvert (Elstner, 1987; Mader et al., 1980). Hydrogen peroxide, which has the longest half-life, provides a good estimate of the relative active oxygen level in the system. There is an opinion that elicitor- or pathogen-stimulated accumulation of H_2O_2 comes only from SOD-catalysed dismutation of superoxide radicals (Auh and Murphy, 1995). SOD and catalase are critical to the immediate level of H_2O_2 since they are involved in production and utilization of the molecule. The existence of multiple molecular forms of SOD, peroxidase, catalase, and other related enzymes and the variation in the activity of these during plant development suggests that each isozyme may have a separate role (Scandalios, 1993).

Specific peroxidase isoenzymes recognized as PR proteins in tobacco (Stintzi et al., 1993), that are identical to or homologous with a lignin-forming peroxidase, have been classified as PR-9. Peroxidases represent another component of an early response system in plants to pathogen attack (Mader and Fussi, 1982; Mader et al., 1980). The products of these enzymes, in the presence of a suitable hydrogen donor and hydrogen peroxide, can have direct antimicrobial and antiviral effects (Van Loon and Callow, 1983). The extracellular location of peroxidase isozymes stimulated during pathogen attack (Birecka et al., 1975), and their affinity for substrates involved in lignification, as well as the capacity of peroxidases to form hydrogen peroxide (Ride, 1975), suggest that peroxidase isozymes may also be involved in the formation of barrier substances which limit the extent of pathogen spread. Elicitation of peroxidase activity and lignin biosynthesis was observed in resistant pepper cell suspension cultures treated with the pathogen *Phytophthora capsici*, but not in susceptible cells (Egea et al., 2001). The release of superoxide and free radical intermediates during lignin polymerization (Grisebach, 1981) may be involved in restricting the growth of both fungal and bacterial pathogens (Klement, 1982; Ride, 1975; Tiburzy and Reisner, 1990). For example, antibacterial components active against *Xanthomonas oryzae* pv. *oryzae* were isolated from rice leaves and found to be lignin precursors (Reimers and Leach, 1991).

6.3.3 Thaumatins and Thaumatin-like Proteins (Osmotins)

The PR-5 family of proteins are referred to as thaumatins or thaumatin-like proteins due to their close or more varying sequence similarity to the intensely sweet protein thaumatin from *Thaumatococcus danielli* (Musthurkrishnan et al., 2001). PR-5 proteins have been shown to inhibit the growth of fungi *in vitro*, causing leakage of cytoplasmic material from ruptured hyphae (Vigers et al., 1991; Niderman et al., 1995). Two proteins highly induced by *Ascochyta rabiei* during infection of chickpea (*Cicer arietinum* L.) were identified as PR-5a and PR-5b proteins (Hanselle et al., 2001). An osmotin-like protein (OLP), which is a member of the thaumatin-like proteins, was purified from the seeds of *Benincasa hispida* (Shih et al., 2001). The homology of thaumatin-like PR-5 proteins with a bifunctional α-amylase/trypsin inhibitor from maize seeds (Richardson et al., 1987) suggests that these proteins could also play a role in protection against phytophagous insects.

Further characterization efforts to cloning and studying the expression of gene encoding osmotins resulted in demonstration that it is highly regulated by ABA and involved in adaptation to osmotic stress (Singh et al., 1987; 1989). Antifungal activity of osmotins appeared to be nonspecific against the cell wall from many fungi, although it is involved in permeabilization of plasma membrane to kill the cells (Abad et al., 1996). Although osmotins were antifungal to many strains of fungi, studies conducted using yeast strains with various resistance to this protein indicated that fungal cell wall proteins, encoded by *PIR* genes, are determinants of resistance to antifungal PR-5 proteins (Yun et al., 1997). Resistance to osmotin in yeast model system appeared to be strongly dependent on the natural polymorphism of the *SSD1* gene where it functions as post-transcriptional regulator of gene expression, cell wall biogenesis, and composition and deposition of PIR proteins (Ibeas et al., 2001). Deposition of such proteins as the fungal cell wall constituents that block the action of osmotin against *Aspergillus nidulans* requires the activity of G-protein mediated signaling pathway, and *A. nidulans* strains mutated to interfere this pathway also demonstrated increased tolerance to SDS, reduced cell wall porosity and increased chitin content in the cell wall (Coca et al., 2000). Further studies using yeast indicated that osmotins indeed have certain target molecules in the cell wall and several cell wall mannoproteins can bind to immobilized osmotin, suggesting that their polysaccharide constituent determines osmotin binding, demonstrating a causal relationship between cell surface phosphomannan and susceptibility of a yeast strain to osmotin (Ibeas et al., 2000). Overexpression of yeast glycoprotein in a plant pathogenic fungus *Fusarium oxysporum* f. sp. *nicotianae*, which is susceptible to osmotin, increased resistance to this antifungal protein and virulence in the fungal pathogen (Narasimhan et al., 2003), further indicating that osmotin plays a role in overall plant defenses against fungal pathogens.

6.3.4 Proteinase Inhibitors

The PR-6 proteins have been shown to be protease inhibitors (reviewed by Green and Ryan, 1972), and include wound-inducible proteinase inhibitors implicated in resistance to insect attack (Lawton et al., 1993). Proteinase inhibitor genes in *Nicotania glutinosa* that are induced in response to wounding as well as infection by TMV have been identified (Choi et al., 2000). Proteinase inhibitors have been shown to confer protection against a variety of insect and nematode pests when expressed in transgenic plants. For example, resistance to the cyst nematode *Globodera tabacum* (Urwin et al., 2002) and tobacco budworm (*Helothis virescens*, Pulliam et al., 2001) was conferred by proteinase inhibitors expressed in transgenic tobacco, and resistance to the potato cyst nematode *Globodera pallida* was conferred by a proteinase inhibitor expressed in transgenic potato (Urwin et al., 2001). Proteinase inhibitor proteins in plants may play roles other than protection against phytophagous insects and nematodes. Phloem-localized proteinase inhibitor proteins have been identified in *Solanum americanum*, which may be involved in regulating proteolysis in the phloem sieve elements (Xu et al., 2001).

Proteins induced under conditions of heat and drought stress have been found to have a putative proteinase inhibitor activity (Satoh et al., 2001), and it may be possible that PR-6 proteins have protective activity against abiotic stresses as well.

6.3.5 Ribonucleases

A recombinant white lupin *PR-10a* gene expressed in *Escherichia coli* exhibited a ribonucleolytic activity against several RNA preparations, including lupin root total RNA providing the first direct evidence of this enzymatic activity in a PR protein (Bantignies et al., 2000). Salicylate-inducible PR-10 genes from apple (*Apa*) were found to be also induced by wounding, ethephon, and exposure to virulent and avirulent fungi (Poupard et al., 2003). A PR-10 protein in western white pine that was associated with acclimation to cold was present in higher amounts in healthy pine needles than in infected ones, suggesting this protein may be involved in protecting frost-damaged plant tissues from pathogen attack (Yu et al., 2000). A similar protein in Douglas fir was found to increase during overwintering of plants but was not associated with acclimation to cold, and may accumulate in response to pathogen infection (Ekramoddoulah et al., 2000). A PR-10 protein was found to be induced in response to ozone and drought stress in birch, and its induction coincided with the formation of visible necrotic lesions and yellowing of leaves (Paakkonen et al., 1998). Ocatin, a member of the PR-10 family that is found in oca roots (*Oxalis tuberosa* Mol.) inhibits the growth of bacteria and fungi *in vitro*, and is expressed only in the pith and outer peel of the tuber, indicating a role in protecting tubers from pathogen attack (Flores et al., 2002). PR-10 genes that accumulate after pathogen attack have also been found in rice (McGee et al., 2001), sorghum (Lo et al., 1999), and alfalfa (Borsics and Lados, 2002). High sequence similarity between ribonuclease from ginseng callus cultures and fungus-elicited PR-proteins in parsley further indicates that at least some of the intracellular PR-proteins are ribonucleases (Moiseyev et al., 1994).

6.3.6 Thionins

The PR-13 family consists of thionins, small (5000 Da) sulfur-rich plant proteins that exert toxicity in various biological systems by destroying membranes (Bohlmann, 1994). They are synthesized as preproteins and secreted into vacuoles, protein bodies, and the plant cell wall, and may be subsequently released upon pathogen infection, and display antifungal and antibacterial activity *in vitro* (Bohlmann, 1994; Terras et al., 1995) The expression of thionins in transgenic plants has been found to protect against pathogenic bacteria in rice (Iwai et al., 2002) and *A. thaliana* (Epple et al., 1997), and thionin concentrations in cell walls have been found to be higher in disease-resistant cultivars of barley and wheat (Ebrahim-Nesbat et al., 1994). *Arabidopsis* mutants constitutively expressing the thionin (*cet*) gene *Thi2.1* showed spontaneous formation of necrotic lesions and an upregulation in the PR-1 gene, reactions that are associated with a salicylate-dependent induced systemic resistance response (Nibbe et al., 2002). Nonspecific resistance to snow moulds and other fungi has been likened to the expression of γ-thionin in winter wheat (Gaudet et al., 2003).

6.4 Patterns of Expression of Chitinases, Glucanases and Peroxidases in Multigenic Resistant and Induced Resistant Plants

In this section, the manner in which hydrolytic and antioxidant enzymes are expressed in plants, which express multigenic resistance and plants in which systemic resistance has been induced, will be compared. Three plant systems (tobacco, tomato, and cabbage) will be discussed in some detail, while work in other plant systems will be mentioned briefly at the end.

6.4.1 Tobacco

Resistance to *Peronospora tabacina* (blue mold) in tobacco is considered to be due to a few genes acting in an additive fashion (Rufty, 1989). Several breeding lines, which have been developed by the use of intraspecific hybridization of wild *Nicotania* species *to N. tabacum* by Rufty (1989) were used for studying the role of preformed hydrolytic enzymes in tobacco. Results from SDS-PAGE and Western blot analyses consistently revealed the presence of chitinase and β-1,3-glucanase isozymes prior to pathogen attack, as well as an earlier induction of isozyme accumulation following attack, in the resistant lines (Tuzun et al., 1997). Enzyme activity assays closely correlated with the Western blot analysis (Robertson, 1995).

Induced systemic resistance to *Peronospora tabacina* (blue mold) occurs naturally under field conditions (i.e., in plants not inoculated by human beings) (Tuzun et al., 1992). Inoculation of tobacco with *Peronospora tabacina* spores or tobacco mosaic virus (TMV) resulted in the induction of systemic resistance against a variety of pathogens (McIntyre et al., 1981) and the accumulation of β-1,3-glucanase and chitinase isozymes prior to foliar inoculation (Tuzun et al., 1989; Ye et al., 1990; Pan et al., 1991,1992). Similar results were observed for tobacco inoculated with viruses, PGPR, or various chemical inducers (Maurhofer et al., 1994, Schneider and Ullrich, 1994; Lusso and Kuć, 1995). Increases in lysozyme, peroxidase, polyphenol oxidase, and phenylalanine ammonium lyase activity, correlated with the induction of ISR, have also been reported (Ye et al., 1990; Schneider and Ullrich, 1994). Inhibition of fungal pathogen growth was found to precede host cell necrosis in induced tobacco, and it is thought that this might be due to the production of defense response elicitors by hydrolytic enzymes (Ye et al., 1992).

Elevated constitutive expression of an endochitinase gene from *Trichoderma viride* in tobacco and potato resulted in significant protection against multiple fungal pathogens (Lorito et al., 1998). Reduced levels of anionic peroxidase, however, did not result in reduced lignification in transgenic tobacco (Lagrimini et al., 1997).

6.4.2 Tomato

Tomato breeding lines and several plant introductions of *Lycopersicon* spp. have already been identified in several studies with heritable foliar resistance to the early blight pathogen *Alternaria solani*, conferred by the presence of multiple genes (Barksdale and Stoner, 1973; Gardner, 1988; Maiero et al., 1990; Maiero and Ng, 1989; Nash and Gardner, 1988). These studies also suggest that expression of a resistant phenotype in a given individual relies on various genetic interactions of an additive and/or epistatic nature. All tomato breeding lines resistant to *A. solani* were found to express significantly higher consitutive levels of chitinase and β-1,3-glucanase isozymes than susceptible plants (Lawrence et al., 1996, 2000). The same 30 kDa chitinase isozyme expressed to a high level in resistant lines was also found to accumulate more rapidly, and to significantly higher levels, in the resistant lines than in the susceptible ones during pathogenesis (Lawrence et al., 1996). The resistant tomato lines expressing elevated levels of chitinase and β-1,3-glucanase isozymes are also able to produce a greater number of, or more effective, elicitors of the hypersensitive response from *A. solani* cell walls than susceptible tomato lines (Lawrence et al., 2000). It is thought that the higher constitutive expression of hydrolytic enzymes might therefore contribute to disease resistance to *A. solani* via the more rapid and greater production of oligosaccharide elicitors upon contact with the pathogen, that in turn activate other defense mechanisms. More rapid accumulation of chitinases in resistant plants during incompatible tomato–pathogen interactions have also been observed *in planta* by other researchers (Benhamou et al., 1990). Two genes encoding basic chitinases, which accumulate during pathogenesis in tomato have been sequenced, and the promoter region of one of these genes cloned (Baykal and Tuzun, unpublished data). The manner in which the gene is regulated is currently being determined.

Tomato plants immunized with β-amino butyric acid (BABA) accumulated β-1,3-glucanase and chitinase (Cohen et al., 1994), while tomato plants immunized with 4-hydroxybenzoic hydrazide, saliclylic hydrazide, or 2-furoic acid accumulated an acidic peroxidase (Miyazawa et al., 1998). Interestingly, this peroxidase was not produced as a result of pathogenesis or wounding, suggesting that different kinds of inducing agents may have different effects on plant physiology. Enkerli et al. (1993) reported correlations between increased tomato chitinase activity, but not β-1,3-glucanase activity, with induction of resistance. Similarly, correlations between induced resistance in tomato and increased production of various antifungal proteins or activity of peroxidases, but not β-1,3-glucanases, have been reported (Anfoka and Buchenauer, 1997). Treatment of tomato roots with the mycoparasite *Pythium oligandrum* (Benhamou et al., 1997), chitosan and *Bacillus pumilis* (Benhamou et al., 1998) or with benzothiadiazole (Behnamou and Belanger, 1998) was able to trigger and amplify plant defense responses to infection with a fungal pathogen, including the deposition of newly formed barriers containing callose and phenolic compounds.

6.4.3 Cabbage

A high level of resistance to the black rot pathogen, *Xanthomonas campestris* pv. *campestris* (XCC), was observed decades ago in the cabbage cultivars Early Fuji and Hugenot (Bain, 1952), and the heritable nature of this resistance was found to involve one major and several modifying genes (Bain, 1955). Cabbage varieties demonstrated to be resistant to a virulent strain of XCC have been found to constitutively express higher levels of the chitinase-lysozyme isozyme CH2 than susceptible cabbage varieties (Dodson et al., 1993). The level of CH2 expression was correlated with the extent of black rot disease resistance. Acidic protein extraction and denaturing electrophoresis identified at least 12 acid-extractable proteins which accumulated in both black-rot resistant and susceptible varieties following XCC infection; however, accumulation was early and more pronounced in the resistant varieties (Tuzun et al., 1997). The chitinase-lysozyme CH2, as well as peroxidase and superoxide dismutase isozymes, accumulate more rapidly and to a greater extent following inoculation with XCC than susceptible varieties (Dodson et al., 1993; Gay and Tuzun, 2000b). Increases in chitinase, lysozyme, peroxidase, and superoxide dismutase activities have also been correlated with increased expression of these isozymes. Higher peroxidase activity in the hydathodal fluids of black rot-resistant cabbage varieties than in susceptible ones was related to increased suppression of XCC growth in the hydathodal fluids (Gay and Tuzun, 2000a). Localized accumulations of peroxidase may function to protect plants against XCC infection, since this pathogen initially invades cabbage via the hydathodes (Staub and Williams, 1972).

Incompatible interactions with *X. campestris* pv. *vesicatoria* and a less pathogenic strain of XCC were sufficient to induce systemic resistance in cabbage against pathogenic isolates of XCC under both greenhouse and field conditions (Jetiyanon, 1994). Immunized plants produced chitinase/lysozyme, β-1,3-glucanase, osmotin, and other pathogenesis-related proteins earlier and in greater quantities than did nonimmunized plants (Tuzun et al., 1997).

6.4.4 Other Systems

Higher constitutive expression of chitinases and/or glucanases in disease-resistant plants relative to susceptible ones has also been noted in barley (Ignatius et al., 1994), grape (Busam et al., 1997), and potato (Wegener et al., 1996). Increases in the expression or activity of chitinase and/or β-1,3-glucanase isozymes in disease resistant plants after pathogen challenge have been reported in barley (Ignatius et al., 1994), pea (Vad et al., 1991), and wheat (Liao et al., 1994; Siefert et al., 1996, Kemp et al., 1999). Chitinase expression increased in wilt-resistant cotton plants following infection by *Verticillium dahliae*, but β-1,3-glucanase expression did not (Cui et al., 2000). Two wheat genes that encode proteins PR-1.1 and PR-1.2, expression of which was induced upon infection with either compatible or incompatible isolates of the fungal pathogen *Erysiphe graminis*, were identified (Molina et al., 1999). Two new PR-4 family proteins (named wheat win3 and wheat win4) showing distinct antifungal activity were identified from wheat (Caruso et al., 2001). A similar protein has been identified from bean leaves, with similarity to PR1 like protein and glucanase, and a thaumatin like protein (Del Campillo and Lewis, 1992). Thionins, defensins, PR-like chitinases, thaumatin like proteins isolated from wheat, barley, sorghum, oats, and maize have antifungal activity (Hejgaard et al., 1991; Vigers et al., 1991).

Increases in the expression and activity of chitinases, β-1,3-glucanases and/or peroxidases after the induction of ISR has also been reported in cotton (Dubery and Slater, 1996), wheat (Liao et al., 1994; Siefert et al., 1996), rice (Manandhar et al., 1999), coffee (Guzzo and Martins, 1996), grape (Busam et al., 1997), cucumber (Schneider and Ullrich, 1994; Ju and Kuć, 1995; Dalisay and Kuć, 1995a,b), bean (Dann et al., 1996; Xue et al., 1998), pepper (Hwang et al., 1997), chestnut (Schafleitner and Wilhelm, 1997), *Cotoneaster watereri* (Mosch and Zeller, 1996), and *Stylosanthes guianensis* (Brown and Davis, 1992). Kogel et al. (1994) reported that ISR in barley is associated with increases in PR-1, peroxidase and chitinase proteins, but not β-1,3-glucanase. Although it appears to be a rather specific case, ISR induced in radish by *Pseudomonas fluorescens* has yet to be explained since no pathogenesis-related proteins accumulate and no changes in cell wall composition occur (Steijl et al., 1999).

Chitosanases, chitinases and β-1,3-glucanases were observed to accumulate in infected spruce seedlings (Sharma et al., 1993), and in the vicinity of the pathogenic fungus in infected spruce and pine (Asiegbu et al., 1999). These observations indicate that the defense responses of gymnosperms are similar to those of angiosperms. Induced resistance to pathogenic fungi in mature Norway spruce trees was found to be localized to the immunized bough rather than being systemic throughout the plant, but this was attributed to the size of the plant rather than a fundamental difference in induced resistance mechanisms (Krokene et al., 1999).

6.5 Regulation in PR Gene Expression

How pathogen infection leads to PR gene expression is as yet not well understood. Some of this is due to the fact that PR gene expression appears to be induced by environmental stimuli (e.g., cold stress, ultraviolet light) as well as developmental cues. Interestingly, the systemic induction of BiP, a lumenal binding protein in tobacco that is required for the normal induction of PR gene expression, occurs prior to the induction of PR genes (Jelitto-van Dooren et al., 1999)

It is possible that some PR proteins, specifically hydrolytic enzymes, act to stimulate an appropriately rapid or intense defense response by amplifying the concentration of nonspecific elicitors that go on to stimulate defense responses, including the production of more hydrolytic PR proteins. A role for hydrolytic

PR proteins as defense response signal amplifiers would be consistent with their generally rapid induction in response to stressful pathogenic infections.

6.5.1 Elicitors of PR Gene Expression

A variety of microbially-produced surface or secreted molecules have been identified as elicitors of PR gene induction, including oligosaccharides, oligochitin and oligoglucan fragments, extracellular glycoproteins and peptides, lipopolysaccharide from *Burkholderia cepacia*, and Avr proteins derived from bacterial and fungal pathogens (Boller and Felix 1996; Coventry and Dubery, 2001).

Avr proteins are specific elicitors, meaning that a pathogen expressing that product is recognized by a host plant expressing the corresponding resistance gene (R), which then activates the disease resistance mechanisms of the host (Staskawicz et al., 1995; Bent, 1996). Activation of tomato PR genes by the *Avr9* gene product of *Cladosporium fulvum* (Wubben et al., 1994), and activation of barley PR gene expression by *Rhychosporium* NIP1 Avr protein (Rohe et al., 1995) provide examples of defense responses induced by specific elicitors. Nonspecific elicitors (i.e., elicitors other than Avr proteins) also seem to be detected by receptors, which then stimulate a defense response. The receptor of a soybean β-glucan elicitor (GE) that induced phytoalexin biosynthesis was identified (Umemoto et al., 1997), and a parsley glycoprotein secreted by *Phytophthora sojae* has been shown to elicit ion channel openings and expression of defense genes including PR genes (Nurnberger et al., 1994; Ligterink et al., 1997).

6.5.2 Activation of PR Gene Expression

Secondary signal molecules such as reactive oxygen species, salicylic acid, ethylene, and jasmonates have been shown to induce PR gene expression (Delledone et al., 1998; Durner et al., 1998; Ecker, 1995). However, it is uncertain whether this induction requires secondary messengers. Proteinase inhibitor (Pin) of tomato is inhibited by ethylene and jasmonates, whereas these secondary signal molecules enhance tobacco osmotin, PR1, and tomato Pin2 (O'Donnell et al., 1996; Farmer et al., 1994). Using an inducible gene expression system, McNellis et al. (1998) directly expressed the AvrRpt2 protein in *Arabidopsis* and stimulated a hypersensitive response as well as induction of PR-1 gene expression.

A DNA microarray analysis of gene-expression changes in *Arabidopsis thaliana*, under 14 different ISR-inducing or ISR-repressing conditions, was used to derive groups of genes with common regulation patterns (regulons). A common promoter element in genes of the PR-1 regulon that binds members of a plant-specific transcription factor family was identified (Maleck et al., 2000). The promoter regions of two peach β-1,3-glucanase genes, designated *PpGns1* and *PpGns2*, identified to be highly expressed upon exposure to *Xanthomonas campestris* pv. pruni, contain elements similar to the *cis*-regulatory elements present in different stress-induced plant genes (Thimmapuram et al., 2001). Receptor-mediated recognition of *Phytophthora sojae* may be achieved through a 13 amino acid peptide sequence (Pep-13) present within an abundant cell wall transglutaminase, which initiates a defense response that includes the transcriptional activation of genes encoding pathogenesis-related (PR) proteins (Kroj et al., 2003). Identification of *cis* regulatory elements mediating pathogen-induced PR gene expression suggested that the regulation primarily occurs at the level of transcription. The *cis* regulatory elements include GCC box (AGCCGCC), W box (TTGACC or TGAC-[N]x-GTCA), MRE-like sequence (A[A/C]C[A/T]A[A/C]C), G box (CACGTG), SA responsive element (SARE, TTCGACCTCC), and a parsley 11bp element mediating PR2 gene expression. Of these only the GCC box and the W box are extensively studied.

GCC box initially denoted as an ethylene responsive element has been identified in the promoter of a number of basic PR genes (Hart et al., 1993). The absence of GCC box in other ethylene responsive genes suggests that GCC box may be associated with the defense response mediated by ethylene. The GCC box confers ethylene-induced transcription of tobacco *gln2* and *PRB-1b* genes. Also, a 140 bp fragment that contains the GCC box from osmotin promoter is necessary to confer responsiveness of osmotin to various stimuli (Ragothama et al., 1997). Thus, GCC box might be a point of cross talk between various signal transduction pathways.

The promoter of an *Arabidopsis* basic PR1-like gene, AtPRB1, establishes organ-specific expression pattern and responsiveness to ethylene and methyl jasmonate (Santamaria et al., 2001). Identification of GCC box binding proteins (EREBP1-4) containing a conserved domain responsible for binding to the GCC box suggests that ethylene further induces the expression of the EREBP genes. Homologues of the EREBP genes have been identified from several species (Kitajama et al., 2000). An *Arabidopsis* EREBP homolog (AtERF1), which acts downstream of EIN3 (a component of the ethylene signaling pathway), has been identified to activate PR gene expression (Chao et al., 1997; Solano et al., 1998). EIN3 has subsequently been shown to be a transcriptional regulator of AtERF1.

A *Glycine max* gene encoding the ethylene-responsive element-binding protein 1 (GmEREBP1) has been shown to have differential expression during soybean cyst nematode infection (Mazarei et al., 2002). Three tomato Pto interacting proteins (Pti), with homology to the tobacco EREBPs have been identified, and shown to bind the GCC box of the tobacco *gln2* gene (Zhou et al., 1997). This suggests that Pti4/5/6 and EREBPs act in the R gene pathway. Ptis have also been shown to be highly regulated proteins. Pto kinase interacts directly with Pti4/5/6, and phosphorylates Pti4 protein specifically, to enhance the ability of Pti4 to activate expression of GCC-box PR genes in tomato (Gu et al., 2000). While *Pti4* is constitutively expressed and shows increased accumulation on infection, *Pti5* transcript is induced only upon infection. Pti4 also responds to mechanical and osmotic stress. A protein kinase that regulates the expression of PR genes has also been identified from rice. The rice mitogen-activated protein kinase (OsMAPK5) has been shown to negatively modulate PR gene expression (PR1 and PR10) and broad-spectrum disease resistance (Xiong and Yang, 2003). Analysis of the *Arabidopsis* PDF1.2 promoter shows a GCC box, and that the promoter confers pathogen and jasmonate responsiveness. This demonstrates that AtPDF1.2 gene is a target for EREBP/Pti class of transcription factors (Wu et al., 2002). Expression of *Pti4*, *Pti5*, or *Pti6* in *Arabidopsis* activated the expression of the salicylic acid-regulated genes *PR1* and *PR2* (Gu et al., 2002).

The W box has been identified in parsley *PR1* and *PR2*, tobacco chitinase, asparagus *PR1*, potato *PR10a*, and maize *Prms*. Promoter deletion analysis showed that the W box is required for the elicitor-induced response of these PR genes. The W box has also been identified in other pathogen responsive genes suggesting a wider role for this *cis* element (Rushton and Somssich, 1998; Somssich, 2003). Rushton et al. (1996) identified a family of parsley proteins, which bind the W box, activating the expression of genes containing the W box. The *Arabidopsis* WRKY protein ZAP1 can activate a W box indicating that ZAP1 is capable of *trans*-activating W box containing genes (De Pater et al., 1996). The parsley *WRKY1* gene has been shown to bind the W box elements and act as a transcriptional activator (Eulgem et al., 1999).

Efforts are on to analyze the regulation of gene expression mediated by other regulatory elements identified to modulate PR gene expression. In *Arabidopsis*, NPR1 was originally discovered as a key regulatory protein that functions downstream of SA in the ISR. Upon induction of ISR, NPR1 activates PR-1 gene expression by physically interacting with a subclass of basic leucine zipper protein transcription factors (TGA/OBF family of transcription factors) that bind to promoter sequences required for SA-inducible PR gene expression (Chao et al., 1997; Zhou et al., 2000; Van Wees et al., 2000). In addition, analysis of the *Arabidopsis* mutant *npr1*, which is impaired in SA signal transduction, revealed that the antagonistic effect of SA on JA signaling requires the NPR1. Nuclear localization of NPR1 indicating that cross-talk between SA and JA is modulated through a novel function of NPR1 in cytosol (Spoel et al., 2003). A negative regulator of ISR, sni1 (suppressor of *npr1*) has been identified in a suppressor screen of *npr1* mutant (Li et al., 1999). Epistatic analysis has identified *cpr5* and *cpr1* as genes acting upstream of SA production and the *npr1* and *cpr6* downstream of SA production (Clarke et al., 1998; Dong, 1998). It also shows that *cpr5* is a negative regulator of the hypersensitive response, and *cpr1* is a negative regulator of SA biosynthesis. In addition the classical PR genes, defensin (PDF1.2) gene and thionin (Thi2.1) are constitutively expressed in *cpr5* and *cpr6* mutants. In contrast, the *cpr1* mutant accumulates only the classical PR genes and not the PDF1.2. A double mutant (npr1: cpr5) accumulated the PDF and not the PR genes, suggesting that expression of PDF is independent of NPR1. Thus, the activation of *Arabidopsis* defense genes appears to follow two separate pathways: an NPR1 dependent pathway for PR1, PR2, and PR5, and an NPR1 independent pathway for PDF and Thi2. Genetic analysis also suggests that CPR6 may be responsible for the crosstalk between SA mediated signaling pathway and the jasmonates/ethylene mediated signaling pathway (Clarke et al., 1998). It appears that overexpression of regulatory genes for induced systemic resistance that results in broad spectrum of resistance (Cao and Dong, 1998) also involves accumulation of PR proteins. A protein identified as the silencing element binding factor (SEBF) that binds elements in the promoter region of potato

PR10a was shown to act as a transcriptional repressor of PR10a expression (Boyle and Brisson, 2001). Elicitor-induced activation of the potato PR-10a requires the binding of the nuclear factor PBF-2 (PR-10a binding factor 2) to an ERE (elicitor response element) in the promoter region, and thus acts as a transcriptional regulator (Desveaux et al., 2000).

6.6 Conclusion

Timely accumulation of PR proteins during pathogenesis can be suggested as a part of defense mechanisms in plants against pathogens and pests. Some of these proteins may have a different role in plant metabolism and/or may just occur there as a part of regulatory systems overall happening during the plant–pathogen interactions. Specific isozymes of the hydrolytic enzymes, on the other hand, which demonstrate differential activity toward the substrate during the release of elicitor molecules from the pathogens may have been evolved as a part of defense mechanisms in "naturally resistant plants". Such isozymes may be bred into the resistant lines of crop varieties act as recognition mechanism to initiate the whole battery of defense mechanisms. It is also clear that some of PR-proteins such as osmotins and hydrolytic enzymes have a direct involvement in reduction of pathogenesis as evidenced by genetic studies as well as microscopic observations. However, it is important to recognize that plant defense mechanisms are complex and more than one factor is involved in the successful existence of plant species over the centuries under the abundance of numerous organisms that can be potentially harmful to plants. Nevertheless pathogenesis is an exception, and is a result of failure of many pathways to be activated in a timely manner. PR proteins are certainly there for a reason, whether they are a part of a major defense mechanisms or not, according to the inducer, they are a part of induced systemic resistance and more studies will further show that they may be the reason of successful breeding efforts, which we have been doing over the centuries to breed disease resistant varieties carrying more than one gene for resistance.

References

Abad, L.R., D'urzo, M.P., Liu, D., Narasimhan, M.L., Reuveni, M., Zhu, J.K., Niu, X., Singh, N., Hasegava, P.M., and Bressan, R.A. 1996. Antifungal activity of tobacco osmotin has specificity and involves plasma membrane permeabilization. *Plant Sci.* 118:11–23.

Anand, A., Zhou, T., Trick, H.N., Gill, B.S., Bockus, W.W., and Muthukrishnan, S. 2003. Greenhouse and field testing of transgenic wheat plants stably expressing genes for thaumatin-like protein, chitinase and glucanase against *Fusarium graminearum. J. Exp. Bot.* 54(384):1101–1111.

Anfoka, G., and Buchenauer, H. 1997. Systemic acquired resistance in tomato against *Phytophthora infestans* by pre-inoculation with tobacco necrosis virus. *Physiol Mol. Plant Pathol.* 1: 351–370.

Antoniw, J.F., Ritter, C.E., Pierpoint, W.S., and van Loon L.C. 1980. Comparison of three pathogenesis-related proteins from plants of two cultivars of tobacco infected with TMV. *J. Gen. Virol.* 47:79–87.

Asiegbu, F.O., Kacprzak, M., Daniel, G., Johansson, M., Stenlid, J., and Mañka M. 1999. Biochemical interactions of conifer seedling roots with *Fusarium* spp. *Can. J. Microbiol.* 45:923–935.

Auh, C.K., and Murphy, T.M. 1995. Plasma membrane redox enzyme is involved in the synthesis of O_2^- and H_2O_2 by *Phytophthora* elicitor-stimulated rose cells. *Plant Physiol.* 107:1241–1247.

Bain, D.C. 1952. Reaction of *Brassica* seedlings to black rot. *Phytopathology* 42:497–500.

Bain, D.C. 1955. Resistance of cabbage to black rot. *Phytopathology* 45:35–37.

Bantignies, B., Seguin, J., Muzac, I., Dedaldechamp, F., Gulick, P., and Ibrahim, R. 2000. Direct evidence for ribonucleolytic activity of a PR-10-like protein from white lupin roots. *Plant Mol. Biol.* 42(6):871–881.

Barksdale, T.H., and Stoner, A.K. 1973. Segregation for horizontal resistance to tomato early blight. *Plant Dis. Rep.* 57:964–965.

Benhamou, N. 1992. Ultrastructural detetion of β-1,3-glucans in tobacco root tissues infected by *Phytophthora parasitica* var. *nicotianae* using a gold-complexed tobacco β-1,3-glucanase. *Physiol. Mol. Plant Pathol.* 92:1108–1120.

Benhamou, N., and Belanger, R. 1998. Benzothiadiazole-mediated induced resistance *to Fusarium oxysporum* f. sp. *radicis-lycopersici* in tomato. *Plant Physiol.* 118:1203–1212.

Benhamou, N., Joosten, M.H.A.J., and de Wit P.J.G.M. 1990. Subcellular localization of chitinase and of its potential substrate in tomato root tissues infected by *Fusarium oxysporum* f. sp. *radicis-lycopersici*. *Plant Physiol.* 92:1108–1120.

Benhamou, N., Kloepper, J.W., and Tuzun, S. 1998. Induction of resistance against *Fusarium* wilt of tomato by combination of chitosan with an endophytic bacterial strain: ultrastructure and cytochemistry of the host response. *Planta* 204:153–168.

Benhamou, N., Rey, P., Chérif, M., Hockenhull, J., and Tirilly, Y. 1997. Treatment with the mycoparasite *Pythium oligandrum* triggers induction of defense-related reactions in tomato roots when challenged with *Fusarium oxysporum* f. sp. *radicis-lycopersici*. *Phytopathology* 87:108–122.

Bent, A.F. 1996. Plant Disease Resistance Genes: Function Meets Structure. *Plant Cell* 8(10):1757–1771.

Birecka, H., Catalfamo, J.L. and Urban, P. 1975. Cell wall and protoplast isoperoxidases in tobacco plants in relation to mechanical injury and infection with tobacco mosaic virus. *Plant Physiol.* 55:611–619.

Bohlmann, H. 1994. The role of thionins in plant protection. *Crit. Rev. Plant Sci.* 13:1–16.

Boller, T. 1985. Induction of hydrolases as a defense reaction against pathogens In *Cellular and Molecular Biology of Plant Stress,* eds. J. Key, and T. Kosuge, pp. 247–262. New York: Alan R. Liss, Inc.

Boller, T., and Felix, G. 1996. Olfaction in Plants: specific perception of common microbial molecules, In *Biology of Plant-Microbe Interactions Stacey,* eds. B.G, Mullin, and P.Gresshoff. St. Paul, MN: International Society for Molecular Plant-Microbe Interactions.

Borowicz, V.A. 1997. A fungal root symbiont modifies plant resistance to an insect herbivore. *Oecologia* 112:534–542.

Borsics, T., and Lados, M. 2002. Dodder infection induces the expression of a pathogenesis-related gene of the family PR-10 in alfalfa. *J. Exp. Bot.* 53(375):1831–1832.

Boyle, B., and Brisson, N. 2001. Repression of the defense gene PR-10a by the single-stranded DNA binding protein SEBF. *Plant Cell.* 13(11):2525–2537.

Broekaert, W.F., Terras, F.R., Cammue, B.P., and Osborn, R.W. 1995. Plant defensins: novel antimicrobial peptides as components of the host defense system. *Plant Physiol.* 108(4): 1353–1358.

Brown, A.E., and Davis, R.D. 1992. Chitinase activity in *Stylosanthes guianensis* systemically protected against *Colletotrichum gloeosporoides*. *J. Phytopathol.* 136:247–256.

Busam, G., Kassemeyer, H.H., and Matern, U. 1997. Differential expression of chitinases in *Vitis vinifera* L. responding to systemic acquired resistance activators or fungal challenge. *Plant Physiol.* 115:1029–1038.

Chao, H., Li, X., and Dong, X. 1998. Generation of broad-spectrum disease resistance by overexpression of an essential regulatory gene in systemic acquired resistance. *Proc. Natl. Acad. Sci.* 95(11):6531–6536.

Caruso, C., Nobile, M., Leonardi, L., Bertini, L., Buonocore, V., and Caporale, C. 2001. Isolation and amino acid sequence of two new PR-4 proteins from wheat. *J. Protein Chem.* 20(4):327–335.

Chao, Q., Rothenberg, M., Solano, R., Roman, G., Terzaghi, W., and Ecker, J.R. 1997. Activation of the ethylene gas response pathway in *Arabidopsis* by the nuclear protein Ethylene-Insensitive3 and related proteins. *Cell* 89(7):1133–1144.

Choi, D., Park, J.A., Seo, Y.S., Chun, Y.J., and Kim, W.T. 2000. Structure and stress-related expression of two cDNAs encoding proteinase inhibitor II of *Nicotiana glutinosa* L. *Biochim. Biophys. Acta* 1492:211–215.

Clarke, J.D., Aarts, N., Feys, B.J., Dong, X., and Parker, J.E. 1998. Constitutive disease resistance requires EDS1 in the *Arabidopsis* mutants cpr1 and cpr6 and is partially EDS1-dependent in cpr5. *Plant J.* 26(4):409–420.

Cocca, M.A., Damsz, B., Yun, D.J., Hasegawa, P.M., Bressan, R.A., and Narasimhan, M.L. 2000. Heterotrimeric G-proteins of a filamerntous fungus regulate cell wall composition and susceptibility to a PR-5 protein. *Plant J.* 22:61–69.

Cohen, Y., Niderman, T., Mosinger, E., and Fluhr, R. 1994. β-aminobutyric acid induces the accumulation of pathogenesis-related proteins in tomato (*Lycopersicon esculentum* L.) plants and resistance to late blight infection caused by *Phytophthora infestans*. *Plant Physiol.* 104:59–66.

Cote, F., Cutt, J.R., Asselin, A., and Klessig, D.F. 1991. Pathogenesis-related acidic beta-1,3-glucanase genes of tobacco are regulated by both stress and developmental signals. *Mol. Plant Microbe Interact.* 4(2):173–181.

Coventry, H.S., and Dubery, I.A. 2001. Lipopolysaccharides from *Burkholderia cepacia* contribute to an enhanced defense capacity and the induction of pathogenesis-related proteins in *Nicotiana tabacum*. *Physiol. Mol. Plant Pathol.* 58:149–158.

Cui, Y., Bell, A.A., Joost, O., and Magill, C. 2000. Expression of potential defense genes in cotton. *Physiol. Mol. Plant Pathol.* 56:25–31.

Dann, E.K., Meuwly, P., Métraux, J.P., and Deverall B.J. 1996. The effect of pathogen inoculation or chemical treatment on activities of chitinase and β-1,3-glucanase and acumulation of salicylic acid in leaves of green bean, *Phaseolus vulgaris* L. *Physiol Mol. Plant Pathol.* 49:307–319.

Dalisay, R.F., and Kuć, J.A. 1995a. Persistence of reduced penetration *by Colletotrichum lagenarium* into cucumber leaves with induced systemic resistance and its relation to enhanced peroxidase and chitinase activities. *Physiol. Mol. Plant Pathol.* 47:329–338.

Dalisay, R.F., and Kuć, J.A. 1995b. Persistence of induced resistance and enhanced peroxidase and chitinase activities in cucumber plants. *Physiol. Mol. Plant Pathol.* 47:315–327.

Davis, J.M., Wu, H., Cooke, J.E., Reed, J.M., Luce, K.S., and Michler, C.H. 2002. Pathogen challenge, salicylic acid, and jasmonic acid regulate expression of chitinase gene homologs in pine. *Mol. Plant Microbe Interact.* 15(4):380–387.

De Jong, A.J., Cordewener, J., Lo Schiavo, F., Terzi, M., Vandekerckhove, J., van Kammen, A., and De Vries, S.C. 1992. A carrot somatic embryo mutant is rescued by chitinase. *Plant Cell.* 4(4):425–433.

Del Campillo, E., and Lewis, L.N. 1992. Identification and kinetics of accumulation of proteins induced by ethylene in bean abscission zones. *Plant Physiol.* 98:955–963.

Delledonne, M., Xia, Y., Dixon, R.A., and Lamb, C. 1998. Nitric oxide functions as a signal in plant disease resistance. *Nature* 394(6693):585–588.

De Pater, S., Greco, V., Pham, K., Memelink, J., and Kijne, J. 1996. Characterization of a zinc- dependent transcriptional activator from *Arabidopsis*. *Nucleic Acids Res.* 24(23):4624–4631.

Desveaux, D., Despres, C., Joyeux, A., Subramaniam, R., and Brisson, N. 2000. PBF-2 is a novel single-stranded DNA binding factor implicated in PR-10a gene activation in potato. *Plant Cell.* 12(8):1477–1489.

Ding, C.K., Wang, C.Y., Gross, K.C., and Smith, D.L. 2002. Jasmonate and salicylate induce the expression of pathogenesis-related protein genes and increase resistance to chilling injury in tomato fruit. Planta 214(6):895–901.

Dodson, K.M., Shaw, J.J., and Tuzun, S. 1993. Purification of a chitinase/lysozye isozyme (CHL2) that is constitutively expressed in cabbage varieties resistant to black rot. (Abstract A28) *Phytopathology* 83:1335.

Dong, X. 1998. SA, JA, ethylene and disease resistance in plants. *Curr. Opin. Plant Biol.*1:316–323.

Dubery, L.A., and Slater, V. 1997. Induced defence responses in cotton leaf disks by elicitors from *Verticillium dahliae*. Phytochemistry 44:1429–1434.

Durner, J., Wendehenne, D., and Klessig, D.F. 1998. Defense gene induction in tobacco by nitric oxide, cyclic GMP, and cyclic ADP-riboseProc *Natl. Acad. Sci. USA* 95(17):10328–10333.

Ebrahim-Nesbat, F., Rohringer, R., and Heitefuss, R. 1994. Effect of rust infection on cell walls of barley and wheat; Immunocytochemistry using anti-barley thionin as a probe. *J. Phytopathol.* 141:38–44.

Ecker, J.R. 1995. The ethylene signal transduction pathway in plants. *Science.* 268(5211):667–675.

Egea, C., Ahmed, A.S., Candela, M., and Candela, M.E. 2001. Elicitation of peroxidase activity and lignin biosynthesis in pepper suspension cells by *Phytophthora capsici. J. Plant Physiol.* 158:151–158.

Ekramoddoullah, A.K.M., Yu, X., Sturrock, R., Zamani, A., and Taylor, D. 2000. Detection and seasonal expression pattern of a pathogenesis-related protein (PR-10) in Douglas-fir (*Pseudotsuga menziesii*) tissues. *Physiol. Plant.* 110:240–247.

Elstner, E.F. 1987. Metabolism of active oxygen species. In *The Biochemistry of Plants*, ed. D.D. Davies, Vol. 11, pp. 253–315. San Diego CA: Academic Press.

Enkerli, J., Gisi, U. and Mosinger, E. 1993. Systemic acquired resistance to *Phythophthora infestans* in tomato and the role of pathogenesis-related proteins. *Physiol. Mol. Plant Pathol.* 43:161–171.

Epple, P., Apel, K., and Bohlmann, H. 1995. An *Arabidopsis thaliana* thionin gene is inducible via a signal transduction pathway different from that for pathogenesis-related proteins. *Plant Physiol.* 109:813–820.

Epple, P., Apel, K., and Bohlmann, H. 1997. Overexpression of an endogenous thionin enhances resistance of *Arabidopsis* against *Fusarium oxysporum*. *Plant Cell* 9 (4):509–520.

Eulgem, T., Rushton, P.J., Schmelzer, E., Hahlbrock, K., and Somssich, I.E. 1999. Early nuclear events in plant defence signalling: rapid gene activation by WRKY transcription factors. *EMBO J.* 18(17):4689–4699.

Farmer, E.E., Caldelari, D., Pearce, G., Walker-Simmons, M.K., and Ryan, C.A. 1994. Diethyldithiocarbamic acid inhibits the octadecanoid signaling pathway for the wounding induction of proteinase inhibitors in tomato leaves. *Plant Physiol.* 106:337–346.

Flores, T., Alape-Giron, A., Flores-Diaz, M., and Flores, H.E. 2002. Ocatin. A novel tuber storage protein from the andean tuber crop oca with antibacterial and antifungal activities. *Plant Physiol.* 128(4):1291–302.

Fought, L., and Kuć, J. 1996. Lack of specificity in plant extracts and chemicals as inducers of systemic resistance in cucumber plants to anthracnose. *J. Phytopathol.* 144:1–6.

Fujibe, T., Saji, H., Arakawa, K., Yabe, N., Takeuchi, Y., and Yamamoto, K.T. 2004. A methyl viologen-resistant mutant of arabidopsis, which is allelic to ozone- sensitive rcd1, is tolerant to supplemental ultraviolet-B irradiation. *Plant Physiol.* 134:1–11.

García-Olmedo, F., Molina, A., Segura, A., and Moreno, M. 1995. The defensive role of nonspecific lipid-transfer proteins in plants. *Trends Microbiol.* 3:72–74.

Gardner R.G. 1988. NC EBR-1 and NC EBR-2 early blight resistant tomato breeding lines. *Hortscience* 23:779–781.

Gaudet, D.A., LaRohe, A., Frick, M., Davoren, J., Puchalski, B., and Ergon, Å. 2000. Expression of plant defence-related (PR-protein) transcripts during hardening and de-hardening of winter wheat. *Physiol. Mol. Plant Pathol.* 57:15–24.

Gaudet, D.A., LaRohe, A., Frick, M., Huel, R., and Puchalski, B. 2003. Cold-induced expression of plant defensin and lipid transfer protein transcripts in winter wheat *Physiol. Plantarum* 117(2):195–205.

Gay, P.A., and Tuzun, S. 2000a. Involvement of a novel peroxidase isozyme and lignification in hydathodes in resistance to black rot disease in cabbage. *Can. J. Bot.* 78:1144–1149.

Gay P.A., and Tuzun, S. 2000b. Temporal and spatial assessment of defense responses in resistant and susceptible cabbage varieties during infection with *Xanthomonas campestris* pv. *campestris*. *Physiol. Mol. Plant Pathol.* 57:201–210.

Green, T.R., and Ryan, C.A. 1972. Wound-induced proteinase inhibitor in plant leaves: a possible defense mechanism against insects. *Science* 175: 776–777.

Grisebach, H. 1981. Lignins. In *The Biochemistry of Plants*, Vol. 7: *Secondary Plant Products*, ed. E.E.Conn., pp. 457–478. New York: Academic Press.

Gu, Y.Q., Yang, C., Thara, V.K., Zhou, J., and Martin, G.B. 2000. Pti4 is induced by ethylene and salicylic acid, and its product is phosphorylated by the Pto kinase. *Plant Cell* 12(5):771–786.

Gu, Y.Q., Wildermuth, M.C., Chakravarthy, S., Loh, Y.T., Yang, C., He, X., Han, Y., and Martin, G.B. 2002. Tomato transcription factors pti4, pti5, and pti6 activate defense responses when expressed in Arabidopsis. *Plant Cell.* 14(4):817–831.

Guzzo, S.D., and Martins, E.M.F. 1996. Local and systemic induction of β-1,3-glucanase and chitnase in coffee leaves protected against *Hemileia vastatrix* by *Bacillus thuringiensis*. J *Phytopathology* 144:449–454.

Ham, K.S., Kauffman, S., Albetsheim, P., and Darvill, A.G. 1991. A soybean pathogenesis-related protein with β-1,3-glucanase activity releases phytoalexin elicitor-active heat-stable fragments from fungal walls. *Mol. Plant Microbe Interact* 4:545–552.

Hanselle, T., Ichinoseb, Y., and Barz, W. 2001. Biochemical and molecular biological studies on infection (*Ascochyta rabiei*)-induced thaumatin-like proteins from chickpea plants (*Cicer arietinum* L.). *Z Naturforsch* [C] 56(11–12):1095–1107.

Hart, C.M., Nagy, F., and Meins, F. Jr. 1993. A 61 bp enhancer element of the tobacco beta-1,3-glucanase B gene interacts with one or more regulated nuclear proteins. *Plant Mol. Biol.* 21(1):121–131.

Hejgaard, J., Jacobsen, S., and Svendsen, I. 1991. Two antifungal thaumatin-like proteins from barley grain. *FEBS Lett.* 291(1):127–131.

Hennig, J., Dewey, R.E., Cutt, J.R., and Klessig D.F. 1993. Pathogen, salicylic acid and developmental dependent expression of a beta-1,3-glucanase/GUS gene fusion in transgenic tobacco plants. *Plant J.* 19934(3):481–493.

Hoffland, E., Pieterse, C.M., Bik, L., and van Pelt, J.A. 1995. Induced systemic resistance in radish is not associated with accumulation of pathogenesis-related proteins. *Physiol. Mol. Plant Pathol.* 46:309–320.

Horvath, D.M., Huang, D.J., and Chua N.H. 1998. Four classes of salicylate-induced tobacco genes. *Mol. Plant Microbe Interact.* 11(9):895–905.

Hwang, S.K., Sunwoo, J.Y., Kim, Y.K., and Kim, B.S. 1997. Accumulation of beta-1,3-glucanase and chitinase isoforms, and salicylic acid in the DL-β-amino-n-butyric acid-induced resistance response of pepper stems to *Phytophthora capsici*. *Physiol. Mol. Plant Pathol.* 51:305–322.

Ibeas, J.I., Lee, H., Damsz, B., Prasad, T.P., Pardo, J.M., Hasagawa, P.M., Bressan, R., and Narasimhan, M.L. 2000. Fungal cell phosphomannans facilitate the toxic activity of a plant PR-5 protein. *Plant J.* 23:375–383.

Ibeas, J.I., Yun, D.-J., Damsz, B., Narasimhan, M.L., Uesono, Y., Ribas, J.C., Lee, H., Hagesawa, P.M., and Pardo, J.M. 2001. Resistance to the PR-5 protein osmotin in the model fungus *Saccharomyces cerevisiae* is mediated by the regulatory effects of SSD1 on cell wall composition. *Plant J.* 25:271–280.

Ignatius, S.M.J., Chopra, R.K., and Muthukrishnan, S. 1994. Effects of fungal infection and wounding on the expression of chitinases and β-1,3-glucanases in near-isogenic lines of barley. *Physiol. Plantarum* 90:584–592.

Iwai, T., Kaku, H., Honkura, R., Nakamura, S., Ochiai, H., Sasaki, T., and Ohashi, Y. 2002. Enhanced resistance to seed-transmitted bacterial diseases in transgenic rice plants overproducing an oat cell-wall-bound thionin. *Plant Microbe Interact.* 15(6): 515–21.

Jelitto-Van Dooren, E.P., Vidal, S., and Denecke, J. 1999. Anticipating endoplasmic reticulum stress. A novel early response before pathogenesis-related gene induction. *Plant Cell* 11(10):1935–1944.

Jetiyanon, K. 1994. Immunization of cabbage for long-term resistance to black rot. M. Sc. thesis, Auburn University, Auburn, AL.

Ju, C., and Kuć, J. 1995. Purification and characterization of an acidic β-1,3-glucanase from cucumber and its relationship to systemic disease resistance induced *by Colletotrichum lagenarum* and tobacco necrosis virus. *Mol. Plant Microbe Interact* 8:899–905.

Karban, R., and Kuć, J. 1999. Induced resistance against pathogens and herbivores: an overview. In *Induced Plant Defenses against Pathogens and Herbivores,* eds. A.A. Agrawal and E. Bent, pp. 1–16. St. Paul, MN: APS Press Kassanis, B., Gianinazzi, S., and White, R.F. 1974. A possible explanation of the resistance of virus infected tobacco to second infection. *J. Gen. Virol.* 23:11–16.

Kauffman, S., Legrand, M., Geoffrey, P., and Fritig, B. 1987. Biological function of pathogenesis-related proteins of tobacco have 1,3-β-glucanase activity. *EMBO J.* 6, 3209–3212.

Keen, N.T., and Yoshikawa, M. 1983. β-1,3-endoglucanases from soybean release elicito-active carbohydrates from fungus cell walls. *Plant Physiol.* 71:460–465.

Kemp, G., Botha, A.M., Kloppers, F.J., and Pretorius, Z.A. 1999. Disease development and β-1,3-glucanase expression following leaf rust infection in resistant and susceptible near-isogenic wheat seedlings. *Physiol. Mol. Plant Pathol.* 55:45–52.

Kitajima, S., Koyama, T., Ohme-Takagi, M., Shinshi, H., and Sato, F. 2000. Characterization of gene expression of NsERFs, transcription factors of basic PR genes from *Nicotiana sylvestris*. *Plant Cell Physiol.* 41(6):817–824.

Klement, Z. 1982. Hypersensitivity. In *Phytopathogenic Prokaryotes*, eds. M.S. Mount and G.H. Lacy, Vol. 2, pp. 149–157. London: Academic Press.Kogel, K.H., Beckhove, U., Dreschers, J., Münch, S., and Rommé, Y. 1994. Acquired resistance in barley: the resistance mechanism induced by 2,6-dichloroisonicotinic acid is a phenocopy of a genatically based mechanism governing race-specific powdery mildew resistance. *Plant Physiol.* 106:1269–1277.

Kroj, T., Rudd, J.J., Nurnberger, T., Gabler, Y., Lee, J., and Scheel, D. 2003. Mitogen-activated protein kinases play an essential role in oxidative burst-independent expression of pathogenesis-related genes in parsley. *J. Biol. Chem.* 278(4):2256–2264.

Krokene, P., Christiansen, E., Solheim, H., Franceschi, V.R., and Berryman, A.A. 1999. Induced resistence to pathogenic fungi in Norway spruce. *Plant Physiol.* 121: 565–569.

Kuranda, M.J., and Robbins, P.W. 1991. Chitinase is required for cell separation during growth of *Saccharomyces cerevisiae. J. Biol. Chem.* 266(29):19758–19767.

Lagrimini, L.M., Burkhart W., Moyer, M., and Rothstein, S. 1987. Molecular cloning of complementary DNA encoding the lignin-forming peroxidase from tobacco: molecular analysis and tissue-specific expression. *Proc. Natl. Acad. Sci. USA* 84:7542–7546.

Lagrimini, L.M., Gingas, V., Finger, F., Rothstein, S., and Liu, T.T.Y. 1997. Characterization of antisense transformed plants deficient in the tobacco anionic peroxidase. *Plant Physiol.* 114:1187–1196.

Lawrence, C.B., Joosten, M.H.A.J., and Tuzun, S. 1996. Differential induction of pathogenesis-related proteins in tomato by *Alternaria solani* and the association of a basic chitinase isozyme with resistance. *Physiol. Mol. Plant Pathol.* 48:361–377.

Lawrence, C.B., Singh, N.P., Qiu, J., Gardner, R.G., and Tuzun, S. 2000. Constitutive hydrolytic enzymes are associated with polygenic resistance of tomato to *Alternaria solani* and may function as an elicitor release mechanism.*Physiol. Mol. Plant Pathol.* 57:211–220.

Lawton, K., Uknes, S., Friedrich, L., Gaffney, T., Alexander, D., Goodman, R., Metraux, J.-P, Kessman H., Ahl Goy P., Gut Rella, M., Ward, E., and Ryals, J. 1993. In *The Molecular Biology of Systemic Acquired Resistance, in Mechanisms of Plant Defense Responses,* eds. B. Frittig, and M. Legrand, p. 422. Dordrecht, The Netherlands: Kluwer Academic Publishing

Legrand, M., Kauffmann, S., Geoffroy, P., and Fritig, B. 1987. Biological function of "pathogenesis-related" proteins: four tobacco pathogenesis related proteins are chitinases. *Proc. Natl. Acad. Sci. USA* 84:6750–6759.

Li, X., Zhang, Y., Clarke, J.D., Li, Y., and Dong, X. 1999. Identification and cloning of a negative regulator of systemic acquired resistance, SNI1, through a screen for suppressors of npr1-1. *Cell* 98(3):329–339.

Liao, Y.C., Krutzaler, F., Fischer, R., Reisner, J.J., and Tizbury, R. 1994. Characterization of a wheat class Ib chitinase gene differentially induced in isogenic lines by infection *with Puccinia graminis. Plant Sci.* 103:177–187.

Ligterink, W., Kroj, T., zur Nieden, U., Hirt, H., and Scheel, D. 1997. Receptor-mediated activation of a MAP kinase in pathogen defense of plants. *Science* 276(5321):2054–2057.

Linthorst, H.J., Danhash, N., Brederode, F.T., van Kan, J.A., De Wit, P.J., and Bol, J.F. 1991. Tobacco and tomato PR proteins homologous to win and pro-hevein lack the "hevein" domain. *Mol. Plant Microbe Interact.* 4(6):586–592.

Lintle, M., and van der Westhuizen, A.J. 2002. Glycoproteins from Russian wheat aphid infested wheat induce defence responses. *Z Naturforsch* [C] 57(9–10):867–873.

Lo, S.C., Hipskind, J.D., and Nicholson, R.L. 1999. cDNA cloning of a sorghum pathogenesis-related protein (PR-10) and differential expression of defense-related genes following inoculation with *Cochiobolus heterostrophus* or *Colletotrichum sublineolum. Mol.Plant Microbe Interact.* 12:479–489.

Lorito, M., Woo, S.L., Fernandez, I.G., Colucci, G., Harman, G.E., Pintor-Toro, J.A., Filippone, E., Muccifora, S., Lawrence, C.B., Zonia, A., Tuzun, S., and Scala, F. 1998. Genes from mycoparasitic fungi as a source for improving plant resistance to fungal pathogens. *Proc. Natl. Acad. Sci. USA* 95:7860–7865.

Lotan, T., Ori, N., and Fluhr, R. 1989. Pathogenesis-related proteins are developmentally regulated in tobacco flowers. *Plant Cell* 1(9):881–887.

Lusso, M., and Kuć, J. 1995. Evidence for transcriptional regulation of β-1,3-glucanase as it relates to induced systemic resistance of tobacco to blue mold. *Mol. Plant Microbe Interact.* 8:473–475.

Mader, M., and Fussi, R. 1982. Role of peroxidase in lignification of tobacco cells: regulation by phenolic compounds. *Plant Physiol.* 70:1132–1134.

Mader, M., Ungemach, J. and, Schloss, P. 1980. The role of peroxidase isoenzyme groups of *Nicotiana tabacum* in hydrogen peroxide formation. *Planta* 147: 467–470.

Maiero, M., Ng, T.J., and Barksdale, T.H. 1990. Inheritance of collar rot resistance in the tomato breeding lines C1943 and NC EBR-2. *Phytopathology* 80:1365–1368.

Maiero, M., and Ng, T.J. 1989. Genetic analyses of early blight resistance in tomatoes. *HortScience* 24:221–226.

Maleck, K., Levine, A., Eulgem, T., Morgan, A., Schmid, J., Lawton, K.A., Dangl, J.L., and Dietrich, R.A. 2000. The transcriptome of *Arabidopsis thaliana* during systemic acquired resistance. *Nat. Genet.* 26(4):403–410.

Manandhar, H.K., Mathur, S.B., Smedegaard-Petersen, V., and Thordal-Christensen, H. 1999. Accumulation of transcripts for pathogenesis-related proteins and peroxidase in rice plants triggered by *Pyricularia oryzae, Bipolaris sorokinana* and UV. light. *Physiol. Mol. Plant Pathol.* 55:289–295.

Mauch, F., Mauch-Mani, B., and Boller, T. 1988. Antifungal hydrolases in pea tissue. II. Inhibition of fungal growth by combinations of chitinases and β-1,3-glucanases. *Plant Physiol.* 88:936–942.

Maurhofer, M., Hase, C., Meuwly, P., Métraux, J.P., and Défago, G. 1994. Induction of systemic resistance of tobacco to tobacco necrosis virus by the root-*colonizing Pseudomonas fluorescens* strain CHA0: Influence of the *gacA* gene and of pyoverdine production. *Phytopathology* 84:139–146.

Mazarei, M., Puthoff, D.P., Hart, J.K., Rodermel, S.R., and Baum, T.J. 2002. Identification and characterization of a soybean ethylene-responsive element-binding protein gene whose mRNA expression changes during soybean cyst nematode infection. *Mol. Plant Microbe. Interact.* 15(6):577–586.

McGee, J.D., Hamer, J.E., and Hodges, T.K. 2001. Characterization of a PR-10 pathogenesis-related gene family induced in rice during infection with *Magnaporthe grisea*. *Mol. Plant Microbe Interact.* 14(7):877–886.

McIntyre, J.L., Dodds, J.A., and Hare, J.D. 1981. Effects of localized infections of *Nicotania tabacum* by tobacco mosaic virus on systemic resistance against diverse pathogens and an insect. *Phytopathology* 71:297–30.

McNellis, T.W., Mudgett, M.B., Li, K., Aoyama, T., Horvath, D., Chua, N.H. and Staskawicz, B.J. (1998). Glucocorticoid-inducible expression of a bacterial avirulence gene in transgenic Arabidopsis induces hypersensitive cell death. *Plant J.* 14(2):247–257.

Melchers, L.S., Apotheker-de Groot, M., van der Knaap, J.A., Ponstein, A.S., Sela-Buurlage, M.B., Bol, J.F., Cornelissen, B.J.C., van den Elzen, P.J.M., and Linthorst, H.J.M. 1994. A new class of tobacco chitinases homologous to bacterial exo-chitinases displays antifungal activity. *Plant J.* 5:469–480.

Memelink, J., Linthorst, H.J., Schilperoort, R.A., and Hoge, J.H. 1990. Tobacco genes encoding acidic and basic isoforms of pathogenesis-related proteins display different expression patterns. *Plant Mol. Biol.* 14(2):119–126.

Métraux, J.-P., Streit, L., and Staub, T.H. 1988. A pathogenesis-related protein in cucumber is a chitinase. *Physiol. Mol. Plant Pathol.* 33:1–9.

Metraux, J.P., Burkhart, W., Moyer, M., Dincher, S., Middlesteadt, W., Williams, S., Payne, G., Carnes M., and Ryals, J. 1989. Isolation of a complementary DNA encoding a chitinase with structural homology to a bifunctional lysozyme/chitinase. *Proc. Natl. Acad. Sci. USA* 86:896–900.

Miyazawa, J., Kawabata, T., and Ogasawara, N. 1998. Induction of an acidic isozyme of peroxidse and acquired resistance to wilt disease in response to treatment of tomato roots with 2-furoic acid, 4-hydroxybenzoic hydrazide or salicylic hydrazide. *Physiol. Mol. Plant Pathol.* 52:115–126.

Moiseyev, G.P., Beintema, J.J., Fedoreyeva, L.I., and Yakovlev, G.I. 1994. High sequence similarity between a ribonuclease from ginseng calluses and fungus-elicited proteins from parsley indicates that intracellular pathogenesis-related proteins are ribonucleases. *Planta* 193(3):470–472.

Molina, A., Gorlach, J., Volrath, S., and Ryals, J. 1999. Wheat genes encoding two types of PR-1 proteins are pathogen inducible, but do not respond to activators of systemic acquired resistance. *Mol. Plant Microbe Interact.* 12(1):53–58.

Mosch, J., and Zeller, W. 1996. Further studies on plant extracts with a resistance induction effect against *Erwinia amylovora*. *Acta Hort.* 411:361–366.

Muthukrishnan, S., Liang, G.H., Trick, H.N., and Gill, B.S. (2001). Pathogenesis-related proteins and their genes in cereals. *Plant Cell, Tissue and Organ Culture.* 64:93–114.

Narasimhan, M.L., Lee, H., Damsz, B., Singh, N.K., Ibeas, J.I., Matsumoto, T.K., Woloshuk, C.P., and Brassan, R. (2003). Overexpression of a cell wall glycoprotein in *Fusarium oxysporum* increases virulence and resistance to a PR-5 protein. *Plant J.* 36:390–400.

Nash, A.F., and Gardner, R.G. 1988. Tomato early blight resistance in a breeding line derived from *Lycopersicon hirsutum* PI 126445. *Plant Dis.* 72:206–209.

Nelson, R.R. 1978. Genetics of horizontal resistance to plant diseases. *Annu. Rev. Plant Pathol.* 16:359–378.

Nibbe, M., Hilpert, B., Wasternack, C., Miersch, O., and Apel, K. 2002. Cell death and salicylate- and jasmonate-dependent stress responses in Arabidopsis are controlled by single cet genes. *Planta* 216(1):120–128.

Niderman, T., Genetet, I., Bruyere, T., Gees, R., Stintzi, A., Legrand, M., Fritig, B., and Mosinger, E. 1995. Pathogenesis-related PR-1 proteins are antifungal. Isolation and characterization of three 14-kilodalton proteins of tomato and of a basic PR-1 of tobacco with inhibitory activity against *Phytophthora infestans. Plant Physiol.* 108(1): 17–27.

Norman, C., Vidal, S., and Palva, E.T. 1999. Oligogalacturonide-mediated induction of a gene involved in jasmonic acid synthesis in response to the cell-wall-degrading enzymes of the plant pathogen *Erwinia carotovora. Mol. Plant Microbe Interact.* 12:640–644.

Nurnberger, T., Nennstiel, D., Jabs, T., Sacks, W.R., Hahlbrock, K., and Scheel, D. (1994). High affinity binding of a fungal oligopeptide elicitor to parsley plasma membranes triggers multiple defense responses. *Cell.* 78(3):449–460.

O'Donnell, P.J., Calvert, C., Atzorn, R., Wasternack, C., Leyser, H.M.O., and Bowles, D.J. 1996. Ethylene as a signal mediating the wound response of tomato plant. *Science* 274(5294):1914–1917.

Okushima, Y., Koizumi, N., Kusano, T., and Sano, H. 2000. Secreted proteins of tobacco cultured BY2 cells: identification of a new member of pathogenesis-related proteins. *Plant Mol. Biol.* 42:479–488.

Pääkkönen, E., Seppänen, S., Holopainen, T., Kokko, H., Kärenlampi, S., Kärenlampi, L., and Kangasjärvi, J. 1998. Induction of genes for the stress proteins PR-10 and PAL in relation to growth, visible injuries and stomatal conductance in birch (*Betula pendula*) clones exposed to ozone and/or drought. *The New Phytologist* 138 (2):295–305.

Pan, S.Q., Ye, X.S., and Kuć, J. 1991. Association of β-1,3-glucanase activity and isoform pattern with systemic resistance to blue mould in tobacco induced by stem injection with *Peronospora tabacina* or leaf inoculation with tobacco mosaic virus. *Physiol. Mol. Plant Pathol.* 39:25–39.

Pan, S.Q., Ye, X.S., and Kuć, J. 1992. Induction of chitinases in tobacco plants systemically protected against blue mold by *Peronospora tabacina* or tobacco mosaic virus. *Phytopathology* 82:119–123.

Pieterse, C.M., van Wees, S.C., Hoffland, E., van Pelt, J.A., and van Loon, L.C. 1996. Systemic resistance in *Arabidopsis* induced by biocontrol bacteria is independent of salicylic acid accumulation and pathogenesis-related gene expression. *Plant Cell* 8(8): 1225–1237.

Ponstein, A.S., Bres-Vloemans, S.A., Sela-Buurlage, M.B., van den Elzen, P.J., Melchers, L.S., and Cornelissen, B.J. 1994. A novel pathogen- and wound-inducible tobacco (*Nicotiana tabacum*) protein with antifungal activity. *Plant Physiol.* 104(1):109–118.

Poupard, P., Parisi, L., Campion, C., Ziadi, S., and Simoneau, P. 2003. A wound-and ethephon-inductible PR-10 gene subclass from apple is differentially expressed during infection with a compatible and an incompatible race of *Venturia inaequalis. Physiol. Mol. Plant Pathol.* 62:3–12.

Pulliam, D.A., Williams, D.L., Broadway, R., and Stewart, Jr. C.N. 2001. Isolation and characterization of a serine proteinase inhibitor cDNA from cabbage and its antibiosis in transgenic tobacco plants. *Plant Cell Biotech. Mol. Biol.* 2(1,2):19–32.

Raghothama, K.G., Maggio, A., Narasimhan, M.L., Kononowicz, A.K., Wang, G., D'Urzo, M.P., Hasegawa, P.M., and Bressan, R.A. 1997. Tissue-specific activation of the osmotin gene by ABA, C2H4 and NaCl involves the same promoter region. *Plant Mol. Biol.* 34(3):393–402.

Redolfi, P. 1983. Occurrence of pathogenesis related (b) and similar proteins in different plant species. *Neth. J. Plant Pathol.* 89:245–256.

Regalado, A.P., Pinheiro, C., Vidal, S., Chaves, I., Ricardo, C.P.P., and Rodrigues-Pousada, C. 2000. The *Lupinus albus* class-III chitinase gene, *IF-3*, is constitutively expressed in vegetative organs and developing seeds. *Planta* 210:543–550.

Reimers, P.J., and Leach, J.E. 1991. Race-specific resistance to *Xanthomonas oryzae* pv. *oryzae* conferred by bacterial blight resistance gene *xa-10* in rice (*Oryza sativa*) involves accumulation of a lignin-like substance in host tissues. *Physiol. Mol. Plant Pathol.* 38: 39–55.

Richardson, M., Valdes-Rodriguez, S., and Blanco-Labra, A. 1987. A possible function for thaumatin and a TMV induced protein suggested by homology to a maize inhibitor. *Nature* 327:432–434.

Ride, J.P. 1975. Lignification in wounded wheat leaves in response to fungi and its possible role in resistance. *Physiol. Plant Pathol.* 5:125–134.

Robertson, T.L. 1995. Hydrolytic enzyme accumulation in tobacco lines resistant *to Peronospora tabacina*, the causal agent of blue mold. M.Sc. thesis, Auburn University, Auburn AL.

Rohe, M., Gierlich, A., Hermann, H., Hahn, M., Schmidt, B., Rosahl, S., and Knogge, W. 1995. The race-specific elicitor, NIP1, from the barley pathogen, *Rhynchosporium secalis,* determines avirulence on host plants of the Rrs1 resistance genotype. *EMBO J.* 14(17):4168–4177.

Rufty, R.C. 1989. Genetics of host resistance to tobacco blue mold. In *Blue Mold of Tobacco,* ed. W.E. McKeen, pp. 141–164. St. Paul, MN: APS Press.

Rushton, P.J., Torres, J.T., Parniske, M., Wernert, P., Hahlbrock, K., and Somssich, I.E. 1996. Interaction of elicitor-induced DNA-binding proteins with elicitor response elements in the promoters of parsley PR1 genes. *EMBO. J.* 15(20):5690–5700.

Rushton, P.J., and Somssich, I.E. 1998. Transcriptional control of plant genes responsive to pathogens. *Curr. Opin. Plant Biol.* 1(4):311–315.

Santamaria M., Thomson C.J., Read N.D., and Loake G.J. 2001. The promoter of a basic PR1-like gene, AtPRB1, from *Arabidopsis* establishes an organ-specific expression pattern and responsiveness to ethylene and methyl jasmonate. *Plant Mol. Biol.* 47(5):641–652.

Satoh, H., Uchida, A., Nakayama, K. and Okada, M. 2001. Water-soluble chlorophyll protein in *Brassicaceae* plants is a stress-induced chlorophyll-binding protein. *Plant Cell Physiol.* 42(9):906–11.

Scandalios, J.G. 1993. Oxygen stress and superoxide dismutases. *Plant Physiol.* 101: 7–12.

Schafleitner, R., and Wilhelm, E. 1997. Effect of virulent and *hypovirulent Cryphonectria parasitica* (Murr.) Barr on the intercellular pathogen related proteins and on total protein pattern of chestnut (*Castanea stativa* Mill.). *Physiol. Mol. Plant Pathol.* 45: 57–67.

Schneider S., and Ullrich, W.R. 1994. Differential induction of resistance and enhanced enzyme activities in cucumber and tobacco caused by treatment with various abiotic and biotic inducers. *Physiol. Mol. Plant Pathol.* 43:57–67.

Sela-Buurlage M. B., Ponstein A.S. Bres-Vloemans S.A., Melchers L.S., van der Elzen P.M.J., and Cornelissen, B.J.C. 1993. Only specific tobacco chitinases and β-1,3-glucanases exhibit antifungal activity. *Plant Physiol.* 101:857–863.

Sharma, P., Børja D., Stougaard P., and Lönneborg A. 1993. PR-proteins accumulating in spruce roots infected with a pathogenic *Pythium* sp. isolate include chitinases, chitosanases and β-1,3-glucanases. *Physiol. Mol. Plant Pathol.* 45:291–304.

Shih, C.T., Wu, J., Jia S., Khan, A.A., Ting, K.H., and Shih, D.S. 2001. Purification of an osmotin-like protein from the seeds of *Benincasa hispida* and cloning of the gene encoding this protein. *Plant Sci.* 160(5):817–826.

Shin, R., Park, C.J., An, J.M., and Paek, K.H. 2003. A novel TMV-induced hot pepper cell wall protein gene (CaTin2) is associated with virus-specific hypersensitive response pathway. *Plant Mol. Biol.* 51(5):687–701.

Shinshi, H., Neuhaus, J., Ryals, J., and Meins, F. 1990. Structure of a tobacco endochitinase gene: evidence that different chitinase genes can arise by transposition of sequences encoding a cysteine-rich domain. *Plant Mol. Biol.* 14:357–368.

Siefert F., Thalmair, M., Langbartels, C., Sandermann, H., and Grossman, K. 1996. Epoxiconazole-induced stimulation of the antifungal hydrolases chitinase and β-1,3-glucanase in wheat. *Plant Growth Reg.* 20:279–286.

Singh, N.K., Bracker, C.A., Hasegava, P.M., Handa, A.K., Buckel, S., Hermodson, M.A., Pfankoch, E., Regnier, F.E., and Bressan, R.A. 1987. Characterization of osmotin. A thaumatin-like protein associated with osmotic adaptation of cultured tobacco cells. *Plant Physiol.* 79:126–137.

Singh, N.K., Nelson, D.E., Kuhn, D., Hasegawa, P.M., and Bressan, R.A. 1989. Molecular cloning of osmotin and regulation of its expression by ABA and adaptation to low water potential. *Plant Physiol.* 90:1096–1101.

Simmonds, N.W. 1991. Genetics of horizontal resistance to diseases of crops. *Biol. Rev.* 66:189–241.

Solano, R., Stepanova, A., Chao, Q. and Ecker, J.R. 1998. Nuclear events in ethylene signaling: a transcriptional cascade mediated by Ethylene-Insensitive3 and Ethylene-Response-Factor1. *Genes Dev.* 12(23): 3703–3714.

Somssich, I.E. 2003. Closing another gap in the plant SAR puzzle. *Cell*, 113(7):815–816.

Somssich, I.E., Schmelzer, E., Bollmann, J., and Hahlbrock, K. 1986. Rapid activation by fungal elicitor of genes encoding "pathogenesis-related" proteins in cultured parsley cells. *Proc. Natl. Acad. Sci. USA* 83:2427–2430.

Spoel, S.H., Koornneef, A., Claessens, S.M.C., Korzelius, J.P., Van Pelt, J.A., Mueller, M.J., Buchala, A.J., Metraux, J.-P., Brown, R., Kazan, K., Van Loon, LC., Dong, X., and Pieterse, C.M.J. 2003. NPR1 modulates cross-talk between salicylate- and jasmonate-dependent defense pathways through a novel function in the cytosol. *Plant Cell* 15:760–770.

Staskawicz, B.J., Ausubel, F.M., Baker, B.J., Ellis, J.G., and Jones, J.D. 1995. Molecular genetics of plant disease resistance. *Science* 268(5211):661–667.

Staub, T., and Williams, P.H. 1972. Factors influencing black rot lesion development in resistant and susceptible cabbage. *Phytopathology* 62:722–728.

Steijl, H., Niemann, G.J., and Boon, J.J. 1999. Changes in chemical composition related to fungal infection and induced resistance in carnation and radish investigated by pyrolysis mass spectrometry. *Physiol Mol. Plant Pathol.* 55:297–311.

Stintzi, A., Heitz, T., Prasad, V., Wiedemann-Merdinoglu, S., Kauffmann, S., Geoffroy, P., Legrand, M., and Fritig, B. 1993. Plant 'pathogenesis-related' proteins and their role in defense against pathogens. *Biochimie.* 75(8):687–706.

Tang, X., Frederick, R.D., Zhou, J., Halterman, D.A., Jia, Y., and Martin, G.B. 1996. Initiation of plant disease resistance by physical interaction of AvrPto and Pto kinase. *Science* 274(5295):2060–2063.

Terras, F.R., Eggermont, K., Kovaleva, V., Raikhel, N.V., Osborn, R.W., Kester, A., Rees, S.B., Torrekens, S., Van Leuven, F., and Vanderleyden, J. 1995. Small cysteine-rich antifungal proteins from radish: their role in host defense. *Plant Cell.* 7(5): 573–588.

Thimmapuram, J., Ko, T.S., and Korban, S.S. 2001. Characterization and expression of beta-1,3-glucanase genes in peach. *Mol. Genet. Genomics.* 265(3):469–479.

Tiburzy, R., and Reisener, H.J. 1990. Resistance of wheat to *Puccinia graminis* f. sp. *tritici:* association of the hypersensitive reaction with the cellular accumulation of lignin- like material and callose. *Physiol. Mol. Plant Pathol.* 36:109–120.

Ton, J., Pieterse, C.J.M., and van Loon, L.C. 1999. Identification of a locus in *Arabidopsis* controlling both the expression of a rhizobacteria-mediated induced systemic resistance (ISR) and basal resistance against *Pseudomonas syringae* pv. *tomato*. *Mol. Plant Microbe Interact.* 12:911–918.

Tuzun, S., Gay, P.A., Lawrence, C.B., Robertson, T.B., and Sayler, R.J. 1997. Biotech-nological applications of inheritable and inducible resistance to diseases in plants. In *Technology Transfer of Plant Biotechnology,* ed. P.M.Gresshoff, pp. 25–40. New York: CRC Press.

Tuzun, S., Juarez, J., Nesmith, W.C., and Kuć, J. 1992. Induction of systemic re-sistance in tobacco against metalaxyl-tolerant strains of *Peronospora tabacina* and the natural occurrence of the phenomenon in Mexico. *Phytopathology* 82:425–429.

Tuzun, S., and Kloepper, J. 1995. Practical application and implementation of induced resistance. In *Induced Resistance to Disease in Plants: Developments in Plant Pathology,* eds. R. Hammerschmidt, and J. Kuć, Vol. 4, pp. 152–168. Boston: Kluwer Academic Publishers.

Tuzun, S., and Kuć, J. 1991. Plant immunization: an alternative approach to pesticides for control of plant diseases in the greenhouse and field. In *The Biological Control of Plant Diseases,* pp. 30–40. Taipei, Taiwan: FFTC Book Series No. 42.

Tuzun, S., Nesmith, W., Ferriss, R.S., and Kuć, J. 1986. Effects of stem injections *with Peronospora tabacina* on growth of tobacco and protection against blue mold in the field. *Phytopathology* 76:938–941.

Tuzun, S., Rao, M.N., Vogeli, U., Schardl, S.L., and Kuć, J. 1989. Induced systemic resis-tance to blue mold: early induction and accumulation of β-1,3-glucanases, chitinases and other pathogenesis-related proteins (b-proteins) in immunized tobacco. *Phytopathology* 79:979–983.

Umemoto, N., Kakitani, M., Iwamatsu, A., Yoshikawa, M., Yamaoka, N., and Ishida, I. 1997. The structure and function of a soybean beta-glucan-elicitor-binding protein. *Proc. Natl. Acad. Sci USA* 94(3):1029–1034.

Urwin, P.E., Troth, K.M., Zubko, E.I., and Atkinson, H.J. 2001. Effective transgenic resis-tance to *Globodera pallida* in potato field trials. *Mol. Breed.* 8(1):95–101.

Urwin, P.E., Lilley, C.J., and Atkinson, H.J. 2002. Ingestion of double-stranded RNA by preparasitic juvenile cyst nematodes leads to RNA interference. *Mol. Plant Microbe Interact.* 15(8):747–752.

Vad, K., Mikkelsen, J.D., and Collinge, D.B. 1991. Induction, purification and charac-terization of chitinase isolated from pea leaves inoculated with *Ascochyta pisi. Planta* 184:24–29.

Van Loon, L.C., and Van Kammen, A. 1970. Polyacrylamide disc electrophoresis of the soluble leaf proteins from Nicotiana tabacum var. "Samsun" and "Samsun NN" II. Changes in protein constitution after infection with tobacco mosaic virus. *Virology* 40:199–205.

Van Loon, L.C. 1975. Polyacrylamide disc electrophoresis of the soluble leaf proteins from Nicotiana tabacum var. "Samsun" and "Samsun NN" III. Influence of temperature and virus strain on changes induced by tobacco mosaic virus. *Physiol. Plant Pathol.* 6:289–299.

Van Loon, L.C. 1982. Regulation of changes in proteins and enzymes associated with active defense against virus infection. In *Active Defense Mechanisms in Plants,* ed. R.K.S. Wood, pp. 247–273. New York: Plenum Press.

Van Loon, L.C., and Callow, J.A. 1983. Transcription and translation in the diseased plant. In *Biochemical Plant Pathology,* ed. J.A. Callow, pp. 385–414. Chichester, UK: John Wiley & Sons.

Van Loon, L.C. 1997. Induced plant resistance in plants and the role of pathogenesis-related proteins. *Eur. J. Plant Pathol.* 103:753–765.

Van Loon, L.C., and Van Strien, E.A. 1999. The families of pathogenesis-related proteins, their activities, and comparative analysis of PR-1 type proteins. *Physiol. Mol. Plant Pathol.* 55:85–97.

Van Loon, L.C., Bakker, P.A.H.M., and Pieterse, C.M.J. 2002. Prospects and challenges for practical application of rhizobacteria-mediated induced systemic resistance. In *Induced Resistance in Plants Against Insects and Diseases,* eds. A. Schmitt, and B. Mauch-Mani, IOBC/wprs Bulletin 25(6):75–82.

Van Wees, S.C.M., De Swart, E.A.M., Van Pelt, J.A., Van Loon, L.C., and Pieterse, C.M.J. 2000. Enhancement of induced disease resistance by simultaneous activation of salicylate- and jasmonate-dependent defense pathways in *Arabidopsis thaliana. Proc. Natl. Acad. Sci. USA* 97:8711–8716.

Vera, P., and Conejero, V. 1988. Pathogenesis-related proteins of tomato. P-69 as an alkaline endoproteinase. *Plant Physiol.* 87:58–63.

Vigers, A.J., Roberts, W.K., Selitrennikoff, C.P. 1991. A new family of plant antifungal proteins. *Mol. Plant Microbe Interact.* 4:315–323.

Vigers, A.J., Roberts, W.K., and Selitrennikoff, C.P. 1991. A new family of plant antifungal proteins. *Mol. Plant Microbe Interact.* 4:315–323.

Vogeli-Lange, R., Frundt, C., Hart, C.M., Nagy, F., and Meins, F. Jr. 1994. Developmental, hormonal, and pathogenesis-related regulation of the tobacco class I beta-1,3-glucanase B promoter. *Plant Mol. Biol.* 25(2):299–311.

Wegener, C.S., Bartling, S., Olsen, O., Weber, J., and von Wertstein, D. 1996. Pectate lyase in transgenic potatoes confers pre-activation of defence against *Erwinia carotovora. Physiol. Mol. Plant Pathol.* 49:359–376.

Wei,Y., Zhang, Z., Andersen, C.H., Schmelzer, E., Gregersen, P.L., Collinge, D.B., Smedegaard-Petersen, V., and Thordal-Christensen, H. 1998. An epidermis/papilla-specific oxalate oxidase-like protein in the defence response of barley attacked by the powdery mildew fungus. *Plant Mol. Biol.* 36:101–112.

Wei, Z.M., and Beer, S.V. 1996. Harpin from *Erwinia amylovora* induces plant resistance. *Acta Hort.* 411:223–225.

Wu K., Tian L., Hollingworth J., Brown, D.C., and Miki, B. 2002. Functional analysis of tomato Pti4 in Arabidopsis. *Plant Physiol.* 128(1):30–37.

Wubben, J.P., Joosten, M.H., and De Wit, P.J. 1994. Expression and localization of two *in planta* induced extracellular proteins of the fungal tomato pathogen *Cladosporium fulvum. Mol. Plant Microbe Interact.* 7(4):516–524.

Xiong, L., and Yang, Y. 2003. Disease resistance and abiotic stress tolerance in rice are inversely modulated by an abscisic acid-inducible mitogen-activated protein kinase. *Plant Cell.* 15(3):745–759.

Xu, Y., Chang, P., Liu, D., Narasimhan, M.L., Raghothama, K.G., Hasegawa, P.M., and Bressan, R.A. 1994. Plant defense genes are synergistically induced by ethylene and methyl jasmonate. *Plant Cell.* 6(8):1077–1085.

Xu, Z.F., Qi, W.Q., Ouyang, X.Z., Yeung, E., and Chye, M.L. 2001. A proteinase inhibitor II of *Solanum americanum* is expressed in phloem. *Plant Mol. Biol.* 47(6):
727–738.

Xue, L., Charest, P.M., and Jabaji-Hare, S.H. 1998. Systemic induction of peroxidases,
1,3- beta-glucanases, chitinases, and resistance in bean plants by binucleate *Rhizoctonia*
species. *Phytopathology* 88:359–365.

Ye, X.S., Järlfors, S., Tuzun, S., Pan, S.Q., and Kuć, J. 1992. Biochemical changes in cell
walls and cellular responses of tobacco leaves related to systemic resistance to blue mold
(*Peronospora tabacina*) induced by tobacco mosaic virus. *Can. J. Bot.* 70:49–57.

Ye, X.S., Pan, S.Q., and Kuć, J. 1990. Association of pathogenesis-related proteins and
activities of peroxidase, β-1,3-glucanase and chitinase with systemic induced resistance
to blue mould of tobacco but not to systemic tobacco mosaic virus. *Physiol. Mol. Plant
Pathol.* 36:523–531.

Yu, X., Ekramoddoullah, A.K.M., and Misra, S. 2000. Characterization of Pin m III cDNA
in western white pine. *Tree Physiol.* 20:663–671.

Yun, D-J, Zhao, Y., Pardo, J.M., Narasihman, M.L., Damsz, B., Lee, H., Abad, L.R., D'urzo,
M.P., Hasegawa, P.M., and Bressan, R.A. 1997. Stress proteins on the yeast cell surface
determine resistance to osmotin, a plant antifungal protein. *Proc. Natl. Acad. Sci. USA*
94:7082–7087.

Zhang, J., and Kirkham, M.B. 1994. Drought-stress-induced changes in activities of superoxide dismutase, catalase, and peroxidase in wheat species. *Plant Cell Physiol.* 35:785–
791.

Zhang, Z., Collinge, D.B., and Thordal-Christensen, H. 1995. Germin-like oxalate oxidase,
a H2O2-producing enzyme, accumulates in barley attacked by the powdery mildew
fungus. *Plant J.* 8:139–145.

Zhou, J., Tang, X., and Martin, G.B. 1997. The Pto kinase conferring resistance to tomato
bacterial speck disease interacts with proteins that bind a *cis*-element of pathogenesis-
related genes. *EMBO J.* 16:3207–3218.

Zhou, J.M., Trifa, Y., Silva, H., Pontier, D., Lam, E., Shah, J., and Klessig, D.F. 2000. NPR1
differentially interacts with members of the TGA/OBF family of transcription factors
that bind an element of the PR-1 gene required for induction by salicylic acid. *Mol. Plant
Microbe Interact.* 13(2):191–202.

7

Chemical Signals in Plant Resistance: Salicylic Acid

CHRISTIANE NAWRATH, JEAN-PIERRE MÉTRAUX, AND THIERRY GENOUD

7.1 Introduction: Systemic Acquired Resistance and Salicylic Acid

Plants are defended against pathogens by constitutive and inducible barriers. Induced resistance is expressed locally at the site of infection as well as in uninfected parts of infected plants. Induced defense responses to pathogens were already described in the first half of the 20th century (Carbone and Arnaudi, 1930; Chester, 1933; Gäumann, 1946). Some decades later, the phenomenon of induced resistance extending beyond the infected sites of a plant was studied in detail in tobacco and cucumber (Madamanchi and Kuć, 1991; Ross, 1966). The classical experimental system consists of a plant infected on the lower leaf with a necrotizing pathogen that induces a resistance response in the upper leaf toward the same or other pathogens. This resistance is referred to as systemic acquired resistance (SAR) and occurs in many di- and monocotyledonous species (Sticher et al., 1997).

The broad systemic response to pathogens and the transmission of a systemic signal are both spectacular and intriguing features of SAR. The induction of SAR by pathogens is a complex process. Elicitors released at the site of infection are recognized by corresponding plant receptors; this leads to modifications in ion homeostasis, production of reactive oxygen species (ROS), and numerous phosphorylation events (Dangl and Jones, 2001). These changes activate a signaling network leading to transcriptional events involved in various aspects of local and SAR responses. A putative signal released from the infected leaf moves to other parts of the plant where it induces defense reactions. Interestingly, besides localized infection by pathogens, colonization of roots with nonpathogenic bacteria can also induce resistance in leaves (Pieterse and van Loon, 1999; Van Loon et al., 1998). Furthermore, localized viral infections can lead to the systemic induction of post-transcriptional gene silencing, a defense mechanism to subsequent viral infections (Waterhouse et al., 2001). Environmental factors such as light or UV irradiation can also have an important impact on SAR (Genoud et al., 2002; Islam et al., 1998; Mercier et al., 2001).

SAR and its broad spectrum of protection inspired researchers to use this phenomenon for novel approaches in plant protection. For instance, nonantibiotic molecules were identified that can induce SAR on various plants under field conditions (Friedrich et al., 1996; Görlach et al., 1996; Métraux et al., 1991). The molecular responses induced during SAR also became an important target for many groups. For example, a set of proteins termed pathogenesis-related or PR-proteins and their associated genes were discovered that are locally and systemically induced in response to elicitors (Van Loon and Van Strien, 1999). Some of these PRs have antibacterial or antifungal activities, indicating a role in pathogen defense. The number of defense-related genes is much wider than originally thought, as shown by genome-wide analyses (Maleck et al., 2000; Schenk et al., 2000). Exogenously applied salicylic acid (SA) was first shown in tobacco to induce PRs and to protect against tobacco mosaic virus. Later, SA was found in plants after pathogen infection, locally and systemically, making SA an endogenous signal for SAR (reviewed in Sticher et al., 1997).

SA is found in many species and can regulate such diverse physiological processes such as thermogenesis, flowering or defense against pathogens (reviewed in Raskin, 1992). Strong correlations were found between induced resistance and endogenous SA accumulation in plant tissue after a localized pathogen infection (reviewed in Sticher et al., 1997). Further support for the importance of SA for SAR came from studies with mutants and transgenic plants that exhibit altered levels of SA. In general, plants with low endogenous SA are impaired in SAR. Conversely, mutants with constitutive high levels of SA exhibit increased tolerance to pathogens (reviewed in Métraux and Durner, 2002). Besides SA, other endogenous molecules have been identified as signals involved in the activation of resistance responses that are SA-independent. These compounds include octadecanoic acid derivatives such as jasmonic acid (JA), methyl jasmonate (MeJA), 12-oxo-phytodienoic acid (OPDA), and ethylene (ET). Interestingly, it was shown in *Arabidopsis thaliana* that SA-dependent responses can provide resistance to a defined spectrum of pathogens only (such as *Peronospora parasitica* or *Pseudomonas syringae*) while JA- and ET-dependent resistance responses seem to operate against another group (*Alternaria brassicicola, Botrytis cinerea*) (Thomma et al., 1998). Thus, a pathogen attack does not trigger a central SA-dependent cascade of reactions leading to the activation of a single set of resistance mechanisms but rather activates a complex network dependent on multiple signals, of which SA is one (Thomma et al., 1998, 2001). Some branches of this network crosstalk with each other, or interfere with pathways triggered by environmental stimuli such as light (Genoud et al., 2002). This increases the flexibility of the network to optimize the defensive reactions of the plant to a given environment. A digital approach based on Boolean logic was proposed to represent such a complex network (Genoud et al., 2001, 2002).

This chapter will focus on our state of knowledge on the biosynthesis and metabolism of SA, the various roles of SA in defense responses, SA-dependent signaling, and the SA-induced defense signaling network.

7.2 Biosynthesis and Metabolism of Salicylic Acid

Several studies have shown that SA derives from the shikimate-phenylpropanoid pathway (reviewed in Sticher et al., 1997). Depending on the species or tissues, two routes from phenylalanine to SA have been described that differ at the hydroxylation of the aromatic ring. Phenylalanine ammonia lyase (PAL) converts phenylalanine (Phe) to cinnamic acid (CA) that can be hydroxylated to form *ortho*-coumaric acid followed by oxidation of the side chain to yield SA. Alternatively, SA results from an oxidation of the side chain of CA to form benzoic acid (BA) that is hydroxylated in the *ortho* position (reviewed in Sticher et al., 1997). In tobacco, SA was postulated to be synthesized from free BA (Yalpani et al., 1993), and recent results indicate that benzoyl glucose, a conjugated form of BA, is the direct precursor of SA (Chong et al., 2001). In cucumber, potato, and rice SA is likely to derive from phenylalanine via CA and BA but the exclusive role of this route in pathogen-induced SA was never fully assessed (reviewed in Sticher et al., 1997).

Arabidopsis thaliana also produces SA locally and systemically after pathogen infection or treatment with UV-C light (Nawrath and Métraux, 1999; Summermatter et al., 1995). In *Arabidopsis*, inhibitor studies with 2-aminoindan-2-phosphonic acid (AIP), an inhibitor of PAL, indicate that the biosynthetic pathway of SA is derived from Phe and CA. AIP-treated plants have lower amounts of SA and are susceptible to *P. parasitica* (Mauch-Mani and Slusarenko, 1996). The SA-induction deficient (*sid1* and *sid2*) mutants are unable to accumulate SA and to express SAR after an infection (Nawrath and Métraux, 1999). The *sid2* mutation was localized to a gene, *ICS*, encoding isochorismate synthase (ICS) (Wildermuth et al., 2001). ICS1 includes a chorismate-binding domain. It shares 57% amino acid identity with a *Catharanthus roseus* ICS (Van Tegelen et al., 1999) and 20% identity with the bacterial ICS, and both proteins have confirmed biochemical activities (Serino et al., 1995; Wildermuth et al., 2001). The ICS1 gene is induced locally and systemically upon localized pathogen infection (Wildermuth et al., 2001). This demonstrates that SA produced by ICS is required for SAR in *Arabidopsis*. An explanation is now needed to explain the discrepancy between these results from studies with AIP-treated plants (Mauch-Mani and Slusarenko, 1996). ESTs for ICS have been annotated in soybean and tomato, making it likely that many higher plants produce pathogen-induced SA from isochorismate (Wildermuth et al., 2001). The presence of a plastid transit peptide and cleavage site in the *ICS1* gene indicates a plastid-localized synthesis of SA. Possibly, the SA pathway in *Arabidopsis* might share common ancestry with prokaryotic endosymbionts (Wildermuth et al., 2001). The presence of W-box elements in the promoter of *ICS1* suggests that WRKY transcription factors may regulate the response to pathogens or stress (Eulgem et al., 2000). The *ICS1* promoter also includes a binding site for Myb transcription factors that regulate genes for plant defense and associated secondary metabolism (Bender and Fink, 1998; Yang and Klessig, 1996). Interestingly, neither bZIP nor NF-κB motifs, typically required for the induction of *PR1* by SA, were found in the promoter of *ICS1,* suggesting a SA-independent regulation after pathogen infection (Cao et al., 1997; Ryals et al., 1997). Indeed, wild-type expression levels of *ICS1* are observed in SA-depleted NahG plants (Wildermuth et al., 2001). Therefore, the expression of *ICS1* is likely to be under the control of a signal other than SA.

Although the site of action of SA is not known, evidence from transgenic plants expressing the *NahG* gene in the cytoplasm (Delaney et al., 1994) supports either a cytoplasmic location or at least a traffic of SA through this compartment. Interestingly, another SA-induction deficient mutant, *eds5/sid1*, was used to identify a membrane protein homologous to the bacterial multidrug and toxin extrusion (MATE) proteins (Brown et al., 1999; Nawrath et al., 2002). MATEs have recently been reported in *Arabidopsis* (Brown et al., 1999; Debeaujon et al., 2001; Diener et al., 2001; Nawrath et al., 2002). It will now be very interesting to learn more on the nature of the substrate(s) transported by EDS5/SID1.

The relative importance of CA- and ICS-derived SA for the induction of SAR needs to be investigated, since the isochorismate pathway might not be unique for *Arabidopsis* (Wildermuth et al., 2001). If both pathways really coexist in a same species, specific stimuli might selectively induce SA by one or the other pathway. In *Arabidopsis*, virulent or avirulent pathogens, ozone stress, or callus formation lead to high levels of SA while wild-type levels of SA are observed in *sid2* mutants that have an inactivated ICS (Nawrath and Métraux, 1999). This supports a unique ICS-derived pathway for pathogen, ozone and callus-induced SA formation. Possibly, wild-type basal levels of SA might derive from the CA pathway. Another source of the basal levels of SA was proposed to result from the action of a second *ICS* gene (*ICS2*), the transcripts of which remain undetected in infected or uninfected leaves of *Arabidopsis* (Wildermuth et al., 2001). Clearly, the function and regulation of CA- and ICS-derived SA needs to be clarified in *Arabidopsis* and other species where CA was proposed as a main precursor for pathogen-induced SA.

SA is also present as a conjugate, either in methylated, hydroxylated, or glycosylated form. In tobacco, volatile methyl salicylate (MeSA) is produced from SA after infection. Interestingly, MeSA can induce defense reactions upon conversion to SA (Seskar et al., 1997; Shulaev et al., 1997). It was proposed to be additive to SA for signaling within a plant and to act as a signal for communication between plants (Shulaev et al., 1997). In tobacco, a predominant and stable SA metabolite is SA-2-*O*-β-D-glucoside (SAG). The ester glucoside (GSA) was also found in tobacco (Enyedi et al., 1992). GSA was observed to accumulate rapidly and transiently after SA application (Lee and Raskin, 1998). GSA was proposed to protect the plant against phytotoxicity of high SA levels, while SAG might represent a slow release form of SA (Lee and Raskin, 1999). A UDP:glucose:SA glucosyltransferase (SAGTase) was isolated from tobacco and oats that can form both SAG and GSA (Edwards, 1994; Lee and Raskin, 1999). The tobacco SAG-Tase has a broad specificity for simple phenolics and its mRNA is rapidly induced upon SA treatment or inoculation with incompatible pathogens (Lee and Raskin, 1999).

7.3 Salicylic Acid and Its Various Roles in Resistance

Endogenous SA, or application of SA, or functional analogs such as BTH (benzo-(1,2,3)-thiadiazole-7-carbothioic acid S-methyl ester; BION®, ACTIGARD®) and INA (2,6-dichloroisonicotinic acid) induce the expression of a set of PR-proteins such as PR1, PR2, and PR5, the expression of which correlates with resistance (Métraux et al., 1991, Uknes et al., 1992, Ward et al., 1991). Interestingly, while some PRs have an antimicrobial activity *in vitro* and were proposed to act similarly *in planta* (reviewed in Punja, 2001) the biological function of PR1, one of the best markers for SAR, is still unknown. Some situations were also described where the induction of some PRs could be dissociated from the action of SA (Nawrath and Métraux 1999; Schaller et al., 2000). In *Arabidopsis* undergoing SAR, 31 genes linked to SAR cluster together with *PR1* (Maleck et al., 2000). This typical defense gene expression pattern is lost in SA-degrading NahG plants (Delaney et al., 1994; Gaffney et al., 1993; Maleck et al., 2000), as well as in mutants blocked in SA biosynthesis (Nawrath and Métraux, 1999). So far, it was tacitly assumed that NahG plants are only affected in SA accumulation. Several studies indicate more complex modifications that could in some cases influence the interpretation of the phenotype observed in NahG plants (Cameron, 2000; Lieberherr et al., 2003; Nawrath and Métraux 1999; Heck et al., 2003; Van Wees and Glazebrook, 2003).

SA also promotes or inhibits cell death depending on the plant pathogen inter-action, environmental conditions, and genetic background of the plant cell (Green-berg et al., 2000). In *Arabidopsis*, many mutants with constitutive high PR1 ex-pression and enhanced resistance form spontaneously HR-like lesions (Dietrich et al., 1994; Greenberg, et al., 1994; Weymann et al., 1995). In some mutants, SA-accumulation and SAR gene expression are only necessary for disease resis-tance, but not for lesion formation, i.e., in *lsd2* and *lsd4* (Hunt et al., 1997). In other mutants, expression of the *NahG* gene suppresses lesion formation as well as disease resistance, e.g., in *lsd6*, *lsd7*, and *ssi1* (Weymann et al., 1995; Shah et al., 1999; Greenberg et al., 2000). SA-dependent cell death has also been ob-served in tobacco expressing the *Cf-9* gene of tomato together with the avirulence gene *Avr9* of *P. syringae* (Hammond-Kosack et al., 1998) as well as in soybean cell cultures infected with avirulent *P. syringae* pv. *glycinea* (Shirasu et al., 1997). In TMV-infected tobacco, the expression of NahG delays the development of the HR (Mur *et al.*, 1997) and attenuates the oxidative burst after inoculation with avirulent bacteria (Mur et al., 2000).

SA-dependent cell death may also be caused by cellular dysfunction associated with superoxide production (Broderson et al., 2002; Jabs et al., 1996; Kliebenstein et al., 1999). For example, superoxide production leads to runaway cell death in the *lsd1* mutant. This might be caused by a defect in the GATA-type transcription factor LSD1 that activates the expression of a Cu/Zn superoxide dismutase (Dietrich et al., 1997). In the *snc1* mutant, an unknown additional factor besides SA was found to be needed for cell death (Li et al., 2001). In some *Arabidopsis* mutants the lesion formation are uncoupled from SA production and SAR. This is the case in *dnd1*, *dnd2, and hrl1* that do not develop HR-like lesions while SA accumulation and SAR remain intact (Yu et al., 1998). In other *Arabidopsis* mutants, e.g., the *acd5* and *ddl1* mutants, SA accumulation, cell death, and disease resistance are uncoupled from each other (Greenberg et al., 2000; Pilloff et al., 2002). For example, SA or BTH induces cell death leading to an increased susceptibility to *P. syringae* and endogenous SA accumulation does not lead to SAR in *acd5* (Greenberg et al., 2000).

The prominent effect of SA on gene expression led many investigators to study its molecular mode of action. SA is unlikely to interact directly with a target site at the promoter of induced genes. Therefore, a search for protein binding sites with high affinity for SA led to the enzyme catalase (Chen et al., 1993). Binding and associated inactivation of catalase was proposed to increase intracellular H_2O_2 that could activate defense gene expression or act as an antimicrobial barrier at the site of invasion (Chen et al., 1993). This catalase inhibition hypothesis was seriously questioned (reviewed in Mauch-Mani and Métraux, 1998). SA was proposed to affect the redox status of the cells. The ability of SA to form free radicals upon inhibition of heme-containing enzymes such as peroxidase or catalase led to the "free radical" hypothesis of SA action (Durner and Klessig, 1995, 1996). Phenolic free radicals can be potent initiators of lipid peroxidation, the products of which might activate defense reactions (Farmer et al., 1998). It remains to be demonstrated that sufficient free radicals are produced in the correct time and space frames to induce defense responses. A novel protein was also found to exhibit high affinity for SA, but its relevance for the induction of SA-dependent resistance has never been completely assessed (Du and Klessig, 1997).

Another aspect of the molecular action of SA is based on its possible involvement in phosphorylation cascades. MAP kinases (MAPKs) typically compose modules of signaling equivalent to the bacterial signal-integrating phosphorelays, which are characterized by a sequence of reversible phosphorylations of the MAPK by MAPK kinases (MAPKK), subsequent to the phosphorylation of MAPKK by MAPKK kinases (MAPKKK) (Nürnberger and Scheel, 2001; Romeis et al., 2000; Wrzaczek and Hirt, 2001; Zhang and Klessig, 2001). The three successive phosphorylation events are locally assisted by a scaffold protein (see for instance Xing et al., 2002), that may also contribute to precisely target the signaling (amplifier) module to a specific location in the cell. In eukaryotes such as yeast, this type of signal transduction apparatus acts in combination with specific receptors (such as trimeric-G-coupled receptors) in the transmission of external stimuli and can be the site of crosstalk modulation by a different perceptive pathway. In plants, SA induces the activity of a protein kinase (referred to as SA-induced protein kinase, SIPK) belonging to the MAP kinase family (Zhang and Klessig, 1997). SIPK was proposed to initiate or be part of a more complex signaling cascade for the induction of defense reactions. In tobacco, the MAPKK NtMEK2 activates SIPK. This is followed by a hypersensitive reaction (HR)-like cell death and activation of the expression of 3-hydroxy-3-methylglutaryl CoA reductase (HMGR) and L-phenylalanine ammonia-lyase (PAL), two genes encoding key enzymes of the biosynthesis of defense-related phenolics (Yang et al., 2001). Unexpectedly, SA is not involved in the NtMEK2-mediated activation of HR (Yang et al., 2001), indicating the existence of alternative signaling cascades for SA. The existence of different MAPK cascades was also inferred from the study of the flagellin cascade in *Arabidopsis* (Asai et al., 2002). In *Arabidopsis*, H_2O_2 activates the MAPKKK ANP1 that activates the SIPK analogs AtMPK3 and AtMPK6, apparently without the implication of SA (Kovtun et al., 1998). In summary, while the activation of MAPKs by SA has been reported in some instances, many studies suggest that kinase cascades can operate without SA. Presumably, such signaling cascades would precede downstream defense responses, some of which are SA-dependent.

A possible molecular action of SA was also considered in relation to priming. This hypothesis proposes that SAR-derived signals prime or condition the plant tissue to react with a faster and more intense induction of defense reactions after an infection. Support for a role of SA in priming was first obtained in elicitor-treated cultured parsley cells (Conrath et al., 2002). The defense responses that can be primed by SA or functional analogs include the oxidative burst, the HR, the production of phenolic compounds, lignin-like polymers or phytoalexins, or the expression of defense-related genes (Conrath et al., 2002). Priming has also been observed in whole plants. *Arabidopsis* pretreated with pathogens or BTH shows an increase in the sensitivity to *P. syringae*-induced activation of the PAL gene and callose deposition, two reactions that are not induced by BTH alone (Kohler et al., 2002). Priming by BTH and pathogen infection for resistance to *P. syringae* requires the activity of the *NIM1/NPR1* gene (Kohler et al., 2002). Interestingly, in *Arabidopsis* the BTH-primed PAL expression and callose deposition could also be induced after wounding or infiltration of leaves with water, indicating that priming may be a point of crosstalk between the response to pathogens and wounding or osmotic stress (Kohler et al., 2002). The nonprotein amino acid β-aminobutyric acid (BABA) protects *Arabidopsis* from infection with *Peronospora parasitica*. BABA acts by potentiating the tissue to a stronger deposition of callose-containing papillae at the fungal infection sites. In response to infection with virulent *P. syringae*, the effect of BABA manifests itself by a potentiation of the induction of PR1 (Zimmerli et al., 2000). Interestingly, the effect of BABA against P. *parasitica* is independent of the SA, JA, and ethylene signaling pathways, whereas BABA potentiation to *P. syringae* is dependent on SA signaling (Zimmerli et al., 2000). Future experiments should elucidate the molecular mode of action of SA in priming of defense responses.

The involvement of SA as a systemic mobile signal was also repeatedly explored. Since SA was detected in the phloem sap, it was initially proposed as the primary signal for SAR that moves from the infected to the uninfected parts of the plant (Malamy et al., 1990; Métraux et al., 1990). However, grafting and leaf excision experiments indicate that while SA is a necessary component for the induction of local and systemic resistance, it is not the primary mobile signal exported from the infected leaf to other parts of the plant (reviewed in Mauch-Mani and Métraux, 1998). Radiolabeling experiments showed that SA synthesized after inoculation can be transported from the infected to the upper leaves by the phloem before resistance was detectable (Mölders et al., 1996; Shulaev et al., 1995). These results might not be incompatible: SA produced in high amounts at infection sites could be translocated together with another primary mobile signal and induce resistance in the distal leaves. Progress in the search for a phloem-mobile signal was recently made using the *Arabidopsis dir1-1* mutant defective in systemic but not in local induced resistance. *DIR1* encodes a putative apoplastic lipid-transfer protein (Maldonado et al., 2002). Analyses of phloem exudates indicate that *dir1-1* plants are missing an essential mobile signal. The authors propose that DIR1 interacts with a lipid-derived molecule to promote long distance signaling.

SA was also found to be involved in the signal transduction pathway for virus resistance. In tobacco or in *Arabidopsis*, SA inhibits the replication or the movement of several RNA viruses, independently of SA-induced PR proteins (Chivasa et al., 1997; Murphy et al., 1999; Murphy and Carr, 2002; Naylor et al., 1998; Wong et al., 2002). In tobacco and *Arabidopsis*, SA-mediated resistance can be induced by cyanide and the mitochondrial electron transport inhibitor antimycin A (AA) or inhibited by salicylhydroxamide acid, suggesting a role of the mitochondrial alternative oxidase (AOX) in virus resistance by an action on the level of ROS in the cell (Maxwell et al., 1999; Murphy and Carr, 2002). AA, H_2O_2, and SA disrupt the normal cytochrome-dependent functions of the mitochondria, lowering the ATP levels and increasing the formation of ROS and AOX (Maxwell et al., 1999; Maxwell et al., 2002). AOX is also induced by pathogen attack, indicating that the same mechanism may act after virus infection (Simons et al., 1999). In addition, plant cells treated with the AA, SA, and H_2O_2 specifically express genes that are involved in programmed cell death. This supports the hypothesis that mitochondria transduce intracellular stress signals to the nucleus, leading to altered defense gene expression (Maxwell et al., 2002).

7.4 Regulation of the SA-Dependent Pathway Leading to PR-Gene Expression

An important element of the signal transduction pathway linking SA to defense responses is the ankyrin-repeat containing protein NPR1 (NON-expressor of PR)/NIM1 (NON-immunity) (Ryals et al., 1997; Cao et al., 1997). NPR1 function is essential for the induction of SAR by pathogens or SAR-inducers, for disease limitation after infection with virulent pathogens as well as for priming (Conrath et al., 2002). Race-specific resistance is modified by *NPR1* in some cases only (Cao et al., 1997; Delaney et al., 1995; Rate and Greenberg, 2001; Rairdan and Delaney, 2002). *NPR1* was found to control certain SA-dependent processes related to cell death and cell growth (Vanacker et al., 2001; Greenberg, 2000). In addition, NPR1 can act in a SA-independent pathway leading to ISR (Pieterse et al., 1998).

NPR1 is localized in the cytoplasm in the absence of SA and locates to the nucleus in the presence of SA, where it may act as transcriptional coactivator in a protein complex (Kinkema et al., 2000; Weigel et al., 2001). NPR1 interacts with members of the TGA family of β-ZIP transcription factors (Deprès et al., 2000; Fan and Dong, 2002; Zhang et al., 1999; Zhou et al., 2000) that may regulate SAR positively or negatively (Lebel et al., 1998; Pontier et al., 2001). However, not all NPR1-dependent genes that consistently cluster with *PR1* in microarray experiments have TGA factor binding sites. In fact, the WRKY factor binding site is the overrepresented promoter element in the *PR1* gene cluster (Maleck et al., 2000).

The *NPR1* gene is induced after pathogen infection or SA treatment via SA-inducible members of the family of WRKY DNA-binding proteins (Robatzek and Somssich, 2002; Yu et al., 2001). Overexpression of the WRKY18 transcription factor leads to a constitutive increase of PR-protein expression that causes detrimental effects to plant growth (Chen and Chen, 2002; Robatzek and Somssich, 2002). In contrast, overexpression of *NPR1* itself leads to enhanced resistance to *P. syringae* and *P. parasitica* without leading to constitutive PR protein expression and detrimental effects (Cao et al., 1998; Friedrich et al., 2001). *NPR1* overexpression also results in an enhanced effectiveness of fungicides making concepts for combination of transgenic and chemical approaches for durable resistance attractive (Friedrich et al., 2001). Interestingly, overexpression of the *Arabidopsis NPR1* gene in rice leads to rice blast resistance, indicating that the signal transduction pathway of disease resistance is conserved between monocots and dicots (Chern et al., 2001). The search for suppressors of *NPR1/NIM1* identified the novel nucleus-localized SNI1 protein that may act as a negative regulator of SAR in wild-type plants (Li et al., 1999).

Several positive regulators of the SA-dependent pathway have been identified, such as EDS1, PAD4, NDR1, and EDS4. EDS1 and PAD4 are two proteins of unknown function containing a lipase-domain that are essential for the resistance to *P. syringae* and *P. parasitica* mediated by proteins of the TIR-NB-LRR resistance proteins (Falk et al., 1999; Feys et al., 2001; Jirage et al., 1999). The regulation of SA accumulation might require an interaction of EDS1 with PAD4 (Feys and Parker, 2000). EDS1 is necessary for the transcriptional regulation of PAD4 and both proteins are necessary for the expression of *EDS5* leading to accumulation of SA after pathogen attack and exposure to UV-C light (Nawrath and Métraux, 1999, Zhou et al., 1998). The expression of *EDS1* and *PAD4* can also be upregulated by SA; and a positive feedback loop was postulated to amplify the SA pathway (Falk et al., 1999; Jirage et al., 1999).

NDR1, a small protein containing a membrane-spanning domain, is required for resistance mediated by most R-genes of the CC-NB-LRR class (Century et al., 1997; Aarts et al., 1998). Thus, NDR1 defines a different pathway than EDS1. NDR1 contributes quantitatively to resistance depending on the respective R-gene. For example, the ability to induce cell death depends strongly on NDR1 when the RPS2 pathway is triggered; this dependence is weaker when the RPM1 pathway is activated (Century et al., 1997; Tornero et al., 2002). A link between ROS and SA production was observed in the *ndr1* mutant: SA accumulation and SAR are impaired in *ndr1* after inoculation with *P. syringae* carrying the *avrRpt2* gene, or after treatment with ROS (Shapiro and Zhang, 2001).

Negative regulators of the SA-pathway may be identified among the large number of mutants that have constitutive PR1 or PR2 expression, high levels of SA, and an increased resistance to virulent strains of *P. syringae* and *P. parasitica*. In general, these mutants are smaller than wild-type plants and many of them also develop spontaneously HR-like lesions, as reviewed in Métraux and Durner, (2004). For example, CPR proteins act at the beginning of the SA-signaling cascade upstream of EDS1 and PAD4 and regulate defense pathway in different ways, i.e., the dwarfism may be dependent on SA, as in *cpr1*, or independent of SA, as in *cpr6* (Clarke et al., 2000; Jirage et al., 2001; Clarke et al., 2001). CPR5 also acts in the senescence pathway as well as in trichome development and has thus a very pleiotropic effect, possibly leading to plant defense only indirectly (Bowling et al., 1997; Boch et al., 1998; Kirik et al., 2001; Yoshida et al., 2002).

EDR1, a MAPKKK of the CTR1 family, is likely to function at the top of a MAP kinase cascade that negatively regulates SA-inducible defense response upstream of EDS1, PAD4 and NPR1 (Frye et al., 1998; 2001). Since the *edr1* mutant does not exhibit constitutive *PR1* expression, EDR1 might be a regulator of the priming response (Conrath et al., 2002).

7.5 The Integration of Salicylic Acid in a Network of Signal Processing

Besides SA, the phytohormones jasmonic acid (JA) and ethylene (ET) are two of the most important signaling molecules involved in defense-related responses. They are also involved in the expression of wound-responsive (WR) genes, some of which are likely to have protective properties against microbial infection. JA and ET mediate a variety of pathways that exhibit multiple forms of crosstalk interactions (reviewed in Pieterse and van Loon, 1999; Genoud and Métraux, 1999; Feys and Parker, 2000; Pieterse et al., this volume). For example, a concomitant activation of the JA and ET pathways is required in *Arabidopsis* for the induction of the antifungal plant defensin gene *PDF1.2* (Penninckx et al., 1998). The SA pathway also exhibits different types of crosstalks with the JA/ET pathways (reviewed in Reymond and Farmer, 1998; Genoud and Métraux, 1999; Genoud et al., 2001). The *Arabidopsis cpr5* and *cpr6* mutants, which have elevated levels of SA and express SAR constitutively, also express marker genes from the JA pathway (Bowling et al., 1997; Clarke et al., 1998). CPR5 and CPR6 regulate resistance through distinct pathways, and SA-mediated, NPR1-independent resistance involves components of the JA/ET-mediated pathways (Clarke et al., 2000). Similarly, the *ssi1* mutation, which bypasses the requirement of NPR1 for SAR function, makes the expression of PDF1.2 SA-dependent (Shah et al., 1999). Also, in *Arabidopsis*, the *eds4* and *pad4* mutations cause reduced SA levels in plants that exhibit a heightened response to inducers of JA-dependent gene expression (Gupta et al., 2000). Another form of crosstalk was observed in the *hrl1* mutant, where the expression of *PDF1.2* is rendered partially *NPR1-* and SA/BTH-dependent. In *hrl1*, ET plays an essential role for the systemic expression of *PR1* and resistance to *P. syringae*, and an impairment in JA-signaling leads to exaggerated cell death and strong dwarfism (Devadas and Raina, 2002). In addition, a MAP kinase activity of *Arabidopsis* (MPK4) has recently been shown to control the repression of SAR. In the mutant *mpk4* plants, SAR is dependent on elevated SA levels, but is independent of NPR1. Interestingly, the activation of the JA-responsive genes *PDF1.2* and *THI2.1* was blocked in *mpk4* expressing NahG, suggesting the requirement of MPK4 in JA-responsive gene expression (Petersen et al., 2000).

Plants integrate information simultaneously received from various environmental stimuli, and from the fluctuating context of their organ-specific activities, developmental stage, and metabolic status. The plasticity in the response of the plant to its environment and to internal cues is also achieved through the use of alternative signaling pathways (Genoud and Métraux, 1999). For instance, SA-induced resistance to *P. syringae* is compromised in *eds4 Arabidopsis* plants when grown at 22°C and 85% relative humidity, but not when grown at 23°C and 50% relative humidity (Gupta et al., 2000). Interestingly, several targets of nitric oxide (NO) in animals, including guanylate cyclase and MAPKs (e.g., SIPK), are also modulated by NO in plants. This observation suggests that a crosstalk exists between a potential NO-signaling pathway and the SA pathway (Klessig et al., 2000).

Data from microarray analysis have recently proven to be invaluable to characterize *Arabidopsis* plants in the context of different environmental and developmental scenarios. Using a microarray prepared with 2,375 expressed sequence tags (ESTs) with a biased representation of putative defense-associated and regulatory genes, Schenk et al. (2000) characterized their expression levels in the plant after inoculation with an incompatible fungal pathogen, or treatment with SA, methyl-jasmonate (Me-JA) (a biologically active JA derivative), or ET. A substantial change in the steady-state abundance of 705 mRNAs was observed, out of which 169 genes were regulated by multiple treatments, with the largest number of coinduced or corepressed genes being responsive both to SA and Me-JA. In a recent study, Chen et al. (2002) confirm that SA- and JA/ET-pathways interact diversely (positively and negatively) to induce the expression or repression of transcription factors in *Arabidopsis* upon infection with bacterial pathogens (of the *Pseudomonas* species). In a related experiment, Maleck et al. (2000) examined transcriptional changes associated with the induction or maintenance of SAR by using a DNA microarray representing approximately 7,000 genes. Gene activity patterns were compared under 14 different SAR-inducing or SAR-repressing conditions; 413 ESTs exhibited differential expression equal to or greater than 2.5-fold in at least two SAR-relevant samples. Two different algorithms were used to generate a hierarchical "clustergram" and "self-organizing maps" (SOMs) to define groups of coregulated genes (Maleck et al., 2000). For instance, a molecular marker for the PR1 gene clustered in SOM c1, which contained 45 ESTs (from a maximum of 31 genes), suggesting that the genes in this regulon function in SAR. Significantly, these genes showed a unique expression profile, being strongly activated in secondary SAR tissue and dependent on NIM1/NPR1/SAI1. Furthermore, the only cis-acting regulatory element present in all known promoters from SOM c1 is the binding site for WRKY transcription factors (W boxes: TTGAC). The authors proposed that NIM1/NPR1/SAI1 may mediate a WRKY-dependent derepression of PR1 regulon genes, or alternatively, that it may drive early expression of a subset of WRKY proteins that subsequently regulate other WRKY-dependent SAR target genes.

Such microarray-based studies illustrate the power of this technique for the analysis of complex signal transduction networks. Clearly, as this and other type of large-scale approaches are further exploited to elucidate the mechanisms controlling gene expression, it is necessary to simultaneously develop appropriate computational-based systems that will enable accurate integration and representation of the increasing amount of data being generated (Genoud et al., 2001).

It is also known that a crosstalk between the light signal transduction and the PR gene signaling pathways occurs in several plants. For instance, recent studies with *Arabidopsis* and maize mutants developing spontaneously HR lesions, and transgenic tomato expressing the R gene *Pto*, have suggested that light critically influences the formation of defensive cell death in plants (Dietrich et al., 1994; Martienssen, 1997; Tang, et al., 1999). Moreover, the light hypersensitive mutant of *Arabidopsis* (*psi2*) produces HR-like lesions and increased PR1 expression on leaves at high intensity of red light (Genoud et al., 1998). This indicates that a crosstalk exists between red light and/or far-red light perception and PR expression signaling pathways. The *psi2* mutant also exhibits a light-fluence-dependent amplification of SA-induced *PR1* gene expression.

We have confirmed the observations that light regulates sensitivity to SA by scoring the expression of PR genes in mutants containing no detectable phyA and B proteins (*phyA-phyB* double mutants; Genoud et al., 2002). In these plants, the expression of the PR genes elicited by either SA or BTH is strongly reduced, and the mutant's resistance to an ecotype-competent pathogen of the *Pseudomonas* group was significantly attenuated. In addition, the measured SA levels in the different mutants indicate that the endogenous level of SA is not modified by light, further suggesting that phytochrome activity modulates the perception of SA.

Other environmental stimuli have been linked to the control of SA production (i.e., they may modulate the SA-pathway upstream of SA production). In tobacco, ultraviolet (UV)-C light or ozone mimic the effect of necrotizing pathogens, inducing a transient increase in SA, in both exposed and unexposed leaves of the plants (Yalpani et al., 1994). This accumulation of SA is paralleled by a higher production of SA conjugate, also by the activation of a benzoic acid 2- hydroxylase, and by an accumulation of PR1. In correlation, an elevated SAR to a subsequent challenge with tobacco mosaic virus (TMV) has been observed. Hence, UV light, ozone fumigation, and TMV activate common, or redundant, signaling pathways leading to SA and PR-protein accumulation and SAR. As partial confirmation of these results, both UV-C and ozone treatment strongly induce the accumulation of SA and SA-conjugate in *Arabidopsis* (Nawrath and Métraux, 1999). Ozone- and superoxide-induced ROS and cell death are differently controlled by JA and ET, as shown in a description of an ozone-sensitive mutant of *Arabidopsis* (*rcd1*;

Overmyer et al., 2000). ET perception and signaling promote ozone-activated cell death while JA signaling might be responsible for the lesion containment. Thus, JA, ET, and SA might contribute to the response of plants submitted to high ozone exposure.

In barley, SA and aspirin were found to induce the accumulation of glycine betaine, an osmoprotectant produced in response to cold, drought, and osmotic stress (Jagendorf and Takabe, 2001), and SA added to the hydroponic growth solution of young maize plants under normal growth conditions provides protection against subsequent low-temperature stress. This last effect might result from the induction of antioxidative enzymes that lead to chilling resistance (Janda et al., 1999). In tobacco cells, two MAPKs, identified as SIPKs (SA-induced protein kinase) are activated in response to salt-induced hyperosmotic stress. One of these SIPKs is a 40 kD protein, that is specific for the hyperosmotic stress and is Ca^{2+}-and abscisic acid (ABA)-independent (Hoyos and Zhang, 2000), therefore the MAP kinase system could play the role of connecting the salt- and the SA-pathway. The interaction between ABA and SA is likely to differ depending on the branches of the pathways that interact, and also in function of the plant species. For instance, ABA suppresses the SA-dependent defense in tomato (Audenaert et al., 2002) and determines the basal susceptibility to *B. cinerea*. In the reaction controlling the protection against heat-stress in *Arabidopsis*, both ABA and SA (together with ET) have been shown to induce protective antioxidants (Larkindale and Knight, 2002). This has been observed in physiological experiments where ABA-insensitive mutant *abi1*, ethylene-insensitive mutant *etr1*, and SA-deficient plant NahG presented a reduction in heat-shock-induced antioxidant production with a correlated decrease in survival. The application of SA, of an ET generating substance, or ABA, have been shown to stimulate the survival of plants exposed to heat-shock; since calcium mimics this effect, Larkindale and Knight (2002) suggest that these crosstalks might be regulated by calcium signals.

7.6 Conclusions and Perspectives

Research on the role of SA in plants has witnessed a steady increase in interest since the first publications on the possible role of SA in the regulation of SAR in the early 1990s. Since then, the number of yearly publications on SA research has followed an increase that does not appear to slow down. This results from a wide recognition of the fundamental role of SA in plant defense and many aspects of its complex mode of action are keenly investigated.

Turning toward the future, breakthroughs will include the identification and characterization of additional signaling components in the SA pathway. For example, one target of research will be the regulatory process that controls the local and distal levels of SA. Another target will undoubtedly be the mode of action of SA itself, its putative binding site and the responses thereof. The response of plants to pathogens is far from a linear cascade of events but constitutes a complex network that integrates information from the internal and external plant environment. The exploration of the properties of this network will be another major area of investigation. This approach will combine results of genome-wide expression analysis, proteomics, metabolomics, mutant studies, as well as bioinformatics. We foresee that computer simulations will be increasingly used to obtain a comprehensive overview of the results.

The advances in this fundamental knowledge will also have an important impact on agronomy. Discoveries of novel genes involved in various aspects of resistance will direct the conventional selection procedures toward new varieties with improved properties. Expression of such genes under inducible promoters will eventually allow the regulation of pathways for various defense reactions alone or in combination. The results obtained from studies of the network of resistance will establish the parameters to be taken into account and to be optimized in order to induce resistance using chemical inducers. Selection of biocontrol bacterial strains that enhance induced resistance of the plant will also profit from the knowledge on the network of information operating in the plant during interactions with pathogens. In summary, research on SA and plant defense will undoubtedly undergo very exciting developments both in our understanding of the related molecular and physiological processes, as well as in the direct or indirect application of this knowledge in agronomy.

References

Aarts, N., Metz, M., Holub, E., Staskawicz, B.J., Daniels, M.J., and Parker, J.E. 1998. Different requirements for EDS1 and NDR1 by disease resistance genes define at least two R gene-mediated signaling pathways in *Arabidopsis*. *Proc. Natl. Acad. Sci. USA.* 95:10306–10311.

Asai, T., Tena, G., Plotnikova, J., Willmann, M.R., Chiu, W.L., Gomez-Gomez, L., Boller, T., Ausubel, F.M., and Sheen, J. 2002. MAP kinase signaling cascade in *Arabidopsis* innate immunity. *Nature* 415:977–983.

Audenaert, K., De Mezer, G.B., and Höfte, M.M. 2002. Abscisic acid determines basal susceptibility of tomato to *Botrytis cinerea* and suppresses salicylic acid-dependent signaling mechanisms. *Plant Physiol.* 128:491–501.

Bender, J., and Fink, G.R. 1998. A Myb homologue, ATR1, activates tryptophan gene expression in *Arabidopsis*. *Proc. Natl. Acad. Sci. USA* 95:5655–5660.

Boch, J., Verbsky, M.L., Robertson, T.L., Larkin, J.C., and Kunkel, B.N. 1998. Analysis of resistance gene-mediated defense responses in *Arabidopsis thaliana* plants carrying a mutation in CPR5. *Mol. Plant Microbe Interact.* 11:1196–1206.

Bowling, S.A., Clarke, J.D., Liu, Y.D., Klessig, D.F., and Dong, X.N. 1997. The *cpr5* mutant of *Arabidopsis* expresses both NPR1-dependent and NPR1-independent resistance. *Plant Cell* 9:1573–1584.

Brodersen, P., Petersen, M., Pike, H.M., Olszak, B., Skov, S., Odum, N., Jorgensen, L.B., Brown, R.E., and Mundy, J. 2002. Knockout of *Arabidopsis* ACCELERATED-CELL-DEATH11 encoding a sphingosine transfer protein causes activation of programmed cell death and defense. *Genes Dev.* 16:490–502.

Brown, M.H., Paulsen, I.T., and Skurray, R.A. 1999. The multidrug efflux protein NorM is a prototype of a new family of transporters. *Mol. Microbiol.* 31:394–395.

Cameron, R.K. 2000. Salicylic acid and its role in plant defense responses: what do we really know? *Physiol. Mol. Plant Pathol.* 56:91–93.

Cao, H., Li, X., and Dong, X.N. 1998. Generation of broad-spectrum disease resistance by overexpression of an essential regulatory gene in systemic acquired resistance. *Proc. Natl. Acad. Sci. USA* 95:6531–6536.

Cao, H., Bowling, S.A., Gordon, A.S., and Dong, X.N. 1994. Characterization of an *Arabidopsis* mutant that is non-responsive to inducers of systemic acquired resistance. *Plant Cell* 6:1583–1592.

Cao, H., Glazebrook, J., Clarke, J.D., Volko, S., and Dong, X.N. 1997. The *Arabidopsis* NPR1 gene that controls systemic acquired resistance encodes a novel protein containing ankyrin repeats. *Cell* 88:57–63.

Carbone, D., and Arnaudi C. 1930. L'immunità nelle piante. Monografie dell'Istituto Sieroterapico Milanese, Milano.

Century, K.S., Shapiro, A.D., Repetti, P.P., Dahlbeck, D., Holub, E., and Staskawicz, B.J. 1997. NDR1, a pathogen-induced component required for *Arabidopsis* disease resistance. *Science* 278:1963–1965.

Chen, C.H., and Chen, Z.X. 2002. Potentiation of developmentally regulated plant defense response by AtWRKY18, a pathogen-induced *Arabidopsis* transcription factor. *Plant Physiol.* 129:706–716.

Chen, Z.X., Silva, H., and Klessig, D.F. 1993. Active oxygen species in the induction of plant systemic acquired resistance by salicylic acid. *Science* 262:1883–1886.

Chen, W., Provart, N.J., Glazebrook, J., Katagiri, F., Chang, H.-S., Eulgem, T., Mauch, F., Luan, S., Zou, G., Whitham, S.A., Budworth, P.R., Tao, Y., Xie, Z., Chen, X., Lam, S., Kreps, J.A., Harper, J.F., Si-Ammour, A., Mauch-Mani, B., Heinlein, M., Kobayashi, K., Hohn, T., Dangl, J.L., Wang, X., and Zhu, T. 2002. Expression profile matrix of *Arabidopsis* transcription factor genes suggests their putative functions in response to environmental stresses. *Plant Cell* 14:559–574.

Chern, M.S., Fitzgerald, H.A., Yadav, R.C., Canlas, P.E., Dong, X.N., and Ronald, P.C. 2001. Evidence for a disease resistance pathway in rice similar to the NPR1-mediated signaling pathway in *Arabidopsis*. *Plant J.* 27:101–113.

Chester, K.S. 1933. The problem of acquired physiological immunity in plants. *Q. Rev. Biol.* 8:275–324.

Chivasa, S., Murphy, A.M., Naylor, M., and Carr, J.P. 1997. Salicylic acid interferes with tobacco mosaic virus replication via a novel salicylhydroxamic acid-sensitive mechanism. *Plant Cell* 9:547–557.

Chong, J., Pierrel, M.A., Atanassova, R., Werck-Reithhart, D., Fritig, B., and Saindrenan, P. 2001. Free and conjugated benzoic acid in tobacco plants and cell cultures induced accumulation upon elicitation of defense responses and role as salicylic acid precursors. *Plant Physiol.* 125:318–328.

Clarke, J.D., Liu, Y.D., Klessig, D.F., and Dong, X.N. 1998. Uncoupling PR gene expression from NPR1 and bacterial resistance: characterization of the dominant *Arabidopsis cpr6-1* mutant. *Plant Cell* 10:557–569.

Clarke, J.D., Volko, S.M., Ledford, H., Ausubel, F.M., and Dong, X.N. 2000. Roles of salicylic acid, jasmonic acid, and ethylene in cpr-induced resistance in *Arabidopsis*. *Plant Cell* 12:2175–2190.

Clarke, J.D., Aarts, N., Feys, B.J., Dong, X.N., and Parker, J.E. 2001. Constitutive disease resistance requires EDS1 in the *Arabidopsis* mutants *cpr1* and *cpr6* and is partially EDS1-dependent in *cpr5*. *Plant J.* 26:409–420.

Clough, S.J., Fengler, K.A., Yu, I.C., Lippok, B., Smith, R.K., and Bent, A.F. 2000. The *Arabidopsis dnd1* "defense no death" gene encodes a mutated cyclic nucleotide-gated ion channel. *Proc. Natl. Acad. Sci. USA* 97:9323–9328.

Conrath, U., Pieterse, C.M.J., and Mauch-Mani, B. 2002. Priming in plant-pathogen interactions. *Trends Plant Sci.* 7:210–216.

Dangl, J.L., and Jones, J.D.G. 2001. Plant pathogens and integrated defense responses to infection. *Nature* 411:826–833.

Debeaujon, I., Peeters, A.J.M., Leon-Kloosterziel, K.M., and Koornneef, M. 2001. The *TRANSPARENT TESTA12* gene of *Arabidopsis* encodes a multidrug secondary transporter-like protein required for flavonoid sequestration in vacuoles of the seed coat endothelium. *Plant Cell* 13:853–871.

Delaney, T.P., Uknes, S., Vernooij, B., Fiedrich, L., Weymann, K., Negrotto, D., Gaffney, T., Gut-Rella, M., Kessmann, H., and Ward, E. 1994. A central role of salicylic acid in plant disease resistance. *Science* 266:1247–1249.

Delaney, T.P., Friedrich, L., and Ryals, J.A. 1995. *Arabidopsis* signal-transduction mutant defective in chemically and biologically induced disease resistance. *Proc. Natl. Acad. Sci. USA* 92:6602–6606.

Desprès, C., DeLong, C., Glaze, S., Liu, E., and Fobert, P.R. 2000. The *Arabidopsis* NPR1/NIM1 protein enhances the DNA binding activity of a subgroup of the TGA family of bZIP transcription factors. *Plant Cell* 12:279–290.

Devadas, S.K., and Raina, R. 2002. Preexisting systemic acquired resistance suppresses hypersensitive response-associated cell death in *Arabidopsis hrl1* mutant. *Plant Physiol.* 128:1234–1244.

Diener, A.C., Gaxiola, R.A., and Fink, G.R. 2001. *Arabidopsis ALF5*, a multidrug efflux transporter gene family member, confers resistance to toxins. *Plant Cell* 13:1625–1637.

Dietrich, R.A., Delaney, T.P., Uknes, S.J., Ward, E.J., Ryals, J.A., and Dangl, J.L. 1994. *Arabidopsis* mutants simulating disease resistance response. *Cell* 77:565–578.

Dietrich, R., Richberg, M., Morel, J.B., Dangl, J., and Jabs, T. 1997. An *Arabidopsis* mutant with impaired cell death control. *Eur. J. Cell Biol.* 72:6–16.

Du, H., and Klessig, D.F. 1997. Identification of a soluble, high-affinity salicylic acid-binding protein in tobacco. *Plant Physiol.* 113:1319–1327.

Durner, J., and Klessig, D.F. 1995. Inhibition of ascorbate peroxidase by salicylic acid and 2,6-dichloroisonicotinic acid, two inducers of plant defense responses. *Proc. Natl. Acad. Sci. USA* 92:11312–11316.

Durner, J., and Klessig, D.F. 1996. Salicylic acid is a modulator of tobacco and mammalian catalases. *J. Biol. Chem.* 271:28492–28501.

Edwards, R. 1994. Conjugation and metabolism of salicylic acid in tobacco. *J. Plant Physiol.* 143:609–614.

Enyedi, A.J., Yalpani, N., Silverman, P., and Raskin, I. 1992. Localization, conjugation, and function of salicylic-acid in tobacco during the hypersensitive reaction to tobacco mosaic- virus. *Proc. Natl. Acad. Sci. USA* 89:2480–2484.

Eulgem, T., Rushton, P.J., Robatzek, S., and Somssich, I.E. 2000. The WRKY superfamily of plant transcription factors. *Trends Plant Sci.* 5:199–206.

Falk, A., Feys, B.J., Frost, L.N., Jones, J.D.G., Daniels, M.J., and Parker, J.E. 1999. EDS1, an essential component of R gene-mediated disease resistance in *Arabidopsis* has homology to eukaryotic lipases. *Proc. Natl. Acad. Sci. USA* 96:3292–3297.

Fan, W., and Dong, X.N. 2002. *In vivo* interaction between NPR1 and transcription factor TGA2 leads to salicylic acid-mediated gene activation in *Arabidopsis*. *Plant Cell* 14:1377–1389.

Farmer, E.E., Weber, H., and Vollenweider, S. 1998. Fatty acid signaling in *Arabidopsis*. *Planta* 206:167–174.

Feys, B.J., and Parker, J.E. 2000. Interplay of signaling pathways in plant disease resistance. *Trends Genet.* 16:449–455.

Feys, B.J., Moisan, L.J., Newman, M.A., and Parker, J.E. 2001. Direct interaction between the *Arabidopsis* disease resistance signaling proteins, EDS1 and PAD4. *EMBO J.* 20:5400–5411.

Friedrich, L., Lawton, K., Ruess, W., Masner, P., Specker, N., Rella, M.G., Meier, B., Dincher, S., Staub, T., Uknes, S., Métraux, J.-P., Kessmann, H., and Ryals, J. 1996. A benzothiadiazole derivative induces systemic acquired resistance in tobacco. *Plant J.* 10:61–70.

Friedrich, L., Lawton, K., Dietrich, R., Willits, M., Cade, R., and Ryals, J. 2001. NIM1 overexpression in *Arabidopsis* potentiates plant disease resistance and results in enhanced effectiveness of fungicides. *Mol. Plant Microbe Interact.* 14:1114–1124.

Frye, C.A., and Innes, R.W. 1998. An *Arabidopsis* mutant with enhanced resistance to powdery mildew. *Plant Cell* 10:947–956.

Frye, C.A., Tang, D.Z., and Innes, R.W. 2001. Negative regulation of defense responses in plants by a conserved MAPKK kinase. *Proc. Natl. Acad. Sci. USA* 98:373–378.

Gäumann, E. 1946. Pflanzliche Infektionslehre, Birkhäuser, Basel.

Gaffney, T., Friedrich, L., Vernooij, B., Negrotto, D., Nye, G., Uknes, S., Ward, E., Kessmann, H., and Ryals, J. 1993. Requirement of salicylic acid for the induction of systemic acquired resistance. *Science* 261:754–756.

Genoud, T., and Métraux, J.P. 1999. Crosstalks in plant cell signaling: structure and function of the genetic network. *Trends Plant Sci.* 4:503–507.

Genoud, T., Millar, A.J., Nishizawa, N., Kay, S.A., Schäfer, E., Nagatani, A., and Chua, N.-H. 1998. An *Arabidopsis* mutant hypersensitive to red and far-red light signals. *Plant Cell* 10:889–904.

Genoud, T., Trevino Santa Cruz, M.B., and Métraux, J.P. 2001. Numeric simulation of plant signaling networks. *Plant Physiol.* 126:1430–1437.

Genoud, T., Buchala, A.J., Chua, N.H., and Métraux, J.-P. 2002. Phytochrome signaling modulates the SA-perceptive pathway in *Arabidopsis*. *Plant J.* 31:87–95.

Görlach, J., Volrath, S., Knaufbeiter, G., Hengy, G., Beckhove, U., Kogel, K.H., Oostendorp, M., Staub, T., Ward, E., Kessmann, H., and Ryals, J. 1996. Benzothiadiazole, a novel class of inducers of systemic acquired resistance, activates gene expression and disease resistance in wheat. *Plant Cell* 8:629–643.

Greenberg, J.T., Guo, A.L., Klessig, D.F., and Ausubel, F.M. 1994. Programmed cell death in plants: A pathogen-triggered response activated coordinately with multiple defense functions. *Cell* 77:551–563.

Greenberg, J.T., Silverman, F.P., and Liang, H. 2000. Uncoupling salicylic acid-dependent cell death and defense-related responses from disease resistance in the *Arabidopsis* mutant *acd5*. *Genetics* 156:341–350.

Gupta, V., Willits, M.G., and Glazebrook, J. 2000. *Arabidopsis thaliana* EDS4 contributes to salicylic acid (SA)- dependent expression of defense responses: evidence for inhibition of jasmonic acid signaling by SA. *Mol. Plant Microbe Interact.* 13:503–511.

Hammond-Kosack, K.E., Tang, S.J., Harrison, K., and Jones, J.D.G. 1998. The tomato *Cf-9* disease resistance gene functions in tobacco and potato to confer responsiveness to the fungal avirulence gene product Avr9. *Plant Cell* 10:1251–1266.

Heck, S., Grau, T., Buchala, A., Métraux J.P., and Nawrath C. 2003. Genetic evidence that expression of NahG modifies defence pathways independent of salicylic acid biosynthesis in the *Arabidopsis–Pseudomonas syringae* pv. *tomato*. *Plant J.* 36:342–352.

Hoyos, M.E., and Zhang, S.Q. 2000. Calcium-independent activation of salicylic acid-induced protein kinase and a 40-kilodalton protein kinase by hyperosmotic stress. *Plant Physiol.* 122:1355–1363.

Hunt, M.D., Delaney, T.P., Dietrich, R.A., Weymann, K.B., Dangl, J.L., and Ryals, J.A. 1997. Salicylate-independent lesion formation in *Arabidopsis lsd* mutants. *Mol. Plant Microbe Interact.* 10:531–536.

Islam, S.Z., Honda, Y., and Arase, S. 1998. Light-induced resistance of broad bean against *Botrytis cinerea. J. Phytopathol.-Phytopathol. Z.* 146:479–485.

Jabs, T., Dietrich, R.A., and Dangl, J.L. 1996. Extracellular superoxide is necessary and sufficient for runaway cell death in an *Arabidopsis* mutant. *Plant Physiol.* 111:76–86.

Jagendorf, A.T., and Takabe, T. 2001. Inducers of glycinebetaine synthesis in barley. *Plant Physiol.* 127:1827–1835.

Janda, T., Szalai, G., Tari, I., and Paldi, E. 1999. Hydroponic treatment with salicylic acid decreases the effects of chilling injury in maize (*Zea mays* L.) plants. *Planta* 208:175–180.

Jirage, D., Tootle, T.L., Reuber, T.L., Frost, L.N., Feys, B.J., Parker, J.E., Ausubel, F.M., and Glazebrook, J. 1999. *Arabidopsis thaliana PAD4* encodes a lipase-like gene that is important for salicylic acid signaling. *Proc. Natl. Acad. Sci. USA* 96:13583–13588.

Jirage, D., Zhou, N., Cooper, B., Clarke, J.D., Dong, X.N., and Glazebrook, J. 2001. Constitutive salicylic acid-dependent signaling in *cpr1* and *cpr6* mutants requires PAD4. *Plant J.* 26:395–407.

Kinkema, M., Fan, W.H., and Dong, X.N. 2000. Nuclear localization of NPR1 is required for activation of PR gene expression. *Plant Cell* 12:2339–2350.

Kirik, V., Bouyer, D., Schobinger, U., Bechtold, N., Herzog, M., Bonneville, J.M., and Hulskamp, M. 2001. CPR5 is involved in cell proliferation and cell death control and encodes a novel transmembrane protein. *Curr. Biol.* 11:1891–1895.

Klessig, D.F., Durner, J., Noad, R., Navarre, D.A., Wendehenne, D., Kumar, D., Zhou, J.M., Shah, J., Zhang, S., Kachroo, P., Trifa, Y., Pontier, D., Lam, E., and Silva, H. 2000. Nitric oxide and salicylic acid signaling in plant defense. *Proc. Natl. Acad. Sci. USA* 97:8849–55.

Kliebenstein, D.J., Dietrich, R.A., Martin, A.C., Last, R.L., and Dangl, J.L. 1999. LSD1 regulates salicylic acid induction of copper zinc superoxide dismutase in *Arabidopsis thaliana. Mol. Plant Microbe Interact.* 12:1022–1026.

Kohler, A., Schwindling, S., and Conrath, U. 2002. Benzothiadiazole-induced priming for potentiated responses to pathogen infection, wounding, and infiltration of water into leaves requires the *NPR1/NIM1* gene in *Arabidopsis. Plant Physiol.* 128:1046–1056.

Kovtun, Y., Chiu, W.L., Zeng, W.K., and Sheen, J. 1998. Suppression of auxin signal transduction by a MAPK cascade in higher plants. *Nature* 395:716–720.

Larkindale, J., and Knight, M.R. 2002. Protection against heat stress-induced oxidative damage in *Arabidopsis* involves calcium, abscisic acid, ethylene, and salicylic acid. *Plant Physiol.* 128:682–695.

Lebel, E., Heifetz, P., Thorne, L., Uknes, S., Ryals, J., and Ward, E. 1998. Functional analysis of regulatory sequences controlling PR1 gene expression in *Arabidopsis. Plant J.* 16:223–233.

Lee, H.I., and Raskin, I. 1998. Glucosylation of salicylic acid in *Nicotiana tabacum* cv. *Xanthi*-nc. *Phytopathology* 88:692–697.

Lee, H.I., and Raskin I. 1999. Purification, cloning, and expression of a pathogen inducible UDP-glucose:salicylic acid glucosyltranferase from tobacco. *J. Biol. Chem.* 274:36637–36642.

Lieberherr, D., Wagner,U., Dubuis P.-H., Métraux, J.P., and Mauch, F. 2003. The rapid induction of glutathione *S*-transferases *AtGSTF2* and *AtGSTF6* by avirulent *Pseudomonas syringae* is the result of combined salicylic acid and ethylene signalling. *Plant Cell Physiol.*, 44:750–757.

Li, X., Zhang, Y.L., Clarke, J.D., Li, Y., and Dong, X.N. 1999. Identification and cloning of a negative regulator of systemic acquired resistance, SNII1, through a screen for suppressors of *npr1-1*. *Cell* 98:329–339.

Li, X., Clarke, J.D., Zhang, Y.L., and Dong, X.N. 2001. Activation of an EDS1-mediated R-gene pathway in the *snc1* mutant leads to constitutive, NPR1-independent pathogen resistance. *Mol. Plant Microbe Interact.* 14:1131–1139.

Madamanchi, N.R., and Kuć, J. 1991. Induced systemic resistance in plants. In *The Fungal Spore and Disease Initiation in Plants and Animals*, eds. G.T. Cole, and H.C. Hoch, pp. 347–362. New York: Plenum Press.

Malamy, J., Carr, J.P., Klessig, D.F., and Raskin, I. 1990. Salicylic acid—a likely endogenous signal in the resistance response of tobacco to viral infection. *Science* 250:1002–1004.

Maldonado, A.M., Doerner, P., Dixon, R.A., Lamb, C.J., and Cameron, R.K. 2002. A putative lipid transfer protein involved in systemic resistance signaling in *Arabidopsis*. *Nature* 419:399–403.

Maleck, K., Levine, A., Eulgem, T., Morgan, A., Schmid, J., Lawton, K.A., Dangl, J.L., and Dietrich, R.A. 2000. The transcriptome of *Arabidopsis thaliana* during systemic acquired resistance. *Nat. Genet.* 26:403–410.

Martienssen, R. 1997. Cell death: fatal induction in plants. *Curr. Biol.* 7:R534–R537.

Mauch-Mani, B., and Slusarenko, A.J. 1996. Production of salicylic acid precursors is a major function of phenylalanine ammonia-lyase in the resistance of *Arabidopsis* to *Peronospora parasitica Plant Cell* 8:203–212.

Mauch-Mani, B., and Métraux, J.-P. 1998. Salicylic acid and systemic acquired resistance to pathogen attack. *Ann. Bot.* 82:535–540.

Maxwell, D.P., Wang, Y., and McIntosh, L. 1999. The alternative oxidase lowers mitochondrial reactive oxygen production in plant cells. *Proc. Natl. Acad. Sci. USA* 96:8271–8276.

Maxwell, D.P., Nickels, R., and McIntosh, L. 2002. Evidence of mitochondrial involvement in the transduction of signals required for the induction of genes associated with pathogen attack and senescence. *Plant J.* 29:269–279.

Mercier, J., Baka, M., Reddy, B., Corcuff, R., and Arul, J. 2001. Shortwave ultraviolet irradiation for control of decay caused by *Botrytis cinerea* in bell pepper: Induced resistance and germicidal effects. *J. Am. Soc. Horticult. Sci.* 126:128–133.

Métraux, J.-P., and Durner, J. 2004. The role of salicylic acid and nitric oxide in programmed cell death and induced resistance. In *Molecular Ecotoxicology of Plants*, ed. H. Sanderman. Berlin: Springer. p. 111–150.

Métraux, J.-P., Signer, H., Ryals, J., Ward, E., Wyss-Benz, M., Gaudin, J., Raschdorf, K., Schmid, E., Blum, W., and Inverardi, B. 1990. Increase in salicylic acid at the onset of systemic acquired resistance in cucumber. *Science* 250:1004–1006.

Métraux, J.-P., Ahl Goy, P., Staub, T., Speich, J., Steinemann, A., Ryals, J., and Ward, E. 1991. Induced systemic resistance in cucumber in response to 2,6-dichloro-isonicotinic acid and pathogens. In *Advances in molecular Genetics of Plant-Microbe Interactions*, eds. H. Hennecke, and D.P.S. Verma, Vol I, pp. 432–439. The Netherlands: Kluwer Academic Publishers.

Mölders, W., Buchala, A., and Métraux, J.-P. 1996. Transport of salicylic acid in tobacco necrosis virus-infected cucumber plants. *Plant Physiol.* 112:787–792.

Mur, L.A.J., Bi Y.M., Darby, R.M., Firek, S., Draper, J. 1997. Compromising early salicylic acid accumulation delays the hypersensitive response and increases viral dispersal during lesion establishment in TMV-infected tobacco. *Plant J.* 12:1113–1126.

Mur, L.A.J., Brown, I.R., Darby, R.M., Bestwick, C.S., Bi, Y.M., Mansfield, J.W., Draper, J. 2000. A loss of resistance to avirulent bacterial pathogens in tobacco is associated with the attenuation of a salicylic acid- potentiated oxidative burst. *Plant J.* 23:609–621.

Murphy, A.M., and Carr, J.P. 2002. Salicylic acid has cell-specific effects on tobacco mosaic virus replication and cell-to-cell movement. *Plant Physiol.* 128:552–563.

Murphy, A.M., Chivasa, S., Singh, D.P., and Carr, J.P. 1999. Salicylic acid-induced resistance to viruses and other pathogens: a parting of the ways? *Trends Plant Sci.* 4: 155–160.

Nawrath, C., and Métraux, J.-P. 1999. Salicylic acid induction-deficient mutants of *Arabidopsis* express *PR-2* and *PR-5* and accumulate high levels of camalexin after pathogen inoculation. *Plant Cell* 11:1393–1404.

Nawrath, C., Heck, S., Parinthawong, N., and Métraux, J.-P. 2002. EDS5, an essential component of salicylic acid-dependent signaling for disease resistance in *Arabidopsis*, is a member of the MATE transporter family. *Plant Cell* 14:275–286.

Naylor, M., Murphy, A.M., Berry, J.O., and Carr, J.P. 1998. Salicylic acid can induce resistance to plant virus movement. *Mol. Plant Microbe Interact.* 11:860–868.

Nürnberger, T., and Scheel, D. 2001. Signal transmission in the plant immune response. *Trends Plant Sci.* 6:372–379.

Overmyer, K., Tuominen, H., Kettunen, R., Betz, C., Langebartels, C., Sandermann, H., and Kangasjarvi, J. 2000. Ozone-sensitive *Arabidopsis rcd1* mutant reveals opposite roles for ethylene and jasmonate signaling pathways in regulating superoxide-dependent cell death. *Plant Cell* 12:1849–1862.

Penninckx, I.A.M.A., Thomma, B.P.H.J., Buchala, A., Métraux, J.-P., and Broekaert, W.F. 1998. Concomitant activation of jasmonate and ethylene response pathways is required for induction of a plant defensin gene in *Arabidopsis*. *Plant Cell* 10:2103–2113.

Petersen, M., Brodersen, P., Naested, H., Andreasson, E., Lindhart, U., Johansen, B., Nielsen, H.B., Lacy, M., Austin, M.J., Parker, J.E., Sharma, S.B., Klessig, D.F., Martienssen, R, Mattsson, O., Jensen, A.B., and Mundy, J. 2000. *Arabidopsis* MAP kinase 4 negatively regulates systemic acquired resistance. *Cell* 103:1111–1120.

Pieterse, C.M.J., van Wees, S.C.M., van Pelt, J.A., Knoester, M., Laan, R., Gerrits, N., Weisbeek, P.J., and van Loon, L.C. 1998. A novel signaling pathway controlling induced systemic resistance in *Arabidopsis*. *Plant Cell* 10:1571–1580.

Pieterse, C.M.J., and van Loon, L.C. 1999. Salicylic acid-independent plant defense pathways. *Trends Plant Sci.* 4:52–58.

Pilloff, R.K., Devadas, S.K., Enyedi, A., and Raina, R. 2002. The *Arabidopsis* gain-of-function mutant *dll1* spontaneously develops lesions mimicking cell death associated with disease. *Plant J.* 30:61–70.

Pontier, D., Miao, Z.H., and Lam, E. 2001. Trans-dominant suppression of plant TGA factors reveals their negative and positive roles in plant defense responses. *Plant J.* 27: 529–538.

Punja, Z.K. 2001. Genetic engineering of plants to enhance resistance to fungal pathogens— a review of progress and future prospects. *Can. J. Plant Pathol.* 23:216–235.

Rairdan, G.J., and Delaney, T.P. 2002. Role of salicylic acid and NIM1/NPR1 in race-specific resistance in *Arabidopsis*. *Genetics* 161:803–811.

Raskin, I. 1992. Role of salicylic acid in plants. *Annu. Rev. Plant Physiol. Plant Mol. Biol.* 43:439–463.

Rate, D.N., and Greenberg, J.T. 2001. The *Arabidopsis aberrant growth and death2* mutant shows resistance to *Pseudomonas syringae* and reveals a role for NPR1 in suppressing hypersensitive cell death. *Plant J.* 27:203–211.

Reymond, P., and Farmer, E.E. 1998. Jasmonate and salicylate as global signals for defense gene expression. *Curr. Opin. Plant Biol.* 1:404–411.

Robatzek, S., and Somssich, I.E. 2002. Targets of AtWRKY6 regulation during plant senescence and pathogen defense. *Genes Dev.* 16:1139–1149.

Rogers, E.E. and Ausubel, F.M. 1997. *Arabidopsis* enhanced disease susceptibility mutants exhibit enhanced susceptibility to several bacterial pathogens and alterations in PR1 gene expression. *Plant Cell* 9:305–316.

Romeis, T., Piedras, P., and Jones, J.D.G. 2000. Resistance gene-dependent activation of a calcium-dependent protein kinase in the plant defense response. *Plant Cell* 12:803–815.

Ross, A.F. (1966). Systemic effects of local lesion formation. In *Viruses of Plants,* eds. A.B.R. Beemster, and J. Dijkstra, pp. 127–150. Amsterdam: North-Holland Publishing.

Ryals, J., Weymann, K., Lawton, K., Friedrich, L., Ellis, D., Steiner, H. Y., Johnson, J., Delaney, T. P., Jesse, T., Vos, P., and Uknes, S. 1997. The *Arabidopsis* NIM1 protein shows homology to the mammalian transcription factor inhibitor I kappa B. *Plant Cell* 9:425–439.

Schaller, A., Roy, P. and Amrhein, N. 2000. Salicylic acid-independent induction of pathogenesis-related gene expression by fusicoccin. *Planta* 210(4):599–606.

Schenk, P.M., Kazan, K., Wilson, I., Anderson, J.P., Richmond, T., Somerville, S.C., and Manners, J. M. 2000. Coordinated plant defense responses in *Arabidopsis* revealed by microarray analysis. *Proc. Natl. Acad. Sci. USA* 97:11655–11660.

Serino, L., Reimmann, C., Baur, H., Beyeler, M., Visca, P., and Haas, D. 1995. Structural genes for salicylate biosynthesis from chorismate in *Pseudomonas aeruginosa. Mol. Gen. Genet.* 249:217–228.

Seskar, M., Shulaev, V., and Raskin, I. 1997. Production of methyl salicylate in pathogen-inoculated tobacco plants. *Plant Physiol.* 114:1147–1147.

Shah, J., Kachroo, P., and Klessig, D.F. 1999. The *Arabidopsis ssi1* mutation restores pathogenesis-related gene expression in *npr1* plants and renders defensin gene expression salicylic acid dependent. *Plant Cell* 11:191–206.

Shapiro, A.D., and Zhang, C. 2001. The role of NDR1 in avirulence gene-directed signaling and control of programmed cell death in *Arabidopsis. Plant Physiol.* 127:1089–1101.

Shirasu, K., Nakajima, H., Rajasekhar, V.K., Dixon, R.A., and Lamb, C. 1997. Salicylic acid potentiates an agonist-dependent gain control that amplifies pathogen signals in the activation of defense mechanisms. *Plant Cell* 9:261–270.

Shulaev, V., Leon, J., and Raskin, I. 1995. Is salicylic acid a translocated signal of systemic acquired resistance in tobacco? *Plant Cell* 7:1691–1701.

Shulaev, V., Silverman, P., and Raskin, I. 1997. Airborne signaling by methyl salicylate in plant pathogen resistance. *Nature* 385:718–721.

Simons, B.H., Millenaar, F.F., Mulder, L., Van Loon, L.C., and Lambers, H. 1999. Enhanced expression and activation of the alternative oxidase during infection of *Arabidopsis* with *Pseudomonas syringae* pv *tomato. Plant Physiol.* 120:529–538.

Sticher, L., MauchMani, B., and Métraux, J.P. 1997. Systemic acquired resistance. *Annu. Rev. Phytopathol.* 35:235–270.

Summermatter, K., Sticher, L., and Métraux, J.-P. 1995. Systemic responses in *Arabidopsis thaliana* infected and challenged with *Pseudomonas syringae* pv *syringae. Plant Physiol.* 108:1379–1385.

Tang, X., Xie, M., Kim, Y.J., Zhou, J., Klessig, D.F., and Martin, G.B. 1999. Overexpression of *Pto* activates defense responses and confers broad resistance. *Plant Cell* 11:15–29.

Thomma, B.P.H.J., Eggermont, K., Penninckx, I.A.M.A., Mauch-Mani, B., Vogelsang, R., Cammue, B.P.A., and Broekaert, W.F. 1998. Separate jasmonate-dependent and salicylate-dependent defense- response pathways in *Arabidopsis* are essential for resistance to distinct microbial pathogens. *Proc. Natl. Acad. Sci. USA* 95:15107–15111.

Thomma, B.P.H.J., Penninckx, I.A.M.A., Broekaert, W.F., and Cammue, B.P.A. 2001. The complexity of disease signaling in *Arabidopsis. Curr. Opin. Immunol.* 13:63–68.

Tornero, P., Merritt, P., Sadanandom, A., Shirasu, K., Innes, R.W., and Dangl, J.L. 2002. RAR1 and NDR1 contribute quantitatively to disease resistance in *Arabidopsis*, and their relative contributions are dependent on the R Gene assayed. *Plant Cell* 14:1005–1015.

Uknes, S., Mauchmani, B., Moyer, M., Potter, S., Williams, S., Dincher, S., Chandler, D., Slusarenko, A., Ward, E., and Ryals, J. 1992. Acquired-Resistance in *Arabidopsis. Plant Cell* 4:645–656.

Vanacker, H., Lu, H., Rate, D.N., and Greenberg, J. 2001. A role for salicylic acid and NPR1 in regulating cell growth in *Arabidopsis. Plant J.* 28:209–216.

Van Loon, L.C., Bakker, P.A.H.M., and Pieterse, C.M.J. 1998. Systemic resistance induced by rhizosphere bacteria. *Annu. Rev. Phytopathol.* 36:453–483.

Van Loon, L.C., and Van Strien, E.A. 1999. The families of pathogenesis-related proteins, their activities, and comparative analysis of PR1 type proteins. *Physiol. Mol. Plant Pathol.* 55:85–97.

Van Tegelen, L.J.P., Moreno, P.R.H., Croes, A.F., Verpoorte, R., and Wullems, G.J. 1999. Purification and CDNA cloning of isochorismatesynthase from elicited cultures of catharanthus roseus. *Plant Physiol.* 119:705–712.

Van Wees, S.C.M., and Glazebrook, J. 2003. Loss of non-host resistance of *Arabidopsis NahG* to *Pseudomonas syringae* pv. *phaseolica* is due to degradation products of salicylic acid. *Plant J.* 33:733–742.

Ward, E.R., Uknes, S.J., Williams, S.C., Dincher, S.S., Wiederhold, D.L., Alexander, D.C., Ahlgoy, P., Métraux, J.-P., and Ryals, J.A. 1991. Coordinate gene activity in response to agents that induce systemic acquired-resistance. *Plant Cell* 3:1085–1094.

Waterhouse, P.M., Wang, M.B., and Lough, T. 2001. Gene silencing as an adaptive defense against viruses. *Nature* 411:834–842.

Weigel, R.R., Bauscher, C., Pfitzner, A.J.P., and Pfitzner, U.M. 2001. NIMIN-1, NIMIN-2 and NIMIN-3, members of a novel family of proteins from *Arabidopsis* that interact with NPR1/NIM1, a key regulator of systemic acquired resistance in plants. *Plant Mol. Biol.* 46:143–160.

Weymann, K., Hunt, M., Uknes, S., Neuenschwander, U., Lawton, K., Steiner, H.Y., and Ryals, J. 1995. Suppression and restoration of lesion formation in *Arabidopsis lsd* mutants. *Plant Cell* 7:2013–2022.

Wildermuth, M.C., Dewdney, J., Wu, G., and Ausubel, F.M. 2001. Isochorismate synthase is required to synthesize salicylic acid for plant defense. *Nature* 417:571–571.

Wong, C.E., Carson, R.A.J., and Carr, J.P. 2002. Chemically induced virus resistance in *Arabidopsis thaliana* is independent of pathogenesis-related protein expression and the NPR1 gene. *Mol. Plant Microbe Interact.* 15:75–81.

Wrzaczek, M., and Hirt, H. 2001. Plant MAP kinase pathways: how many and what for? *Biol. Cell* 93:81–87.

Xing, T., Ouellet, T., and Miki, B.L. 2002. Towards genomic and proteomic studies of protein phosphorylation in plant-pathogen interactions. *Trends Plant Sci.* 7:224–230.

Yalpani, N., Leon, J., Lawton, M.A., and Raskin, I. 1993. Pathway of salicylic acid biosynthesis in healthy and virus-inoculated tobacco. *Plant Physiol.* 103:315–321.

Yalpani, N., Enyedi, A.J., Leon, J., and Raskin, I. 1994. Ultraviolet-light and ozone stimulate accumulation of salicylic-acid, pathogenesis-related proteins and virus-resistance in tobacco. *Planta* 193:372–376.

Yang, K.Y., Liu, Y.D., and Zhang, S.Q. 2001. Activation of a mitogen-activated protein kinase pathway is involved in disease resistance in tobacco. *Proc. Natl. Acad. Sci. USA* 98:741–746.

Yang, Y.O., and Klessig, D.F. 1996. Isolation and characterization of a tobacco mosaic virus-inducible myb oncogene homolog from tobacco. *Proc. Natl. Acad. Sci. USA* 93:14972–14977.

Yoshida, S., Ito, M., Nishida, I., and Watanabe, A. 2002. Identification of a novel gene *HYS1/CPR5* that has a repressive role in the induction of leaf senescence and pathogen-defense responses in *Arabidopsis thaliana*. *Plant J.* 29:427–437.

Yu, D.Q., Chen, C.H., and Chen, Z.X. 2001. Evidence for an important role of WRKY DNA binding proteins in the regulation of *NPR1* gene expression. *Plant Cell* 13:1527–1539.

Yu, I.C., Parker, J., and Bent, A.F. 1998. Gene-for-gene disease resistance without the hypersensitive response in *Arabidopsis dnd1* mutant. *Proc. Natl. Acad. Sci. USA* 95:7819–7824.

Zhang, S.Q., and Klessig, D.F. 1997. Salicylic acid activates a 48-kD MAP kinase in tobacco. *Plant Cell* 9:809–824.

Zhang, S.Q., and Klessig, D.F. 2001. MAPK cascades in plant defense signaling. *Trends Plant Sci.* 6:520–527.

Zhang, Y.L., Fan, W.H., Kinkema, M., Li, X., and Dong, X.N. 1999. Interaction of NPR1 with basic leucine zipper protein transcription factors that bind sequences required for salicylic acid induction of the PR1 gene. *Proc. Natl. Acad. Sci. USA* 96:6523–6528.

Zhou, J.M., Trifa, Y., Silva, H., Pontier, D., Lam, E., Shah, J., and Klessig, D.F. 2000. NPR1 differentially interacts with members of the TGA/OBF family of transcription factors that bind an element of the PR1 gene required for induction by salicylic acid. *Mol. Plant Microbe Interact.* 13:191–202.

Zhou, N., Tootle, T.L., Tsui, F., Klessig, D.F., and Glazebrook, J. 1998. PAD4 functions upstream from salicylic acid to control defense responses in *Arabidopsis*. *Plant Cell* 10:1021–1030.

Zimmerli, L., Jakab, C., Métraux, J.P., and Mauch-Mani, B. 2000. Potentiation of pathogen-specific defense mechanisms in *Arabidopsisg* by β-aminobutyric acid. *Proc. Natl. Acad. Sci. USA* 97:12920–12925.

8

Signaling in Plant Resistance Responses: Divergence and Cross-Talk of Defense Pathways

CORNÉ M.J. PIETERSE, ANDREAS SCHALLER, BRIGITTE MAUCH-MANI, AND UWE CONRATH

8.1 Introduction

Plants possess inducible defense mechanisms to protect themselves against attack by microbial pathogens and herbivorous insects. The endogenous signaling molecules salicylic acid, ethylene, and jasmonic acid, and the peptide messenger systemin play important roles in the regulation of these induced defense responses. Disease resistance of plants can also be induced by chemical agents, such as 2,6-dichloroisonicotinic acid, benzothiadiazole, and the nonprotein amino acid β-aminobutyric acid. In most cases, these chemical agents mimic or ingeniously make use of the same pathways that are activated by the endogenous defense signals. This review is focussed on the current state of research on signal transduction pathways involved in induced resistance against pathogens and insects. Recent advances in induced resistance research revealed that the signaling pathways involved are interconnected, resulting in overlap, synergism, and antagonism between the different signal transduction pathways. Divergence and crosstalk of pathways in defense response signaling provide the plant with flexibility and the opportunity for fine-tuning of resistance responses, thereby enabling it to cope with different forms of stress more efficiently.

8.2 Salicylic Acid Induces Systemic Resistance Responses

Over the past decade it became increasingly clear that the endogenous signal salicylic acid (SA) serves multiple roles in plants. For example, SA is involved in the regulation of cell growth (Vanacker et al., 2001), flowering, and thermogenesis (for reviews, see Malamy and Klessig, 1992; Raskin, 1992; Klessig and Malamy, 1994; Shah and Klessig, 1999). SA also plays a crucial role in plant defense against pathogens by affecting lesion formation (Weymann et al., 1995) and by activating induced disease resistance (Dempsey et al., 1999; Shah and Klessig, 1999; Nawrath et al., in this volume). The latter is variously referred to as systemic acquired resistance (SAR) or induced systemic resistance (ISR). Although these terms are synonymous (Hammerschmidt et al., 2001), we refer to the SA pathway-dependent, induced disease resistance as SAR. SAR is characterized by a long-lasting resistance against a broad spectrum of pathogens both at the initial infection site and in the distal, uninoculated organs. The most compelling evidence for the important role of SA in the onset of SAR comes from studies with transgenic tobacco and *Arabidopsis* plants expressing the *NahG* gene from *Pseudomonas putida*. This gene encodes a salicylate hydroxylase, which destroys the SA signal by converting it to catechol. Upon pathogen attack, *NahG* transgenic tobacco and *Arabidopsis* plants do not accumulate enhanced levels of SA nor do they establish SAR (Gaffney et al., 1993; Delaney et al., 1994). The SAR state is activated by many microbes that cause tissue necrosis but it can also be induced by exogenous application of SA or its functional analogs 2,6-dichloroisonicotinic acid (INA) and benzothiadiazole (BTH) (Ryals et al., 1996; Sticher et al., 1997; Dempsey et al., 1999).

The onset of SAR is associated with an early increase in endogenous SA levels and with the immediate expression of a specific set of so-called *SAR* genes, some of which encode pathogenesis-related (PR) proteins (Ryals et al., 1996; Sticher et al., 1997; Dempsey et al., 1999). While it is known that some PR proteins display antimicrobial activity (Van Loon and Van Strien, 1999), their actual role in SAR is still unclear and can depend on the plant-pathogen system. In fact, a strict correlation between increased accumulation of PR proteins before challenge pathogen attack and SAR has not always been observed. To gain a better understanding of the mechanisms that contribute to SAR, it is necessary, therefore, to study further defense-associated cellular events that are induced faster or to a greater extent in attacked, SAR-protected plants. Such events include the activation of defense-related genes other than those encoding PR proteins, and the deposition of callose (Kohler et al., 2002).

In addition to *SAR* gene expression, SAR is also associated with priming (sensitizing) which enhances the plant's capacity for the rapid and effective activation of cellular defense responses, that are induced only upon contact with a (challenging) pathogen (Kuć, 1987; Katz et al., 1998; Conrath et al., 2002). These responses include hypersensitive cell death (Mittler and Lam, 1996), cell wall fortification (Hammerschmidt and Kuć, 1982; Stumm and Gessler, 1986; Schmele and Kauss, 1990), the production of reactive oxygen species (Doke et al., 1996), and the activation of defense-related genes (Ryals et al., 1996; Sticher et al., 1997).

The role of SA in *PR* gene expression as a part of SAR is discussed by Nawrath et al. and will therefore not be discussed here in detail. This section of our review will rather focus on the progress made in elucidating the role of SA in priming for potentiated activation of cellular defense responses.

8.2.1 Salicylic Acid-Induced Priming in a Cell Culture Model System

Over the past 13 years, it has been reported that a pretreatment of parsley cell cultures with low doses of the SAR inducers SA, INA, and BTH did not directly induce various assayed, cellular defense responses (Kauss et al., 1992a; 1993; Kauss and Jeblick, 1995; Thulke and Conrath, 1998; Katz et al., 1998; 2002). Yet, a preincubation with the SAR inducers primed the cells for potentiated (augmented) activation of defense responses, that were subsequently induced by otherwise noninducing doses of an elicitor from *Phytophthora sojae* cell walls (Kauss et al., 1992a; 1993; Kauss and Jeblick, 1995; Thulke and Conrath, 1998; Katz et al., 1998; 2002). The potentiated responses include the early oxidative burst (Kauss and Jeblick, 1995), a rapidly induced K^+/pH response (Katz et al., 2002), the incorporation into the cell wall of various phenolics and a lignin-like polymer (Kauss et al., 1993), and the secretion of antimicrobial coumarin phytoalexins resulting from an enhanced activity of coumarin biosynthetic enzymes (Kauss et al., 1992a) and augmented expression of some of the genes encoding these enzymes (Kauss et al., 1992a; 1993; Katz et al., 1998; Thulke and Conrath, 1998). In a similar manner, in soybean suspension cells, physiological concentrations of SA strongly augmented defense gene activation, H_2O_2 accumulation, and the hypersensitive necrosis response (HR) that was induced by treatment with avirulent *Pseudomonas syringae* pv. *glycinea* (Shirasu et al., 1997). However, since the SA-mediated potentiation of defense responses in soybean cells did not depend on prolonged pre-treatment with SA, this mechanism of regulation obviously differs from the time-dependent priming in cultured parsley cells. Together, the observations made with parsley and soybean suspension cells revealed that plant cell cultures can be suitable model systems for studying the SA-, INA-, and BTH-induced priming for potentiated activation of cellular plant defense responses.

8.2.2 Salicylic Acid Serves a Dual Role in the Activation of Defense Responses

While elucidating the influence of SA and BTH on the activation of defense-related genes in the parsley cell culture, it became obvious that the inducer's effect on gene activation depends on the gene that is being monitored (Katz et al., 1998; Thulke and Conrath, 1998). One set of genes, such as those encoding anionic peroxidase and mannitol dehydrogenase, was found to be directly induced by relatively low concentrations of the two SAR inducers tested (Katz et al., 1998; Thulke and Conrath, 1998). A second set of parsley defense-related genes, including those encoding phenylalanine ammonia-lyase (PAL), 4-coumarate:CoA ligase, intracellular PR-10 proteins and a hydroxyproline-rich glycoprotein, was only faintly responsive to the treatment with relatively low concentrations of SA or BTH. Yet, already at low inducer concentrations, these genes displayed SA- and BTH-dependent potentiation of their expression following treatment with a low elicitor dose (Katz et al., 1998; Thulke and Conrath, 1998). For instance, more than

0.5 mmolar SA was required to activate *PAL* using only SA, whereas as little as 0.01 mmolar SA greatly potentiated the activation of the *PAL* gene by an otherwise faintly inducing elicitor concentration (Thulke and Conrath, 1998). These results revealed a dual role for SAR inducers in the activation of plant defense responses: a direct one in the immediate induction of certain defense genes at higher inducer concentrations, and an indirect one which requires only low doses of the inducers to prime for potentiated activation of another class of defense genes. As the potentiation by SA and BTH of both elicited *PAL* gene expression and coumarin secretion strongly depended on an extended preincubation period, the SAR inducers are assumed to mediate a time-dependent response that shifts the cells on the alert (Katz et al., 1998; Thulke and Conrath, 1998). Whether this shift includes the proposed synthesis of cellular factors with crucial roles in the coordination and expression of cellular defense responses remained uncertain.

Similar observations to those made in parsley have been reported for cowpea seedlings (Latunde-Dada and Lucas, 2001). The BTH-mediated SAR response of cowpea is associated with rapid and transient increases in the activity of PAL and chalcone isomerase followed by accelerated accumulation of kievitone and phaseollidin phytoalexins in infected hypocotyls. These responses were not observed in induced, uninoculated tissues, suggesting that the protection of cowpea seedlings by BTH is mediated via potentiation of early defense mechanisms (Latunde-Dada and Lucas, 2001). In cucumber hypocotyls with INA-induced SAR (Fauth et al., 1996), and in wounded soybean tissue (Graham and Graham, 1994), potentiation was also detected for the development of elicitation competency. Whether the enhanced induction of elicitation competency is based on a similar priming mechanism to the one described above for parsley cells is unclear.

8.2.3 Activators of SAR Induce Priming in Arabidopsis

In *Arabidopsis*, BTH directly activates *PR-1* and primes the plants for potentiated *PAL* gene expression induced by phytopathogenic *Pseudomonas syringae* pv. *tomato* (*Pst*) (Kohler et al., 2002). BTH-induced priming also augments both *PAL* gene activation and callose deposition induced by either mechanically wounding the leaves with forceps or infiltrating them with water (Kohler et al., 2002). These observations with *Arabidopsis* not only confirm the above described dual role for SAR inducers in the activation of cellular plant defense responses, they also suggest that priming might be common to several signaling pathways, mediating crosstalk between pathogen defense and wound or osmotic stress responses (see below).

Intriguingly, when SAR was biologically induced by previous infection of *Arabidopsis* with an avirulent strain of *Pst*, there was potentiated activation of both the *PAL* and the *PR-1* gene upon challenge infection with virulent *Pst* (Cameron et al., 1999; Van Wees et al., 1999; Kohler et al., 2002). Priming is thus likely to play an important role not only in chemically induced but also in pathogen-activated SAR of plants. The same conclusion was drawn from studies with SA-primed transgenic tobacco plants displaying potentiated expression of chimeric

Asparagus officinalis PR-1::GUS and *PAL-3::GUS* defense genes after wounding or pathogen attack (Mur et al., 1996). The *Arabidopsis edr1* mutant constitutively displays enhanced resistance to *Pst* (strain DC3000) and to the fungal pathogen *Erisyphe cichoracearum* (Frye and Innes, 1998). Interestingly, *edr1* differs from other enhanced disease resistance mutants because it shows no constitutive expression of *PR-1* and *PR-2*, although transcripts of both of these genes accumulate after pathogen attack. This finding, and the fact that *edr1* shows stronger expression of defense responses, such as the HR and callose deposition, after infection strongly suggest an involvement of EDR1 in priming. *EDR1* codes for a putative mitogen-activated protein kinase kinase kinase (MAPKKK) and mediates disease resistance via SA-inducible defense responses (Frye et al., 2001). Future mutational approaches in *Arabidopsis* are expected to yield more genes that play a role in priming.

The *Arabidopsis npr1* mutant (also known as *nim1* or *sai1*) accumulates wild-type levels of SA when treated with avirulent pathogens but is unable to mount biologically or chemically induced SAR (Cao et al., 1994; Delaney et al., 1995; Shah et al., 1997). Interestingly, the potentiation by BTH-priming of both *Pst*-induced *PAL* gene activation and wound- or water infiltration-induced *PAL* gene expression and callose deposition are absent in *npr1* (Kohler et al., 2002). The *Arabidopsis cpr1* and *cpr5* mutants, on the other hand, which express constitutive SAR in the absence of a pretreatment with SAR inducers (Bowling et al., 1994; 1997), are permanently primed for potentiated *PAL* gene activation by *Pst* infection and for augmented *PAL* gene expression and callose deposition upon wounding or water infiltration (Kohler et al., 2002). Constitutive priming in *cpr1* and *cpr5* could be due to the expression of a multiplicity of defense-related genes in these plants, or the activation of other stress response mechanisms besides SAR (Boch et al., 1998; Clarke et al., 2001), although these possibilities remain remote. More likely, however, the enhanced levels of SA in *cpr1* and *cpr5* (Bowling et al., 1994; 1997) cause a permanently primed (alarm) state. Because of constitutive priming, *cpr1* and *cpr5* might be able to rapidly and effectively induce their various cellular defense mechanisms, thus leading to enhanced resistance to pathogens, wounding, or water infiltration (Kohler et al., 2002). In this context it is noteworthy that the constitutively enhanced pathogen resistance of another *Arabidopsis* mutant, *cpr5-2*, has been ascribed to the potentiated induction of the *PR-1* gene upon infection with virulent *Pseudomonas syringae* strains (Boch et al., 1998). There is evidence that a null *eds1* mutation suppresses the disease resistance of both *cpr1* and *cpr6* but only partially that of *cpr5*, indicating a different requirement of *CPR* genes for EDS1 (Clarke et al., 2001). EDS1 also likely plays a role in priming in connexion with PAD4 (Jirage et al., 2001). Although both proteins act upstream of pathogen-induced SA accumulation, their expression can be potentiated by SA-pre-treatment of the plants. It has been proposed that EDS1 is involved in the amplification of defense responses, possibly by associating with PAD4 (Feys et al., 2001).

The strong correlation between the presence of SAR and priming supports the conclusion that priming is an important mechanism for SAR in plants. This assumption is further substantiated by the close correlation between the ability of various chemicals to induce SAR against tobacco mosaic virus (TMV) in tobacco (Conrath et al., 1995) and their capability to prime for potentiated *PAL* expression induced by either elicitor treatment in parsley cells (Katz et al., 1998; Thulke and Conrath, 1998) or *Pst* infection, wounding, or water infiltration in *Arabidopsis* plants (Kohler et al., 2002). In addition, in *NahG*-transgenic tobacco plants that are unable to establish SA-mediated priming, both the onset of the HR and the activation of an active oxygen-responsive chimeric *Asparagus officinalis PR-1::GUS* reporter gene were significantly delayed when infected with avirulent *Pseudomonads*. The attenuation of priming and the loss of potentiated production of active oxygen species were accompanied by a lack of resistance to the bacteria (Mur et al., 2000). Furthermore, overexpressing the disease resistance gene *PTI5* in tomato potentiates pathogen-induced defense gene expression and enhances the resistance to *Pst* (He et al., 2001). Finally, a complete or partial inactivation of the MLO protein was shown to prime young barley seedlings for potentiated induction of defense responses associated with enhanced resistance against powdery mildew (Büschges et al., 1997).

8.3 Jasmonic Acid and Ethylene: Important Signals in Plant Defense Responses

Apart from SA, the defense signaling molecules jasmonic acid (JA) and ethylene (ET) have also been implicated in the regulation of resistance responses. In many cases, infection by microbial pathogens and attack by herbivorous insects was shown to be associated with enhanced production of these phytohormones and a concomitant activation of distinct sets of defense-related genes (De Laat and Van Loon, 1981; Gundlach et al., 1992; Peña-Cortés et al., 1993; Mauch et al., 1994; Reymond et al., 2000; Schenk et al., 2000). Compelling evidence for a role of JA and ET in disease resistance came from genetic analyses of mutants and transgenic plants that are affected in the biosynthesis or perception of these compounds. In many plant–pathogen interactions, JA and ET appeared to be involved in local and/or systemic induction of defense responses.

8.3.1 Genetic Evidence for a Role of Jasmonic Acid and Ethylene in Pathogen Resistance

Genetic evidence of a role for JA in plant defense came particularly from analyses of *Arabidopsis* mutants affected in the biosynthesis or perception of JA. The JA-response mutant *coi1* displays enhanced susceptibility to the necrotrophic fungi *Alternaria brassicicola* and *Botrytis cinerea* (Thomma et al., 1998), and the bacterial soft-rot pathogen *Erwinia carotovora* (Norman-Setterblad et al., 2000). Another JA-insensitive *Arabidopsis* mutant, *jar1*, allows enhanced growth of *Pst* in the leaves (Pieterse et al., 1998). These findings demonstrate that JA-dependent defense responses contribute to the basal resistance of *Arabidopsis* against different microbial pathogens. Furthermore, both *jar1* and the *fad3 fad7 fad8* triple mutant of *Arabidopsis*, which is deficient in the biosynthesis of the JA precursor linolenic acid, exhibit susceptibility to normally nonpathogenic soilborne *Pythium* spp. (Staswick et al., 1998; Vijayan et al., 1998), indicating that JA also plays a role in nonhost resistance. A role for JA in defense against herbivorous insects is indicated by the observation that the *Arabidopsis fad3 fad7 fad8* mutant exhibited extremely high mortality after attack by larvae of the common saprophagous fungal gnat, *Bradysia impatiens* (McConn et al., 1997). Furthermore, a JA-deficient tomato mutant, *def-1*, was found to be compromised in the wound-inducible expression of defense genes and resistance to *Manduca sexta* larvae (Howe et al., 1996).

The role of ET in plant resistance seems more ambiguous. In some cases, ET is involved in disease resistance, whereas in other cases it is associated with symptom development. For instance, several ET-insensitive mutants of *Arabidopsis* have been reported to exhibit enhanced susceptibility to *B. cinerea* (Thomma et al., 1999), *Pst* (Pieterse et al., 1998), and *E. carotovora* (Norman-Setterblad et al., 2000), indicating that ET-dependent defense responses contribute to basal resistance against these pathogens. A similar phenomenon was observed in tomato and soybean mutants with reduced sensitivity to ET, which developed more severe symptoms when infected by the fungal pathogens *B. cinerea* (Díaz et al., 2002), *Septoria glycinea*, or *Rhizoctonia solani* (Hoffman et al., 1999). In addition, ET-insensitive tobacco plants transformed with the mutant ET receptor gene *etr1* from *Arabidopsis* displayed susceptibility to the normally nonpathogenic oomycete *Pythium sylvaticum* (Knoester et al., 1998). Thus, ET obviously also plays a role in nonhost resistance. In other cases, reduced ET sensitivity was associated with disease tolerance. For example, ET-insensitive tomato genotypes allowed growth of virulent *Pst* and *Xanthomonas campestris* pv. *vesicatoria* to levels similar to those in wild-type tomato plants, but developed less severe disease symptoms (Lund et al., 1998; Ciardi et al., 2000). A similar phenomenon was found in the ET-insensitive *ein2* mutant of *Arabidopsis*, which displayed increased tolerance to virulent *Pst* and *X. campestris* pv. *campestris* (Bent et al., 1992). In addition, soybean mutants with reduced sensitivity to ET developed disease symptoms similar or less-severe than those in the wild type when infected with the bacterial pathogen *P. syringae* pv. *glycinea* or the oomycete *Phytophthora sojae* (Hoffman et al., 1999). In these interactions, ET is clearly involved in symptom development, rather than in disease resistance.

The dual role of ET in plant defense might reflect its involvement in various physiological processes in the plant. ET plays an important role in senescence (Abeles et al., 1992) and lesion development of hypersensitively reacting plant tissues (Knoester et al., 2001). Since necrotrophic pathogens feed on dead cells, both functions of ET might be favorable for the development of disease caused by such types of pathogens. Biotrophic pathogens, in contrast, need living cells to complete their life cycle. Thus, the same functions of ethylene might help to restrict these types of pathogens. Support for this hypothesis comes from experiments with hypersensitively reacting *Arabidopsis* plants. On the one hand, the hypersensitively responding tissue was more susceptible to infection by the necrotrophic fungi *B. cinerea* and *Sclerotinia sclerotiorum*, but, on the other hand, inhibited the growth of biotrophic pathogens (Govrin and Levine, 2000).

8.3.2 Jasmonic Acid- and Ethylene-Mediated Induced Defenses against Pathogens

Besides their role in basal resistance, JA and ET also function as key regulators in induced defense responses that act systemically to enhance resistance against subsequent pathogen attack. For instance, infection of *Arabidopsis* with the fungal pathogen *A. brassicicola* results in local and systemic activation of the *PDF1.2* gene, encoding a plant defensin with anti-fungal properties. Mutant analysis revealed that *PDF1.2* gene expression is regulated through a JA- and ET-dependent signaling pathway that functions independently of SA (Penninckx et al., 1996; 1998). Another example comes from studies on the interaction between the bacterial pathogen *E. carotovora* and its host plants tobacco and *Arabidopsis*. Infection of leaves of these plants with *E. carotovora*, or treatment of the leaves with elicitors of this pathogen, activated an SA-independent systemic resistance and a set of defense-related genes that differs from that induced upon exogenous application of SA (Vidal et al., 1997; Norman-Setterblad et al., 2000). Interestingly, most of the *E. carotovora*-induced genes appeared to be regulated by JA and ET.

Another type of JA/ET-dependent induced pathogen resistance is triggered by selected strains of nonpathogenic rhizosphere bacteria. Strains that were isolated from naturally disease-suppressive soils, mainly fluorescent *Pseudomonas* spp., were found to promote plant growth by suppressing soilborne pathogens. This biological control activity is effective under field conditions (Zehnder et al., 2001) and in commercial greenhouses (Leeman et al., 1995), and can be the result of competition for nutrients, siderophore-mediated competition for iron, antibiosis, or secretion of lytic enzymes (Bakker et al., 1991). Some of the biological control strains reduce disease through a plant-mediated mechanism that is phenotypically similar to pathogen-induced SAR, as the induced resistance is systemically activated and is effective against various types of pathogens. This type of induced disease resistance is referred to here as rhizobacteria-mediated induced systemic resistance (ISR) (Van Loon et al., 1998; Pieterse et al., 2002). In *Arabidopsis*, rhizobacteria-mediated ISR activated by *Pseudomonas fluorescens* WCS417r and *Pseudomonas putida* WCS358r has been shown to function independently of SA and *PR* gene activation (Pieterse et al., 1996; Van Wees et al., 1997). Instead, rhizobacteria-mediated ISR signaling requires JA and ET, because *Arabidopsis* mutants impaired in their ability to respond to either of these two phytohormones are unable to express ISR (Pieterse et al., 1998; Ton et al., 2001; 2002a). The state of rhizobacteria-mediated ISR is not only independent of *PR* gene expression, but is also not associated with the activation of other known defense-related genes (Van Wees et al., 1999). Upon challenge with a pathogen, however, ISR-expressing plants show enhanced expression of certain JA- and ET-responsive genes such as *AtVSP, PDF1.2,* and *HEL* (Van Wees et al., 1999; Hase and Pieterse, unpublished observations), suggesting that ISR-expressing tissue is primed to activate specific JA- and ET-inducible genes faster and/or to a higher level upon pathogen attack. As mentioned above, the priming phenomenon has already been observed in other processes in plants responding to stress signals and is regarded to enhance the plant's ability to defend itself against different types of biotic or abiotic stress (Conrath et al., 2002).

8.3.3 Priming of Defense Responses During Rhizobacteria-Mediated ISR

Although expression of rhizobacteria-mediated ISR in *Arabidopsis* requires an intact response to both JA and ET (Pieterse et al., 1998), the analysis of local and systemic levels of these plant hormones revealed that ISR is not associated with changes in the production of these signals (Pieterse et al., 2000). This finding suggests that ISR is based on an enhanced sensitivity to these plant hormones rather than on an increase in their production. If this is true, ISR-expressing plants are primed to react faster or more strongly to JA and ET produced after pathogen attack.

The hypothesis that ISR may be based on an enhanced sensitivity to JA is supported by the finding that the expression of the JA-inducible gene *AtVSP* was potentiated in ISR-expressing leaves after challenge with *Pst* (Van Wees et al., 1999). In the same study, the expression of several other JA-responsive genes was tested as well, but these failed to show an enhanced expression level in ISR-expressing leaves, suggesting that ISR in *Arabidopsis* is associated with potentiation of a specific set of JA-responsive genes. Potentiation of defense responses by JA has been reported in other systems as well. For instance, pre-treatment with methyl jasmonate potentiates the elicitation of various phenylpropanoid defense responses in parsley suspension cell cultures (Kauss et al., 1992b) and primes them for enhanced induction of the early oxidative burst (Kauss et al., 1994). Moreover, JA potentiates the expression of the *PR-1* gene in rice and the level of resistance against *Magnaporthe grisea* induced by low doses of INA (Schweizer et al., 1997).

The role of ethylene in priming is more complex. After treatment with a saturating dose of 1 millimolar of the ethylene precursor 1-aminocyclopropane-1-carboxylate (ACC), ISR-expressing plants emit significantly more ethylene than ACC-treated control plants (Pieterse et al., 2000). Evidently, the capacity to convert ACC to ethylene is increased in ISR-expressing plants. Because in infected tissues, ACC levels rapidly increase as a result of pathogen-induced ACC synthase activity, the enhanced ACC-converting capacity of ISR-expressing plants likely primes the plant for a faster or greater production of ethylene upon pathogen attack. In *Pst*-infected *Arabidopsis* plants induced for ISR, the production of ET was indeed enhanced during the first 24 hours after infection compared to uninduced plants (Hase and Pieterse, unpublished observations). Interestingly, exogenous application of ACC has been shown to induce resistance against *Pst* in *Arabidopsis*

(Pieterse et al., 1998). Therefore, a faster or greater production of ET in the initial phase of infection might contribute to the enhanced resistance against this pathogen.

8.4 Systemins: Peptide Signals in The Systemic Wound Response

In the early 1970s, Green and Ryan (1972) observed an accumulation of proteinase inhibitors (PIs) in tomato and potato plants after herbivore-induced or mechanical wounding in both the injured leaves and undamaged parts of the plants. In this landmark study, Green and Ryan (1972) suggested this systemic reaction to be an inducible defense response directed against herbivorous insects. It is now clear that the systemic wound response is not limited to proteinase inhibitors but rather includes a large number of proteins which may contribute, directly or indirectly, to enhanced insect resistance in many plant species (Constabel, 1999; Reymond et al., 2000; Ryan, 2000; Walling, 2000). The wound response in the *Solanaceae* attracted considerable attention over the past 30 years and has developed into a model system of long-distance signaling in plants. Much effort has been devoted to the identification of a hypothetical wound signal that is generated at the site of injury, transmitted throughout the aerial parts of the plant, and capable of inducing the expression of defense genes in undamaged tissues. Physical stimuli such as hydraulic waves that result from the release of xylem tension upon wounding or action and variation potentials have been implicated in the wound signal transduction process, as well as chemical signaling molecules including JA, ET, abscisic acid, oligogalacturonides (OGAs), and systemins. The activity of these signals and their contribution to long-distance signal transduction has been covered in several reviews (Schaller and Ryan, 1995; Bowles, 1998; Ryan, 2000; de Bruxelles and Roberts, 2001; León et al., 2001) and is also discussed by Korth and Thompson (this volume). This section will instead focus on systemins, their discovery, activity, and signaling properties.

8.4.1 Systemins in Different Plant Species

The systemic wound response of tomato plants is characterized by the accumulation of a large number of defense proteins (systemic wound response proteins, SWRPs) (Ryan, 2000). A search for the hypothetical signaling molecule(s) that allows tomato plants to respond systemically to a local stimulus (i.e., wounding), led to the identification of the first plant peptide with a signaling function in 1991 (Pearce et al., 1991; Ryan, 1992). A 18-amino-acid peptide was isolated from the leaves of tomato plants on the basis of its ability to induce the expression of SWRPs using a sensitive bioassay. The peptide was named "systemin" to emphasize its central role as an inducing compound and the systemic nature of the response (Pearce et al., 1991). Based on the systemin amino acid sequence, the cDNA and gene of prosystemin were cloned, and found to encode a systemin precursor of

200 amino acids (McGurl and Ryan, 1992; McGurl et al., 1992). The systemin sequence is found close to the C-terminus of the precursor. There is a single gene for prosystemin in the haploid tomato genome from which two different polypeptides are derived by differential splicing of the pre-mRNA. The polymorphism is located in the nonsystemin portion of the polypeptides and does not seem to affect their wound signaling properties (Li and Howe, 2001). Highly similar prosystemins have been identified in closely related plant species (potato, bell pepper, and black nightshade) exhibiting 73–88 % identity with the tomato sequence, but not outside the family of *Solanaceae* (Constabel et al., 1998). Homology-based approaches failed to identify prosystemin in the more distantly related tobacco. A search for tobacco signaling molecules functionally related to tomato systemin identified two 18-amino-acid peptide inducers of PI synthesis in tobacco leaves (Pearce et al., 2001). The two peptides are derived from a single precursor protein of 165 amino acids. The precursor of the tobacco systemins is not homologous to the previously identified prosystemins from other *Solanaceae* but contains sequence motifs present also in hydroxyproline-rich cell wall glycoproteins (Pearce et al., 2001). Likewise, tobacco systemins themselves bear no structural similarity to tomato systemin. Therefore, systemins are now considered to represent a structurally diverse group of polypeptides that are produced in injured plants and function as signaling molecules in the activation of defense genes (Pearce et al., 2001). Systemic responses to herbivore attack have been documented in more than 100 plant species (Karban and Baldwin, 1997). It will be interesting to see which proteins exert systemin function in these plants and whether or not further distinct proteins have evolved to perform systemin's signaling function. In the following discussion of systemin activity and signaling we will focus on the properties of the tomato peptide, presently the only one that has been thoroughly investigated.

8.4.2 The Activity of Tomato Systemin

A wealth of physiological data point toward a role for systemin as a signal molecule in the wound signal transduction pathway in tomato plants. In addition to *SWRP* gene expression, the synthetic tomato peptide triggers physiological reactions that are characteristic to the wound response. Changes in plasma membrane permeability are among the earliest cellular responses to treatment with systemin and oligogalacturonide elicitors of the wound response. The influx of calcium and protons and the efflux of potassium and chloride ions lead to an increase in the cytoplasmic free calcium concentration, intracellular acidification, depolarization of the plasma membrane, and alkalinization of the apoplast (Felix and Boller, 1995; Thain et al., 1995; Moyen and Johannes, 1996; Moyen et al., 1998; Schaller, 1998). These early events are essentially indistinguishable from those triggered in plant cells after pathogen recognition or elicitation (Conrath et al., 1991; Ebel and Mithöfer, 1998; Scheel, 1998; Katz et al., 2002). In both wound and pathogen defense responses, these ion fluxes were shown to be necessary and sufficient for the subsequent activation of defense genes (Fukuda, 1996; Jabs et al., 1997; Schaller and Oecking, 1999; Blume et al., 2000; Schaller and Frasson, 2001).

Both wounding and systemin stimulate the accumulation of calmodulin as well as of polygalacturonase, phospholipase, and protein kinase activities which may all contribute to the transduction of the wound signal in tomato (Conconi et al., 1996; Stankovic and Davies, 1997; Stratmann and Ryan, 1997; Bergey and Ryan, 1999; Bergey et al., 1999; Narváez-Vásquez et al., 1999; Chico et al., 2002) and other plant species (e.g., Seo et al., 1995; Vian et al., 1996; Lee et al., 1997; Rojo et al., 1998; Seo et al., 1999; Dhondt et al., 2000; Jonak et al., 2000; Wang et al., 2000; Ishiguro et al., 2001). Both wounding and systemin stimulate the synthesis and transient accumulation of JA (Peña-Cortés et al., 1993; Doares et al., 1995a), another inducer of defense gene expression (Farmer and Ryan, 1990; Farmer et al., 1991). This finding places the octadecanoid pathway for JA biosynthesis downstream of both wounding and systemin in the signaling pathway that leads to the expression of wound-responsive genes (Farmer and Ryan, 1992). Consistently, a rapid and transient induction of JA biosynthetic enzymes is observed after wounding or systemin treatment and is followed by a delayed and more sustained induction of SWRPs with a direct role in deterring insect herbivores (Ryan, 2000; Strassner et al., 2002). The production of ET is triggered by wounding and systemin treatment (Felix and Boller, 1995; O'Donnell et al., 1996), and both ET and JA were shown to be required for *SWRP* gene activation (O'Donnell et al., 1996). Finally, a local and systemic production of H_2O_2 was observed in tomato plants upon wounding and systemin treatment and was shown to depend on a functional octadecanoid pathway. Hence, a role for H_2O_2 as a second messenger downstream of JA was proposed (Orozco-Cardenas and Ryan, 1999; Orozco-Cárdenas, 2000).

8.4.3 The Role of Tomato Systemin in Wound Signal Transduction

The activities elucidated for tomato systemin are essentially consistent with a model of wound signaling originally proposed by Farmer and Ryan (1992). According to this model, systemin is released from prosystemin as a consequence of wounding, translocated throughout the aerial parts of the plant, and then interacts with a cell-surface receptor in the target tissue. This interaction results in the activation of a lipase, which releases linolenic acid from membrane lipids to serve as a substrate of the octadecanoid pathway for the biosynthesis of JA which, in turn, activates defense genes (Farmer and Ryan, 1992). The model was later refined to account for the requirement of ET for *SWRP* gene activation (O'Donnell et al., 1996), the defense signaling activity of oxylipins other than JA (Stintzi et al., 2001), the action of H_2O_2 as a second messenger downstream of JA (Orozco-Cárdenas, 2000), and the involvement of ion fluxes across the plasma membrane and reversible protein phosphorylation in wound signaling (Schaller, 1999; Ryan, 2000; Schaller, 2001). Important support for this model includes the characterization of a cell-surface binding site for systemin exhibiting characteristics of a functional systemin receptor (Meindl et al., 1998; Scheer and Ryan, 1999; Stratmann et al., 2000; Scheer and Ryan, 2002), as well as data derived from the analysis of transgenic and mutant tomato plants. Transgenic tomato plants in which the expression of prosystemin was suppressed by the antisense RNA technology were impaired in both the wound-induced accumulation of PIs and resistance to insect larvae demonstrating an absolute requirement of prosystemin for the activation of the wound response in tomato plants (McGurl et al., 1992; Orozco-Cardenas et al., 1993). In a converse manner, constitutive accumulation of SWRPs was observed in tomato plants overexpressing prosystemin under control of the constitutive CaMV 35S promoter (McGurl et al., 1994). Extragenic suppressors of the *35S::prosystemin*-mediated SWRP accumulation were identified and characterized, demonstrating that wounding and systemin induce defense gene expression through a common signaling pathway (Howe et al., 1996; Howe and Ryan, 1999). Surprisingly, when ectopically expressed, prosystemin appears to be sufficient to trigger defense gene activation and, thus, wounding is no longer required. In prosystemin-overexpressing plants, untimely processing of prosystemin may occur, or the ectopic expression of prosystemin even alleviates the need for processing, as full-length prosystemin was shown to be as active as systemin in the induction of *SWRP* gene expression when supplied to tomato plants via the transpiration stream (Dombrowski et al., 1999; Vetsch et al., 2000). Grafting experiments were performed using *35S::prosystemin*-expressing plants as the root stock and wild-type tomato as the scion. SWRPs were found to accumulate in the scion, demonstrating that the overexpression of prosystemin is sufficient to generate a graft-transmissible signal for defense gene activation. Similarly, addition of systemin or prosystemin to wound sites on leaves of prosystemin antisense plants caused *SWRP* gene activation in the distal unwounded leaves (Dombrowski et al., 1999). These observations are consistent with systemin itself being the mobile signal. (Pro)systemin-induced synthesis of another, as yet unidentified signaling molecule, however, cannot be excluded.

A microarray comprising 235 cDNAs was used to analyze the relative changes in gene expression in wounded and distal, unwounded leaves of tomato plants. While transcripts for SWRPs with direct defense function (i.e., the "late" defense genes, e.g., those for PIs; Ryan, 2000) accumulated to high levels in both tissues, the coordinate induction of genes for octadecanoid pathway enzymes dedicated to JA biosynthesis ("early" defense genes; Ryan, 2000) was observed locally but not systemically (Strassner et al., 2002). In this study, JA and its precursor 12-oxophytodienoic acid accumulated in the damaged, but not in distal, leaves of wounded plants (Strassner et al., 2002) which is consistent with previous reports of limited systemic JA accumulation (Bowles, 1998; Rojo et al., 1999; Ziegler et al., 2001). Hence, synthesis and accumulation of JA in systemic leaves do not seem to be required for defense gene activation. However, this does not necessarily imply that systemic *SWRP* gene activation, as suggested by Bowles (1998), is JA-independent: A recent study showed that systemic wound signaling requires the capacity to synthesize JA in the wounded leaf, whereas the ability to perceive JA is required in the systemic leaves. Elegant grafting experiments were performed using tomato mutants that either fail to synthesize (*spr-2*; Howe and Ryan, 1999) or perceive (*jai-1*; Li et al., 2001) the JA signal. When grafted plants were wounded below the graft junction, activation of the wound response in the scion depended on the ability to perceive JA. Wound- or (pro)systemin-induced activation of the JA biosynthetic pathway, on the other hand, was required in the lower part of the plant for the generation of a graft-transmissible signal, but not for defense gene activation in the scion (Li et al., 2002). The data suggest that the activity of (pro)systemin is required in the wounded leaf to promote the production of a systemic signal, possibly JA or another octadecanoid-derived molecule. Therefore, the model of wound signaling originally proposed by Farmer and Ryan (1992) may describe local rather than systemic wound signal transduction events. Thirty years after the initial report on the phenomenon (Green and Ryan, 1972), the identity of the systemic signal molecule in the wound response of plants is still unclear.

Another level of complexity is added by the cell-type-specific expression of genes involved in the wound response, which is certainly highly relevant for the processes leading to both local and systemic activation of defense genes. "Early genes", i.e., those rapidly induced after wounding, including those encoding prosystemin and some of the JA biosynthetic enzymes, are expressed in vascular bundles (Jacinto et al., 1997; Kubigsteltig et al., 1999; Hause et al., 2000), whereas "late genes", i.e., those for SWRPs with a direct role in plant defense, are expressed in palisade and adjacent spongy mesophyll cells (Shumway et al., 1976; Walker-Simmons and Ryan, 1977; Ryan, 2000). The temporally and spatially separated expression of the two classes of genes led to the suggestion that wound-signaling events may initially be activated in the vascular bundles to produce second messengers (octadecanoids, OGAs, H_2O_2) that will then induce defense gene expression in mesophyll cells (Orozco-Cárdenas, 2000; Ryan, 2000). Some of the second messengers may exert their effects over long distances and contribute to systemic signal transduction.

Further work is needed to precisely understand systemin action and function in tomato plants. Obviously, these studies will have to be extended to other plant species, particularly to *Arabidopsis*. The plethora of signaling mutants available in *Arabidopsis* will be useful to advance our understanding of the complexity of wound signal transduction as well as the interaction of the systemin signal transduction pathway with other defense signaling pathways (see below).

8.5 β-Aminobutyric Acid Activates Resistance Responses

β-Aminobutyric acid (BABA) is a nonprotein amino acid, which is only rarely found in nature. BABA has been described as part of a small, 9-kilodalton proteinaceous inhibitor of trypsin and microbial serine proteinases isolated from *Yersinia pseudotuberculosis* (Burtseva and Kofanova, 1996). In addition, BABA was found in root exudates of tomato plants grown in solarized soil (Gamliel and Katan, 1992). Despite its rare occurrence, BABA is an interesting compound. This is because of its close structural similarity to a highly bioactive substance, the neurotransmitter GABA, whose natural occurrence is well documented in plants (Shelp et al., 1999). Also, BABA is a potent inducer of acquired disease resistance (Jakab et al., 2001). Applied as either a soil drench or foliar spray, BABA has a broad spectrum of activity against viruses, bacteria, oomycetes, fungi, and nematodes (Jakab et al., 2001). This wide range of activity supports a role for BABA as an inducer of acquired disease resistance, especially since the substance was shown not to be directly toxic to microorganisms (reviewed by Jakab et al., 2001). As BABA is highly water-soluble it is readily taken up by plant roots and then distributed throughout the plant (Cohen and Gisi, 1994; Jakab et al., 2001).

8.5.1 β-Aminobutyric Acid-Induced Priming

Depending on the method of application, mild phytotoxic effects of BABA have been observed. BABA has been sprayed on leaves, injected into stems of plants, supplied via petiole dip, or applied as a soil drench to the root system. When applied as a foliar spray to tobacco plants, BABA, and to a lesser extent α-aminobutyric acid, but not GABA, were phytotoxic at a concentration of 100 μg ml^{-1} (ca. 1 mmolar) (Cohen, 1994). Small necrotic lesions started to form on treated leaves two days after spraying. A rapid induction of necrotic lesions in tobacco was also observed by Siegrist et al. (2000) after foliar treatment with 10 mmolar BABA. Localized necrosis was accompanied by the formation of reactive oxygen species, lipid peroxidation, callose deposition around the lesions, and an increase in the SA content of the leaves (Siegrist et al., 2000). No such effects were observed in plants treated with GABA, even at concentrations as high as 2000 μg ml^{-1}(ca. 20 mmolar) (Cohen, 1994; Siegrist et al., 2000).

In *Arabidopsis*, spraying BABA onto leaves also leads to the formation of small necrotic lesions and to an accumulation of *PR* gene transcripts, with a pattern that is similar to the one observed when SA is used to induce resistance. However, when supplied via the root system, BABA concentrations sufficient to induce resistance, do not induce defense gene expression in *Arabidopsis* (Zimmerli et al., 2000). This observation suggests that the induction of resistance by BABA in *Arabidopsis* is not primarily based on a previous accumulation of defense gene transcripts. Rather, an additional mechanism of resistance induction seems to be present in BABA-treated *Arabidopsis* plants. This conclusion is supported by the observation that BABA induces resistance against the oomycete *Peronospora parasicita* in wild-type *Arabidopsis* plants as well as in plants that are impaired in defense gene expression (Zimmerli et al., 2000), such as the *npr1, jar1,* or *etr1* mutants (Bleeker et al., 1988; Staswick et al., 1992; Cao et al., 1994), and *NahG* transgenic *Arabidopsis* plants (Delaney et al., 1994). In this case, resistance is independent of the presence of SA and *PR* or other defense gene activation. Common to the BABA-mediated defense mechanism observed in the different mutant and wild-type plants is a more rapid and stronger deposition of callose-containing papillae at the site of infection by *P. parasitica* (Zimmerli et al., 2000). BABA primes *Arabidopsis* to effectively react to *P. parasitica* infection with papillae deposition, thus making further defense responses obsolete since ingress by *P. parasitica* has already been stopped at this point. Interestingly, a similar observation was made with *NahG* tobacco challenged with downy mildew: there was no difference in the protection by BABA between *NahG* and wild-type plants (Cohen et al., 2000). It is probable that also in this case priming for potentiated induction of SA-independent defense mechanisms is responsible for the observed protection.

When BABA-pretreated *Arabidopsis* plants are challenged with a virulent strain of *Pst*, priming becomes apparent as a strong potentiation of *PR-1* gene expression (Zimmerli et al., 2000). In this case, the induction kinetics are very similar to those observed in response to avirulent *Pst* (Zimmerli et al., 2000). In the interaction between *Arabidopsis* and *Pst*, priming by BABA is dependent on an intact SA signaling pathway, but independent on a functioning JA/ET pathway as evident from experiments with the same defense response mutants as described above (Zimmerli et al., 2000). Interestingly, in the *Arabidopsis-B. cinerea* interaction, it is *PR-1* that again shows strongly potentiated expression (Zimmerli et al., 2001) and not *PDF1.2* (Thomma et al., 1998) that is commonly thought to play a role in defense against *B. cinerea*.

In contrast to other inducers of SAR, such as SA or BTH (Kohler et al., 2002), BABA itself does not induce *PR* gene expression (Zimmerli et al., 2000). Using BABA, it is possible therefore to clearly separate priming and defense gene activation. This will greatly facilitate the future analysis of priming phenomena in induced resistance.

8.6 Cross-Talk Between Signaling Pathways

Over the past years, evidence has accumulated indicating that the SA-, JA-, ET-, and systemin-dependent defense pathways can affect each other, either positively or negatively. Although the observed pathway interactions vary between species and the type of attacker used, it is becoming increasingly clear that cross-talk between signaling pathways is important for the plant to fine-tune its defense responses. For example, JA and ET have been shown to act synergistically in the activation of genes encoding defense-related plant proteins, such as PIs and defensins (O'Donnell et al., 1996; Penninckx et al., 1998). Moreover, JA and ET have been shown to support the action of SA resulting in enhanced *PR* gene expression (Lawton et al., 1994; Xu et al., 1994; Schweizer et al., 1997). On the other hand, SA, INA, and BTH suppress JA-dependent defense gene expression (Doherty et al., 1988; Peña-Cortés et al., 1993; Bowling et al., 1997; Niki et al., 1998; Fidantsef et al., 1999; Van Wees et al., 1999), possibly through inhibition of JA biosynthesis and action (Peña-Cortés et al., 1993; Doares et al., 1995b; Harms et al., 1998). Consistent with this, Preston et al. (1999) demonstrated that TMV-infected tobacco plants displaying SAR are unable to express normal JA-mediated wound responses, probably due to inhibition of JA signaling by increased SA levels resulting from the TMV infection. Also, in the *Arabidopsis ssi2* mutant, the SA-dependent signaling pathway is constitutively activated, while JA-dependent signaling is suppressed (Kachroo et al., 2001). Conversely, in pathogen-inoculated *NahG* plants, which are unable to accumulate significant SA levels, expression of the JA/ET-responsive defensin gene *PDF1.2* was at least twofold higher than in wild-type plants (Penninckx et al., 1996). Inhibitory effects of salicylates on ET biosynthesis have also been reported (reviewed by Shah and Klessig, 1999). Thus, activating the SA pathway confers resistance to a broad spectrum of microbial pathogens but, at the same time, may have detrimental effects on the JA/ET-dependent signal transduction mechanism that confers resistance against insects and certain groups of pathogens.

An additional level of antagonistic regulation of wound- and pathogen-induced defense responses is provided by the proton electrochemical gradient across the plasma membrane. In tomato and tobacco plants, activation of the plasma membrane H^+-ATPase by the fungal toxin fusicoccin (FC) induces the accumulation of both basic and acidic PR proteins (Fukuda, 1996; Roberts and Bowles, 1999; Schaller and Oecking, 1999; Frick and Schaller, 2002). Also, expression of a bacterial proton pump induced a lesion mimic phenotype, activated multiple defense responses, and increased the resistance to microbial pathogens in transgenic tobacco and potato plants (Mittler et al., 1995; Abad et al., 1997; Rizhsky and Mittler, 2001). In addition to activating pathogen defense responses, the hyperpolarization of the plasma membrane by FC-treatment resulted in a suppression of wound-, systemin-, OGA-, and JA-induced *SWRP* gene expression (Doherty and Bowles, 1990; Schaller, 1999; Frick and Schaller, 2002). Both the activation of pathogen response genes and the repression of wound-induced genes by FC were shown to be at least partly independent of SA, as they (i) were incompatible with the timing of FC-induced SA accumulation in tomato leaves, and (ii) occurred under conditions of inhibited SA biosynthesis (Schaller et al., 2000). Furthermore, FC induced *PR* gene expression in *NahG* tobacco and tomato plants, i.e. plants unable to accumulate significant amounts of SA (Schaller et al., 2000; Frick and Schaller, 2002).

While activation of the H^+-ATPase induced *PR* gene expression and *SWRP* gene supression, inhibitors of the plasma membrane H^+-ATPase activity and ionophores that dissipate the proton electrochemical gradient induced *SWRP* genes in tomato (Schaller and Oecking, 1999; Schaller and Frasson, 2001). Octadecanoid-dependent signaling was also triggered by the ion-channel-forming peptide alamethicin (Engelberth et al., 2001), and, more generally, a role for the pore-forming properties of elicitors in the induction of defense responses has been discussed (Klüsener and Weiler, 1999). Apparently, wound and pathogen defense signaling pathways are differentially affected by changes in the proton electrochemical gradient. Therefore, the plasma membrane H^+-ATPase may act as a switch activating either wound or pathogen defense responses.

Several studies have provided evidence for trade-offs between SA-dependent pathogen resistance and JA-dependent insect resistance, indicating that the activation of a particular defense mechanism can reduce the resistance to certain groups of pathogens or herbivorous insects. For instance, Moran (1998) demonstrated that SAR in cucumber against *Colletotrichum orbiculare* was associated with reduced resistance against feeding by spotted cucumber beetles *(Diabrotica undecimpunctata howardi)* and enhanced reproduction of melon aphids *(Aphis gossypii)*. A similar phenomenon was observed by Preston et al. (1999) who demonstrated that TMV-infected tobacco plants induced for SAR display higher sensitivity to tobacco hornworm (*Manduca sexta*) grazing when compared with noninduced control plants. Furthermore, it has been shown that transgenic tobacco plants with reduced SA levels, caused by silencing of the *PAL* gene, exhibit reduced SAR against TMV but enhanced herbivore-induced resistance to *Heliothis virescens* larvae (Felton et al., 1999). In a converse manner, *PAL*-overexpressing tobacco displays a strong reduction of herbivore-induced insect resistance, while TMV-induced SAR was enhanced in these plants.

The SAR inducer BTH has in some cases also been shown to reduce insect resistance. Exogenous application of BTH to tomato plants enhanced the level of resistance against *Pst*, but improved the suitability of tomato for feeding by leaf chewing larvae of the corn earworm (*Helicoverpa zea*) (Stout et al., 1999). A similar phenomenon was observed by Thaler et al. (1999) who reported compromised resistance to the beet armyworm (*Spodoptera exigua*) upon application of BTH to field-grown tomato plants. In most cases, the reduced insect resistance of SAR-expressing plants could be attributed to the inhibition of JA production by either BTH or increased SA levels.

8.7 Concomitant Activation of Induced Disease Resistance Mechanisms

Though negative interactions between the SA- and JA/ET-dependent signal transduction pathways have clearly been shown, other studies argue against such a negative relationship. A genetic screen for the isolation of *Arabidopsis* signal transduction mutants that constitutively express the JA/ET-responsive *THI2.1* gene yielded two mutants, which showed concomitant induction of both the SA- and the JA-dependent signaling pathways (Hilpert et al., 2001). The finding that some gene transcripts which increase after *A. brassicicola* infection of *Arabidopsis* leaves also accumulate upon treatment with SA, JA, and ET, also points to an overlap of the different signaling pathways, at least in *Arabidopsis* (Schenk et al., 2000). In this context, it is worthwhile to mention that a pre-treatment with systemin was shown to prime tomato cell suspension cultures for augmented induction of the H_2O_2 burst induced by the addition of OGAs or water (Stennis et al., 1998). In a similar manner, preincubating cultured parsley cells with JA potentiated the subsequent activation of phenylpropanoid defense responses by a *P. sojae* cell wall elicitor (Kauss et al., 1992b). Also, priming *Arabidopsis* plants with BTH (Kohler et al., 2002) or BABA (Jakab et al., 2001) enhanced the subsequent induction of defense responses against biotic and abiotic stresses. Thus, priming likely represents a molecular mechanism at which the systemin, JA/ET, BABA, and SA signaling pathways merge.

Failure to demonstrate a negative relationship between signaling mechanisms was also reported on the level of pathogen or insect resistance. For instance, inoculating lower leaves of tobacco plants with TMV does not affect the growth of tobacco aphid (*Myzus nicotianae*) populations (Ajlan and Potter, 1992). In a similar manner, there is no negative effect of BTH application on the population growth of whiteflies (*Bemisia argentifolii*) and leaf miners (*Liriomyza* spp.) (Inbar et al., 1998). Interestingly, Stout et al. (1999) have demonstrated that inoculation of tomato leaves with *Pst* induced resistance to both *Pst* and the corn earworm (*Helicoverpa zea*) in distal parts of the *Pst*-inoculated plants. Conversely, feeding by *H. zea* induced resistance against both *Pst* and *H. zea*. A nice demonstration of simultaneous pathogen and insect resistance in the field was provided by Zehnder et al. (2001). The authors observed that rhizobacteria-mediated ISR of cucumber against insect-transmitted bacterial wilt disease, caused by *Erwinia tracheiphila*, was associated with reduced feeding of the cucumber beetle vector. It appeared that induction of ISR was associated with reduced concentrations of cucurbitacin, a secondary metabolite and powerful feeding stimulant for cucumber beetles. Induction of rhizobacteria-mediated ISR against *E. tracheiphila* was also effective in the absence of beetle vectors, suggesting that ISR protects cucumber against bacterial wilt not only by reducing beetle feeding and pathogen transmission, but also through induction of defense responses that act against the bacterial pathogen. These observations indicate that negative interactions between induced pathogen- and insect resistance are by no means general.

The question of whether SA- and JA/ET-dependent resistance against microbial pathogens can be expressed simultaneously was recently addressed by Van Wees et al. (2000). In *Arabidopsis*, SA-dependent, necrosis-triggered SAR and JA/ET-dependent, rhizobacteria-mediated ISR are each effective against various pathogens, although their spectrum of effectiveness partly diverges (Ton et al., 2002b). Both SAR and ISR are effective against *Pst*. Simultaneous activation of both types of induced resistance resulted in an additive effect on the level of induced protection against this pathogen. In *Arabidopsis* genotypes that are blocked in either SAR or ISR, this additive effect was absent. Moreover, induction of ISR did not affect expression of the SAR marker gene *PR-1* in plants expressing SAR. Together, these observations demonstrate that the signaling pathways involved in both types of induced resistance can be compatible and that there is not necessarily significant cross-talk between them. Therefore, combining SAR and ISR can provide an attractive tool for improving disease control in plants.

References

Abad, M.S., Hakimi, S.M., Kaniewski, W.K., Rommens, C.M.T., Shulaev, V., Lam, E., and Shah, D.M. 1997. Characterization of acquired resistance in lesion-mimic transgenic potato expressing bacterio-opsin. *Mol. Plant Microbe Interact.* 10:635–645.

Abeles, F.B., Morgan, P.W., and Saltveit, M.E.J. 1992. *Ethylene in Plant Biology*. San Diego, CA: Academic Press.

Ajlan, A.M., and Potter, D.A. 1992. Lack of effect of tobacco mosaic virus-induced systemic acquired resistance on arthropod herbivores in tobacco. *Phytopathology* 82:647–651.

Bakker, P.A.H.M., Van Peer, R., and Schippers, B. 1991. Suppression of soil-borne plant pathogens by fluorescent pseudomonads: mechanisms and prospects. In *Biotic Interactions and Soil-Borne Diseases*, eds. A.B.R. Beemster, G.J. Bollen, M. Gerlagh, M.A. Ruissen, B. Schippers, and A. Tempel, pp. 217–230. Amsterdam, The Netherlands: Elsevier Scientific Publishers.

Bent, A.F., Innes, R.W., Ecker, J.R., and Staskawicz, B.J. 1992. Disease development in ethylene-insensitive *Arabidopsis thaliana* infected with virulent and avirulent *Pseudomonas* and *Xanthomonas* pathogens. *Mol. Plant Microbe Interact.* 5:372–378.

Bergey, D.R., and Ryan, C.A. 1999. Wound- and systemin-inducible calmodulin gene expression in tomato leaves. *Plant Mol. Biol.* 40:815–823.

Bergey, D.R., Orozco-Cardenas, M., de Moura, D.S., and Ryan, C.A. 1999. A wound- and systemin-inducible polygalacturonase in tomato leaves. *Proc. Natl. Acad. Sci. USA* 96:1756–1760.

Bleecker, A.B., Estelle, M.A., Somerville, C., and Kende, H. 1988. Insensivity to ethylene conferred by a dominant mutation in *Arabidopsis thaliana*. *Science* 241:1086–1089.

Blume, B., Nürnberger, T., Nass, N., and Scheel, D. 2000. Receptor-mediated increase in cytoplasmic free calcium required for activation of pathogen defense in parsley. *Plant Cell* 12:1425–1440.

Boch, J., Verbsky, M.L., Robertson, T.L., Larkin, J.C., and Kunkel, B.N. 1998. Analysis of resistance gene-mediated defense responses in *Arabidopsis thaliana* plants carrying a mutation in *CPR5*. *Mol. Plant Microbe Interact.* 12:1196–1206.

Bowles, D. 1998. Signal transduction in the wound response of tomato plants. *Philos. Trans. R. Soc.* 353:1495–1510.

Bowling, S.A., Guo, A., Cao, H., Gordon, A.S., Klessig, D.F., and Dong, X. 1994. A mutation in *Arabidopsis* that leads to constitutive expression of systemic acquired resistance. *Plant Cell* 6:1845–1857.

Bowling, S.A., Clarke, J.D., Liu, Y., Klessig, D.F., and Dong, X. 1997. The *cpr5* mutant of *Arabidopsis* expresses both NPR1-dependent and NPR1-independent resistance. *Plant Cell* 9:1573–1584.

Burtseva, T.I., and Kofanova, N.N. 1996. Inhibitor of trypsin and microbial serine proteinases isolated from *Yersinia pseudotuberculosis*. *Biochemistry (Moskow)* 61:1335–1341.

Büschges, R., Hollricher, K., Panstruga, R., Simons, G., Wolter, M., Frijters, A., Van Daelen, R., Van der Lee, T., Diergaarde, P., Groenendijk, J., Töpsch, S., Vos, P., Salamini, F., and Schulze-Lefert, P. 1997. The barley *MLO* gene: a novel control element of plant pathogen resistance. *Cell* 88:695–705.

Cameron, R.K., Paiva, N.L., Lamb, C.J., and Dixon, R.A. 1999. Accumulation of salicylic acid and *PR-1* gene transcripts in relation to the systemic acquired resistance (SAR) response induced by *Pseudomonas syringae* pv. *tomato* in *Arabidopsis*. *Physiol. Mol. Plant Pathol.* 55:121–130.

Cao, H., Bowling, S.A., Gordon, A.S., and Dong, X. 1994. Characterization of an *Arabidopsis* mutant that is nonresponsive to inducers of systemic acquired resistance. *Plant Cell* 8:1583–1592.

Chico, J.M., Raíces, M., Téllez-Iñón, M.T., and Ulloa, R.M. 2002. A calcium-dependent protein kinase is systemically induced upon wounding in tomato plants. *Plant Physiol.* 128:256–270.

Ciardi, J.A., Tieman, D.M., Lund, S.T., Jones, J.B., Stall, R.E., and Klee, H.J. 2000. Response to *Xanthomonas campestris* pv. *vesicatoria* in tomato involves regulation of ethylene receptor gene expression. *Plant Physiol.* 123:81–92.

Clarke, J.D., Aarts, N., Feys, B.J., Dong, X., and Parker, J.E. 2001. Constitutive disease resistance requires EDS1 in the *Arabidopsis* mutants *cpr1* and *cpr6* and is partially EDS1-dependent in *cpr5*. *Plant J.* 26:409–420.

Cohen, Y. 1994. 3-aminobutyric acid induces systemic resistance against *Peronospora tabacina*. *Physiol. Mol. Plant Pathol.* 44:273–288.

Cohen, Y., and Gisi, U. 1994. Sytemic translocation of ^{14}C-DL-3-aminobutyric acid in tomato plants in relation to induced resistance against *Phytophthora infestans. Physiol. Mol. Plant Pathol.* 45:441–446.

Cohen, Y., Ovadia, A., and Oka, Y. 2000. Is induced-resistance reversible? The BABA case. (Abstract), *First International Symposium on Induced Resistance to Plant Diseases*, Island of Corfu, Greece, 22–27 May 2000, p. 68.

Conconi, A., Miquel, M., Browse, J.A., and Ryan, C.A. 1996. Intracellular levels of free linolenic and linoleic acids increase in tomato leaves in response to wounding. *Plant Physiol.* 111:797–803.

Conrath, U., Jeblick, W., and Kauss, H. 1991. The protein kinase inhibitor, K-252a, decreases elicitor-induced Ca^{2+} uptake and K^+ release, and increases coumarin synthesis in parsley cells. *FEBS Lett.* 279:141–144.

Conrath, U., Chen, Z., Ricigliano, J.R., and Klessig, D.F. 1995. Two inducers of plant defense responses, 2,6-dichloroisonicotinic acid and salicylic acid, inhibit catalase activity in tobacco. *Proc. Natl. Acad. Sci. USA* 92:7143–7147.

Conrath, U., Pieterse; C.M.J., and Mauch-Mani, B. 2002. Priming in plant-pathogen interactions. *Trends Plant Sci.* 5:210–216.

Constabel, C.P. 1999. A survey of herbivory-inducible defensive proteins and phytochemicals. In *Induced Plant Defenses Against Pathogens and Herbivores. Biochemistry, Ecology, and Agriculture,* eds. A.A. Agrawal, S. Tuzun, and E. Bent, pp. 137–166. St. Paul, MN: APS Press.

Constabel, P.C., Yip, L., and Ryan, C.A. 1998. Prosystemin from potato, black nightshade, and bell pepper: primary structure and biological activity of predicted systemin polypeptides. *Plant Mol. Biol.* 36:55–62.

de Bruxelles, G.L., and Roberts, M.R. 2001. Signals regulating multiple responses to wounding and herbivores. *Crit. Rev. Plant Sci.* 20:487–521.

De Laat, A.M.M., and Van Loon, L.C. 1981. Regulation of ethylene biosynthesis in virus-infected tobacco leaves. I. Determination of the role of methionine as a precursor of ethylene. *Plant Physiol.* 68: 256–260.

Delaney, T.P., Uknes, S., Vernooij, B., Friedrich, L., Weymann, K., Negrotto, D., Gaffney, T., Gut-Rella, M., Kessmann, H., Ward, E., and Ryals, J. 1994. A central role of salicylic acid in plant disease resistance. *Science* 266:1247–1250.

Delaney, T.P., Friedrich, L., and Ryals, J. 1995. *Arabidopsis* signal transduction mutant defective in chemically and biologically induced disease resistance. *Proc. Natl. Acad. Sci. USA* 92:6602–6606.

Dempsey, D.A, Shah, J., and Klessig, D.F. 1999. Salicylic acid and disease resistance in plants. *Crit. Rev. Plant Sci.* 18:547–575.

Dhondt, S., Geoffroy, P., Stelmach, B.A., Legrand, M., and Heitz, T. 2000. Soluble phospholipase A2 activity is induced before oxylipin accumulation in tobacco mosaic virus-infected tobacco leaves and is contributed by patatin-like enzymes. *Plant J.* 23:431–440.

Díaz, J., Ten Have, A., and Van Kan, J.A.L. 2002. The role of ethylene and wound signaling in resistance of tomato to *Botrytis cinerea. Plant Physiol.* 129:1314–1315.

Doares, S.H., Syrovets, T., Weiler, E.W., and Ryan, C.A. 1995a Oligogalacturonides and chitosan activate plant defensive genes through the octadecanoid pathway. *Proc. Natl. Acad. Sci. USA* 92:4095–4098.

Doares, S.H., Narváez-Vásquez, J., Conconi, A, and Ryan, C.A. 1995b. Salicylic acid inhibits synthesis of proteinase inhibitors in tomato leaves induced by systemin and jasmonic acid. *Plant Physiol.* 108:1741–1746.

Doherty, H.M., and Bowles, D.J. 1990. The role of pH and ion transport in oligosaccharide-induced proteinase inhibitor accumulation in tomato plants. *Plant Cell Environ.* 13:851–855.

Doherty, H.M., Selvendran, R.R., and Bowles, D.J. 1988. The wound response of tomato plants can be inhibited by aspirin and related hydroxy-benzoic acids. *Physiol. Mol. Plant Pathol.* 33:377–384.

Doke, N., Miura, Y., Sanchez, L.M., Park, H.J., Noritake, T., Yoshiola, H., and Kawakita, K. 1996. The oxidative burst protects plants against pathogen attack: mechanism and role as an emergency signal for plant bio-defence: a review. *Gene* 179:45–51.

Dombrowski, J.E., Pearce, G., and Ryan, C.A. 1999. Proteinase inhibitor-inducing activity of the prohormone prosystemin resides exclusively in the C-terminal systemin domain. *Proc. Natl. Acad. Sci. USA* 96:12947–12952.

Ebel, J., and Mithöfer, A. 1998. Early events in the elicitation of plant defence. *Planta* 206:335–348.

Engelberth, J., Koch, T., Schueler, G., Bachmann, N., Rechtenbach, J., and Boland, W. 2001. Ion channel-forming alamethicin is a potent elicitor of volatile biosynthesis and tendril coiling. Cross talk between jasmonate and salicylate signaling in lima bean. *Plant Physiol.* 125:369–377.

Farmer, E.E., and Ryan, C., A. 1990. Interplant Communication: Airborne methyl jasmonate induces synthesis of proteinase inhibitors in plant leaves. *Proc. Natl. Acad. Sci. USA* 87:7713–7716.

Farmer, E.E., and Ryan, C.A. 1992. Octadecanoid precursors of jasmonic acid activate the synthesis of wound-inducible proteinase inhibitors. *Plant Cell* 4:129–134.

Farmer, E.E., Johnson, R.R., and Ryan, C.A. 1991. Regulation of expression of proteinase inhibitor genes by methyl jasmonate and jasmonic acid. *Plant Physiol.* 98:995-1002.

Fauth, M., Merten, A., Hahn, M.G., Jeblick, W., and Kauss, H. 1996. Competence for elicitation of H_2O_2 in hypocotyls of cucumber is induced by breaching the cuticle and is enhanced by salicylic acid. *Plant Physiol.* 110:347–354.

Felix, G., and Boller, T. 1995. Systemin induces rapid ion fluxes and ethylene biosynthesis in *Lycopersicon peruvianum* cells. *Plant J.* 7:381–389.

Felton, G.W., Korth, K.L., Bi, J.L., Wesley, S.V., Huhman, D.V., Mathews, M.C., Murphy, J.B., Lamb, C., and Dixon, R.A. 1999. Inverse relationship between systemic resistance of plants to microorganisms and to insect herbivory. *Curr. Biol.* 9:317–320.

Feys, B.J., Moisan, L.J., Newmann, M.A., and Parker, J.E. 2001. Direct interaction between the *Arabidopsis* disease resistance signaling proteins, EDS1 and PAD4. *EMBO J.* 20:5400–5411.

Fidantsef, A.L., Stout, M.J., Thaler, J.S., Duffey, S.S., and Bostock, R.M. 1999. Signal interactions in pathogen and insect attack: expression of lipoxygenase, proteinase inhibitor II, and pathogenesis-related protein P4 in the tomato, *Lycopersicon esculentum*. *Physiol. Mol. Plant. Pathol.* 54:97–114.

Frick, U.B., and Schaller, A. 2002 cDNA microarray analysis of fusicoccin-induced changes in gene expression in tomato plants. *Planta* 216:83–94.

Frye, C.A., and Innes, R.W. 1998. An *Arabidopsis* mutant with enhanced resistance to powdery mildew. *Plant Cell* 10:947–956.

Frye, C.A., Tang, D., and Innes, R.W. 2001. Negative regulation of defense responses in plants by a conserved MAPKK kinase. *Proc. Natl. Acad. Sci. USA* 98:373–378.

Fukuda, Y. 1996. Coordinated activation of chitinase genes and extracellular alkalinization in suspension-cultured tobacco cells. *Biosci. Biotech. Biochem.* 60:2005–2010.

Gaffney, T., Friedrich, L., Vernooij, B., Negrotto, D., Nye, G., Uknes, S., Ward, E., Kessmann, H., and Ryals, J. 1993. Requirement of salicylic acid for the induction of systemic acquired resistance. *Science* 261:754–756.

Gamliel, A., and Katan, J. 1992. Influence of seed and root exudates on fluorescent Pseudomonas and fungi in solarized soil. *Phytopathology* 82:320–327.

Govrin, E.M., and Levine, A. 2000. The hypersensitive reaction facilitates plant infection by the necrotrophic fungus *Botrytis cinerea*. *Curr. Biol.* 10:751–757.

Graham, T.L., and Graham, M.Y. 1994. Wound-associated competency factors are required for the proximal cell responses of soybean to the *Phytophthora sojae* wall glucan elicitor. *Plant Physiol.* 105:571–578.

Green, T.R., and Ryan, C.A. 1972. Wound-induced proteinase inhibitor in plant leaves: A possible defense mechanism against insects. *Science* 175:776–777.

Gundlach, H., Mueller, M.J., Kutchan, T.M., and Zenk, M.H. 1992. Jasmonic acid is a signal transducer in elicitor-induced plant cell cultures. *Proc. Natl. Acad. Sci. USA* 89:2389–2393.

Hammerschmidt, R., and Kuć, J. 1982. Lignification as a mechanism for induced systemic resistance in cucumber. *Physiol. Plant Pathol.* 20:61–71.

Hammerschmidt, R., Métraux, J.P., and Van Loon, L.C. 2001. Inducing resistance: a summary of papers presented at the *First International Symposium on Induced Resistance to Plant Diseases*, Corfu, May 2000. *Eur. J. Plant Pathol.* 107:1–6.

Harms, K., Ramirez, I., and Peña-Cortés, H. 1998. Inhibition of wound-induced accumulation of allene oxide synthase transcripts in flax leaves by aspirin and salicylic acid. *Plant Physiol.* 118:1057–1065.

Hause, B., Stenzel, I., Miersch, O., Maucher, H., Kramell, R., Ziegler, J., and Wasternack, B. 2000. Tissue-specific oxylipin signature of tomato flowers: allene-oxide cyclase is highly expressed in distinct flower organs and vascular bundles. *Plant J.* 24:113–126.

He, P., Warren, R.F., Zhao, T., Shan, L., Zhu, L., Tang, X., and Zhou, J.-M. 2001. Overexpression of *PTI5* in tomato potentiates pathogen-induced defense gene expression and enhances disease resistance to *Pseudomonas syringae* pv. *tomato*. *Mol. Plant Microbe Interact.* 14:1453–1457.

Hilpert, B., Bohlmann, H., Op den Camp, R., Przybyla, D., Miersch, O., Buchala, A., and Apel, K. 2001. Isolation and characterization of signal transduction mutants of *Arabidopsis thaliana* that constitutively activate the octadecanoid pathway and form necrotic microlesions. *Plant J.* 26:435–446.

Hoffman, T., Schmidt, J.S., Zheng, X., and Bent, A.F. 1999. Isolation of ethylene-insensitive soybean mutants that are altered in pathogen susceptibility and gene-for-gene disease resistance. *Plant Physiol.* 119:935–949.

Howe, G.A., and Ryan, C.A. 1999. Suppressors of systemin signaling identify genes in the tomato wound response pathway. *Genetics* 153:1411–1421.

Howe, G.A., Lightner, J., Browse, J., and Ryan, C.A. 1996. An octadecanoid pathway mutant (JL5 of tomato is compromised in signaling for defense against insect attack. *Plant Cell* 8:2067–2077.

Inbar, M., Doostdar, H., Sonoda, R.M., Leibee, G.L., Mayer, R.T. 1998. Elicitors of plant defensive systems reduce insect densities and disease incidence. *J. Chem. Ecol.* 24:135–149.

Ishiguro, S., Kawai-Oda, A., Nishida, I., and Okada, K. 2001. The *defective in anther dehiscence1* gene encodes a novel phospholipase A1 catalyzing the initial step of jasmonic acid biosynthesis, which synchronizes pollen maturation, anther dehiscence, and flower opening in *Arabidopsis*. *Plant Cell* 13:2191–2209.

Jabs, T., Tschöpe, M., Colling, C., Hahlbrock, K., and Scheel, D. 1997. Elicitor-stimulated ion fluxes and O_2^- from the oxidative burst are essential components in triggering defense gene activation and phytoalexin biosynthesis in parsley. *Proc. Natl. Acad. Sci. USA* 94:4800–4805.

Jacinto, T., McGurl, B., Franceschi, V., Delano-Freier, J., and Ryan, C.A. 1997. Tomato prosystemin promoter confers wound-inducible, vascular bundle-specific expression of the β-glucuronidase gene in transgenic tomato plants. *Planta* 203:406–412.

Jakab, G., Cottier, V., Toquin, V., Rigoli, G., Zimmerli L., Métraux, J.P., and Mauch-Mani, B. 2001. β-aminobutyric acid-induced resistance in plants. *Eur. J. Plant Pathol.* 107: 29–37.

Jirage, D., Zhou, N., Cooper, B., Clarke, J.D., Dong, X., and Glazebrook, J. 2001. Constitutive salicylic acid-dependent signaling in *cpr1* and *cpr6* mutants requires PAD4. *Plant J.* 26:395–407.

Jonak, C., Beisteiner, D., Beyerly, J., and Hirt, H. 2000. Wound-induced expression and activation of WIG, a novel glycogen synthase kinase 3. *Plant Cell* 12:1467–1475.

Kachroo, P., Shanklin, J., Shah, J., Whittle, E.J., and Klessig, D.F. 2001. A fatty acid desaturase modulates the activation of defense signaling pathways in plants. *Proc. Natl. Acad. Sci. USA* 98:9448–9453.

Karban, R., and Baldwin, I.T. 1997. *Induced Responses to Herbivory.* Chicago: University of Chicago Press.

Katz, V.A., Thulke, O.U., and Conrath, U. 1998. A benzothiadiazole primes parsley cells for augmented elicitation of defense responses. *Plant Physiol.* 117:1333–1339.

Katz, V.A., Fuchs, A., and Conrath, U. 2002. Pretreatment with salicylic acid primes parsley cells for enhanced ion transport following elicitation. *FEBS Lett.* 520:53–57.

Kauss, H., and Jeblick, W. 1995. Pretreatment of parsley suspension cultures with salicylic acid enhances spontaneous and elicited production of H_2O_2. *Plant Physiol.* 108:1171–1178.

Kauss, H., Theisinger-Hinkel, E., Mindermann, R., and Conrath, U. 1992a. Dichloroisonicotinic and salicylic acid, inducers of systemic acquired resistance, enhance fungal elicitor responses in parsley cells. *Plant J.* 2:655–660.

Kauss, H., Krause, K., and Jeblick, W. 1992b. Methyl jasmonate conditions parsley suspension cells for increased elicitation of phenylpropanoid defense responses. *Biochem. Biophys. Res. Comm.* 189:304–308.

Kauss, H., Franke, R., Krause, K., Conrath, U., Jeblick, W., Grimmig, B., and Matern, U. 1993. Conditioning of parsley *Petroselinum crispum*) suspension cells increases elicitor-induced incorporation of cell wall phenolics. *Plant Physiol.* 102:459–466.

Kauss, H., Jeblick, W., Ziegler, J., and Krabler, W. 1994. Pretreatment of parsley (*Petroselinum crispum* L.) suspension cultures with methyl jasmonate enhances elicitation of activated oxygen species. *Plant Physiol.* 105:89–104.

Klessig, D.F., and Malamy, J. 1994. The salicylic acid signal in plants. *Plant Mol. Biol.* 26:1439–1485.

Klüsener, B., and Weiler, E.W. 1999. Pore-forming properties of elicitors of plant defense reactions and cellulolytic enzymes. *FEBS Lett.* 459:263–266.

Knoester, M., Van Loon, L.C., Van den Heuvel, J., Hennig, J., Bol, J.F., and Linthorst, H.J.M. 1998. Ethylene-insensitive tobacco lacks nonhost resistance against soil-borne fungi. *Proc. Natl. Acad. Sci. USA* 95:1933–1937.

Knoester, M., Linthorst, H.J.M., Bol, J.F., and Van Loon, L.C. 2001. Involvement of ethylene in lesion development and systemic acquired resistance in tobacco during the hypersensitive reaction to tobacco mosaic virus. *Physiol. Mol. Plant Pathol.* 59:45–57.

Kohler, A., Schwindling, S., and Conrath, U. 2002. Benzothiadiazole-induced priming for potentiated responses to pathogen infection, wounding, and infiltration of water into leaves requires the *NPR1/NIM1* gene in *Arabidopsis*. *Plant Physiol.* 128:1046–1056.

Kubigsteltig, I., Laudert, D., and Weiler, E.W. 1999. Structure and regulation of the *Arabidopsis thaliana* allene oxide synthase gene. *Planta* 208:463–471.

Kuć, J. 1987. Translocated signals for plant immunization. *Ann. N.Y. Acad. Sci.* 494:221–223.

Latunde-Dada, A.O., and Lucas, J.A. 2001. The plant defense activator acibenzolar-S-methyl primes cowpea [*Vigna unguiculata* (L.) Walp.] seedlings for rapid induction of resistance. *Physiol. Mol. Plant Pathol.* 58:199–208.

Lawton, K.A., Potter, S.L., Uknes, S., and Ryals, J. 1994. Acquired resistance signal transduction in *Arabidopsis* is ethylene independent. *Plant Cell* 6:581–588.

Lee, S., Suh, S., Kim, S., Crain, R.C., Kwak, J.M., Nam, H.-G., and Lee, Y. 1997. Systemic elevation of phosphatidic acid and lysophospholipid levels in wounded plants. *Plant J.* 12:547–556.

Leeman, M., Van Pelt, J.A., Hendrickx, M.J., Scheffer, R.J., Bakker, P.A.H.M., and Schippers, B. 1995. Biocontrol of fusarium wilt of radish in commercial greenhouse trials by seed treatment with *Pseudomonas fluorescens* WCS374. *Phytopathology* 85:1301–1305.

León, J., Rojo, E., and Sánchez-Serrano, J.J. 2001. Wound signalling in plants. *J. Exp. Bot.* 52:1–9.

Li, L., and Howe, G.A. 2001. Alternative splicing of prosystemin pre-mRNA produces two isoforms that are active as signals in the wound response pathway. *Plant Mol. Biol* 46:409–419.

Li, L., Li, C., and Howe, G.A. 2001. Genetic analysis of wound signaling in tomato. Evidence for a dual role of jasmonic acid in defense and female fertility. *Plant Physiol.* 127:1414–1417.

Li, L., Li, C., Lee, G.I., and Howe, G.A. 2002. Distinct roles for jasmonate synthesis and action in the systemic wound response of tomato. *Proc. Natl. Acad. Sci. USA* 99:6416–6421.

Lund, S.T., Stall, R.E., and Klee, H.J. 1998. Ethylene regulates the susceptible response to pathogen infection in tomato. *Plant Cell* 10:371–382.

Malamy, J., and Klessig, D.F. 1992. Salicylic acid and plant disease resistance. *Plant J.* 2:643–654.

Mauch, F., Hadwiger, L.A., and Boller, T. 1994. Ethylene: symptom, not signal for the induction of chitinase and β-1,3-glucanase in pea pods by pathogens and elicitors. *Plant Physiol.* 76:607–611.

McConn, J., Creelman, R.A., Bell, E., Mullet, J.E., and Browse, J. 1997. Jasmonate is essential for insect defense in *Arabidopsis*. *Proc. Natl. Acad. Sci. USA* 94:5473–5477.

McGurl, B., and Ryan, C.A. 1992. The organization of the prosystemin gene. *Plant Mol. Biol.* 20:405–409.

McGurl, B., Pearce, G., Orozco-Cardenas, M., and Ryan, C.A. 1992. Structure, expression and antisense inhibition of the systemin precursor gene. *Science* 255:1570–1573.

McGurl, B., Orozco-Cardenas, M., Pearce, G., and Ryan, C.A. 1994. Overexpression of the prosystemin gene in transgenic tomato plants generates a systemic signal that constitutively induces proteinase inhibitor synthesis. *Proc. Natl. Acad. Sci. USA* 91:9799-9802.

Meindl, T., Boller, T., and Felix, G. 1998. The plant wound hormone systemin binds with the N-terminal part to its receptor but needs the C-terminal part to activate it. *Plant Cell* 10:1561–1570.

Mittler, R., and Lam, E. 1996. Sacrifice in the face of foes: pathogen-induced programmed cell death in plants. *Trends Microbiol.* 4:10–15.

Mittler, R., Shulaev, V., and Lam, E. 1995. Coordinated activation of programmed cell death and defense mechanisms in transgenic tobacco plants expressing a bacterial proton pump. *Plant Cell* 7:29–42.

Moran, P. 1998. Plant-mediated interactions between insects and fungal plant pathogen and the role of chemical responses to infection. *Oecologia* 115:513–530.

Moyen, C., and Johannes, E. 1996. Systemin transiently depolarizes the tomato mesophyll cell membrane and antagonizes fusicoccin-induced extracellular acidification of mesophyll tissue. *Plant Cell Environ.* 19:464–470.

Moyen, C., Hammond-Kosack, K.E., Jones, J., Knight, M.R., and Johannes, E. 1998. Systemin triggers an increase of cytoplasmic calcium in tomato mesophyll cells: Ca^{2+} mobilization from intra- and extracellular compartments. *Plant Cell Environ.* 21:1101–1111.

Mur, L.A.J., Brown, I.R., Darby, R.M., Bestwick, C.S., Bi, Y.-M., Mansfield, J.W., and Draper, J. 1996. Salicylic acid potentiates defense gene expression in tissue exhibiting acquired resistance to pathogen attack. *Plant J.* 9:559–571.

Mur, L.A., Brown, I.R., Darby, R.M., Bestwick, C.S., Bi, Y.-M., Mansfield, J.W., Draper, J. 2000. A loss of resistance to avirulent bacterial pathogens in tobacco is associated with the attenuation of a salicylic acid-potentiated oxidative burst. *Plant J.* 23:609–621.

Narváez-Vásquez, J., Florin-Christensen, J., and Ryan, C.A. 1999. Positional specificity of a phospholipase A activity induced by wounding, systemin, and oligosaccharide elicitors in tomato leaves. *Plant Cell* 11:2249–2260.

Niki, T., Mitsuhara, I., Seo, S., Ohtsubo, N., and Ohashi, Y. 1998. Antagonistic effect of salicylic acid and jasmonic acid on the expression of pathogenesis-related (PR) protein genes in wounded mature tobacco leaves. *Plant Cell Physiol.* 39:500–507.

Norman-Setterblad, C., Vidal, S., and Palva, T.E. 2000. Interacting signal pathways control defense gene expression in *Arabidopsis* in response to cell wall-degrading enzymes from *Erwinia carotovora*. *Mol. Plant Microbe Interact.* 13:430–438.

O'Donnell, P.J., Calvert, C., Atzorn, R., Wasternack, C., Leyser, H.M.O., and Bowles, D.J. 1996. Ethylene as a signal mediating the wound response of tomato plants. *Science* 274:1914–1917.

Orozco-Cárdenas, M.L. 2000. Hydrogen peroxide acts as a second messenger for the induction of defense genes in tomato plants in response to wounding, systemin, and methyl jasmonate. *Plant Cell* 13:179–191.

Orozco-Cardenas, M., and Ryan, C.A. 1999. Hydrogen peroxide is generated systemically in plant leaves by wounding and systemin via the octadecanoid pathway. *Proc. Natl. Acad. Sci. USA* 96:6553–6557.

Orozco-Cardenas, M., McGurl, B., and Ryan, C.A. 1993. Expression of an antisense prosystemin gene in tomato plants reduces the resistance toward *Manduca sexta* larvae. *Proc. Natl. Acad. Sci. USA* 90:8273–8276.

Pearce, G., Strydom, D., Johnson, S., and Ryan, C.A. 1991. A polypeptide from tomato leaves induces wound-inducible proteinase inhibitor proteins. *Science* 253:895–898.

Pearce, G., Moura, D.S., Stratmann, J., and Ryan, C.A. 2001. Production of multiple plant hormones from a single polyprotein precursor. *Nature* 411:817–820.

Peña-Cortés, H., Albrecht, T., Prat, S., Weiler, E.W., and Willmitzer, L. 1993. Aspirin prevents wound-induced gene expression in tomato leaves by blocking jasmonic acid biosynthesis. *Planta* 191:123–128.

Penninckx, I.A.M.A., Eggermont, K., Terras, F.R.G., Thomma, B.P.H.J., De Samblanx, G.W., Buchala, A., Métraux, J.-P., Manners, J.M., and Brokaert, W.F. 1996.

Pathogen-induced systemic activation of a plant defensin gene in *Arabidopsis* follows a salicylic acid-independent pathway. *Plant Cell* 8:2309–2323.

Penninckx, I.A.M.A., Thomma, B.P.H.J., Buchala, A., and Métraux, J.-P. 1998. Concomitant activation of jasmonate and ethylene response pathways is required for induction of a plant defensin gene in *Arabidopsis*. *Plant Cell* 10:2103–2113.

Pieterse, C.M.J., Van Wees, S.C.M., Hoffland, E., Van Pelt, J.A., and Van Loon, L.C. 1996. Systemic resistance in *Arabidopsis* induced by biocontrol bacteria is independent of salicylic acid accumulation and pathogenesis-related gene expression. *Plant Cell* 8:1225–1237.

Pieterse, C.M.J., Van Wees, S.C.M., Van Pelt, J.A., Knoester, M., Laan, R., Gerrits, H., Weisbeek, P.J., and Van Loon, L.C. 1998. A novel signaling pathway controlling induced systemic resistance in *Arabidopsis*. *Plant Cell* 10:1571–1580.

Pieterse, C.M.J., Van Pelt, J.A., Ton, J., Parchmann, S., Mueller, M.J., Buchala, A.J., Métraux, J.-P., and Van Loon, L.C. 2000. Rhizobacteria-mediated induced systemic resistance (ISR) in *Arabidopsis* requires sensitivity to jasmonate and ethylene but is not accompanied by an increase in their production. *Physiol. Mol. Plant Pathol.* 57:123-134.

Pieterse, C.M.J., Van Wees, S.C.M., Ton, J., Van Pelt, J.A., and Van Loon, L.C. 2002. Signaling in rhizobacteria-induced systemic resistance in *Arabidopsis thaliana*. *Plant Biol.* 4:535–544.

Preston, C.A., Lewandowski, C., Enyedi, A.J., and Baldwin, I.T. 1999. Tobacco mosaic virus inoculation inhibits wound-induced jasmonic acid-mediated responses within but not between plants. *Planta* 209:87–95.

Raskin, I. 1992. Role of salicylic acid in plants. *Annu. Rev. Plant Physiol. Plant Mol. Biol.* 43:439–463.

Reymond, P., Weber, H., Damond, M., and Farmer, E.E. 2000. Differential gene expression in response to mechanical wounding and insect feeding in *Arabidopsis*. *Plant Cell* 12:707–719.

Rizhsky, L., and Mittler, R. 2001. Inducible expression of bacterio-opsin in transgenic tobacco and tomato plants. *Plant Mol. Biol.* 46:313–323.

Roberts, M.R., and Bowles, D.J. 1999. Fusicoccin, 14-3-3 proteins, and defense responses in tomato plants. *Plant Physiol.* 119:1243–1250.

Rojo, E., Titarenko, E., León, J., Berger, S., Vancanneyt, G., and Sánchez-Serrano, J.J. 1998. Reversible protein phosphorylation regulates jasmonic acid-dependent and -independent wound signal transduction pathways in *Arabidopsis thaliana*. *Plant J.* 13:153-165.

Rojo, E., León, J., and Sánchez-Serrano, J.J. 1999. Cross-talk between wound signalling pathways determines local versus systemic gene expression in *Arabidopsis thaliana*. *Plant J.* 20:135–142.

Ryals, J.A., Neuenschwander, U.H., Willits, M.G., Molina, A., Steiner, H.-Y., and Hunt, M.D. 1996. Systemic acquired resistance. *Plant Cell* 8:1809–1819.

Ryan, C.A. 1992. The search for the proteinase inhibitor-inducing factor, PIIF. *Plant Mol. Biol.* 19:123–133.

Ryan, C.A. 2000. The systemin signaling pathway: differential activation of plant defensive genes. *Biochim. Biophys. Acta* 1477:112–121.

Schaller, A. 1998. Action of proteolysis-resistant systemin analogues in wound signalling. *Phytochemistry* 47:605–612.

Schaller, A. 1999. Oligopeptide signalling and the action of systemin. *Plant Mol. Biol.* 40:763–769.

Schaller, A. 2001. Bioactive peptides as signal molecules in plant defense, growth, and development. In *Bioactive Natural Products,* ed. Atta-Ur-Rahman, Vol. 25, pp. 367–411. Amsterdam, The Netherlands: Elsevier.

Schaller, A., and Frasson, D. 2001. Induction of wound response gene expression in tomato leaves by ionophores. *Planta* 212:431–435.

Schaller, A., and Oecking, C. 1999. Modulation of plasma membrane H^+-ATPase activity differentially activates wound and pathogen defense responses in tomato plants. *Plant Cell* 11:263–272.

Schaller, A., and Ryan, C.A. 1995. Systemin—a polypeptide defense signal in plants. *BioEssays* 18:27–33.

Schaller, A., Roy, P., and Amrhein, N. 2000. Salicylic acid-independent induction of pathogenesis-related gene expression by fusicoccin. *Planta* 210:599–606.

Scheel, D. 1998. Resistance response physiology and signal transduction. *Curr. Opin. Plant Biol.* 1:305–310.

Scheer, J.M., and Ryan, C.A. 1999. A 160 kDa systemin receptor on the cell surface of *Lycopersicon peruvianum* suspension cultured cells: kinetic analyses, induction by methyl jasmonate and photoaffinity labeling. *Plant Cell* 11:1525–1535.

Scheer, J.M., and Ryan, C.A. 2002. The systemin receptor SR160 from *Lycopersicon perivianum* is a member of the LRR receptor kinase family. *Proc. Natl. Acad. Sci. USA* 99:9585–9590.

Schenk, P.M., Kazal, K., Wilson, I., Anderson, J.P., Richmond, T., Somerville, S.C., and Manners, J.M. 2000. Coordinated plant defense responses in *Arabidopsis* revealed by microarray analysis. *Proc. Natl. Acad. Sci.* 97:11655–11660.

Schmele, I., and Kauss, H. 1990. Enhanced activity of the plasma membrane localized callose synthase in cucumber leaves with induced resistance. *Physiol. Mol. Plant Pathol.* 37:221–228.

Schweizer, P., Buchala, A., and Métraux, J.-P. 1997. Gene expression patterns and levels of jasmonic acid in rice treated with the resistance inducer 2,6-dichloroisonicotinic acid. *Plant Physiol.* 115:61–70.

Seo, S., Okamoto, M., Seto, H., Ishizuka, K., Sano, H., and Ohashi, Y. 1995. Tobacco MAP kinase: A possible mediator in wound signal transduction pathways. *Science* 270:1988–1992.

Seo, S., Sano, H., and Ohashi, Y. 1999. Jasmonate-based wound signal transduction requires activation of WIPK, a tobacco miotogen-activated protein kinase. *Plant Cell Physiol.* 11:289–298.

Shah, J., and Klessig, D.F. 1999. Salicylic acid: signal perception and transduction. In *Biochemistry and Molecular Biology of Plant Hormones,* eds. P.P.J. Hooykaas, M.A. Hall, and K.R. Libbenga, pp. 513–541. Amsterdam, The Netherlands: Elsevier Science.

Shah, J., Tsui, F., and Klessig, D.F. 1997. Characterization of a salicylic acid-insensitive mutant *(sai1* of *Arabidopsis thaliana,* identified in a selective screen utilizing the SA-inducible expression of the *TMS2* gene. *Mol. Plant Microbe Interact.* 10:69–78.

Shelp, B.J., Bown, A.W., and McLean, M.D. 1999. Metabolism and function of gamma-aminobutyric acid. *Trends Plant Sci.* 4:446–452.

Shirasu, K., Nakajima, H., Rajasekhar, K., Dixon, R.A., and Lamb, C. 1997. Salicylic acid potentiates an agonist-dependent gain control that amplifies pathogen signals in the activation of defense mechanisms. *Plant Cell* 9:261–270.

Shumway, L.K., Yang, V.V., and Ryan, C.A. 1976. Evidence for the presence of proteinase inhibitor I in vacuolar bodies of plant cells. *Planta* 129:161–165.

Siegrist, J., Orober, M., and Buchenauer, H. 2000. β-aminobutyric acid-mediated enhancement of resistance in tobacco to tobacco mosaic virus depends on the accumulation of salicylic acid. *Physiol. Mol. Plant. Pathol.* 56:95–106.

Stankovic, B., and Davies, E. 1997. Intercellular communication in plants: electrical stimulation of proteinase inhibitor gene expression in tomato. *Planta* 202:402–406.

Staswick, P.E., Su, W.P., and Howell, S.H. 1992. Methyl jasmonate inhibition of root growth and induction of a leaf protein are decreased in an *Arabidopsis thaliana* mutant. *Proc. Natl. Acad. Sci. USA* 89:6837–6840.

Staswick, P.E., Yuen, G.Y., and Lehman, C.C. 1998. Jasmonate signaling mutants of *Arabidopsis* are susceptible to the soil fungus *Pythium irregulare*. *Plant J.* 15:747–754.

Stennis, M.J., Chandra, S., Ryan, C.A., and Low, P. 1998. Systemin potentiates the oxidative burst in cultured tomato cells. *Plant Physiol.* 117:1031–1036.

Sticher, L., Mauch-Mani, B., and Métraux, J.-P. 1997. Systemic acquired resistance. *Annu. Rev. Phytopath.* 35:235–270.

Stintzi, A., Weber, H., Reymond, P., Browse, J., and Farmer, E.E. 2001. Plant defense in the absence of jasmonic acid: the role of cyclopentenones. *Proc. Natl. Acad. Sci. USA* 98:12837–12842.

Stout, M.J., Fidantsef, A.L., Duffey, S.S., and Bostock, R.M. 1999. Signal interactions in pathogen and insect attack: systemic plant-mediated interactions between pathogens and herbivores of the tomato, *Lycopersicon esculentum*. *Physiol. Mol. Plant Pathol.* 54:115–130.

Strassner, J., Schaller, F., Frick, U.B., Howe, G.A., Weiler, E.W., Amrhein, N., Macheroux, P., and Schaller, A. 2002. Characterization and cDNA-microarray expression analysis of 12-oxophytodienoate reductases reveals differential roles for octadecanoid biosynthesis in the local versus the systemic wound response. *Plant J.* 32:585–601.

Stratmann, J.W., and Ryan, C.A. 1997. Myelin basic protein kinase activity in tomato leaves is induced systemically by wounding and increases in response to systemin and oligosaccharide elicitors. *Proc. Natl. Acad. Sci. USA* 94:11085–11089.

Stratmann, J., Scheer, J., and Ryan, C.A. 2000. Suramin inhibits initiation of defense signaling by systemin, chitosan, and a β-glucan elicitor in suspension-cultured. *Lycopersicon peruvianum* cells. *Proc. Natl. Acad. Sci. USA* 97:8862–8867.

Stumm, D., and Gessler, C. 1986. Role of papillae in the induced systemic resistance of cucumbers against *Colletotrichum lagenarium*. *Physiol. Mol. Plant Pathol.* 29:405–410.

Thain, J.F., Gubb, I.R., and Wildon, D.C. 1995. Depolarization of tomato leaf cells by oligogalacturonide elicitors. *Plant Cell Environ.* 18:211–214.

Thaler, J.S., Fidantsef, A.L., Duffey, S.S., and Bostock, R.M. 1999. Trade-offs in plant defense against pathogens and herbivores: a field demonstration of chemical elicitors of induced resistance. *J. Chem. Ecol.* 25:1597–1609.

Thomma, B.P.H.J., Eggermont, K., Penninckx, I.A.M.A., Mauch-Mani, B., Vogelsang, R., Cammue, B.P.A., and Broekaert, W.F. 1998. Separate jasmonate-dependent and salicylate-dependent defense-response pathways in *Arabidopsis* are essential for resistance to distinct microbial pathogens. *Proc. Natl. Acad. Sci. USA*95:15107–15111.

Thomma, B.P.H.J., Eggermont, K., Tierens, K.F.M., and Broekaert, W.F. 1999. Requirement of functional *ethylene-insensitive 2* gene for efficient resistance of *Arabidopsis* to infection by *Botrytis cinerea*. *Plant Physiol.* 121:1093–1102.

Thulke, O.U., and Conrath, U. 1998. Salicylic acid has a dual role in the activation of defense-related genes in parsley. *Plant J.* 14:35–42.

Ton, J., Davison, S., Van Wees, S.C.M., Van Loon, L.C., and Pieterse, C.M.J. 2001. The *Arabidopsis ISR1* locus controlling rhizobacteria-mediated induced systemic resistance is involved in ethylene signaling. *Plant Physiol.* 125:652–661.

Ton, J., De Vos, M., Robben, C., Buchala, A.J., Métraux, J.-P., Van Loon, L.C., and Pieterse, C.M.J. 2002a. Characterisation of *Arabidopsis* enhanced disease susceptibility mutants that are affected in systemically induced resistance. *Plant J.* 29:11–21.

Ton, J., Van Pelt, J.A., Van Loon, L.C., and Pieterse, C.M.J. 2002b. Differential effectiveness of salicylate-dependent and jasmonate/ethylene-dependent induced resistance in *Arabidopsis. Mol. Plant Microbe Interact.* 15:27–34.

Van Loon, L.C., and Van Strien, E.A. 1999. The families of pathogenesis-related proteins, their activities, and comparative analysis of PR-1 type proteins. *Physiol. Mol. Plant Pathol.* 55:85–97.

Van Loon, L.C., Bakker, P.A.H.M., and Pieterse, C.M.J. 1998. Systemic resistance induced by rhizosphere bacteria. *Annu. Rev. Phytopathol.* 36:453–483.

Van Wees, S.C.M., Pieterse, C.M.J., Trijssenaar, A., Van 't Westende, Y.A.M., Hartog, F., and Van Loon, L.C. 1997. Differential induction of systemic resistance in *Arabidopsis* by biocontrol bacteria. *Mol. Plant Microbe Interact.* 10:716–724.

Van Wees, S.C.M., Luijendijk, M., Smoorenburg, I., van Loon, L.C., and Pieterse, C.M.J. 1999. Rhizobacteria-mediated induced systemic resistance (ISR) in *Arabidopsis* is not associated with a direct effect on known defense-genes but stimulates the expression of the jasmonate-inducible gene *ATVSP* upon challenge. *Plant Mol. Biol.* 41:537–549.

Van Wees, S.C.M., De Swart, E.A.M., Van Pelt, J.A., Van Loon, L.C., and Pieterse, C.M.J. 2000. Enhancement of induced disease resistance by simultaneous activation of salicylate- and jasmonate-dependent defense pathways in *Arabidopsis thaliana. Proc. Natl. Acad. Sci. USA* 97:8711–8716.

Vanacker, H., Lu, H., Rate, D.N., and Greenberg, J.T. 2001. A role for salicylic acid and NPR1 in regulating cell growth in *Arabidopsis. Plant J.* 28:209–216.

Vetsch, M., Janzik, I., and Schaller, A. 2000. Characterization of prosystemin expressed in the baculovirus/insect cell system reveals biological activity of the systemin precursor. *Planta* 211:91–91.

Vian, A., Henry-Vian, C., Schantz, R., Ledoigt, G., Frachisse, J.-M., Desbiez, M.-O., and Julien, J.-L. 1996. Is membrane potential involved in calmodulin gene expression after external stimulation in plants? *FEBS Lett.* 380:93–96.

Vidal, S., Ponce de León, I., Denecke, J., and Palva, T.E. 1997. Salicylic acid and the plant pathogen *Erwinia carotovora* induce defense genes via antagonistic pathways. *Plant J.* 11:115–123.

Vijayan, P., Shockey, J., Levesque, C.A., Cook, R.J., and Browse, J. 1998. A role for jasmonate in pathogen defense of *Arabidopsis. Proc. Natl. Acad. Sci. USA* 95:7209–7214.

Walker-Simmons, M.K., and Ryan, C.A. 1977. Immunological identification of proteinase inhibitors I and II in isolated leaf vacuoles. *Plant Physiol.* 60:61–63.

Walling, L.L. 2000. The myriad plant responses to herbivores. *J. Plant Growth Regul.* 19:195–216.

Wang, C., Zien, C.A., Afitlhile, M., Welti, R., Hildebrand, D.F., and Wang, X. 2000. Involvement of phospholipase D in wound-induced accumulation of jasmonic acid in *Arabidopsis. Plant Cell* 12:2237–2246.

Weymann, K., Hunt, M., Uknes, S., Neuenschwander, U., Lawton, K., Steiner, H.-Y., and Ryals, J. 1995. Suppression and restauration of lesion formation in *Arabidopsis lsd* mutants. *Plant Cell* 7:2013–2022.

Xu, Y., Chang, P.-F- L., Liu, D., Narasimhan, M.L., Raghothama, K.G., Hasegawa, P.M., and Bressan, R.A. 1994. Plant defense genes are synergistically induced by ethylene and methyl jasmonate. *Plant Cell* 6:1077–1085.

Zehnder, G.W., Murphy, J.F., Sikora, E.J., and Kloepper, J.W. 2001. Application of rhizobacteria for induced resistance. *Eur. J. Plant Pathol.* 107:39–50.

Ziegler, J., Keinänen, M., and Baldwin, I. 2001. Herbivore-induced allene oxide synthase transcripts and jasmonic acid in *Nicotiana attenuata*. *Phytochemistry* 58:729–738.

Zimmerli, L., Jakab, G., Métraux, J.P., and Mauch-Mani, B. 2000. Potentiation of pathogen-specific defense mechanisms in *Arabidopsis* by beta-aminobutyric acid. *Proc. Natl. Acad. Sci. USA* 97:12920–12925.

Zimmerli, L., Métraux, J.P., and Mauch-Mani, B. 2001. β-aminobutyric acid-induced protection of *Arabidopsis* against the necrotrophic fungus *Botrytis cinerea*. *Plant Physiol.* 126:517–523.

9

The Relationship Between Basal and Induced Resistance in *Arabidopsis*

Jurriaan Ton, Corné M.J. Pieterse, and L.C. Van Loon

9.1 Introduction

Plants are constantly exposed to potentially pathogenic microorganisms. They possess an extensive array of passive and active defense mechanisms, and only a small proportion of microorganisms are capable of infecting the plant and causing disease. Plant resistance can be broadly defined as the plant's ability to suppress or retard the damaging activity of a pathogen. The most common type of resistance is nonhost resistance. This type of resistance protects the plant entirely from infection by most potential pathogens, and is manifested as an inability of the pathogen to cause disease upon contact with any individual of a particular plant species. In such an interaction, the pathogen is nonpathogenic.

If certain individuals within populations of the species are susceptible to some races of a pathogen but resistant to other races of the same pathogen, the interaction usually follows a gene-for-gene relationship. In compatible interactions, the pathogen is able to colonize the plant and cause disease. In contrast, in incompatible interactions, the pathogen is capable of initiating infection, but rapidly arrested at the site of infection. The resulting race-specific or vertical resistance is generally controlled by a single dominant resistance (R) gene in the host, which encodes a product that either directly or indirectly recognizes the product of a matching dominant effector (avirulence; *Avr*) gene expressed by the pathogen. Usually, this early recognition of the so-called avirulent pathogen gives rise to a hypersensitive response (HR). The HR involves a range of active defense mechanisms, including a form of programmed cell death at the site of infection.

Accumulation of anti-microbial compounds, fortification of cell walls, and expression of defense-related genes in the surrounding tissue all contribute to inhibit further colonization of the plant by the pathogen (Hammond-Kosack and Jones, 1996). If the invading pathogen does not carry an *Avr* gene that is recognized by the host, the plant fails to activate a HR. However, these so-called virulent pathogens can still be restrained by nonspecific defenses that can afford various levels of protection. This type of resistance is not well defined, but is generally referred to

197

as polygenic, multigenic, horizontal, or basal resistance, and acts in slowing down the rate of disease development.

Besides primary resistance responses, plants can express an enhanced defensive capacity after being exposed to certain biotic or abiotic stimuli. This resistance is commonly referred to as induced resistance. In this chapter, we focus on the relationship between basal resistance and induced resistance.

9.2 Signal Compounds Involved in Primary Disease Resistance

The plant hormones salicylic acid (SA), jasmonic acid (JA), and ethylene (ET) have repeatedly been implicated in the regulation of resistance responses. In many cases, infection by both avirulent and virulent pathogens is associated with enhanced production of these regulators, and exogenous application of these compounds often results in an enhanced level of resistance (Boller, 1991; Dempsey et al., 1999; Pieterse et al., 1996, 1998, 2000; Thomma et al., 2000). Moreover, blocking the response to either of these signals can render plants more susceptible to certain pathogens or even insects (Delaney et al., 1994; Knoester et al., 1998; McConn et al., 1997; Staswick et al., 1998; Stout et al., 1999; Thomma et al., 1998; Ton et al., 2001; Van Wees et al., 1999). A central role for SA became apparent with the use of NahG transformants. Transgenic NahG plants constitutively express the bacterial *NahG* gene, encoding salicylate hydroxylase, which converts SA into catechol. Tobacco and *Arabidopsis* NahG plants show enhanced disease susceptibility to a broad range of oomycetous, fungal, bacterial, and viral pathogens (Delaney et al., 1994; Kachroo et al., 2000). Recently, a screen based on impaired accumulation of SA after pathogen infection resulted in the identification of two *Arabidopsis* mutants that are affected in pathogen-induced biosynthesis of SA (Nawrath and Métraux, 1999). Both mutants, *sid1* and *sid2*, displayed enhanced susceptibility to the virulent pathogens *Pseudomonas syringae* pv. *tomato* and *Peronospora parasitica*, demonstrating the importance of SA in the basal resistance against both the bacterial and the oomycetous pathogen. Mutants *sid1* and *sid2* are allelic with enhanced disease susceptibility mutants *eds5* (Nawrath and Métraux, 1999) and *eds16* (Wildermuth et al., 2001), respectively, which were characterized as having enhanced susceptibility to a virulent strain of the bacterial pathogen *Xanthomonas campestris* pv. *raphani* (Rogers and Ausubel, 1997) and the mildew fungus *Erysiphe orontii* (Wildermuth et al., 2001).

Evidence for the role of JA in pathogen resistance came predominantly from analyses of *Arabidopsis* mutants affected in the biosynthesis of, or responsiveness to, JA. The JA-response mutant *coi1* has been documented as displaying enhanced susceptibility to the necrotrophic fungi *Alternaria brassicicola* and *Botrytis cinerea* (Thomma et al., 1998), and the bacterial leaf pathogen *Erwinia carotovora* pv. *carotovora* (Norman-Setterblad et al., 2000). Furthermore, *coi1* was reported to also exhibit an altered level of basal resistance against *P. syringae* (Feys et al., 1994). Another JA-insensitive mutant of *Arabidopsis*, *jar1*, allows enhanced levels of growth of virulent *P. syringae* pv. *tomato* in the leaves (Pieterse et al., 1998). This clearly demonstrates that JA-dependent defenses contribute to basal resistance against these pathogens. Furthermore, both the *jar1* mutant and the *fad3, fad7, fad8* triple mutant, which is defective in JA biosynthesis, exhibit susceptibility to normally nonpathogenic soilborne oomycetes of the genus *Pythium* (Staswick et al., 1998; Vijayan et al., 1998). These findings indicate that JA plays a role in nonhost resistance against these oomycetes. Besides involvement in resistance responses against microbial pathogens, JA also contributes to basal resistance against insects (McConn et al., 1997).

The role of ET in plant resistance seems more ambiguous. In some cases ET promotes disease development, whereas in other cases it is associated with disease resistance. For instance, ET-insensitive tomato genotypes allowed wild-type levels of growth of virulent *P. syringae* pv. *tomato* and *X. campestris* pv. *vesicatoria*, but symptoms of disease were less severe (Ciardi et al., 2000; Lund et al., 1998). In these cases, ET clearly regulates symptom expression rather than plant resistance. In *Arabidopsis* the ET-insensitive mutant *ein2-1* was found to be compromised in disease development due to infection by *P. syringae* pv. *tomato* and *X. campestris* pv. *vesicatoria* (Bent et al., 1992). However, various ET-insensitive genotypes of *Arabidopsis* also allow enhanced levels of growth of *P. syringae* pv. *tomato* and *X. campestris* pv. *vesicatoria* (Pieterse et al., 1998; Ton et al., 2002c), indicating that ET also contributes to basal resistance. Knoester et al. (1998) reported that ET-insensitive tobacco transformed with the mutant ET receptor gene *etr1-1* from *Arabidopsis* (Tetr tobacco), displayed susceptibility to the normally nonpathogenic oomycete *Pythium sylvaticum*. This demonstrates that, like JA, ET plays a role in nonhost resistance against *Pythium*. Furthermore, several ET-insensitive mutants of *Arabidopsis* exhibit enhanced disease susceptibility to *B. cinerea* (Thomma et al., 1999), *Plectosphaerella cucumerina* (Berrocal-Lobo et al., 2002) and *E. carotovora* (Norman-Setterblad et al., 2000). These observations indicate that ET-dependent defenses contribute to basal resistance against these pathogens.

SA, JA, and ET not only regulate basal and nonhost resistance responses, but are also instrumental in boosting defense reactions in race-specific resistance. All three regulators are strongly increased during the hypersensitive reaction (e.g., Pieterse et al., 2000) and induce the expression of several defense-related genes (Maleck et al., 2000). However, plant genotypes that are impaired in the production of, or the responsiveness to, SA, JA, or ET are still capable of expressing an HR (Delaney et al., 1994; Knoester et al., 1998; Vijayan et al., 1998), indicating that *R* gene-dependent resistance is still functional in the absence of any of these regulators.

Depending on the host-pathogen interaction, SA, JA, and ET appear to be differentially involved in basal resistance or nonhost resistance. In *Arabidopsis*, some pathogens have been shown to be resisted predominantly through SA-dependent pathways, i.e., *P. parasitica* and turnip crinkle virus (TCV), whereas others are resisted predominantly through JA- and ET-dependent resistance mechanisms, i.e., *A. brassicicola*, *B. cinerea*, and *E. carotovora*. Table 9.1 summarizes the data demonstrating differential involvement of SA, JA, and ET in basal

TABLE 9.1. Differential involvement of salicylic acid (SA), jasmonic acid (JA), and ethylene (ET) in the regulation of basal resistance in different plant–pathogen interactions.

Plant species	Pathogen	Signals involved in basal resistance[a]			Reference
		SA	JA	ET	
Arabidopsis	*Peronospora parasitica*	+	0	0	Delaney et al. (1994); Thomma et al. (1998)
Arabidopsis	Turnip crinkle virus	+	0	0	Kachroo et al. (2000)
Tobacco	Tobacco mosaic virus	+	n.d.	0	Delaney et al. (1994); Knoester et al. (1998)
Tobacco	*Phytophthora infestans*	+	n.d.	n.d.	Delaney et al. (1994)
Tobacco	*Cercospora nicotianae*	+	n.d.	n.d.	Delaney et al. (1994)
Arabidopsis	*Xanthomonas campestris*	+	+	+	Rogers and Ausubel (1997); Ton et al. (2002c)
Arabidopsis	*Pseudomonas syringae*	+	+	+	Delaney et al. (1994); Pieterse et al. (1998)
Arabidopsis	*Plectosphaerella cucumerina*	+	+	+	Berrocal-Lobo et al. (2002)
Arabidopsis	*Pythium irregulare*	n.d.	+	n.d.	Staswick et al. (1998)
Arabidopsis	*Pythium mastophorum*	n.d.	+	n.d.	Vijayan et al. (1998)
Tobacco	*Pythium sylvaticum*	n.d.	n.d.	+	Knoester et al. (1998)
Arabidopsis	*Alternaria brassicicola*	0	+	+	Thomma et al. (1998, 1999)
Arabidopsis	*Botrytis cinerea*	0	+	+	Thomma et al. (1998, 1999)
Arabidopsis	*Erwinia carotovora*	0	+	+	Norman-Setterblad et al. (2000)

[a]Based on the enhanced susceptibility of transgenics/mutants of *Arabidopsis* and tobacco, impaired in the accumulation of, or responsiveness to, a particular hormone.
+: transgenic/mutant displaying enhanced disease susceptibility compared to wild-type plants.
0: transgenic/mutant displaying the same level of basal resistance as wild-type plants.
n.d.: not determined.

resistance. The information presented is based on enhanced susceptibility phenotypes of transgenics or mutants of *Arabidopsis* and tobacco that are impaired either in the accumulation of, or in the responsiveness to, any of these signal compounds.

9.3 Induced Disease Resistance

9.3.1 Biologically and Chemically Induced Resistance

Plants also possess adaptive defense mechanisms to counteract pathogen or insect attack. Upon appropriate stimulation, plants are capable of developing an enhanced defensive capacity, commonly referred to as induced resistance. The state of induced resistance depends either on defensive compounds that are produced as a result of the induction treatment, and/or on a more rapid and stronger activation of extant defense mechanisms upon challenge inoculation with a pathogen. The latter mechanism is variously referred to as "priming", "sensitization", or "potentiation". In either case, the resistance-inducing agent can predispose the plant to better resist subsequent pathogen attack. Induced resistance is nonspecific in being effective against a wide range of pathogens, and is typically characterized by both a restriction of pathogen growth and a reduction in disease symptoms compared to noninduced plants infected by the same pathogen (Hammerschmidt, 1999).

Induced resistance triggered by biological agents can be subdivided into two broad categories. The classical type of biologically induced resistance is variously referred to as systemic acquired resistance (SAR) or induced systemic resistance (ISR), and occurs in distal plant parts after localized infection by mainly necrosis-inducing pathogens. Although the two terms are synonymous (Hammerschmidt et al., 2001), for convenience we refer to this type of induced resistance as SAR. Ross (1961) was the first to provide a detailed physiological characterization of the SAR phenomenon. He demonstrated that tobacco plants that reacted hypersensitively to tobacco mosaic virus (TMV) developed an enhanced resistance in the noninoculated upper leaves against subsequent infection by TMV or tobacco necrosis virus. Over the years, SAR has been documented as an effective defense response in various plant species against a broad range of pathogens (Kuć, 1982; Ryals et al., 1996; Sticher et al., 1997). The expression of SAR is associated with the transcriptional activation of genes encoding pathogenesis-related proteins (PRs; Van Loon, 1997) and the accumulation of these proteins. For this reason, *PR* mRNAs or PR-proteins are generally taken as markers for the enhanced resistance state of SAR (Kessmann et al., 1994; Ryals et al., 1996).

The second type of biologically induced resistance develops systemically in response to colonization of plant roots by selected strains of nonpathogenic rhizobacteria. In 1991, two research groups independently demonstrated that rhizosphere-colonizing *Pseudomonas* spp. have the potential to enhance the resistance of the host plant (Van Peer et al., 1991; Wei et al., 1991). This type of induced resistance is generally not associated with the expression of *PR* genes. In order to distinguish this type of induced resistance from pathogen-induced SAR, the term rhizobacteria-mediated ISR was introduced (Pieterse et al., 1996). Rhizobacteria-mediated ISR has been demonstrated in different plant species under conditions in which the rhizobacteria remained spatially separated from the challenging pathogen (Van Loon et al., 1998), demonstrating that the phenomenon is plant-mediated.

A variety of chemicals have been shown to induce resistance as well. Several of these compounds are activators of the SAR response. For instance, SA, 2,6-dichloroisonicotinic acid (INA) and benzothiadiazole (BTH) induce the same set of *PR* genes that is induced upon biological induction of SAR. Moreover, their action involves signaling steps that are also required for the expression of SAR (Lawton et al., 1996; Uknes et al., 1992; Ward et al., 1991). However, the nonprotein amino acid β-aminobutyric acid (BABA) appears to act in a different manner, as this compound has been reported to induce resistance without concomitant expression of *PR* genes (Cohen and Gisi, 1994; Zimmerli et al., 2000). The mode of action of BABA seems to be based on the priming of basal resistance mechanisms that act specifically against the attacking pathogen. Thus, BABA treatment results in an enhanced expression of SA-dependent basal defenses if the plant is invaded by *Pseudomonas syringae*, but stimulates SA-independent callose accumulation upon infection with *Peronospora parasitica* (Zimmerli et al., 2000).

9.3.2 SAR: Triggering and Signaling

SA was first suggested to be involved in SAR signaling based on the observation that exogenously applied SA induced resistance associated with the accumulation of PRs (Uknes et al., 1992; Van Loon and Antoniw, 1982; Ward et al., 1991; White, 1979). Furthermore, both Malamy et al. (1990) and Métraux et al. (1990) observed a strong accumulation of SA in the infected leaves of hypersensitively reacting tobacco and of cucumber with limited fungal infection, respectively. In the noninfected plant parts there was a delayed and weaker accumulation of SA that correlated with the development of SAR. Conclusive evidence for a key role of SA in SAR came from analysis of SA-nonaccumulating NahG plants. Both tobacco and *Arabidopsis* plants expressing the *NahG* gene were found to be blocked in the expression of pathogen-induced SAR, indicating that endogenous accumulation of SA is essential for SAR signaling (Gaffney et al., 1993; Lawton et al., 1995; Figure 9.1). The observation that mutants *sid1* and *sid2* of *Arabidopsis*, which are both affected in pathogen-inducible biosynthesis of SA, are equally impaired in the expression of SAR against *P. parasitica* (Nawrath and Métraux, 1999) supports this conclusion.

Initially, SA was also considered a candidate for the systemically transported SAR signal. Apart from the earlier observations that accumulation of SA preceded the development of SAR and *PR*-gene expression in noninoculated plant parts (Malamy et al., 1990; Métraux et al., 1990), Shulaev et al. (1995) reported that ^{18}O-containing SA molecules that had been synthesized locally in the infected leaf, were transported systemically throughout the plant. However, grafting experiments with tobacco strongly suggested that SA is not the systemically transported signal. Vernooij et al. (1994) demonstrated that a nontransformed scion grafted on a TMV-infected SA-nonaccumulating NahG rootstock expressed SAR, whereas a NahG scion grafted on a TMV-infected nontransformed rootstock failed to develop SAR. Similar results were obtained by graftings between nontransformed tobacco plants and transgenics exhibiting epigenetic cosuppression of the *PAL* gene encoding phenylalanine ammonia-lyase, causing a strongly reduced biosynthesis of SA (Pallas et al., 1996). Indeed, Smith-Becker et al. (1998) demonstrated that upon primary infection of a single cucumber leaf, the accumulation of SA in phloem fluids was preceded by a transient increase in PAL activity in the stems and petioles. These results suggested that SA is synthesized de novo in stems and petioles in response to an early mobile signal from the inoculated leaf. Even though SA is transported within the plant, it is unlikely to act as the transported SAR signal.

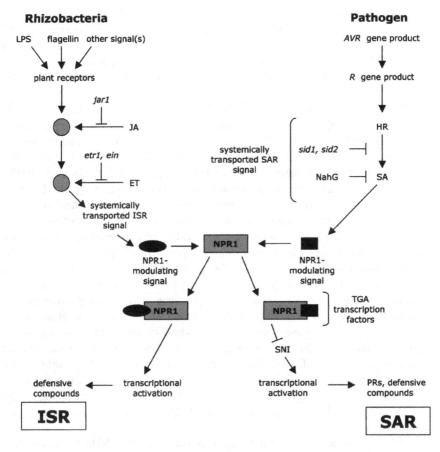

Rhizobacteria   **Pathogen**

FIGURE 9.1. Proposed model for the signal transduction network controlling rhizobacteria-mediated induced systemic resistance (ISR) and pathogen-induced systemic acquired resistance (SAR) in *Arabidopsis thaliana*. *Pseudomonas fluorescens* WCS417r-mediated ISR is controlled by a pathway that is dependent on responsiveness to jasmonic acid (JA) and ethylene (ET), whereas pathogen-induced SAR is controlled by a pathway that depends on accumulation of salicylic acid (SA). Both pathways require the defense regulatory protein NPR1 that differentially regulates SA- and JA/ET dependent defense mechanisms, depending on the pathway that is activated upstream of it (Pieterse et al., 1998). LPS: lipopolysaccharide; NahG: salicylate hydroxylase; PRs: pathogenesis-related proteins; SNI: transcriptional repressor of SAR genes (Li et al., 1999); TGA transcription factors: family of transcription factors interacting with SA-induced NPR1 (Després et al., 2000).

Recently, it was found that ET-insensitive Tetr tobacco plants (Knoester et al., 1998) develop less SAR, and concomitantly accumulate lower amounts of SA and fail to express *PR* genes in the plant parts distal from primary TMV infection. Moreover, grafting experiments demonstrated that a Tetr scion grafted on a TMV-infected nontransformed rootstock expressed SAR, whereas a nontransformed scion on a TMV-infected Tetr rootstock did not (Verberne et al., 2003). Because the Tetr plants produce copious amounts of ET (Van Loon, unpublished results), ET itself cannot act as the mobile signal. These results clearly show that ET plays a promotive role in the generation or translocation of the mobile SAR signal.

Another essential mediator of the SAR signaling pathway is the defense regulatory protein NPR1. A screen for mutants in *Arabidopsis* that failed to exhibit increased expression of a *BGL2* (*PR-2*)-β-glucuronidase (GUS) reporter gene in response to SA treatment yielded the *npr1* mutant (Cao et al., 1994). Since then, several mutant screens based on impaired SAR expression (Delaney et al., 1995), reduced SA-induced *PR* gene expression (Shah et al., 1997), or enhanced disease susceptibility (Glazebrook et al., 1996) all resulted in the identification of mutants allelic to *npr1*, illustrating the broad involvement of NPR1 in plant defense. In *npr1* plants, no induced resistance was evident after pretreatment with SA or its functional analogue INA, indicating that NPR1 functions downstream of the accumulation of SA in the SAR signaling pathway (Cao et al., 1994; Figure 9.1).

Clues as to the molecular basis of NPR1 function came from analysis of its predicted protein sequence, showing the presence of ankyrin repeats, a protein motif that is known to mediate protein–protein interactions (Cao et al., 1997; Ryals et al., 1997). By use of the yeast two-hybrid system for identifying protein–protein interactions, the NPR1 protein was recently demonstrated to interact with members of the TGA family of transcription factors (Després et al., 2000; Zhang et al., 1999; Zhou et al., 2000; Figure 9.1). A subset of these transcription factors showed specific binding to a promoter element within the *PR-1* gene, suggesting a link between NPR1 and the transcriptional activation of *PR-1* genes during the onset of SAR.

A further factor implicated in the regulation of SAR is the SNI1 protein. This factor was identified by a mutant screen for genetic suppressors of the *npr1* mutation (Li et al., 1999). The resulting recessive *sni1* mutant showed restored SAR expression and *PR-1* transcription in response to treatment with INA, indicating SNI1 functions as a negative regulator in the establishment of SAR. It was proposed that SNI1 acts as a transcriptional repressor of SAR that can be counteracted by NPR1 after activation of the SA-dependent SAR pathway. Thereupon, the transcription factors of the TGA family would be allowed to activate the expression of *PR-1* and other genes involved in the establishment of SAR (Figure 9.1).

9.3.3 Rhizobacteria-Mediated ISR: Bacterial Determinants

Rhizobacteria are present in large numbers on the root surface, where plant exudates and lysates provide nutrients (Lynch and Whipps, 1991). Many rhizobacterial strains can suppress soilborne diseases by antagonizing the pathogen (Bakker et al., 1991; Wei et al., 1996). Thus, in order to prove experimentally that resistance is induced by specific rhizobacterial strains, the pathogen and the rhizobacteria must remain spatially separated to prevent direct antagonistic interactions. During the early interaction between ISR-inducing rhizobacteria and the host plant, the rhizobacteria must produce one or more ISR-eliciting compounds that are perceived by the plant at the root surface. Under iron-limiting conditions, certain rhizobacterial strains can produce SA as an iron-scavenging siderophore (Meyer et al., 1992; Visca et al., 1993). Elicitation of ISR in tobacco by *Pseudomonas fluorescens* strain CHA0 might be fully explained by the bacterial production of SA, because treatment of plant roots with CHA0 bacteria triggered accumulation of SA-inducible PRs in the leaves (Maurhofer et al., 1994). Furthermore, transformation of *P. fluorescens* strain P3 with the SA-biosynthetic gene cluster from CHA0 strongly improved the ISR-inducing capacity of P3 (Maurhofer et al., 1998).

Another strain that was suggested to elicit ISR by production of SA, is *Pseudomonas aeruginosa* 7NSK2. A SA-deficient mutant of 7NSK2 failed to induce systemic resistance in bean and tobacco, whereas two mutants affected in either pyoverdin or pyochelin siderophores were still capable of inducing resistance (De Meyer and Höfte, 1997). Moreover, root bacterization of NahG tobacco plants with the wild-type strain failed to induce resistance against TMV, suggesting that 7NSK2-mediated ISR is dependent on bacterially produced SA (De Meyer et al., 1999a). Indeed, SA-negative mutants of 7NSK2 lost the capacity to induce resistance in tomato to *Botrytis cinerea*. However, SA is used by the bacterium to produce the siderophore pyochelin, and pyochelin together with the bacterially produced antibiotic pyocyanin are now taken to be responsible for the induction of resistance through the generation of highly reactive hydroxyl radicals that cause cell damage (Audenaert et al., 2002).

Although these examples demonstrate that rhizobacteria-mediated ISR can be mediated by bacterially produced SA, resulting in the activation of the SA-dependent SAR pathway, other ISR-inducing rhizobacteria have been demonstrated to activate a SA-independent pathway (Iavicoli et al., 2003; Pieterse et al., 1996, 1998; Press et al., 1997; Ryu et al., 2003; Van Wees et al., 1997; Yan et al., 2002), implying involvement of other bacterial factors (Figure 9.1). So far, various structural and metabolic compounds have been implicated in the elicitation of rhizobacteria-mediated ISR (Van Loon et al., 1998). Purified outer membrane lipopolysaccharides (LPS), pseudobactin-type siderophores, antibiotics, and flagella of some nonpathogenic *Pseudomonas* strains have been shown to induce systemic resistance in selected plant species (Iavicoli et al., 2003; Leeman et al., 1995a; Van Peer and Schippers, 1992; Van Wees et al., 1997; Bakker, unpublished results). Putative receptors for the bacterial LPS have not been characterized in plants. Therefore, the molecular mechanisms behind the perception of LPS as related to ISR signaling remain unclear. Bacteria do possess specific receptors for uptake of iron-containing pseudobactin siderophores, but those are not well characterized at the protein level and their involvement in the induction of ISR has not been demonstrated. In contrast, plants have been shown to possess a sensitive perception system for bacterial flagellins (Felix et al., 1999). Recently, a flagellin receptor of *Arabidopsis* was characterized as a receptor kinase sharing structural and functional homology with known plant resistance genes (Gomez-Gomez and

Boller, 2000). These results suggest that the perception of bacterial flagella can result directly in elicitation of a defense-signaling pathway. Although exogenous application of purified LPS, siderophores, or flagella can induce systemic resistance in radish and *Arabidopsis* (Leeman et al., 1995a; Van Peer and Schippers, 1992; Van Wees et al., 1997), bacterial mutants lacking these determinants were still able to elicit ISR in *Arabidopsis* (Van Wees et al., 1997; Bakker, unpublished results). This indicates that several determinants can be involved in the elicitation of rhizobacteria-mediated ISR (Figure 9.1).

9.3.4 Rhizobacteria-Mediated ISR: A Genetic Interaction Between the Rhizobacterium and the Host

ISR-inducing rhizobacteria show little specificity in their colonization of roots of different plant species (Van Loon et al., 1998). However, the ISR-inducing rhizobacterial strains *Pseudomonas putida* WCS358r and *P. fluorescens* WCS374r act differentially on different plant species: *Arabidopsis* is responsive to WCS358r, whereas radish and carnation are not (Leeman et al., 1995b; Van Peer, 1990; Van Peer and Schippers, 1992; Van Wees et al., 1997). Conversely, radish is responsive to WCS374r, whereas *Arabidopsis* is not. *P. fluorescens* strain WCS417r has the ability to elicit ISR in both plant species. These findings indicate that ISR requires a specific interaction between the plant and the nonpathogenic rhizobacterium, which must depend on specific genetic traits of both the rhizobacterium and the host plant. Thus, elicitation of ISR appears to be quite specific with regard to both the host species and the rhizobacterial strain.

In the past decade, the introduction of *Arabidopsis* as a model species has provided many new tools for investigating molecular and genetic aspects of plant–pathogen interactions. To unravel the genetic and molecular basis of rhizobacterially-mediated induced systemic resistance, an *Arabidopsis*-based assay system was developed (Pieterse et al., 1996) in which strain WCS417r was adopted as the ISR-inducing agent, and the agent of bacterial speck disease, *P. syringae* pv. *tomato* strain DC3000, was used as the challenging pathogen (Whalen et al., 1991). Moreover, to compare ISR to pathogen-induced SAR, a HR-eliciting strain of *P. syringae* pv. *tomato*, carrying the avirulence gene *avrRpt2*, was used as an inducer to elicit SAR.

When comparing three *Arabidopsis* accessions, Van Wees et al. (1997) found that the accessions Columbia (Col-0) and Landsberg *erecta* (L*er*) were responsive to induction of ISR by WCS417r, as evidenced by a reduction in symptoms of bacterial speck and multiplication of the pathogen upon inoculation with *P. syringae* pv. *tomato*. In contrast, accession RLD1 failed to develop WCS417r-mediated ISR against this pathogen. Root colonization of RLD1 by WCS417r was of the same order as on Col-0 and L*er*, indicating that RLD1 supports growth of WCS417r bacteria in the rhizosphere, but that within the species *Arabidopsis thaliana* genetic variation is present for ISR inducibility by WCS417r. When seven additional *Arabidopsis* accessions were tested for their ability to express WCS417r-mediated ISR and avirulent *P. syringae*-induced SAR, all displayed normal levels of pathogen-induced SAR. However, only six ecotypes were capable of expressing WCS417r-mediated ISR, whereas accession Ws-0, like RLD1, was not. This WCS417r-nonresponsive phenotype of both RLD1 and Ws-0 was associated with an increased susceptibility to *P. syringae* pv. *tomato* infection. The F_1 progenies of crosses between ISR-noninducible accessions and inducible accessions (Col-0 × RLD1, RLD1 × Col-0, Ws-0 × Col-0, Ws-0 × L*er*) were fully capable of expressing ISR and exhibited a relatively high level of basal resistance, similar to that of their WCS417r-responsive parent. This indicated that the potential to express ISR and the relatively high level of basal resistance against *P. syringae* pv. *tomato* are both inherited as dominant traits.

Analysis of the F_2 and F_3 progeny of a Col-0 × RLD1 cross revealed that the potential to express ISR and basal resistance against *P. syringae* pv. *tomato* cosegregate in a 3:1 fashion, implying that both resistance mechanisms are monogenically determined and genetically linked. Neither the responsiveness to WCS417r, nor the relatively high level of basal resistance against *P. syringae* pv. *tomato* were complemented in the F_1 progeny of crosses between RLD1 and Ws-0, indicating that RLD1 and Ws-0 are both affected in the same locus. This locus, designated *ISR1*, controls both expression of ISR and basal resistance against *P. syringae* pv. *tomato*. Thus, the naturally occurring variation in both ISR inducibility and basal resistance is based on differences at the *ISR1* locus. The observed association between ISR and basal resistance against *P. syringae* pv. *tomato* suggests that rhizobacteria-mediated ISR against *P. syringae* pv. *tomato* in *Arabidopsis* requires the presence of a single dominant gene that functions in the basal resistance response against *P. syringae* pv. *tomato* infection (Ton et al., 1999).

The accessions RLD1 and Ws-0 also failed to express WCS417r-mediated ISR against the bacterium *Xanthomonas campestris* pv. *armoraciae* and the oomycete *P. parasitica*. However, the level of basal resistance against these pathogens was increased relative to the ISR-inducible accession Col-0, rather than decreased. (Ton et al., 2002b). Neither the *ISR1*, nor the *isr1* genotypes developed ISR against turnip crinkle virus (TCV), indicating that WCS417r-mediated ISR is ineffective against the virus. In contrast, both *ISR1* and *isr1* genotypes were capable of expressing SAR against all pathogens tested, indicating that SAR functions independently of the *ISR1* locus.

9.4 Differential Signaling In Pathogen-Induced SAR and Rhizobacteria-Mediated ISR

9.4.1 SA-Independent ISR Requires Responsiveness to Jasmonate and Ethylene

The existence of an SA-independent pathway controlling ISR was first demonstrated in *Arabidopsis* when Pieterse et al. (1996) found that WCS417r-mediated ISR was fully maintained in NahG plants, and not associated with the transcriptional activation of genes encoding SA-inducible PRs. Further studies revealed that treatment of the roots with WCS417r bacteria failed to trigger ISR in the JA-insensitive *jar1* or in the ET-insensitive *etr1* mutants, indicating that the JA and ET response pathways are essential for the establishment of this type of ISR (Pieterse et al., 1998, 2002). Moreover, using methyl jasmonate (MeJA) and the ET precursor 1-aminocyclopropane-1-carboxylate (ACC) as chemical activators of the ISR pathway, it was demonstrated that JA functions upstream of ET in the ISR signaling pathway (Pieterse et al., 1998; Figure 9.1).

An SA-independent but JA- and ET-dependent pathway was also established for the induction of systemic resistance by *Bacillus pumilus* SE34 and *P. fluorescens* 89B61 in tomato against late blight, caused by *Phytophthora infestans* (Yan et al., 2002). By contrast, induction of systemic resistance in *Arabidopsis* by *P. fluorescens* CHA0 to *Peronospora parasitica* was blocked in the *eir1* mutant, but not in the ethylene-insensitive *etr1* or *ein2* mutants (Iavicoli et al., 2003). Whereas induction of resistance in *Arabidopsis* to *P. syringae* pv. *tomato* by *Serratia marcescens* 90–166 required JA or ET signaling, SA-independent induction by *P. fluorescens* 89B61 did not, and induction by *B. pumilus* T4 was independent of NPR1 (Ryu et al., 2003). These observations indicate variations in the requirement for elicitation of ISR by different bacterial strains in *Arabidopsis*.

To further investigate the roles of JA and ET in ISR signaling, the levels of these signaling molecules were determined in plants upon root bacterization with WCS417r. Both systemically and at the site of application of the bacteria, JA content and the level of ET evolution remained unaltered upon ISR induction (Knoester et al., 1999; Pieterse et al., 2000). Also, *LOX2*-cosuppressed S-12 plants, that are blocked in lipoxygenase-mediated production of JA after wounding (Bell et al., 1995) and pathogen infection (Pieterse et al., 2000), were normally responsive to bacterial induction treatments (Pieterse et al., 2000), indicating that ISR can be expressed in the absence of increased JA levels. These data suggest that the JA and ET dependency of ISR is not based on an enhancement of JA and ET production, but rather on an enhanced sensitivity to these hormones.

Since modulation of ET sensitivity in ET-response mutants of *Arabidopsis* results in an altered level of basal expression of ET-responsive genes (Knoester et al., 1999), increased expression would be expected if ISR-expressing plants have enhanced sensitivity to JA and ET. However, when Van Wees et al. (1999) analyzed a large set of known, well-characterized defense-related genes of *Arabidopsis* upon induction by WCS417r, none of these defense-related genes were up-regulated in roots or leaves of ISR-expressing plants. Furthermore, a differential screening from a cDNA library representing mRNAs of ISR-expressing leaves did not result in the identification of genes that were significantly up-regulated upon induction of ISR (Van Wees, 1999). Thus, WCS417r-mediated ISR, unlike pathogen-induced SAR, is neither associated with major changes in *PR* gene expression, nor with changes in the expression of JA- and ET-inducible genes. An alternative explanation for the JA- and ET-dependency of ISR could be that basal levels of both hormones are required for priming the plant to be conducive to ISR signaling (Figure 9.1).

To further elucidate the role of ET in the ISR signaling pathway, Knoester et al. (1999) tested several well-characterized *Arabidopsis* mutants that are disturbed in different steps of the ET-response pathway. None of these mutants expressed ISR upon treatment of the roots with WCS417r, demonstrating that all known components of the ET signaling pathway are required for the expression of ISR. Mutant *eir1*, which is insensitive to ET in the roots only, did not develop ISR after application of WCS417r to the roots, but did after application to the leaves. Based on this observation it was postulated that ET signaling is required at the site of application of the inducer, suggesting that, similar to SAR in tobacco (Knoester et al., 2001; Verberne et al., 2003), ET is involved in the generation or translocation of the systemically transported signal (Figure 9.1). The finding that JA signaling functions upstream of ET signaling in the ISR pathway (Pieterse et al., 1998) implies that JA signaling is required at the site of WCS417r application as well (Figure 9.1). However, these findings do not rule out the possibility that components of the JA and ET response are also required for the expression of ISR in tissues distant from the site of application of the inducing bacterium.

9.4.2 The Dual Role of NPR1 in Induced Resistance

Although the signaling pathways controlling WCS417r-mediated ISR and pathogen-induced SAR clearly differ, both pathways share at least one common signaling component. Pieterse et al. (1998) reported that the *npr1* mutant of *Arabidopsis* is not only impaired in the expression of SAR, but also fails to express ISR after treatment of the roots with WCS417r bacteria. This demonstrated that NPR1 is required for the establishment of both SA-dependent SAR and JA-and ET-dependent ISR. Elucidation of the sequence of ISR signaling events revealed that NPR1 functions downstream of the JA and ET response in the ISR pathway, indicating that NPR1 regulates the activation of both SA-dependent defense-related genes and so far unidentified JA- and ET-dependent defense components (Pieterse et al., 1998). Thus, NPR1 differentially regulates either SA- or JA/ET-dependent defense responses, depending on the pathway that is activated upstream of it (Figure 9.1). Recently, Van Wees et al. (2000) demonstrated that simultaneous activation of SAR and ISR results in an enhanced level of protection against *P. syringae* pv. *tomato*. In addition, it was demonstrated that simultaneous activation of both responses is not associated with enhanced levels of *NPR1* transcription. Thus, the constitutive level of NPR1 is sufficient for the expression of both defense responses.

Further evidence suggesting a regulatory function of NPR1 in SA-independent defense responses came from a genetic study by Clarke et al. (1998). A screen for mutants in transgenic *Arabidopsis* constitutively expressing the *BGL2*-GUS reporter gene yielded the identification of the dominant *cpr6* mutant. This mutant possessed enhanced levels of SA in combination with enhanced pathogen resistance and increased constitutive expression of both SA- and JA-responsive genes. The enhanced resistance of *cpr6* against *P. syringae* pv. *maculicola* was abolished in the *cpr6, npr1* double mutant, despite unaltered constitutive expression of

SA-inducible *PR* genes. This not only indicates that *PR* genes can be controlled in a NPR1-independent manner, but also illustrates that *cpr6*-mediated resistance, like WCS417r-mediated ISR, is controlled through an NPR1-dependent pathway that is not associated with SA-inducible *PR* gene expression.

9.5 Induced Resistance in Relation to Basal Resistance

9.5.1 Induced Resistance is Expressed as an Enhancement of Basal Resistance

The enhanced defensive capacity of plants expressing induced resistance can be based on physiological and biochemical changes in response to the resistance-inducing treatment, or on mechanisms that are expressed only after pathogen challenge of the induced tissues. In the case of SAR, accumulation of PRs is triggered as a result of the inducing treatment. Certain PRs that are synthesized de novo upon SAR induction have antifungal activity. However, the contribution of PRs to induced resistance remains uncertain (Van Loon, 1997). PRs may contribute to resistance against oomycetes, fungi, or bacteria by their hydrolytic action on pathogen cell walls, but it is difficult to envisage a function in viral resistance. Despite several attempts in the case of WCS417r-mediated ISR, metabolic changes before challenge inoculation with a pathogen have not been identified. This suggests that the enhanced defensive capacity of plants expressing induced resistance is largely based on increased post-challenge defense responses. Indeed, the plant may become sensitized to activate appropriate defense mechanisms faster and more strongly upon infection with a challenging pathogen.

Examples of primed expression of defense mechanisms have been reported for both SAR and ISR. Notably, these mechanisms also operate in noninduced plants, but they occur at lower frequency, intensity, or at a later stage during pathogen attack (Hammerschmidt, 1999). For example, noninduced cucumber plants upon infection with *Colletotrichum lagenarium* develop papillae at the sites of attempted penetration of the fungus. These papillae contain callose and lignin, which are thought to act as a barrier to pathogen penetration. In induced plants, the enhanced resistance was associated with a faster formation of significantly more papillae at the sites of appressoria formation than in noninduced plants. Moreover, the papillae in induced plants contained higher amounts of callose and lignin (Hammerschmidt and Kuć, 1982; Kovats et al., 1991). Likewise, *Arabidopsis* plants pre-treated with the chemical inducer BABA and subsequently challenged with *P. parasitica* displayed intensified deposition of callose-rich papillae at the sites of pathogen penetration (Zimmerli et al., 2000). This suggests that the induced resistance against *C. lagenarium* and *P. parasitica* is realized through a primed expression of papilla formation, a mechanism that also determines the level of basal resistance against these pathogens. A stimulation of other defense mechanisms, such as accumulation of hydroxyproline-rich glycoproteins and increased peroxidase activity was also observed (Hammerschmidt, 1999). Similarly, challenge-inoculated carnation plants expressing rhizobacteria-mediated ISR against *Fusarium oxysporum* f.sp. *dianthi*, accumulated phytoalexins earlier and to a greater extent than noninduced plants (Van Peer et al., 1991). In all these examples, the induced resistance appeared as a faster and stronger expression of defense mechanisms that also contributed to the basal resistance of noninduced plants. These findings suggest that the enhanced defensive capacity of plants expressing induced resistance is largely based on enhanced expression of extant basal defense mechanisms.

Induced resistance as an enhancement of extant basal resistance would imply that plant genotypes differing in genetically determined basal resistance could differ in the extent to which induced resistance can be expressed. Indeed, in carnation WCS417r-mediated ISR against Fusarium wilt was considerably more effective in the moderately resistant cultivar Pallas than in the susceptible cultivar Lena (Van Peer et al., 1991). An apparently opposite relationship was described by Liu et al. (1995), who reported that *P. putida* 89B-27-mediated ISR in cucumber against *Colletotrichum orbiculare* was expressed in three susceptible cultivars, but not in a resistant one. This result could be interpreted in the sense that in the already highly resistant cultivar defenses could not be further enhanced upon induction of ISR. However, a correlation between induced resistance and basal resistance is not always apparent. For example, both susceptible and moderately resistant radish cultivars were capable of expressing rhizobacteria-mediated ISR against Fusarium wilt (Leeman et al., 1995b).

9.5.2 Induced Resistance as Primed Expression of SA-Dependent or JA/ET-Dependent Defenses

Interestingly, SA, JA, and ET have all been implicated in the regulation of priming of defense responses (Conrath et al., 2002). For instance, parsley cells pretreated with either JA, SA, or its functional analogues, showed primed accumulation of active oxygen species, secretion of cell wall phenolics, accumulation of coumarin phytoalexins, and *PAL* gene expression upon treatment with the *Pmg* elicitor of *Phytophthora megasperma* f.sp. *glycinea* (Katz et al., 1998; Kauss et al., 1992, 1993, 1994; Thulke and Conrath, 1998). Notably, in intact plants these defense responses all contribute to local resistance responses after primary pathogen attack. In tobacco, it was demonstrated that SAR-expressing plants showed primed *PR-10* and *PAL* gene expression upon infection with different pathogenic pseudomonads (Mur et al., 1996). In *Arabidopsis*, Lawton et al. (1994) showed that plants preexposed to ET were sensitized to SA-induced *PR-1* gene expression, suggesting that ET potentiates defense mechanisms that contribute to SAR. Indeed, ET-insensitive tobacco plants expressing the mutant *etr1-1* gene of *Arabidopsis* showed a reduced SAR response (Knoester et al., 2001).

Analysis of mutants and transgenics, particularly in *Arabidopsis* and tobacco, revealed that signaling pathways controlling basal resistance are often involved also in induced resistance responses. For instance, as described in Section 9.2, SA-nonaccumulating NahG plants of both *Arabidopsis* and tobacco exhibit enhanced susceptibility to a variety of pathogens (Delaney et al., 1994). At the same time, they are affected in the expression of pathogen-induced SAR (Gaffney et al., 1993; Lawton et al., 1995). A similar correlation was found for *Arabidopsis* plants mutated in the *NPR1* gene. Those mutants are not only blocked in the expression of pathogen-induced SAR (Cao et al., 1994; Delaney et al., 1995) and WCS417r-mediated ISR (Pieterse et al., 1998), but their level of basal resistance is also lower against *P. syringae* and *P. parasitica* (Delaney et al., 1995; Glazebrook et al., 1996). Similarly, the JA-insensitive *jar1* mutant and the ET-insensitive *etr1* mutants are affected in the expression of WCS417r-mediated ISR (Pieterse et al., 1998), and concurrently allow tenfold higher levels of growth of *P. syringae* pv. *tomato* in the leaves than wild-type plants upon primary infection.

Phenotypically, mutants *jar1* and *etr1* strongly resemble the *isr1* phenotype of accessions RLD1 and Ws-0. Therefore, we considered the possibility that eco-types RLD1 and Ws-0 are impaired in either JA or ET signaling, and whether the *ISR1* locus might be involved. Compared to the ISR-inducible accession Col-0, accessions RLD1 and Ws-0 were not affected in JA-induced inhibition of root growth or expression of the JA-responsive vegetative storage protein gene *AtVSP*, suggesting that the *ISR1* locus is not involved in JA signaling. However, RLD1 and Ws-0 were affected in their ET-dependent triple response and showed re-duced expression of the ET-responsive hevein gene *HEL,* and the plant defensin gene *PDF1.2* after exogenous application of ACC. Moreover, in contrast to Col-0, both RLD1 and Ws-0 did not develop resistance against *P. syringae* pv. *tomato* after treatment of the leaves with ACC. Analysis of the F_2 and F_3 progeny of a cross between Col-0 (*ISR1/ISR1*) and RLD1 (*isr1/isr1*) revealed that the re-duced sensitivity to ET cosegregates with the recessive alleles of the *ISR1* locus (Ton et al., 2001). These results indicated that the *ISR1* locus encodes a novel component of the ET-response pathway, which is required for the expression of rhizobacteria-mediated ISR. Hence, the observed association between ISR-noninducibility and reduced basal resistance against *P. syringae* pv. *tomato* in the *Arabidopsis* accessions RLD1 and Ws-0 can be attributed to a reduced sensitivity to ET.

Thus, in many cases there seems to be a correlation between the presence of a certain level of basal resistance and the capacity of a plant to develop in-duced resistance. Nevertheless, various ET-insensitive mutants of *Arabidopsis* are unaffected in their SAR response (Knoester et al., 1999; Lawton et al., 1994, 1995). Interestingly, upon challenge inoculation with *P. syringae* pv. *tomato*, SAR-expressing *Arabidopsis* plants showed a primed expression of SA-inducible *PR* genes (Cameron et al., 1999; Van Wees et al., 1999), whereas ISR-expressing *Arabidopsis* plants displayed a primed expression of the JA-inducible *AtVSP* gene (Van Wees et al., 1999). These results clearly indicate that both types of induced resistance are associated with priming of different defense responses. Therefore, it is tempting to speculate that SAR is achieved through a primed expression of SA-dependent basal defenses, whereas WCS417r-mediated ISR is achieved through a primed expression of JA/ET-dependent basal resistance. A model is schematically represented in Figure 9.2.

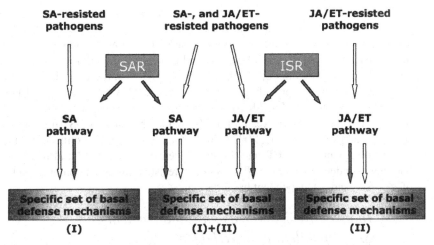

FIGURE 9.2. Model explaining pathogen-induced SAR and rhizobacteria-mediated ISR as a primed expression of basal defense mechanisms. SA-dependent basal defense mechanisms (I) are primed in SAR-induced plants. Consequently, infection of SAR-expressing tissue triggers a faster and stronger activation of SA-dependent defense mechanisms, resulting in an effective protection against pathogens that are resisted through SA-dependent basal resistance, i.e., *P. parasitica* and TCV. Conversely, pathogen infection of plants pretreated with ISR-inducing WCS417r bacteria results in priming of JA/ET-dependent basal defense mechanisms (II). Accordingly, ISR-expressing tissues show a faster and stronger expression of JA/ET-dependent defense mechanisms upon infection, resulting in an effective protection against pathogens that are resisted through JA/ET-dependent basal resistance, i.e., *A. brassicicola*. Pathogens that are resisted through a combination of SA and JA/ET-dependent basal resistance, i.e., *P. syringae* and *X. campestris*, are sensitive to both SAR and ISR (I) + (II).

9.5.3 Impaired Induced Resistance as a Result of Reduced SA-Dependent or JA/ET-Dependent Basal Resistance

Because of the association between induced resistance and basal resistance, a collection of *Arabidopsis eds* mutants with enhanced disease susceptibility to pathogenic *P. syringae* bacteria (Glazebrook et al., 1996; Volko et al., 1998) was screened for their potential to express rhizobacteria-mediated ISR and pathogen-induced SAR against *P. syringae* pv. *tomato*. Out of 11 *eds* mutants tested, *eds4-1*, *eds8-1*, and *eds10-1* were nonresponsive to induction of ISR by WCS417r, whereas mutants *eds5-1* and *eds12-1* were nonresponsive to induction of SAR (Ton et al., 2002a). While *eds5-1* is known to be allelic to *sid1*, and blocked in the synthesis of SA (Nawrath and Métraux, 1999), further analysis of *eds12-1* revealed that the SAR-impaired phenotype of this mutant is caused by a reduced sensitivity to SA. Analysis of the ISR-impaired *eds* mutants revealed that they are insensitive to induction of resistance by MeJA (*eds4-1*, *eds8-1*, and *eds10-1*) or ACC (*eds4-1* and *eds10-1*). Moreover, *eds4-1* and *eds8-1* showed reduced expression of the *PDF1.2* gene after treatment with MeJA and ACC, which was associated with a reduced sensitivity to either ET (*eds4-1*) or MeJA (*eds8-1*). Although blocked in rhizobacteria-, MeJA-, and ACC-induced protection, mutant *eds10-1* showed normal responsiveness to both MeJA and ACC. Together, these results indicated that EDS12 is required for SAR and acts downstream of SA, whereas EDS4, EDS8, and EDS10 are required for ISR and act in either the JA response (EDS8), the ET response (EDS4), or downstream of the JA and ET response (EDS10) in the ISR signaling pathway (Ton et al., 2002a). Together, these results not only confirm the dual involvement of JA, ET, and SA in induced resistance and basal resistance, but they also demonstrate that *P. syringae* is resisted through a combined action of JA-, ET-, and SA-dependent basal resistance.

9.5.4 Induced Resistance as an Enhancement of SA-Dependent or JA/ET-Dependent Basal Resistance

Over the past years, plant genotypes affected in SA, JA, or ET signaling have been linked repeatedly to enhanced disease susceptibility to specific pathogens and even insects (Delaney et al., 1994; Knoester et al., 1998; McConn et al., 1997; Staswick et al., 1998; Vijayan et al., 1998). Evidence is accumulating that SA-, JA-, and ET-dependent defenses contribute to basal resistance against different pathogens. JA- and ET-insensitive *Arabidopsis* genotypes exhibit enhanced susceptibility to necrotrophic pathogens, i.e., *A. brassicola* and *B. cinerea*, indicating that basal resistance against these pathogens is, at least in part, conferred by JA- and ET-dependent defenses (Thomma et al., 1998, 1999). Conversely, genotypes impaired in SA accumulation exhibit enhanced susceptibility to predominantly biotrophic pathogens, i.e., *P. parasitica* and TCV (Kachroo et al., 2000; Nawrath and Métraux, 1999; Thomma et al., 1998), indicating that these pathogens are predominantly resisted through SA-dependent defenses.

Because SA is a key regulator of SAR, whereas JA and ET sensitivity are required for ISR, SAR and ISR might also be differentially effective against different pathogens. Indeed, the fungus *A. brassicicola*, which is resisted through JA/ET-dependent basal defenses, was inhibited considerably in plants expressing WCS417r-mediated ISR, whereas expression of SAR induced by either INA or avirulent *P. syringae* was ineffective against this pathogen (Ton et al., 2002c). Conversely, *P. parasitica* and TCV, which are both resisted through predominantly SA-dependent basal defenses, were strongly inhibited by the expression of SAR, while ISR yielded only weak and no protection, respectively. SAR induced by avirulent *P. syringae* and ISR triggered by WCS417r bacteria were equally effective against *P. syringae* pv. *tomato* and *X. campestris* pv. *armoraciae*, which are resisted through a combination of SA-, JA-, and ET-dependent basal defenses (Ton et al., 2002c). Thus, ISR is predominantly effective against pathogens that are resisted through basal defenses that are activated by a JA/ET-dependent mechanism, whereas SAR is more effective against pathogens that are resisted through SA-dependent basal defenses.

As discussed in Section 9.5.2, *Arabidopsis* genotypes affected in JA/ET-dependent basal resistance against *P. syringae* are impaired in WCS417r-mediated ISR, whereas genotypes affected in SA-dependent basal resistance against *P. syringae* are impaired in avirulent pathogen-induced SAR. In accordance with the earlier notion that induced disease resistance is an enhancement of genetically determined basal resistance by which extant defense mechanisms are expressed earlier and to higher levels (Van Loon, 1997), these results strongly suggest that WCS417r-mediated ISR involves an enhancement of JA- and ET-dependent basal resistance, whereas SAR constitutes an enhancement of SA-dependent basal resistance. Consequently, pathogens such as *P. syringae* and *X. campestris*, which are resisted through a combined action of SA-dependent and JA/ET-dependent basal defenses, are sensitive to both SAR and ISR (Pieterse et al., 1998; Ton et al., 2002c). As a result, *Arabidopsis* RLD1 and Ws-0 can still enhance their defensive capacity through the expression of SAR, even though they have lost their ability to express ISR.

9.6 Combining SAR and ISR as a Method to Improve Biocontrol of Plant Diseases

Van Wees et al. (2000) demonstrated that combined treatment of *Arabidopsis* with ISR-inducing WCS417r and SAR-inducing avirulent *P. syringae* results in an enhanced level of induced protection against *P. syringae* pv. *tomato*. Moreover, the resistance of the constitutively SAR-expressing mutant *cpr1* could be increased further by treatment of the roots with ISR-inducing WCS417r bacteria. This indicates that the JA/ET-dependent ISR pathway and the SA-dependent SAR pathway act additively on the level of protection against this pathogen. *X. campestris* pv. *armoraciae* is also resisted through a combined action of JA/ET-dependent and SA-dependent defense pathways (Ton et al., 2002c). Therefore, one can predict that simultaneous activation of SAR and ISR will result in an enhanced level of protection against *X. campestris* pv. *armoraciae* as well. Indeed, recent observations confirmed that simultaneous activation of WCS417r-mediated ISR and avirulent *P. syringae*-induced SAR conferred enhanced protection against this bacterium (Van Pelt and Pieterse, unpublished results). Additionally, SAR and ISR seem to confer differential protection against different types of pathogens. Thus, combining SAR and ISR can protect the plant against a wider spectrum of pathogens, and even result in an additive level of induced protection against pathogens that are resisted through both the JA/ET- and the SA-dependent pathways. This additive action of pathogen-induced SAR and WCS417r-mediated ISR in resistance against pathogenic bacteria is at variance with the reported antagonism between SA-dependent SAR and JA-induced resistance against insects (Stout et al., 1999). However, this apparent discrepancy can be explained by the fact that induced resistance against insects depends on a signaling pathway requiring enhanced accumulation of JA (McConn et al., 1997), whereas WCS417r-mediated ISR is dependent on sensitivity to JA and ET rather than elevated levels of these regulators (Pieterse et al., 2000).

Biological control of plant diseases is still in its infancy, because the level of protection and its consistency are generally not sufficient to compete with conventional methods of disease control. One approach to improve the efficacy and consistency of biological control against soilborne pathogens is to apply combinations of antagonistic microorganisms with different mechanisms of action (De Boer, 2000). Alternatively, microorganisms can be engineered to express disease-suppressive traits constitutively at high levels. Manipulation of the plants by introducing race-specific *R* genes into plants is another attractive approach, because it renders the plant completely resistant to a pathogen. However, resistance based on gene-for-gene resistance, offers protection against only a single pathogen, and the pathogen can overcome the resistance by mutation. Transgenic approaches to engineer durable and broad-spectrum resistance are promising, but still under development. Our findings that the combination of SAR and ISR confers protection against a wider spectrum of pathogens and results in enhanced levels of protection against specific bacterial pathogens (Van Wees et al., 2000), offers great potential for integrating both forms of induced resistance in future agricultural practices.

The chemical plant activator BION suppresses plant diseases through BTH-mediated activation of the SAR response (Friedrich et al., 1996; Lawton et al., 1996). Nevertheless, SAR does not protect the plant against necrotrophic pathogens such as *A. brassicicola* and *B. cinerea* (Thomma et al., 1998). Furthermore, BION has been reported to reduce plant growth and seed set under field conditions (Heil et al., 2000). By contrast, resistance-inducing rhizobacteria can improve plant growth under field conditions. This rhizobacteria-mediated growth promotion results mainly from the antagonistic activity against soilborne pathogens and other deleterious microorganisms (Kloepper et al., 1980; Schippers et al., 1987). Furthermore, resistance-inducing rhizobacteria, in general, do not solely induce resistance through JA/ET-dependent ISR. Some rhizobacteria appear to activate the SAR response through production of SA at the root surface (De Meyer et al., 1999b; De Meyer and Höfte, 1997; Maurhofer et al., 1994, 1998). Agricultural inoculants containing combinations of selected ISR-inducing rhizobacteria and SA-producing rhizobacteria could have an advantage in that three disease-suppressive mechanisms, i.e., microbial antagonism, ISR- and SAR-action, are combined. Therefore, activation of both SAR and ISR through rhizobacterial treatments offers not only great potential for improving the efficacy and consistency of biocontrol with plant growth-promoting rhizobacteria, but would also broaden its spectrum of effectiveness. Furthermore, elucidation of the molecular mechanisms underlying ISR and SAR may lead to the identification of key regulatory components that could be engineered to constitutive expression in crop plants, in order to enhance their level of basal resistance against a broad spectrum of pathogens.

References

Audenaert, K., Pattery, T., Cornelis, P., and Höfte, M. 2002. Induction of systemic resistance to *Botrytis cinerea* in tomato by *Pseudomonas aeruginosa* 7NSK2: role of salicylic acid, pyochelin, and pyocyanin. *Mol. Plant Microbe Interact.* 15:1147–1156.

Bakker, P.A.H.M., Van Peer, R., and Schippers, B. 1991. Suppression of soil-borne plant pathogens by fluorescent pseudomonads: mechanisms and prospects. In *Biotic Interactions and Soil-Borne Diseases,* eds. A.B.R. Beemster, G.J. Bollen, M. Gerlagh, M.A. Ruissen, B. Schippers, and A. Tempel, pp. 217–230. Amsterdam, The Netherlands: Elsevier Scientific Publishers.

Bell, E., Creelman, R.A., and Mullet, J.E. 1995. A chloroplast lipoxygenase is required for wound-induced accumulation of jasmonic acid in *Arabidopsis. Proc. Natl. Acad. Sci. USA* 92:8675–8679.

Bent, A.F., Innes, R.W., Ecker, J.R., and Staskawicz, B.J. 1992. Disease development in ethylene-insensitive *Arabidopsis thaliana* infected with virulent and avirulent *Pseudomonas* and *Xanthomonas* pathogens. *Mol. Plant Microbe Interact.* 5:372–378.

Berrocal-Lobo, M., Molina, A., and Solano, R. 2002. *ETHYLENE-RESPONSE-FACTOR1* in *Arabidopsis* confers resistance to several necrotrophic fungi. *Plant J.* 29:23–33.

Boller, T. 1991. Ethylene in pathogenesis and disease resistance. In *The Plant Hormone Ethylene,* eds. A.K. Mattoo, and J.C. Suttle, pp. 293–314. Boca Raton: CRC Press.

Cameron, R.K., Paiva, N.C., Lamb, C.J., and Dixon, R.A. 1999. Accumulation of salicylic acid and *PR-1* gene transcripts in relation to the systemic acquired resistance (SAR) response by *Pseudomonas syringae* pv. *tomato* in *Arabidopsis. Physiol. Mol. Plant Pathol.* 55:121–130.

Cao, H., Bowling, S.A., Gordon, A.S., and Dong, X. 1994. Characterization of an *Arabidopsis* mutant that is nonresponsive to inducers of systemic acquired resistance. *Plant Cell* 6:1583–1592.

Cao, H., Glazebrook, J., Clarke, J.D., Volko, S., and Dong, X. 1997. The *Arabidopsis NPR1* gene that controls systemic acquired resistance encodes a novel protein containing ankyrin repeats. *Cell* 88:57–63.

Ciardi, J.A., Tieman, D.M., Lund, S.T., Jones, J.B., Stall, R.E., and Klee, H.J. 2000. Response to *Xanthomonas campestris* pv. *vesicatoria* in tomato involves regulation of ethylene receptor gene expression. *Plant Physiol.* 123:81–92.

Clarke, J.D., Liu, Y., Klessig, D.F., and Dong, X. 1998. Uncoupling *PR* gene expression from NPR1 and bacterial resistance: characterization of the dominant *Arabidopsis cpr6-1* mutant. *Plant Cell* 10:557–569.

Cohen, Y., and Gisi, U. 1994. Systemic translocation of ^{14}C-DL-3-aminobutyric acid in tomato plants in relation to induced resistance against *Phytophthora infestans. Physiol. Mol. Plant Pathol.* 45:441–456.

Conrath, U., Pieterse, C.M.J., and Mauch-Mani, B. 2002. Priming in plant–pathogen interactions. *Trends Plant Sci.* 7:210–216.

De Boer, M. 2000. Combining *Pseudomonas* strains to improve biological control of fusarium wilt in radish. Ph. D. thesis, Utrecht University, The Netherlands.

De Meyer, G., Audenaert, K., and Höfte, M. 1999a. *Pseudomonas aeruginosa* 7NSK2-induced systemic resistance in tobacco depends on *in planta* salicylic acid accumulation, but is not associated with *PR-1a* expression. *Eur. J. Plant Pathol.* 105:513–517.

De Meyer, G., Capieau, K., Audenaert, K., Buchala, A., Métraux, J.-P., and Höfte, M. 1999b. Nanogram amounts of salicylic acid produced by the rhizobacterium *Pseudomonas aeruginosa* 7NSK2 activate the systemic acquired resistance pathway in bean. *Mol. Plant-Microbe Interact.* 12:450–458.

De Meyer, G., and Höfte, M. 1997. Salicylic acid produced by the rhizobacterium *Pseudomonas aeruginosa* 7NSK2 induces resistance to leaf infection by *Botrytis cinerea* on bean. *Phytopathology* 87:588–593.

Delaney, T.P., Friedrich, L., and Ryals, J.A. 1995. *Arabidopsis* signal transduction mutant defective in chemically and biologically induced disease resistance. *Proc. Natl. Acad. Sci. USA* 92:6602–6606.

Delaney, T.P., Uknes, S., Vernooij, B., Friedrich, L., Weymann, K., Negrotto, D., Gaffney, T., Gut-Rella, M., Kessmann, H., Ward, E., and Ryals, J. 1994. A central role of salicylic acid in plant disease resistance. *Science* 266:1247–1250.

Dempsey, D.A., Shah, J., and Klessig, D.F. 1999. Salicylic acid and disease resistance in plants. *Crit. Rev. Plant Sci.* 18:547–575.

Després, C., DeLong, C., Glaze, S., Liu, E., and Fobert, P.R. 2000. The *Arabidopsis* NPR1/NIM1 protein enhances the DNA binding activity of a subgroup of the TGA family of bZIP transcription factors. *Plant Cell* 12:279–290.

Felix, G., Duran, J.D., Volko, S., and Boller, T. 1999. Plants have a sensitive perception system for the most conserved domain of bacterial flagellin. *Plant J.* 18:265–276.

Feys, B.J.F., Benedetti, C.E., Penfold, C.N., and Turner, J.G. 1994. *Arabidopsis* mutants selected for resistance to the phytotoxin coronatine are male sterile, insensitive to methyl jasmonate, and resistant to a bacterial pathogen. *Plant Cell* 6:751–759.

Friedrich, L., Lawton, K., Ruess, W., Masner, P., Specker, N., Gut-Rella, M.G., Meier, B., Dincher, S., Staub, T., Métraux, J.-P., Kessmann, H., and Ryals, J. 1996. A benzothiadiazole derivate induces systemic resistance in tobacco. *Plant J.* 10:61–70.

Gaffney, T., Friedrich, L., Vernooij, B., Negrotto, D., Nye, G., Uknes, S., Ward, E., Kessmann, H., and Ryals, J. 1993. Requirement of salicylic acid for the induction of systemic acquired resistance. *Science* 261:754–756.

Glazebrook, J., Rogers, E.E., and Ausubel, F.M. 1996. Isolation of *Arabidopsis* mutants with enhanced disease susceptibility by direct screening. *Genetics* 143:973–982.

Gomez-Gomez, L., and Boller, T. 2000. FLS2: an LRR receptor-like kinase involved in the perception of the bacterial elicitor flagellin in *Arabidopsis*. *Mol. Cell* 5:1003–1012.

Hammerschmidt, R. 1999. Induced disease resistance: how do induced plants stop pathogens? *Physiol. Mol. Plant Pathol.* 55:77–84.

Hammerschmidt, R., and Kuć, J. 1982. Lignification as a mechanism for induced systemic resistance in cucumber. *Physiol. Plant Pathol.* 17:61–71.

Hammerschmidt, R., Métraux, J.-P., and Van Loon, L.C. 2001. Inducing resistance: a summary of papers presented at the First International Symposium on Induced Resistance to Plant Diseases, Corfu, May 2000. *Eur. J. Plant Pathol.* 107:1–6.

Hammond-Kosack, K.E., and Jones, J.D.G. 1996. Resistance gene-dependent plant defense responses. *Plant Cell* 8:1773–1791.

Heil, M., Hilper, A., Kaiser, W., and Linsenmair, K.E. 2000. Reduced growth and seed set following chemical induction of pathogen defence: does systemic acquired resistance (SAR) incur allocation costs? *J. Ecol.* 88:645–654.

Iavicoli, A., Boutet, E., Buchala, A., and Métraux, J.P. 2003. Induced systemic resistance in *Arabidopsis thaliana* in response to root inoculation with *Pseudomonas fluorescens* CHA0. *Mol. Plant Microbe Interact.* 16:851–858.

Kachroo, P., Yoshioka, K., Shah, J., Dooner, K.D., and Klessig, D.F. 2000. Resistance to turnip crinkle virus in *Arabidopsis* is regulated by two host genes and is salicylic acid dependent but NPR1, ethylene, and jasmonate independent. *Plant Cell* 12:677–690.

Katz, V.A., Thulke, O.U., and Conrath, U. 1998. A benzothiadiazole primes parsley cells for augmented elicitation of defense responses. *Plant Physiol.* 117:1333–1339.

Kauss, H., Franke, R., Krause, K., Conrath, U., Jeblick, W., Grimmig, B., and Matern, U. 1993. Conditioning of parsley (*Petroselium crispum*) suspension cells increases elicitor-induced incorporation of cell wall phenolics. *Plant Physiol.* 102:459–466.

Kauss, H., Jeblick, W., Ziegler, J., and Krabler, W. 1994. Pretreatment of parsley (*Petroselinum crispum*) L. suspension cultures with methyl jasmonate enhances elicitation of activated oxygen species. *Plant Physiol.* 105:89–104.

Kauss, H., Theisinger-Hinkel, E., Mindermann, R., and Conrath, U. 1992. Dichloroisonicotinic and salicylic acid, inducers of systemic acquired resistance, enhance fungal elicitor responses in parsley cells. *Plant J.* 2:655–660.

Kessmann, H., Staub, T., Ligon, J., Oostendorp, M., and Ryals, J. 1994. Activation of systemic acquired disease resistance in plants. *Eur. J. Plant Pathol.* 100:359–369.

Kloepper, J.W., Leong, J., Teintze, M., and Schroth, M.N. 1980. Enhanced plant growth by siderophores produced by plant growth-promoting rhizobacteria. *Nature* 286:885–886.

Knoester, M., Linthorst, H.J.M., Bol, J.F., and Van Loon, L.C. 2001. Involvement of ethylene in lesion development and systemic acquired resistance in tobacco during the hypersensitive reaction to tobacco mosaic virus. *Physiol. Mol. Plant Pathol.* 59:45–57.

Knoester, M., Pieterse, C.M.J., Bol, J.F., and Van Loon, L.C. 1999. Systemic resistance in *Arabidopsis* induced by rhizobacteria requires ethylene-dependent signaling at the site of application. *Mol. Plant Microbe Interact.* 12:720–727.

Knoester, M., Van Loon, L.C., Van den Heuvel, J., Hennig, J., Bol, J.F., and Linthorst, H.J.M. 1998. Ethylene-insensitive tobacco lacks nonhost resistance against soil-borne fungi. *Proc. Natl. Acad. Sci. USA* 95:1933–1937.

Kovats, K., Binder, A., and Hohl, H.L. 1991. Cytology of induced systemic resistance of tomato to *Colletotrichum lagenarium*. *Planta* 183:484–490.

Kuć, J. 1982. Induced immunity to plant disease. *Bioscience* 32:854–860.

Lawton, K., Weymann, K., Friedrich, L., Vernooij, B., Uknes, S., and Ryals, J. 1995. Systemic acquired resistance in *Arabidopsis* requires salicylic acid but not ethylene. *Mol. Plant Microbe Interact.* 8:863–870.

Lawton, K.A., Friedrich, L., Hunt, M., Weymann, K., Delaney, T., Kessmann, H., Staub, T., and Ryals, J. 1996. Benzothiadiazole induces disease resistance in *Arabidopsis* by activation of the systemic acquired resistance signal transduction pathway. *Plant J.* 10:71–82.

Lawton, K.A., Potter, S.L., Uknes, S., and Ryals, J. 1994. Acquired resistance signal transduction in *Arabidopsis* is ethylene independent. *Plant Cell* 6:581–588.

Leeman, M., Van Pelt, J.A., Den Ouden, F.M., Heinsbroek, M., Bakker, P.A.H.M., and Schippers, B. 1995a. Induction of systemic resistance against fusarium wilt of radish by lipopolysaccharides of *Pseudomonas fluorescens*. *Phytopathology* 85:1021–1027.

Leeman, M., Van Pelt, J.A., Den Ouden, F.M., Heinsbroek, M., Bakker, P.A.H.M., and Schippers, B. 1995b. Induction of systemic resistance by *Pseudomonas fluorescens* in radish cultivars differing in susceptibility to fusarium wilt, using a novel bioassay. *Eur. J. Plant Pathol.* 101:655–664.

Li, X., Zhang, Y., Clarke, J.D., Li, Y., and Dong, X. 1999. Identification and cloning of a negative regulator of systemic acquired resistance, SNI1, through a screen for suppressors of *npr1-1*. *Cell* 98:329–339.

Liu, L., Kloepper, J.W., and Tuzun, S. 1995. Induction of systemic resistance in cucumber against angular leaf spot by plant growth-promoting rhizobacteria. *Phytopathology* 85:1064–1068.

Lund, S.T., Stall, R.E., and Klee, H.J. 1998. Ethylene regulates the susceptible response to pathogen infection in tomato. *Plant Cell* 10:371–382.

Lynch, J.M., and Whipps, J.M. 1991. Substrate flow in the rhizosphere. In *The Rhizosphere and Plant Growth*, eds. D.L. Keister, and P.B. Cregan, pp. 15–24. Dordrecht, The Netherlands: Kluwer.

Malamy, J., Carr, J.P., Klessig, D.F., and Raskin, I. 1990. Salicylic acid: a likely endogenous signal in the resistance response of tobacco to viral infection. *Science* 250:1002–1004.

Maleck, K., Levine, A., Eulgem, T., Morgan, A., Schmid, J., Lawton, K.A., Dangl, J.L., and Dietrich, R.A. 2000. The transcriptome of *Arabidopsis thaliana* during systemic acquired resistance. *Nature Genet.* 26:403–410.

Maurhofer, M., Hase, C., Meuwly, P., Métraux, J.P., and Défago, G. 1994. Induction of systemic resistance to tobacco necrosis virus by the root-colonizing *Pseudomonas fluorescens* strain CHA0: influence of the *gacA* gene and pyoverdine production. *Phytopathology* 84:139–146.

Maurhofer, M., Reimmann, C., Schmidli-Sacherer, P., Heeb, S.D., and Défago, G. 1998. Salicylic acid biosynthesis genes expressed in *Pseudomonas fluorescens* strain P3 improve the induction of systemic resistance in tobacco against tobacco necrosis virus. *Phytopathology* 88:678–684.

McConn, J., Creelman, R.A., Bell, E., Mullet, J.E., and Browse, J. 1997. Jasmonate is essential for insect defense in *Arabidopsis*. *Proc. Natl. Acad. Sci. USA* 94:5473–5477.

Métraux, J.P., Signer, H., Ryals, J., Ward, E., Wyss-Benz, M., Gaudin, J., Raschdorf, K., Schmid, E., Blum, W., and Inverardi, B. 1990. Increase in salicylic acid at the onset of systemic acquired resistance in cucumber. *Science* 250:1004–1006.

Meyer, J.-M., Azelvandre, P., and Georges, C. 1992. Iron metabolism in *Pseudomonas*: Salicylic acid, a siderophore of *Pseudomonas fluorescens* CHA0. *Biofactors* 4:23–27.

Mur, L.A.J., Naylor, G., Warner, S.A.J., Sugars, F.M., White, R.F., and Draper, J. 1996. salicylic acid potentiates defence gene expression in tissue exhibiting acquired resistance to pathogen attack. *Plant J.* 9:559–571.

Nawrath, C., and Métraux, J.P. 1999. Salicylic acid induction-deficient mutants of *Arabidopsis* express PR-2 and PR-5 and accumulate high levels of camalexin after pathogen inoculation. *Plant Cell* 11:1393–1404.

Norman-Setterblad, C., Vidal, S., and Palva, T.E. 2000. Interacting signal pathways control defense gene expression in *Arabidopsis* in response to cell wall-degrading enzymes from *Erwinia carotovora*. *Mol. Plant Microbe Interact.* 13:430–438.

Pallas, J.A., Paiva, N.L., Lamb, C., and Dixon, R.S. 1996. Tobacco plants epigenetically suppressed in phenylalanine ammonia-lyase expression do not develop systemic acquired resistance in response to infection by tobacco mosaic virus. *Plant J.* 10:281–293.

Pieterse, C.M.J., Van Pelt, J.A., Ton, J., Parchmann, S., Mueller, M.J., Buchala, A.J., Métraux, J.P., and Van Loon, L.C. 2000. Rhizobacteria-mediated induced systemic resistance (ISR) in *Arabidopsis* requires sensitivity to jasmonate and ethylene but is not accompanied by an increase in their production. *Physiol. Mol. Plant Pathol.* 57:123–134.

Pieterse, C.M.J., Van Wees, S.C.M., Ton, J., Van Pelt, J.A., and Van Loon, L.C. 2002. Signalling in rhizobacteria-induced systemic resistance in *Arabidopsis thaliana*. *Plant Biol.* 4:535–544.

Pieterse, C.M.J., Van Wees, S.C.M., Hoffland, E., Van Pelt, J.A., and Van Loon, L.C. 1996. Systemic resistance in *Arabidopsis* induced by biocontrol bacteria is independent of salicylic acid accumulation and pathogenesis-related gene expression. *Plant Cell* 8:1225–1237.

Pieterse, C.M.J., Van Wees, S.C.M., Van Pelt, J.A., Knoester, M., Laan, R., Gerrits, H., Weisbeek, P.J., and Van Loon, L.C. 1998. A novel signaling pathway controlling induced systemic resistance in *Arabidopsis. Plant Cell* 10:1571–1580.

Press, C.M., Wilson, M., Tuzun, S., and Kloepper, J.W. 1997. Salicylic acid produced by *Serratia marcescens* 91-166 is not the primary determinant of induced systemic resistance in cucumber or tobacco. *Mol. Plant Microbe Interact.* 10:761–768.

Rogers, E.E., and Ausubel, F.M. 1997. *Arabidopsis* enhanced disease susceptibility mutants exhibit enhanced susceptibility to several bacterial pathogens and alterations in *PR-1* gene expression. *Plant Cell* 9:305–316.

Ross, A.F. 1961. Systemic acquired resistance induced by localized virus infections in plants. *Virology* 14:340–358.

Ryals, J., Weymann, K., Lawton, K., Friedrich, L., Ellis, D., Steiner, H.Y., Johnson, J., Delaney, T.P., Jesse, T., Vos, P., and Uknes, S. 1997. The *Arabidopsis* NIM1 protein shows homology to the mammalian transcription factor inhibitor IκB. *Plant Cell* 9:425–439.

Ryals, J.A., Neuenschwander, U.H., Willits, M.G., Molina, A., Steiner, H.-Y., and Hunt, M.D. 1996. Systemic acquired resistance. *Plant Cell* 8:1808–1819.

Ryu, C.M., Hu, C.H., Reddy, M.S., and Kloepper, J.W. 2003. Different signaling pathways of induced resistance by rhizobacteria in *Arabidopsis thaliana* against two pathovars of *Pseudomonas syringae. New Phytol.* 160:413–420.

Schippers, B., Bakker, A.W., and Bakker, P.A.H.M. 1987. Interactions of deleterious and beneficial rhizosphere micoorganisms and the effect of cropping practices. *Annu. Rev. Phytopathol.* 115:339–358.

Shah, J., Tsui, F., and Klessig, D.F. 1997. Characterization of a salicylic acid-insensitive mutant (*sai1*) of *Arabidopsis thaliana*, identified in a selective screen utilizing the SA-inducible expression of the *tms2* gene. *Mol. Plant Microbe Interact.* 10:69–78.

Shulaev, V., Leon, J., and Raskin, I. 1995. Is salicylic acid a translocated signal of systemic acquired resistance in tobacco? *Plant Cell* 7:1691–1701.

Smith-Becker, J., Marois, E., Huguet, E.J., Midland, S.L., Sims, J., and Keen, N.T. 1998. Accumulation of salicylic acid and 4-hydroxybenzoic acid in phloem fluids of cucumber during systemic acquired resistance is preceded by a transient increase in phenylalanine ammonia-lyase activity in petioles and stems. *Plant Physiol.* 116:231–238.

Staswick, P.E., Yuen, G.Y., and Lehman, C.C. 1998. Jasmonate signaling mutants of *Arabidopsis* are susceptible to the soil fungus *Pythium irregulare. Plant J.* 15:747–754.

Sticher, L., Mauch-Mani, B., and Métraux, J.-P. 1997. Systemic acquired resistance. *Annu. Rev. Phytopathol.* 35:235–270.

Stout, M.J., Fidantsef, A.L., Duffey, S.S., and Bostock, R.M. 1999. Signal interactions in pathogen and insect attack: systemic plant-mediated interactions between pathogens and herbivores of the tomato, *Lycopersicon esculentum. Physiol. Mol. Plant Pathol.* 54:115–130.

Thomma, B.P.H.J., Eggermont, K., Broekaert, W.F., and Cammue, B.P.A. 2000. Disease development of several fungi on *Arabidopsis* can be reduced by treatment with methyl jasmonate. *Plant Physiol. Biochem.* 38:421–427.

Thomma, B.P.H.J., Eggermont, K., Penninckx, I.A.M.A., Mauch-Mani, B., Vogelsang, R., Cammue, B.P.A., and Broekaert, W.F. 1998. Separate jasmonate-dependent and salicylate-dependent defense-response pathways in *Arabidopsis* are essential for resistance to distinct microbial pathogens. *Proc. Natl. Acad. Sci. USA* 95:15107–15111.

Thomma, B.P.H.J., Eggermont, K., Tierens, K.F.M., and Broekaert, W.F. 1999. Requirement of functional *ethylene-insensitive 2* gene for efficient resistance of *Arabidopsis* to infection by *Botrytis cinerea*. *Plant Physiol*. 121:1093–1102.

Thulke, O.U., and Conrath, U. 1998. Salicylic acid has a dual role in the activation of defense-related genes in parsley. *Plant J*. 14:35–42.

Ton, J., Davison, S., Van Wees, S.C.M., Van Loon, L.C., and Pieterse, C.M.J. 2001. The *Arabidopsis ISR1* locus controlling rhizobacteria-mediated induced systemic resistance is involved in ethylene signaling. *Plant Physiol*. 125:652–661.

Ton, J., De Vos, M., Robben, C., Buchala, A.J., Métraux, J.P., Van Loon, L.C., and Pieterse, C.M.J. 2002a. Characterisation of *Arabidopsis* enhanced disease susceptibility mutants that are affected in systemically induced resistance. *Plant J*. 29:11–21.

Ton, J., Pieterse, C.M.J., and Van Loon, L.C. 1999. Identification of a locus in *Arabidopsis* controlling both the expression of rhizobacteria-mediated induced systemic resistance (ISR) and basal resistance against *Pseudomonas syringae* pv. *tomato*. *Mol. Plant Microbe Interact*. 12:911–918.

Ton, J., Van Pelt, J.A., Van Loon, L.C., and Pieterse, C.M.J. 2002b. The *Arabidopsis ISR1* locus is required for rhizobacteria-mediated induced systemic resistance against different pathogens. *Plant Biol*. 4:224–227.

Ton, J., Van Pelt, J.A., Van Loon, L.C., and Pieterse, C.M.J. 2002c. Differential effectiveness of salicylate-dependent and jasmonate/ethylene-dependent induced resistance in *Arabidopsis*. *Mol. Plant Microbe Interact*. 15:27–34.

Uknes, S., Mauch-Mani, B., Moyer, M., Potter, S., Williams, S., Dincher, S., Chandler, D., Slusarenko, A., Ward, E., and Ryals, J. 1992. Acquired resistance in *Arabidopsis*. *Plant Cell* 4:645–656.

Van Loon, L.C. 1997. Induced resistance and the role of pathogenesis-related proteins. *Eur. J. Plant Pathol*. 103:753–765.

Van Loon, L.C., and Antoniw, J.F. 1982. Comparison of the effects of salicylic acid and ethephon with virus-induced hypersensitivity and acquired resistance in tobacco. *Neth. J. Plant Pathol*. 88:237–256.

Van Loon, L.C., Bakker, P.A.H.M., and Pieterse, C.M.J. 1998. Systemic resistance induced by rhizosphere bacteria. *Annu. Rev. Phytopathol*. 36:453–483.

Van Peer, R. 1990. Microbial interactions and plant responses in soilless cultures—root colonization by pseudomonads: mechanisms, plant responses and effects on fusarium wilt. Ph. D. thesis, Utrecht University, The Netherlands.

Van Peer, R., Niemann, G.J., and Schippers, B. 1991. Induced resistance and phytoalexin accumulation in biological control of fusarium wilt of carnation by *Pseudomonas* sp. strain WCS417r. *Phytopathology* 91:728–734.

Van Peer, R., and Schippers, B. 1992. Lipopolysaccharides of plant growth-promoting *Pseudomonas* sp. strain WCS417r induce resistance in carnation to fusarium wilt. *Neth. J. Plant Pathol*. 98:129–139.

Van Wees, S.C.M. 1999. Rhizobacteria-mediated induced systemic resistance in *Arabidopsis*: signal transduction and expression. Ph. D. thesis, Utrecht University, The Netherlands.

Van Wees, S.C.M., De Swart, E.A.M., Van Pelt, J.A., Van Loon, L.C., and Pieterse, C.M.J. 2000. Enhancement of induced disease resistance by simultaneous activation of salicylate- and jasmonate-dependent defense pathways in *Arabidopsis thaliana*. *Proc. Natl. Acad. Sci. USA* 97:8711–8716.

Van Wees, S.C.M., Luijendijk, M., Smoorenburg, I., Van Loon, L.C., and Pieterse, C.M.J. 1999. Rhizobacteria-mediated induced systemic resistance (ISR) in *Arabidopsis* is not associated with a direct effect on expression of known defense-related genes but stimulates the expression of the jasmonate-inducible gene *Atvsp* upon challenge. *Plant Mol. Biol.* 41:537–549.

Van Wees, S.C.M., Pieterse, C.M.J., Trijssenaar, A., Van 't Westende, Y.A.M., Hartog, F., and Van Loon, L.C. 1997. Differential induction of systemic resistance in *Arabidopsis* by biocontrol bacteria. *Mol. Plant Microbe Interact.* 10:716–724.

Verberne, M.C., Hoekstra, J., Bol, J.F., and Linthorst, H.J.M. 2003. Signaling of systemic acquired resistance in tobacco depends on ethylene perception. *Plant J.* 35:27–32.

Vernooij, B., Friedrich, L., Morse, A., Reist, R., Kolditz-Jawhar, R., Ward, E., Uknes, S., Kessmann, H., and Ryals, J. 1994. Salicylic acid is not the translocated signal responsible for inducing systemic acquired resistance but is required in signal transduction. *Plant Cell* 6:959–965.

Vijayan, P., Shockey, J., Levesque, C.A., Cook, R.J., and Browse, J. 1998. A role for jasmonate in pathogen defense of *Arabidopsis*. *Proc. Natl. Acad. Sci. USA* 95:7209–7214.

Visca, P., Ciervo, A., Sanfilippo, V., and Orsi, N. 1993. Iron-regulated salicylate synthesis by *Pseudomonas* spp. *J. Gen. Microbiol.* 139:1995–2001.

Volko, S.M., Boller, T., and Ausubel, F.M. 1998. Isolation of new *Arabidopsis* mutants with enhanced disease susceptibility to *Pseudomonas syringae* by direct screening. *Genetics* 149:537–548.

Ward, E.R., Uknes, S.J., Williams, S.C., Dincher, S.S., Wiederhold, D.L., Alexander, D.C., Ahl-Goy, P., Métraux, J.P., and Ryals, J.A. 1991. Coordinate gene activity in response to agents that induce systemic acquired resistance. *Plant Cell* 3:1085–1094.

Wei, G., Kloepper, J.W., and Tuzun, S. 1991. Induction of systemic resistance of cucumber to *Colletotichum orbiculare* by select strains of plant-growth promoting rhizobacteria. *Phytopathology* 81:1508–1512.

Wei, G., Kloepper, J.W., and Tuzun, S. 1996. Induced systemic resistance to cucumber diseases and increased plant growth by plant growth-promoting rhizobacteria under field conditions. *Phytopathology* 86:221–224.

Whalen, M.C., Innes, R.W., Bent, A.F., and Staskawicz, B.J. 1991. Identification of *Pseudomonas syringae* pathogens of *Arabidopsis* and a bacterial locus determining avirulence on both *Arabidopsis* and soybean. *Plant Cell* 3:49–59.

White, R.F. 1979. Acetylsalicylic acid (aspirin) induces resistance to tobacco mosaic virus in tobacco. *Virology* 99:410–412.

Wildermuth, M.C., Dewdney, J., Wu, G., and Ausubel, F.M. 2001. Isochorismate synthase is required to synthesize salicylic acid for plant defence. *Nature* 414:562–565.

Yan, Z., Reddy, M.S., Ryu, C.M., McInroy, J.A., Wilson, M., and Kloepper, J.W. 2002. Induced systemic protection against tomato late blight elicited by plant growth-promoting rhizobacteria. *Phytopathology* 92:1329–1333.

Zhang, Y., Fan, W., Kinkema, M., Li, X., and Dong, X. 1999. Interaction of NPR1 with basic leucine zipper protein transcription factors that bind sequences required for salicylic acid induction of the *PR-1* gene. *Proc. Natl. Acad. Sci. USA* 96:6523–6528.

Zhou, J.M., Trifa, Y., Silva, H., Pontier, D., Lam, E., Shah, J., and Klessig, D.F. 2000. NPR1 differentially interacts with members of the TGA/OBF family of transcription factors that bind an element of the *PR-1* gene required for induction by salicylic acid. *Mol. Plant-Microbe Interact.* 13:191–202.

Zimmerli, L., Jakab, G., Métraux, J.-P., and Mauch-Mani, B. 2000. Potentiation of pathogen-specific defense mechanisms in *Arabidopsis* by β-aminobutyric acid. *Proc. Natl. Acad. Sci. USA* 97:12920–12925.

10

Induced Systemic Resistance Mediated by Plant Growth-Promoting Rhizobacteria (PGPR) and Fungi (PGPF)

ELIZABETH BENT

10.1 Definitions of PGPR and PGPF

10.1.1 PGPR

Plant growth-promoting rhizobacteria, or PGPR, are a heterogenous group of non-pathogenic bacteria that are associated with plant roots (colonizing either the root itself, or the rhizosphere), and mediate improvements in plant growth or health. While it is quite possible for a bacterium to benefit plant growth or health while colonizing the phyllosphere (e.g., Bashan and de-Bashan, 2002), and the phyllosphere colonizers could in theory influence plant defense responses, this discussion will be restricted to soil-inhabiting rhizobacteria.

False impressions can easily be generated by considering any PGPR as part of a uniform group of organisms interacting similarly with plants. The classification of different bacteria as "PGPR" *does not* reflect a biological similarity between these bacteria: PGPR vary from one another quite radically in taxonomy, in physiology, and in their interactions with plants. When attempting to compare literature reports, it is invaluable to take into account whether the organisms under study are in any way biologically similar (e.g., producing a similar compound, or belonging to the same phylogenetic group). It is very frustrating to read reports in which the authors assume that, since one bacterium classified as a PGPR produces a particular plant response, that *all* bacteria classified as PGPR must effect this same response.

In reality, PGPR interact with their host plants by a variety of mechanisms, and most PGPR probably employ more than one of these mechanisms, either simultaneously, or at different times under different conditions. Also, despite the name, PGPR do not always promote plant growth. A bacterium that promotes the growth of one plant may have no effect, or a deleterious effect, upon the growth of other plants, and a bacterium that promotes the growth of a given plant under one set of environmental conditions may have no effect, or a deleterious effect, on the same plant under different conditions (Tuzun and Bent, 1999, and references therein). The term "PGPR" should therefore be considered an operational rather than an absolute term, which describes the effect of a bacterium on a given range of plant hosts under a given range of environmental conditions only, as previously suggested in Bent and Chanway (1998).

Organisms identified as PGPR have diverse taxonomy (Glick, 1995), and include Firmicutes or Gram-positive bacteria (e.g., members of the Actinomycetales, including *Frankia* and *Streptomyces*, and Bacilli, including *Bacillus* and *Paenibacillus*), as well as Gram-negative organisms in various subdivisions of the Proteobacteria: Rhizobiaceae (*Rhizobium, Bradyrhizobium*), Rhodospirillaceae (*Azospirillum*), and Acetobacteraceae (*Acetobacter*) in the α-Proteobacteria; members of the Burkholderia group (*Burkholderia*) in the β-Proteobacteria, and members of the Enterobacteriaceae (*Enterobacter, Pantoea, Serratia*) and Pseudomonaceae (*Pseudomonas, Flavimonas*) in the γ-Proteobacteria.

Some PGPR form symbiotic structures with plants (e.g., rhizobial or actinorhizal nodules) while others are "associative", and live freely in the rhizosphere soil, the root surface, or even the interior of the root itself (Glick, 1995; Sturz et al., 2000).

10.1.2 PGPF

The defintion of plant growth-promoting fungi, or PGPF, is similar to that of PGPR except that the organisms in question are fungi (here including true fungi as well as oomycetes) rather than bacteria. While mycorrhizal fungi are known to improve the growth of plants and affect the expression of plant defense responses (Lambais and Mehdy, 1995; Peterson and Farquhar, 1994; Ruiz-Lozano et al., 1999; Sirrenberg et al., 1995), a comprehensive discussion of the interactions between mycorrhizal fungi and plants is beyond the scope of this chapter. Our definition of PGPF, therefore, is limited to nonsymbiotic saprotrophic fungi that live freely in rhizosphere soil or on the plant root surface.

The same caveats identified above for PGPR hold for PGPF: not every organism identified as a PGPF will improve plant growth under all conditions, or in association with all plant hosts (e.g., Ousley et al., 1993). The term "PGPF" is a convenient but artificial category, not an indication of any real biological similarity between organisms classified as PGPF, and when comparing the results of different studies, the fact that the organisms under question may be radically different from one another, or differ in their interactions with plants, must always be kept in mind. As with PGPR, it is helpful to keep in mind any phylogenetic or taxonomic similarities between PGPF reported in the literature when comparing reports.

Characterized fungi reported in the literature as PGPF primarily include ascomycetes (*Penicillium, Trichoderma, Fusarium, Phoma, Gliocladium*) and oomycetes (*Pythium, Phythophthora*). Interestingly, some reported PGPF are non-pathogenic or hypovirulent strains of phytopathogenic fungi (Table 10.2).

10.2 How PGPR and PGPF Interact with Plants to Improve Growth

There are a variety of ways in which PGPR and PGPF, here discussed together, may improve the growth or health of plants. A detailed discussion of each of these mechanisms is beyond the scope of this chapter, and the reader is referred to reviews by Buchenauer (1998), Glick (1995), and Whipps (2001) for more information.

Mechanisms of plant growth promotion include increasing plant nutrient acquisition, modification of plant growth and development, modification of the soil environment to promote plant growth, and biocontrol of plant pathogens. Biocontrol can be via direct mechanisms, where the pathogen itself is attacked, or via indirect ones, where plant defense responses against the pathogen are induced.

10.2.1 Plant Growth Promotion

Nitrogen-fixing rhizobial and actinorhizal nodules can increase plant uptake of nitrogen, and PGPR that assist in the formation of rhizobial nodules and the vigor of activity within them have also been identified (Srinivasan et al., 1996; Tokala et al., 2002). Free-living nitrogen-fixing bacteria that colonize the rhizosphere or interior tissues of plants may also improve plant growth by providing nitrogen (Sevilla et al., 2001), and this may be especially important in nutrient-limiting environments. Mycorrhizal infection can improve plant uptake of water as well as nutrients, phosphorus in particular (Peterson and Farquhar, 1994), and PGPR have been identified which assist mycorrhizal fungi in colonizing plants (Garbaye, 1994). Siderophore-overproducing mutants of a metal-tolerant soil bacterium were found to help plants overcome growth inhibition by heavy metals in soil, most likely by providing the plant with iron (Burd et al., 2000). Saprophytic PGPF can improve the nutrient supply also: for example, phosphate-solubilizing fungi have been identified which promote plant growth (Whitelaw et al., 1999).

Plant growth-altering hormones such as auxin, cytokinins, or giberellins, which can alter root morphology and stimulate growth, are known to be produced by rhizobacteria (Costacurta and Vanderleyden, 1995; Patten and Glick, 1996, 2002) as well as rhizofungi (Furukawa et al., 1996). PGPR may also produce enzymes that degrade the precursors of plant growth-inhibiting hormones such as ethylene, indirectly enhancing plant growth (Glick, 1995).

PGPR and PGPF may also improve plant growth indirectly, via alterations to the structure of rhizosphere soil, which benefit the plant. Exopolysaccharide-producing PGPR have been found to significantly increase rhizosphere soil aggregation and the volume of soil macropores, resulting in increased water and fertilizer availability to inoculated sunflowers (Alami et al., 2000). Desertified soils, in which the soil structure has been degraded, contain a greater number of hydrostable soil aggregates after inoculation with PGPR and fungi, and this improvement in soil structure may assist natural plant communities in recolonizing these soils (Requena et al., 2001).

10.2.2 Disease Control

Perhaps the most research on plant-growth promoting microorganisms has been devoted to determining how they can be used to protect plants from disease. Pathogen control by PGPR may involve the production of antimicrobial enzymes, antibiotics, predation, or it may occur via the systemic induction of plant defense responses (ISR) (Buchenauer, 1998; Whipps, 2001). Phyllosphere as well as rhizosphere bacteria have also been shown to successfully control pathogens via niche exclusion (Bashan and de-Bashan, 2002; Buchenauer, 1998). Bacteria may employ more than one mechanism simultaneously to control pathogens.

Pathogen control by PGPF may also occur via niche exclusion, antibiosis, predation, mycoparasitism, and ISR induction (Shivanna et al., 1996; Mauchline et al., 2002; Whipps, 2001). Hypovirulent pathogen isolates containing double-stranded RNA (dsRNA) may also control more virulent isolates via anastomosis, in which dsRNA conferring hypovirulence is transferred to the virulent isolate (e.g., Batten et al., 2000).

Fungi may employ more than one control mechanism simultaneously. For example, a nonpathogenic strain of *Fusaruim oxysporum* was found to control *Pythium ultimum* via a combination of ISR, antibiosis, and mycoparasitism (Benhamou et al., 2002), and *Trichoderma* isolates, known to act directly on pathogens as biocontrol agents, have been also found capable of inducing systemic resistance (de Meyer et al., 1998).

10.3 The Difference Between ISR and Direct Biological Control

It is important to draw a clear distinction between direct mechanisms of biological control, in which the PGPR/F acts directly upon the pathogen, and indirect mechanisms that require the induction of plant defense responses. This distinction is not always understood: there have been recent reports in which the authors conclude that biocontrol agents acted via induction of systemic resistance in plants, when alternate explanations for the reduction in disease symptoms of inoculated plants, such as antibiosis or niche exclusion, were not tested. The criteria for distinguishing between biocontrol agents that act via direct or via indirect (ISR) mechanisms have been thoroughly described elsewhere (van Loon et al., 1998).

It is also important to distinguish between ISR and race-specific, gene-for-gene types of interactions. ISR can be a nonspecific phenomenon, in which a variety of nonspecific elicitors can stimulate the plant's innate, and already existing, defenses. The plant does not acquire new defense mechanisms during the process of stimulation; ISR makes use of the plant's existing set of defense responses. Ton et al. (1999) provide an excellent illustration of this principle: ecotypes of *Arabidopsis thaliana* which exhibited greater innate, or "basal" susceptibility to *P. syringae* pv. *tomato*, also failed to develop ISR after treatment with *P. fluorescens* WCS417r, a bacterium known to elicit this response in other ecotypes of *A. thaliana*. A genetic association was observed between basal resistance and ability to develop ISR, supporting the idea that plants with a more effective set of resistance responses (or greater basal resistance) will be able to muster a more effective ISR response than plants with a less effective set of resistance responses. This pattern was also observed in cucumbers inoculated with *Pseudomonas* isolates (Arndt et al., 1998), and in a variety of other plant systems where pathogenesis-related (PR) proteins are constitutively expressed at higher levels in cultivars expressing greater basal resistance to a given pathogen (Tuzun and Bent, 1999, Vleeshouwers, 2000).

10.3.1 Plant Nutrition and Improved Resistance to Disease

Since the defensive mechanisms activated in plants by a variety of plant-beneficial microorganisms are still largely unknown (e.g., van Wees et al., 1999) or remain unstudied (Tables 10.1, 10.2), the following possibility should be mentioned. In determining whether a PGPR/F inoculant can induce systemic resistance in a plant, direct biocontrol interactions between PGPR/F and the pathogens used must be ruled out, but there is no requirement to directly observe the induction of a plant defense mechanism. It is sufficient to observe that the inoculated plants have improved resistance to the disease, and that this effect cannot be explained by alternate biocontrol mechanisms, for the phenomenon to be labeled "ISR". Plant-beneficial microorganisms, by improving plant nutrition or the rate or extent of plant growth, might improve plant resistance to or tolerance of pathogens or herbivores *without* the direct induction of any known plant defensive response. Fertilization of plants is known to improve plant tolerance of disease and herbivory (Goncalves et al., 2000; Matichenkov et al., 2000), and induced resistance to herbivory in soybean by an arbuscular mycorrhizal fungus was attributed to improved plant nutrition, rather than induction of any plant defense mechanisms (Borowicz, 1997). The effect of improved fertilization on disease resistance may be pathogen specific, however, and not provide consistent results against different pathogens (Ellis et al., 2000).

10.4 PGPR and PGPF-Mediated ISR

10.4.1 Explanation of Terminology Used in this Chapter

There are many reports of ISR/SAR induced by rhizosphere organisms in which the defensive mechanism for the resistance is unknown (Tables 10.1, 10.2). Recent research also indicates that there are more than two biochemical pathways by which induced resistance can be activated (e.g., Bostock et al., 2001; Dong and Beer, 2000; Mayda et al., 2000a,b; Zimmerli et al., 2000, Ryu et al., 2003). Moreover, since the mechanisms by which many PGPR or PGPF mediate ISR have never actually been studied, I find it would be impossible to discuss this topic without some generic term that means only resistance in plants which is inducible and systemic.

I will use "ISR" as the generic term. To distinguish between different established or hypothetical mechanisms that produce ISR, I will use a prefix suggesting a

TABLE 10.1. Overview of literature reporting PGPR, which induce systemic resistance in plants.

PGPR designation	Plant host(s)	Challenge organism(s)	Defense response in plant[a]	Reference
Pseudomonas fluorescens 89-B-61, *Serratia marcesens* 90-166	Cucumber (*Cucumis sativus*), tobacco (*Nicotania tabacum*)	*Colletotrichum orbiculare, Fusarium oxysporum* f. sp. *cucumerinum, Pseudomonas syringae* pv. *tabaci* (tobacco)		Liu et al. (1995a,b); Press et al. (1997, 2001)
P. fluorescens 89-B-61, *S. marcesens* 90-166, *Bacillus pumilis* SE34, T4, *B. pasteurii* C-9	Tobacco (*N. tabacum*)	*Peronospora tabacina*		Zhang et al. (2002)
P. fluorescens 89-B-61, *S. marcesens* 90-166, *Bacillus pumilis* SE34, T4	Thale cress (*Arabidopsis thaliana*)	*Pseudomonas syringae* pv. *tomato, P. syringae* pv. *maulicola*		Ryu et al. (2003)
P. fluorescens WCS374, WCS417	Radish (*Raphanus sativus*)	*F. oxysporum* f. sp. *raphani*		Leeman et al. (1995, 1996)
P. fluorescens WCS374r, WCS417r	Thale cress (*A. thaliana*)	*P. syringae* pv. *tomato* DC3000	Primes conversion of ACC to ethylene (ethylene could be involved in defense responses?)	Hase et al. (2003)
P. fluorescens WCS417r	Carnation (*Dianthus caryophyllus*)	*F. oxysporum* f. sp. *dianthi*	Phytoalexin accumulation	van Peer et al. (1991)
P. fluorescens WCS417r	Radish (*R.. sativus*), carnation (*D. caryophyllus*)	*F. oxysporum* f. sp. *raphani, F. oxysporum* f. sp. *dianthi*	Changes in cell wall composition	Steijl (1999)
P. fluorescens WCS417r, WCS358r	Thale cress (*A. thaliana*)	*F. oxysporum* f. sp. *raphani, P. syringae* pv. *tomato*		van Wees et al., 1997, 1999; Pieterse et al., 1996
P. fluorescens WCS417r	Thale cress (*A. thaliana*)	*Alternaria brassicicola, Xanthomonas campestris* pv. *armoraciae, P. syringae* pv.*tomato*		Ton et al (1999, 2001, 2002)
P. fluorescens WCS417r	Tomato (*Lycoperscion esculentum*)	*F. oxysporum* f. sp. *lycopersici*		Duijff et al. (1998)

Bacterial strain	Plant	Pathogen	Response/Mechanism	Reference
P. aeruginosa 7NSK2	Bean (*Phaseolus vulgaris*)	*Botrytis cinerea*	Induction of phenylalanine ammonia-lyase, SA accumulation	de Meyer et al. (1999)
P. fluorescens WCS417, P. aeruginosa KMPCH	Bean (*P. vulgaris*)	*Colletotrichum lindemuthianum*		Bigirimana and Hofte (2002)
P. aeruginosa KMPCH	Bean (*P. vulgaris*)	*B. cinerea*	HR and phytoalexin accumulation	de Meyer et al. (1998)
Nine *Pseudomonas* sp. strains and two *Serratia* sp. strains	Bean (*P. vulgaris*)	Fusarium solani pv. phaseoli		Hynes et al. (1994)
P. fluorescens S97	Bean (*P. vulgaris*)	*Pseudomonas syringae* pv. *syringae*		Alström (1991)
Pseudomonas. putida BTP1, M3	Cucumber (*C. sativus*)	*Pythium aphanidermatum*	Phytoalexin accumulation	Ongena et al. (2000)
Pseudomonas WB1, WB15, WB52	Cucumber (*C. sativus*)	*Pythium ultimum, Rhizoctonia solani*	Callose deposition	Arndt et al. (1998)
P. fluorescens CHA0	Tobacco (*N. tabacum*)	Tobacco necrosis virus	Accumulation of PR proteins	Maurhofer et al. (1994)
P. fluorescens CHA0, EP1, Pf1, P. putida KKM1	Sugarcane (*Saccharum* sp. hybrids)	*Colletotrichum falcatum*	Enhanced lignification and activity of defense proteins	Viswanathan and Samiyappan (2002a,b)
P. fluorescens Pf1, FP7	Rice (*Oryza sativa*)	*R. solani*	Accumulation of chitinase and peroxidase isozymes	Nandakumar et al. (2001)
P. fluorescens 7-14, P. putida Vl4i	Rice (*O. sativa*)	*Magnaporthe grisea*	Accumulation of SA	Krishnamurthy and Gnanamanickam (1998)
P. fluorescens WR8-3, WR9-11, P. putida WR9-16	Watermelon (*Citrullus lanatus* cv. Geumchon)	*Didymella bryoniae*		Lee et al. (2001)
P. fluorescens Blight Ban A506, Pantoea agglomerans C9-1	Apple (*Malus domestica*)	*Erwinia amylovora*		Momol et al. (1999)
Pantoea agglomerans E278Ar	Radish (*Raphanus sativus*)	*X. campestris* pv. *armoraciae*		Han et al. (2000)
Bacillus subtilis FZB-G	Tomato (*Lycopersicon esculentum*)	*Fusarium oxysporum* f. sp. *lycopersici*		Gupta et al.. (2000)
B. subtilis AF 1	Groundnut (*Arachis hypogaea*)	*Aspergillus niger*	Increased lipoxygenase activities, production of antifungal hydroperoxides	Sailaja et al. (1997)
B. pumilis SE 34, Serratia plymuthica R1GC4, P. fluorescens 63-28	tomato (*L. esculentum*), pea (*Pisum sativum*), cucumber (*C. sativus*)	*Fusarium oxysporum* f. sp. *lycopersici, P. ultimum*	Callose and cellulose deposition, accumulation of phytoalexins and PR proteins	Benhamou et al. (1996, 1998, 2000); M'Piga et al. (1997)

TABLE 10.1. (Continued)

PGPR designation	Plant host(s)	Challenge organism(s)	Defense response in plant[a]	Reference
P. fluorescens 63-28, *P. corrugata* 13	Cucumber (*C. sativus*)	*Pythium aphanidermatum*	Induction of peroxidases and enzymes involved in lignification	Chen et al. (1998, 2000)
B. pumilis SE 34, SE 52, INR7, *S. marcesens* 90-166	Loblolly pine (*Pinus taeda*)	*Cronartium quercum* f. sp. *fusiforme*		Enebak and Carey (2000)
Paenibacillus polymyxa B2	thale cress (*A. thaliana*)	*Erwinia carotovora*	Induced changes in drought stress, SA-ISR and JA-ISR-related genes	Timmusk and Wagner (1999)
Bacillus thuringiensis Berliner	coffee (*Coffea arabica*)	*Hemileia vastatrix*	Accumulation of PR proteins	Guzzo and Martins (1996)
B. pumilis INR7, *B. subtilis* GB03, *Curtobacterium flaccumfaciens* ME1, alone or in mixtures	Cucumber (*C. sativus*)	*C. orbiculare, P. syringae* pv. *lachrymans, Erwinia tracheiphila,* alone or in mixtures		Raupach and Kloepper (1998)
Bacillus amyloliquefaciens EXTN-1	tobacco (*N. tabacum*), thale cress (*A. thaliana*), cucumber (*C. sativus*)	Pepper mild mottle virus (tobacco, cress), *C. orbiculare* (cucumber)	Accumulation of PR and defense-related gene transcripts (tobacco, cress), callose deposition (cucumber)	Ahn et al. (2002); Jeun et al. (2001)
B. pumilis SE34, *B. subtilis* IN937b, *B. amyloliquifaciens* IN937a, *Kluyvera cryocrescens* IN114	Tomato (*L. esculentum*)	Cucumber mosaic cucumovirus, tomato mottle virus (not IN114)		Zehnder et al. (2000); Murphy et al. (2000)
B. pumilis INR-7, T4, *Flavomonas oryzihabitans* INR-5, *P. putida* 89-B-61, *S. marcesens* 90-166	Cucumber (*C. sativus*)	Cucumber beetle (*Acalymma vittata*), *Erwinia tracheiphila* (T4, 90-166 only)	Decreased levels of cucurbitacin (INR-7, INR-5)	Zhender et al. (1997a,b, 2000)
Rhizobium elti G12, *Agrobacterium radiobacter* G12, *Bacillus sphaericus* B43	Potato (*Solanum tuberosum*)	Potato cyst nematode (*Globodera pallida*)		Reitz et al. (2000); Hasky-Günther et al. (1998)
Pseudomonas-like sp. P29, P80, *Bacillus cereus* B1	White clover (*Trifolium repens*)	Clover cyst nematode (*Heterodera trifolii*)		Kempster et al. (2001)

[a] This column summarizes positive reports of plant defense-related reactions that were induced or augmented by PGPR inoculation. Negative results (i.e., reports that a substance or reaction is *not* involved), or reports of optimal conditions or plant or bacterial genotypes required for ISR, are not included.

TABLE 10.2. Overview of literature reporting PGPF (fungi and oomycetes) which induce systemic resistance in plants.

PGPF designation	Plant host(s)	Challenge organism(s)	Defense response in plant[a]	Reference
Trichoderma GT3-2, Fusarium GF18-3, Penicillium GP17-2, Phoma GS8-2, sterile fungus GU23-3	Cucumber (Cucumis sativus)	Colletotrichum orbiculare, Fusarium oxysporum f. sp. cucumerinum, Pseudomonas syringae pv. lachrymans	Superoxide generation, induction of lignification	Koike et al. (2001)
Phoma GS8-2, sterile fungus GU23-3, and uncharacterized PGPF GS8-1, GS8-2, GU21-2	Cucumber (C. sativus)	C. orbiculare		Meera et al. (1995)
Nonpathogenic Fusarium oxysporum Fo47	Cucumber (C. sativus)	Pythium ultimum	Production of physical barriers to infection, production of antimicrobial compounds	Benhamou et al. (2002)
Nonpathogenic F. oxysporum isolates	Watermelon (Citrullus lanatus), tomato (Lycopersicon esculentum)	F. oxysporum f. sp. niveum, F. oxysporum f. sp. lycopersici		Larkin et al. (1996); Larkin and Fravel (1999)
Nonpathogenic F. oxysporum isolates	Asparagus (Asparagus officinalis)	Fusarium oxysporum f. sp. asparagi	Hypersensitive response, lignification, increased defense protein production, production of antifungal metabolites	He et al. (2002)
Penicillium janczewskii	Cotton (Gossypium barbadense), melon (Cucumis melo L. var. reticulatus)	Rhizoctonia solani	Hypersensitive response and elevated defense protein activity (melon), increased peroxidase activity (both)	Madi and Katan (1998)
Trichoderma harizanum T39	Tomato (L. esculentum), lettuce (Lactuca sativa), pepper (Capsicum annuum), bean (Phaseolus vulgaris), tobacco (Nicotania tabacum)	Botrytis cinerea		De Meyer et al. (1998)
Nonpathogenic Alternaria cucumarina, Cladosporium fulvum	Cucumber (C. sativus)	Sphaerotheca fuliginea		Reuveni and Reuveni (2000)
Nonpathogenic, binucleate Rhizoctonia sp.	Bean (P. vulgaris)	Rhizoctonia solani, Colletotrichum lindemuthianum	Systemic increase in PR proteins, accumulation of phenolics, increased lignification	Xue et al. (1998)
Nonpathogenic Phytophthora cryptogea	Tomato (L. esculentum)	F. oxysporum f. sp. lycopersici		Attitalla et al. (2001)
Aureobasidium pullulans	Apple (Malus domestica cv. Red Delicious)	Botrytis cinerea, Penicillium expansum	Increases in PR proteins (but control may be due to biocontrol rather than ISR)	Ippolito et al. (2000)

[a] This column summarizes positive reports of plant defense-related reactions that were induced or augmented by PGPF inoculation. Reports of optimal conditions, or plant or bacterial genotypes required for ISR, are not included.

compound involved in the biochemical response. I feel this scheme is simple to understand and allows for the discussion of many different potential mechanisms, as well as reports of ISR where the biochemical pathway or mechanism involved in induction is unknown.

10.4.2 Overview of Known Pathways Involved in Microbially-Stimulated ISR

There are at least three, and potentially more, interconnected biochemical mechanisms by which ISR can be activated in plants. I will focus here on those mechanisms, which are or could be stimulated by rhizosphere microorganisms. By "mechanism" I refer to the entire biochemical pathway involved in recognition of the stimulus and generation of the response in the plant. It should be noted that multiple mechanisms can share part of the same biochemical "pathway" if receptors for different elicitors activate the same signaling cascades at some point. The pathways involved in the ISR phenomenon and their interconnections are reviewed by Nawrath et al. and Pieterse et al. (Chapter 7 and 8 of this volume).

It should also be noted that these mechanisms may not function in every plant species or variety. ISR in peanut (*Arachis hypogaea* L.), for example, was induced by β-aminobutyric acid (BABA) but not by a variety of PGPR inoculants or commonly used chemical elicitors, including salicylic acid, methyl jasmonate, and ethylene (Zhang et al., 2001).

Salicylate-mediated, salicylate-dependent or "classical" ISR (here defined as SA-ISR; also sometimes defined in recent works as "systemic acquired resistance" or "SAR") was the first mechanism identified. It is sometimes also referred to as "pathogen-mediated" ISR, since the phenomenon was first observed on plants inoculated with plant pathogens, but nonpathogenic organisms may also stimulate SA-ISR (see Section 10.4.4). SA-ISR typically involves the accumulation of pathogenesis-related (PR) proteins and the induction of a hypersensitive response.

Jasmonate-mediated or jasmonate-dependent ISR (JA-ISR) is less well characterized; it is not associated with PR protein accumulation or a hypersensitive response, but appears to involve changes to plant secondary metabolism, resulting in the accumulation of phytoalexins in at least some plants. Since the first organisms discovered to induce this set of responses were rhizobacteria, JA-ISR has generally been associated with PGPR, but this association is misleading since it gives the impression that *all* PGPR that elicit ISR do so via the JA-ISR mechanism exclusively, which is not the case. This mechanism is discussed further in Section 10.4.5.

The above example of peanut plants in which ISR cannot be elicited by jasmonate or salicylate, but instead by BABA (Zhang et al., 2001), illustrate the existence of a third ISR mechanism. This mechanism is induced by aminobutyric acids (Jakab et al., 2001; Zimmerli et al., 2000). I am not aware of any evidence that plant-beneficial microorganisms use this mechanism to activate ISR in plants, although this is possible in theory. I will therefore not discuss this mechanism further.

Neither jasmonate, salycilate, nor aminobutyric acids appear to be involved in a fourth, potentially distinct, mechanism of ISR, in which the associated responses differ from those seen in JA-ISR and SA-ISR. This mechanism appears to be induced by the onset of cellular insensitivity to auxin (Mayda et al., 2000a,b). The potential for involvement of auxin in microbially mediated ISR is discussed in Section 10.4.6.

There are additional reports of ISR mechanisms induced by riboflavin (Dong and Beer, 2000), a modified antiviral protein (Zoubenko et al., 2000) or ceramides (Bostock et al., 2001), via which microorganisms could in theory induce disease resistance should they produce a sufficiently similar inducing substance, but it is beyond the scope of this chapter to discuss these.

10.4.3 Specificity and Induction of More than One ISR Mechanism by Microorganisms

There may be some specificity in the ability of PGPR to induce an ISR response, although the basis for such specificity is currently unknown: for example, different strains of *Pseudomonas fluorescens* were found to induce ISR in radish or in *Arabidopsis*, but not both plants (van Wees et al., 1997), and while a variety of PGPR (*Bacillus pumilis, Serratia marcesens*, and *Pseudomonas fluorescens*) were found to induce resistance in *NahG* (salicylate-deficient) plants, some of these same strains were also found to function via pathways deficient in jasmonic acid, or ethylene signaling, or in plants deficient in *nprl*, previously thought to be required for ISR mediated by nonpathogenic rhizobacteria (Ryu et al., 2003). There is also no theoretical reason why a single organism (or, as happens more often in nature, a consortia of organisms) cannot induce resistance via *more* than one ISR pathway, either by stimulating different responses in different plant hosts, or by stimulating different responses in the same plant under varying conditions. An ISR-inducing PGPR was found to induce changes in *Arabidopsis* drought stress-related genes, as well as genes relating to the SA-ISR and JA-ISR pathways (Timmusk and Wagner, 1999), suggesting that biotic and abiotic stress responses may be linked, and that one organism may possess the ability to induce more than one ISR pathway.

10.4.4 PGPR that Activate SA-ISR

SA-ISR is still commonly thought to be restricted to necrotrophic phytopathogenic fungi and bacteria, despite the fact that there are several reports of PGPR that induce systemic resistance by SA-ISR (van Loon et al., 1998; Table 10.1). Because of the strong linkage between this pathway and the presence of necrotic pathogens, or metabolites of necrotic pathogens, that induce an oxidative burst and a hypersensitive response, it has been hypothesized that the nonpathogenic bacteria that do induce plant defenses via this pathway may have evolved, or acquired genes from pathogenic organisms (Tuzun and Bent, 1999). This is not a unique idea, Arndt et al. (1998) described an ISR-inducing strain of *Pseudomonas* that could "imitate infections of soilborne pathogens" in tomato, and Reitz et al. (2000) suggest that the induction of PR proteins by PGPR strain *P. fluorescens* CHA0 is due to "stress" caused by this organism on the plant.

Typical hallmarks of SA-ISR include the systemic accumulation of salicylic acid and a variety of PR protein isoforms, including chitinases, β-1,3-glucanases, and thaumatin-like proteins, increased lignification or callose deposition, the production of phytoalexins or phenolic antimicrobial secondary compounds, and increased expression of enzymes associated with active oxygen species, lignification, or plant secondary metabolism (Kobayashi et al., 1995; Hammerschmidt and Smith-Becker, 1999).

Accumulation of salicylate or PR proteins in response to PGPR inoculation has been described in several plant-PGPR systems (Zdor and Anderson, 1992; Table 10.1) along with the strengthening of physical barriers to infection and the accumulation of antifungal substances (Table 10.1). The latter responses may be temporally separated from the onset of PR protein induction and occur prior to PR protein accumulation (Benhamou et al., 1996, 1998, 2000; M'Piga et al., 1997).

Harpins produced by bacterial plant pathogens are known to elicit SA-ISR (Dong et al., 1999; Strobel et al., 1996). Tuzun and Bent (1999) speculated that nonpathogenic rhizobacteria that induce ISR may express harpin-like proteins that cause microscopic necrotic lesions and so stimulate the SA-ISR response. Since then, conserved type III secretion system genes, similar to the *hrp* cluster in plant pathogens, have been reported in PGPR, including *Rhizobium* sp. and *Pseudomonas fluorescens* (Preston et al., 2001). The nature of hypersensitive responses mediated by PGPR and pathogens seems to differ, which may explain why some PGPR can induce resistance via SA-ISR yet do not cause symptoms of disease: HR mediated by *P. fluorescens* were slower, required at least tenfold more cells, and were induced differently in different tissues, compared to HR mediated by the pathogen *P. syringae* (Preston et al., 2001). Preston et al. (2001) speculated that type III secretion systems may play broadly conserved roles in plant-microbe interactions, and may help nonpathogens as well as pathogens to live intimately with plants.

A variety of stress conditions, including exposure to salicylate, inhibits the production of the OmpF porin in *Escherichia coli* (Ramani and Boyake, 2001). OmpC and OmpF porins function as nonselective pores in the outer membrane of *E. coli* through which small hydrophilic molecules can diffuse, with the channel diameter of OmpC being slightly smaller. A decrease in OmpF expression would therefore result in generally smaller channels available for nonselective diffusion, and increased protection against the entry of larger molecules that are more likely to be toxic to the cell. It has been suggested that the ability of a bacterium to colonize plant tissues and the rhizosphere is influenced by its sensitivity to phytoalexins (Hynes et al., 1994), and it is tempting to speculate that regulation of porin size in response to plant defense signals may help Gram negative ISR-inducing PGPR to survive plant defense responses.

10.4.5 PGPR that Activate JA-ISR

JA-ISR is elicited by jasmonic acid and its derivatives, as well as by ethylene, and is implicated in systemic wound responses (Staswick and Lehman, 1999). Plant responsiveness to jasmonate and ethylene is required for the JA-ISR response to be generated in *Arabidopsis thaliana* (Knoester et al., 1999; Ton et al., 2001, Pieterse et al., 1998). Ethylene is not produced by at least some of the PGPR that are known to induce the JA-ISR response, and ethylene levels in the vicinity of induced plants do not always rise (Knoester et al., 1999). However, *Arabidopsis* roots show an increased ability to convert 1-aminocyclopropane-1-carboxylate (ACC) to ethylene after treatment with *Pseudomonas fluorescens* (both strains inducing ISR and not), suggesting that strains of this bacterium may prime plants to produce greater quantities of ethylene upon pathogen infection (Hase et al., 2003). The phytopathogenic fungus *Botrytis cinerea* also produces ethylene, both in vitro and on tomato fruit (Cristescu et al., 2002), and it is possible that nonpathogenic organisms may also produce ethylene. Rather than acting as an elicitor, ethylene may play a regulatory role, and modify plant defense responses (those regulated by JA, and others) according to particular circumstances. Salicylate, for example, was shown to enhance the expression of genes regulated by both ethylene and jasmonic acid in *Arabidopsis*, while suppressing the expression of genes regulated by jasmonic acid alone (Norman-Setterblad et al., 2000).

Ethylene production in higher plants requires ACC synthase activity, and ACC synthase expression is induced by auxin (Yi et al., 1999). As described previously, rhizobacterial inoculation has been shown to result in enhanced ability of *Arabidopsis* to convert ACC to ethylene, although the mechanism by which this occurs is unclear (Hase et al., 2003). Ethylene also regulates auxin levels: nitrilase is a key enzyme involved in auxin biosynthesis, and a gene encoding a nitrilase-like protein was found to strongly bind an ethylene-responsive element-binding protein (Xu et al., 1998). Interestingly, the expression of two ACC synthase genes in lupin increased in response to wounding (Bekman et al., 2000) and the expression of a particular ACC synthase gene in mung bean also increased continuously in response to 24-epibrassinolide (BR), an active brassinosteroid, until 24 hours after treatment (Yi et al., 1999). BR is known to promote auxin-induced ethylene production (Yi et al., 1999). Are brassiniosteroids and auxins part of a JA-ISR defense mechanism, controlling the level of expression and timing of this particular response in different plant tissues via their regulation of ethylene production? Many PGPR have been identified which produce or degrade auxin or auxin precursors, or affect auxin levels within plants (Patten and Glick, 1996). In addition, the ability of microorganisms to produce auxins in the rhizosphere will vary with environmental factors (e.g., available tryptophan levels), leaving open the possibility that microbially-mediated plant defense induction that requires the auxin production may not function in all environments or on all plant types. The elicitation of JA-ISR by strains of *Pseudomonas fluorescens* has been linked to the production of lipopolysaccharides and siderophores, but these elicitors do not fully account for ISR elicitation by these strains (van Wees et al., 1997). It would be interesting to determine if rhizobacteria that stimulate JA-ISR can also produce auxin, and under what circumstances.

The only plant defense responses known to be activated by JA-ISR are increases in phytoalexin production (van Peer et al., 1991) and alterations of the composition of lignin that seem to retard pathogen ingress (Steijl et al., 1999). No pathogenesis-related proteins appear to be induced, and there is apparently no hypersensitive response nor, to our knowledge, induction of enzymes related to lignification or plant secondary metabolism, although this possibility may not have been ade-quately explored (Pieterse et al., 1996; Reitz et al., 2001; van Wees et al., 1999). The transient accumulation of a single jasmonate-inducible transcript (*Atvsp*) has been noted in *Arabidopsis* treated with a PGPR known to activate JA-ISR (van Wees et al., 1999), but it is not clear that this is a defense response. AtVsp is a vegetative storage protein (VSP) in *Arabidopsis*; such proteins accumulate in the vacuoles of young leaves and developing reproductive structures, and serve a nutritional function by acting as a storage form for amino acids. VSPs are induced in older plant parts upon wounding (Berger et al., 1995), but do not appear to have any direct role in plant defenses. It should be possible for rhizobacteria that stimulate even transient accumulations of ethylene in plant tissues to also elicit the expression of at least some PR proteins, however, since ethylene can increase the expression of a variety of defense-related genes, including osmotin, chitinases, β-1,3-glucanases, thaumatin-like proteins, and protein kinases (del Campillo and Lewis, 1992; Xu et al., 1998).

JA-ISR induces responses within plants that appear to activate only a subset of available plant defenses (i.e., phytoalexin accumulation and lignification). This appears to explain why JA-ISR has been found to be a less effective mechanism for protection against disease in *Arabidopsis* than SA-ISR, which elicits a broad array of defenses (Ton et al., 2002). The presence of multiple disease resistance mechanisms in plants may reflect plant defenses geared toward different pathogen strategies, as well as some measure of functional redundancy.

10.4.6 Can PGPR Stimulate ISR by Modifying Auxin Levels?

Mayda et al. (2000a,b) have described an interesting model for another, apparently independently regulated, defense response induction pathway, which may help ex-plain how induced resistance in plants against viruses occurs in some plant–virus interactions. The tomato *CEV-1* gene is an anionic peroxidase induced during com-patible viral infections, but not during incompatible infections. *CEV-1* expression is also not induced by salicylate, methyl jasmonate, or ethylene, or wounding, and is therefore unlikely to be involved in the typical SA-ISR or JA-ISR responses. *CEV-1* is rapidly induced when connections between plant cells are broken in normal plants, and is also induced in auxin-insensitive tomato mutants. It is hypothesized that *CEV-1* is up-regulated via the induction of plant cell insensitivity to auxin, im-posed upon plants during compatible viral infections, and that auxin itself does not induce this gene (Mayda et al., 2000a). A *CEV-1* recessive mutant (*dth9*) was found to be more susceptible to fungal and bacterial infections, although salicylic acid metabolism and expression of PR genes remained normal, and was insensitive to exogenously applied auxin (Mayda et al., 2000b). *CEV-1*, and similarly-regulated genes, could participate in a defense response that is controlled by changes in the ability of plant cells to perceive auxin, a phenomenon which is linked to auxin homeostatic mechanisms (Leyser, 2002).

One mechanism for auxin signal transduction involves the targeted degradation of transcriptional regulators that participate in complex and competing systems, modulating the expression of a wide variety of genes (Leyser, 2002), including, probably, defense-related genes. Links between auxin metabolism and plant defense responses have already been identified earlier in this review, the most obvious of these being the link between ethylene and auxin. Auxin and ethylene each regulate levels of the other in plant tissues (Xu et al., 1998; Yi et al., 1999), and ethylene, as outlined previously, is known to be involved in JA-ISR. Auxin-activated gene transcripts were also found to accumulate in tobacco upon inoculation with compatible and incompatible bacterial pathogens (Froissard et al., 1994), and auxins negatively regulated the expression of defense-related genes in tobacco and carrot (Jouanneau et al., 1991; Ozeki et al., 1990). In pepper, an auxin-repressed protein was among a variety of genes induced by pathogen infection (Jung and Hwang, 2000).

Could there be more than one ISR mechanism in plants that is controlled by auxins, one where the role of auxin is to control ethylene levels, which, in turn, control the expression of defense-related genes, and other(s) where auxins themselves control these responses, perhaps serving as negative regulators? If there are multiple, auxin-regulated ISR mechanisms (which may or may not interact, although it seems probable that they would), is it possible for bacterial strains which produce or degrade auxins or their precursors—either from a location exterior to the plant, such as in the rhizosphere, or from a location *within* the plant, as in the case of naturally occurring, nonpathogenic endophytic microorganisms—to manipulate these mechanisms? It has often been noted that treatment of plants with biological or chemical elicitors of ISR can produce, in addition to resistance, significant increases in plant growth and yield (Tuzun and Bent, 1999). Given that plants must make an energy investment in their plant defenses, this result is counterintuitive. However, if known plant growth stimulants such as auxins are involved in plant defense responses, increases in plant growth under most conditions would be expected.

10.4.7 PGPF-Mediated ISR

Most of the PGPF studied seem to stimulate the SA-ISR pathway in plants, judging solely from the reports of the activation of defense responses (e.g., PR protein induction) normally linked to this pathway (Table 10.2). As this pathway is most closely related to ISR induced by pathogens, and many of the PGPF found to induce systemic resistance are themselves nonvirulent forms of plant pathogens (Table 10.2), this is perhaps to be expected. It is a mistake to link SA-ISR responses with fungi or with pathogens only, however, SA-ISR can also be induced by nonpathogenic, mycorrhizal fungi (Cordier et al., 1998; Lambais and Mehdy, 1995), bacterial pathogens (Preston et al., 2001), and PGPR (Table 10.1). It is also important to realize that not all necrotrophic pathogens are able to induce ISR, even when they induce defense reactions (Govrin and Levine, 2002). If the defense reactions induced by these pathogens are, for whatever reason, wrongly timed or of insufficient extent to contain a particular pathogen, there will of course be disease.

Given that many fungi produce auxins, or auxin precursors, it is possible that PGPF could stimulate plant defenses via an auxin-regulated ISR pathway. Interestingly, the PGPF *Penicillium janczewskii* and its sterile culture filtrate were both able to induce ISR and to alter cotton root development (Madi and Katan, 1998). Whether auxin or auxin precursors were involved in these phenomena was not determined, but the alterations in root development were consistent with the observed effects of microbially-supplied auxin (Patten and Glick, 2002).

10.5 Microbial Elicitors of PGPR- and PGPF-Mediated ISR

10.5.1 Elicitor Production by Microorganisms May Vary with their Physiological Status

Whether a bacterium can function to induce systemic resistance in plants will probably relate to its physiological status. Different phase culture filtrates from *Bacillus subtilis* strain FZB-G differed in their ability to activate ISR against *Fusarium oxysporum* f.sp. *radicis-lycopersici* in tomato: stationary phase filtrates were fungitoxic, and did not induce resistance, while the opposite was observed for transition phase filtrates (Gupta et al., 2000). The effect of *B. subtilis* FZB-G on tomato resistance will therefore depend upon the physiological state of the bacterial cells, a variable often overlooked by researchers studying PGPR-plant interactions. The composition of the growth medium used to produce or deliver a bacterial or fungal biocontrol inoculant, and therefore the physiological state of the inoculant, can affect its ability to control pathogens (Fuchs et al., 2000; Hoitink and Boehm, 1999; Ousley et al., 1993), and it is not inconceivable that PGPR/F that induce ISR are similarly influenced.

10.5.2 Avr Elicitors

A variety of established or putative elicitors of ISR are produced by bacteria and fungi. Specific defense response elicitors are those involved in gene-for-gene interactions, where an inducing organism (i.e., an incompatible pathogen) expresses an avirulence (*avr*) gene, the product of which is detected by a plant possessing a resistance (*R*) gene, triggering a defense response. The same defensive responses are presumably triggered by specific as by nonspecific elicitors, but it is not clear how ISR and Avr-R interactions are related. Avr products do not seem to act in the same fashion as known ISR elicitors, since the elicitor Avr9 did not induce systemic resistance, and sometimes enhanced pathogen growth, when applied to tomato or transgenic canola expressing the *Cf-9* resistance gene (Hennin et al., 2001).

10.5.3 Oligosaccharides and Peptides

Nonspecific elicitors are perhaps more interesting, in that they are broadly conserved and less likely to be overcome by pathogen mutation, and that all the elicitors implicated in PGPR- and PGPF-mediated ISR seem to be of this type. Oligosaccharide and peptide elicitors derived from fungal cell walls can elicit plant defense responses, as reviewed by Hahn (1996). These elicitors need not be derived from virulent phytopathogenic fungi: sterile culture filtrates of the PGPF *Penicillium janczewskii* induced ISR in melon, cotton, and tobacco (Madi and Katan, 1998), and filtrates in which various PGPF were grown contained a high molecular weight (> 12,000 Da) elicitor able to induce defense responses in cucumber and tobacco (Koike et al., 2001). Chitin, which is present not only in fungal cell walls but in arthropod exoskeletons and nematode egg membranes, has long been known to stimulate ISR (El Ghaouth et al., 1994) and has been used by itself or in combination with agricultural inoculants for this purpose.

10.5.4 Lipopolysaccharides

Lipopolysaccharide (LPS) is present on the outer membrane of Gram-negative bacteria, and it consists of a lipid moiety linked to a polysaccharide that contains a conserved core region and a variable antigenic (*o*-antigen) region (Freer, 1985). Gram-positive organisms do not produce LPS. Bacterial LPS is known to affect plant defense responses, including the expression of PR protein genes, synthesis of antimicrobial compounds, and the hypersensitive response (Dow et al., 2000). Crude cell wall extracts as well as purified LPS from *P. fluorescens* strains WCS374 and WCS417 were able to induce resistance in radish while similar preparations from *P. putida* WCS358, or from mutants of WCS374 and WCS417 lacking an o-antigenic side chain, did not (Leeman et al., 1995). In contrast, outer membrane fragments derived from *P. fluorescens* WCS417r have been shown to induce resistance in radish and carnation, but this resistance was also observed when fragments were prepared from a mutant of WCS417r lacking an o-antigenic side chain (van Wees et al., 1997). LPS from *Rhizobium elti* strain G12 was also found to induce resistance in potato to potato cyst nematode (Reitz et al., 2000).

10.5.5 Siderophores

Siderophores are low molecular weight, iron-sequestering compounds produced by bacteria under iron-limiting conditions. It was thought for some time that control of pathogens by several PGPR depended upon competition for iron, and that PGPR-produced siderophores reduced the amount of iron available to pathogens (de Weger et al., 1988). It has since been demonstrated that ISR mediated by at least some PGPR depends upon siderophore production, or the level of iron availability in the rhizosphere (de Meyer and Hofte, 1998; Leeman et al., 1996). ISR has been induced in radish by purified pseudobactins (fluorescent siderophores), as well as by concentrations of SA as low as 1 ng (de Meyer et al., 1999), which may also be produced under low-iron conditions and used as a siderophore by *Pseudomonas* (de Meyer and Hofte, 1997; Leeman et al., 1996). For *P. aeruginosa* 7NSK2, SA production is essential for ISR induction (de Meyer and Hofte, 1998), but for *Serratia marcesens* strain 90-166, SA or pseudobactin production is not required for ISR induction (Press et al., 1997). However, siderophores may help protect *S. marcesens* 90-166 against activated oxygen species and facilitate its colonization of the root interior (Press et al., 2001).

10.5.6 Flagellins, Harpins, and Other Bacterial Proteins

Bacterial flagellin is also a potent elicitor of plant defense responses in *Arabidopsis thaliana* (Gómez- Gómez et al, 1999) and tomato (Felix et al., 1999). Plants may be able to distinguish between flagellin from different sources: peptides corresponding to conserved eubacterial flagellin domains produced a response (oxidative burst, callose deposition, and production of PR proteins) in *A. thaliana*, while peptides corresponding to these regions from *Agrobacterium tumefaciens* and *Rhizobium meliloti* were inactive (Gómez- Gómez et al., 1999). Heat-killed cells and culture filtrates of *Bacillus sphaericus* were found to induce resistance in potato to potato cyst nematode (Hasky-Günther et al., 1998), and it is possible that peptidoglycan or flagellin, sheared from cell surfaces during centrifugation, are responsible for this.

Other bacterial proteins, including those encoded by *hrp* clusters in both pathogenic and nonpathogenic strains of bacteria (Dong et al., 1999; Preston et al., 2001; Strobel et al., 1996) have been found to elicit ISR. A bacterial proton pump from *Halobacterium halobium* that was constitutively expressed in potatoes also elicited ISR, but this was probably due to effects on cell membrane polarization, which mimicked early events in plant defense responses (Abad et al., 1997).

10.5.7 Elicitins and Mycotoxins

Elicitins are low molecular weight peptides produced by oomycete fungi, including all analyzed species of the plant pathogen *Phytophthora* (Keller et al., 1996) and the mycoparasite *Pythium oligandrum* (Benhamou et al., 2001). Elicitins are not virulence factors, but rather avirulence factors, since the most virulent organisms are those which produce little or no elicitin (Keller et al., 1996). Elicitins can induce ISR, and different elicitins may vary in their ability to stimulate plant defense reactions. For example, resistance induced by a basic elicitin, cryptogein, induced both necrosis in tobacco leaves and the transcription of a variety of defense-related genes, while resistance induced by an acidic elicitin, capsicein, was not accompanied by visible necrosis but still induced the transcription (albeit to a lesser extent) of the same genes (Keller et al., 1996). Cryptogein was also found to increase the extent of apoplastic RNase activity, which, in turn, was found to be sufficient to reduce infection of tobacco by *Phytophthora parasitica* (Galiana et al., 1997). Proteinaceous ISR elicitors may also be produced by non-oomycete fungi, for example, a *Fusarium oxysporum* 24 kDa protein was found to induce ethylene production and varying defense responses in different weed species (Jennings et al., 2000).

Mycotoxins are produced by virulent fungi, and include the AAL-toxins and fumonisins, groups of structurally related sphingosine analogs that are produced by *Alternaria alternata* f. sp. *lycopersici* and *Fusarium moniliforme*, respectively. Both kinds of toxins have been found to induce cell death in plants, apparently by disrupting ceramide synthesis (Bostock et al., 2001). Treatment of plant roots with ceramide has been found to induce ceramide accumulation in leaves, as well as systemic resistance (Bostock et al., 2001).

10.5.8 Detection of Nonspecific Elicitors by Plants

Plant receptor-like kinases that bind peptidoglycan, a polymer found in the cell walls of bacteria, or chitin have been identified (Shiu and Bleecker, 2001), as well as plant plasma membrane proteins that have a high binding affinity for chitin fragments (Okada et al., 2002), plant proteins that are rapidly phosphorylated in response to flagellin or chitin (Peck et al., 2001), and calmodulin isoforms that are activated by nonspecific fungal elicitors (Heo et al., 1999). These discoveries may help explain how the detection of nonspecific elicitors can stimulate a plant defense response.

10.6 Spectrum of PGPR- and PGPF-Mediated ISR Activity

PGPR-induced systemic resistance has been observed on a wide variety of plants, including monocots, dicots, and gymnosperms, in response to several types of pathogens or herbivores (Table 10.1). I am not aware of any studies performed with plants that belong outside these categories (e.g., mosses and ferns), although it would be interesting from an evolutionary perspective to know if the more ancient plant forms can express induced systemic resistance.

The majority of reports I was able to find focus on plant pathogenic fungi or bacteria, but PGPR-induced resistance to viruses, nematodes, and herbivorous insects has also been reported (Table 10.1). For example, *Bacillus sphaericus*, *Agrobacterium radiobacter*, and *Rhizobium elti* can induce ISR against the potato cyst nematode, *Globodera pallida* (Hasky-Günther et al., 1998; Reitz et al., 2001). Inoculation with *Bacillus pumilis* strain INR-7 can decrease levels of the feeding stimulant cucurbitacin in cucumber leaves, resulting in reduced feeding on these plants by cucumber beetles (Zhender et al., 1997a).

PGPF-induced resistance has likewise been observed on a variety of angiosperms in response to various plant pathogens (Table 10.2). Mycorrhizal fungi, although not included in the definition of PGPF, have also been observed to induce defense responses in angiosperms (Cordier et al., 1998; Lambais and Mehdy,

1995) and gymnosperms (Sylvia and Sinclair, 1983; Strobel and Sinclair, 1991; Salzer et al., 1996).

10.7 Effects of the Environment and Other Microorganisms on PGPR- and PGPF-Mediated ISR

In a natural environment, PGPR and PGPF exist in the midst of a wide variety of other micro- and macroorganisms, some of which may themselves exert effects on plant defense responses, and all of which can be influenced by the host plant, soil- and climate-related factors, and other nearby vegetation. These factors can only be briefly outlined here.

Microbial rhizosphere communities have been observed to vary between soil and plant types (Catellan et al., 1998; Weland et al., 2001; Latour et al., 1996; Kuske et al., 2002; Timonen et al., 1998), in response to crop rotations (Vargas-Ayala et al., 2000) and the addition of organic soil amendments (Zhang et al., 1996; Bent and Topp, unpublished observations). The microbial rhizosphere community can vary with depth (Kuske et al., 2002), location along the root surface, and the nutritional status of the plant, which will affect the composition of root exudates upon which microorganisms feed (Yang and Crowley, 2000). Precipitation will affect the distribution of microorganisms in the rhizosphere, as percolating water in the soil can flush bacteria off roots and down into the soil (Mawdsley and Burns, 1994). Earthworms have been shown to facilitate the movement of bacteria within soils, and by providing an environment in which plasmid transfer between bacteria can readily occur, may increase the rate of gene transfer between bacteria in natural soils (Daane et al., 1997). The physicochemical properties of a soil will also have a profound influence on microbial metabolism, and the ability of a microorganism to produce compounds by which it interacts with plants or with other microorganisms. For example, the production of siderophores and antibiotics by *Pseudomonas fluorescens* CHA0 was found to be modulated by such factors as phosphate availability, the ratio of carbon sources to nutrients, the presence of soluble cobalt, molybdenum or zinc, and the composition of available carbon sources (Duffy and Defago, 1999).

It is not always easy to identify which microorganisms in the environment are affecting plant growth or metabolism. "Non-culturable" soil microorganisms that cannot be cultured using traditional techniques appear to make up a majority of the organisms present in soils, based upon analyses of rDNA extracted from soils (Amann et al., 1995). Such non-culturable organisms may be so because they are only able to grow in mixed cultures, as has recently been demonstrated for non-culturable marine bacteria (Kaeberlein et al., 2002), and an obligately biotrophic mycorrhizal fungus that requires a bacterium to grow in vitro (Hildebrandt et al, 2002). The presence of other bacterial rhizosphere colonists has been shown to improve the ability of some bacterial strains to colonize specific root microsites (Bent et al., 2002), and microorganisms in soil or on plant roots will naturally exist in biofilms, which change in chemical and microbial composition over time.

Direct metabolic interactions between members of a biofilm community have been reported (Moller et al., 1998). The production of quorum sensing signal molecules that regulate and coordinate the activity of individuals within a given bacterial species has been known for some time (Pierson et al., 1998). Bacteria can respond to quorum sensing-like molecules produced by other rhizobacteria (Steidle et al., 2001), by plants (Teplitski et al., 2000), and even destroy the quorum-sensing molecules produced by other bacterial species (Dong et al., 2002). Bacterial activity in the rhizosphere can therefore be altered directly by plants or other microorganisms via quorum-sensing molecules.

Mycorrhizal fungi are known to induce plant defense responses directly, as outlined in the previous section. The presence of nonhost plants, or their root exudates, can sometimes prevent mycorrhizal fungi from colonizing plants they would otherwise be able to infect (Fontenla et al., 1999), and root exudates containing allelopathic, phytotoxic compounds can prevent other plant species from establishing in the vicinity of the producer (e.g., Yamane et al., 1992). The infection of roots with mycorrhizal fungi will alter the composition of root exudates, and therefore the community of rhizosphere microorganisms (Belimov et al., 1999), and can even increase the number of rhizosphere protozoa (Jentscke et al., 1995). Protozoa feed upon rhizosphere bacteria and can alter their spatial distribution upon surfaces (Lawrence and Snyder, 1998), their taxonomic and functional diversity (Bonkowski, 2002) and appear to induce physiological responses in the bacteria, which remain uneaten (Kandeler et al., 1999). Protozoa can also influence plant growth directly via a mechanism that is unrelated to the release of nutrients from bacteria during grazing (Jentscke et al., 1995).

Grazing of roots or foliage by herbivores can also alter the composition of the microbial rhizosphere community, by altering the quantity or quality of root exudates or plant litter (Bardgett et al., 1998; Denton et al., 1999).

PGPR field trials can be quite variable in their results, and it is consistently hypothesized that other soil microorganisms may be interfering with the ability of the PGPR inoculants to adequately colonize plant roots or interact with plants (Bent and Chanway, 1998; Bent et al., 2000, and references therein). This interference may be due to an inability to adequately colonize the plant root, or alternatively, due to the alteration of growth-promoting or signaling molecules produced by PGPR and PGPF by other rhizosphere microorganisms. The effects of PGPR or PGPF on plant defense responses under natural conditions are therefore very likely to be affected by the presence of other organisms in the plant's environment, which may include other plants, protists, earthworms, insects, herbivorous animals, fungi, and bacteria. This is in addition to climactic and soil factors that can alter the physiology of plants, and potentially affect interactions between PGPR or PGPF and their host plants.

Still, it may be possible to manipulate soil microbial communities so that plant defenses are stimulated and plant growth and health improved. Disease-suppressive soils may contain microorganisms that disrupt the disease cycle in a variety of ways, including direct attacks on the pathogen or utilization of substrates the pathogen requires to locate host roots (Yin et al., in press). While ISR has not been clearly identified with a disease-suppressive soil to our knowledge, ISR can be induced by compost amendments (Hoitink and Boehm, 1999; Zhang et al., 1996, 1998), as well as aqueous extracts of compost, or autoclaved compost amended with a biocontrol agent (Hoitink and Boehm, 1999; Zhang et al., 1998). In each of these cases, the induction of systemic resistance was attributed to the presence of compost-related microorganisms.

10.8 Conclusion

PGPR and PGPF interact with plants in complex and numerous ways, especially under natural conditions where each organism is part of a dynamic consortium that fluctuates in response to environmental biotic and abiotic stimuli. Systemic resistance in plants is induced via several different mechanisms by PGPR, and possibly by several in PGPF, although less research has been conducted on PGPF-mediated ISR. These mechanisms are likely to interact at some point with the host plant's hormonal balances, especially since both bacteria and fungi are capable of synthesis of various plant hormones such as auxin. More details of these interesting plant–microbe interactions will be elucidated as research progresses, and I hope to see more work in the near future conducted on the mechanisms governing microbially-mediated ISR as well as the role(s) of typically growth-related phytohormones (auxins, cytokinins) in the ISR phenomenon. Emerging technology, such as oligonucleotide fingerprinting of rRNA genes (OFRG; Valinsky et al., 2002a,b) will enable the detailed study of rhizosphere and endophytic consortia in natural soils and in the interior of the plant. Using OFRG, it will become possible to determine the conditions under which natural or artificially produced microbial consortia tend to flourish and induce ISR in particular plants. It may one day be possible to engineer ISR-stimulating soils containing stable populations of ISR-inducing microbial consortia for particular crops, via the strategic addition of substrates, inocula, or other soil treatments.

References

Abad, M.S., Hakimi, S.M., Kaniewki, W.K., Rommens, C.M.T., Shulaev, V., Lam, E., and Shah, D.M. 1997. Characterization of acquired resistance in lesion-mimic transgenic potato expressing bacterio-opsin. *Mol. Plant Microbe Interact.* 10:635–645.

Agrawal, A.A., Tuzun, S., and Bent, E. 1999. Editor's note on terminology. In *Induced Plant Defenses Against Pathogens and Herbivores: Biochemistry, Ecology and Agriculture*, eds. Agrawal, A.A., Tuzun, S., and Bent, E., p. ix. St. Paul, MN: American Phytopathological Society.

Ahn, I.P., Park, K., and Kim, C.H. 2002. Rhizobacteria-induced resistance perturbs viral disease progress and triggers defense-related gene expression. *Molecules and Cells* 13:302–308.

Alami, Y., Achouak, W., Marol, C., and Heulin, T. 2000. Rhizosphere soil aggregation and plant growth promotion of sunflowers by an exopolysaccharide-producing Rhizobium sp. strain isolated from sunflower roots. *Appl. Environ. Microbiol.* 66:3393–3398.

Alström, S. 1991. Induction of disease resistance in common bean susceptible to halo blight bacterial pathogen after seed bacterization with rhizosphere pseudomonads. *J. Gen. Appl. Microbiol.* 37:495–501.

Amann, R.I., Ludwig, W., and Schleifer, K.-H. 1995. Phylogenetic identification and in situ detection of individual microbial cells without cultivation. *Microbiol. Rev.* 59:143–169.

Arndt, W., Kolle, C., and Buchenauer, H. 1998. Effectiveness of fluorescent pseudomonads on cucumber and tomato plants under practical conditions and preliminary studies on the mode of action of the antagonists. *Z. Pflanzenkrank. Pflanzenschutz* 105:198–215.

Attitalla, I.H., Johnson, P., Brishammar, S., and Quintanilla, P. 2001. Systemic resistance to Fusarium wilt in tomato induced by *Phytopththora cryptogea. J. Phytopathol.* 149:373–380.

Bardgett, R.D., Wardle, D.A., and Yeates, G.W. 1998. Linking above-ground and below-ground interactions: how plant responses to foliar herbivory influence soil organisms. *Soil Biol. Biochem.* 30:1867–1878.

Bashan, Y., and de-Bashan, L.E. 2002. Protection of tomato seedlings against infection by *Pseudomonas syringae* pv. tomato by using the plant growth-promoting bacterium *Azospirillum brasiliense. Appl. Environ. Microbiol.* 68:2637–2643.

Batten, J.S., Scholthof, K-B.G., Lovic, B.R., Miller, M.E., and Martyn, R.D. 2000. Potential for biocontrol of Monosporascus root rot/vine decline under greenhouse conditions using hypovirulent isolates of *Monosporascus cannonballus. Eur. J. Plant Pathol.* 106:639–649.

Bekman, E.P., Saibo, N.J.M., di Cataldo, A., Regalado, A.P., Ricardo, C.P., and Rodrigues-Pousada, C. 2000. Differential expression of four genes encoding 1-aminocyclopropane-1-carboxylate synthase in *Lupinus albus* during germination, and in response to indole-3-acetic acid and wounding. *Planta* 211:633–672.

Belimov, A.A., Sebrennikova, N.V., and Stepanok, V.V. 1999. Interaction of associative bacteria and an endomycorrhizal fungus with barley upon dual inoculation. *Microbiology* 68:122–126.

Benhamou, N., Belanger, R.R., Rey, P., and Tirilly, Y. 2001. Oligandrin, the elicitin-like protein produced by the mycoparasite *Pythium oligandrum*, induces systemic resistance to *Fusarium* crown and root rot in tomato plants. *Plant Physiol. Biochem.* 39:681–698.

Benhamou, N., Gagné, S., Le Quéré, D., and Dehbi, L. 2000. Bacterial-mediated induced resistance in cucumber: beneficial effect of the endophytic bacterium *Serratia plymuthica* on the protection against infection by *Pythium ultimum. Phytopathology* 90:45–56.

Benhamou, N., Garand, C., and Goulet, A. 2002. Ability of nonpathogenic *Fusarium oxysporum* strain Fo47 to induce resistance against *Pythium ultimum* infection in cucumber. *Appl. Environ. Microbiol.* 68:4044–4060.

Benhamou, N., Kloepper, J.W., Quadt-Hallmann, A., and Tuzun, S. 1996. Induction of defense-related ultrastructural modifications in pea root tissues inoculated with endophytic bacteria. *Plant Physiol.* 112:919–929.

Benhamou, N., Kloepper, J.W., and Tuzun, S. 1998. Induction of resistance against Fusarium wilt of tomato by combination of chitosan with an endophytic bacterial strain: ultrastructure and cytochemistry of the host response. *Planta* 204:153–168.

Bent, E., Breuil, C., Chanway, C.P., and Enebak, S. 2002. Surface colonization of lodgepole pine (*Pinus contorta* var. *latifolia* [Dougl. Engelm.]) roots *by Pseudomonas fluorescens* and *Paenibacillus polymyxa* under gnototbiotic conditions. *Plant Soil* 240:187–196.

Bent, E., and Chanway, C.P. 1998. The growth-promoting effects of an endophytic rhizobacterium on lodgepole pine are partially inhibited by the presence of other rhizobacteria. *Can. J. Microbiol.* 44: 980–988.

Bent, E., Tuzun, S., Chanway, C.P., and Enebak, S. 2000. Alterations in plant growth and in root hormone levels of lodgepole pines inoculated with rhizobacteria. *Can. J. Microbiol.* 47:793–800.

Berger, S., Bell, E., Sadka, A., and Mullet, J.E. 1995. *Arabidopsis thaliana AtVsp* is homologous to soybean *VspA* and *VspB*, genes encoding vegetative storage protein acid phosphatases, and is regulated similarly by methyl jasmonate, wounding, sugars, light and phosphate. *Plant Mol. Biol.* 27:933–942.

Bigirimana, J., and Hofte, M., 2002. Induction of systemic resistance to *Colletotrichum lindemuthianum* in bean by a benzothiadiazole derivative and rhizobacteria. *Phytoparasitica* 30:159–168.

Bonkowski, M. 2002. Protozoa and plant growth: trophic links and mutualism. *Eur. J. Protistol.* 37:363–365.

Borowicz, V.A. 1997. A fungal root symbiont modifies plant resistance to an insect herbivore. *Oecologia* 112:534–542.

Bostock, R.M., Karban, R., Thaler, J.S., Weyman, P.D., and Gilchrist, D. 2001. Signal interactions in induced resistance to pathogens and insect herbivores. *Eur. J. Plant Pathol.* 107:103–111.

Buchenauer, H. 1998. Biological control of soil-borne diseases by rhizobacteria. *Z. Pflanzenkrank. Pflanzenschutz* 105:329–348.

Burd, G.I., Dixon, D.G., and Glick, B.R. 2000. Plant growth-promoting bacteria that decrease heavy metal toxicity in plants. *Can. J. Microbiol.* 46:237–245.

Catellan, A.J., Hartel, P.G., and Fuhrman, J.J. 1998. Bacterial composition in the rhizosphere of nodulating and non-nodulating soybean. *Soil Sci. Soc. Am. J.* 62:1549–1555.

Chen, C., Belanger, R.R., Benhamou, N., and Paulitz, T.C. 1998. Induced systemic resistance ISR by *Pseudomonas* spp. Impairs pre- and post-infection development of *Pythium aphanidermatum* on cucumber roots. *Eur. J. Plant Pathol.* 104:877–886.

Chen, C., Belanger, R.R., Benhamou, N., and Paulitz, T.C. 2000. Defense enzymes induced in cucumber roots by treatment with plant growth-promoting rhizobacteria PGPR and *Pythium aphanidermatum. Physiol. Mol. Plant Pathol.* 56:13–23.

Cordier, C., Pozo, M.J., Barea, J.M., Gianinazzi, S., and Gianinazzi-Pearson, V. 1998. Cell defense responses associated with localized and systemic resistance to *Phytophthora parasitica* induced in tomato by an arbuscular mycorrhizal fungus. *Mol. Plant Microbe Interact.* 11:1017–1028.

Costacurta, A., and Vanderleyden, J. 1995. Synthesis of phytohormones by plant-associated bacteria. *Crit. Rev. Microbiol.* 21:1–18.

Cristescu, S.M., de Martinis, D., te Lintel Hekkert, S., Parker, D.H., and Harren, F.M.J. 2002. Ethylene production by *Botrytis cinerea* in vitro and in tomatoes. *Appl. Environ. Microbiol.* 68:5342–5350.

Daane, L.L., Molina, J.A.E., and Sadowsky, M.J. 1997. Plasmid transfer between spatially separated donor and recipient bacteria in earthworm-containing soil microcosms. *Appl. Environ. Microbiol.* 63:679–686.

de Meyer, G., Bigirimana, J., Elad, Y., and Hofte, M. 1998. Induced systemic resistance in *Trichoderma harizanum* T39 biocontrol of *Botrytis cinerea. Eur. J. Plant Pathol.* 104:279–286.

de Meyer, G., Capieau, K., Audenaert, K., Buchala, A., Metraux, J.P., and Hofte, M. 1999. Nanogram amounts of salicylic acid produced by the rhizobacterium *Pseudomonas aeruginosa* 7SNK2 activate the systemic acquired resistance pathway in bean. *Mol. Plant Microbe Interact.* 12:450–458.

de Meyer, G., and Hofte, M. 1997. Salicylic acid produced by the rhizobacterium *Pseudomonas aeruginosa* 7SNK2 induces resistance to leaf infection by *Botrytis cinerea* on bean. *Phytopathology* 87:588–593.

de Weger, L.A., van Arendonk, J.J.C.M., Recourt, K., van der Hofstad, G.A.J.M., Weisbeek, P.J., and Lugtenberg, B. 1988. Siderophore-mediated uptake of Fe^{3+} by the plant growth-stimulating *Pseudomonas putida* strain WCS358 and by other rhizosphere microorganisms. *J. Bacteriol.* 170:4693–4698

del Campillo, E., and Lewis, L.N. 1992. Identification and kinetics of accumulation of proteins induced by ethylene in bean abscission zones. *Plant Physiol.* 98:955–961.

Denton, C.S., Bardgett, R.D., Cook, R., and Hobbs, P.J. 1999. Low amounts of root herbivory positively influence the rhizosphere microbial community in a temperate grassland soil. *Soil Biol. Biochem.* 31:155–165.

Dong, H., and Beer, S.V. 2000. Riboflavin induces disease resistance in plants by activating a novel signal transduction pathway. *Phytopathology* 90:801–811.

Dong, H., Delaney, T.P., Bauer, D.W., and Beer, S.V. 1999. Harpin induces disease resistance in *Arabidopsis* through the systemic acquired resistance pathway mediated by salicylic acid and the NIM1 gene. *Plant J.* 20:207–215.

Dong, Y.-H., Gusti, A.R., Zhang, Q., Xu, J.-L., and Zhang, L.-H. 2002. Identification of quorum-quenching *N*-acyl homoserine lactonases from *Bacillus* species. *Appl. Environ. Microbiol.* 68:1754–1759.

Dow, M., Newman, M.-A., and von Roepenack, E. 2000. The induction and modulation of plant defense responses by bacterial lipopolysaccharides. *Annu. Rev. Phytopathol.* 38:241–261.

Duffy, B.K., and Defago, G. 1999. Environmental factors modulating antibiotic and siderophore biosynthesis by *Pseudomonas fluorescens* biocontrol strains. *Appl. Environ. Microbiol.* 65:2429–2438.

Duijff, B.J., Pouhair, D., Olivain, C., Alabouvette, C., and Lemanceau, P. 1998. Implication of systemic induced resistance in the suppression of Fusarium wilt of tomato by *Pseudomonas fluorescens* WCS417r and by nonpathogenic *Fusaruim oxysporum* Fo47. *Eur. J. Plant Pathol.* 104:903–910.

El Ghaouth, A., Arul, J., Wilson, C., and Benhamou, N. 1994. Ultrastructural and cytochemical aspects of the effect of chitosan on decay of bell pepper fruit. *Physiol. Mol. Plant Pathol.* 44:417–432.

Ellis, H., Jeger, M.J., and van Beusichem, M.L. 2000. Effect of nitrogen supply rate on disease resistance in tomato depends upon the pathogen. *Plant Soil* 218:239–347.

Enebak, S.A., and Carey, W.A. 2000. Evidence for induced systemic protection to fusiform rust in loblolly pine by plant growth-promoting rhizobacteria. *Plant Dis.* 84:306–308.

Felix, G., Duran, J.D., Volko, S., and Boller, T. 1999. Plants have a sensitive perception system for the conserved domain of bacterial flagellin. *Plant J.* 18:265–276.

Fontenla, S., Garcia-Romera, I., and Ocampo, J.A. 1999. Negative influence of non-host plants on the colonization of *Pisum sativum* by the arbuscular mycorrhizal fungus *Glomus mosseae*. *Soil Biol. Biochem.* 31:1591–1597.

Freer, J.H. 1985. Illustrated guide to the anatomy of the bacterial cell envelope. In *Immunology of the Bacterial Cell Envelope*, eds. D.E.S. Stewart-Tull, and M. Davies, pp. 355–383. New York: John Wiley & Sons.

Froissard, D., Gough, C., Czernic, P., Schneider, M., Toppan, A., Roby, D., and Marco, Y. 1994. Structural organization of *str*246C and *str*246N, plant defense-related genes from *Nicotania tabacum*. *Plant Mol. Biol.* 26:515–521.

Fuchs, J.-G., Moënne-Loccoz, Y., and Défago, G. 2000. The laboratory medium used to grow biocontrol *Pseudomonas* sp. Pf153 influences its subsequent ability to protect cucumber from black root rot. *Soil Biol. Biochem.* 32:421–424.

Furukawa, T., Koga, J., Adachi, T., Kishi, K., and Syono, K. 1996. Efficient conversion of L-tryptophan to indole-3-acetic acid and/or tryptophol by some species of *Rhizoctonia*. *Plant Cell Physiol.* 37:899–905.

Galiana, E., Bonnet, P., Conrod, S., Keller, H., Panabieres, F., Ponchet, M., Poupet, A., and Ricci, A. 1997. RNase activity prevents the growth of a fungal pathogen in tobacco leaves and increases upon induction of systemic acquired resistance with elicitin. *Plant Physiol.* 115:1557–1567.

Garbaye, J. 1994. Helper bacteria: a new dimension to the mycorrhizal symbiosis. *New Phytol.* 128:197–210.

Glick, B.R. 1995. The enhancement of plant growth by free-living bacteria. *Can. J. Microbiol.* 41:109–117.

Gómez-Gómez, L., Felix, G., and Boller, T. 1999. A single locus determines sensitivity to bacterial flagellin in *Arabidopsis thaliana*. *Plant J.* 18:277–284.

Goncalves, M.C., de Souza-Dias, J.A.C., Granja, N.P., Furlani, P.R., and Costa, A.S. 2000 Effect of nutrient supply on symptom expression and concentration of potato leafroll virus (PLRV) in *Physalis heterophylla*. *Summa Phytopathol.* 26:9–14.

Govrin, E.M., and Levine, A. 2002. Infection of *Arabidopsis* with a necrotrophic pathogen, *Botrytis cinerea*, elicts various defense responses but does not induce systemic acquired resistance (SAR). *Plant Mol. Biol.* 48:267–276.

Gupta, V.P., Bochow, H., Dolej, S., and Fischer, I. 2000. Plant growth-promoting *Bacillus subtilis* strain as potential inducer of systemic resistance in tomato against *Fusarium* wilt. *Z. Pflanzenkrank. Pflanzenschutz* 107:145–154.

Guzzo, S.D., and Martins, E.M.F. 1996. Local and systemic induction of β-1,3-glucanase and chitinase in coffee leaves protected against *Hemileia vastatrix* by *Bacillus thuringiensis. J. Phytopathol.*144:449–454.

Hahn, M.G. 1996. Microbial elicitors and their receptors in plants. *Annu. Rev. Phytopathol.* 34:387–412.

Hammerschmidt, R., and Smith-Becker, J.A. 1999. The role of salicylic acid in disease resistance. In *Induced Plant Defenses Against Pathogens and Herbivores: Biochemistry, Ecology and Agriculture*, eds. A.A. Agrawal, S. Tuzun, and E. Bent, pp. 37–53. St. Paul, MN: American Phytopathological Society.,

Han, D.Y., Coplin, D.L., Bauer, W.D., and Hoitink, H.A.J. 2000. A rapid bioassay for screening rhizosphere microorganisms for their ability to induce systemic resistance. *Phytopathology* 90:327–332.

Hase, S., Van Pelt, J.A., Van Loon, L.C., and Pieterse, C.M.J. 2003. Colonization of *Arabidopsis* roots by *Pseudomonas fluorescens* primes the plant to produce higher levels of ethylene upon pathogen infection. *Physiol. Mol. Plant Pathol.* 62:219–226.

Hasky-Günther, K., Hoffmann-Hergarten, S., and Sikora, R.A. 1998. Resistance against the potato cyst nematode *Globodera pallida* systemically induced by the rhizobacteria *Agrobacterium radiobacter* (G12) and *Bacillus sphaericus* (B43). *Fund. Appl. Nematol.* 21:511–517.

He, C.Y., Hsiang, T., and Wolyn, D.J. 2002. Induction of systemic disease resistance and pathogen defense responses in *Asparagus officinalis* inoculated with nonpathogenic strains *of Fusaruim oxysporum. Plant Pathol.* 51:225–230.

Hennin, C., Diederichsen, E., and Hofte, M. 2001. Local and systemic resistance to fungal pathogens triggered by an AVR9-mediated hypersensitive response in tomato and oilseed rape carrying the *Cf-9* resistance gene. *Physiol. Mol. Plant Pathol.* 59:287–295.

Heo, W.D., Lee, S.H., Kim, M.C., Kim, J.C., Chung, W.S., Chun, H.J., Lee, K.J., Park, C.Y., Park, H.C., Choi, J.Y., and Cho, M.J. 1999. Involvement of specific calmodulin isoforms in salicylic acid-independent activation of plant disease resistance responses. *Proc. Natl. Acad. Sci. USA* 96:766–771.

Hildebrandt, U., Janetta, K., and Bothe, H. 2002. Towards growth of arbuscular mycorrhizal fungi independent of a plant host. *Appl. Environ. Microbiol.* 68:1919–1924.

Hoitink, H.A.J., and Boehm, M.J. 1999. Biocontrol within the context of soil microbial communities: a substrate-dependent phenomenon. *Annu. Rev. Phytopathol.* 37:427–446.

Hynes, R.K., Hill, J., Reddy, M.S., and Lazarovitz, G. 1994. Phytoalexin production by wounded white bean *Phaseolus vulgaris*) cotyledons and hypocotyls in response to inoculation with rhizobacteria. *Can. J. Microbiol.* 40:548–554.

Ippolito, A., El Ghaouth, A., Wilson, C.L., and Wsiniewski, M. 2000. Control of postharvest decay of apple fruit by *Aureobasidium pullulans* and induction of defense responses. *Postharvest Biol. Technol.* 19:265–272.

Jakab, G., Cottier, V., Toquin, B., Rigoli, G. Zimmerli, L., Metraux, J.P., and Mauch-Mani, B. 2001. B-aminobutyric acid-induced resistance in plants. *Eur. J. Plant Pathol.* 107:29–37.

Jennings, J.C., Apel-Birkhold, P.C., Bailey, B.A., and Anderson, J.D. 2000. Induction of ethylene biosynthesis and necrosis in weed leaves by a *Fusarium oxysporum* protein. *Weed Sci.* 48:7–14.

Jentschke, G., Bonkowski, M., Godbold, D.L., and Scheu, S. 1995. Soil protozoa and forest tree growth:non-nutritional effects and interaction with mycorrhizae. *Biol. Fert. Soils* 20:263–269.

Jeun, Y.C., Park, K., and Kim, C.H. 2001. Different mechanisms of induced systemic resistance and systemic acquired resistance against *Colletotrichum orbiculare* on the leaves of cucumber plants. *Mycobiology* 29:19–26.

Jouanneau, J.P., Lapous, D., and Guern, J. 1991. In plant protoplasts, the spontaneous expression of defense reactions and the responsiveness to exogenous elicitors are under auxin control. *Plant Physiol.* 96: 459–466.

Jung, H.W., and Hwang, B.K. 2000. Isolation, partial sequencing, and expression of pathogenesis-related cDNA genes from pepper leaves infected by *Xanthomonas campestris pv. vesicatoria*. *Mol. Plant. Microbe Interact.* 13:136–142.

Kaeberlein, T., Lewis, K., and Epstein, S.S. 2002. Isolating "uncultivable" microorganisms in pure culture in a simulated natural environment. *Science* 296:1127-1129.

Kandeler, E., Kampichler, C., Joergensen, R.G., and Molter, K. 1999. Effects of mesofauna in a spruce forest on soil microbial communities and N cycling in field mesocosms. *Soil Biol. Biochem.* 31:1783–1792.

Keller, H., Blein, J.P., Bonnet, P., and Ricci, P. 1996. Physiological and molecular characteristics of elicitin-induced systemic acquired resistance in tobacco. *Plant Physiol.* 110:365–376.

Kempster, V.N., Davies, K.A., and Scott, E.S. 2001. Chemical and biological induction of resistance to the clover cyst nematode (*Heterodera trifolii*) in white clover (*Trifolium repens*). *Nematology* 3:35–43.

Knoester, M., Pieterse, C.J.M., Bol, J.F., and van Loon, L.C. 1999. Systemic resistance in *Arabidopsis* induced by rhizobacteria requires ethylene-dependent signaling at the site of application. *Mol. Plant Microbe Interact.* 12:720–727.

Kobayashi, I., Murdoch, L.J., Kunoh, H., and Hardham, A.R. 1995. Cell biology of early events in the plant resistance response to infection by pathogenic fungi. *Can. J. Bot.* 73:S418–S425.

Koike, N., Hyakumachi, M., Kageyama, K., Tsuyumu, S., and Doke, N. 2001. Induction of systemic resistance in cucumber against several diseases by plant growth-promoting fungi: lignification and superoxide generation. *Eur. J. Plant Pathol.* 107:523–533.

Krishnamurthy, K., and Gnanamanickam, S.S. 1998. Induction of systemic resistance and salicylic acid accumulation in *Oryza sativa*, L. in the biological suppression of rice blast cause by treatments with *Pseudomonas* spp. *World J. Microbiol. Biotechnol.* 14:935–937.

Kuske, C.R., Ticknor, L.O., Miller, M.E., Dunbar, J.M., Davis, J.A., Barns, S.M., and Belnap, J. 2002. Comparison of soil bacterial communities in the rhizospheres of three plant species and the interspaces in an arid grassland. *Appl. Environ. Microbiol.* 68:1854–1863.

Larkin, R.P., and Fravel, D.R. 1999. Mechanisms of action and dose-response relationships governing biological control of Fusaruim wilt of tomato by nonpathogenic *Fusarium* spp. *Phytopathology* 89:1152–1161.

Larkin, R.P., Hopkins, D.L., and Martin, F.N. 1996. Suppression of Fusarium wilt of watermelon by nonpathogenic *Fusarium oxysporum* and other microorganisms recovered from disease-suppressive soil. *Phytopathology* 86:812–819.

Lambais, M.R., and Mehdy, M.C. 1995. Differential expression of defense-related genes in arbuscular mycorrhiza. *Can. J. Bot.* 73:S533–S540.

Latour, X., Corberand, T., Laguerre, G., Allard, F., and Lemanceau, P. 1996. The composition of fluorescent pseudomonad populations associated with roots is influenced by plant and soil type. *Appl. Environ. Microbiol.* 62:2449–2456.

Lawrence, J.R., and Snyder, R.A. 1998. Feeding behaviour and grazing impacts of a *Euplotes* sp. on attached bacteria. *Can. J. Microbiol.* 44:623–629.

Lee, Y.H., Lee, W.H., Lee, D.K., and Shim, H.K. 2001. Factors relating to induced systemic resistance in watermelon by plant growth-promoting *Pseudomonas* sp. *Plant Pathol. J.* 17:174–179.

Leeman, M., den Ouden, F.M., van Pelt, J.A., Dirkx, F.P.M, Steijl, H., Bakker, P.A.H.M., and Schippers, B. 1996. Iron availability affects induction of systemic resistance to Fusarium wilt of radish by *Pseudomonas fluorescens*. *Phytopathology* 86:149–155.

Leeman, M., van Pelt, J.A., Den Ouden, F.M. Heinsbroek, M., Bakker, P.A.H.M., and Schippers, B. 1995. Induction of systemic resistance against Fusarium wilt of radish by lipopolysaccharides of *Pseudomonas fluorescens*. *Phytopathology* 85:1021–1027.

Leyser, O. 2002. Molecular genetics of auxin signaling. *Annu. Rev. Plant Biol.* 53:377–398.

Liu, L., Kloepper, J.W., and Tuzun, S. 1995a. Induction of systemic resistance in cucumber against bacterial angular leaf spot by plant growth-promoting rhizobacteria. *Phytopathology* 85:843–847.

Liu, L., Kloepper, J.W., and Tuzun, S. 1995b. Induction of systemic resistance in cucumber against Fusarium wilt by plant growth-promoting rhizobacteria. *Phytopathology* 85:695–698.

Madi, L., and Katan, J. 1998. *Penicillium janczewskii* and its metabolites, applied to leaves, elicit systemic acquired resistance to stem rot caused by *Rhizoctonia solani*. *Physiol. Mol. Plant Pathol.* 53:163–175.

Maurhofer, M., Hase, C., Meuwly, P., Métraux, J.P., and Défago, G. 1994. Induction of systemic resistance of tobacco to tobacco necrosis virus by the root-colonizing *Pseudomonas fluorescens* strain CHA0: influence of the *gacA* gene and of pyoverdine production. *Phytopathology* 84:139–146.

Matichenkov, V.V., Calvert, D.V., and Snyder, G.H. 2000. Prospective of silicon fertilization for citrus in Florida. *Soil Crop Sci. Soc. Florida Proc.* 59:137–141.

Mauchline, T.H., Kerry, B.R., and Hirsch, P.R. 2002. Quantification in soil and the rhizosphere of the nematophagous fungus *Verticillium chlamydosporium* by competitive PCR and comparison with selective plating. *Appl. Environ. Microbiol.* 68:1846–1853.

Mawdsley, J.L., and Burns, R.G. 1994. Root colonization by a *Flavobacterium* species and the influence of percolating water. *Soil Biol. Biochem.* 26:861–870.

Mayda, E., Marqués, C., Conejero, V., and Vera, P. 2000a. Expression of a pathogen-induced gene can be mimicked by auxin insensitivity. *Mol. Plant Microbe Interact*, 13:23–31.

Mayda, E., Mauch-Mani, B., and Vera, P. 2000b. Arabidopsis *dth9* mutation identifies a gene involved in regulating disease susceptibility without affecting salicylic acid-dependent responses. *Plant Cell* 12:2119–2128.

Meera, M.S., Shivanna, M.B., Kageyama, K., and Hyakumachi, M. 1995. Persistence of induced systemic resistance in cucumber in relation to root colonization by plant growth promoting fungal isolates. *Crop Prot.* 14:123–130.

Moller, S., Sternberg, C., Andersen, J.B., Christensen, B.B., Ramos, J.L., Givskov, M., and Molin, S. 1998. In situ gene expression in mixed-culture biofilms: evidence of metabolic interactions between community members. *Appl. Environ. Microbiol.* 64:721–732.

Momol., M.T., Norelli, J.L., Aldwinckle, H.S., and Saygili, H. 1999. Evaluation of biological control agents, systemic acquired resistance inducers and bactericides for the control of fire blight on apple blossom. *Acta Hort.* 489:553–557.

M'Piga, P., Bélanger, R.R., Paulitz, T.C., and Benhamou, N. 1997. Increased resistance to *Fusarium oxysporum* f. sp. *radicis-lycopersici* in tomato plants treated with the endophytic bacterium *Pseudomonas fluorescens* stran 63-28. *Physiol. Mol. Plant Pathol.*50:301–320.

Murphy, J.F., Zhender, G.W., Schuster, D.J., Sikora, E.J., Polston, J.E., and Kloepper, J.W. 2000. Plant growth-promoting rhizobacterial mediated protection in tomato against tomato mottle virus. *Plant Dis.* 84:779–784.

Nandakumar, R., Babu, S., Viswanathan, R., Raguchanger, T., and Samiyappan, R. 2001. Induction of systemic resistance in rice against sheath blight disease by *Pseudomonas fluorescens*. *Soil Biol. Biochem.* 33:603–612.

Norman-Setterblad, C., Vidal, S., and Palva, E.T. 2000. Interacting signal pathways control defense gene expression in *Arabidopsis* in response to cell wall-degrading enzymes from *Erwinia carotovora*. *Mol. Plant Microbe Interact.* 13:430–438.

Okada, M., Matsumura, M., Ito, Y., and Shibuya, N. 2002. High-affinity binding proteins for *N*-acetylchitooligosaccharided elicitor in the plasma membranes from wheat, barley and carrot cells: conserved presence and correlation with the responsiveness to the elicitor. *Plant Cell Physiol.* 43:505–512.

Ongena, M., Daayf, F., Jacques, P., Thonart, P., Benhamou, N., Paulitz, T.C., and Belanger, R.R. 2000. Systemic induction of phytoalexins in cucumber in response to treatments with fluorescent pseudomonads. *Plant Pathol.* 49:523–530.

Ousley, M.A., Lynch, J.M., and Whipps, J.M. 1993. Effect of *Trichoderma* on plant growth: a balance between inhibition and growth promotion. *Microb. Ecol.* 26:277–285.

Ozeki, Y., Komamine, A., and Tanaka, Y. 1990. Induction and repression of phenylalanine ammonia-lyase and chalcone synthase enzyme proteins and mRNAs in carrot cell suspension cultures regulated by 2,4-D. *Physiol. Plant.* 78:400–408.

Patten, C.L., and Glick, B.R. 1996. Bacterial biosynthesis of indole-3-acetic acid. *Can. J. Microbiol.* 42:207–220.

Patten, C.L., and Glick, B.R. 2002. Role of *Pseudomonas putida* indoleacetic acid in development of the host plant root system *Appl. Environ. Microbiol.* 68:3795–3801.

Peck, S.C., Nühse, T.S., Hess, D., Iglesias, A., Meins, F., and Boller, T. 2001. Directed proteomics identifies a plant-specific protein rapidly phosphorylated in response to bacterial and fungal elicitors. *Plant Cell* 13:1467–1475.

Peterson, R.L., and Farquhar, M.L. 1994. Mycorrhizas-integrated development between roots and fungi. *Mycologia* 86:311–326.

Pierson, L.S., Wood, D.W., and Pierson, E.A. 1998. Homoserine lactone-mediated gene regulation in plant-associated bacteria. *Annu. Rev. Phytopathol.* 36:207–225.

Pieterse, C.M.J., van Wees, S.C.M., Hoffland, E., van Pelt, J.A., and van Loon, L.C. 1996. Systemic resistance in *Arabidopsis* induced by biocontrol bacteria is independent of salicylic acid accumulation and pathogenesis-related gene expression. *Plant Cell* 8:1225–1237.

Pieterse, C.M.J., van Wees, S.C.M., van Pelt, J.A., Knoester, M., Laan, R., Gerrits, H., Weisbeek, P.J., and van Loon, L.C. 1998. A novel signaling pathway controlling induced systemic resistance in *Arabidopsis*. *Plant Cell* 10:1571–1580.

Press, C.M., Loper, J.E., and Kloepper, J.W. 2001. Role of iron in rhizobacteria-mediated induced systemic resistance of cucumber. *Phytopathology* 91:593–598.

Press, C.M., Wilson, M., Tuzun, S., and Kloepper, J.W. 1997. Salicylic acid produced by *Serratia marcesens* 90-166 is not the primary determinant of induced systematic resistance in cucumber or tobacco. *Mol. Plant Microbe Interact.* 10:761–768.

Preston, G.M., Bertrand, N., and Ralney, P.B. 2001. Type III secretion in plant growth-promoting *Pseudomonas fluorescens* SBW25. *Mol. Microbiol.* 41(5):999–1014.

Ramani, N., and Boyake, K. 2001. Salicylate inhibits the translation and transcription of *ompF* in *Escherichia coli*. *Can. J. Microbiol.* 47:1053–1057.

Raupach, G.S., and Kloepper, J.W. 1998. Mixtures of plant growth-promoting rhizobacteria enhance control of multiple cucumber pathogens. *Phytopathology* 88:1158–1164.

Reitz, M., Rudolph, K., Schröder, I., Hoffmann-Hergarten, S., Hallmann, J., and Sikora, R.A. 2000. Lipopolysaccharides of *Rhizobium elti* strain G12 act in potato roots as an inducing agent of systemic resistance to infection by the cyst nematode *Globodera pallida*. *Appl. Environ. Microbiol.* 66:3515–3518.

Reitz, M., Hoffmann-Hergarten, S., Hallmann, J., and Sikora, R.A. 2001. Induction of systemic resistance in potato by rhizobacterium *Rhizobium elti* strain G12 is not associated with accumulation of pathogenesis-related proteins and enhanced lignin biosynthesis. *Z. Pflanzenkrank. Pflanzenschutz* 108:11–20.

Requena, N., Perez-Solis, E., Azcon-Aguilar, C., Jeffries, P., and Barea, J.-M. 2001. Management of indigenous plant-microbe symbioses aids restoration of desertified ecosystems. *Appl. Environ. Microbiol.* 67:495–498.

Reuveni, M., and Reuveni, R. 2000. Prior inoculation with non-pathogenic fungi induces systemic resistance to powdery mildew on cucumber plants. *Eur. J. Plant Pathol.* 106:633–638.

Ruiz-Lozano, J.M., Roussel, H., Gianinazzi. S., and Gianinazzi-Pearson, V. 1999. Defense genes are differentially induced by a mycorrhizal fungus and *Rhizobium* sp. in wild-type and symbiosis-defective pea genotypes. *Mol. Plant Microbe Interact.* 12:976–984.

Ryu, C.-M., Hu, C.-H., Reddy, M.S., and Kloepper, J.W. 2003. Different signaling pathways of induced resistance by rhizobacteria in *Arabidopsis thaliana* against two pathovars of *Pseudomonas syringae*. *New Phytol.* 160:413–420.

Sailaja, P.R., Podile, A.R., and Reddanna, R. 1997. Biocontrol strain of *Bacillus subtilis* AF 1 rapidly induces lipoxygenase in groundnut *Arachis hypogaea* L.) compared to crown rot pathogen *Aspergillus niger*. *Eur. J. Plant Pathol.* 104:125–132.

Salzer, P., Hebe, G., Reith, A., Zitterel-Haid, B., Stransky, H., Gaschler, K., and Hager, A. 1996. Rapid reactions of spruce cells to elicitors released from the ectomycorrhizal fungus *Hebeloma crustiniliforme*, and inactivation of these elicitors by extracellular spruce enzymes. *Planta* 198:118–126.

Sevilla, M., Burris, R.H., Gunapala, N., and Kennedy, C. 2001. Comparison of benefit to sugarcane plant growth and $^{15}N_2$ incorporation following inoculation of sterile plants with *Acetobacter diazotrophicus* wild-type and Nif⁻mutant strains. *Mol. Plant Microbe Interact.* 14:358–366.

Shiu, S.H., and Bleecker, A.B. 2001. Plant receptor-like kinase gene family: diversity, function and signaling. *Science's STKE*, http://www.stke.org/cgi/content/full/OC_sigtrans;2001/113/re22

Shivanna, M.B., Meera, M.S., and Hyakumachi, M. 1996. Role of root colonization ability of plant growth promoting fungi in the suppression of take-all and common root rot of wheat. *Crop Prot.* 15:497–504.

Sirrenberg, A., Salzer, P., and Hager, A. 1995. Induction of mycorrhiza-like structures and defence reactions in dual cultures of spruce callus and ectomycorrhizal fungi. *New Phytol.* 130:149–156.

Srinivasan, M., Petersen, D.J., and Holl, F.B. 1996. Influence of indoleacetic-acid-producing *Bacillus* isolates on the nodulation of *Phaseolus vulgaris* by *Rhizobium elti* under gnotobiotic conditions. *Can. J. Microbiol.* 42:1006–1014.

Staswick, P.E. and Lehman, C.C. 1999. Jasmonic acid-signaled responses in plants. In *Induced Plant Defenses Against Pathogens and Herbivores: Biochemistry, Ecology and Agriculture*, eds. A.A. Agrawal, S. Tuzun, and E. Bent, pp. 117–136. St. Paul, MN: American Phytopathological Society.

Steidle, A., Sigl, K., Schuhegger, R., Ihring, A., Schmid, M., Gantner, A. Stoffels, M., Reidel, K., Givskov, M., Hartmann, A., Langebartels, C., and Eberl, L. 2001. Visualization of *N*-acylhomoserine lactone-meidated cell-cell communication between bacteria colonizing the tomato rhizosphere. *Appl. Environ. Microbiol.* 67:5761–5770.

Steijl, H., Niemann, G.J., and Boon, J.J. 1999. Changes in chemical composition related to fungal infection and induced resistance in carnation and radish investigated by pyrolysis and mass spectrometry. *Physiol. Mol. Plant Pathol.* 55:297–311.

Strobel, N.E., Ji, C., Gopalan, S., Kuć, J.A., and He, S.Y. 1996 Induction of systemic acquired resistance in cucumber *by Pseudomonas syringae* pv. *syringae* 61 HrpZPss protein. *Plant J.* 9:431–439.

Strobel, N.E and Sinclair, W.A. 1991. Role of flavanolic wall infusions in the resistance induced by *Laccaria bicolor* to *Fusarium oxysporum* in primary roots of Douglas-fir. *Phytopathology* 81:420–425.

Sturz, A.V., Christie, B.R., and Nowak, J. 2000. Bacterial endophytes: potential role in developing sustainable systems of crop production. *Crit. Rev. Plant Sci.* 19:1–30.

Sylvia, D.M., and Sinclair, W.A. 1983. Phenolic compounds and resistance to fungal pathogens induced in primary roots of Douglas-fir seedlings by the ectomycorrhizal fungus *Laccaria bicolor*. *Phytopathology* 73:390–397.

Teplitski, M., Robinson, J.B., and Bauer, W.D. 2000. Plants secrete substances that mimic bacterial *N*-acyl homoserine lactone signal activities and affect population density-dependent behaviors in associated bacteria. *Mol. Plant Microbe Interact.* 13:637–648.

Timmusk, S., and Wagner, E.G.H. 1999. The plant-growth-promoting rhizobacterium *Paenibacillus polymyxa* induces changes in *Arabidopsis thaliana* gene expression: a possible connection between biotic and abiotic stress responses. *Mol. Plant Microbe Interact.*12:951–959.

Timonen, S., Jorgensen, K.S., Haahtela, K., and Sen, R. 1998. Bacterial community structure at defined locations of *Pinus sylvestris- Suillus bovinus* and *Pinus-sylvestris-Paxillus involutus* mycorrhizospheres in dry pine forest humus and nursery peat. *Can. J. Microbiol.* 44:499–513.

Tokala, R.K., Strap, J.L., Jung, C.M., Crawford, D.L., Salove, M.H., Deobald, L.A., Bailey, J.F., and Morra, M.J. 2002. Novel plant-microbe rhizosphere interaction involving *Streptomyces lydicus* WYEC108 and the pea plant *Pisum sativum. Appl. Environ. Microbiol.* 68:2161–2171.

Ton, J., Davison, S., van Wees, S.C.M., van Loon, L.C., and Pieterse, C.J.M. 2001. The Arabidopsis *ISR1* locus controlling rhizobacteria-mediated induced systemic resistance is involved in ethylene signaling. *Plant Physiol.* 125:652–661.

Ton, J., Pieterse, C.J.M., and van Loon, L.C. 1999. Identification of a locus in *Arabidopsis* controlling both the expression of rhizobacteria-mediated induced systemic resistance (ISR and basal resistance against *Pseudomonas syringae* pv. *tomato. Mol. Plant Microbe Interact.* 12:911–918.

Ton, J., van Pelt, J.A., van Loon, L.C., and Pieterse, C.M.J. 2002. Differential effectiveness of salicylate-dependent and jasmonate/ethylene-dependent induced resistance in *Arabidopsis. Mol. Plant Microbe Interact.* 15:27–34.

Tuzun, S., and Bent, E. 1999. The role of hydrolytic enzymes in multigenic and microbially-induced resistance in plants. In *Induced Plant Defenses Against Pathogens and Herbivores: Biochemistry, Ecology and Agriculture*, eds. A.A. Agrawal, S. Tuzun, and E. Bent, pp. 95–115. St. Paul, MN: American Phytopathological Society.

Valinsky, L., Vedova, G.D., Scupham, A.J., Alvey, S. Figueroa, A. Yin, B., Hartin, R.J. Chrobak, M., Crowley, D.E., Jiang, T., and Borneman, J. 2002a Analysis of bacterial community composition by oligonucleotide fingerprinting of rRNA genes. *Appl. Environ. Microbiol.* 68:3243–3250.

Valinsky, L., Vedova, G.D., Jiang, T., and Borneman, J. 2002b. Oligonucleotide fingerprinting of rRNA genes for analysis of fungal community composition. *Appl. Environ. Microbiol.* 68:5999–6004.

van Loon, L.C., Bakker, P.A.H.M., and C.M.J. Pieterse. 1998. Systemic resistance induced by rhizosphere bacteria. *Annu. Rev. Phytopathol.* 36:453–483.

van Peer, R., Niemann, G.J. and Schippers, B. 1991. Induced resistance and phytoalexin accumulation in biological control of Fusarium wilt of carnation by *Pseudomonas* sp. strain WCS417r. *Phytopathology* 81:728–734.

van Wees, S.C.M., Luijendijk, M., Smoorenburg, I., van Loon, L.C., and Pieterse, C.M.J. 1999. Rhizobacteria-mediated induced systemic resistance ISR in *Arabidopsis* is not associated with a direct effect on expression of known defense-related genes but stimulates the expression of the jasmonate-inducible gene *Atvsp* upon challenge. *Plant Mol. Biol.* 41:537–549.

van Wees, S.C.M., Pieterse, C.M.J., Trisjssenaar, A., Van't Westende, Y.A.M., Hartog, F., and van Loon, L.C. 1997. Differential induction of systemic resistance in Arabidopsis by biocontrol bacteria. *Mol. Plant Microb. Interact.* 10:716–724.

Varagas-Ayala, R., Rodriguez-Kabana, R., Morgan-Jones, G. McInroy, J.A., and Kloepper, J.W. 2000. Shifts in soil microflora induced by velvetbean (*Mucina deeringiana*) in cropping systems to control root-knot nematodes. *Biol. Control* 17:11–22.

Viswanathan, R., and Samiyappan, R. 2002a. Induced systemic resistance by fluorescent pseudomonads against red rot disease of sugarcane caused by *Colletotrichum falcatum. Crop Prot.* 21:1–10.

Viswanathan, R., and Samiyappan, R. 2002b. Role of oxidative enzymes in the plant growth promoting rhizobacteria (PGPR) mediated induced systemic resistance in sugarcane against *Colletotrichum falcatum. Z. Pflanzenkrank. Pflanzenschutz* 109:88–100.

Vleeshouwers, V.G.A.A., van Dooijeweert, W., Govers, F., Kamoun, S., and Colon, L.T. 2000. Does basal PR gene expression in *Solanum* species contribute to non-specific resistance to *Phytophthora infestans? Physiol. Mol. Plant Pathol.* 57:35–42.

Weland, G., Neumann, R., and Backhaus, H. 2001. Variation in microbial communities in soil, rhizosphere and rhizoplane in response to crop species, soil type, and crop development. *Appl. Environ. Microbiol.* 67:5849–5854.

Whipps, J.M. 2001. Microbial interactions and biocontrol in the rhizosphere. *J. Exp. Bot.* 52:487–511.

Whitelaw, M.A., Harden, T.J., and Heylar, K.R. 1999. Phosphate solubilization in solution culture by the soil fungus *Penicillium radicum. Soil Biol. Biochem.* 31:655–665.

Xu, P., Narasimhan, M.L., Samson, T., Coca, M.A., Huh, G.-H., Zhou, J., Martin, G.B. Hasegawa, P.M., and Bressan, R.A. 1998. A nitrilase-like protein interacts with GCC box DNA-binding proteins involved in ethylene and defense responses. *Plant Physiol.* 118:867–874.

Xue, L., Charest, P.M., and Jabaji-Hare, S.H. 1998. Systemic induction of peroxidases, 1,3-β-glucanases, chitinases and resistance in bean plants by binucleate *Rhizoctonia* species. *Phytopathology* 88:359–365.

Yamakazi, H., Sunao, K., Tsuguo, H., and Takeshi, K. 2000. Effect of calcium concentration in nutrient solution on development of bacterial wilt and population of its pathogen *Ralstonia solanacearum* in grafted tomato seedlings. *Soil Sci. Plant Nutrition* 46:535–539.

Yamane, A., Nishimura, H., and Mizutani, J. 1992. Allelopathy of yellow fieldcress *Rorippa sylvestris*): identification and characterization of phytotoxic constituents. *J. Chem. Ecol.* 18:683–691.

Yang, C.H., and Crowley, D.E. 2000. Rhizosphere microbial community structure in relation to root location and plant iron nutritional status. *Appl. Environ. Microbiol.* 66:345–351.

Yi, H.C., Joo, S., Nam, K.H., Lee, J.S., Kang, B.G., and Kim, W.T. 1999. Auxin and brassinosteroid differentially regulate the expression of three members of the 1-aminocyclopropane-1-carboxylase synthase gene family in mung bean (*Vigna radiata* L.). *Plant Mol. Biol.* 41:443–454.

Yin, B., Scupham, A.J., Menge, J.A., and Borneman, J. 2004. Identifying microorganisms which fill a niche similar to that of the pathogen: a new investigative model for biological control research. *Plant Soil.* 259:19–27.

Zdor, R.E., and Anderson, A.J. 1992. Influence of root colonizing bacteria on the defense responses of bean. *Plant Soil* 140:99–107.

Zhang, S., Reddy, M.S., and Kloepper, J.W. 2002. Development of assays for assessing induced systemic resistance by plant growth-promoting rhizobacteria against blue mold of tobacco. *Biol. Control* 23:79–86.

Zhang, S., Reddy, M.S., Kokalis-Burelle, N., Wells, L.W., Nightengale, S.P., and Kloepper, J.W. 2001. Lack of induced resistance in peanut to late blight spot disease by plant growth-promoting rhizobacteria and chemical eliciters. *Plant Dis.* 85:879–884.

Zhang, W., Dick, W.A., and Hoitink, H.A.J. 1996. Compost-induced systemic acquired resistance in cucumber to Pythium root rot and anthracnose. *Phytopathology* 86:1066–1070.

Zhang, W., Han, D.Y., Dick, W.A., Davis, K.R., and Hoitink, H.A.J. 1998. Compost and compost water extract-induced systemic acquired resistance in cucumber and *Arabidopsis. Phytopathology* 88:450–455.

Zhender, G., Kloepper, J., Tuzun, S., Yao, C., Wei. G. Chambliss, O., and Shelby, R. 1997a. Insect feeding on cucumber mediated by rhizobacteria-induced plant resistance. *Entomologia Experimentalis et Applicata* 83:81–85.

Zhender, G., Kloepper, J., Yao, C., and Wei, G. 1997b. Induction of systemic resistance in cucumber against cucumber beetles (Coleoptera: Chrysomelidae) by plant growth-promoting rhizobacteria. *J. Econ. Entomol.* 90:391–396.

Zhender, G.W., Yao, C., Murphy, J.F., Sikora, E.R., and Kloepper, J.W. 2000. Induction of resistance in tomato against cucumber mosaic cucumovirus by plant growth-promoting rhizobacteria. *BioControl* 45:127–137.

Zhender, G.W., Yao, C., Wei, G., and Kloepper, J.W. 2000. Influence of methyl bromide fumigation on microbe-induced resistance in cucumber. *Biocontrol Sci. Technol.* 10:687–693.

Zimmerli, L., Jakab, G., Métraux, J.P., and Mauch-Mani, B. 2000. Potentiation of pathogen-specific defense mechanisms in Arabidopsis by β-aminobutyric acid. *Proc. Natl. Acad. Sci. USA* 97:12920–12925.

Zoubenko, O., Hudak, K., and Tumer, N.E. 2000. A non-toxic pokeweed antiviral protein mutant inhibits pathogen infection via a novel salicylic acid-independent pathway. *Plant Mol. Biol.* 44:219–229.

11

Chemical Signals in Plants: Jasmonates and the Role of Insect-Derived Elicitors in Responses to Herbivores

Kenneth L. Korth and Gary A. Thompson

11.1 Introduction

In any environment where plants grow, one can be quickly awestruck at the complex relationships that occur between plants and insects. Striking variation in the types of responses, on both sides of the interaction, is easily observed at the whole-organism level and these are multiplied many times over when events are considered at the cellular, biochemical, and molecular levels. A multitude of potential responses can occur in a plant following injury by a phytophagous insect. Complex sets of signals modulate a suite of responses, either within a given plant in response to different herbivores or in different plant species in response to similar herbivores (Figure 11.1). Understanding the central themes of plant defense against insects, as well as those responses unique to a particular insect or plant, is key to obtaining an overall picture of plant defenses against herbivores.

Plants are continually vulnerable not only to herbivory but also to environmental stresses and infestation by pathogens. However, preformed responses against potential stresses are not always necessary or effective. Induction of broad-spectrum defenses against a variety of enemies, ranging from viruses to large animals, although potentially difficult, could offer the most efficient means of defense. However, induced responses carry with them metabolic costs and are most valuable when directed at the stress that is of immediate concern. To accomplish this, some plants have evolved highly specific surveillance and response mechanisms for defense against the herbivores and other pests to which they are exposed.

On the face of it, biochemical changes in response to herbivory would clearly seem to benefit the plant, although the ecological advantages of induced resistance are sometimes difficult to measure (Heil and Baldwin, 2002). Nonetheless, herbivory-induced insect resistance has a positive effect by reducing subsequent herbivory and improving overall fitness, if the induction occurs early in the life of the plant (Agrawal, 1998). Furthermore, induced resistance can negatively affect

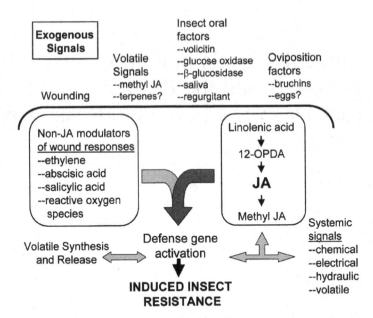

FIGURE 11.1. A schematic representation of the central role of jasmonic acid (JA) and related compounds in the complex and interconnected responses of plants to signals initiated during plant–insect interactions. Exogenous signals can originate from insects, plants, or environmental stresses. Bidirectional arrows indicate that events such as systemic signaling and release of volatile compounds can regulate, and/or be regulated by, activation of defense genes. Overlapping arrows illustrate crosstalk between the signaling cascades mediated by plant-derived factors. Although not discussed in detail in the text, non-JA plant components such as reactive oxygen species (Neill et al., 2001; Orozco-Cárdenas et al., 2001) and salicylic acid (Ozawa et al., 2000) also play an important role in responses to insect herbivores.

herbivore populations (Underwood, 1999). A central role for jasmonic acid and structurally related compounds (hereafter referred to as "jasmonates") as a "master switch" in plant wound responses is well established (Wasternack and Parthier, 1997). Treatment of field-grown plants with jasmonates decreases herbivory, concomitant with increased seed set in wild plants (Baldwin, 1998) or without reducing yield in crop plants (Thaler, 1999). Thus, measurable benefits in fitness and reproduction can offset the metabolic costs to the plant for utilizing induced defenses.

Several excellent reviews have recently focused on the biosynthesis and activity of jasmonates and their roles in defense mechanisms (Creelman and Mullet, 1997; Seo et al., 1997; Weber, 2002). Another recent paper reviewed significant aspects of plant reactions to insect feeding, with emphases on plant fitness and insect-induced responses (Kessler and Baldwin, 2002). This chapter will address several areas of importance in plant responses to wounding, with a focus on the role of jasmonate-mediated signaling. Plant-derived compounds such as ethylene (O'Donnell et al., 1996) and abscisic acid (Hildmann et al., 1992) can serve complementary roles in transduction of wound signals, but will not be covered in this paper. Rather, the discussion will focus on (1) how the jasmonate pathway functions in wound responses, (2) the specificity of plant responses to biotic versus abiotic damage, and (3) the role of insect-derived elicitors in triggering plant defenses.

11.2 Plant Defense Signaling Pathways in Response to Herbivores

11.2.1 The Jasmonate Pathway

Jasmonic acid (JA) serves as an important hormonal regulator of several steps of plant growth and development. Jasmonates also accumulate rapidly in wounded plant tissue (Creelman et al., 1992) and treatment of plants with exogenous jasmonates leads to enhanced defense gene expression and protein accumulation (Farmer and Ryan, 1990; Creelman et al., 1992; Farmer and Ryan, 1992; Farmer et al., 1992). In spite of the long-standing importance attributed to jasmonates, critical questions remain regarding their formation and mode of action.

Following a wound event, lipid membrane components are probably released via the action of a lipase. It was recently shown that induction of a JA-mediated wound response is dependent on the activation of a phospholipase D (Wang et al., 2000). Lipase-activated formation of free linolenic acid can lead ultimately, via the octadecanoid pathway, to the production of 12-oxo-phytodienoic acid (OPDA) and finally to JA and methyl jasmonate (meJA) (reviewed by Howe and Schilmiller, 2002; Weber, 2002). Polyunsaturated fatty acids are converted to hydroperoxy fatty acids by lipoxygenases, to form a class of compounds known as oxylipins, of which jasmonates are the best-studied subclass.

Overexpression of enzymes involved in jasmonate synthesis in transgenic plants illustrates our incomplete understanding of jasmonate-mediated signaling. Activity of allene oxide synthase (AOS) is a key regulatory step in the production of jasmonates (Laudert and Weiler, 1998). Overproduction of this enzyme in transgenic plants leads to elevated levels of jasmonates similar to those in wounded plants, but does not necessarily cause increased expression of jasmonate-inducible genes (Harms et al., 1995). This suggests that in addition to jasmonates, independent signals must be required for some wound-inducible responses. In contrast, transgenic plants overexpressing JA carboxyl methyltransferase, which catalyzes the formation of meJA from JA, have enhanced levels of meJA and do exhibit constitutive expression of some jasmonate-inducible genes (Seo et al., 2001). Several lines of evidence suggest that there is considerable specificity in the lipases, lipoxygenases, and probably other enzymes in the oxylipin pathway producing an "oxylipin signature" and jasmonates that accumulate in response to wounding (Howe and Schilmiller, 2002; Strassner et al., 2002). The levels and ratios of the individual types of oxylipins might constitute an important aspect of how a plant recognizes and responds to a particular stress event. As supporting evidence for this idea, exogenous jasmonates could induce events separately from those induced by endogenously formed jasmonates (Kramell et al., 2000).

11.2.2 The Utility of Jasmonate-Deficient Mutants

Although biochemical approaches have been invaluable in elucidating the biosynthetic pathway leading to jasmonates, mutant lines with impaired jasmonate-mediated signaling provide tools for understanding how plants respond to wounding. An *Arabidopsis* mutant lacking some fatty acid desaturase (*fad*) activities clearly demonstrated a functional role for jasmonates in insect defense. A *fad* triple-mutant line contains negligible amounts of the jasmonate precursor linolenic acid and does not accumulate jasmonates. McConn et al. (1997) found that the *fad* triple mutant was particularly susceptible to the fungal gnat *Bradysia impatiens* and that insect resistance could be rescued with exogenous meJA. Even in the absence of jasmonates, wound-inducible expression of at least one gene encoding a glutathione-S-transferase was observed in *fad* mutants, indicating the presence of a jasmonate-independent wound signal pathway. The same triple-mutant line was later used to show that jasmonates are essential for defense against pathogens in *Arabidopsis* (Vijayan et al., 1998).

The *opr3* mutant of *Arabidopsis* does not convert 12-OPDA to JA but exhibits normal levels of insect and pathogen resistance (Stinzi et al., 2001). Interestingly, *opr3* mutants demonstrate wound-inducible expression of some genes that were thought to be JA-dependent. Precursors of JA apparently function in some of the roles previously thought to be performed by JA and/or meJA, further highlighting the importance of an oxylipin signature in wound responses. Other *Arabidopsis* mutants have been identified with impaired responses to jasmonates, rather than defective jasmonate synthesis (Staswick et al., 1992; Berger et al., 1996). Wound-induced gene expression is greatly reduced in the *coi1* mutant, especially in systemic tissues. However, increased transcript accumulation of some genes occurs in wounded leaves, providing a further indication of jasmonate-independent wound signaling in *Arabidopsis* (Titarenko et al., 1997). The requirement of an ethylene signaling pathway was demonstrated in tomato (O'Donnell et al., 1996). Through combination of *Arabidopsis* mutant genes *in planta*, the respective roles and interdependence of the jasmonate- and ethylene-pathways have been established (Penninckx et al., 1998).

Genetic studies in other plant species are also providing insights into the role of jasmonates in defense. The *def1* mutant line of tomato has low levels of 12-OPDA and JA, and is significantly more susceptible to lepidopteran insect and spider mite damage than wild-type plants (Howe et al., 1996; Li et al., 2002b). Exogenous JA can rescue insect resistance and increase levels of wound-induced gene expression in the mutant. Furthermore, *def1* tomato plants emit less herbivore-induced volatile attractants of predatory insects, indicating a crucial role for jasmonates in both direct and indirect herbivore defenses (Thaler et al., 2002).

11.2.3 Nonchemical Signaling Events

Systemic signaling by jasmonates is thought to occur through the phloem. In tomato plants, biosynthesis and recognition of jasmonates is required for some systemic responses to wounding (Li et al., 2002a). However, the rapid speed of some systemic responses suggests that a chemical translocated in the phloem is not the most likely candidate as a long-distance signal. Electrical signals mediating wound responses, such as the systemic induction of proteinase inhibitors, have been shown to be propagated through the phloem in the absence of translocation (Wildon et al., 1992; Rhodes et al., 1996). The continuous endomembrane system within sieve tubes certainly lends itself to such signaling. The type of wound administered almost certainly plays a role in the type of electrical signal that can be measured. For example, signals in a flame-wounded plant are transmitted differently than those in an electrically stimulated plant (Stankovic and Davies, 1997). In addition, the direction of wound-induced electrical currents in maize roots differs depending on the type of damage. Artificial damage causes a large inward current, whereas natural damage caused by the emergence of lateral roots results in an outward current (Meyer and Weisenseel, 1997).

Hydraulic signals also have been suggested as playing a major role in systemic transmission of wound responses in plants. Alarcon and Malone (1994) used changes in leaf thickness to infer that substantial hydraulic pressure changes occur in plants after herbivory. Hydraulic events were most evident in detached leaves with only minimal changes occurring in intact plants damaged by lepidopteran larvae. Further studies (Malone and Alarcon, 1995) using heat-girdled tomato plants concluded that only hydraulic signaling could explain systemic induction of proteinase inhibitor levels following wounding with heat. Other studies, again using proteinase inhibitor induction in tomato as a marker, suggested that both electrical and hydraulic signals could be involved in systemic responses (Herde et al., 1995). Indeed, changes in electrical potential might result directly from changes in internal pressure of the xylem (Stankovic and Davies, 1998).

Feeding damage by lepidopteran pests on *Nicotiana* plants results in increased JA levels that are much higher than those caused by artificial wounding. This difference is presumed to be due to the presence of oral elicitors in the herbivore (see Section 11.3.2). Schittko et al. (2000) showed that the signal leading to JA amplification spread at rates faster than could be explained by introduction of oral factors during feeding. They concluded that amplification of the increase in JA must spread within the leaf by a fast-traveling mechanism, such as a gaseous, electrical, or hydraulic signal. Therefore, even the levels of jasmonates might be regulated by alternative non-JA signals.

11.3 Plant Responses to Insect-Derived Elicitors

11.3.1 Specificity of Responses to Biotic Damage

In many studies measuring biochemical responses to wounding, plants have been damaged either by cutting or crushing leaf tissue. It was quickly recognized that in contrast to mechanical damage alone, herbivory could enhance accumulation of specific insect-induced proteins (Walker-Simmons et al., 1984). The differential response of plants to insect damage and artificial mechanical damage has been demonstrated by a variety of plant biochemical characters such as phenolic levels (Hartley and Lawton, 1991), volatile compound release (Turlings et al., 1990), amino acid profiles (Tomlin and Sears, 1992), enzyme activities (Stout et al., 1994), and transcript accumulation (Korth and Dixon, 1997). At the whole plant level, regrowth of plant tissue was significantly increased following herbivory as compared to mechanical cutting (Capinera and Roltsch, 1980; Dyer and Bokhari, 1976). The most likely explanations for these insect-specific responses are that physical or chemical aspects unique to herbivory, specifically the type of injury and/or elicitors introduced into the plant, serve as signals differentiating biotic from abiotic damage.

Plant Volatiles Released Following Herbivory

In virtually all plant species examined, feeding by arthropods ranging from mites (Dicke and Sabelis 1988) and caterpillars (Turlings et al., 1990) to beetles (Bolter et al., 1997), can stimulate release of volatile chemicals from leaves. The release of volatiles is systemic, occurring from both damaged and undamaged leaves of wounded plants (Turlings and Tumlinson, 1992). It is also damage-specific. Insect feeding usually results in release of a different profile and quantity of compounds than does artificial wounding (Pichersky and Gershenson, 2002). Herbivory can be mimicked by adding crude insect regurgitant to a wound site eliciting volatile release (Turlings et al., 1990). It was this response that was used to isolate volicitin, an elicitor present in lepidopteran regurgitant (discussed below) (Alborn et al., 1997).

The release of herbivore-induced plant volatiles is of special biological interest because volatiles can serve as attractants to several classes of natural enemies of herbivores (Takabayashi and Dicke, 1996) such as predatory mites (Dicke and Sabelis, 1988) and parasitoid wasps (Turlings et al., 1990). In *Nicotiana attenuata* natural levels of oviposition by herbivores are reduced, while levels of predation by egg-feeding bugs are increased (Kessler and Baldwin, 2001). In an agricultural setting, parasitoid wasp populations were larger near both herbivore-damaged and regurgitant-treated maize plants than near undamaged plants (Ockroy et al., 2001). Induced terpenes might also deter herbivory by mimicking alarm pheromones that repel aphids (Gibson and Pickett, 1983).

Volatile-mediated interactions of plants with other organisms are complex. In addition to playing a role in plant:insect interactions, plant volatiles appear to affect plant pathogens. Elicitors from plant pathogenic fungi have been shown to cause release of induced plant volatiles via a JA-dependent pathway (Piel et al., 1997). Plant volatile profiles from insect-damaged and pathogen-infected leaves differ. Pathogen-induced volatiles reduce growth of in vitro-grown fungi, suggesting that volatiles directly defend against fungal infestation on the leaf surface (Cardoza et al., 2002). Furthermore, volatiles might serve as important plant:plant signals (reviewed in Baldwin et al., 2002; Farmer, 2001).

Several chemical classes of volatiles, including terpenes, sulfides, nitriles, indole and others are released following herbivory (Paré and Tumlinson, 1999; van

Poecke et al., 2001). Green leaf volatiles, namely C_6 alcohols and aldehydes, are released following damage and are derived from lipoxygenase-mediated breakdown of membrane components. Terpenes are produced by either of the two independent biosynthetic pathways, whereas indole and methyl salicylate are products of the shikimate pathway. The release of many of the terpenes and indole is delayed by nearly a full day following initial herbivory, indicating that an induction of biosynthetic pathways might be necessary. Chemical labeling experiments show that increases in levels of terpenes and indole occur via de novo biosynthesis resulting from long-term feeding by caterpillars (Paré and Tumlinson, 1997a). Signal transduction from herbivore damage leading to biosynthesis of volatiles is not well understood. Enzymatic activities and levels of transcripts encoding the proteins in the terpene and indole biosynthetic pathways can increase dramatically in response to herbivory or insect regurgitant (Bouwmeester et al., 1999; Degenhardt and Gershenzon, 2000; Frey et al., 2000; Korth and Dixon, 1997). The induced volatiles themselves might act as a plant–plant signal inducing the expression of some defense genes, including some involved in terpene synthesis (Arimura et al., 2000).

Plant jasmonates serve a crucial role in the production of herbivore-induced volatiles. It is well known that exogenous application of jasmonates leads to increased expression of many genes involved in production of volatiles and to volatile release itself (Hopke et al., 1994). Moreover, responses are specific, possibly due to the type of damage and/or the presence of herbivore-specific elicitors. The addition of exogenous JA causes the release of a volatile profile that is significantly different to that of mite-infested lima bean plants (Dicke et al., 1999). The addition of JA to leaves can lead to volatile profiles similar to those caused by caterpillar damage; whereas addition of both JA and salicylic acid were necessary to achieve a volatile profile similar to that from mite-infested plants (Ozawa et al., 2000). Exogenous JA added to excised leaves will also cause volatile release. However, the timing and profiles of released compounds can differ greatly depending on the assay used (Schmelz et al., 2001), illustrating the high degree of variation dependent on how plants are treated and how elicitors are added. *Arabidopsis thaliana* releases herbivore- and artificial damage-induced volatiles that attract parasitoid wasps (van Poecke et al., 2001). The availability of jasmonate mutants in *Arabidopsis* could provide useful tools to dissect the role of jasmonates in the release of herbivore-induced plant volatiles.

The high degree of specificity in plant recognition of herbivores was well illustrated when it was shown that parasitoid wasps are more attracted to plants that have been fed upon by their preferred host caterpillar, over plants damaged by a closely related caterpillar (De Moraes et al., 1998). This suggests that the amounts and profiles of insect-derived elicitors can cause different plant responses, to the point that wasps can distinguish between plants damaged by different species. Even a single insect species can cause different plant responses depending on the maturity of the caterpillar. Early instar caterpillars (the preferred host for the wasp eggs) cause a higher degree of volatile release that attracts more parasitoids to maize plants than do older caterpillars (Takabayashi et al., 1995). Specificity could be due to oral elicitors since regurgitant from early instar larvae caused a much greater emission of volatiles than regurgitant from older larvae (Takabayashi et al., 1995).

Specificity of Gene Activation and Suppression

Differential gene expression patterns indicate the presence of multiple signaling events responsive to various components of the herbivory complex, such as wounding, jasmonates, or insect elicitors (Pickett and Poppy, 2001). A growing body of data suggests that none of the pathways is completely independent, and that there is a significant degree of overlap between the pathways responding to individual inducing factors.

Feeding by the silverleaf whitefly *Bemisia argentifolii* on squash causes the accumulation of transcripts for the *SLW*1 and *SLW*3 genes, whereas wounding or bacterial infection have no effect on expression of these genes (van de Ven et al., 2000). The *SLW*3 gene is unresponsive to either jasmonate or ethylene, suggesting that a unique pathway regulates its induction. Furthermore, both *SLW*1 and *SLW*3 are induced by water stress, showing that plant responses to herbivory can overlap with responses to abiotic stresses. Related findings were achieved using microarray analysis of *Arabidopsis* responses. Reymond et al. (2000) compared gene expression patterns in response to herbivory by a lepidopteran, water stress, and mechanical damage. Transcripts for one gene, encoding a hevein-like protein, were induced by *Pieris rapae* feeding and not by the other stimuli tested. In general, mechanical wounding resulted in gene expression patterns that were more like those resulting from water stress than from insect damage. Some aspect of insect feeding might therefore suppress the expression of genes that would otherwise be induced by wounding; lepidopteran regurgitant has been shown to suppress the accumulation of some wound-induced transcripts (Schittko et al., 2001). When gene expression patterns responding to lepidopteran damage in *Nicotiana attenuata* were measured, it was found that none of the induced genes shared high similarity with insect-induced genes from the *Arabidopsis* studies (Hermsmeier et al., 2001).

The differences in plant responses to chewing insects compared to sucking insects might be as great or greater than those observed when comparing chewing insects with pathogens. In many ways, plants appear to perceive and respond to sucking/piercing insects with signaling events resembling those observed following pathogen infection (reviewed by Walling, 2000). The overlapping patterns are variable, though, for example, aphid feeding causes gene induction patterns that are unique from those caused by wounding or chemical induction of systemic acquired resistance to pathogens (Moran and Thompson, 2001; Moran et al., 2002). The timing of accumulation patterns of some insect- and wound-responsive transcripts can also differ depending on the type of damage, with insect damage generally causing a more rapid plant response (Korth and Dixon, 1997). This observation might be analogous at some level to plant responses to pathogens, where some defense-related genes accumulate faster in response to virulent pathogen strains than to avirulent strains (e.g., Zhou et al., 1997). In addition to timing differences, transcript levels of gene-family members can accumulate to different ratios when comparing insect damage to artificial damage (Berger et al., 2002).

There is a high degree of overlap in gene expression patterns following plant treatment with inducers of defense genes (Schenk et al., 2000). There is ample evidence for reciprocal interference of salicylic acid- and jasmonate-mediated signaling pathways; however, some studies point to cooperative responses induced by these same compounds (Felton and Korth, 2000; Kunkel and Brooks, 2002). It is most likely that changes in gene expression patterns, along with other physiological responses, are the result of complex interconnected signaling networks that are responsive to a wide variety of biotic and environmental cues.

11.3.2 Characterization of Insect-Derived Signals

During the intimate association of an insect herbivore with its host, any component of the insect that comes into contact with a plant could evoke defense responses. The most likely source of elicitors from herbivores is from oral factors that enter plant tissues during the feeding process. Generally, elicitors introduced during feeding could be derived from either saliva or digestive juices that exit the mouth as regurgitant. Recent discovery of some of these factors is shedding light on the specificity of plant responses to biotic damage, but many insect-specific responses in plants have still not been associated with a defined elicitor(s). To date, the only known specific insect-derived elicitors of plant responses are introduced into plant tissues either during feeding or oviposition. Although elicitors in insect waste such as honeydew or frass have not been identified, it is possible that digestive waste products could be a signal of herbivory. In addition, gall-forming insects have a profound effect on plant cell growth as they commandeer plant cells as a food source or a safe haven; clearly in this case there are signals introduced by the insect that trigger an extreme alteration of normal plant development. Although the idea that insect-derived factors affect plant physiology has been assumed for some time (e.g., Miles, 1968; Dyer and Bokhari, 1976), it is only in recent years that the chemical nature of some of these elicitors has been determined.

Insect-Derived Signals to the Plant

The importance of elicitors in plant recognition of insects was inferred from the studies that showed the release of specific volatile signals from plants following herbivore damage as compared to artificial wounding (Dicke, 1994; Turlings et al., 1995). Application of regurgitant to wound sites also causes differential accumulation of plant phenolics when compared to artificial damage (Hartley and Lawton, 1991). These findings support the idea that specific signals coming from the arthropod indicate biotic damage to the plant.

The first reported insect-derived elicitor of volatile release was an enzyme, a β-glucosidase from the cabbage butterfly, *Pieris brassicae* (Mattiacci et al., 1995). β-glucosidase is abundant in both head extracts and regurgitant and is thought to function directly in the release of volatile compounds through cleavage of preformed glucosides in cabbage leaves. There is no evidence that β-glucosidase directly or indirectly induces other defenses. Indeed, this enzyme seems to be a very general elicitor, as purified almond β-glucosidase and human saliva added to leaf wound sites also cause the differential release of volatile parasitoid attractants (Mattiacci et al., 1995).

Another class of volatile-release elicitors was first demonstrated by the discovery of "volicitin", N-(17-hydroxylinolenoyl)-L-glutamine, from regurgitant of beet armyworm, *Spodoptera exigua* (Alborn et al., 1997). When purified natural or synthetic volicitin is added to wounded leaves or through a cut stem, maize seedlings respond by releasing the same volatile terpenoids and indole released after caterpillar damage (Turlings et al., 1993; 2000). In contrast to β-glucosidase, volicitin probably causes release of volatiles indirectly. Beet armyworm caterpillar feeding and treatment of plants with regurgitant both induce a delayed, de novo biosynthesis of terpenoids and indole (Paré and Tumlinson, 1997a,b).

Volicitin was the first of a series of heat-stable fatty acid–amino acid conjugates (FACs) identified in lepidopteran regurgitant. Structurally similar, but considerably less active FACs are found along with volicitin in beet armyworm regurgitant (Alborn et al., 2000). Nonhydroxylated linolenic and linoleic acids conjugated with either glutamine or glutamate are also abundant in regurgitant of tobacco hornworm (*Manduca sexta*) and a related species (*M. quinquemaculata*) (Halitschke et al., 2001). Linolenic acid is a precursor of jasmonic acid, and its release from membranes by a lipase activity is thought to be one of the earliest steps in JA-mediated wound responses. The structural similarity of the fatty acid component of these regurgitant-based compounds with JA precursors suggests that they function by incorporation into the octadecanoid signaling pathway (Alborn et al., 1997). Addition of JA precursors to leaves is known to trigger proposed defense responses, such as the biosynthesis of proteinase inhibitor proteins in tomato (Farmer and Ryan, 1992). However, comparison of the effects of equivalent amounts of insect-derived FACs with free fatty acids showed that defense-induction activity in regurgitant is not the result of fatty acid moieties feeding into the octadecanoid pathway (Halitschke et al., 2001). This confirms that the FACs produced in insects act to induce plant defense pathways via a specific herbivore-recognition system. Heat-stable low-molecular weight elicitors of plant terpene biosynthesis have also been identified in crude extracts of sweet potato weevil *Cylas formicarius* (Sâto et al., 1981), although the structure of these compounds and their mode of entry into plants remain unknown.

Glucose oxidase (GOX), a salivary insect enzyme, has also been suggested to function as a signal of lepidopteran feeding to plants. GOX converts glucose to gluconic acid and H_2O_2 via oxidation and is the most abundant protein found in the labial salivary glands of a lepidopteran herbivore *Helicoverpa zea* Boddie (Eichenseer et al., 1999). In addition to playing a role in digestion, it might also serve as an elicitor in plants through the production of either gluconic acid or H_2O_2 (Felton and Eichenseer, 1999). Addition of purified GOX to tobacco (*Nicotiana tabacum*) can suppress the production of nicotine, which is an indicator of induced plant defenses (Musser et al., 2002). If *H. zea* larvae are treated to prevent the release of labial gland saliva via ablation of the spinneret at the base of the mouth, then tobacco plants produce significantly more nicotine in response to feeding damage. Furthermore, larvae had lower survival and body weight when fed on tobacco injured by insects with ablated spinnerets as compared to tobacco injured by control insects. This indicates a direct role of labial salivary components, most likely GOX, in induced insect defenses (Musser et al., 2002). Thus, different insect-derived signals can act to either enhance or suppress plant defenses. In the case of β-glucosidase and FACs, these oral factors apparently elicit an enhanced indirect plant defense, through the production of volatiles that attract natural enemies of the herbivore. Introducing insect GOX into plant tissue seems to be a distinct advantage for the herbivore because it suppresses some induced plant defenses.

The gross effects on alteration of plant appearance and physiology are often most noticeable following damage by sucking and piercing insects, such as aphids or whiteflies. Aphids and other phloem-feeding insects typically feed on phloem sap by inserting their highly modified mouthparts or stylets directly into sieve elements of the phloem. In spite of all that is known of the close association of stylets with plant tissue, little is certain regarding the chemical nature of specific elicitors that might be found there. At a feeding site, sucking insects introduce two types of saliva to the plant from the stylet. From the onset of probing, a continuous lipoprotein sheath surrounding the stylet is formed from viscous stylet secretions also referred to as "gelatinous saliva" (Miles, 1999). In addition to the sheath saliva, "watery saliva" is also egested into the plant tissue during both stylet penetration and feeding. The watery saliva contains enzymes, such as proteases, phosphatases, lipases, polyphenol oxidases, and peroxidases, many of which could act in plant signaling events (Walling, 2001). In addition to elicitors of plant defense responses, it is also proposed that toxins can be introduced during insect feeding (e.g., Fouché et al., 1984). Gene induction responses to aphid feeding in *Arabidopsis* are substantially different compared to those caused by artificial wounding. This indicates that there is specific plant recognition of some component, possibly oral elicitors, of the aphid-feeding complex (Moran and Thompson, 2001).

Elicitors Introduced During Oviposition

In addition to elicitors introduced via mouthparts, some insect factors introduced during oviposition can have acute effects on plants. "Bruchins" are a recently identified set of compounds found in both the pea- and cowpea-weevil (Doss et al., 2000). The bruchins from these insects are esterified long-chain α,ω-diols. In resistant pea plants (*Pisum sativum*) carrying the *Np* allele, the presence of bruchins at a weevil oviposition site causes a stimulation of plant cell division leading to neoplastic growth beneath the egg. This serves the plant as a form of induced resistance, because the uncontrolled cell growth effectively lifts the insect egg from the plant surface, and the newly hatched larvae are subsequently deterred from entry into the pea pod. The specificity of this response was clearly shown in experiments where amounts as small as 1 fmol of purified bruchins added to pea pods resulted in neoplastic cell growth.

Numerous other insect species might elicit unique responses in plants following oviposition. The whitebacked planthopper (WBPH, *Sogatella furcifera*) is a serious pest of rice in large regions of Asia. Female WBPH penetrate the epidermis of rice leaves with the ovipositor and lay egg masses in the intercellular air spaces of leaf sheaths or midribs. In some rice cultivars, watery lesions surrounding the egg form within 12 hours following oviposition resulting in the death of WBPH eggs (Suzuki et al., 1996). WBPH resistant varieties of rice have enhanced levels of lesion formation surrounding oviposition sites, whereas susceptible varieties do not form such lesions (Sogawa and Liu, 2001). Factors must be present either in WBPH eggs or are introduced during oviposition that trigger a specific response in resistant varieties of rice. The nature of these potential elicitors is unknown, but they represent another potential form of insect-derived indicators of biotic damage in plants.

Gene-For-Gene Interactions in Plant: Insect Systems

A hypersensitive response (HR) can occur in resistant plant lines when a specific avirulence gene product from a pathogen is recognized by a corresponding resistance (R) gene product in the plant (reviewed by Keen, 1990). The HR is manifested as a localized area of cell death, a lesion that serves to effectively seal off potential advancement of pathogen growth in its host so infection cannot proceed. Genetic studies of both plants and pathogens have confirmed this gene-for-gene interaction in scores of cases involving fungi, bacteria, and viruses, and its importance in developing and maintaining resistant plant varieties has long been recognized.

There are only a few examples of well-characterized gene-for-gene interactions that occur in plants during insect infestation. Among these, the response of wheat varieties to the Hessian fly (*Mayetiola destructor*) was one of the earliest identified. Use of genetic resistance, often conferred by single genes, has been successful in plant breeding strategies designed to control this important and destructive pest. Insect genes conferring survival that are complementary to wheat resistance genes have been identified (Hatchett and Gallun, 1970). Recently, techniques utilizing molecular markers (Stuart et al., 1998) and in situ hybridization (Shukle and Stuart, 1995) have been used to focus on the genetics of the Hessian fly, however the nature of the avirulence gene products is still unknown.

Gene-for-gene interactions have also been shown to occur in other plant:insect systems, such as that of greenbug (*Schizaphis graminum*) on sorghum (Puterka and Peters, 1995), and the rice gall midge (*Orseolia oryzae*) on rice (Bentur et al., 1992; Sardesai et al., 2001). Although its study has been somewhat neglected, the HR has been suggested to be the most common form of plant defense against galling insects (Fernandes and Negreiros, 2001). Whether the HR is controlled at a gene-for-gene level remains to be determined. The low number of demonstrated cases of gene-for-gene complementation in plant responses to insects could partially reflect the difficulties in performing careful genetic studies on many insect herbivore species. Alternatively, few cases of such interactions could exist in nature and plants rely on less-specific means of resistance to deter arthropod pests. Certainly, the majority of herbivore-induced defense responses known in plants, including those covered in this chapter, would fall under the category of broad-spectrum resistance.

11.4 Summary

Among higher plants there are some underlying similarities in plant defense responses against herbivores; a central role for jasmonates seems the key in responses of most plants to wounding. The high degree of phenotypic variation seen among natural plant–insect interactions is perhaps mirrored by a high degree of genetic and chemical variation at the base of whole plant and cellular interactions. With recent genomic and biochemical advances, the specific steps in the biosynthetic pathway leading to jasmonates are now essentially complete (Strassner, 2002). Although inroads have been laid, many questions remain. For example, how are jasmonate-centered pathways affected by other plant- and insect-derived defense signals? Furthermore, how do the specific levels of individual jasmonates and other signals affect plant responses?

The relative ratios of discrete components of blends of volatile signals are known to often be much more important than are the absolute amounts of any one compound. This is the case in plant attraction of both parasitoid wasps and pollinators. Insect oral factor blends with differing amounts of individual insect elicitors might explain the differential responses that can occur when plants are injured by closely related herbivores. Finally, recent studies illustrate the importance of the specificity of jasmonate biosynthetic enzymes leading to varying levels of jasmonates and a related "oxylipin signature". Again, the relative ratios of individual plant compounds could be more important than the actual quantity of any individual compound. Thus, mixtures of insect elicitors, plant-derived signals, and plant volatiles could ultimately manifest their biological specificity because of the relative amounts of their individual components.

Acknowledgments

We gratefully acknowledge research support from USDA-NRI Entomology and Nematology Program, Grant. No. 99-35302-10671 (G.A.T.), the Samuel Roberts Noble Foundation (K.L.K.) and the Arkansas Rice Research and Promotion Board (K.L.K.). We thank Jorge Ayala for helpful comments on the manuscript and Shehan Welihindha for manuscript preparation.

References

Agrawal, A.A. 1998. Induced responses to herbivory and increased plant performance. *Science* 279:1201–1202.

Alarcon, J.-J., and Malone, M. 1994. Substantial hydraulic signals are triggered by leaf-biting insects in tomato. *J. Exp.. Bot.* 45:953–957.

Alborn, H.T., Turlings, T.C.J., Jones, T.H., Stenhagen, G., Loughrin, J.H., and Tumlinson, J.H. 1997. An elicitor of plant volatiles from beet armyworm oral secretions. *Science* 276:945–949.

Alborn, H.T., Jones, T.H., Stenhagen, G.S., and Tumlinson, J.H. 2000. Identification and synthesis of volicitin and related components from beet armyworm oral secretions. *J. Chem. Ecol.* 26:203–220.

Arimura, G., Ozawa, R., Shimoda, T., Nishioka, T., Boland, W., and Takabayashi, J. 2000. Herbivory-induced volatiles elicit defence genes in lima bean leaves. *Nature* 406:512–515.

Baldwin, I.T. 1998. Jasmonate-induced responses are costly but benefit plants under attack in native populations. *Proc. Natl. Acad. Sci. USA* 95:8113–8118.

Baldwin, I.T., Kessler, A., and Halitschke, R. 2002. Volatile signaling in plant–plant–herbivore interactions: what is real? *Curr. Opin. Plant Biol.* 5:351–354.

Bentur, J.S., Pasalu, I.C., and Kalode, M.B. 1992. Inheritance of virulence in rice gall midge (*Orseolia oryzae*). *Indian J. Agric. Sci.* 62:490–493.

Berger, S., Bell, E., and Mullet, J.E. 1996. Two methyl jasmonate-insensitive mutants show altered expression of *AtVsp* in response to methyl jasmonate and wounding. *Plant Physiol.* 111:525–531.

Berger, S., Mitchell-Olds, T., and Stotz, H.U. 2002. Local and differential control of vegetative storage protein expression in response to herbivore damage in *Arabidopsis thaliana*. *Physiol. Plantarum* 114:85–91.

Bolter, C.J., Dicke, M., Van Loon, J.J., Visser, J.H., and Posthumus, M.A. 1997. Attraction of Colorado potato beetle to herbivore-damaged plants during herbivory and after its termination. *J. Chem. Ecol.* 23:1003–1023.

Bouwmeester, H.J., Verstappen, F.W.A., Posthumus, M.A., and Dicke, M. 1999. Spider mite-induced (3S)-(*E*)-nerolidol synthase activity in cucumber and lima bean. The first dedicated step in acyclic C11-homoterpene biosynthesis. *Plant Physiol.* 121:173–180.

Capinera, J.L., and Roltsch, W.J. 1980. Response of wheat seedlings to actual and simulated migratory grasshopper defoliation. *J. Econ. Entomol.* 73:258–261.

Cardoza, Y.J., Alborn, H.T., and Tumlinson, J.H. 2002. *In vivo* volatile emissions from peanut plants induced by simultaneous fungal infection and insect damage. *J. Chem. Ecol.* 28:161–174.

Creelman R.A., and Mullet, J.E. 1997. Biosynthesis and action of jasmonates in plants. *Annu. Rev. Plant Physiol. Plant Mol. Biol.* 48:355–81.

Creelman R.A., Tierney, M.L., and Mullet, J.E. 1992. Jasmonic acid/methyl jasmonate accumulate in wounded soybean hypocotyls and modulate wound gene expression. *Proc. Natl. Acad. Sci. USA.* 89:4938–4941.

De Moraes, C.M., Lewis, W.J., Paré, P.W., Alborn, H.T., and Tumlinson, J.H. 1998. Herbivore-infested plants selectively attract parasitoids. *Nature* 393:570–573.

Degenhardt, J., and Gershenzon, J. 2000. Demonstration and characterization of (*E*)-nerolidol synthase from maize: A herbivore-inducible terpene synthase participating in (3*E*)-4, 8-dimethyl-1,3,7-nonatriene biosynthesis. *Planta* 210:815–822.

Dicke, M. 1994. Local and systemic production of volatile herbivore-induced terpenoids: their role in plant-carnivore mutualism. *J. Plant Physiol.* 143:465–472.

Dicke, M., and Sabelis, M.W. 1988. How plants obtain predatory mites as bodyguards. *Neth. J. Zool.* 38:148–165.

Dicke, M., Gols, R., Ludeking, D., and Posthumus, M.A. 1999. Jasmonic acid and herbivory differentially induce carnivore-attracting plant volatiles in lima bean plants. *J. Chem. Ecol.* 25:1907–1922.

Doss, R.P., Oliver, J.E., Proebsting, W.M., Potter, S.W., Kuy, S., Clement, S.L., Williamson, R.T., Carney, J.R., and DeVilbiss, E.D. 2000. Bruchins: Insect-derived plant regulators that stimulate neoplasm formation. *Proc. Natl. Acad. Sci. USA* 97:6218–6223.

Dyer, M.I., and Bokhari, U.G. 1976 Plant-animal interactions: Studies of the effects of grasshopper grazing on blue grama grass. *Ecology* 57:762–772.

Eichenseer, H., Mathews, M.C., Bi, J.L., Murphy, J.B., and Felton, G.W. 1999. Salivary glucose oxidase: multifunctional roles for *Heliocoverpa zea. Arch. Insect Biochem. Physiol.* 42:99–109.

Farmer, E.E. 2001. Surface-to-air signals. *Nature* 411:854–856.

Farmer, E.E., and Ryan, C.A. 1990. Interplant communication: airborne methyl jasmonate induces synthesis of proteinase inhibitors in plant leaves. *Proc. Natl. Acad. Sci. USA* 87:7713–7716.

Farmer, E.E., and Ryan, C.A. 1992. Octadecanoid precursors of jasmonic acid activate the synthesis of wound-inducible proteinase inhibitors. *Plant Cell* 4:129–134.

Farmer, E.E., Johnson, R.R., and Ryan, C.A. 1992. Regulations of expression of proteinase inhibitor genes by methyl jasmonate and jasmonic acid. *Plant Physiol.* 98:995–1002.

Felton, G.W, and Eichenseer, H. 1999. Herbivore saliva and its effects on plant defense against herbivores and pathogens. In *Induced Plant Defenses Against Pathogens and Herbivores,* eds. A.A. Agrawal, S. Tuzun, and E. Bent, pp. 19–36. St. Paul, MN: American Phytopathological Society.

Felton, G.W., and Korth, K.L. 2000. Trade-offs between pathogens and herbivore resistance. *Curr. Opin. Plant Biol.* 3:309–314.

Fernandes, G.W., and Negreiros, D. 2001. The occurrence and effectiveness of hypersensitive reaction against galling herbivores across host taxa. *Ecol. Entomol.* 26:46–55.

Fouché, A., Verhoeven, R.L., Hewitt, P.H., Walters, M.C., Kriel, C.F., and De Jager, J. 1984. Russian aphid (*Diuraphis noxia*) feeding damage on wheat, related cereals and a *Bromus* grass species. *Tech. Commun. S. Afr. Dept. Agric.* 191:22–33.

Frey, M., Stettner, C. Paré, P.W., Schmelz, E.A., Tumlinson, J.H., and Gierl, A. 2000. An herbivore elicitor activates the gene for indole emission in maize. *Proc. Natl. Acad. Sci. USA* 97:14801–14806.

Gibson, R.W., and Pickett, J.A. 1983. Wild potato repels aphids by release of aphid alarm pheromone. *Nature.* 302:608–609.

Halitschke, R., Schittko, U., Pohnert, G., Boland, W., and Baldwin, I.T. 2001. Molecular interactions between the specialist herbivore *Manduca sexta* (Lepidoptera, Sphingidae) and its natural host *Nicotiana attenuata*. III. Fatty acid-amino acid conjugates in herbivore oral secretions are necessary and sufficient for herbivore-specific plant responses. *Plant Physiol.* 125:711–717.

Harms, K., Atzorn, R., Brash, A., Kühn, H., Wasternack, C., Willmitzer, L., and Peña-Cortés, H. 1995. Expression of a flax allene oxide synthase cDNA leads to increased endogenous jasmonic acid (JA) levels in transgenic potato plants but not to a corresponding activation of JA-responding genes. *Plant Cell.* 7:1645–1654.

Hartley, S.E., and Lawton, J.H. 1991. Biochemical aspects and significance of the rapidly induced accumulation of phenolics in birch foliage. In *Phytochemical Induction by Herbivores*, eds. D.W. Tallamy, and M.J. Raupp, pp. 105–132. New York: John Wiley & Sons.

Hatchett, J.H., and Gallun, R.L. 1970. Genetics of the ability of the Hessian fly, *Mayetiola destructor*, to survive on wheats having different genes for resistance. *Ann. Entomol. Soc. Am.* 63:1400–1407.

Heil, M., and Baldwin, I.T. 2002. Fitness costs of induced resistance: Emerging experimental support for a slippery concept. *Trends Plant Sci.* 7:61–67.

Herde, O., Fuss, H., Peña-Cortés, H., and Fisahn, J. 1995. Proteinase inhibitor II gene expression induced by electrical stimulation and control of photosynthetic activity in tomato plants. *Plant Cell Physiol.* 36:737–742.

Hermsmeier, D., Schittko, U., and Baldwin, I.T. 2001. Molecular interactions between the specialist herbivore *Manduca sexta* (Lepidoptera, Sphingidae) and its natural host *Nicotiana attenuata*. I. Large-scale changes in the accumulation of growth- and defense-related plant mRNAs. *Plant Physiol.* 125:683–700.

Hildmann, T., Ebneth, M., Peña-Cortes, H., Sánchez-Serrano, J.J., Willmitzer, L., and Prat, S. 1992. General roles of abscisic and jasmonic acids in gene activation as a result of mechanical wounding. *Plant Cell*, 4:1157–1170.

Hopke, J., Donath J., Blechert, S., and Boland, W. 1994. Herbivore-induced volatiles: the emission of acyclic homoterpenes from leaves of *Phaseolus lunatus* and *Zea mays* can be triggered by a beta-glucosidase and jasmonic acid. *FEBS Lett.* 352:146–150.

Howe, G.A., and Schilmiller, A.L. 2002. Oxylipin metabolism in response to stress. *Curr. Opin. Plant Biol.* 5:230–236.

Howe, G.A., Lightner, J., Browse, J., and Ryan, C.A. 1996. An octadecanoid pathway mutant (*JL5*) of tomato is compromised in signaling for defense against insect attack. *Plant Cell.* 8:2067–2077.

Keen, N.T. 1990. Gene-for-gene complementarity in plant-pathogen interactions. *Annu. Rev. Genet.* 24:447–463.

Kessler, A., and Baldwin, I.T. 2001. Defensive function of herbivore-induced plant volatile emissions in nature. *Science* 291:2141–2144.

Kessler, A., and Baldwin, I.T. 2002. Plant responses to insect herbivory: The emerging molecular analysis. *Annu. Rev. Plant Biol.* 53:299–328.

Korth, K.L., and Dixon, R.A. 1997. Evidence for chewing insect-specific molecular events distinct from a general wound response in leaves. *Plant Physiol.* 115:1299–1305.

Kramell, R., Miersch, O., Atzorn, R., Parthier, B., and Wasternack, C. 2000. Octadecanoid-derived alteration of gene expression and the "oxylipin signature" in stressed barley leaves: implications for different signaling pathways. *Plant Physiol.* 123:177–187.

Kunkel, B.N., and Brooks, D.M. 2002. Cross talk between signaling pathways in pathogen defense. *Curr. Opin. Plant Biol.* 5:325–331.

Laudert, D., and Weiler, E.W. 1998. Allene oxide synthase: a major control point in *Arabidopsis thaliana* octadecanoid signalling. *Plant J.* 15:675–684.

Li, L., Li, C., Lee, G.I., and Howe, G.A. 2002a. Distinct roles for jasmonate synthesis and action in the systemic wound response of tomato. *Proc. Natl. Acad. Sci. USA* 99:6416–6421.

Li, C., Williams, M.M., Loh, Y.-T., Lee, G.I., and Howe, G.A. 2002b. Resistance of cultivated tomato to cell content-feeding herbivores is regulated by the octadecanoid-signaling pathway. *Plant Physiol.* 130:494–503.

Malone, M., and Alarcon, J.J. 1995. Only xylem-borne factors can account for systemic wound signalling in the tomato plant. *Planta* 196:740–746.

Mattiacci, L., Dicke, M., and Posthumus, M.A. 1995. β-Glucosidase: an elicitor of herbivore-induced plant odor that attracts host-searching parasitic wasps. *Proc. Natl. Acad. Sci. USA* 92:2036–2040.

McConn, M., Creelman, R.A., Bell, E., Mullet, J.E., and Browse, J. 1997. Jasmonate is essential for insect defense in *Arabidopsis*. *Proc. Natl. Acad. Sci. USA* 94:5473–5477.

Meyer, A.J., and Weisenseel, M.H. 1997. Wound-induced changes of membrane volatage, endogenous currents, and ion fluxes in primary roots of maize. *Plant Physiol.* 114:989–998.

Miles, P.W. 1968. Insect secretions in plants. *Annu. Rev. Phytopath.* 6:137–164.

Miles, P.W. 1999. Aphid saliva. *Biol. Rev.* 74:41–85.

Moran, P.J., and Thompson, G.A. 2001. Molecular responses to aphid feeding in *Arabidopsis* in relation to plant defense pathways. *Plant Physiol.* 125:1074–1085.

Moran, P.J., Cheng, Y., Cassell, J.L., and Thompson, G.A. 2002. Gene expression profiling of *Arabidopsis thaliana* in compatible plant-aphid interactions. *Arch. Insect Bioch. Physiol.* 51:182–203.

Musser, R.O., Hum-Musser, S.M., Eichenseer, H., Peiffer, M., Ervin, G., Murphy, J.B., and Felton, G.W. 2002. Caterpillar saliva beats plant defences. *Nature* 416:599–600.

Neill, S.J., Desikan, R., Clarke, A., Hurst, R.D., and Hancock, J.T. 2002. Hydrogen peroxide and nitric oxide as signalling molecules in plants. *J. Exptl. Bot.* 53:1237–1247.

Ockroy, M.L.B., Turlings, T.C.J., Edwards, P.J., Fritzsche-Hoballah, M.E., Ambrosetti, L., Bassetti, P., and Dorn, S. 2001. Response of natural populations of predators and parasitoids to artificially induced volatile emissions in maize plants (*Zea mays* L.). *Agri. Forest Entomol.* 3:201–209.

O'Donnell, P.J., Calvert, C., Atzorn, R., Wasternack, C., Leyser, H.M.O., and Bowles, D.J. 1996. Ethylene as a signal mediating the wound response of tomato plants. *Science* 274:1914–1917.

Orozco-Cárdenas, M.L., Narváez-Vásquez, J., and Ryan, C.A. 2001. Hydrogen peroxide acts as a second messenger for the induction of defense genes in tomato plants in response to wounding, systemin, and methyl jasmonate. *Plant Cell* 13:179–191.

Ozawa, R., Arimura, G., Takabayashi, J., Shimoda, T., and Nishioka, T. 2000. Involvement of jasmonate- and salicylate-related signaling pathways for the production of specific herbivore-induced volatiles in plants. *Plant Cell Physiol.* 41:391–398.

Paré, P.W., and Tumlinson, J.H. 1997a. *De novo* biosynthesis of volatiles induced by insect herbivory in cotton plants. *Plant Physiol.* 114:1161–1167.

Paré, P.W., and Tumlinson, J.H. 1997b. Induced synthesis of plant volatiles. *Nature* 385:30–31.

Paré, P.W., and Tumlinson, J.H. 1999. Plant volatiles as a defense against insect herbivores. *Plant Physiol.* 121:325–331.

Penninckx, I.A.M.A., Thomma, B.P.H.J., Buchala, A., Métraux, J.-P., and Broekaert, W.F. 1998. Concomitant activation of jasmonate and ethylene response pathways is required for induction of a plant defensin gene in *Arabidopsis*. *Plant Cell* 10:2103–2113.

Pichersky, E., and Gershenzon, J. 2002. The formation and function of plant volatiles: perfumes for pollinator attraction and defense. *Curr. Opin. Plant Biol.* 5:237–243.

Pickett, J.A., and Poppy, G.M. 2001. Switching on plant genes by external chemical signals. *Trends Plant Sci.* 6:137–139.

Piel, J., Atzorn, R., Gäbler, R., Kühnemann, F., and Boland, W. 1997. Cellulysin from the plant parasitic fungus *Trichoderma viride* elicits volatile biosynthesis in higher plants via the octadecanoid signaling cascade. *FEBS Lett.* 416:143–148.

Puterka, G.J., and Peters, D.C. 1995. Genetics of greenbug (Homoptera: Aphididae) virulence to resistance in sorghum. *J. Econ. Entomol.* 88:421–429.

Reymond, P., Weber, H., Damond, M., and Farmer, E.E. 2000. Differential gene expression in response to mechanical wounding and insect feeding *Arabidopsis*. *Plant Cell* 12:707–719.

Rhodes, J.D., Thain, J.F., and Wildon, D.C. 1996. The pathway for systemic electrical signal conduction in the wounded tomato plant. *Planta* 200:50–57.

Sardesai, N., Rajyashri, K.R., Behura, S.K., Nair, S., and Mohan, M. 2001. Genetic, physiological and molecular interactions of rice and its major dipteran pest, gall midge. *Plant Cell, Tissue Org. Cult.* 64:115–131.

Satô, K., Uritani, I., and Saito, T. 1981. Characterization of the terpene-inducing factor isolated from the larvae of the sweet potato weevil, *Cylas formicarius* Fabricius (Coleoptera:Brenthidae). *Appl. Ent. Zool.* 16:103–112.

Schenk, P.M., Kazan, K., Wilson, I., Anderson, J.P., Richmond, T., Somerville, S.C., and Manners, J.M. 2000. Coordinated plant defense responses in *Arabidopsis* revealed by microarray analysis. *Proc. Natl. Acad. Sci. USA* 97:11655–11660.

Schittko, U., Preston, C.A., and Baldwin, I.T. 2000. Eating the evidence? *Manduca sexta* larvae can not disrupt specific jasmonate induction in *Nicotiana attenuata* by rapid consumption. *Planta* 210:343–346.

Schittko, U., Hermsmeier, D., and Baldwin, I.T. 2001. Molecular interactions between the specialist herbivore *Manduca sexta* (Lepidoptera, Sphingidae) and its natural host *Nicotiana attenuata*. II. Accumulation of plant mRNAs in response to insect-derived cues. *Plant Physiol.* 125:701–710.

Schmelz, E.A., Alborn, H.T., and Tumlinson, J.H. 2001. The influence of intact-plant and excised-leaf bioassay designs on volicitin- and jasmonic acid-induced sesquiterpene volatile release in *Zea mays*. *Planta* 214:171–179.

Seo, H.S., Song, J.T., Cheong, J.J., Lee, Y.H., Hwang, I., and Lee, J.S. 2001. Jasmonic acid carboxyl methyltransferase: A key enzyme for jasmonate-regulated plant responses. *Proc. Natl. Acad. Sci. USA.* 98:4788–4793.

Seo, S., Sano, H., and Ohashi, Y. 1997. Jasmonic acid in wound signal transduction pathways. *Physiol. Plant.* 101:740–745.

Shukle, R.H., and Stuart, J.J. 1995. Physical mapping of DNA sequences in the Hessian fly, *Mayetiola destructor*. *J. Heredity* 86:1–5.

Sogawa, K., and Liu, G. 2001. Whitebacked planthopper resistance in Chinese japonica rice. In *Rice Research for Food Security and Poverty Alleviation*, eds. S. Peng, and B. Hardy, pp. 393–400. Los Baños, Philippines: International Rice Research Institute.

Stankovic, B., and Davies, E. 1997. Intercellular communication in plants: electrical stimulation of proetienase inhibitor gene expression in tomato. *Planta* 202:402–406.

Stankovic, B., and Davies, E. 1998. The wound response in tomato involves rapid growth and electrical responses, systemically up-regulated transcription of proteinase inhibitor and calmodulin and down-regulated translation. *Plant Cell Physiol.* 39:268–274.

Staswick, P.E., Su, W., and Howell, S.H. 1992. Methyl jasmonate inhibition of root growth and induction of a leaf protein are decreased in an *Arabidopsis thaliana* mutant. *Proc. Natl. Acad. Sci. USA* 89:6837–6840.

Stintzi, A., Weber, H., Reymond, P., Browse, J., and Farmer, E.E. 2001. Plant defense in the absence of jasmonic acid: The role of cyclopentenones. *Proc. Natl. Acad. Sci. USA.* 98:12837–12842.

Stout, M.J., Workman, J., and Duffey, S.S. 1994. Differential induction of tomato foliar proteins by arthropod herbivores. *J. Chem. Ecol.* 20:2575–2594.

Strassner, J., Schaller, F., Frick, U.B., Howe, G.A., Weiler, E.W., Amrhein, N., Macheroux, P., and Schaller, A. 2002. Characterization and cDNA-microarray expression analysis of 12-oxophytodienoate reductases reveals differential roles for octadecanoid biosynthesis in the local versus the systemic wound response. *Plant J.* 32:585–601.

Stuart, J.J., Schulte, S.J., Hall, P.S., and Mayer, K.M. 1998. Genetic mapping of Hessian fly avirulence gene *vH6* using bulked segregant analysis. *Genome.* 41:702–708.

Suzuki, Y., Sogawa, K., and Seino, Y. 1996. Ovicidal reaction of rice plants against the whitebacked planthopper, *Sogatella fucifera* Horvath (Homoptera: Delphacidae). *Appl. Entomol. Zool.* 31:111–118.

Takabayashi, J., and Dicke, M. 1996. Plant-carnivore mutualism through herbivore-induced carnivore attractants. *Trends Plant Sci.* 1:109–113.

Takabayashi, J., Takahashi, S., Dicke, M., and Posthumus, M.A. 1995. Developmental stage of herbivore *Pseudaletia separata* affects production of herbivore-induced synomone by corn plants. *J. Chem. Ecol.* 21:273–287.

Thaler, J.S. 1999. Induced resistance in agricultural crops: Effects of jasmonic acid on herbivory and yield in tomato plants. *Environ. Entomol.* 28:30–37.

Thaler, J.S., Farag, M.A., Paré P.W., and Dicke, M. 2002. Jasmonate-deficient plants have reduced direct and indirect defences against herbivores. *Ecol. Lett.* 5:764–774.

Titarenko, E., Rojo, E., León, J., and Sánchez-Serrano, J.J. 1997. Jasmonic acid-dependent and -independent signaling pathways control wound-induced gene activation in *Arabidopsis thaliana*. *Plant Physiol.* 115:817–826.

Tomlin, E.S., and Sears, M.K. 1992. Effects of Colorado potato beetle and potato leafhopper on amino acid profile of potato foliage. *J. Chem. Ecol.* 18:481–488.

Turlings, T.C.J., and Tumlinson, J.H. 1992. Systemic release of chemical signals by herbivore-injured corn. *Proc. Natl. Acad. Sci. USA* 89:8399–8402.

Turlings, T.C.J., Tumlinson, J.H., and Lewis, W.J. 1990. Exploitation of herbivore-induced plant odors by host-seeking parasitic wasps. *Science* 250:1251–1253.

Turlings, T.C.J., McCall, P.J., Alborn, H.T., and Tumlinson, J.H. 1993. An elicitor in caterpillar oral secretions that induces corn seedlings to emit chemical signals attractive to parasitic wasps. *J. Chem. Ecol.* 19:411–425.

Turlings, T.C.J., Loughrin, J.H., McCall, P.J., Röse, U.S.R., Lewis, W.J., and Tumlinson, J.H. 1995. How caterpillar-damaged plants protect themselves by attracting parasitic wasps. *Proc. Natl. Acad. Sci. USA* 92:4169–4174.

Turlings, T.C.J., Alborn, H.J., Loughrin, J.H., and Tumlinson, J.H. 2000. Volicitin, an elicitor of maize volatiles in oral secretions of *Spodoptera exigua*: isolation and bioactivity. *J. Chem. Ecol.* 26:189–202.

Underwood, N. 1999. The influence of induced plant resistance on herbivore population dynamics. In *Induced Plant Defenses Against Pathogens and Herbivores*, eds. A.A. Agrawal, S. Tuzun, and E. Bent, pp. 211–230. St. Paul, MN: American Phytopathological Society.

van de Ven, W.T.G., LeVesque, C.S., Perring, T.M., and Walling, L.L. 2000. Local and systemic changes in squash gene expression in response to silverleaf whitefly feeding. *Plant Cell.* 12:1409–1423.

Van Poecke, R.M.P., Posthumus, M.A., and Dicke, M. 2001. Herbivore-induced volatile production by *Arabidopsis thaliana* leads to attraction of the parasitoid *Cotesia rubecula*: Chemical, behavioral, and gene-expression analysis. *J. Chem. Ecol.* 27:1911–1928.

Vijayan, P., Shockey, J., Lévesque, C.A., Cook, R.J., and Browse, J. 1998. A role for jasmonate in pathogen defense of *Arabidopsis*. *Proc. Natl. Acad. Sci. USA* 95:7209–7214.

Walker-Simmons, M., Holländer-Czytko, H., Andersen, J.K., and Ryan, C.A. 1984. Wound signals in plants: A systemic plant wound signal alters plasma membrane integrity. *Proc. Natl. Acad. Sci. USA* 81: 3737–3741.

Walling, L.L. 2000. The myriad plant responses to herbivores. *J. Plant Growth Regul.* 19:195–216.

Walling, L.L. 2001. Induced resistance: From the basic to the applied. *Trends Plant Sci.* 6:445–447.

Wang, C., Zien C.A., Afitlhile, M., Welti, R., Hildebrand, D.F., and Wang, X. 2000. Involvement of phospholipase D in wound-induced accumulation of jasmonic acid in *Arabidopsis. Plant Cell* 12:2237–2246.

Wasternack, C., and Parthier, B. 1997. Jasmonate-signalled plant gene expression. *Trends Plant Sci.* 2:302–307.

Weber, H. 2002. Fatty acid-derived signals in plants. *Trends Plant Sci.* 7:217–224.

Wildon, D.C., Thain, J.F., Minchin, P.E.H., Gubb, I.R., Reilly, A.J., Skipper, Y.D., Doherty, H.M., O'Donnell, P.J., and Bowles, D.J. 1992. Electrical signalling and systemic proteinase inhibitor induction in the wounded plant. *Nature* 360:62–65.

Zhou, J., Tang, X., and Martin, G.B. 1997. The Pto kinase conferring resistance to tomato bacterial speck disease interacts with proteins that bind a *cis*-element of pathogenesis-related genes. *EMBO J.* 16:3207–3218.

12

Tree Defenses Against Insects

Erkki Haukioja

12.1 Introduction

Our understanding of individual resistance mechanisms against insects, and especially against pathogens, has improved tremendously during the last decade (Heath and Boller, 2002). This progress has largely been achieved by studying plants (such as cotton, tomato, or *Arabidopsis*) with suitable traits for experimental work. Trees share just the opposite characteristics: long life span, large size, architectural complexity, and an often short and distinct period of leaf growth. Only saplings can be studied effectively under experimental laboratory conditions; it is not clear, however, to what extent the defenses of saplings are similar to those of mature trees. Hence the defenses of woody plants are not well known; presumably, trees employ mechanisms that are widespread in the plant kingdom. The specific features, compared to herbs, also offer opportunities to study mechanisms of plant defense, and their interactions, that are not obvious in short-lived hosts.

In explaining tree resistance against insects, the traditional emphasis in forest entomology has been to pay attention to secondary compounds, such as phenolics and terpenoids, and—especially in the older literature—to foliar proteins and sugars. A major problem is that plant phenolics and terpenoids are not likely to serve for insect defense alone. Nevertheless, they may strongly modify insect performance and consumption, thus linking them to plant defense. Some recent observations indicate that tree defense may involve cascades of defense reactions, induced for instance by products of the octadecanoid pathway or ethylene. What we know less about is to what extent the effects of defensive cascades are based on the specific end products of these pathways, to what extent on interactions of compounds in the pathways with "traditional" defenses such as phenols or with nutritive compounds. In any case, tree defenses are probably fundamentally multigenic.

Variance in individual defense mechanisms, and the relative importance of these mechanisms, is poorly known in all plants, including trees. The use of genetically uniform plants under standardized conditions is likely to eliminate variance in defense expressions that depends on environmental conditions or on plant biochemical variation, leading easily to an overestimation of the role of single mechanisms behind plant resistance. Accordingly, individual, potentially defensive mechanisms may or may not be important under field conditions, where their importance has to be judged against all the factors that contribute to a reduction in damage. This problem is especially relevant in the defense of long-lived plants against insects, simply because the longer the plant lifespan, the more diverse the herbivore types and environmental conditions the plant is likely to encounter. Possible interactions may take place among specific defense cascades, secondary compounds, and nutritive plant traits. All this means that trees need multiple means to handle the diversity of insects and environmental situations.

In this chapter I deal with certain major ecological issues involved in tree defenses, by emphasizing the complications that emerge when we try to include insect traits in the picture (Haukioja, 2003; for another recent review of tree defenses against insects, see Larsson, 2002). My emphasis comes from birch studies, and biases the discussion toward birches and other deciduous trees. I will deal with the defenses of conifer foliage, but not, for instance, with defenses against bark beetles (Raffa, 1991). I will first review the ecological aspects relating to traits creating resistance in woody plants. I will then discuss the problem of how long-lived plants such as trees can remain resistant against short-lived pests that have numerous generations during the lifetime of a single host. Finally, I will discuss some general practical and theoretical problems encountered in research into tree resistance.

12.2 Constitutive versus Induced Resistance

Ecologically the most relevant classification of tree resistance is that of constitutive versus induced resistance. The levels of constitutive defenses do not depend on prior contacts with herbivores. The boundary between constitutive and induced resistance, however, is not sharp. For instance, the efficiency of typical constitutive defenses, such as phenolic compounds, may depend critically on polyphenol oxidases (PPOs) that are activated when insect feeding crushes the leaf cells and allows plant oxidases to react with phenolics (Felton et al., 1992). This leads to the formation of quinones, which are regarded as more toxic than phenols. Although PPOs are known to be inducible by insect damage and jasmonic acid (Constabel et al., 2000; Tscharntke et al., 2001), constitutive PPOs also function as very rapidly inducible defenses.

Constitutive defenses may contribute to the average levels of pest populations, but unlike herbivore-induced responses they are not good candidates in explaining temporal variation in insect population densities (e.g., Larsson, 2002). The most spectacular and economically important effects of forest insects relate to their multiannual population outbreaks or cycles, which often lead to large-scale defoliations. Herbivore-induced responses can in theory modify pest populations, and because of their variable time lags they have gained much attention in woody plant research.

12.2.1 Constitutive Tree Defenses

Constitutive defenses have been studied in forest entomology for a long time, and in particular secondary compounds have received much attention. Defense is usually credited to phenolics or terpenoids, the main groups of secondary compounds in the foliage of deciduous and coniferous trees, respectively. Sugars and proteins may also be important contributors to variance in the success of forest pests, as emphasized in the older German forest-entomological literature (e.g., Schwenke, 1968). Similarly, the stress hypothesis by White (1974) is based on the importance of temporal variance in soluble nitrogen.

Although the levels of constitutive defenses do not depend on previous encounters with insects, the concentrations of constitutive secondary compounds show very dynamic variation, both temporal (Zou and Cates, 1995; Nerg et al., 1994; Habermann, 2000; Riipi et al., 2002) and spatial (Suomela et al., 1995a,b). In birch, the seasonal succession in dominant groups of foliar phenolics takes place very rapidly during leaf development: from phenolic aglycons via galloylglucoses, phenolic glycosides, and ellagitannins to proanthocyanidins (condensed tannins) (Nurmi et al., 1996; Kause et al., 1999b; Salminen et al., 2001; Haukioja, 2003; Valkama et al., 2003). Together with the simultaneous rapid decline in protein contents and the hump-backed seasonal trend in foliar sugars (Riipi et al., 2002), this rapid succession may explain why young and mature leaves represent very different diets for leaf chewers (Haukioja et al., 2002). At different times of leaf development, the same trees may represent such different diets that even specialized species of sawflies experience strongly reduced growth and increased mortality on birch leaves not matching the typical leaf age at that developmental stage of the larva (Martel et al., 2001). These observations suggest that the level of "tree defense" may be an elusive concept.

Compared to studies emphasizing the effects of phenolics on insects (Feeny, 1970; Rossiter et al., 1988, Bryant et al., 1993; Julkunen-Tiitto et al., 1996; Keinänen et al., 1999; Ossipov et al., 2001; Salminen and Lempa, 2002), much less attention has been paid to proteins and sugars as determinants of tree defense and insect performance (see Mattson and Scriber, 1987, Slansky, 1993; Clancy, 1992; Zou and Cates, 1994, Haukioja et al., 2002). Interestingly, different foliar sugars may have different effects on insects. High levels of galactose (Zou and Cates, 1994) and sucrose (Clancy, 1992) may retard insect growth on artificial diets. These sugars also displayed negative correlations with insect growth in birch leaves, contrary to the positive correlations shown by glucose and fructose, even at the time of peaking leaf sugars (Haukioja et al., 2002; Henriksson et al., 2003).

The effects of individual amino acids in tree foliage on insects are not well understood, and seem to show variable correlations with insect growth, perhaps because of rapid changes in the concentrations and ratios of individual foliar amino acids, and high intra-tree variation (Suomela et al., 1995b; Riipi et al., 2002). The stress hypothesis by White (1974) claims that insect outbreaks are caused by the increased availability of soluble nitrogen in stressed trees. There exist no good data to demonstrate the validity of this claim, but in birch we found no consistent correlations between soluble amino acids and the performance of an outbreaking lepidopteran (Haukioja et al., 2002; Henriksson et al., 2003). Actually, in several birch-insect data sets the best correlate with insect growth is the amount of water ingested with consumed leaf mass. Leaf water content was found to be correlated with soluble and cell-wall bound proteins, but correlations between insect growth and these groups of proteins were not as high as those with water (Haukioja et al., 2002; Henriksson et al., 2003).

It is increasingly obvious that defensive secondary compounds do not function independently of the nutritional background of tree leaves (Schopf, 1986, Jensen, 1989; Haukioja et al., 2002). Simpson and Raubenheimer (2001) demonstrated that gallic acid had a strong effect on locusts only when combined with extreme protein:carbohydrate ratios in the diet. Haukioja et al. (2002) found that while 7 out of the 35 individual phenolic compounds displayed a significant correlation with the growth or consumption of a geometrid moth, 17 more compounds showed significant interactive (insect trait * phenolic trait * leaf trait) effects with three seasonally changing leaf traits: water content, sugar:protein ratio, and toughness. In these data sets, the sugar:protein ratio was not a more important covariate than water content or toughness, perhaps because only two (glucose and fructose) of the four dominant birch leaf sugars (Riipi et al., 2002) displayed consistent positive correlations with insect growth.

Another reason relates to the logic of insect feeding (Haukioja, 2003). The goal of insect feeding is to obtain a sufficient amount of those nutritive compounds that limit its performance, in forest insects usually larval growth. On high nutritive diets insects have their demands satisfied with less consumption than on diets containing low concentrations of nutritive compounds (Haukioja et al., 2002). Low consumption on high nutritive diets naturally means a simultaneous limited intake of potentially defensive compounds. On low quality birch leaves, late-season sawfly species strongly increase their consumption (Kause et al., 1999a). This increase in consumption with declining leaf nutrition is truly dramatic: the ratio of leaf mass consumed relative to larval growth (in terms of dry matter) for a lepidopteran larva feeding on the nutritious leaves of mid-June was below 3, but in late summer sawflies it was as high as 15 (Haukioja, 2003).

All in all, foliar nutrient and phenolic contents interact strongly in their effects on insects, just as in mammals (Villalba et al., 2002). Maintaining a low nutritive value sounds like an appealing solution for plant defense, and in evolutionary time it may be so. But it can also easily lead to high loss of leaf mass to adapted herbivore species (Moran and Hamilton, 1980). The critical question is what really sets a limit to further consumption before the insect's nutritive demands are satisfied. Surprisingly, the answer to this question is not well known, presumably for several reasons. First, due to compensatory feeding, the amounts of single allelochemicals or groups of physiologically related allelochemicals consumed may be so high that the insect cannot handle any more. This is especially likely on low nutritive diets (Slansky and Wheeler, 1992). Second, due to low concentrations of growth-limiting compounds, the feeding rate needed may be so high that the insect simply cannot consume and process that much plant mass. This is because moving the leaf mass consumed through the gut demands metabolic work, due to the active construction of peritrophic membranes that separate the gut contents from the gut walls, for example (Barbehenn, 2001). In this case, there may be no single compound responsible for the cessation of feeding; it may simply become impossible to continue feeding. Third, the ratios of different nutritive compounds (particularly carbohydrates vs. proteins) may be so far from optimal that the elimination of nonlimiting nutrients puts an end to insect feeding (Raubenheimer and Simpson, 1994). This is a complex question because different nutritive compounds demand different elimination processes (surplus sugars are respired, proteins are excreted as ammonium), and because of interactive effects between nutrient ratios and secondary compounds (Simpson and Raubenheimer, 2001).

12.2.2 Herbivore-Induced Responses in Trees

Plant responses are regarded as defenses due to their end result, which is reduced damage. The term "induced defense" therefore pools together heterogeneous mechanisms (Karban and Kuć, 1999). Some of these undoubtedly are true defenses, designed by natural selection for that purpose. Induced defenses may also be inclusive and occur even when no specific defense pathways are activated. In other words, the null hypothesis is not that no change in plant quality takes place after herbivore damage. The drying up of partially consumed leaves is an example. Since plants are modular organisms, any damage to hormonally active plant meristems is bound to rearrange resource flows within the plant (Haukioja et al., 1990; Dyer et al., 1991), and this may lead to drastic changes in the suitability of adjacent plant parts for insects. Consistent with the multiple ways whereby induced resistance against insects can emerge is the pronounced variance between the results of individual experiments; for birch see Ruohomäki et al. (1992), for pines Watt (1990), Lyytikäinen (1994), Trewhella et al. (1997), Raffa et al. (1998) and Smits et al. (2001). Unlike many of the mechanisms activated by defense pathways, such as the octadecanoid cascade, responses based on mechanical damage via altered sink-source relations seem to be local (Tuomi et al., 1988, Långström et al., 1990; Henriksson, 2001, Henriksson et al., 2003).

Some herbivore-induced responses may occur immediately after the damage; the long lifespan of trees, however, also allows carry-over responses of damage into foliage produced in the growth season(s) following the damage. Due to variation in the duration of induced responses relative to insect generation time, induced plant responses can cause different and even opposite pressures on insect population densities (Haukioja, 1982; 1990). I therefore deal separately with rapid induced (RIR) and delayed induced resistance (DIR).

Rapid Induced Defenses

RIR is experienced by the same generation of herbivores that triggered the response with adverse effects on the insect. RIR is likely to dampen out fluctuations in insect population density: the higher the insect density, the more likely it becomes that induced responses will be expressed and will detrimentally affect the insect. There are numerous observations of rapid induced systems against insects in the foliage of various woody plants, such as birch (Haukioja and Niemelä, 1977; Edwards and Wratten, 1982; Wratten et al., 1984; Fowler and MacGarvin, 1986; Hanhimäki and Senn, 1992), oak (Schultz and Baldwin, 1982; Rossiter et al., 1988), poplar (Havill and Raffa, 1999), larch (Krause and Raffa, 1992), pine (Litvak and Monson, 1998), fir (Litvak and Monson, 1998), and alder (Seldal et al., 1994; Dolch and Tscharntke, 2000).

A key challenge in explaining rapid induced tree defenses is to understand interactions between specific defense pathways (such as cascades induced by octadecanoid compounds, salicylic acid and by ethylene), phenolic metabolism, and primary leaf nutrients. Changes in tree nutrient levels in connection with RIR are poorly known (Beardmore et al., 2000). An induced resistance response is often characterized by an accumulation of phenolics in deciduous species, and of terpenoids in conifers. The role of jasmonic acid in the increased synthesis of terpenoids has been demonstrated by Martin et al. (2002). Jasmonate treatments did not induce higher levels of phenolic synthesis in alder (Tscharntke et al., 2001) or birch leaves (Ossipov, unpublished observations), but a response was found in poplar sink leaves (Arnold and Schultz, 2002). Since jasmonates are implemental in the activation of plant enzymes, such as PPOs, they may still control the efficacy of phenolic compounds on insects (Constabel et al., 2000; Tscharntke et al., 2001). The cascade induced by ethylene was found to induce increased production of phenolics (Tscharntke et al., 2001), reiterating the old observation in eucalypts (Hillis, 1975).

Tscharntke et al. (2001) demonstrated no increase in predator densities after induction of defense in alder, but, in general, the possible role of parasitoids in RIR in woody plants is not known. This is a promising area, since the spread of herbivore-induced volatiles (Dolch and Tscharnke, 2000) may have an even forest-wide effect on predation.

Delayed Induced Defenses

Multiannual, large-scale insect outbreaks are characteristic of many woody plants. Induced defenses offer a potential explanation for the collapse of peak populations; in order to lead to a low population level, however, they would have to severely impair insect performance over several generations, i.e., lead to delayed negative feedbacks. The increased resistance associated with DIR has been shown to last for one to four years after manual defoliations in birch (Neuvonen and Haukioja, 1991; Kaitaniemi et al., 1999) and after natural defoliation in larch (Benz, 1974; Omlin, 1980). Thus DIR—unlike RIR—offers a potential factor for the decline and the low phase of insect populations after outbreaks of forest insects.

The chemical basis of DIR is not well understood. The simplest explanation is that defoliations reduce tree nutrient reserves. In the years following defoliations the trees actually tend to be low in nutritive compounds, especially nitrogen and water (Valentine et al., 1983; Kaitaniemi et al., 1998). Honkanen and Haukioja (1998) presented a physiological null hypothesis for DIR, based on the observation that primordial leaves within short shoot buds are initiated the year before their flush. Early damage to the leaves supporting the buds weakens the sink strength of the buds, and the following year reduces their ability to draw resources from the common pool. Such leaves are therefore likely to remain small, low in nutrients, and high in phenolics. Levels of phenolics in the leaves of trees defoliated the previous year have been found to be higher than in undefoliated control trees. Most early studies refer to the consequences of manual defoliations (Tuomi et al., 1984; Hartley and Lawton, 1990; Bryant et al., 1993; Ruohomäki et al., 1996; Kaitaniemi et al., 1999), and their relevance to field conditions is not clear. Benz (1974) reported a drop in both consumption and efficiency of larval growth in larch trees that had been naturally defoliated in previous year; these measures were associated with low nitrogen and high fiber contents in needles. Kaitaniemi et al. (1998) demonstrated increased levels of phenolics (and lower levels of sugars and proteins) in trees naturally defoliated by moth larvae during previous years, compared to trees that had been protected from defoliations by pyrethrine spraying. They also found more seasonal overlap between the peaks of hydrolyzable and condensed tannins in birch leaves after natural defoliation (Kaitaniemi et al., 1998).

Whether DIR operates via the third trophic level in trees is not known, and clearly deserves more attention, as does the use of larval-produced rather than manual defoliation. Kaitaniemi and Ruohomäki (2001) experimentally introduced few geometrid larvae on individual trees (leading to negligible consumption at the whole-tree level). The following year they found that larvae disappeared at a significantly faster rate from these trees than from trees without caterpillars the year before. Although the experiment did not reveal whether the larvae disappeared due to predation or dispersal, the study suggests that natural larval damage, unlike manual damage (Tuomi et al., 1984), leads to systemic DIR.

Induced Susceptibility

It is easy to interpret induced resistance as an evolved mechanism of plant defense. The dangers of that approach are indicated by observations showing that plant responses to herbivory may also make further herbivory more likely. It is actually very easy to manipulate trees so as to improve the quality of their foliage for herbivores: one simply breaks the apical dominance by removing the most apical meristems. After that the previously suppressed meristems start to grow actively, becoming rich in nutrients and low in tannins. After mammalian browsing, plants may become better for insects (Danell and Huss-Danell, 1985); insect feeding can lead to a similar outcome (Bryant et al., 1991). Since responses to the breaking of apical dominance are of a very general nature, it is not surprising that induced susceptibility has been found in many types of woody plants, including eucalypts (Landsberg, 1990), birch (Haukioja et al., 1990), oak (Hunter and West, 1990), alder (Williams and Myers, 1984), willows (Hjälten and Price, 1996), and pines (Trewhella et al., 1997; Raffa et al., 1998).

The chemical changes that describe induced susceptibility have not been studied in detail. In birch they seem to be just the opposite of DIR—an increase in nutrients and a decline in phenolics (Danell et al., 1997); better insect performance, however, may also follow even though some chemical traits may indicate poor foliage quality (Wagner and Evans, 1985; Raffa et al., 1998).

Induced susceptibility pinpoints certain unconsidered possibilities with regard to the origin of population cycles. If insects are able to manipulate the plant hormonal system for their own benefit, this may improve the foliage; hypothetically, this could create positive feedbacks into population numbers. This might be reflected in an improved reproductive capacity in insects during the increase phase. A further hypothetical possibility is that during the increase phase of the population, insects could sabotage the expression of defensive cascades and reduce the effects of possible indirect defenses (Dicke and van Poecke, 2002). The sabotage of indirect defenses (Edwards, 1989; Krause and Raffa, 1995; Kaitaniemi et al., 1997) might even contribute to a low rate of parasitism during the years of population build-up, a necessary prerequisite for increasing insect populations.

12.3 Resistance of Long-Lived Hosts Against Short-Lived Pests

Defoliating forest insects have short generation times compared to their hosts. This creates a potential evolutionary dilemma: the numerous generations of herbivores during the lifetime of a single host suggest the insect should easily develop genotypes that exactly match the genotypes of the host (Edmunds and Alstad, 1978), perhaps at the cost of being able to live on other host individuals. In studying the idea of adaptive deme formation, van Zandt and Mopper (1998) found evidence for locally adapted insect populations; however, it did not correlate in any simple way with the herbivore dispersal rate, feeding type, or mode of reproduction.

An obvious answer to the question of why short-lived herbivores are not able to completely breach the genetic resistance system of their long-lived trees is that insects interact with tree phenotypes, not genotypes, and that the same tree genotype is able to express vastly different resistance mechanisms. Induced defenses represent an important source of variance in plant defense (Denno and McClure, 1983), which insects try to overcome by moving around or by compensatory feeding. In birch sawflies these alternatives seem to be mutually exclusive (Kause et al., 1999a).

By no means do induced defenses represent the only source of variability. The architectural complexity of trees also promotes spatial heterogeneity within canopies (Lawton, 1983). In mountain birch, Suomela and Nilson (1994) showed that most of the variance in insect performance in birch canopies was within, not among, trees, mainly between ramets and between branches. Actually the largest component of variance in tree foliage probably is seasonal. Since secondary compounds function interactively with other leaf traits (water content, sugar/protein ratio, toughness) (Schopf, 1986; Haukioja et al., 2002), this creates rich phenotypic variation in foliage quality of the same tree genotypes.

From the viewpoint of insect adaptation to a single host, a critical point is that during its development an insect may encounter successive qualities of leaves from the same tree so different that the tree simply does not represent a single target for the insect to adapt to (Adler and Karban, 1994). This is especially likely in insects that feed on growing leaves, and is obviously less likely in sucking insects. Interestingly, none of the examples of adaptive deme formation in van Zandt and Mopper (1998) referred to foliage-chewing insects.

Pronounced changes in diet quality during the lifetime of an arboreal chewing insect take place during the time of most intense leaf development. Because of the large seasonal variation in the main nutritive, phenolic, and terpenoids compounds, individual insect genotypes may not be optimally adapted to handle each successive defensive trait. The interactive actions of phenolics with nutritive leaf traits (Haukioja et al., 2002) further enhance the challenge of dealing with the foliage of an individual tree, as suggested by the observation that most of the variance in sugars and amino acids fell within and not among trees (Suomela et al., 1995b).

Accordingly, in birch it has been common not to find significant correlations when spring-feeding insects have been repeatedly tested on leaves of the same trees. On young, nutritious leaves the lack of correlation may result from compensatory feeding on poorer quality trees producing similar growth on different host individuals, but also from the rapid developmental switching of main foliar defenses. Still, each larva that survived to pupation on a single host had to be able to handle all the seasonally switching defenses. Ruusila et al. (2005) tested autumnal moth larvae at different instars in the same trees; in accordance with the above logic, they did not find significant difference among trees but did find tree $*$ instar interactions. In other words, different trees were best for different instars. Major switches in tree suitability seemed to take place between the third instar and later ones (i.e., at the time when gallotannin contents declined, Riipi et al., 2002). Brood-specific larval growth, on the other hand, was similar in the fourth and fifth instars (see also Ayres et al., 1987). The results suggest that one component of the defensive strategy of mountain birch against *Epirrita autumnata*, its most important herbivore, may be to minimize the probability of the evolution of well adaptive insect genotypes. Birch leaf quality remains relatively invariant in mid and late season, allowing insects to adapt to certain developmental phases of the host (Hanhimäki et al., 1995; Kause et al., 2001). Martel et al. (2001) found that birch sawflies did manage poorly when they were experimentally exposed to younger (and more nutritious) leaves or to more mature (nutritionally inferior) ones than those that they typically consume. This suggests that mid- to late-season sawfly species were specialized to certain phases of birch leaf development, not to host trees per se.

12.4 Discussion

Low consumption, the final outcome of successful defense, can be achieved by numerous different combinations of defensive solutions, and we lack a holistic understanding of the plant traits that prevent consumption by insects or that curtail it to values less than optimal for insect development. This, however, is the critical problem in explaining how plant defenses lead to the ultimate goal, reduced consumption. It is important to realize that a trait with adverse effects on a herbivore may or may not be an important component of plant defense. A fundamental task in identifying crucial plant defenses is thus to determine the contribution of different plant traits to consumption (the inverse of defense); in other words, we should concentrate on how much these traits explain of the variance (r^2) in consumption. This is more important for a holistic understanding than whether a certain trait makes a defensive contribution (i.e., whether it correlates with consumption), or whether it is detrimental for a herbivore. Scriber and Feeny (1979) demonstrated the importance of water for arboreal insects, and Haukioja et al. (2002) showed that in most cases leaf water content explained more of the variance in growth and to some extent in consumption of a defoliator than any phenolic compound measured.

12.4.1 Possibilities for General Theories

Within the ecological paradigm of plant defense, research on tree defenses has long concentrated on measuring the levels of those compounds that are likely to be defensive, such as phenolics and terpenoids. The most popular general theories (the carbon-nutrient balance and the growth-differentiation balance hypotheses) have stressed the importance of trade-offs between plant growth and defense, and the defense level has often been measured as the concentration of phenolic compounds (see e.g., Herms and Mattson, 1992; Hamilton et al., 2001). In woody plants there are actually strong negative correlations between nutrient levels (high values characterizing rapidly growing plant parts) and total phenolics, or their usually largest component, condensed tannins. This has traditionally been explained by the passive accumulation of carbon in defensive compounds after carbon demands for plant growth have been met. However, the key point may not be in plant growth versus defense, since other types of putative defenses (hydrolyzable tannins, terpenoids) of woody plants do not respond to variable carbon availability in the same way as condensed tannins (Haukioja et al., 1998; Koricheva et al., 1998). Furthermore, correlations between the synthesis of putatively defensive compounds and the growth of other plant tissues may not be generally inversely related, as indicated by an analysis of the accumulation of phenolic compounds in growing birch leaves and shoots (Riipi et al., 2002). As such this does not refute trade-offs at the whole-plant level; the synthesis of leaf phenolics in mountain birch, however, depends on local and not whole-tree carbon resources, as shown by experiments shading whole canopies or individual branches only (Henriksson et al., 2003).

I have elsewhere introduced an alternative, adaptive explanation for the strong trade-off between leaf nutritive quality and its "quantitative" (sensu Feeny, 1975) defenses, such as condensed tannins (Haukioja, 2003). Quantitative defenses are obviously targeted against adapted species of herbivores, which are well equipped to handle single defensive compounds in their host. Quantitative defenses would not be effective against such species if the plant were rich in nutrients; this is because on nutrient-rich diets insects can satisfy their requirements for nutritive compounds with such low consumption that the *amounts* of tannins ingested are inconsequential. Accordingly, quantitative defenses can be effective only if plant nutrient content is low, and if compensatory feeding leads to increased consumption of defensive compounds. This scenario predicts a tight negative correlation between plant nutrient content and quantitative defenses such as condensed tannins. The explanation is not sensitive to the matter of whether the reason for high nutrition is nitrogen fertilization, shade or seasonal schedules.

Quite obviously, possibilities for the creation of new general theories as broad as the carbon-nutrient balance hypothesis are limited (see e.g., Lerdau and Coley, 2002, Nitao et al., 2002, Koricheva, 2002). This is because of the emerging picture of plant defenses as the outcomes of complex and probably idiosyncratic interactions between secondary compounds, nutritive plant traits and specific defense cascades.

12.4.2 What Next?

During recent years specific defense pathways have been shown to operate in trees, such as poplars (Constabel et al., 2000), alder (Tscharntke et al., 2001), and spruce (Martin et al., 2002). In birch as well, strong negative correlations between insect growth and leaf fatty acids of the octadecanoid pathway suggest the importance of this pathway (Haukioja et al., unpublished data). The emerging genetic maps of woody plants and the use of microarrays for trees too will offer new tools for the study of specific defense systems with the same accuracy as for well-known herbs (e.g., Chiron et al., 2000). At that point we will have the tools to start to unravel the specific and relative roles of particular defense pathways, of constitutive secondary compounds, and of nutritive compounds as codeterminants of tree defenses. Several major problems still remain. Experimentation with trees is easy only with saplings, while the most dramatic incidents of herbivory tend to occur in mature or overmature stands (e.g., Ruohomäki et al., 1997). The context-dependence and interactive functions of defenses suggest that generalizations about defenses may be risky, and we have to keep in mind possible species-specific effects of plant defense responses.

Particular interest should be accorded to the idea that trees may utilize indirect defenses, i.e., recruit parasitoids as part of tree defense, presumably via volatile exudates luring parasitoids. This idea is really worth testing; it is well known that trees release huge amounts of volatiles into the atmosphere (Rhoades, 1985; Hakola et al., 1998; Monson and Holland, 2001). Yet a further source of complexity is the possibility that some of these mechanisms may interact with "traditional" plant defenses, producing unanticipated results; dietary phenolics, for instance, may increase larval resistance against viral diseases (Hunter and Schultz, 1993).

References

Adler, F.R., and Karban, R. 1994. Defended fortresses or moving targets? Another model of inducible defenses inspired by military metaphors. *Am. Nat.* 144:813–832.

Arnold, T.M., and Schultz, J.C. 2002. Induced sink strength as a prerequisite for induced tannin biosynthesis in developing leaves of *Populus*. *Oecologia* 130:585–593.

Ayres, M.P., Suomela, J., and MacLean, S.F., Jr. 1987. Growth performance of *Epirrita autumnata* (Lepidoptera: Geometridae) on mountain birch: trees, broods, and tree∗brood interactions. *Oecologia* 74:450–457.

Barbehenn, R.V. 2001. Roles of peritrophic membranes in protecting herbivorous insects from ingested plant allelochemicals. *Arch. Insect Biochem. Physiol.* 47:86–99.

Beardmore, T., Wetzel, S., and Kalous, M. 2000. Interactions of airborne methyl jasmonate with vegetative storage protein gene and protein accumulation and biomass partitioning in Populus plants. *Can. J. For. Res.* 30:1106–1113.

Benz, G. 1974. Negative Rückkoppelung durch Raum- und Nahrungskonkurrenz sowie zyklische Veränderung dr Nahrungsgrundlage als Regelprinzip in der Populationsdynamik des Grauen Lärchenwicklers, *Zeiraphera diniana* (Guenée) (Lep., Tortricidae). *Z. Ang. Ent.* 76:196–228.

Bryant, J.P., Heitkoning, I., Kuropat, P., and Owen-Smith, N. 1991. Effects of severe defoliation on the long-term resistance to insect attack and on leaf chemistry in six woody species of the southern African savanna. *Am. Nat.* 137:50–63.

Bryant, J.P., Reichardt, P.B., Clausen, T.P., and Werner, R.A. 1993. Effects of mineral nutrition on delayed inducible resistance in Alaska paper birch. *Ecology* 74:2072–2084.

Chiron, H., Drouet, A., Lieutier, F., Payer, H.D., Ernst, D., and Sandermann, H. 2000. Gene induction of stilbene biosynthesis in Scots pine in response to ozone treatment, wounding, and fungal infection. *Plant Physiol.* 124:865–872.

Clancy, K.M. 1992. The role of sugars in western spruce budworm nutritional ecology. *Ecol. Entomol.* 17:189–197.

Constabel, C.P., Yip, L., Patton, J.J., and Christopher, M.E. 2000. Polyphenol oxidase from hybrid poplar. Cloning and expression in response to wounding and herbivory. *Plant Physiol.* 124:285–295.

Danell, K., Haukioja, E., and Huss-Danell, K. 1997. Morphological and chemical responses of birch leaves and twigs to winter browsing along a gradient of plant productivity. *Écoscience* 4:296–303.

Danell, K., and Huss-Danell, K. 1985. Feeding by insects and hares on birches earlier affected by moose browsing. *Oikos* 44:75–81.

Denno, R.F., and McClure, M.S. eds. 1983. *Variable Plants and Herbivores*. New York: Academic Press

Dicke, M., and van Poecke, R.M.P. 2002. Signaling in plant-insect interactions: signal transduction in direct and indirect plant defence. In *Plant Signal Transduction*, eds. D. Scheel, and C. Wasternack, pp. 289–316. Oxford: Oxford University Press

Dolch, R., and Tscharntke, T. 2000. Defoliation of alders (*Alnus glutinosa*) affects herbivory by leaf beetles on undamaged neighbours. *Oecologia* 125:504–511.

Dyer, M.I., Acra, M.A., Wang, G.M., Coleman, D.C., Freckman, D.W., McNaughton, S.J., and Strain, B.R. 1991. Source-sink carbon relations in two *Panicum coloratum* ecotypes in response to herbivory. *Ecology* 72:1472–1483.

Edmunds, G.F., and Alstad, D.N. 1978. Coevolution in insect herbivores and conifers. *Science* 199:941.

Edwards, P.J. 1989. Insect herbivory and plant defence theory. In *Towards a More Exact Ecology*, eds. P.J. Grubb, and J.B. Whittaker, pp. 275–297. Oxford: Blackwell. Edwards, P.J., and Wratten, S.D. 1982. Wound-induced changes in palatability in birch (*Betula pubescens* Ehr. ssp. *pubescens*). *Am. Nat.* 120:816–818.

Feeny, P. 1970. Seasonal changes in oak leaf tannins and nutrients as a cause of spring feeding by winter moth caterpillars. *Ecology* 51:565–581.

Feeny, P. 1975. Biochemical coevolution between plants and their insect herbivores. In *Co-Evolution of Animals and Plants*, eds. L.E. Gilbert, and P.H. Raven, pp. 3–19. Austin: University of Texas Press.

Felton, G.W., Donato, K.K., Broadway, R.M., and Duffey, S.S. 1992. Impact of oxidized plant phenolics on the nutritional quality of dietary protein to a noctuid herbivore, *Spodoptera exigua*. *J. Insect Physiol.* 38:277–285.

Fowler, S.V., and MacGarvin, M. 1986. The effects of leaf damage on the performance of insect herbivores on birch, *Betula pubescens*. *J. Anim. Ecol.* 55:565–573.

Habermann, M. 2000. The larch casebearer and its host tree II. Changes in needle physiology of the infested trees. *For. Ecol. Manage.* 136:23–34.

Hakola, H., Rinne, J., and Laurila, T. 1998. The hydrocarbon emission rates of tea-leafed willow (*Salix phylicifolia*), silver birch (*Betula pendula*) and European aspen (*Populus tremula*). *Atmosph. Environ.* 32:1825–1833.

Hamilton, J.G., Zangerl, A.R., DeLucia, E.H., and Berenbaum, M.R. 2001. The carbon-nutrient balance hypothesis: its rise and fall. *Ecol. Lett.* 4:86–95.

Hanhimäki, S., and Senn, J. 1992. Sources of variation in rapidly inducible responses to leaf damage in the mountain birch-insect herbivore system. *Oecologia* 91:318–331.

Hanhimäki, S., Senn, J., and Haukioja, E. 1995. The convergence in growth of foliage-chewing insect species on individual mountain birch trees. *J. Anim. Ecol.* 64:543–552.

Hartley, S.E. and Lawton, J.H. 1990. Damage-induced changes in birch foliage: mechanisms and effects on insect herbivores. In *Population Dynamics of Forest Insects*, eds. A.D. Watt, S.R. Leather, M.D. Hunter, and N.A.C. Kidd, pp. 147–155. Andover, Hampshire: Intercept. Haukioja, E. 1982. Inducible defences of white birch to a geometrid defoliator, *Epirrita autumnata*. In *Insect–Plant Relationships*, eds. J.H. Visser, and A.K. Minks, pp. 199–203. Wageningen, The Netherlands: Pudoc.

Haukioja, E. 1990. Induction of defenses in trees. *Annu. Rev. Entomol.* 36:25–42.

Haukioja, E. 2003. Putting the insect into the birch-insect interaction. *Oecologia* 136:161–168.

Haukioja, E., and Niemelä, P. 1977. Retarded growth of a geometrid larva after mechanical damage to leaves of its host tree. *Ann. Zool. Fennici* 14:48–52.

Haukioja, E., Ossipov, V., Koricheva, J., Honkanen, T., Larsson, S., and Lempa, K. 1998. Biosynthetic origin of carbon-based secondary compounds: cause of variable responses of woody plants to fertilization? *Chemoecology* 8:133–139.

Haukioja, E., Ossipov, V., and Lempa, K. 2002. Interactive effects of leaf maturation and phenolics on consumption and growth of a geometrid moth. *Entomol. Exp. Appl.* 104:125–136.

Haukioja, E., Ruohomäki, K., Senn, J., Suomela, J., and Walls, M. 1990. Consequences of herbivory in the mountain birch (*Betula pubescens* ssp *tortuosa*): importance of the functional organization of the tree. *Oecologia* 82:238–247.

Havill, N.P., and Raffa, K.F. 1999. Effects of elicitation treatment and genotypic variation on induced resistance in *Populus*: impacts on gypsy moth (Lepidoptera: Lymantriidae) development and feeding behavior. *Oecologia* 120:295–303.

Heath, M.C. and Boller, T. 2002. Biotic interactions levels of complexity in plant interactions with herbivores, pathogens and mutualists. *Curr. Opin. Plant Biol.* 5:277–278.

Henriksson, J. 2001. Differential shading of branches or whole trees: survival, growth, and reproduction. *Oecologia* 126:482–486.

Henriksson, J., Haukioja, E., Ossipov, V., Ossipova, S., Sillanpää, S., Kapari, L., and Pihlaja, K. 2003. Effects of host shading on consumption and growth of the geometrid Epirrita autumnata: interactive roles of water, primary and secondary compounds. *Oikos* 103:3–16.

Herms, D.A., and Mattson,W.J. 1992. The dilemma of plants: to grow or defend. *Quart. Rev. Biol.* 67:283–335.

Hillis, W.E. 1975. Ethylene and extraneous material formation in woody tissues. *Phytochemistry* 14:2559–2562.

Hjältén, J., and Price, P.W. 1996. The effects of pruning on willow growth and sawfly population densities. *Oikos* 77:549–555.

Honkanen, T., and Haukioja, E. 1998. Intraplant regulation and plant/herbivore interactions. *Écoscience* 5:470–479.

Hunter, M.D., and Schultz, J.C. 1993. Induced plant defenses breached? Phytochemical induction protects a herbivore from disease. *Oecologia* 94:195–203.

Hunter, M.D., and West, C. 1990. Variation in the effects of spring defoliation on the late season phytophagous insects of *Quercus robur*. In *Population Dynamics of Forest Insects*, eds. A.D. Watt, S.R. Leather, M.D. Hunter, and N.A.C. Kidd, pp. 123–135. Andover, Hampshire: Intercept.

Julkunen-Tiitto, R., Rousi, M., Bryant, J.P., Sorsa, S., Keinänen, M., and Sikanen, H. 1996. Chemical diversity of several Betulaceae species: comparison of phenolics and terpenoids in northern birch stems. *Trees* 11:16–22.

Kaitaniemi, P., Neuvonen, S., and Nyyssönen, T. 1999. Effects of cumulative defoliations on growth, reproduction, and insect resistance in the mountain birch. *Ecology* 80:524–532.

Kaitaniemi, P., and Ruohomäki, K. 2001. Sources of variability in plant resistance against insects: free caterpillars show strongest effects. *Oikos* 95:461–470.

Kaitaniemi, P., Ruohomäki, K., and Haukioja, E. 1997. Consumption of apical buds as a mechanism of alleviating host plant resistance for *Epirrita autumnata* larvae. *Oikos* 78:230–238.

Kaitaniemi, P., Ruohomäki, K., Ossipov, V., Haukioja, E., and Pihlaja, K. 1998. Delayed induced changes in the biochemical composition of host plant leaves during an insect outbreak. *Oecologia* 116:182–190.

Karban, R., and Kuć, J. 1999. Induced resistance against pathogens and herbivores: an overview. In *Induced Plant Defenses Against Pathogens and Herbivores*, eds. A.A. Agrawal, S. Tuzun, and E. Bent, pp. 1–16. St. Paul, MN: APS Press.

Kause, A., Haukioja, E., and Hanhimäki, S. 1999a. Phenotypic plasticity in foraging behavior of sawfly larvae. *Ecology* 80:1230–1241.

Kause, A., Ossipov, V., Haukioja, E., Lempa, K., and Hanhimäki, S. 1999b. Multiplicity of biochemical factors determining quality of growing birch leaves. *Oecologia* 120:102–112.

Kause, A., Saloniemi, I., Morin, J.P., Haukioja, E., Hanhimäki, S., and Ruohomäki, K. 2001. Seasonally varying diet quality and the quantitative genetics of development time and body size in birch feeding insects.*Evolution* 55:1992–2001.

Keinänen, M., Julkunen-Tiitto, R., Mutikainen, P., Walls, M., Ovaska, J., and Vapaavuori, E. 1999. Trade-offs in secondary metabolism: effects of fertilization, defoliation, and genotype on birch leaf phenolics. *Ecology* 80:1970–1986.

Koricheva, J. 2002. Meta-analysis of sources of variation in fitness costs of plant antiherbivore defenses. *Ecology* 83:176–190.

Koricheva, J., Larsson, S., Haukioja, E., and Keinänen, M. 1998. Regulation of woody plant secondary metabolism by resource availability: hypothesis testing by means of meta-analysis. *Oikos* 83:212–226.

Krause, S.C., and Raffa, K.F. 1992. Comparison of insect, fungal, and mechanically induced defoliation of larch—effects on plant productivity and subsequent host susceptibility. *Oecologia* 90:411–416.

Landsberg, J. 1990. Dieback of rural eucalypts: response of foliar dietary quality and herbivory to defoliation. *Aust. J. Ecol.* 15:89–96.

Långström, B., Tenow, O., Ericsson, A., Hellqvist, C., and Larsson, S. 1990. Effects of shoot pruning on stem growth, needle biomass, and dynamics of carbohydrates and nitrogen in Scots pine as related to season and tree age. *Can. J. For. Res.* 20:514–523.

Larsson, S. 2002. Resistance in trees to insects—an overview of mechanisms and interactions. In *Mechanisms and Deployment of Resistance in Trees to Insects*, eds. M.R. Wagner, K.M. Clancy, F. Lieutier, and T.D. Paine, pp. 1–29. Dordrecht, The Netherlands: Kluwer Academic Press.

Lawton, J.H. 1983. Plant architecture and the diversity of phytophagous insects. *Annu. Rev. Entomol.* 28:23–39.

Lerdau, M., and Coley, P.D. 2002. Benefits of the carbon-nutrient balance hypothesis. *Oikos* 98:533–535.

Litvak, M.E., and Monson, R.K. 1998. Patterns of induced and constitutive monoterpene production in conifer needles in relation to insect herbivory. *Oecologia* 114:531–540.

Lyytikäinen, P. 1994. Effects of natural and artificial defoliations on sawfly performance and foliar chemistry of Scots pine saplings. *Ann. Zool. Fennici* 31:307–318.

Martel, J., Hanhimäki, S., Kause, A., and Haukioja, E. 2001. Diversity of birch sawfly responses to seasonally atypical diets across a seasonal gradient of larval feeding periods. *Entomol. Exp. Appl.* 100:301–309.

Martin, D., Tholl, D., Gershenzon, J., and Bohlmann, J. 2002. Methyl jasmonate induces traumatic resin ducts, terpenoid resin biosynthesis, and terpenoid accumulation in developing xylem of Norway spruce stems. *Plant Physiol.* 129:1003–1018.

Mattson, W.J., and Scriber, J.M. 1987. Nutritional ecology of insect folivores of woody plants: nitrogen, water, fiber, and mineral considerations. In *Nutritional Ecology of Insects, Mites, Spiders and Related Invertebrates*, eds. F. Slansky, Jr., and J.G. Rodriquez, pp. 105–146. New York: John Wiley.

Monson, R.K., and Holland, E.A. 2001. Biospheric trace gas fluxes and their control over troposhperic chemistry. *Annu. Rev. Ecol. Syst.* 32:547–576.

Moran, N., and Hamilton, W.D. 1980. Low nutritive quality as defence against herbivores. *J. Theor. Biol.* 86:247–254.

Nerg, A., Kainulainen, P., Vuorinen, M., Hanso, M., Holopainen, J.K., and Kurkela, T. 1994. Seasonal and geographical variation of terpenes, resin acids and total phenolics in nursery grown seedlings of Scots pine (*Pinus sylvestris* L.). *New Phytol.* 128:703–713.

Neuvonen, S., and Haukioja, E. 1991. The effects of inducible resistance in host foliage on birch-feeding herbivores. In *Phytochemical Induction by Herbivores*, eds. D.W. Tallamy, and M.J. Raupp, pp. 277–291. New York: John Wiley & Sons.

Nitao, J.K., Hamilton, J.G., Zangerl, A.R., DeLucia, E.H., and Berenbaum, M. 2002. CNB: requiescat in pace? *Oikos* 98:539–545.

Nurmi, K., Ossipov, V., Haukioja, E., and Pihlaja, K. 1996. Variation of total phenolic content and low-molecular-weight phenolics in foliage of the mountain birch trees (*Betula pubescens* ssp. *tortuosa*). *J. Chem. Ecol.* 22:2033–2050.

Omlin, F.X. 1980. Aspekte zum wechselweisen Beziehungsgefüge Lärchenwickler (*Zeiraphera diniana* GN.) und Lärche (*Larix decidua* MILL.). 1. Teil: Einflussnahme des Lärchenwicklers auf die Lärche. *Mitt. Schweiz. Entomol. Ges.* 53:379–399.

Ossipov, V., Haukioja, E., Hanhimäki, S., and Pihlaja, K. 2001. Phenolic and phenolic-related factors as determinants of suitability of mountain birch leaves to an herbivorous insect. *Biochem. Syst. Ecol.* 29:223–240.

Raffa, K.F. 1991. Induced defensive reactions in conifer-bark beetle systems. In *Phytochemical Induction by Herbivores*, eds. D.W. Tallamy, and M.J. Raupp, pp. 245–276. New York: John Wiley.

Raffa, K.F., Krause, S.C., and Reich, P.B. 1998. Long-term effects of defoliation on red pine suitability to insects feeding on diverse plant tissues. *Ecology* 79:2352–2364.

Raubenheimer, D., and Simpson, S.J. 1994. The analysis of nutrient budgets. *Funct. Ecol.* 8:783–791.

Rhoades, D.F. 1985. Offensive-defensive interactions between herbivores and plants: their relevance to herbivore population dynamics and community theory. *Am. Nat.* 125:205–238.

Riipi, M., Ossipov, V., Lempa, K., Haukioja, E., Koricheva, J., Ossipova, S., and Pihlaja, K. 2002. Seasonal changes in birch leaf chemistry: are there trade-offs between leaf growth and accumulation of phenolics? *Oecologia* 130:380–390.

Rossiter, M.C., Schultz, J.C., and Baldwin, I.T. 1988. Relationships among defoliation, red oak phenolics, and gypsy moth growth and reproduction. *Ecology* 69:267–277.

Ruohomäki, K., Chapin, F.S., III, Haukioja, E., Neuvonen, S., and Suomela, J. 1996. Delayed inducible resistance in mountain birch in response to fertilization and shade. *Ecology* 77:2302–2311.

Ruohomäki, K., Hanhimäki, S., Haukioja, E., Iso-Iivari, L., Neuvonen, S., Niemelä, P., and Suomela, J. 1992. Variability in the efficacy of delayed inducible resistance in mountain birch. *Entomol. Exp. Appl.* 62:107–115.

Ruohomäki, K., Virtanen, T., Kaitaniemi, P., and Tammaru, T. 1997. Old mountain birches at high altitudes are prone to outbreaks of *Epirrita autumnata* (Lepidoptera: Geometridae). *Environ. Entomol.* 26:1096–1104.

Ruusila, V., Morin, J.P., van Oik, T., Saloniemi, I., Ossipov, V., and Haukioja, E. 2005. A short-lived herbivore on a long lived host: tree resistance to herbivory depends of leaf age. *Oikos* 108:99–104.

Salminen, J.-P., and Lempa, K. 2002. Effects of hydrolysable tannins on an herbivorous insect: fate of individual tannins in insect digestive tract. *Chemoecology* 12:203–211.

Salminen, J.P., Ossipov, V., Haukioja, E., and Pihlaja, K. 2001. Seasonal variation in the content of hydrolysable tannins in leaves of *Betula pubescens*. *Phytochemistry* 57:15–22.

Schopf, R. 1986. The effect of secondary needle compounds on the development of phytophagous insects. *For. Ecol. Manage.* 15:55–64.

Schultz, J.C., and Baldwin, I.T. 1982. Oak leaf quality declines in response to defoliation by gypsy moth larvae. *Science* 217:149–151.

Scriber, J.M., and Feeny, P. 1979. Growth of herbivorous caterpillars in relation to feeding specialization and to the growth form of their food plants. *Ecology* 60:829–850.

Schwenke, W. 1968. Neue Hinweise auf eine Abhängigkeit der Vermehrung blatt- und nadelfressender Forstinsekten vom Zuckergehalt ihrer Nahrung. *Z. Ang. Ent.* 61:365–369.

Seldal, T., Dybwad, E., Andersen, K., and Högstedt, G. 1994. Wound-induced proteinase inhibitors in grey alder (*Alnus incana*): a defence mechanism against attacking insects. *Oikos* 71:239–245.

Simpson, S.J., and Raubenheimer, D. 2001. The geometric analysis of nutrient-allelochemical interactions: a case study using locusts. *Ecology* 82:422–439.

Slansky, F., Jr. 1993. Nutritional ecology: the fundamental quest for nutrients. In *Caterpillars. Ecological and Evolutionary Constraints on Foraging*, eds. N.E. Stamp, and T.M. Casey, pp. 29–91. New York: Chapman and Hall.

Slansky, F., Jr., and Wheeler, G.S. 1992. Caterpillars' compensatory feeding response to diluted nutrients leads to toxic allelochemical dose. *Entomol. Exp. Appl.* 65:171–186.

Smits, A., Larsson, S., and Hopkins, R. 2001. Reduced realised fecundity in the pine looper *Bupalus piniarius* caused by host plant defoliation. *Ecol. Entomol.* 26:417–424.

Suomela, J., Kaitaniemi, P., and Nilson, A. 1995a. Systematic within-tree variation in mountain birch leaf quality for a geometrid, *Epirrita autumnata*. *Ecol. Entomol.* 20:283–292.

Suomela, J., and Nilson, A. 1994. Within-tree and among-tree variation in growth of *Epirrita autumnata* on mountain birch leaves. *Ecol. Entomol.* 19:45–56.

Suomela, J., Ossipov, V., and Haukioja, E. 1995b. Variation among and within mountain birch trees in foliage phenols, carbohydrates and amino acids, and in growth of *Epirrita autumnata* larvae. *J. Chem. Ecol.* 21:1421–1446.

Trewhella, K.E., Leather, S.R., and Day, K.R. 1997. Insect induced resistance in lodgepole pine: Effects on two pine feeding insects. *J. Appl. Entomol.* 121:129–136.

Tscharntke, T., Thiessen, S., Dolch, R., and Boland, W. 2001. Herbivory, induced resistance, and interplant signal transfer in *Alnus glutinosa*. *Biochem. Syst. Ecol.* 29:1025–1047.

Tuomi, J., Niemelä, P., Haukioja, E., Sirén, S., and Neuvonen, S. 1984. Nutrient stress: an explanation for plant anti-herbivore responses to defoliation. *Oecologia* 61:208–210.

Tuomi, J., Niemelä, P., Rousi, M., Sirén, S., and Vuorisalo, T. 1988. Induced accumulation of foliage phenols in mountain birch: branch response to defoliation? *Am. Nat.* 132:602–608.

Valentine, H.T., Wallner, W.E., and Wargo, P.M. 1983. Nutritional changes in host foliage during and after defoliation, and their relation to the weight of gypsy moth pupae. *Oecologia* 57:298–302.

Valkama, E., Salminen, J.P., Koricheva, J., and Pihlaja, K. 2003. Comparative analysis of leaf trichome structure and composition of epicuticular flavonoids in Finnish birch species. *Ann. Bot.* 91:643–655.

van Zandt, P.A., and Mopper,S. 1998. A meta-analysis of adaptive deme foundation in phytophagous insect populations. *Am. Nat.* 152:595–604.

Villalba, J.J., Provenza, F.D., and Bryant, J.P. 2002. Consequences of the interaction between nutrients and plant secondary metabolites on herbivore selectivity: benefits or detriments for plants? *Oikos* 97:282–292.

Wagner, M.R., and Evans, P.D. 1985. Defoliation increases nutritional quality and allelochemics of pine seedlings. *Oecologia* 67:235–237.

Watt, A.D. 1990. The consequences of natural, stress-induced and damage-induced differences in tree foliage on the population dynamics of the pine beauty moth. In *Population Dynamics of Forest Insects*, eds. A.D. Watt, S.R. Leather, M.D. Hunter, and N.A.C. Kidd, pp. 157–168. Andover, Hampshire: Intercept.

White, T.C.R. 1974. A hypothesis to explain outbreaks of looper caterpillars, with special reference to populations of *Selidosema suavis* in a plantation of *Pinus radiata* in New Zealand. *Oecologia* 16:279–301.

Williams, K.S., and Myers, J.H. 1984. Previous herbivore attack of red alder may improve food quality for fall webworm larvae. *Oecologia* 63:166–170.

Wratten, S.D., Edwards, P.J., and Dunn, I. 1984. Wound-induced changes in the palatability of *Betula pubescens* and *Betula pendula*. *Oecologia* 61:372–375.

Zou, J., and Cates, R.G. 1994. Role of Douglas fir (*Pseudotsuga menziesii*) carbohydrates in resistance to budworm (*Choristoneura occidentalis*). *J. Chem. Ecol.* 20:395–405.

Zou, J.P., and Cates, R.G. 1995. Foliage constitutients of Douglas-fir (*Pseudotsuga menziesii* (Mirb)) Franco (Pinaceae)—their seasonal variation and potential role in Douglas-fir resistance adn silvicultural management. *J. Chem. Ecol.* 21:387–402.

13

The Role of Terpene Synthases in the Direct and Indirect Defense of Conifers Against Insect Herbivory and Fungal Pathogens

Dezene P.W. Huber and Jörg Bohlmann

13.1 Introduction

Insect herbivory is a strong selective pressure that plants must cope with on several different levels. First, plants are non-mobile, so they cannot escape their highly mobile antagonists in space. Instead they must defend themselves in the place in which they are growing. Second, many plants live for several years, and some trees live for hundreds or even thousands of years, whereas many insect herbivores have generation times ranging from one or two years to as short as months or even days. Thus, in order to maximize lifetime reproductive output many plants must withstand numerous successive waves of newly eclosed and rapidly evolving insect herbivores during the course of their lives. Third, most plants are plagued not with merely one, but with numerous different species of pest insects, each insect species potentially attacking a different portion or developmental stage of a given plant (Bernays and Chapman, 1994) and each insect species likely differing in phenotypic traits related to resistance or tolerance to plant defenses (Thompson, 1994).

In order to survive these pressures, and indeed in order to thrive to an extent that allows successful reproduction, plants have developed complex defenses that are both physical and chemical in nature (Karban and Baldwin, 1997). Of the classes of compounds involved in chemical defense, and often in physical defense as well, terpenoids are the most numerous. Humans are quite familiar with terpenoids as compounds that determine many characteristic aromas and fragrances of fruits, vegetables, and spices, and terpenoids are commonly used as natural aromas and fragrances in foodstuffs, perfumes, and household and industrial cleaning agents, and air fresheners. Because of their inherent biological activities in other organisms, terpenoids of plant origin have been widely used in traditional and modern medicine (e.g., the diterpenoid anticancer drug taxol, or the sesquiterpenoid

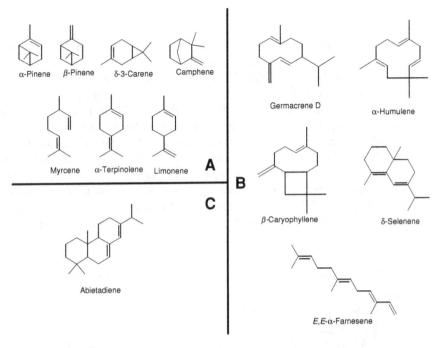

FIGURE 13.1. A very small representation of the diversity of known plant terpenoids. Monoterpenoids (A), sesquiterpenoids (B), and a diterpenoid (C).

antimalaria drug artemisinine). The outcomes of many interactions of plants with insects or pathogens can depend a great deal on the quantity and identity of terpenoids produced by plants and on the particulars of the genetic and physiological regulation and of the biosynthetic capabilities of terpenoid synthases (TPS), the enzymes in plants that catalyze the final reactions that result in the often diverse array (Figure 13.1) of terpenoids (Bohlmann et al., 1998b; Cane, 1999; Wise and Croteau, 1999; MacMillan and Beale, 1999; Davis and Croteau, 2000; Greenhagen and Chappell, 2001).

Multigenic terpenoid-based defense in plants shows complexity at the level of organization and evolution of large *TPS* gene families (Bohlmann et al., 1998b; Aubourg et al., 2002), at the level of *TPS* gene expression, at the level of unique enzyme mechanisms of multi-product and single-product TPS enzymes (Cane, 1999; Wise and Croteau, 1999; MacMillan and Beale, 1999; Greenhagen and Chappell, 2001), and at the level of changes in plant histology related to delivery of terpenoid defense compounds in specialized anatomical structures (Martin et al., 2002). In addition, the effects of plant terpenoid defenses on target organisms, potential pathogens and herbivores, are complex and varied. In this review we intend to describe and unravel a portion of this complexity and demonstrate the effectiveness of multigenic, terpenoid-based defenses of plants.

13.2 Terpenoid Biosynthesis

The precursors of terpenoid biosynthesis are derived from the condensation of two or more five-carbon units of isopentenyl diphosphate (IPP) and the IPP isomer, dimethylallyl diphosphate (DMAPP) (Figure 13.2). IPP and DMAPP are each synthesized via one of the two separate pathways and each of the pathways seems to be localized to a separate compartment in the cell (Figure 13.2). Enzymes in the mevalonate pathway are found in the cytoplasm whereas enzymes in the 2-C-methyl-D-erythritol 4-phosphate pathway (MEP) are found in the plastids (Lange et al., 2001). Condensation of one, two, or three units of IPP with DMAPP by the catalytic action of geranyl diphosphate synthase, farnesyl diphosphate synthase, or geranyl geranyl diphosphate synthase results in the formation of geranyl diphosphate (GPP), farnesyl diphosphate (FPP), and geranyl geranyl diphosphate (GGPP), respectively (Koyama and Ogura, 1999). Monoterpene synthases

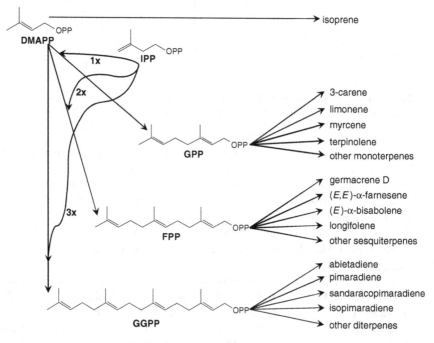

FIGURE 13.2. The substrates used by various TPS are derived from the enzyme-catalyzed condensation of dimethylallyl diphosphate (DMAPP) with one, two, or three isopentenyl diphosphate (IPP) units. The resulting geranyl diphosphate (GPP), farnesyl diphosphate (FPP), and geranyl geranyl diphosphate (GGPP) are the substrates for monoterpene synthases, sesquiterpene synthases, and diterpene synthases, respectively. Via the action of many different TPS, each substrate may be converted into any of a large number of possible products. DMAPP may also be converted, via isoprene synthase, into isoprene.

(mono-TPS) utilize GPP as a substrate to synthesize monoterpenoids (C_{10}), sesquiterpene synthases (sesqui-TPS) utilize FPP as a substrate to synthesize sesquiterpenoids (C_{15}), and diterpene synthases (di-TPS) utilize GGPP as a substrate to synthesize diterpenoids (C_{20}) (Bohlmann et al., 1998b). Like the reactions in the mevalonate and MEP pathways, the reactions catalyzed by prenyltransferases, mono-TPS, di-TPS, and sesqui-TPS are also localized to the plastids or the cytoplasm (McGarvey and Croteau, 1995; Croteau et al., 2000). The terpenoids synthesized by the TPS are often further modified by oxidations or additions of functional groups.

13.3 Terpene Synthases

Most TPS catalyze the formation of one or more major products and several additional minor products. The formation of multiple products from a single substrate in the active site of TPS is based on reaction mechanisms that involve formation of enzyme-bound, highly reactive carbocation intermediates and subsequent intramolecular electrophilic attacks at double bonds, rearrangements and ultimately quenching by proton elimination, addition of water, or readdition of the diphosphate group (Davis and Croteau, 2000) (Figure 13.3). However, several highly specialized single-product TPS have also been described. As an example for a typical multiproduct TPS, one sesqui-TPS isolated from grand fir (*Abies grandis*) produces mainly δ-selinene and (*E,E*)-germacrene B, and also produces 32 other sesquiterpenes (Steele et al., 1998a). The same study describes a second sesqui-TPS that produces mainly γ-humulene and sibirene and also produces 50 other sesquiterpenes. Typical single product enzymes are the mono-TPS, myrcene synthase, from grand fir (Bohlmann et al., 1997) and *epi*-aristolochene synthase, a sesqui-TPS of phytoalexin formation, in species of tobacco (Facchini and Chappell 1992; Bohlmann et al., 2002). There are many other examples of both single product TPS (e.g., Wildung and Croteau, 1996; Bennett et al., 2002) and multiple product TPS (e.g., Bohlmann et al., 1999; Bohlmann et al., 2000; Lu et al., 2002; Lücker et al., 2002; Fäldt et al., 2003a,b; Hölscher et al., 2003). In many cases plausible reaction schemes have been proposed for single-product and multiple-product enzymes (e.g., Steele et al., 1998a; Bohlmann et al., 1999, 2000; Hölscher et al., 2003; Fäldt et al., 2003b). While one TPS may synthesize many products, separate enantiomers of the same terpenoid product seem to be produced primarily by separate enzymes (Schmidt et al., 1998; Phillips et al., 2003). Thus, plants generally seem to have separate enzymes, and multiple corresponding genes, to produce various enantiomers of specific terpenoids (Phillips et al., 2003).

In addition to the fact that single TPS can synthesize multiple terpenoid products from simple precursors, plants contain multiple *TPS* genes each with differing function (Figure 13.4). To date, 11 *TPS* cDNAs have been functionally characterized from *A. grandis*, one di-TPS, three sesqui-TPS and seven mono-TPS (Stofer Vogel et al., 1996; Bohlmann et al., 1997, 1998a; 1999). A similar number of

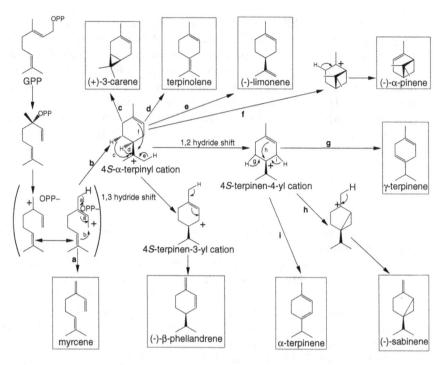

FIGURE 13.3. A plausible mechanism, proposed by Fäldt et al. (2003b), for monoterpene biosynthesis by (+)-3-carene synthase from Norway spruce, *Picea abies*. While (+)-3-carene, and to a lesser extent terpinolene, are the major products, a number of other monoterpene products are formed via a highly reactive carbocation intermediate and subsequent electrophilic attacks and rearrangement of bonds. GPP is geranyl diphosphate.

functionally diverse *TPS* genes have recently been characterized from two other conifer systems, Norway spruce (*Picea abies*) and Sitka spruce (*P. sitchensis*) (Fäldt et al., 2003b; Byun McKay et al., 2003; Martin and Bohlmann, unpublished observations). The comparative, functional analysis of large families of *TPS* genes in these closely related conifer species has shed some new light on the evolution of structure–function relatedness in gymnosperm *TPS*, some of which appear to represent archetype plant *TPS* genes (Trapp and Croteau, 2001b). Analysis of the angiosperm *Arabidopsis thaliana* genome has revealed 30 *TPS* genes (*AtTPS*) that are likely to be functional and involved in the synthesis of terpenoids used in plant defense or in other activities, such as endogenous plant signaling and communication, along with two additional *AtTPS* genes involved in biosynthesis of gibberellic acid (Aubourg et al., 2002). However, much work needs to be completed on this newly emerging system for the study of terpenoids, as only 5 of the 30 *AtTPS* have been functionally characterized to date (Bohlmann et al., 2000; Chen et al., 2003; Fäldt et al., 2003a) and the role of terpenoids in *A. thaliana* biology is very poorly understood (Fäldt et al., 2003a; Chen et al., 2003; van Poecke et al., 2001). If *Abies*, *Picea*, and *Arabidopsis* are good indicators, it seems

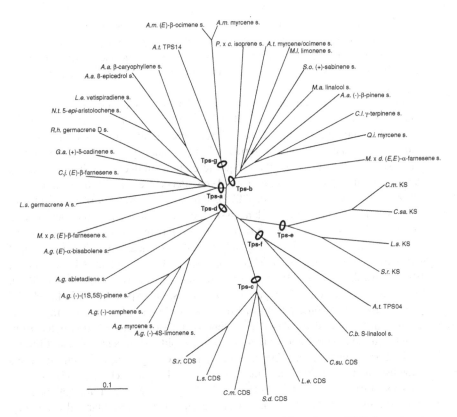

FIGURE 13.4. Phylogenetic tree of representative terpenoid synthases (TPS) from the seven TPS subfamilies. Most angiosperm sesquiterpene synthases and all angiosperm diterpene synthases are in *TPS-a*; angiosperm monoterpene synthases with the RRX$_8$W motif are in *TPS-b*; copalyl diphosphate synthases (CDS) are in *TPS-c*; gymnosperm TPS are in *TPS-d*; kaurene synthases (KS) are in *TPS-e*, a linalool synthase from *Clarkia breweri* and the *Arabidopsis* AtTPS04 are in *TPS-f*; the known *Antirrhinum majus* monoterpene synthases (two of three shown in this tree) and *Arabidopsis* AtTPS14 are in *TPS-g*. Taxon abbreviations are as follows: *A.a.* is *Artemisia annua*; *A.g.* is *Abies grandis*; *A.m.* is *Anitrrhinum majus*; *A.t.* is *Arabidopsis thaliana*; *C.b.* is *Clarkia breweri*; *C.j.* is *Citrus junos*; *C.l.* is *Citrus limon*; *C.m.* is *Cucurbita maxima*; *C.sa.* is *Cucumis sativus*; *C.su.* is *Croton sublyratus*; *G.a.* is *Gossypium arboretum*; *L.e.* is *Lycopersicon esculentum*; *L.s.* is *Lactuca sativa*; *M.a.* is *Mentha aquatica*; *M. × d.* is *Malus* domestica; *M.l.* is *Mentha longifolia*; *M. × p.* is *Mentha × piperita*; *N.t.* is *Nicotiana tabacum*; *P. × c.* is *Populus × canescens*; *Q.i.* is *Quercus ilex*; *R.h.* is *Rosa hybrida*; *S.d.* is *Scoparia dulcis*; *S.o.* is *Salvia officinalis*; *S.r.* is *Stevia rebaudiana*. Sequences were obtained from GenBank and from other published material. The alignment was completed with ClustalX and the tree was visualized with TreeView.

that plants in general contain large and functionally diverse *TPS* gene families. Much current effort directed at functional *TPS* gene discovery in other species will soon serve to test this hypothesis. This effort is being accelerated both by data mining from plant genome sequencing projects and from full-length cDNA projects.

The multiple *TPS* genes present in a given plant's genome may be constitutively expressed, or expression may instead be induced by exogenous factors. Analyses of the terpenoid components of constitutive, defense related conifer resin has revealed the presence of numerous terpenoids, presumably produced via constitutive expression of *TPS* genes and then sequestered in resin ducts or other specialized anatomical structures (Erdtman et al., 1968; Persson et al., 1996; Sjödin et al., 1996; Fäldt, 2000; Martin et al., 2002). Because the products of constitutively expressed TPS are present in plant tissue on a continual basis, they are among the first defenses that an herbivore or an invading pathogen will encounter. Constitutive terpenoids contribute to plant chemical characteristics that act to repel herbivores, modify their behavior, or otherwise defend the plant (Chararas et al., 1982; Edwards et al., 1993; Trapp and Croteau, 2001a; Wallin and Raffa, 2000). In contrast to constitutively expressed *TPS* genes, the expression of induced *TPS* genes is increased some period of time after a plant's first contact with an antagonist. Evidence of increased *TPS* gene expression can be extrapolated from changes in the quantity and/or identity of terpenoid components of resin (Tomlin et al., 2000; Martin et al., 2002, 2003b), from changes in TPS activity levels (Lewinsohn et al., 1991; Funk and Croteau, 1994; Martin et al., 2002), or from changes in transcript levels of individual *TPS* genes (Steele et al., 1998b; Fäldt et al., 2003b; Byun McKay et al., 2003) after insect attack, mechanical wounding, or treatment with chemical elicitors.

In summary, terpenoid-based defenses in conifers are complex and considerable potential exists for substantial multiple gene-based fine-tuning of the defense by the plant in response to different threats. First, terpenoid biosynthesis proceeds by at least two separate pathways, each involving a number of enzymes that are the products of multiple genes. Regulation of transcription of any of these genes, or regulation of activity of the resulting enzymes, plays a considerable role in the final amount and variety of terpenoids produced (Fäldt et al., 2003b; Martin et al., 2002, 2003b). Because monoterpenoid and diterpenoid biosynthesis are, at least in part, separated from sesquiterpenoid biosynthesis, a responding tree can upregulate the pathway leading to one group of compounds independent of the other (Martin et al., 2002, 2003b). Second, the TPS enzymes that catalyze the final committed step toward various terpenoid products often produce more than one major terpenoid product and usually produce a number of minor products as well (e.g., Steele et al., 1998a; Fäldt et al., 2003b). Third, the few plants that have been intensively studied to date, e.g., *A. grandis* and *Picea abies*, have a large number of functionally diverse *TPS* genes. And fourth, differential regulation of numerous *TPS* genes, and of the genes in the preceding biosynthetic pathways, can result in changes in the composition of the final mixture of terpenoids and the effect can be quite complex due to the one-substrate/multiple-product nature of TPS.

13.4 Spruce Defense: A Case Study of Multigenic Terpenoid Defense

Among the gymnosperms, the terpenoid-based responses of spruce to herbivory by the white pine weevil (*Pissodes strobi*) and the regulation of terpenoid biosynthesis accumulation and emission has been one of the most comprehensively studied to date (Alfaro et al., 2002). *P. strobi* feeding on white spruce (*Picea glauca*) results in the formation of an induced resin that is richer in monoterpenes and diterpene acids than the constitutive resin of the same plant (Tomlin et al., 2000). In several species of spruce, feeding and oviposition by *P. strobi* (Alfaro 1995; Byun McKay et al., 2003), inoculation with insect-associated fungi (Franceschi et al., 2000; Nagy et al., 2000), or simulated weevil damage (Tomlin et al., 1998; Franceschi et al., 2002; Martin et al., 2002; Byun McKay et al., 2003) also induce dramatic histological changes. In particular, herbivory causes the formation of conspicuous resin ducts in the developing xylem. The induced resin ducts are important for the delivery of induced resin components to the site of herbivore damage.

Recent work has shown that the octadecanoid molecule, methyl jasmonate (MeJA), plays a substantial role in the induction of *TPS* genes and in defense-related changes in terpenoid biochemistry and anatomy in spruce (Franceschi et al., 2002; Fäldt et al., 2003b; Martin et al., 2002, 2003b). Topical application of MeJA on spruce trees results in changes in *mono-TPS* and *di-TPS* gene expression and TPS enzyme activities, resin chemistry, terpenoid volatile release, and stem histology that are very similar to changes following arthropod herbivore feeding (Franceschi et al., 2002; Fäldt et al., 2003b; Martin et al., 2002, 2003b). Application of MeJA on *P. abies* causes the formation of traumatic resin ducts in the developing xylem of spruce stems (Franceschi et al., 2002; Martin et al., 2002), similar to those seen after weevil feeding on *P. glauca* (Alfaro, 1995) and *P. sitchensis* (Byun McKay et al., 2003). MeJA also increases monoterpene and diterpene content in the wood (xylem) but not substantially in the bark (phloem), while sesquiterpenes are not affected, indicative of differential effects on *TPS* gene expression in different tissues. Total mono-TPS and di-TPS activities, but not sesqui-TPS activities, increase in the wood after MeJA treatment while GPP synthase, FPP synthase, and GGPP synthase activities only increase transiently in the wood and not at all in the bark (Martin et al., 2002), likely as a result of differential gene expression for enzymes involved in the middle and late stages of terpenoid biosynthesis. Indeed, the terpenoid content of the MeJA-induced *P. abies* developing xylem can be correlated to increased TPS transcript accumulation (Fäldt et al., 2003b). Northern analyses of MeJA-treated trees using mono-TPS- and di-TPS-based probes reveal a strong increase in transcript level at two days following MeJA treatment for mono-TPS with elevated transcript levels remaining up to at least 16 days after treatment. An evaluation of di-TPS transcript levels reveals a parallel increase in accumulation (Fäldt et al., 2003b). Similarly, increased levels of weevil-induced monoterpenes in *P. sitchensis* stems can be correlated with increased levels of mono-TPS transcripts after herbivory or mechanical wounding (Byun McKay et al., 2003).

MeJA treatment also affects the terpenoid content of the foliage of Norway spruce, though the effect is not as pronounced as in the stem (Martin et al., 2002, 2003b). While the terpenoid content of the foliage does not increase strongly, the emission of terpenoids from the foliage increases dramatically after MeJA treatment and volatile terpenoids are released on a diurnal cycle, with most of the volatiles being released during the light period (Martin et al., 2003b). MeJA treatment causes a dramatic shift in the composition of terpenoid volatiles released from the foliage. While sapling trees emitted mainly monoterpene hydrocarbons on a constitutive basis, the MeJA-induced tissues released large amounts of oxygenated monoterpenes, e.g., linalool, and sesquiterpenes, e.g., farnesene and bisabolene, in addition to *de novo* emission of the nonterpenoid methyl salicylate (Martin et al., 2003b). In a previous study, Litvak and Monson (1998) also found that while mono-TPS activity increases in a number of conifer species in response to foliage damage caused by herbivores or artificial damage, total monoterpene content of the needles does not increase. Measurements of monoterpenoid volatile release from MeJA-induced needles of Norway spruce indicates that an increase in mono-TPS and sesqui-TPS activity is necessary to replace the terpenoids lost to volatilization (Martin et al., 2003b). Indeed, MeJA-induced needles showed large increases in the level of enzyme activities for linalool synthase and for sesquiterpene synthases (Martin et al., 2003b). This subset of TPS enzymes and their genes were not induced in stem tissues by the same treatment (Martin et al., 2002), supporting further the concept of differential constitutive and induced expression of members of a large and functionally diverse *TPS* gene family in different organs or tissues of spruce. Quantitative and qualitative changes of monoterpenoids, sesquiterpenoids, and methyl salicylate released as volatiles from herbivore-damaged or MeJA-treated foliage may serve to play a role in attracting predators of spruce herbivores in a form of indirect defense (Martin et al., 2003b) similar to analogous situations in angiosperm systems (Paré and Tumlinson, 1997a,b; Kessler and Baldwin, 2001). The maintenance of terpenoid levels in the needles may also be important in defense against defoliators, as seen in other systems (Chen et al., 2002).

In summary, the regulation of terpene biosynthesis resulting in the formation of induced, defensive terpenoids in spruce seems to be controlled by a large family of *TPS* genes that are differentially regulated in bark, xylem, and foliage of spruce, potentially involving multiple signaling pathways. The octadecanoid signaling compound, MeJA, induces *TPS* gene expression (Fäldt et al., 2003b) and terpenoid biosynthesis in spruce such that terpenoid build-up and histological changes affecting delivery of the compounds is different in different plant tissues (Martin et al., 2002, 2003b), at different times of the day (Martin et al., 2003b), and in accordance with different herbivores which attack different portions of the tree such as stem boring weevils and bark beetles or foliage-feeding Lepidoptera and Hymenoptera. The complexity of terpenoid-based defense thus becomes evident as a combined result of the effects of multiple TPS genes, some capable of producing more than one product, acting on substrates produced from two biosynthetic pathways and potentially regulated by multiple signaling pathways, including the octadecanoid pathway cascades that are often initiated by herbivore threats (Farmer and Ryan,

1992; Reymond and Farmer, 1998). In addition, dramatic histological events of resin duct formation are induced by herbivory, likely the result of the expression of many interacting genes controlling complex changes in xylem differentiation (Franceschi et al., 2002; Martin et al., 2002). These anatomical changes serve to augment the effects of the overall terpenoid-based defense by allowing efficient, rapid, and targeted delivery of the terpenoids to herbivore-attacked tissues. The many genes involved in the process allow the final terpenoid composition of the resulting resins or volatile mixtures to be differentially fine-tuned in spruce at different times of the day, at different points following herbivore attack, in different plant tissues, and in response to different herbivores.

13.5 Multigenic Terpenoid Defenses and Communications in Conifer—Bark Beetle—Pathogen Interactions

The multigenic plant terpenoid defenses are at least part of the evolutionary success of long-lived conifer species. It is plausible that the interactions of conifers with insects, e.g., bark beetles and insect-vectored fungal pathogens, have shaped the evolution not only of large and functionally diverse *TPS* gene families in conifers, but have also influenced many other aspects of terpenoid-based communications and defenses in these plant—insect—fungus complexes (Seybold et al., 2000). Bark beetles are among the most important mortality agents of conifers and other trees in nature (Wood, 1963). Over the past four decades much work has been completed on the biological interactions between these insects and their host trees, particularly in regard to host-derived terpenoids and bark beetle terpenoid pheromones (Seybold et al., 2000). The wealth of biological information gleaned in previous research about the relationship between these insects and their host serves to create questions that make this system very amenable to new molecular and genomics approaches.

While some bark beetles and related ambrosia beetles utilize fallen trees, many species rely at least partially, and often solely, on standing, living trees. Because conifers possess potent, chemical-based defenses and because these defenses are likely to be strong in viable trees, bark beetles have evolved to overcome the defenses. The primary strategy utilized by many tree-killing species is mass aggregation (Rudinsky, 1962). Upon encountering a suitable host tree and commencing feeding, bark beetles synthesize (Seybold and Tittiger, 2003) and release powerful aggregation pheromones (Borden, 1985). Many of these insect pheromones are terpenoids that are derived at least in part from oxidation of host tree monoterpene hydrocarbons (Byers, 1995; Seybold et al., 2000), but apparently also by *de novo* terpene synthesis in insects or associated fungi (Martin et al., 2003a; Seybold and Tittiger, 2003). Conspecific beetles respond to the released aggregation pheromone and to host-derived volatiles (Byers, 1995) released from damaged tree tissue by the boring activities of insects already on the tree, and arrive to attack the tree in large numbers. The swarms of attacking insects burrow through the outer bark and mine the phloem tissue in their feeding and egg-laying activities. The combined effects of large numbers of beetles and their progeny mining the phloem and the growth of symbiotic fungi vectored by the attacking beetles (Whitney, 1982; Paine et al., 1997; Solheim and Krokene, 1998a,b) serve to quickly weaken the defense response of the attacked tree and often ultimately serve to kill it.

Although the beetles' strategy of mass attack is often successful, healthy trees are able to withstand large aggregations. Raffa and Berryman (1983) found that lodgepole pine (*Pinus contorta*) is able to withstand the attacks up to approximately 40 pairs of beetles/m^2 of bark surface. Up to that point sufficient resources are present in the tree to surround both the nascent galleries and their builders with toxic terpenoids or to actually fill the galleries with terpenoid-based resin. Beetles trapped in the resin flow are not able to signal conspecifics to aggregate and thus the attack ceases. It is evident that the efficacy of the terpenoid defense of the tree has its foundation in the speed and quality of the response of the tree to the attacking insects and the growing insect-associated fungi. If the response is too slow, or if the compounds involved are ineffective, the insects could attract enough conspecifics to ultimately doom the tree. In turn, the speed and quality of the response has a multipart foundation based on the defense gene expression, terpenoid biochemistry, anatomy, histology, and physiological condition of the tree.

Successful defense by a host tree against aggregating bark beetles is dependent upon resin terpenoid characteristics. In *P. contorta* the physical characteristics of the resin do not seem to play as great a role in resistance as does the actual overall quantity of resin produced (Raffa and Berryman, 1982). However, in some other *Pinus* spp., the physical characteristics of the resin, such as viscosity and time to crystallization, seem to be the most important factors in resisting bark beetle attack (Hodges et al., 1979). The physical characteristics of the resin are, to a large extent, governed by the relative ratios of monoterpenoids, sesquiterpenoids, and diterpenoids present. Because there are two biosynthetic pathways involved in the formation of representatives of these classes (Lange et al., 2001), genetic control of the final resin composition is possible at that level. In addition, the regulated increased expression of representative genes of one class of *TPS* genes over another, or different timing of expression, produces resin with particular characteristics. The timing of expression of different classes of *TPS* genes and enzymes seems to play a role in the formation of *A. grandis* traumatic resin after wounding (Steele et al., 1998b) and in the formation of traumatic resin and volatile emissions in *P. abies* (Martin et al., 2002, 2003b; Miller and Bohlmann, unpublished observations). Much more work in this area is required in order to fully understand the multigenic regulation of resin formation involved in terpenoid biosynthesis and defense.

A number of studies have shown that bark beetles and their associated fungi respond to particular terpenoids or particular blends of terpenoids. For instance, a number of terpenoids have a toxic effect on the fir engraver (*Scolytus ventralis*) feeding on *A. grandis*. Limonene, in particular, was inhibitory to the beetle-vectored fungus (Raffa et al., 1985). Limonene on its own is toxic to the spruce beetle (*Dendroctonus rufipennis*) but the eastern larch beetle (*D. simplex*) was susceptible to limonene as well as myrcene and β-phellandrene (Werner, 1995). Another study focused on the Douglas-fir beetle (*Dendroctonus pseudotsugae*), which prefers to infest Douglas fir (*Pseudotsuga menziesii*) to its alternate host, western larch (*Larix occidentalis*). The difference in preference may be partly due to the different levels of the monoterpenoid 3-carene in larch and Douglas fir (Reed et al., 1986). A study on the postlanding behavior of *Ips pini* has shown that the behavior of the insect is mainly regulated by the overall level of terpenoids, but that a number of behavioral effects could be particularly attributed to α-pinene and β-pinene concentrations (Wallin and Raffa, 2000).

Interaction of hosts with different bark beetle species also plays a role in the evolution and variation of resin biochemistry in conifers. Ponderosa pine (*P. ponderosa*) with historical exposure to the mountain pine beetle (*Dendroctonus ponderosae*) contained a dissimilar diversity of resin compounds than populations of the same species of pines historically affected by the western pine beetle (*D. brevicomis* (Sturgeon and Mitton, 1986). Such a variation in resin terpenoid composition is expected because different beetles react differently to individual terpenoids. It should be noted that while certain compounds may sometimes be important in the preferences of bark beetles to certain conifers, a large statistical study designed to test the preference of nine species of bark beetles to ten conifers indicated that the overall mixture of terpenoids in the essential oil of a tree is a more important factor influencing beetle choice than the presence or absence of any group of particular terpenoids (Chararas et al., 1982). In support of this view are various studies which provide evidence that mechanical wounding (Ruel et al., 1998) or inoculations with fungi associated with bark beetles (Lieutier *et al.* 1989; Viiri et al., 2001) result in overall increased resin flow and an increase in the content of a number of terpenoids simultaneously. Although much of the published research into bark beetle responses to terpenoids has focused on one or a few terpenoids at a time, the results that point to complex chemical responses of trees to herbivory highlight the need for concerted attempts to unravel the joint effects of conifer species-specific terpenoid blends on their specialist bark beetles.

13.6 Conclusion

The results described in this brief review provide fertile ground for future analyses of the presence or absence or sequence variation of certain *TPS* genes in particular species or within populations of conifers and of overall expression patterns of *TPS* and other genes in the pathways affecting terpenoid biosynthesis and ultimately resin composition, volatile emission, and defense against insects and pathogens in conifers. The multigene *TPS* families in conifers can be explored for development of SNPs and other markers potentially associated with variation in defenses and resistance to insect pests or pathogens in conifers. Such work ought to be carried out in terms of what is already known about tree biochemical, anatomical, and physiological responses, terpenoid toxicity, insect behavior before and during colonization of host trees, and differential host preferences. An approach that takes into account the multigenic nature of terpenoid defenses in conifers will doubtless shed more light on the coevolutionary relationship between bark beetles and their host trees, and will provide new tools for monitoring and managing these economically and ecologically important conifer—insect interactions.

Acknowledgments

Due to the limited size of this review, in comparison with the burgeoning scope of the literature in this field, we apologize to all of the researchers and authors whose work we were unable to cite.

Note Added in Proof

Since we submitted this review for publication, several articles have been published which are not included; therefore, we would like to draw the attention of the reader to the following three recent papers relevant to this chapter:

(1) Martin, D.M., Fäldt, J., and Bohlmann, J. 2004. Functional characterization of nine Norway spruce TPS genes and evolution of gymnosperm terpene synthases of the TPS-d subfamily. *Plant Physiol.* 135:1908–1927.
(2) Miller, B., Madilao, L.L., Ralph, S., and Bohlmann, J. 2005. Insect-induced conifer defense. White pine weevil and methyl jasmonate induce traumatic resinosis, de novo formed volatile emissions, and accumulation of terpenoid synthase and putative octadecanoid pathway transcripts in Sitka spruce. *Plant Physiol.* 137:369–382.
(3) Hudgins, J.W. and Franceschi, V.R. 2004. Methyl jasmonate-induced ethylene production is responsible for conifer phloem defense responses and reprogramming of stem cambial zone for traumatic resin duct formation. *Plant Physiol.* 135:2134–2149.

References

Alfaro, R.I. 1995. An induced defense reaction in white spruce to attack by the white pine weevil, *Pissodes strobi. Can. J. Forensic Res.* 25:1725–1730.

Alfaro, R.I., Borden, J.H., King, J.N., Tomlin, E.S., McIntosh, R.L., and Bohlmann, J. 2002. Mechanisms of resistance in conifers against shoot infesting insects. The case of the white pine weevil *Pissodes strobi* (Peck) (Coleoptera: Curculionidae). In *Mechanisms and Deployment of Resistance to Insects*, eds. M.R. Wagner, K.M. Clancy, F. Lieutier, and T.D. Paine, pp. 105–130. Boston, MA: Kluwer Academic Publishers.

Aubourg, S., Lecharny, A., and Bohlmann, J. 2002. Genomic analysis of the terpenoid synthase (*AtTPS*) gene family of *Arabidopsis thaliana. Mol. Genet. Genomics* 267:730–745.

Bennett, M.H., Mansfield, J.W., Lewis, M.J., and Beale, M.H. 2002. Cloning and expression of sesquiterpene synthase genes from lettuce (*Lactuca sativa* L.). *Phytochemistry* 60:255–261.

Bernays, E.A., and Chapman, R.F. 1994. *Host-Plant Selection by Phytophagous Insects.* Boston, MA: Kluwer Academic Publishers.

Bohlmann, J., Steele, C.L., and Croteau, R. 1997. Monoterpene synthases from grand fir (*Abies grandis*): cDNA isolation, characterization, and functional expression of myrcene synthase, (−)-(4S)-limonene synthase, and (−)-(1S,5S)-pinene synthase. *J. Biol. Chem.* 272:21784–27792.

Bohlmann, J., Crock, J., Jetter, R., and Croteau, R. 1998a. Terpenoid-based defenses in conifers: cDNA cloning, characterization, and functional expression of wound-inducible (*E*)-alpha-bisabolene synthase from grand fir (*Abies grandis*). *Proc. Natl. Acad. Sci.* 95:6756–6761.

Bohlmann, J., Gilbert Meyer-Gauen, G., and Croteau, R. 1998b. Plant terpenoid synthases: molecular biology and phylogenetic analysis. *Proc. Natl. Acad. Sci.* 95:4126–4133.

Bohlmann, J., Phillips, M., Ramachandiran, V., Katoh, S., and Croteau, R. 1999. cDNA cloning, characterization, and functional expression of four new monoterpene synthase members of the *Tpsd* gene family from grand fir (*Abies grandis*). *Arch. Biochem. Biophys.* 368:232–243.

Bohlmann, J., Martin, D., Oldham, N.J., and Gershenzon, J. 2000. Terpenoid secondary metabolism in *Arabidopsis thaliana*: cDNA cloning, characterization, and functional expression of a myrcene/(*E*)-β-ocimene synthase. *Arch. Biochem. Biophys.* 375:261–269.

Bohlmann, J., Stauber, E.J., Krock B., Oldham, N.J., Gershenzon J., and Baldwin, I.T. 2002. Gene expression of 5-epi-aristolochene synthase and formation of capsidiol in roots of *Nicotiana attenuata* and *N. sylvestris. Phytochemistry* 60:109–116.

Borden J.H. 1985. Aggregation pheromones. In *Comprehensive Insect Physiology, Biochemistry and Pharmacology*, eds. G.A. Kerkut, and L.I. Gilbert, Vol. 9, pp. 257–285. Oxford, UK: Pergamon Press.

Byers, J.A. 1995. Host-tree chemistry affecting colonization in bark beetles. In *Chemical Ecology of Insects 2*, eds. R.T. Cardé, and W.J. Bell, pp. 154–213. New York: Chapman and Hall.

Byun McKay, S.A., Hunter, W.L., Godard, K.A., Wang, S.X., Martin, D.M., Bohlmann, J., and Plant, A.L. 2003. Insect attack and wounding induce traumatic resin duct development and gene expression of (−)-pinene synthase in Sitka spruce. *Plant Physiol.* 133:368–378.

Cane, D.E. 1999. Sesquiterpene biosynthesis: cyclization mechanisms. In *Comprehensive Natural Products Chemistry*, ed. D.E. Cane, Vol. 2, pp.155–200. Oxford: Pergamon.

Chararas, C., Revolon, C., Feinberg, M., and Ducauze, C. 1982. Preference of certain Scolytidae for different conifers: a statistical approach. *J. Chem. Ecol.* 8:1093–1110.

Chen, Z., Kolb, T.E., and Clancy, K.M. 2002. The role of monoterpenes in resistance of Douglas fir to western spruce budworm defoliation. *J. Chem. Ecol.* 28:897–920.

Chen, F., Tholl, D., D'Auria, J.C., Farooq, A., Pichersky, E., and Gershenzon, J. 2003. Biosynthesis and emission of terpenoid volatiles from Arabidopsis flowers. *Plant Cell* 15:481–494.

Croteau, R., Kutchan, T.M., and Lewis, N.G. 2000. Natural products (secondary metabolites). In *Biochemistry and Molecular Biology of Plants*, eds. B.B. Buchanan, W. Gruissem, and R.L. Jones, pp. 1250–1318. Rockville, MD: American Society of Plant Physiologists.

Davis, E.M., and Croteau, R. 2000. Cyclization enzymes in the biosynthesis of monterpenes, sesquiterpenes, and diterpenes. *Topics Curr. Chem.* 209:53–95.

Edwards, P.B., Wanjura, W.J., and Brown, W.V. 1993. Selective herbivory by Christmas beetles in response to intraspecific variation in *Eucalyptus* terpenoids. *Oecologia* 95: 551–557.

Erdtman, H., Kimland, B., Norin, T., and Daniels P.J.L. 1968. The constituents of the "pocket resin" from Douglas fir *Pseudotsuga menziesii* (Mirb.) Franco. *Acta Chem. Scand.* 22:938–942.

Facchini, P.J., and Chappell, J. 1992. A gene family for an elicitor-induced sesquiterpene cyclase in tobacco. *Proc. Natl. Acad. Sci. USA* 89:11088–11092.

Fäldt, J. 2000. Volatile constituents in conifers and conifer-related wood-decaying fungi. Ph.D. thesis. Royal Institute of Technology, Department of Chemistry, Organic Chemistry. Stockholm, Sweden.

Fäldt, J., Arimura, G-i., Gershenzon, J., Takabayashi, J., and Bohlmann, J. 2003a. Functional identification of *AtTPS03* as (E)-β-ocimene synthase: a new monoterpene synthase catalyzing jasmonate- and wound-induced volatile formation in *Arabidopsis thaliana*. *Planta* 216:745–751.

Fäldt, J., Martin, D., Miller, B., Rawat, S., and Bohlmann, J. 2003b. Traumatic resin defense in Norway spruce (*Picea abies*): methyl jasmonate-induced terpene synthase gene expression, and cDNA cloning and functional characterization of (+)-3-carene synthase. *Plant Mol. Biol.* 51:119–133.

Farmer, E.E., and Ryan C.A.1992. Octadecanoid precursors of jasmonic acid activate the synthesis of wound inducible proteinase inhibitors. *Plant Cell* 4:129–134.

Franceschi, V.R., Krokene, P., Krekling, T., and Christiansen, E. 2000. Phloem parenchyma cells are involved in local and distant defense responses to fungal inoculation or bark beetle attack in Norway spruce (Pinaceae). *Am. J. Bot.* 87:314–326.

Franceschi, V.R., Krekling, T., and Christiansen, E. 2002. Application of methyl jasmonate on *Picea abies* (Pinaceae) stems induces defense-related responses in phloem and xylem. *Am. J. Bot.* 89:578–586.

Funk, C., and Croteau, R. 1994. Diterpenoid resin acid biosynthesis in conifers - characterization of two cytochrome P450-dependent monooxygenases and an aldehyde dehydrogenase involved in abietic acid biosynthesis. *Arch. Biochem. Biophys.* 308:258–266.

Greenhagen, B., and Chappell, J. 2001. Molecular scaffolds for chemical wizardry: Learning nature's rules for terpene cyclases. *Proc. Natl. Acad. Sci. USA* 98:13479–13481.

Hodges, J.D., Elam, W.W., Watson, W.F., and Nebeker, T.E. 1979. Oleoresin characteristics and susceptibility of 4 southern pines to southern pine beetle, *Dendroctonus frontalis* (Coleoptera: Scolytidae), attacks. *Can. Entomol.* 111:889–896.

Hölscher, D.J., Williams, D.C., Wildung, M.R., and Croteau, R. 2003. A cDNA clone for 3-carene synthase from *Salvia stenophylla*. *Phytochemistry* 62:1081–1086.

Karban, R., and Baldwin, I.T. 1997. *Induced Responses to Herbivory*. Chicago: University of Chicago Press.

Kessler, A., and Baldwin, I.T. 2001. Defensive function of herbivore-induced plant volatile emissions in nature. *Science* 291:2141–2144.

Koyama, T., and Ogura K. 1999. Isopentenyl diphosphate isomerase and prenyl transferases. In *Comprehensive Natural Products Chemistry*, ed. D.E. Cane, Vol. 2, pp. 69–96. Oxford: Pergamon. Lange, B.M., Ketchum, R.E.B., and Croteau, R.B. 2001. Isoprenoid biosynthesis. Metabolite profiling of peppermint oil gland secretory cells and application to herbicide target analysis. *Plant Physiol.* 127:305–314.

Lewinsohn, E., Gijzen, M., and Croteau, R. 1991. Defense-mechanisms of conifers - differences in constitutive and wound-induced monoterpene biosynthesis among species. *Plant Physiol.* 96:44–49.

Lieutier, F., Cheniclet, C., and Garcia, J. 1989. Comparison of the defense reactions of *Pinus pinaster* and *Pinus sylvestris* to attacks by two bark beetles (Coleoptera: Scolytidae) and their associated fungi. *Environ. Entomol.* 18:228–234.

Litvak, M.E., and Monson, R.K. 1998. Patterns of induced and constitutive monoterpene production in conifer needles in relation to insect herbivory. *Oecologia* 114:531–540.

Lu, S., Xu, R., Jia, J-W., Pang, J., Matsuda, S.P.T., and Chen, X-Y. 2002. Cloning and functional characterization of a β-pinene synthase from *Aretemisia annua* that shows a circadian pattern of expression. *Plant Physiol.* 130:477–486.

Lücker, J., El Tamer, M.K., Schwab, W., Verstappen, F.W.A., van der Plas, L.H.W., Bouwmeester, H.J., and Verhoeven, H.A. 2002. Monoterpene biosynthesis in lemon (*Citrus limon*): cDNA isolation and functional analysis of four monoterpene synthases. *Eur. J. Biochem.* 269:3160–3171.

MacMillan, J., and Beale, M.H. 1999. Diterpene biosynthesis. In *Comprehensive Natural Products Chemistry*, ed. D.E. Cane, Vol. 2, pp. 217–244. Oxford: Pergamon.

Martin, D., Tholl, D., Gershenzon, J., and Bohlmann, J. 2002. Methyl jasmonate induces traumatic resin ducts, terpenoid resin biosynthesis, and terpenoid accumulation in developing xylem of Norway spruce stems. *Plant Physiol.* 129:1003–1018.

Martin, D., Bohlmann, J., Gershenzon, J., Francke, W., and Seybold, S.J. 2003a. A novel sex-specific and inducible monoterpene synthase activity associated with a pine bark beetle, the pine engraver, *Ips pini. Naturwissen.* 90:173–179.

Martin, D.M., Gershenzon, J., and Bohlmann, J. 2003b. Induction of volatile terpene biosynthesis and diurnal emission by methyl jasmonate in foliage of Norway spruce. *Plant Physiol.* 132:1586–1599.

McGarvey, D.J., and Croteau, R. 1995. Terpenoid metabolism. *Plant Cell* 7:1015–1026.

Nagy, N.E., Franceschi, V.R., Solheim, H., Krekling, T., and Christiansen, E. 2000. Wound-induced traumatic resin duct development in stems of Norway spruce (Pinaceae): Anatomy and cytochemical traits. *Am. J. Bot.* 87:302–313.

Paine, T.D., Raffa, K.F., and Harrington, T.C. 1997. Interactions among scolytid bark beetles, their associated fungi, and live host conifers. *Ann. Rev. Entomol.* 42:179–206.

Paré, P.W., and Tumlinson, J.H. 1997a. De novo biosynthesis of volatiles induced by insect herbivory in cotton plants. *Plant Physiol.* 114:1161–1167.

Paré, P.W., and Tumlinson, J.H. 1997b. Induced synthesis of plant volatiles. *Nature* 385:30–31.

Persson, M., Sjödin, K., Borg-Karlson, A.K., Norin, T., and Ekberg, I. 1996. Relative amounts and enantiomeric compositions of monoterpene hydrocarbons in xylem and needles of *Picea abies. Phytochemistry* 42:1289–1297.

Phillips, M.A., Wildung, M.R., Williams, D.C., Hyatt, D.C., and Croteau, R. 2003. cDNA isolation, functional expression, and characterization of (−)-α-pinene synthase and (+)-α-pinene synthase from loblolly pine (*Pinus taeda*): stereocontrol in pinene biosynthesis. *Arch. Biochem. Biophys.* 411: 267–276.

Raffa, K.F., and Berryman, A.A. 1982. Physiological differences between lodgepole pines, *Pinus contorta* var. *latifolia*, resistant and susceptible to the mountain pine beetle, *Dendroctonus ponderosae*, and associated microorganisms. *Environ. Entomol.* 11:486–492.

Raffa, K.F., and Berryman, A.A. 1983. The role of host plant resistance in the colonization behavior and ecology of bark beetles (Coleoptera: Scolytidae). *Ecol. Monogr.* 53:27–50.

Raffa, K.F., Berryman, A.A., Simasko, J., Teal, W., and Wong, B.I. 1985. Effects of grand fir, *Abis grandis*, monoterpenes on the fir engraver, *Scolytus ventralis* (Coleoptera: Scolytidae), and its symbiotic fungus. *Environ. Entomol.* 14:552–556.

Reed, A.N., Hanover, J.W., and Furniss, M.M. 1986. Douglas-fir and western larch: chemical and physical properties in relation to Douglas-fir beetle attack. *Tree Physiol.* 1:277–287.

Reymond, P., and Farmer, E.E. 1998. Jasmonate and salicylate as global signals for plant defense. *Curr. Opin. Plant Biol.* 1:404–411.

Rudinsky, J.A. 1962. Ecology of Scolytidae. *Ann. Rev. Entomol.* 7:327–348.

Ruel, J.J., Ayres, M.P., and Lorio, P.L., Jr. 1998. Loblolly pine responds to mechanical wounding with increased resin flow. *Can. J. For. Res.* 28:596–602.

Schmidt, C.O., Bouwmeester, H.J., de Kraker, J-W., and König, W.A. 1998. Biosynthesis of (+)- and (−)-germacrene D in *Solidago canadensis*: isolation of two enantioselective germacrene D synthases. *Angew. Chem. Int. Ed.* 37:1400–1402.

Seybold, S.J., Bohlmann, J., and Raffa, K.F. 2000. Biosynthesis of coniferophagous bark beetle pheromones and conifer isoprenoids: Evolutionary perspective and synthesis. *Can. Entomol.* 132:697–753.

Seybold, S.J., and Tittiger, C. 2003. Biochemistry and molecular biology of *de novo* isoprenoid pheromone production in the Scolytidae. *Annu. Rev. Entomol.* 48:425–453.

Sjödin, K., Persson, M., Borg-Karlson, A-K., and Torbjörn H. 1996. Enantiomeric compositions of monoterpene hydrocarbons in different tissues of four individuals of *Pinus sylvestris*. *Phytochemistry* 41:439–445.

Solheim, H., and Krokene, P. 1998a. Growth and virulence of *Ceratocystis rufipenni* and three blue-stain fungi isolated from the Douglas-fir beetle. *Can. J. Bot.* 76:1763–1769.

Solheim, H., and Krokene, P. 1998b. Growth and virulence of mountain pine beetle associate blue-stain fungi, *Ophiostoma clavigerum* and *Ophiostoma montium*. *Can. J. Bot.* 76:561–566.

Steele, C.L., Crock, J., Bohlmann, J., and Croteau, R. 1998a. Sesquiterpene synthases from grand fir (*Abies grandis*): comparison of constitutive and wound-induced activities, and cDNA isolation, characterization, and bacterial expression of δ-selinene synthase and γ-humulene synthase. *J. Biol. Chem.* 273:2078–2089.

Steele, C.L., Katoh, S., Bohlmann, J., and Croteau, R. 1998b. Regulation of oleoresinosis in grand fir (*Abies grandis*): differential transcriptional control of monoterpenes, sesquiterpene, and diterpene synthase genes in response to wounding. *Plant Physiol.* 116:1497–1504.

Stofer Vogel, B., Wildung, M.R., Vogel, G., and Croteau, R. 1996. Abietadiene synthase from Grand Fir (*Abies grandis*). cDNA isolation, characterization, and bacterial expression of a bifunctional diterpene cylase involved in resin acid biosynthesis. *J. Biol. Chem.* 271:23262–23268.

Sturgeon, K.B., and Mitton, J.B. 1986. Biochemical diversity of ponderosa pine, *Pinus ponderosa*, and predation by bark beetles (Coleoptera: Scolytidae). *J. Econ. Entomol.* 79:1064–1068.

Thompson, J.N. 1994. *The Coevolutionary Process.* Chicago: University of Chicago Press.

Trapp, S., and Croteau, R. 2001a. Defensive resin biosynthesis in conifers. *Ann. Rev. Plant Physiol. Plant Mol. Biol.* 52:689–724.

Trapp, S.C., and Croteau, R.B. 2001b. Genomic organization of plant terpene synthases and molecular evolutionary implications. *Genetics* 158:811–832.

Tomlin, E.S., Alfaro, R.I., Borden, J.H., and Fangliang, H. 1998. Histological response of resistant and susceptible white spruce to simulated white pine weevil damage. *Tree Physiol.* 18:21–28.

Tomlin, E.S., Antonejevic, E., Alfaro, R.I., and Borden J.H. 2000. Changes in volatile terpene and diterpene resin acid composition of resistant and susceptible white spruce leaders exposed to simulated white pine weevil damage. *Tree Physiol.* 20:1087–1095.

van Poecke, R.M.P., Posthumus, M.A., and Dicke, M. 2001. Herbivore-induced volatile production by *Arabidopsis thaliana* leads to attraction of the parasitoid *Cotesia rubecula*: chemical, behavioral, and gene-expression analysis. *J. Chem. Ecol.* 27:1911–1928.

Viiri, H., Annila, E., Kitunen, V., and Niemaelä, P. 2001. Induced responses in stilbenes and terpenes in fertilized Norway spruce after inoculation with blue-stain fungus, *Ceratocystis polonica*. *Trees* 15:112–122.

Wallin, K.F., and Raffa, K.F. 2000. Influences of host chemical and internal physiology on the multiple steps of postlanding host acceptance behavior of *Ips pini* (Coleoptera: Scolytidae). *Environ. Entomol.* 29:442–453.

Werner, R.A. 1995. Toxicity and repellency of 4-allylanisole and monoterpenes from white spruce and tamarack to the spruce beetle and eastern larch beetle (Coleoptera: Scolytidae). *Environ. Entomol.* 24:372–379.

Whitney, H.S. 1982. Relationships between bark beetles and symbiotic organisms. In *Bark Beetles in North American Conifers: A System for the Study of Evolutionary Biology*, eds. J.B. Mitton, and K.B. Sturgeon, pp. 183–211. Austin, TX: University of Texas Press.

Wildung, M.R., and Croteau, R. 1996. A cDNA clone for taxadiene synthase, the diterpene cyclase that catalyzes the committed step of taxol biosynthesis. *J. Biol. Chem.* 271:9201–9204.

Wise, M.L., and Croteau, R. 1999. Monoterpene biosynthesis. In *Comprehensive Natural Products Chemistry*, ed. D.E. Cane, Vol. 2, pp. 97–154. Oxford: Pergamon.

Wood, S.L. 1963. A revision of the bark beetle genus *Dendroctonus* Erichson (Coleoptera: Scolytidae). *Great Basin Nat.* 23:1–117.

14

Mechanisms Involved in Plant Resistance to Nematodes

ERIN BAKKER, ROBERT DEES, JAAP BAKKER, AND ASKA GOVERSE

14.1 Introduction

Coevolution between nematodes and plants gave rise to obligatory plant parasites. Though representing a small minority of species within the phylum Nematoda, the plant parasitic nematodes receive ample attention, mainly because they are a major yield-limiting factor in crops such as potato, beet, cereals, soybean, and tomato.

When obligatory plant parasitic nematodes are considered, a number of different feeding strategies can be discriminated. Ectoparasitic nematodes like *Trichodorus* spp. feed on rhizodermis cells of different plants, whereas endoparasitic nematodes like root-knot (*Meloidogyne* spp.) and cyst nematodes (*Heterodera* spp. and *Globodera* spp.) establish a permanent feeding site inside the plant root. Once a feeding site is induced, the nematode fully depends on it for growth and development. This durable strategy is successful: endoparasites invade a wide range of plant species and in agriculture they reside among the most persistent and harmful nematodes.

The endoparasitic nematodes can be controlled by crop rotation, chemical soil disinfestation, and resistant cultivars. However, the broad host-range of the root-knot nematodes and the extreme persistence of the cyst nematodes in the absence of a host compromise the usefulness of crop rotation. Cysts can survive in the soil for over 10 years. The chemical control methods of sedentary plant parasitic nematodes involve very unspecific and extremely harmful pesticides, and due to increasing concern regarding environmental issues and stricter governmental regulations, this method has practically been abandoned. Therefore, resistant cultivars are becoming increasingly important and the scientific studies on the underlying genes and resistance mechanisms are of great interest.

14.2 Nematode Resistance

In this chapter, we focus on the mechanisms involved in plant resistance to root-knot and cyst nematodes. The intimate interaction between endoparasitic nematodes and their host plants has been extensively studied over the years to allow the development of durable crop protection strategies (Williamson and Hussey, 1996; Jung et al., 1998; Williamson, 1998, 1999; Bakker, 2002). The recent mapping and cloning of *R* genes that confer resistance to endoparasitic nematodes is a major contribution to the elucidation of the genetic and molecular mechanisms that underlie nematode resistance. In Table 14.1, an overview is given of the nematode resistance genes that are currently mapped in or isolated from agronomically important crops like potato, tomato, pepper, beet, soybean, and cereals or their wild relatives.

14.2.1 Nematode Resistance in Potato

Several years ago, geneticists struggled with the complex genetics of the heterozygous tetraploid potato *Solanum tuberosum* and its wild tetraploid relatives. However, once these species were converted to diploid plants by pseudogamy (Hermsen and Verdenius, 1973) potato genetics had made a major leap forward. Today, many nematode *R* genes have been mapped in potato (reviewed by Gebhardt and Valkonen, 2001). The genes mostly confer resistance to potato cyst nematode species, but one locus (*Rmc1*) confers resistance to the root-knot nematode species *M. chitwoodi* (Rouppe van der Voort et al., 1999).

The first nematode resistance gene isolated from potato is *Gpa2* (Van der Vossen et al., 2000). This single dominant *R* gene confers resistance to the potato cyst nematode *Globodera pallida* and was identified from a complex locus in potato by map-based cloning. *Gpa2* belongs to the class of genes that contain a leucine zipper domain, a nucleotide binding site, and a leucine rich repeat domain (LZ-NBS-LRR). The LZ and NBS are thought to be involved in signal transduction, while the LRR domain is most likely responsible for *R* gene specificity (Ellis et al., 2000). *Gpa2* is highly similar to *Rx1* (93% nucleotide identity and 88% amino acid homology), which confers resistance to a completely unrelated pathogen viz. potato virus X.

Besides *Gpa2* and *Rx1*, the complex locus on chromosome XII contains two other resistance gene homologues (RGHs): *RGH1*, a putative *R* gene with unknown specificity, and RGH3, probably a pseudogene with a truncated effector domain (Van der Vossen et al., 2000). This *R* gene locus originates from the wild potato subspecies *S. tuberosum* ssp. *andigena* and has been studied extensively as an introgression segment in the diploid *S. tuberosum* ssp. *tuberosum* clone SH83-92-488 (SH).

Study of the three *S. tuberosum* ssp. *tuberosum* haplotypes from two diploid potato clones (SH and RH89-039-16 (RH)) revealed nine additional *Gpa2/Rx1* homologues (Bakker et al., 2003). The RGHs were identified with a specific primer pair based on conserved motifs of the LRR domain from *Gpa2* and *Rx1*. Sequence analysis of the RGHs revealed that they are highly similar to *Gpa2* and *Rx1* with sequence identities ranging from 93% to 95%. A modified AFLP method was used to facilitate the genetic mapping of the RGHs. They are all located in the *Gpa2/Rx1* cluster on chromosome XII.

Several other quantitative or single dominant resistance loci have been located in potato. The quantitative trait locus (QTL) *Gro1.4* is located on chromosome III

TABLE 14.1. Overview of mapped loci that confer resistance to root-knot and cyst nematodes in major crops.

Gene	Pathogen	SD or QTL	Origin	cloned	chromosome	Reference
Gro1.4	G. rostochiensis	QTL	Solanum spegazzinii	no	III	Kreike et al. (1996)
Gpa4	G. pallida	QTL	S. tuberosum ssp. tuberosum	no	IV	Bradshaw et al. (1998)
Gpa	G. pallida	QTL	S. spegazzinii	no	V	Kreike et al. (1996)
Gpa5	G. pallida	QTL	Solanum spp.[a]	no	V	Rouppe van der Voort et al. (2000)
Grp1	G. pal/G. ros	QTL	Solanum spp.[a]	no	V	Rouppe van der Voort et al. (1998)
H1	G. rostochiensis	SD	S. tuberosum ssp. andigena	no	V	Gebhardt et al. (1993); Pineda et al. (1993)
GroVI	G. rostochiensis	SD	S. vernei	no	V	Jacobs et al. (1996)
Gro1	G. rostochiensis	SD	S. spegazzinii	no	VII	Ballvora et al. (1995); Barone et al. (1990)
Gpa6	G. pallida	QTL	Solanum spp.[3)	no	IX	Rouppe van der Voort et al. (2000)
Gro1.2	G. rostochiensis	QTL	S. spegazzinii	no	X	Kreike et al. (1996)
Gro1.3	G. rostochiensis	QTL	S. spegazzinii	no	XI	Kreike et al. (1996)
Gpa2	G. pallida	SD	S. tuberosum ssp. andigena	YES	XII	Rouppe van der Voort et al. (1997); Van der Vossen et al. (2000)
Rmc1	M. chitwoodi	SD	S. bulbocastanum	no	XI	Brown et al. (1996)
Hero	G. rostochiensis	SD	Lycopersicon pimpernellifolium	YES	11	Ernst et al. (2002); Ganal et al. (1995)
Mi-1	M. spp.	SD	L. peruvianum	YES	6	Aarts et al. (1998); Milligan et al. (1998); Vos et al. (1998)
Mi-3	M. inc/M. jav	SD	L. peruvianum	no	12	Yaghoobi et al. (1995)
Mi-LA2157	M. incognita/ M. javanica	SD	L. peruvianum	no	6	Veremis et al. (1999)
Me3	M. spp.	SD	Capsicum annuum	no	9	Djian-Caporalino et al. (2001)
RHG1	H. glycines	bigenic	Glycine max	no	G	Concibido et al. (1997); Meksem et al. (2002)

RHG4	H. glycines	bigenic	G. max	no	A	Meksem et al. (2002)
Hs1^{pro1}	H. schachtii	SD	Beta procumbens	YES	I	Cai et al. (1997); Lange et al. (1993)
Hs1^{pat1}	H. schachtii	SD	B. patellaris	no	I	Lange et al. (1993)
Hs1^{web1}	H. schachtii	SD	B. webbiana	no	I	Kleine et al. (1998)
Hs2^{web7}	H. schachtii	SD	B. webbiana	no	VII	Kleine et al. (1998)
Hs2^{pro7}	H. schachtii	SD	B. procumbens	no	VII	Lange et al. (1993)
Hs3^{web8}	H. schachtii	SD	B. webbiana	no	VIII	Kleine et al. (1998)
Ha1	H. avenae	SD	Hordeum vulgare	no	2	Barr et al. (1998)
Ha2	H. avenae	SD	H. vulgare	no	2	Kretschmer et al. (1997)
Ha3	H. avenae	SD	H. vulgare	no	2	Barr et al. (1998)
Ha4	H. avenae	SD	H. vulgare	no	5	Barr et al. (1998)
Cre1	H. avenae	SD	Triticum aestivum	no	2B	Slootmaker et al. (1974)
Cre3	H. avenae	?	T. aestivum	no	2D	Eastwood et al. (1994)
Cre5(=CreX)	H. avenae	QTL	Ae. ventricosa/T. ventricosum	no	2AS	Jahier et al. (2001)
Cre6	H. avenae	SD	Ae. ventricosa/T. ventricosum	no	5N^{v}	Ogbonnaya et al. (2001)
CreF	H. avenae	SD	T. aestivum	no	7HL	Paull et al. (1998)
CreR	H. avenae	SD	Secale cereale	no	6R	Asiedu et al. 1990)

SD = single dominant R gene

QTL = quantitative trait locus

[a] Mapped in (tetraploid) clone that is an interspecific hybrid between S. tuberosum and several wild potato species including S. vernei, S. vernei ssp. ballsii, S. oplocense, and S. tuberosum ssp. Andigena.

of *S. spegazzinii* and confers resistance to *G. rostochiensis* (Kreike et al., 1996). *Gpa4* on chromosome IV is a QTL from *S. tuberosum* ssp. *tuberosum* that confers resistance to *G. pallida*. *Gro1*, *Gro1.2*, and *Gro1.3* on chromosomes VII, X, and XI, respectively, confer resistance to *G. rostochiensis* and originate from *S. spegazzinii* (Ballvora et al., 1995; Barone et al., 1990; Kreike et al., 1996). *Rmc1* is a single dominant gene on chromosome XI of *S. bulbocastanum* (Brown et al., 1999) and is the only gene mapped that confers resistance to a root-knot nematode (*M. chitwoodi*) in potato. On chromosome V, five resistance loci (*Gpa*, *Gpa5*, *Grp1*, *GroVI*, and *H1*) have been mapped. *Gpa* is a QTL from *S. spegazzinii* and confers resistance to *G. pallida* (Bradshaw et al., 1998). *Gpa5* confers a multigenic resistance to *G. pallida* coacting with *Gpa6*, which is present on chromosome IX (Rouppe van der Voort et al., 2000). *Gpa5*, *Gpa6*, and *Grp1* are all mapped in an interspecific hybrid between *S. tuberosum* and several wild species including *S. vernei*, *S. vernei* ssp. *ballsii*, *S. oplocense*, and *S. tuberosum* ssp. *andigena* and therefore the exact origins of these genes are unknown. *Grp1* is a single dominant gene conferring resistance to both *G. pallida* and *G. rostochiensis* (Rouppe van der Voort et al., 1998). *GroVI* and *H1* are also single dominant genes. They both confer resistance to *G. rostochiensis*. *GroVI* originates from *S. vernei* (Jacobs et al., 1996) and *H1* from *S. tuberosum* ssp. *andigena* (Gebhardt et al., 1993; Pineda et al., 1993).

Over the past decades, nematode resistance has been successfully introgressed into cultivars of several crops. The *H1* gene is a good example; since its discovery in 1955 (Huijsman, 1955) it has been used in many commercially available cultivars. In the United Kingdom, the potato cyst nematode *Globodera rostochiensis* caused enormous losses in potato yields until *H1* was successfully introgressed from the wild potato subspecies *S. tuberosum* ssp. *andigena*. Even today, after many decades of use, the gene is very effective against *Globodera rostochiensis*. This makes it one of the most durable resistance genes known. However, it is noted that most *G. rostochiensis* populations in the United Kingdom have been replaced by *G. pallida*, which is virulent for *H1* (Evans, 1993). At the moment a map-based cloning approach is being employed to isolate the *H1* gene (Bakker et al., unpublished data).

Almost all nematode *R* genes in potato are located in regions where other *R* genes are also present (reviewed by Gebhardt and Valkonen, 2001). This applies not only for single dominant genes, but also for QTLs. Theoretically, the quantitative behavior of these resistance traits can be caused by partial resistance of the host plant. Another option is that quantitative resistance is also mediated by *R* genes, but that the potato cyst nematode population used to screen for resistance consist of a mixture of virulent and avirulent genotypes. The latter explanation complies with the locations of most QTLs, which are often linked to *R* genes.

14.2.2 Nematode Resistance in Tomato and Pepper

The high genome synteny of potato species results in the colocalization of *R* genes mapped in different species (Gebhardt and Valkonen, 2001). This synteny, however,

is not confined to potato. Related genera like tomato and pepper also have a high
R gene synteny. Grube et al. (2000) observed that 48 out of 84 R genes studied
are located within 15 cM of R gene positions in other genera. Tomato and potato
genomes have an especially high overall synteny (Tanksley et al., 1992). Many
nematode R genes in potato (*Gro1.4*, *Gpa4*, *Gpa6*, *Gro1.2*, *Gro1.3*, and *Gpa2*)
colocalize with R genes in tomato (*Hero* and *Mi1*) (Grube et al., 2000). Although
the pepper genome is much more differentiated from tomato than potato through
rearrangements (Livingstone et al., 1999) almost all nematode R genes in potato
and tomato (*Gpa2* as the single exception) colocalize with R genes in pepper
(Grube et al., 2000).

In tomato and pepper, several R genes have been mapped that confer resistance
to root-knot nematode species. All root-knot nematode R genes described so far
in these two crops are inherited as single dominant genes. Usually, root-knot nematode resistances are tested with population composed of genetically identical
individuals, as opposed to many potato cyst nematode resistances in potato for
which a genetic mixture is used. The fact that all root-knot nematode loci represent single dominant resistances supports the idea that the quantitative behavior of
the potato cyst nematode R genes is caused by the mixture of virulent and avirulent
genotypes in the potato cyst nematode populations. Additionally, one potato cyst
nematode R gene (*Hero*) conferring resistance to G. *rostochiensis* and partial resistance to G. *pallida*. *Hero* is a single dominant gene that is located on chromosome
4 on an introgression segment originating from *Lycopersicon pimpinellifolium*
(Ganal et al., 1995). The gene has recently been cloned (Ernst et al., 2002) and is
part of a cluster of 14 highly homologues NBS-LRR genes that are present in a
118 kb region.

The first root-knot nematode resistance gene was cloned from the *Mi1* locus on
chromosome 6 of L. *peruvianum* (Milligan et al., 1998; Vos et al., 1998). Three
resistance gene homologues were identified in a 52 kb region including two genes
that were transcribed (*Mi-1.1* and *Mi-1.2*) and one pseudogene. Complementation
assays showed that the single dominant *Mi-1.2* gene was sufficient to confer resistance to root-knot nematodes. *Mi-1.2* belongs to the NBS-LRR super family of
R genes and is most homologous to the *Prf* gene from tomato that is required for
resistance to *Pseudomonas syringae* (Salmeron et al., 1996). The protein contains
a coiled coil (CC) motif and an extra long domain at the N-terminus. Interestingly, *Mi-1.2* does not only confer resistance to root-knot nematodes, but also to a
completely unrelated organism, the potato aphid *Macrosiphum euphorbiae* (Rossi
et al., 1998; Vos et al., 1998).

Other root-knot nematode resistance genes include *Mi3* (Yaghoobi et al.,
1995), *Mi5* (Veremis and Roberts, 1996), and *Mi-LA2157* (Veremis et al., 1999).
Mi-LA2157 is located on the same chromosome as *Mi1*. Both genes are derived
from L. *peruvianum*, although not from the same accession. The main difference
between *Mi1* and *Mi-LA2157* is the heat-stability of *Mi-LA2157*. At temperatures
above 32°C *Mi-LA2157* can still confer resistance (Veremis et al., 1999), while
Mi1 cannot (Dropkin, 1969; Holtzmann, 1963). *Mi3* also becomes nonfunctional
at high temperatures and is located on chromosome 12 of L. *peruvianum* (accession

PI126443-1MH) (Yaghoobi et al., 1995). *Mi5* is tightly linked to *Mi3* and originates from the same accession of *L. peruvianum*. However, the two genes differ in heat-stability: while *Mi3* cannot confer resistance at 32°C, *Mi5* can. Interestingly, the pepper root-knot nematode *R* gene *Me3* (Djian-Caporalino et al., 2001) and the potato cyst nematode *R* gene *Gpa2* (Van der Vossen et al., 2000) are present roughly in the same region as *Mi3* on the homoeologous chromosomes of pepper (9) and potato (XII). After extensive marker studies, however, it is unclear whether the three genes are part of orthologous clusters (Djian-Caporalino et al., 2001). Maybe when all three genes are cloned, sequence comparison will reveal more information on this issue. On the same chromosome two other genes are located in different accessions of pepper (PM217 and CM334, whereas *Me3* is carried by PM687) that confer resistance to *Meloidogyne* spp. (Djian-Caporalino, personal communication). The three genes can be distinguished by their different response patterns in root cells upon nematode infection (Djian-Caporalino et al., 2002).

14.2.3 Nematode Resistance in Beet

The first nematode *R* gene was cloned from beet (Cai et al., 1997). This gene, *Hs1^{pro-1}*, is a single dominant gene that confers resistance to the beet cyst nematode *Heterodera schachtii*. The gene product consists of a leucine-rich domain and a putative membrane-spanning domain. This protein structure corresponds with the *Cf-9* resistance gene product from tomato, although no significant sequence homology was detected.

Sugar beet plants carrying the *Hs1^{pro-1}* locus often suffer from tumor formation on leaves and root systems and from the occurrence of a so called "multi-top" phenotype (Sandal et al., 1997). The introgression segment was reduced in an attempt to obtain resistant plants lacking these negative traits. In two plants that still have the resistant phenotype, the segment was reduced to 35% and 17% of the original segment. However, molecular analysis showed that both plants lost the *Hs1^{pro-1}* gene. This indicates that another nematode resistance gene is present at the introgression segment (Sandal et al., 1997).

Resistance to beet cyst nematode is not present in the gene pool of *Beta vulgaris*. All resistances present in cultivars are the result of interspecific crosses using resistant wild beet species of the section Procumbentes (*B. patellaris, B. procumbens* and *B. webbiana*) (Kleine et al., 1998). These interspecific crosses can result in three types of nematode resistant sugar beet plants: (1) chromosome additions (2n = 19) (De Jong et al., 1985; Loptien, 1984; Speckman et al., 1985), (2) fragment additions (2n = 18 + f) (Brandes et al., 1987; De Jong et al., 1986), and (3) introgressed segments into the sugar beet genome (2n = 18) (Sandal et al., 1997). Within the wild beet species at least three resistance genes located on three different chromosomes were distinguished: *Hs1* on the homoeologous chromosomes I of each species (*Hs1^{pro-1}, Hs1^{pat-1}, Hs1^{web-1}*), *Hs2* on the homoeologous chromosomes VII of *B. procumbens* and *B. webbiana* (*Hs2^{pro-7}, Hs2^{web-7}*), and *Hs3* on chromosome VIII of *B. webbiana* (*Hs3^{web-8}*) (Kleine et al., 1998). Interestingly, three resistance genes originating from the wild beets

B. procumbens ($Hs1^{pro-1}$) and *B. webbiana* ($Hs1^{web-1}$, $Hs2^{web-7}$) that have been transferred to sugar beet were all mapped to chromosome IV (Heller et al., 1996) of *B. vulgaris*, although they originate from chromosomes I and VII of *B. procumbens* and *B. webbania*.

14.2.4 Nematode Resistance in Soybean

In soybean, resistance gene homologues were identified at the *Rhg1* and *Rhg4* loci that confer digenic resistance to the soybean cyst nematode *Heterodera glycines* Ichinohe race 3 (Hg_0). The genes are located on two different linkage groups (LG) *viz. Rhg1* on LG G and *Rhg4* on LG A2 (Meksem et al., 2001). The candidate genes show high homology to the bacterial resistance gene *Xa21* and an *Arabidopsis* receptor-like kinase (RLK) gene family. The proteins consist of three functional domains including 12 extracellular LRRs, a transmembrane domain and a kinase domain (Meksem et al., 2002).

Only a few sources of resistance to the soybean cyst nematode are known: PI 88788, PI 437.654, Peking, PI 90763, and PI 209332 (Meksem et al., 2001). Most cultivars are derived from PI 88788 because of other interesting agricultural traits (Skorupska et al., 1994). In all soybean accessions, resistance is controlled by a few genes (Caldwell et al., 1960; Matson and Williams, 1965; Myers and Anand, 1991; Rao-Arelli, 1994; Rao-Arelli et al., 1992). In contrast to the *Rhg1/Rhg4* resistance, however, interaction of one dominant and one or more recessive genes are suggested (Rao-Arelli et al., 1992). The phenomenon of colocalizing *R* genes does not occur only in the Solanaceae. An interesting example in soybean is the strong (but not complete) association of the soybean cyst nematode *R* gene *Rhg1* with a locus conferring resistance to sudden death syndrome (SDS) caused by the fungus *Fusarium solani* f. sp. *phaseoli* (Chang et al., 1997).

14.2.5 Nematode Resistance in Cereals

Root-knot nematodes are not a major problem in cereals and therefore no root-knot nematode *R* genes have been mapped. The cereal cyst nematode, however, is a major problem in cereal cultivation and hence, much effort has been put into screening and breeding for resistances against cereal cyst nematode (*Heterodera avenae*). *Ha1*, *Ha2*, and *Ha3* are all single dominant genes mapped on chromosome 2 of barley (*Hordeum vulgare*) (Barr et al., 1998; Kretschmer et al., 1997). *Ha4* is also a single dominant barley gene, but this one is mapped on chromosome 5 (Barr et al., 1998). Furthermore, three genes have been mapped in *Triticum aestivum* (common bread wheat) namely *Cre1* on chromosome 2B (Slootmaker et al., 1974), *Cre3* on chromosome 2D (Eastwood et al., 1994), and *CreF* on chromosome 7HL (Paull et al., 1998). Other genes have been mapped in *Aegilops ventricosa* (a wild grass species) and in *Secale cereale* (rye): *Cre5* and *Cre6* are mapped on chromosomes 2AS (Jahier et al., 2001) and $5N^v$ (Ogbonnaya et al., 2001) of *Ae. ventricosa*, respectively, whereas *CreR* is mapped on chromosome 6R of *S. cereale* (Asiedu et al., 1990). Although none of these resistance genes has been cloned, NBS-LRR like sequences have been isolated from the *Cre3* region of wheat (Lagudah et al., 1997).

14.3 Resistance Mechanisms

14.3.1 The Compatible Plant-Nematode Interaction

To understand the response mediated by nematode resistance genes, knowledge of the compatible plant–nematode interaction is indispensable (for reviews about nematode parasitism and feeding cell development see Davis et al. (2000) and Goverse et al. (2000), respectively). Preparasitic juveniles from both cyst and root-knot nematodes hatch from eggs and migrate through the soil in search of a suitable host plant. After penetration of the rhizodermis, the root-knot nematode migrates intercellularly whereas the cyst nematode migrates intracellularly through the root. Migration stops when a cell is encountered that is suitable as a starting point for feeding site formation.

Cyst and root-knot nematodes are able to manipulate plant cells for their own benefit. Regardless of the nematode–host plant combination, the mechanisms of feeding site induction are similar among sedentary endoparasitic nematodes. A multinuclear cell complex is formed inside the root on which the nematode fully depends for nutrition. The nematode-exploited plant cells are metabolically highly active and adapted to withdraw large quantities of nutrient solutions from the vascular system of the host plant (Jones and Northcote, 1972).

Upon feeding, the parasitic juvenile develops into an adult after three molts. The shape of adult females changes from vermiform to saccate, lemon-shaped or spherical. Cyst nematodes are obligately sexual, and eggs will only be produced upon fertilization by the vermiform and mobile males. Within the egg, the first stage juvenile molts and the resulting infective juvenile will hatch in the presence of a suitable host plant. Though sexual reproduction occurs in some *Meloidogyne* species present in the temperate regions, important pathogens such as *M. incognita*, *M. javanica*, and *M. arenaria* reproduce via mitotic parthenogenesis.

14.3.2 The Incompatible Plant-Nematode Interaction

In resistant plants, feeding cell initiation and development are arrested, resulting in starvation of the nematode. Two major types of resistance responses can be distinguished based on extensive microscopic observations of several incompatible plant–nematode interactions (Table 14.2). The first type is characterized by a rapid hypersensitive response (HR) resulting in necrosis of the feeding site within a couple of days post infection, whereas the second type blocks the development of the feeding cell in a late stage of the infection process.

The HR, which is a common defense mechanism to a wide variety of pathogens including endoparasitic nematodes (Table 14.2a), is accompanied by an oxidative burst resulting in the production of H_2O_2 (Waetzig et al., 1999) and the accumulation of phenylpropanoid compounds (Robinson et al., 1988). For *Mi-1*,

TABLE 14.2 Resistance responses upon cyst and root-knot nematode infection.
14.2a. Induction of an HR upon infection (dpi = 0–5)

Host plant	Nematode	R gene	Reference
Tomato	*M. incognita*	*Mi1*	Paulson and Webster (1972); Riggs and Winstead (1959)
Potato	*G. rostochiensis*	*H1*	Rice et al. (1985)
Soybean	*H. glycines*	?	Mahalingam and Skorupska (1996)
Soybean	*H. glycines*	?	Endo (1991); Kim et al. (1987); Ross (1958)
Pepper	*M. incognita*	*Me3*	Bleve-Zacheo et al. (1998)
Arabidopsis	*H. glycines*	Non-host	Grundler et al. (1997)
Wheat	*H. avenae*	?	Bleve-Zacheo et al. (1995)

14.2b. Arrest of feeding cell development (dpi > 14).

Host plant	Nematode	R gene	Reference
Wheat	*H. avenae*	?	Williams and Fisher (1993)
Barley	*H. avenae*	*Ha2, Ha3*	Seah et al. (2000)
Sugar beet	*H. schachtii*	$Hs1^{pro-1}$	Yu (1984)
Pepper	*M. incognita*	*Me1*	Bleve-Zacheo et al. (1998)
Common bean	*H. glycines*	?	Becker et al. (1999)
Tobacco	*M. incognita M. acrita*	?	Powell (1962)
Potato	*G. pallida*	*Gpa2*	Goverse et al. (2001)
Potato	*G. pallida G. rostochiensis*	*Grp1*	Rice et al. (1987)

necrosis occurs in cells surrounding the migratory track and the initial feeding cell (Paulson and Webster, 1972; Riggs and Winstead, 1959). But in case of the *H1* gene, a small feeding cell is induced that becomes encapsulated by a layer of necrotic cells (Rice et al., 1985). Hence, further expansion of the feeding site is prevented. However, these small feeding cells are still able to provide sufficient nutrients for the development of adult males. For cyst nematodes, sex is epigenetically determined in the first week of feeding site development (Trudgill, 1967). Under favorable conditions, the nematode becomes female and in case food is limited it becomes a male. Also for root-knot nematodes, several studies suggest that, even for obligatory parthenogenetic species, under less favorable conditions more males are formed than usual (reviewed by Triantaphyllou, 1973). This phenomenon explains the development of increased numbers of adult males on plants that contain nematode resistance genes.

In contrast, the late resistance response is characterized by the degeneration of the feeding site around two weeks post infection and the absence of an HR. This resistance mechanism is observed for a wide variety of incompatible plant–nematode interactions (Table 14.2b). Initially, no clear differences can be observed between the compatible and the incompatible interaction. The nematode is able to establish a functional feeding site that allows the development of males and females. In later stages, however, differences in morphology of the feeding cells are observed. In resistant plants, the proliferation of the feeding cell is arrested, resulting in less dense cytoplasm and more vacuoles compared to feeding cells induced in a susceptible plant. Moreover, it is often observed that the connection between the feeding cell and the vascular tissue is less pronounced. These features indicate that the metabolic activity of the feeding cell is reduced as a result of the resistance response. Finally, the cells adjacent to the feeding site become necrotic followed by the degradation of the feeding cell itself. This resistance mechanism results in the limitation of nutrients in a late stage of nematode development— after sex determination—and explains the presence of relatively large numbers of females on plants that show such a slow defense response. However, the food supply is finally insufficient, resulting in the arrest of female development and reproduction.

14.3.3 Activation of a Resistance Response by Nematode R Genes

Both types of responses can be mediated by single dominant *R* genes that confer resistance to either cyst or root-knot nematodes. For example, *H1* and *Mi-1* trigger a rapid HR upon infection, whereas *Gpa2* and *Mel* result in a slow resistance response (Table 14.2). The nematode *R* genes that were recently cloned share structural motifs with *R* genes that confer resistance to other plant pathogens like bacteria, fungi, oomycetes, and viruses. This shows that nematode *R* genes are part of the plant survey system that results in the activation of a defense response upon infection. Therefore, it is assumed that common signal transduction pathways are involved in nematode resistance. Functional analysis of nematode *R* genes will increase our understanding about their role in the activation of a resistance response.

Interestingly, *Mi-1* confers resistance to two completely unrelated organisms *viz.* root-knot nematodes and aphids (Rossi et al., 1998; Vos et al., 1998). This suggests that mechanisms of nematode and aphid resistance have similar signal transduction pathways. Another remarkable example of *R* genes that mediate resistance to two distinct pathogens are the closely related genes *Gpa2* and *Rx1*, which confer resistance to the potato cyst nematode *G. pallida* and the potato virus X, respectively (Van der Vossen et al., 2000, Bendahmane et al., 1999). However, the activation of *Gpa2* results in a slow resistance response, whereas *Rx1* results in extreme resistance, i.e., a very quick resistance response without a visible HR. The intriguing question is how such highly homologous *R* genes are able to induce such apparently different defense mechanisms. This could be due to differences in the protein structure such as the absence of an acidic tail for *Gpa2*, or differences in the pathology of the plant–pathogen interactions. Conservation of the LZ-NBS domains of *Gpa2* and *Rx1* suggests that common signal transduction pathways are involved in nematode and virus resistance (Van der Vossen et al., 2000). Domain swaps between the two genes are currently investigated to test this hypothesis (Dees and Goverse, unpublished data).

Extensive structure–function analysis has resulted in a model for *Mi-1* as a regulatory protein. Chimeric gene constructs between *Mi-1* (= *Mi-1.2*) and *Mi-1.1*—a homologue that does not confer nematode resistance—showed that specific regions in the LRR domain are involved in the induction of a cell death response. This is suppressed by an intramolecular interaction between the N-terminal domain of the protein and the LRR domain (Hwang et al., 2000; Williamson, 2002). Moreover, mutagenesis studies in tomato (*Mi-1*/*Mi-1*) resulted in the identification of *rme1*, a gene involved in aphid and root-knot nematode resistance. This gene is specifically required for *Mi-1* signaling and is not required for *I2* mediated resistance to *Fusarium oxysporum* (Martinez de Ilarduya et al., 2001).

The homology between the soybean resistance gene candidates *Rhg1* and *Rhg4* and an *Arabidopsis* receptor-like kinase gene family allows studying their role in signal transduction. From interaction studies, it is suggested that an active heterodimer of *Rhg1* and *Rhg4* is required for nematode resistance (Meksem et al., 2002). This interaction between RHG1 and RHG4 most likely accounts for the digenic resistance response mediated by these two single dominant *R* genes in soybean.

14.3.4 Specificity of Nematode R Genes

Although cyst and root-knot nematodes use similar mechanisms to parasitize their host plants, no resistance is found to be effective against both types of nematodes. Some *R* genes like *Mi-1* recognize several root-knot nematode species, whereas others like *Gpa2* confer resistance to specific populations of one nematode species. It is unknown how the plant recognizes the parasite upon infection of the roots. As for other *R* genes, it is assumed that the LRR domain is indirectly or directly involved in pathogen recognition and that this domain determines the specificity of the resistance response. This is supported by the finding that the highly homologous *R* genes *Gpa2* and *Rx1* show most variation in this part of the protein. The ratio between nonsynonymous and synonymous amino acid substitutions was larger than 1 for the solvent-exposed residues of the LRR region (Van der Vossen et al., 2000). This suggests that these parts of the LRR domain are subject to diversifying selection and pathogen recognition.

The *Mi-1* gene product contains an LRR domain that recognizes the avirulence gene product of root-knot nematodes and the potato aphid (Rossi et al., 1998; Vos et al., 1998). It is very unlikely that aphids and nematodes have an avirulence gene product in common, although the feeding behaviors of both pathogens share the withdrawing of nutrients from the vascular bundle. This is the first example of dual specificity of an *R* gene to unrelated organisms. Previously, it was reported that *RPM1* was able to recognize two different Avr products from the bacterium *P. syringae* (Grant et al., 1995).

Moreover, it is assumed that *R* gene products are part of large protein complexes (Dangl and Jones, 2001). The N-terminal domain of Mi-1 shows homology with a PCI domain that consist of an α-helix of about 200 amino acid residues, mostly located at the extreme C-terminus of the protein and present in subunits of multiprotein complexes (Hofmann and Bucher, 1998). This indicates that Mi-1 may be part of a multiprotein regulator complex and could serve as a structural scaffold protein by interacting with other proteins (Vos et al., 1998). As such, a model is proposed in which the Avr-product from aphids and nematodes could interact with a specific receptor, and then trigger a signal transduction cascade via the Mi-1 protein (Rossi et al., 1998) instead of a direct interaction with the *R* protein.

14.3.5 Recognition of Nematode Avirulence Gene Products

Proteins playing a pivotal role in parasitism are detected by the plant surveillance system. However, no nematode avirulence gene product (Avr-product) has been identified yet. If an HR is induced in cells adjacent to the migratory track, it is likely that the invading nematode is recognized in an early stage of the infection process. If, however, feeding cell development is arrested in a resistant plant, recognition of the nematode could occur in a later stage. During invasion, close contact occurs between plant cells and the nematode surface coat. In addition, secretions are released by the amphids, sense organs involved in chemotaxis. Salivary proteins are secreted both for enzymatic degradation of the cell walls during migration and for the induction of a feeding cell. Recently, the genomes of two near isogenic virulent and avirulent lines of *M. incognita* were analyzed using AFLP. One of the polymorphic fragments present in the virulent line and absent in the avirulent line *map*-1 encodes for a secretory protein that is produced by the amphids (Semblat et al., 2001).

The structure of Gpa2 and Mi-1 suggest that these proteins are localized in the cytoplasm of the plant cell. The avirulence gene product is expected to be colocalized with the *R* gene product. Hence, nematode secretions that are injected into the feeding cell might interact with this class of *R* genes. In case of *Rhg1*, *Rhg4*, and *Hs1^{pro1}*, the LRR domain is predicted to be localized in the extracellular space. Therefore, the putative Avr-products from the nematodes will most probably originate from the nematode surface coat, the amphids, or the subventral oesophageal glands.

14.3.6 Induced Systemic Resistance

Besides the classic "gene-for-gene" type of resistances, some studies have been devoted to a more general type of resistance: the so called Induced Systemic Resistance or ISR (reviewed by (Ramamoorthy et al., 2001). The effect of rhizobacteria on nematode infections has been shown for the cereal cyst nematode *H. avenae* and the root-knot nematode *M. arenaria* (Sikora, 1992), the root-knot nematode *M. incognita* on tomato (Santhi and Sivakumar, 1997) and on cotton (Hallmann et al., 1999), the potato cyst nematode *G. pallida* (Hasky et al., 1998), the beet cyst nematode *H. shachtii* (Neipp and Becker, 1999), the clover cyst nematode *H. trifolii* (Kempster et al., 2001), and the free living plant parasitic nematode *Bursaphelenchus xylophilus* causing pine wilt disease (Kosaka et al., 2001).

It has been shown that bacterial surface components such as lipopolysacharides (LPS) and more specifically the O-antigen play an important role in ISR (Leeman et al., 1995; Van Peer et al., 1991; Van Wees et al., 1997). However, the same LPS did not work for different plant species, suggesting that the effect is host specific with other LPS fractions than the O-antigen involved in recognition, such as the core-region and/or the lipid A fraction (Van Wees et al., 1997). For ISR to *G. pallida* it has been demonstrated that the core-region is the main elicitor, whereas the lipid A factor plays a much more modest role in the mechanism (Reitz et al., 2002).

14.4 Perspectives

In the last decade, significant progress has been made in the mapping and cloning of nematode resistance genes. The use of these *R* genes and their homologues in extensive structure–function analyses will increase our knowledge about the mechanisms that underlie disease resistance in plants. In case of sedentary endoparasitic nematodes, it will be interesting to investigate how the development of a multicellular feeding cell complex is inhibited by the induction of a resistance response. The identification of increasing numbers of nematode resistance gene homologues from different plant species will facilitate studying of the molecular evolution of *R* genes. Moreover, comparative analyses of *R* gene clusters in different plant species may allow the development of PCR-based cloning strategies for disease resistance genes from related wild species. Finally, a major challenge for the near future will be the identification of the corresponding avirulence products from the nematode which is not only important for studying signal transduction processes, but also for understanding the coevolution between nematodes and plants.

References

Aarts, M.G.M., Te Lintel Hekkert, B., Holub, E.B., Beynon, J.L., Stiekema, W.J., and Pereira, A. 1998. Identification of R-gene homologous DNA fragments genetically linked to disease resistance loci in *Arabidopsis thaliana*. *Mol. Plant Microbe Interact.* 11:251–258.

Asiedu, R., Fisher, J.M., and Driscoll, C.J. 1990. Resistance to *Heterodera avenae* in the rye genome of Triticale. *Theor. Appl. Genet.* 79:331–336.

Bakker, E., Butterbach, P., Rouppe van der Voort, J., Van der Vossen, E., Van Vliet, J., Bakker, J., and Goverse, A. 2003. Genetic and physical mapping of homologues of the virus resistance gene *Rx1* and the cyst nematode resistance gene *Gpa2* in potato. *Theor. Appl. Genet.* 106:1524–1531.

Bakker, J. 2002. Durability of resistance against potato cyst nematodes. *Euphytica* 124: 157–162.

Ballvora, A., Hesselbach, J., Niewohner, J., Leister, D., Salamini, F., and Gebhardt, C. 1995. Marker enrichment and high-resolution map of the segment of potato chromosome VII harboring the nematode resistance gene *Gro1*. *Mol. Gen. Genet.* 249:82–90.

Barone, A., Ritter, E., Schachtschabel, U., Debener, T., Salamini, F., and Gebhardt, C. 1990. Localization by restriction-fragment-length-polymorphism mapping in potato of a major dominant gene conferring resistance to the potato cyst nematode *Globodera rostochiensis*. *Mol. Gen. Genet.* 224:177–182.

Barr, A.R., Chalmers, K.J., Karakousis, A., Kretschmer, J.M., Manning, S., Lance, R.C.M., Lewis, J., Jeffries, S.P., and Langridge, P. 1998. RFLP mapping of a new cereal cyst nematode resistance locus in barley. *Plant Breed.* 117:185–187.

Becker, W.F., Ferraz, S., and Monteiro Da Silva, E.A. 1999. Histopathological alterations in the roots of common beans (*Phaseolus vulgaris*) infected by *Heterodera glycines*. *Nematol. Brasileira* 23:34–46.

Bendahmane, A., Kanyuka, K., and Baulcombe, D.C. 1999. The *Rx* gene from potato controls separate virus resistance and cell death responses. *Plant Cell* 11:781–791.

Bleve-Zacheo, T., Bongiovanni, M., Melillo, M.T., and Castagnone-Sereno, P. 1998. The pepper resistance genes Me1 and Me3 induce differential penetration rates and temporal sequences of root cell ultrastructural changes upon nematode infection. *Plant Sci.* 133:79–90.

Bleve-Zacheo, T., Melillo, M.T., Andres, M., Zacheo, G., and Romero, M.D. 1995. Ultrastructure of initial response of graminaceous roots to infection by *Heterodera avenae*. *Nematologica* 41:80–97.

Bradshaw, J.E., Meyer, R.C., Milbourne, D., McNicol, J.W., Phillips, M.S., and Waugh, R. 1998. Identification of AFLP and SSR markers associated with quantitative resistance to *Globodera pallida* (Stone) in tetraploid potato (*Solanum tuberosum* subsp. *tuberosum*) with a view to marker-assisted selection. *Theor. Appl. Genet.* 97:202–210.

Brandes, A., Jung, C., and Wricke, G. 1987. Nematode resistance derived from wild beet and its meiotic stability in sugar beet. *Plant Breed.* 99:56–64.

Brown, C.R., Yang, C.P., Mojtahedi, H., Santo, G.S., and Masuelli, R. 1996. RFLP analysis of resistance to Columbia root-knot nematode derived from *Solanum bulbocastanum* in a BC2 population. *Theor. Appl. Genet.* 92:572–576.

Brown, N.R., Noble, M.E.M., Endicott, J.A., and Johnson, L.N. 1999. The structural basis for specificity of substrate and recruitment peptides for cyclin-dependent kinases. *Nat. Cell Biol.* 1:438–443.

Cai, D., Kleine, M., Kifle, S., Harloff, H.-J., Sandal, N.N., Marcker, K.A., Klein-Lankhorst, R.M., Salentijn, E.M.J., Lange, W., Stiekema, W.J., Wyss, U., Grundler, F.M.W., and Jung, C. 1997. Positional cloning of a gene for nematode resistance in sugar beet. *Science* 275:832–834.

Caldwell, B.E., Brim, C.A., and Ross, J.P. 1960. Inheritance of resistance of soybeans to the cyst nematode, *Heterodera glycines*. *Agron. J.* 52:635–636.

Chang, S.J.C., Doubler, T.W., Kilo, V.Y., AbuThredeih, J., Prabhu, R., Freire, V., Suttner, R., Klein, J., Schmidt, M.E., Gibson, P.T., and Lightfoot, D.A. 1997. Association of loci underlying field resistance to soybean sudden death syndrome (SDS) and cyst nematode (SCN) race 3. *Crop Sci.* 37:965–971.

Concibido, V.C., Lange, D.A., Denny, R.L., Orf, J.H., and Young, N.D. 1997. Genome mapping of soybean cyst nematode resistance genes in 'Peking', PI 90763, and PI 88788 using DNA markers. *Crop Sci.* 37:258–264.

Dangl, J.L., and Jones, J.D.G. 2001. Plant pathogens and intergrated defence responses to infection. *Nature* 411:826–833.

Davis, E.L., Hussey, R.S., Baum, T.J., Bakker, J., and Schots, A. 2000. Nematode parasitism genes. *Ann. Rev. Phytopathol.* 38:365–396.

De Jong, J.H., Speckman, G.J., de Bock, T.S.M., Lange, M., and van Voorst, A. 1986. Alien chromosome fragments conditioning resistance to beet cyst nematode in diploid descendants from monosomic additions of *Beta procumbens* to *Beta vulgaris*. *Can. J. Genet. Cytol.* 28:439–443.

De Jong, J.H., Speckman, G.J., de Bock, T.S.M., and van Voorst, A. 1985. Monosomic additions with resistance to beet cyst nematode obtained from hybrids of *Beta vulgaris* and wild type *Beta* species of the section Patellares.II. Comparative analysis of alien chromosome. *Zeitschrift Pflanzenzucht* 95:84–94.

Djian-Caporalino, C., Lefebvre, V., Palloix, A., and Abad, P. 2002. Characterisation and high-resolution genetic mapping of root-knot nematode resistance genes in pepper (*Capsicum annuum* L.): comparison with the tomato and potato nematode resistance gene locations. *Nematology* 4:229.

Djian-Caporalino, C., Pijarowski, L., Fazari, A., Samson, M., Gaveau, L., O'Byrne, C., Lefebvre, V., Caranta, C., Palloix, A., and Abad, P. 2001. High-resolution genetic mapping of the pepper (*Capsicum annuum* L.) resistance loci *Me-3* and *Me-4* conferring heat-stable resistance to root-knot nematodes (*Meloidogyne* spp.). *Theor. Appl. Genet.* 103:592–600.

Dropkin, V.H. 1969. The necrotic reaction of tomato and other hosts resistant to *Meloidogyne*: reversal by temperature. *Phytopathology* 59:1632–1637.

Eastwood, R.F., Lagudah, E.S., and Appels, R. 1994. A directed search for DNA-sequences tightly linked to cereal cyst-nematode resistance genes in *Triticum tauschii. Genome* 37:311–319.

Ellis, J., Dodds, P., and Pryor, T. 2000. Structure, function and evolution of plant disease resistance genes. *Curr. Opin. Plant Biol.* 3:278–284.

Endo, B.Y. 1991. Ultrastructure of initial responses of susceptible and resistant soybean roots to infection by *Heterodera glycines. Rev. Nematol.* 14:73–94.

Ernst, K., Kumar, A., Kriseleit, D., Kloos, D.-U., Phillips, M.S., and Ganal, M.W. 2002. The broad-spectrum potato cyst nematode resistance gene (*Hero*) from tomato is the only member of a large gene family of NBS-LRR genes with an unusual amino acid repeat in the LRR region. *Plant J.* 31:127–136.

Evans, K. 1993. New approaches for potato cyst-nematode management. *Nematropica* 23:221–231.

Ganal, M.W., Simon, R., Brommonschenkel, S.H., Arndt, M., Phillips, M.S., Tanksley, S.D., and Kumar, A. 1995. Genetic mapping of a wide spectrum nematode resistance gene (*Hero*) against Globodera rostochiensis in tomato. *Mol. Plant Microbe Interact.* 8:886–891.

Gebhardt, C., Mugniery, D., Ritter, E., Salamini, F., and Bonnel, E. 1993. Identification of RFLP markers closely linked to the *H1* gene conferring resistance to Globodera rostochiensis in potato. *Theor. Appl. Genet.* 85:541–544.

Gebhardt, C., and Valkonen, J.P.T. 2001. Organization of genes controlling disease resistance in the potato genome. *Ann. Rev. Phytopathol.* 39:79–102.

Goverse, A., Bakker, E., Smant, G., Sandbrink, H., Van der Vossen, E., and Bakker, J. 2001. R-gene homologs in potato confer resistance against distinct pathogens:nematodes and viruses. *Phytopathology* 91:S147.

Goverse, A., Engler, J.D., Verhees, J., van der Krol, S., Helder, J., and Gheysen, G. 2000. Cell cycle activation by plant parasitic nematodes. *Plant Mol. Biol.* 43:747–761.

Grant, M.R., Godiard, L., Straube, E., Ashfield, T., Lewald, J., Sattler, A., Innes, R.W., and Dangl, J.L. 1995. Structure of the *Arabidopsis Rpm1* gene enabling dual- specificity disease resistance. *Science* 269:843–846.

Grube, R.C., Radwanski, E.R., and Jahn, M. 2000. Comparative genetics of disease resistance within the Solanaceae. *Genetics* 155:873–887.

Grundler, F.M.W., Sobczak, M., and Lange, S. 1997. Defence responses of *Arabidopsis thaliana* during invasion and feeding site induction by the plant-parasitic nematode *Heterodera glycines*. *Physiol. Mol. Plant Pathol.* 50:419–429.

Hallmann, J., Rodriguez, K.R., and Kloepper, J.W. 1999. Chitin-mediated changes in bacterial communities of the soil, rhizosphere and within roots of cotton in relation to nematode control. *Soil Biol Biochem.* 31:551–560.

Hasky, G.K., Hoffmann, H.S., and Sikora, R.A. 1998. Resistance against the potato cyst nematode Globodera pallida systemically induced by the rhizobacteria Agrobacterium radiobacter (G12) and *Bacillus sphaericus* (B43). *Fund. Appl. Nematol.* 21:511–517.

Heller, R., Schondelmaier, J., Steinrucken, G., and Jung, C. 1996. Genetic localization of four genes for nematode (*Heterodera schachtii* Schm) resistance in sugar beet (*Beta vulgaris* L.). *Theor. Appl. Genet.* 92:991–997.

Hermsen, J.G.T., and Verdenius, J. 1973. Selection from *Solanum tuberosum* Group Phureja of genotypes combining high-frequency haploid induction with homozygosy for embryospot. *Euphytica* 22:244–259.

Hofmann, K., and Bucher, P. 1998. The PCI domain: a common theme in three multiprotein complexes. *TIBS* 23:204–205.

Holtzmann, O.V. 1963. Effect of soil temperature on resistance of tomato roor-knot nematode (*Meloidogyne incognita*). *Phytopathology* 55:990–992.

Huijsman, C.A. 1955. Breeding for resistance in the potato eelworm. II. Data on the inheritance in *andigenum-tuberosum* crosses obtained in 1954. *Euphytica* 4:133–140.

Hwang, C.F., Bhakta, A.V., Truesdell, G.M., Pudlo, W.M., and Williamson, V.M. 2000. Evidence for a role of the N terminus and leucine-rich repeat region of the *Mi* gene product in regulation of localized cell death. *Plant Cell* 12:1319–1329.

Jacobs, J.M.E., van Eck, H.J., Horsman, K., Arens, P.F.P., Verkerk-Bakker, B., Jacobsen, E., Pereira, A., and Stiekema, W.J. 1996. Mapping of resistance to the potato cyst nematode *Globodera rostochiensis* from the wild potato species *Solanum vernei*. *Mol. Breed.* 2:51–60.

Jahier, J., Abelard, P., Tanguy, A.M., Dedryver, F., Rivoal, R., Khatkar, S., and Bariana, H.S. 2001. The *Aegilops ventricosa* segment on chromosome 2AS of the wheat cultivar 'VPM1' carries the cereal cyst nematode resistance gene *Cre5*. *Plant Breed.* 120:125–128.

Jones, M.G.K., and Northcote, D.H. 1972. Nematode-induced syncytium:a multinucleate transfer cell. *J. Cell Sci.* 10:789–809.

Jung, C., Daguang, C., and Kleine, M. 1998. Engineering nematode resistance in crop species. *TIPS* 3:266–271.

Kempster, V.N., Davies, K.A., and Scott, E.S. 2001. Chemical and biological induction of resistance to the clover cyst nematode (Heterodera trifolii) in white clover (Trifolium repens). *Nematology* 3:35–43.

Kim, Y.H., Riggs, R.D., and Kim, K.S. 1987. Structural changes associated with resistance of soybean to *Heterodera glycines*. *J. Nematol.* 19:177–187.

Kleine, M., Voss, H., Cai, D., and Jung, C. 1998. Evaluation of nematode-resistant sugar beet (*Beta vulgaris* L.) lines by molecular analysis. *Theor. Appl. Genet.* 97:896–904.

Kosaka, H., Aikawa, T., Ogura, N., Tabata, K., and Kiyohara, T. 2001. Pine wilt disease caused by the pine wood nematode: The induced resistance of pine trees by the avirulent isolates of nematode. *Eur. J. Plant Pathol.* 107:667–675.

Kreike, C.M., Kok-Westeneng, A.A., Vinke, J.H., and Stiekema, W.J. 1996. Mapping of QTLs involved in nematode resistance, tuber yield and root development in *Solanum* sp. *Theor. Appl. Genet.* 92:463–470.

Kretschmer, J.M., Chalmers, K.J., Manning, S., Karakousis, A., Barr, A.R., Islam, A., Logue, S.J., Choe, Y.W., Barker, S.J., Lance, R. C. M., and Langridge, P. 1997. RFLP mapping of the *Ha2* cereal cyst nematode resistance gene in barley. *Theor. Appl. Genet.* 94:1060–1064.

Lagudah, E.S., Moullet, O., and Appels, R. 1997. Map-based cloning of a gene sequence encoding a nucleotide binding domain and a leucine-rich region at the *Cre3* nematode resistance locus of wheat. *Genome* 40:659–665.

Lange, W., Muller, J., and Debock, T.S.M. 1993. Virulence in the beet cyst-nematode (*Heterodera-schachtii*) versus some alien genes for resistance in beet. *Fund. Appl. Nematol.* 16:447–454.

Leeman, M., Van, P.J.A., Den, O.F.M., Heinsbroek, M., Bakker, P.A.H., and Schippers, B. 1995. Induction of systemic resistance against fusarium wilt of radish by lipopolysaccharides of Pseudomonas fluorescens. *Phytopathology* 85:1021–1027.

Livingstone, K.D., Lackney, V.K., Blauth, J.R., Van Wijk, R., and Jahn, M.K. 1999. Genome mapping in *Capsicum* and the evolution of genome structure in the Solanaceae. *Genetics* 152:1183–1202.

Loptien, H. 1984. Breeding nematode resistant beets. I. Development of resistant alien crosses between *Beta vulgaris* L. and wild species of the section Patellares. *Zeitschrift Pflanzenzucht* 92:208–220.

Mahalingam, R., and Skorupska, H.T. 1996. Cytological expression of early response to infection by *Heterodera glycines* Ichinohe in resistant PI 437654 soybean. *Genome* 39:986–998.

Martinez de Ilarduya, O.M., Moore, A.E., and Kaloshian, I. 2001. The tomato *Rme1* locus is required for *Mi-1*-mediated resistance to root-knot nematodes and the potato aphid. *Plant J.* 27:417–425.

Matson, A.L., and Williams, L.F. 1965. Evidence of fourth gene for resistance to the soybean cyst nematode. *Crop Sci.* 5:477.

Meksem, K., Jamai, A., Ruben, E., Triwitayakorn, K., Ishihara, H., Zobrist, K., and Lightfoot, D.A. 2002. Resistance to soybean cyst nematode:an *Xa21*-like gene family that requires possible demerisation for signal transduction. *Nematology* 4: 147.

Meksem, K., Pantazopoulos, P., Njiti, V.N., Hyten, L.D., Arelli, P.R., and Lightfoot, D.A. 2001. 'Forrest' resistance to the soybean cyst nematode is bigenic:saturation mapping of the *Rhg1* and *Rhg4* loci. *Theor. Appl. Genet.* 103:710–717.

Milligan, S.B., Bodeau, J., Yaghoobi, J., Kaloshian, I., Zabel, P., and Williamson, V. 1998. The root-knot nematode resistance gene *Mi* from tomato is a member of the leucine zipper, nucleotide binding, leucine-rich repeat family of plant genes. *Plant Cell* 10:1307–1319.

Myers, G.O., and Anand, S.C. 1991. Inheritance of resistance and genetic-relationships among soybean plant introductions to races of soybean cyst nematode. *Euphytica* 55:197–201.

Neipp, P.W., and Becker, J.O. 1999. Evaluation of biocontrol activity of rhizobacteria from Beta vulgaris against Heterodera schachtii. *J. Nematol.* 31:54–61.

Ogbonnaya, F.C., Seah, S., Delibes, A., Jahier, J., Lopez-Brana, I., Eastwood, R.F., and Lagudah, E.S. 2001. Molecular-genetic characterisation of a new nematode resistance gene in wheat. *Theor. Appl. Genet.* 102:623–629.

Paull, J.G., Chalmers, K.J., Karakousis, A., Kretschmer, J.M., Manning, S., and Langridge, P. 1998. Genetic diversity in Australian wheat varieties and breeding material based on RFLP data. *Theor. Appl. Genet.* 96:435–446.

Paulson, R.E., and Webster, J.M. 1972. Ultrastructure of the hypersensitive reaction in roots of tomato, *Lycopersicon esculentum* L., to infection by the root-knot nematode, *Meloidogyne incognita*. *Physiol. Plant Pathol.* 2:227–234.

Pineda, O., Bonierbale, M.W., and Plaisted, R.L. 1993. Identification of RFLP markers linked to the *H1* gene conferring resistance to the potato cyst nematode *Globodera rostochiensis*. *Genome* 36:152–156.

Powell, N.T. 1962. Histological basis of resistance to root-knot nematodes in flue-cured tobacco. *Phytopathology* 52:25.

Ramamoorthy, V., Viswanathan, R., Raguchander, T., Prakasam, V., and Samiyappan, R. 2001. Induction of systemic resistance by plant growth promoting rhizobacteria in crop plants against pests and diseases. *Crop Prot.* 20:1–11.

Rao-Arelli, A.P. 1994. Inheritance of resistance to *Heterodera glycines* race-3 in soybean accessions. *Plant Dis.* 78:898–900.

Rao-Arelli, A.P., Anand, S.C., and Wrather, J.A. 1992. Soybean resistance to soybean cyst nematode race-3 is conditioned by an additional dominant gene. *Crop Sci.* 32:862–864.

Reitz, M., Oger, P., Meyer, A., Niehaus, K., Farrand, S.K., Hallmann, J., and Sikora, R.A. 2002. Importance of the O-antigen, core-region and lipid A of rhizobial lipopolysaccharides for the induction of systemic resistance in potato to Globodera pallida. *Nematology* 4:73–79.

Rice, S.L., Leadbeater, B.S.C., and Stone, A.R. 1985. Changes in cell structure in roots in resistance potatoes parasitized by potato cyst-nematodes. I. Potatoes with resistance gene *H1* derived from *Solanum tuberosum* ssp. *andigena*. *Physiol. Plant Pathol.* 27:219–234.

Rice, S.L., Stone, A.R., and Leadbeater, B.S.C. 1987. Changes in cell structure in roots of resistant potatoes parasitized by potato cyst nematodes. 2. Potatoes with resistance derived from *Solanum vernei*. *Physiol. Mol. Plant Pathol.* 31:1–14.

Riggs, R.D., and Winstead, C.R. 1959. Studies on resistance in tomato to root-knot nematodes and on the occurence of pathogenic biotypes. *Phytopathology* 49:716–724.

Robinson, M.P., Atkison, H.J., and Perry, R.N. 1988. The association and partial characterisation of a fluorescent hypersensitive response of potato roots to the potato cyst nematodes *Globodera rostochiensis* and *G. pallida*. *Rev. Nematol.* 11:99–107.

Ross, J.P. 1958. Host-parasite relationship of the soybean cyst nematode in resistant soybean roots. *Phytopathology* 48:578–579.

Rossi, M., Goggin, F.L., Milligan, S.B., Kaloshian, I., Ullman, D.E., and Williamson, V.M. 1998. The nematode resistance gene *Mi* of tomato confers resistance against the potato aphid. *Proc. Natl. Acad. Sci. USA* 95:9750–9754.

Rouppe van der Voort, J., Van der Vossen, E., Bakker, E., Overmars, H., Van Zandvoort, P., Hutten, R., Klein-Lankhorst, R., and Bakker, J. 2000. Two additive QTLs conferring broad-spectrum resistance in potato to *Globodera pallida* are localized on resistance gene clusters. *Theor. Appl. Genet.* 101:1122–1130.

Rouppe van der Voort, J., Wolters, P., Folkertsma, R., Hutten, R., Van Zandvoort, P., Vinke, H., Kanyuka, K., Bendahmane, A., Jacobsen, E., Janssen, R., and Bakker, J. 1997. Mapping of the cyst nematode resistance locus *Gpa2* in potato using a strategy based on comigrating AFLP markers. *Theor. Appl. Genet.* 95:874–880.

Rouppe van der Voort, J.N.A.M., Janssen, G.J.W., Overmars, H., Van Zandvoort, P.M., Van Norel, A., Scholten, O.E., Janssen, R., and Bakker, J. 1999. Development of a PCR-based selection assay for root-knot nematode resistance (Rmc1) by a comparative analysis of the Solanum bulbocastanum and S. tuberosum genome. *Euphytica* 106:187–195.

Rouppe van der Voort, J.R., Lindeman, W., Folkertsma, R., Hutten, R., Overmars, H., van der Vossen, E., Jacobsen, E., and Bakker, J. 1998. A QTL for broad-spectrum resistance to cyst nematode species (*Globodera* spp.) maps to a resistance gene cluster in potato. *Theor. Appl. Genet.* 96:654–661.

Salmeron, J.M., Oldroyd, G.E.D., Rommens, C.M.T., Scofield, S.R., Kim, H.S., Lavelle, D.T., Dahlbeck, D. and Staskawicz, B.J. 1996. Tomato *Prf* is a member of the leucine-rich repeat class of plant disease resistance genes and lies embedded within the *Pto* kinase gene cluster. *Cell* 86:123–133.

Sandal, N.N., Salentijn, E.M.J., Kleine, M., Cai, D.G., Arens-de Reuver, M., Van Druten, M., de Bock, T.S.M., Lange, W., Steen, P., Jung, C., Marcker, K., Stiekema, W.J., and Klein-Lankhorst, R.M. 1997. Backcrossing of nematode-resistant sugar beet: a second nematode resistance gene at the locus containing *Hs1*$^{(pro-1)}$? *Mol. Breed.* 3:471–480.

Santhi, A., and Sivakumar, C.V. 1997. Biocontrol potential of Pseudomonas fluorescens (Migula) against root-knot nematode, Meloidogyne incognita (Kofoid and White, 1919) Chitwood, 1949 on tomato. *J. Biol. Control* 9:113–115.

Seah, S., Miller, C., Sivasithamparam, K., and Lagudah, E.S. 2000. Root responses to cereal cyst nematode (*Heterodera avenae*) in hosts with different resistance genes. *New Phytol.* 146:527–533.

Semblat, J.P., Rosso, M.N., Hussey, R.S., Abad, P., and Castagnone-Sereno, P. 2001. Molecular cloning of a cDNA encoding an amphid-secreted putative avirulence protein from the root-knot nematode *Meloidogyne incognita*. *Mol. Plant Microbe Interact.* 14:72–79.

Sikora, R.A. 1992. Management of the antagonistic potential in agricultural ecosystems for the biological-control of plant parasitic nematodes. *Ann. Rev. Phytopathol.* 30:245–270.

Skorupska, H.T., Choi, I.S., Raoarelli, A.P., and Bridges, W.C. 1994. Resistance to soybean cyst-nematode and molecular polymorphism in various sources of Peking soybean. *Euphytica* 75:63–70.

Slootmaker, L.A.J., Lange, W., Jochemsen, G., and Schepers, J. 1974. Monosomic analysis in bread wheat of resistance to cereal root eelworm. *Euphytica* 23:497–503.

Speckman, G.J., de Bock, T.S.M., and de Jong, J.H. 1985. Monosomic additions with resistance to beet cyst nematode obtained from hybrids of *Beta vulgaris* and wild *Beta* species of the section Patellaris. *Zeitschrift Pflanzenzucht* 95:74–83.

Tanksley, S.D., Ganal, M.W., Prince, J.P., De Vicente, M.C., Bonierbale, M.W., Broun, P., Fulton, T.M., Giovannoni, J.J., Grandillo, S., Martin, G.B., Messeguer, R., Miller, J.C., Miller, L., Paterson, A.H., Pineda, O., Roder, M.S., Wing, R.A., Wu, W., and Young, N.D. 1992. High-density molecular linkage maps of the tomato and potato genomes. *Genetics* 132:1141–1160.

Triantaphyllou, A.C. 1973. Environmental sex differentiation of nematodes in relation to pest management, In *Anual Review of Phytopathology*, pp. 441–462. Palo Alto, CA: Annual Reviews Inc.

Trudgill, D.L. 1967. The effect of environment on sex determination in *Heterodera rostochiensis*. *Nematologica* 13:263–272.

Van der Vossen, E., Rouppe van der Voort, J., Kanyuka, K., Bendahmane, A., Sandbrink, H., Baulcombe, D., Bakker, J., Stiekema, W., and Klein-Lankhorst, R. 2000. Homologues of a single resistance-gene cluster in potato confer resistance to distinct pathogens:a virus and a nematode. *Plant J.* 23:567–576.

Van Peer, R., Niemann, G.J., and Schippers, B. 1991. Induced resistance and phytoalexin accumulation in biological control of Fusarium wilt of carnation by Pseudomonas sp. strain WCS417r. *Phytopathology* 81:728–734.

Van Wees, S.C.M., Pieterse, C.M.J., Trijssenaar, A., Van T Westende, Y.A.M., Hartog, F., and Van Loon, L.C. 1997. Differential induction of systemic resistance in Arabidopsis by biocontrol bacteria. *Mol. Plant Microbe Interact.* 10:716–724.

Veremis, J.C., and Roberts, P.A. 1996. Identification of resistance to *Meloidogyne javanica* in the *Lycopersicon peruvianum* complex. *Theor. Appl. Genet.* 93:894–901.

Veremis, J.C., Van Heusden, A.W., and Roberts, P.A. 1999. Mapping a novel heat-stable resistance to *Meloidogyne* in *Lycopersicon peruvianum*. *Theor. Appl. Genet.* 98:274–280.

Vos, P., Simons, G., Jesse, T., Wijbrandi, J., Heinen, L., Hogers, R., Frijters, A., Groenendijk, J., Diergaarde, P., Reijans, M., Fierens-Osterenk, J., De Both, M., Peleman, J., Liharska, T., Hontelez, J., and Zabeau, M. 1998. The tomato *Mi-1* gene confers resistance to both root-knot nematodes and potato aphids. *Nat. Biotech.* 16:1365–1369.

Waetzig, G.H., Sobczak, M., and Grundler, F.M.W. 1999. Localization of hydrogen peroxide during the defence response of *Arabidopsis thaliana* against the plant-parasitic nematode *Heterodera glycines*. *Nematology* 1:681–686.

Williams, K.J., and Fisher, J.M. 1993. Development of *Heterodera avenae* Woll and host cellular- responses in susceptible and resistant wheat. *Fund. Appl. Nematol.* 16:417–423.

Williamson, V.M. 1998. Root-knot nematode resistance genes in tomato and their potential for future use. *Annu. Rev. Phytopathol.* 36:277–293.

Williamson, V.M. 1999. Plant nematode resistance genes. *Curr. Opin. Plant Biol.* 2:327–331.

Williamson, V.M. 2002. Functional analysis of the *Mi-1* gene. *Nematology* 4:147.

Williamson, V.M., and Hussey, R.S. 1996. Nematode pathogenesis and resistance in plants. *Plant Cell* 8:1735–1745.

Yaghoobi, J., Kaloshian, I., Wen, Y., and Williamson, V.M. 1995. Mapping a new nematode resistance locus in *Lycopersicon peruvianum*. *Theor. Appl. Genet.* 91:457–464.

Yu, M.H. 1984. Resistance to *Heterodera schachtii* in patellares section of the genus *Beta*. *Euphytica* 33:633–640.

15

Mechanisms Involved in Induced Resistance to Plant Viruses

Androulla Gilliland, Alex M. Murphy, Chui Eng Wong, Rachael A.J. Carson, and John P. Carr

15.1 Introduction and Definitions

During the coevolution of plants and their pathogens, the pathogens have developed a wide variety of strategies to infect and exploit their hosts. In response to this pressure, plants have countered by deploying a range of defense mechanisms. Some of these are conceptually simple, for example, defenses based on physical barriers such as the cell wall or cuticle, or resistance engendered by preexisting antimicrobial compounds (Osbourn, 1996). However, certain resistance mechanisms, most particularly those that are inducible, are complex in nature and have proved to be more difficult to understand.

Inducible resistance mechanisms can be triggered by exposure to pathogenic and nonpathogenic organisms, as well as by certain abiotic stimuli and chemicals. Some of these resistance responses are only local in extent, for example, the synthesis of phytoalexins (weak, broad spectrum antibiotics: Kuć, 1995; Hammerschmidt, 1999), or by localized programmed cell death that occurs close to the sites of pathogen penetration (see Section 15.2). However, several other inducible resistance mechanisms are expressed systemically, i.e., throughout the entire plant. The nomenclature used to describe inducible systemic resistance phenomena can be confusing and is not used in a uniform way throughout the literature. While originally synonymous, in recent literature systemic resistance induced by necrotizing pathogens via a salicylic acid-dependent mechanism is often referred to as "systemic acquired resistance" (SAR), while the term "induced systemic resistance" (ISR) has become a common term used to describe resistance mechanisms induced by nonpathogenic microbes via a jasmonic acid/ethylene dependent mechanism (see Section 15.2.3).

In general, SAR/ISR phenomena confer protection against a broad spectrum of pathogens, even though the initial induction of resistance may depend on a highly specific plant–pathogen interaction (Section 15.2.1.1). Thus, early studies showed SAR induced by one virus was effective against unrelated viruses (Ross, 1966) or even that SAR induced by a fungus could inhibit infection by a virus (Bergstrom et al., 1982). Similarly, in the case of ISR, resistance against pathogenic fungi can be induced by inoculation of roots with nonpathogenic rhizobacteria (Pieterse et al., 2001; Zehnder et al., 2001; Ton et al., 2002).

In recent years, it has become clear that there is yet another systemic resistance phenomenon in plants: RNA silencing. In contrast to SAR and ISR, RNA silencing is highly specific with respect both to its induction and activity. RNA silencing (Section 15.5) is a homology-based RNA degradation mechanism that probably occurs in all eucaryotes, including plants (Grant, 1999; Baulcombe, 2001; Voinnet, 2001). In plants, it appears to function, at least in part, as a defense mechanism against viruses (Waterhouse et al., 1999, 2001; Voinnet, 2001).

In this chapter, our aim is to review current knowledge of induced resistance mechanisms, specifically with respect to resistance against viruses. Viruses pose a distinct challenge to the plant. Unlike the cellular pathogens (fungi, oomycetes, and bacteria) all viruses are acellular obligate intracellular parasites that must replicate in intimate association with specific components of the host cell (Hull, 2002). As a consequence, most of the inducible defense factors discovered so far, particularly if they are extracellular or targeted against pathogen cell structure or function, have no impact on the virus life cycle. In addition to reviewing our rather scant knowledge of antiviral factors and mechanisms there will also be a discussion of the signal transduction networks that regulate resistance induction and how these might coordinate resistance against diverse pathogens. Moreover, we will address one of the most challenging areas in the study of plant virus resistance, namely, how the "classical" resistance phenomena of SAR and ISR may be coordinated with, or related to, RNA silencing.

15.2 The Hypersensitive Response and the Triggering of Broad Spectrum Induced Resistance to Pathogens: SAR and ISR

15.2.1 Pathogen-Induced Resistance: The Hypersensitive Response and SAR

Resistance to a pathogen is often accompanied by a response known as the hypersensitive response (HR): the rapid, localized death of cells at the infection site. The HR can occur in resistant plants in response to viruses, as well as bacteria, fungi, or nematodes (Goodman and Novacky, 1994; Kombrink and Schmelzer, 2001). In the best-understood systems, the occurrence of the HR depends upon the possession by the plant and invader of corresponding resistance (R) and avirulence (Avr) genes, respectively, also known as a "gene-for-gene" interaction (Flor, 1971). According to the modern conceptualization of the gene-for-gene interaction, which is based on recent progress in the isolation and functional analysis of R and Avr genes, R gene products are believed (or in some cases known) to act as receptor molecules that directly or indirectly detect specific elicitors, which are the direct or indirect products of the pathogen's *Avr* gene (Bergelson et al., 2001; Dangl and Jones, 2001).

In gene-for-gene interactions involving viruses, viral gene products identified as elicitors capable of triggering the HR include replication proteins (Padgett et al., 1997; Abbink et al., 1998; Erickson et al., 1999), viral capsid proteins (Culver and Dawson, 1989; Culver et al., 1994; Bendahmane et al., 1995), and viral movement proteins (Weber and Pfitzner, 1998). Plant *R* genes conferring hypersensitivity to a number of pathogens have been identified and isolated in sufficient numbers to allow their classification into several distinct families (Jones, 2000; Dangl and Jones, 2001). Unfortunately, relatively few virus-specific *R* genes have been examined so far making it difficult to draw any wide-ranging conclusions about any specific or unique properties that might distinguish them from others. Virus-specific *R* genes include the *N* resistance gene from *Nicotiana*, that confers hypersensitivity to *Tobacco mosaic virus* (TMV) and almost all other tobamoviruses (Dinesh-Kumar et al., 1995), as well as the *HRT* gene from *Arabidopsis*, which is required for the HR exhibited by plants of the Dijon ecotype infected with the carmovirus *Turnip crinkle virus* (TCV) (Cooley et al., 2000; Kachroo et al., 2000).

Plant Cell Death and Resistance

The HR is a correlative feature of many, but not all, resistance interactions controlled by *R* genes (Dangl et al., 1996; Shirasu and Schulze-Lefert, 2000). Conceivably, the cell death reaction seen in the HR may inhibit replication of certain pathogens or deprive them of nutrients. However, investigators now consider this to be a simplistic view and that a more important role for the HR is in the generation of signals that cause local and systemic changes in the plant. Perhaps this is why a local HR is often associated with the onset of systemic resistance (Dangl et al., 1996; Pennell and Lamb, 1997; Birch et al., 2000). Of course, cell death as necrosis can also occur in pathogen-infected susceptible plants, but this form of cell death is distinctly different from the HR (see below).

It is generally thought that the HR is a form of programmed cell death (PCD). PCD can be defined as cell death resulting from a complex set of genetically controlled physiological and morphological processes. These result in the selective destruction of cells that can be expended (Pennell and Lamb, 1997; Birch et al., 2000). It differs from necrosis, which is caused by microbial toxins or injury, and is not regulated and limited by the plant (Pennell and Lamb, 1997; Birch et al., 2000). It has been suggested that the HR is comparable to the form of animal PCD known as apoptosis. Features shared by the HR and apoptosis include the activation of complex signaling networks, changes in ion fluxes, the generation of reactive oxygen species, and changes in protein phosphorylation (reviewed and discussed in detail by Birch et al., 2000; Heath, 2000; Gilchrist, 1998; Pennell and Lamb, 1997, and by Watanabe and Lam (Chapter 1 of this volume)). Several recent results are consistent with the idea that the HR may be similar to apoptosis. For example, a number of groups have reported cysteine protease activity during the HR, which may be indicative of caspase-type activity (reviewed by Birch et al., 2000; Lam and del Pozo, 2000). Furthermore, Bax, an animal PCD effector protein, induced plant cell death when expressed from a viral vector (Lacomme and Santa Cruz, 1999). However, there is at present no evidence that plant genomes encode either caspases or Bax-type proteins.

Whatever mechanism of PCD is used during HR, the key question we need to answer is whether or not host cell death is required for resistance to viruses. On one hand, there is often a correlative association between host cell death in the HR and resistance, and recent evidence indicates that certain natural products produced during the HR (e.g., scopoletin: Chong et al., 2002) may have antiviral activity. However, there is no direct evidence showing that cell death is an absolute requirement for the limitation of virus multiplication, at least during the period immediately following the appearance of the HR. This point is exemplified by the results of three studies that investigated the interaction of *NN* genotype *Nicotiana* plants with TMV. In the first of these examples, Weststeijn (1981) exploited the temperature-sensitive nature of the HR and TMV localization in tobacco containing the *N* resistance gene to show that increased temperature could be used to facilitate the escape of virus from lesions for up to 12 days after HR appearance. More recently, Santa Cruz and colleagues (Wright et al., 2000) used genetically engineered TMV expressing green fluorescent protein (GFP) to confirm that virus remained in living cells at the periphery of the HR lesion for several days following the appearance of the HR in *NN* genotype *Nicotiana edwardsonii*. In the third example, cell death was inhibited by growing *NN* genotype tobacco plants in a low oxygen atmosphere, but TMV remained localized in these plants (Mittler et al., 1996).

Studies in other systems indicate at the genetic level that cell death and resistance can be entirely separate phenomena. For example, an examination of the factors controlling the induction of the HR in cowpea by *Cucumber mosaic virus* (CMV) showed that specific and distinct amino acids within the viral RNA polymerase sequence were responsible for the induction of virus localization and the elicitation of cell death (Kim and Palukaitis, 1997). Furthermore, cell death and resistance induction triggered by *Cauliflower mosaic virus* (CaMV) in *Nicotiana* species are controlled by separate host genes (Cole et al., 2001). Taken with the results of the *N* gene/TMV system, these findings strongly suggest that cell death alone is not responsible for virus localization during or after the HR.

SAR and Salicylic Acid

The induction of HR during a resistance response often results in the induction of SAR. Probably the best-known example of this effect in plant–virus interactions is the response of tobacco plants possessing the *N* resistance gene. Ross (1961a,b) showed that inoculation of these plants with TMV resulted in an enhanced degree of resistance to a second inoculation with the virus. This was manifested as the formation of smaller and fewer necrotic lesions not only in tissue close to the primary lesion (Ross 1961a) but also on uninoculated parts of the plant (Ross, 1961b). Subsequent work showed that SAR induced by one pathogen could confer resistance to unrelated pathogens (Ross 1966; Bergstrom et al., 1982; Naylor et al., 1998). We now know that the HR and SAR are coordinated and controlled by a complex signal transduction network.

Salicylic acid (SA) plays a central role in the signal transduction pathway that results in SAR. Indeed, SA has repeatedly been shown to accumulate to high levels in both primary inoculated tissue as well as in distal tissue displaying SAR (Malamy et al., 1990; Métraux et al., 1990, and also reviewed by Dempsey et al., 1999). In fact, the signaling pathways leading to SAR are dependent on endogenous accumulation of SA. If this is blocked by engineering plants to express the salicylate degrading enzyme SA-hydroxylase (*nahG*-transgenic plants: Gaffney et al., 1993; Delaney et al., 1994; Mur et al., 1997), or by mutation of SA biosynthetic genes (Nawrath and Métraux, 1999), plants are not able to express SAR and are more susceptible to pathogens. Due, in part, to conflicting results obtained from grafting experiments with different *nahG*-transgenic tobacco lines (Vernooij et al., 1994; Darby et al., 2000) it is not at all clear whether SA is the translocated signal responsible for establishing SAR throughout the plant. Thus, the nature of the mobile SAR-inducing signal remains to be established, although recent work with the *Arabidopsis dir*-1-1 mutant suggests that a lipid-derived molecule may be the signal (Maldonado et al., 2002).

A prominent feature of the HR and the induction of SAR is the synthesis by the plant of pathogenesis-related (PR) proteins. These are a highly diverse group of proteins that have been classified into a number of families (van Loon and van Strien, 1999). Some PR proteins are induced by SA, while others are regulated by other factors such as ethylene or jasmonic acid (see Section 15.2.3), or combinations of factors (Schenk et al., 2000).

Several of the PR proteins have been shown to have direct antimicrobial activity. For example, those with chitinase or β-1,3-glucanase activity can cause the breakdown of fungal cell walls (Mauch et al., 1988; Rauscher et al., 1999; Schlumbaum et al., 1986). Some PR proteins, including the most-studied PR protein, PR-1, are known to have antimicrobial properties but their mode of action remains unknown (Alexander et al., 1993; Niderman et al., 1995). So far, none of the PR proteins examined to date have been shown to have any clear antiviral activity (see Section 15.3.3). Paradoxically, overexpression of extracellular β-1,3-glucanase PR proteins (PR-2 family; van Loon and van Strien, 1999) can promote viral movement due to the increased breakdown of callose around the plasmodesmata (Bucher et al., 2001). Nevertheless, because PR proteins accumulate so abundantly, they are routinely used as a general marker for the induction of SAR (Ward et al., 1991; Kessmann et al., 1994).

In *Arabidopsis* the NPR1 protein plays a key role downstream of SA in the induction of many *PR* genes and in the establishment of SAR against fungal and bacterial pathogens (Cao et al., 1994; Glazebrook et al., 1996; Delaney et al., 1995; Shah *et al.*, 1997). NPR1 is a 65 kDa protein containing ankyrin repeats (Cao et al., 1997; Ryals et al., 1997) that interacts directly with members of the TGA/OBF family of transcription factors (Zhou et al., 2000). Two of these NPR1-interacting transcription factors have now been found to bind the crucial element of the *PR-1* promoter (Zhou et al., 2000) that has previously been shown to be involved in SA-induction of resistance (Lebel et al., 1998). However, as will be discussed later (Sections 15.3.2 and 15.3.3), SA-induced resistance to viruses does not require PR-protein expression and nor is it dependent on the activity of NPR1.

15.2.2 Chemically Induced Resistance

Many different classes of chemicals can induce some form of resistance (Kuć, 2001). However, in searches for agronomically useful molecules that could mimic the action of SA to induce resistance against viruses and other pathogens, two chemicals have been studied in greatest detail: benzo(1,2,3)-thiadiazole-7-carbothioic acid S-methyl ester ("Bion" or "Actigard") and INA (2,6-dichloroisonicotinic acid) (Friedrich et al., 1996; Lawton et al., 1996; Gorlach et al., 1996). Of these, Bion has been deployed commercially in several countries as a plant protectant or growth-promoting chemical (Oostendorp et al., 2001). Other chemicals, which are the subject of intense investigation are BABA (ß-aminobutyric acid) and Oryzemate (probenazole). BABA protects against a wide range of pathogens, including viruses. However, it has not been resolved whether or not it acts via a stimulation of the SA pathway (Jakab et al., 2001). In contrast, probenazole-induced resistance has been shown to require SA and NPR1 activity (Yoshioka et al., 2001). None of these chemicals show direct antimicrobial activity *in vitro*, but activate resistance to the same range of pathogens as biotic inducers of SAR (Oostendorp et al., 2001; Jakab et al., 2001).

Of course, SA itself and aspirin (acetylsalicylic acid), when applied onto TMV-resistant or susceptible tobacco leaves trigger resistance responses (White, 1979; White et al., 1983). It was this early work that prompted investigation of SA levels in pathogen-infected plants (Malamy et al., 1990; Metraux et al., 1990) and laid the groundwork for many subsequent studies of signal transduction in SAR induction. In susceptible plants resistance is characterized by a delay in the onset of disease symptoms and by a decreased yield of virus (Chivasa et al., 1997; Naylor et al., 1998; White et al., 1983).

15.2.3 Nonpathogenic Rhizobacterial-Induced Resistance

The plant hormones jasmonic acid (JA) and ethylene (ET) have emerged as important signals in the induction of ISR (van Loon et al., 1998). ISR is a relatively recently discovered induced resistance phenomenon triggered by colonization of plant roots by some nonpathogenic bacteria (van Loon et al., 1998). Depending on the colonizing bacterium, ISR can be mediated either by SA or independently of SA (by JA and/or ET), both via NPR 1 (Pieterse et al., 2001; Ton et al., 2002) (see Figure 15.1). Recent work using the ISR-stimulating rhizobacterium *Pseudomonas fluorescens* WCS417r in *Arabidopsis* showed that effective resistance could be induced against the fungus *Alternaria brassicicola*, the bacterium *Xanthomonas campestris*, and the oomycete *Peronospora parasitica* (Ton et al., 2002). Interestingly, using the same ISR-inducing bacterial strain they showed that ISR induced

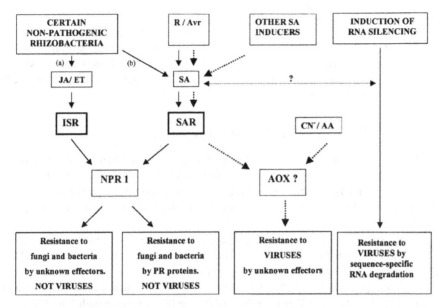

FIGURE 15.1. Model of some of the pathways leading to resistance to viruses and other pathogens. The dashed arrows highlight the pathways we think are specific to virus resistance based on evidence from tobacco and *Arabidopsis*. This pathway leads to resistance to viruses using salicylic acid (SA), and potentially alternative oxidase (AOX), as pathway intermediates. The virus specific signaling pathway is activated independently of the other resistance pathways by specific recognition of a virus or by other inducers of SA, as well as cyanide (CN⁻) and antimycin A (AA). Recent evidence also indicates possible interaction(s) between virus resistance mediated by RNA silencing and SA-mediated signaling pathways (dotted double headed arrow).

The solid arrows highlight signaling pathways leading to resistance to other pathogens such as fungi and bacteria. This can be through an *R/Avr* recognition SA-dependent signaling pathway or through a nonpathogenic rhizobacterial-induced systemic resistance (ISR) signaling pathway, which itself can operate through the signaling molecules (a) jasmonate (JA) and ethylene (ET) or (b) SA. The pathways leading to resistance to bacteria and fungi operate through the induction of the *Npr 1* gene, which leads to the expression of pathogenesis-related (PR) proteins.

via the JA/ET pathway was not effective against TCV (Ton et al., 2002), indicating that a separate signaling pathway is involved (see Figure 15.1).

15.3 SAR and SA-Induced Resistance to Viruses

Extensive reprogramming of both primary and secondary plant metabolism and gene expression levels is initiated during a resistance response (Hammond-Kosack and Jones, 1996; Kombrink and Schmelzer, 2001; Dixon, 2001). Some of the host protein changes are directly involved in the resistance response, for example, biosynthesis of phytoalexins and SA, but some of these may only be secondary to the defense response and it is difficult with our present knowledge to distinguish between these roles (Hull, 2002). However, signaling pathways and resistance mechanisms, that are known (or thought to be), involved in virus resistance will be highlighted in this section.

15.3.1 The Biology of Plant Viruses

Viruses pose a distinct problem to plant defenses. Since they are dependent on host factors for their replication and movement through the plant (Hull, 2002), the potential targets for plant defense mechanisms are for the most part unique to viruses and distinct from those that could be useful in defense against bacteria and fungi.

Although some plant viruses utilize negative-sense single- or double-stranded RNA, most plant viruses possess genomes consisting of positive-sense (i.e., mRNA sense) single-stranded (ss) RNA. These viruses replicate and in some cases synthesize "sub-genomic" mRNA in the cytoplasm of host cells using an RNA-dependent RNA polymerase (RdRp) complex consisting of proteins encoded by the virus plus factors seconded from the host cell (Buck, 1996; Hull, 2002). There are fewer groups of DNA viruses that infect plants, although some of the diseases they cause can be serious (Hull, 2002). The two best-studied groups of plant DNA viruses are the geminiviruses and the caulimoviruses. Geminiviruses possess circular, single-stranded DNA genomes and replicate in the host nucleus using a host DNA polymerase (Hanley-Bowdoin et al., 1999; Hull, 2002). The caulimoviruses are double-stranded DNA pararetroviruses; that is, they encode a reverse transcriptase that allows them to replicate via an RNA intermediate (Hohn and Futterer, 1997; Hull, 2002).

Viruses such as TMV can enter a plant cell through small wounds caused by abrasion but many other types of virus are introduced into the plant by other organisms acting as vectors. Inside the cell, the virus uncoats, replicates, and begins the process of local, cell-to-cell, movement. Most viruses produce one or more movement proteins that mediate transfer of viral RNA (or in some cases entire virus particles) between neighboring cells via cytoplasmic connections called plasmodesmata (Carrington, 1999; Heinlein, 2002). Eventually, the virus reaches the host's vascular system and can begin moving systemically. Although the process by which viruses enter the vasculature is poorly understood, it is known that most viruses move in the phloem tissue, the elements of which are responsible for translocation of carbohydrates and other metabolites around the plant (Leisner and Turgeon, 1993; Nelson and van Bel, 1998). Viruses are translocated preferentially toward young leaves, where they unload from the veins and begin to invade the surrounding tissue (Oparka and Santa Cruz, 2000).

In principle, any of the stages in the viral infection process—entry, replication, intercellular movement, and systemic movement—could be targets of induced resistance.

15.3.2 A Virus-Specific Signaling Pathway Occurring Downstream of SA: A Role for Alternative Oxidase?

SAR against viruses is mediated by a distinct pathway that, downstream of SA, splits away from the branch leading to NPR1-regulated PR protein induction (see Figure 15.1). The initial evidence for this was based on pharmacological experiments. In tobacco, SA-induced resistance to the replication of TMV and PVX and to the movement of CMV was inhibited by salicylhydroxyamic acid (SHAM) (Chivasa et al., 1997; Naylor et al., 1998). However, SHAM did not prevent SA-induced synthesis of PR proteins or prevent SA-induced resistance to fungal or bacterial pathogens (Chivasa et al., 1997). In later experiments, nonlethal concentrations of antimycin A (AA) or cyanide (CN^-) were found to induce resistance to TMV in susceptible tobacco (Chivasa et al., 1997) and to the related tobamovirus Turnip vein-clearing virus (TVCV) in *Arabidopsis* (Wong et al., 2002), without inducing *PR1* gene expression. Based on this pharmacological evidence, a model was proposed in which the signal transduction pathways involved in virus resistance separate downstream of SA; one branch (sensitive to SHAM) leads to resistance to viruses, the other (SHAM insensitive) to the induction of PR proteins and to bacterial and fungal resistance (see Figure 15.1). Genetic evidence gained using *Arabidopsis* mutants appears to confirm this model. Kachroo et al. (2000) showed that *HRT* gene-mediated resistance to TCV was SA-dependent, but independent of the *NPR1* gene. Wong et al. (2002) found that SA- and AA-mediated resistance to TVCV is also independent of NPR1 activity.

All plants have a mitochondrial alternative oxidase (AOX) that by itself constitutes a branch of the cytochrome path linking ubiquinol oxidation directly to the reduction of oxygen to water but without the concomitant generation of ATP. This branch, which consists of a single component, the alternative oxidase (AOX), is usually referred to as the alternative respiratory pathway (Affourtit et al., 2002). SHAM is an inhibitor of AOX and both AA and CN^- stimulate its activity (Laties, 1982; Moore and Siedow, 1991). AA and CN^- inhibit electron transfer in the cytochrome pathway at complexes III and IV, respectively, thus forcing the engagement of the alternative pathway (see Murphy et al., 1999). The known functions of the alternative pathway are in thermogenesis (heat production) in *Arum* plants and, of more general importance, in the dissipation of reactive oxygen species (Maxwell et al., 1999). The findings that SHAM, AA, and CN^- can affect induced resistance to viruses has led us to suggest that AOX may play another role in plant biology, namely as an element in defensive signal transduction (Figure 15.1).

Additional support for the role of AOX in the transduction of virus-specific defense signals has come from our own work and that of others. Both *Aox* gene expression and AOX protein accumulation are elevated in plant tissue (tobacco and *Arabidopsis*) undergoing the HR, expressing SAR or after treatment with SA or INA (Lennon et al., 1997; Chivasa and Carr, 1998; Lacomme and Roby, 1999; Simons et al., 1999). Additionally, we have found in *Arabidopsis* that the induction of *Aox* gene expression by SA and AA is not dependent upon the presence of a functional *Npr1* gene, consistent with a role in NPR1-independent signaling

(Wong et al., 2002). It is interesting to speculate that AOX may play a role in defensive signaling by modulation of reactive oxygen species levels or redox poise in the mitochondrion. Unfortunately, the current pharmacological and correlative data does not provide conclusive evidence that AOX plays a role(s) in the HR or SA-induced resistance to viruses. We are currently investigating this using a viral gene expression vector (TMV.AOX) and transgenic tobacco and *Arabidopsis* transformed with *Aox* cDNA sequences under the control of the constitutive 35S promoter from CaMV to test the effects of altering AOX levels on virus resistance and the HR. Using similar transgenic plants another group has suggested that AOX cannot be a critical component in SA-induced virus resistance (Ordog et al., 2002). However, we have evidence that other factors, particularly the induction of the SA-regulated RdRp (Section 15.5.2) may obscure the role of AOX in this pathway (Gilliland et al., 2003). Thus, it may be premature to dismiss a potential role for AOX in virus resistance. Rather, our latest evidence suggests that AOX may regulate signaling (possibly via modulation of mitochondrial ROS levels) leading to a subset of resistance responses (Gilliland et al., 2003).

15.3.3 Resistance Against Viruses in SA-Treated and SAR-Expressing Plants

Although it is now evident that SA-induced resistance to viruses lies on a separate branch of the signal transduction pathway to that enhancing resistance to cellular pathogens (Section 15.3.1), we have not yet identified the host gene products responsible for limiting virus spread. In contrast, a number of the components responsible for limiting the spread of fungal and bacterial pathogens have been identified, which include the PR proteins (Section 15.2.1.2). However, none of the currently identified PR proteins have been implicated in virus resistance. Transgenic tobacco plants constitutively expressing one or more PR proteins were still susceptible to TMV (Cutt et al., 1989; Linthorst et al., 1989) but showed enhanced resistance against oomycete and fungal pathogens (Alexander et al., 1993). Recently, an SA-regulated host-encoded RdRp was identified in tobacco and shown to have antiviral properties (Xie et al., 2001). However, it was found that antisense suppression of the gene did not abolish SA-induced resistance to viruses, indicating that it is not essential for induced resistance to viruses, although it could still contribute to resistance (see Section 15.5).

15.3.4 SA can Interfere with Virus Replication

SA treatment can inhibit the accumulation of certain positive sense ssRNA viruses in directly inoculated tissues and/or protoplasts from tobacco and *Arabidopsis* and cowpea (Chivasa et al., 1997; Hooft van Huijsduijnen et al., 1986; Naylor et al., 1998; Murphy and Carr, 2002; Wong et al., 2002). An early study in cowpea protoplasts demonstrated that SA treatment could interfere with Alfalfa mosaic virus (AlMV) replication (Hooft van Huijsduijnen et al., 1986). In TMV-susceptible

FIGURE 15.2. Chemically induced suppression of viral disease symptoms. Untreated or salicylic acid (SA) treated *Arabidopsis thaliana* (ecotype Columbia) plants were inoculated with *Cauliflower mosaic virus* (CaMV) strain Cabb-BJI. The plants were photographed at 13 days postinoculation. The untreated plant displays typical CaMV-induced disease symptoms (leaf distortion and chlorosis on the uninoculated leaves). The SA-treated plant displays suppressed CaMV symptoms. Surviving inoculated leaves are indicated by arrows. Scale bar = 1.5 cm.

tobacco leaf tissue, SA caused a dramatic reduction of TMV RNA accumulation (Chivasa et al., 1997). More specifically, it was also found that the ratio of genomic RNA to coat protein mRNA and the ratio of plus- to minus- sense RNAs were affected by SA, suggesting that SA induces interference with the activity of the TMV RdRp complex (Chivasa et al., 1997; Naylor et al., 1998). Similar effects of SA on TMV RNA accumulation were observed in mesophyll protoplasts generated from SA-treated tobacco plants, demonstrating that in this case, SA-induced resistance is operating at the single cell level (Murphy and Carr, 2002). This shows inhibition of replication, rather than cell-to-cell movement is the principal effect. However, inhibition of virus movement also plays a role in SA-induced resistance to TMV in intact leaf tissue (Section 15.3.6).

    Much of the work on induced resistance has focused on RNA viruses. However, the virus specific defense pathway not only enhances resistance against RNA viruses, but also against a DNA virus, CaMV. Both SA and CN⁻ treatment of *Arabidopsis* delayed the appearance of symptoms caused by CaMV (Figure 15.2 and Table 15.1). SA treatment severely depressed the accumulation of CaMV DNA and probably as a consequence, the 35S and 19S RNA species (Carson, 1999). The SA-induced resistance against CaMV occurred in directly inoculated tissue indicating that replication, or possibly cell-to-cell movement, is being targeted (Carson, 1999). SHAM could also antagonize SA-induced resistance against CaMV (Table 15.1), although it must be pointed out that in many experiments SHAM proved a less consistent inhibitor in *Arabidopsis* than in tobacco (discussed in Wong et al., 2002). This data shows that the virus specific defense pathway can

TABLE 15.1. The effect of SA, CN⁻, and SHAM on CaMV-induced disease symptoms in *Arabidopsis thaliana.*

Treatment[b]	Plants displaying symptoms[a] by:			
	14 dpi	16 dpi	18 dpi	21 dpi
Control	0/18	8/18	14/18	14/18
SA	0/18	0/18	0/18	0/18
CN⁻	0/18	1/18	2/18	3/18
SA + SHAM	0/18	2/18	8/18	10/18
CN⁻ + SHAM	1/18	6/18	12/18	13/18

[a] Systemic symptoms: chlorosis and distortion of noninoculated, and young, emerging leaves (see Figure 15.2), appearing 14–21 days post inoculation (dpi).
[b] Plants were sprayed with water (Control), 1 mM salicylic acid (SA) or 50 μM potassium cyanide (CN⁻), alone or in combination with 1 mM salicylhydroxamic acid (SHAM) (SA + SHAM; CN⁻ + SHAM) for four days prior to inoculation with *Cauliflower mosaic virus* (CaMV).

induce interference with the life cycle of viruses with very different replication and gene expression strategies (Section 15.3.1).

15.3.5 SA Can Inhibit Virus Long-Distance Movement

Not all viruses are affected at the inoculation site in SA-treated plants. CMV can evade SA-induced interference with replication in tobacco (Naylor et al., 1998; Ji and Ding, 2001). However, SA-treated tobacco plants show a marked delay in CMV symptom development. It was found that although CMV could replicate in directly inoculated SA-treated tobacco leaves, its entry into the phloem cells was delayed (Naylor et al., 1998). Similar results were also observed with AlMV (Naylor, 1999). Presumably, SA affects one or more cell types within the vascular bundle in a way that prevents or slows down phloem loading. CMV inoculated onto *N*-gene tobacco expressing SAR due to prior exposure to TMV was also restricted in long-distance movement (Naylor et al., 1998). Mutant CMV that is unable to express the 2b resistance suppressor protein succumbs to SA-induced interference with CMV replication (Ji and Ding, 2001) and this is discussed in further detail in Section 15.5.

15.3.6 SA has Cell-Specific Effects

Whilst work with CMV revealed that SA could target long-distance movement of viruses in addition to their replication, further investigation using viruses expressing GFP unveiled the ability of SA to have distinct effects on the same virus in different cell types (Murphy and Carr, 2002). Treatment of susceptible tobacco with SA restricted TMV expressing GFP (TMV.GFP) to single epidermal cell infection sites. The replication of TMV.GFP in single epidermal cells appeared similar in control as well as SA-treated plants, as judged by GFP fluorescence levels, indicating that SA was inhibiting cell-to-cell movement. Recovery of cell-to-cell movement was achieved to some extent when TMV movement protein (MP) was supplied *in trans* from tobacco plants constitutively expressing TMV-MP (Murphy and Carr, 2002). However, even in TMV-MP transgenic plants that had been treated with SA, TMV.GFP was restricted to the epidermal cell layer and did not appear to move into the mesophyll cell layer beneath. This data demonstrates that SA can inhibit cell-to-cell movement of TMV.GFP in the epidermis, but interferes with TMV.GFP replication in the mesophyll cell (Murphy and Carr, 2002).

15.3.7 Relevance of SA- and Cyanide-Induced Resistance to Plants Expressing R-gene Mediated Resistance and SAR

Most of our experiments on SA-induced resistance to viruses have been carried out in susceptible plants, i.e., plants that do not possess a resistance gene for the challenging virus (e.g., Chivasa et al., 1997; Naylor et al., 1998; Murphy and Carr, 2002; Wong et al., 2002). These studies showed that SA-induced resistance phenomena occur in the absence of any *R/Avr* gene interaction or HR-associated cell death. This tells us that although cell death is not required to restrict virus spread, the *R/Avr*-gene interaction may be necessary for resistance that can completely halt the virus. SA treatment usually impedes virus infection rather than preventing it altogether.

Investigations of TMV in SA-treated susceptible tobacco can also shed some light on why fewer and/or smaller visible necrotic lesions appear on SAR-expressing *N*-gene tobacco plants after inoculation with TMV. TMV.GFP is limited to single-cell infection sites in SA-treated tobacco (Murphy and Carr, 2002). This finding may be significant since the HR mediated by the *N*-gene, unlike many other pathogen-induced cell death phenomena, cannot occur at the single cell level. For example, TMV does not cause necrosis of TMV-infected protoplasts from *N*-gene tobacco (Otsuki et al., 1972) and a movement-deficient TMV.GFP construct that could only infect single epidermal cells did not elicit cell death in an *N*-gene containing host (Wright et al., 2000). These studies indicate that *N*-gene mediated cell death can occur only when the virus has infected a group of cells, although so far, the number of cells that constitute a "doomed quorum" is still not known. Thus, in SA-treated tobacco or SAR-expressing *N*-gene containing tobacco, the reduction in the number of HR lesions produced after challenge with TMV may be due, in part, to the limitation of virus to single cells or groups of cells that are too small in number to trigger the HR.

It has also been demonstrated that the virus-specific signaling pathway (Section 15.3.2) is essential for *N* gene-mediated resistance against TMV (Chivasa and Carr, 1998). When *N* gene-containing tobacco was transformed with the bacterial *nahG* gene (see Section 15.2.1.2), it was found that these transgenic plants could no longer restrict the spread of TMV or TMV-induced necrosis (Bi et al., 1995; Ryals et al., 1995; Mur et al., 1997). Thus, it was proposed that SA is required for virus localization early on in the HR (Mur et al., 1997). It was then found that cyanide treatment restored *N* gene-mediated resistance to TMV in plants expressing SA hydroxylase (Chivasa and Carr, 1998). Thus, the virus specific defense pathway is required for *N* gene-mediated TMV localization as well as for the subsequent establishment of acquired resistance.

15.4 Non-SA-Dependent Chemically Induced Resistance to Viruses

Although chemically induced forms of resistance to plant viruses have been well studied for many years, the plant gene products that are directly responsible for inhibiting one or more aspects of the viral life cycle have remained elusive. Ueki and Citovsky (2002) recently identified the protein mediating an inducible resistance phenomenon that inhibits systemic movement of TVCV in tobacco. In recent years, most studies of inducible resistance to viruses have concentrated on either RNA silencing (Section 15.5) or on resistance mechanisms regulated by SA (Section 15.3). In contrast, Citovsky and co-workers have focused on an inducible resistance phenomenon that is quite distinctive in that it apparently antagonizes the establishment of systemic gene silencing (Ueki and Citovsky, 2001). Furthermore, it is not reliant on salicylic acid-mediated signaling (Citovsky et al., 1998). This resistance to viruses is induced by treatment of plants with nontoxic concentrations of ions of the heavy metal cadmium (Ghoshroy et al., 1998). The protein that Ueki and Citovsky (2002) identified, CdiGRP, is a glycine-rich protein that promotes the accumulation of callose in the vascular tissue. This callose build up might restrict the unloading of viruses out of the phloem and inhibits the systemic virus movement.

15.5 RNA Silencing and Resistance to Viruses

15.5.1 RNA Silencing Mechanisms

RNA silencing, otherwise known as RNA interference, quelling or posttranscriptional gene silencing, is a process which inhibits the expression of homologous genes and occurs in many eukaryotic organisms (Grant, 1999; Baulcombe, 2001; Voinnet, 2001). RNA silencing is a sequence-specific RNA degradation process, affecting all highly homologous sequences in which foreign, overexpressed, or aberrant RNA molecules are targeted for destruction in a sequence-specific manner (Baulcombe, 2001; Voinnet, 2001).

Work with *Arabidopsis* mutants defective in RNA silencing showed that their susceptibility to virus infection was increased (Morrain et al., 2000). These and other findings have lent support to the theory that RNA silencing may have evolved as an intrinsic mechanism to protect cellular organisms from virus infection (reviewed by Waterhouse et al., 1999, 2001; Voinnet, 2001). Most RNA in cells is single-stranded. The occurrence of double-stranded (ds) RNA in the cytoplasm normally occurs most often as a result of the replication cycle of most types of

RNA virus (Buck, 1996; Hull, 2002). Viral dsRNA is an intermediate in replication and is produced by the activity of the virus-encoded RdRp (Section 15.3.1). Thus, the synthesis of dsRNA might signal to the plant cell that it is being invaded by viruses (Baulcombe, 2001). Some wild-type, nontransgenic plants have been shown to recover from infection with certain types of viruses by specifically degrading viral RNA (Covey et al., 1997; Ratcliff et al., 1997). Following recovery, plants were immune to secondary infection by the same or closely related viruses (Ratcliff et al., 1997; 1999).

An important property of RNA silencing in plants is that it spreads systemically (Palauqui et al., 1997; Voinnet and Baulcombe, 1997). Initially, the translocated signal was thought to consist of 21–25 nucleotide dsRNA molecules that accumulated in plants displaying silencing (Hamilton and Baulcombe, 1999; Waterhouse et al., 2001). These short RNAs, now known as short interfering RNAs (siRNAs: Voinnet, 2001), are produced by a recently discovered enzyme termed "dicer", a member of the RNase III family of nucleases. Dicer specifically cleaves dsRNA, producing small nucleotides which then serve as "guide sequences" that target an associated nuclease complex to degrade specific mRNA (Bernstein et al., 2001; Lipardi et al., 2001; Di Serio et al., 2001). The substrate for dicer is dsRNA that is produced by the action of either viral or host RdRp enzymes that are thought to play a key role in the initiation of RNA silencing (reviewed by Ahlquist, 2002). Recently, the role of siRNAs in systemic signaling has been thrown into some doubt by the finding that the potyviral HC-Pro protein, a viral counter-defense protein, can inhibit production of siRNAs without abolishing propagation of the systemic silencing signal (Mallory et al., 2001). However, the role originally proposed for siRNAs in determining the sequence-specificity of silencing is still generally accepted (Hamilton and Baulcombe, 1999).

15.5.2 Are there Connections Between RNA Silencing and Salicylic Acid-Mediated Resistance to Viruses?

Although it might seem that "classical" SAR and RNA silencing are separate and distinct induced antiviral systems, two lines of evidence suggest a connection. The first line of evidence comes from studies of the 2b protein of CMV, a multifunctional protein that influences symptom severity and virus movement (Soards et al., 2002), as well as interfering with host defense (Li and Ding, 2001). The 2b protein of CMV was one of the first viral proteins to be identified as a suppressor of RNA silencing (Brigneti et al., 1998). However, since then several plant (Vance and Vaucheret, 2001) and animal (Li et al., 2002) viruses have been shown to possess a silencing suppressor protein with a variety of properties. In the case of the 2b protein, that can prevent the initiation of gene silencing (Béclin et al., 1998). The silencing suppressor activity appears to be related to its ability to accumulate in the host cell nucleus (Lucy et al., 2000), where it is thought to interfere with defensive gene expression (Mayers et al., 2000).

Ji and Ding (2001) have shown that in addition to preventing the initiation of gene silencing, the 2b protein can interfere with SA-induced resistance to viruses.

They used a mutant of CMV that cannot express functional 2b protein (CMV∆2b) and transgenic *Nicotiana* plants expressing the 2b protein to investigate the effect of the 2b protein on SA-induced resistance to CMV. They found that the 2b protein suppresses SA-induced *Aox* gene expression and that the local accumulation and movement of CMV∆2b is, in contrast to wild-type CMV (Naylor et al., 1998), inhibited in directly inoculated tissue of SA-treated plants (Ji and Ding, 2001). Furthermore, when TMV was modified to express the CMV 2b protein, its replication was not as severely inhibited as wild-type TMV replication in SA-treated tissue (Ji and Ding, 2001). What is the 2b protein doing to plant defense? Possibly, it is disrupting defensive signaling (Guo and Ding, 2002), gene expression (Mayers et al., 2000; Ji and Ding, 2001) or both. Another possibility is that it is preventing SA from priming an RNA silencing mechanism (see below).

The second line of evidence suggesting a connection between RNA silencing and SAR arose from the recent discovery that a tobacco RdRp gene (*NtRdRp1*) was induced by TMV infection, by SA treatment and by treatment with SA analogues (Xie et al., 2001). Plant RdRps are believed to be crucial factors in the induction of many RNA silencing phenomena (Ahlquist, 2002). Certainly, NtRdRp is able to perform an antiviral role as shown when transgenic antisense RdRp1 plants, deficient in SA-inducible NtRdRp activity, were found to be more susceptible to both TMV and PVX than wild-type plants (Xie et al., 2001). Thus, a potential role for SA may be to increase the production of an RdRp required for RNA silencing. It should be noted however, that when transgenic antisense RdRp1 plants were pretreated with SA, they did not resist TMV any better than wild-type plants (Xie et al., 2001). This implies that virus resistance in SA-treated tobacco plants requires additional factors such as those described in Section 15.3.

15.6 Future Work and Perspectives

Viruses are important plant pathogens but our knowledge of viral pathogenesis and host resistance is still relatively primitive. Nevertheless, we are steadily accumulating information, which will help us to further our understanding of viral diseases and perhaps even to control them more effectively in the future. There are many aspects of antiviral resistance mechanisms for which we have little or no information. For example, are ion fluxes and changes in Ca^{2+} homeostasis involved, and indeed are they important in the early stages of the resistance response? Is there a build up of reactive oxygen species in mitochondria of plants infected with viruses and if so, what role does AOX have with respect to these species? The use of *Arabidopsis* mutants has revolutionized our knowledge of induced host resistance mechanisms. Unfortunately, up to now most mutant screens have been devised with fungal or bacterial pathogens in mind or used PR gene promoters to drive reporter gene expression. This has meant that host genes controlling virus-specific signaling may have been overlooked.

Engineering disease resistance in transgenic plants has been used very successfully to control virus diseases, particularly due to the use of pathogen-derived resistance, in which plants are transformed with viral gene sequences (Fitchen and Beachy, 1993; Wilson, 1993). However, pathogen-derived resistance is usually very virus- or strain-specific (Fitchen and Beachy, 1993; Wilson, 1993) and one way in which we could control viral diseases would be through the modification of defensive signal transduction pathways by using either genetic manipulation or chemical intervention. However, to be successful this will require substantial knowledge and understanding of each crop species and corresponding pathogen (Dietrich, 2000; Stuiver and Custers, 2001).

It is now apparent that plants use a complex web of signaling pathways to mount induced resistance responses. There are several parallel pathways, with branching and converging points (Genoud and Métraux, 1999; Møller and Chua, 1999; Nurnberger and Scheel, 2001). Using microarray analysis, Schenk et al. (2000) revealed the existence of a tangled network of regulatory interactions and coordination among the different signaling pathways, most notably between SA and JA pathways which had previously seemed to be antagonistic to each other (Schenk et al., 2000). The ability of NPR1 to participate in both SA-dependent and SA-independent signaling pathways highlights it as an important convergence point and as a mediatior of a range of systemic resistance pathways (Bowler and Fluhr, 2000; Parker, 2000). The resultant type of resistance, whether it is to viruses or other plant pathogens appears to be directly influenced by the nature of the input signal (Parker, 2000) (Figure 15.1). It is now clear that more than one pathway may be involved in SAR against viruses (Section 15.5.2), and so the next challenge is to understand the manner in which virus-specific defensive signaling pathways integrate into the wider network of induced resistance mechanisms in plants.

Acknowledgments

Work in our laboratory has received funding from the Biotechnology and Biological Sciences Research Council, The Leverhulme Trust, The Royal Society, the Broodbank Fund of the University of Cambridge, and the Cambridge University Isaac Newton Trust. We thank Dr Cathy Moore for her critical reading of the manuscript.

References

Abbink, T.E.M., Tjernberg, P.A., Bol, J.F., and Linthorst, H.J.M. 1998. Tobacco mosaic virus helicase domain induces necrosis in *N* gene-carrying tobacco in the absence of virus replication. *Mol. Plant Microbe Interact.* 11:1242–1246.

Affourtit, C., Albury, M.S.W., Crichton, P.G., and Moore, A.L. 2002. Exploring the molecular nature of alternative oxidase regulation and catalysis. *FEBS Lett.* 510:121–126.

Ahlquist, P. 2002. RNA-dependent RNA polymerases, viruses, and RNA silencing. *Science* 296:1270–1273.

Alexander, D., Goodman, R.M., Gutrella, M., Glascock, C., Weymann, K., Friedrich, L., Maddox, D., Ahl-Goy, P., Luntz, T., Ward, E., and Ryals, J. 1993. Increased tolerance to

2 oomycete pathogens in transgenic tobacco expressing pathogenesis-related protein-1a. *Proc. Natl. Acad. Sci. USA* 90:7327–7331.

Baulcombe, D.C. 2001. Diced defence: RNA silencing. *Nature* 409:295–296.

Béclin, C., Berthomé, R., Palauqui, J.C., Tepfer, M., and Vaucheret, H. 1998. Infection of tobacco or *Arabidopsis* plants by CMV counteracts systemic post-transcriptional silencing of nonviral (*trans*)genes. *Virology* 252:313–317.

Bendahmane, A., Kahm, B.A., Dedi, C., and Baulcombe, D.C. 1995. The coat protein of potato virus X is a strain specific elicitor of *Rx*1-mediated virus resistance in potato. *Plant J.* 8:933–941.

Bergelson, J., Kreitman, M., Stahl, E.A., and Tian, D. 2001. Evolutionary dynamics of plant *R*-genes. *Science* 292:2281–2285.

Bergstrom, G.C., Johnson, M.C., and Kuć, J. 1982. Effects of local infection of cucumber by *Colletotrichum lagenarium, Pseudomonas lachrymans* or tobacco necrosis virus on systemic resistance to cucumber mosaic virus. *Phytopathology* 72:922–925.

Bernstein, E., Caudy, A.A., Hammond, S.M., and Hannon, G.J. 2001. Role for a bidentate ribonuclease in the initiation step of RNA interference. *Nature* 409:363–366.

Bi, Y., Kenton, P., Mur, L., Darby, R., and Draper, J. 1995. Hydrogen peroxide does not function downstream of salicylic acid in the induction of PR protein expression. *Plant J* 8:235–245.

Birch, P.R.J., Avrova, A.O., Dellagi, A., Lacomme, C., Santa Cruz, S., and Lyon, G.D. 2000. Programmed cell death in plants in response to pathogen attack. In *Molecular Plant Pathology*, eds. M. Dickinson, and J. Beynon, Vol. 4, pp. 175–197. Sheffield, UK: CRC Press.

Bowler, C., and Fluhr, R. 2000. The role of calcium and activated oxygen as signals for controlling cross-tolerance. *Trends Plant Sci.* 5:241–246.

Brigneti, G., Voinnet, O., Li, W.X., Ji, L.-H., Ding, S.W., and Baulcombe, D.C. 1998. Viral pathogenicity determinants are supressors of transgene silencing in *Nicotiana benthamiana*. *EMBO J.* 17:6739–6746.

Bucher, G.L., Tarina, C., Heinlein, M., Di Serio, F., Meins, F. Jr, Iglesias, V.A. 2001. Local expression of enzymatically active class I beta-1, 3-glucanase enhances symptoms of TMV infection in tobacco. *Plant J.* 28:361–369.

Buck, K.W. 1996. Comparison of the replication of positive stranded RNA viruses of plant and animals. *Adv. Virus Res.* 47:159–251.

Cao, H., Bowling, S.A., Gordon, A.S., and Dong, X.N. 1994. Characterization of an *Arabidopsis* mutant that is nonresponsive to inducers of systemic acquired resistance. *Plant Cell* 6:1583–1592.

Cao, H., Glazebrook, J., Clarke, J.D., Volko, S., and Dong, X.N. 1997. The *Arabidopsis* NPR1 gene that controls systemic acquired resistance encodes a novel protein containing ankyrin repeats. *Cell* 88:57–63.

Carrington, J.C. 1999. Reinventing plant virus movement. *Trends Microbiol.* 8:312–313.

Carson, R.A.J. 1999. The effects of salicylic acid on the responses of plants to heat stress and virus infection. PhD thesis, University of Cambridge.

Chivasa, S., and Carr, J.P. 1998. Cyanide restores *N* gene-mediated resistance to tobacco mosaic virus in transgenic tobacco expressing salicylic acid hydroxylase. *Plant Cell* 10:1489–1498.

Chivasa, S., Murphy, A.M., Naylor, M., and Carr, J.P. 1997. Salicylic acid interferes with tobacco mosaic virus replication via a novel salicylhydroxamic acid-sensitive mechanism. *Plant Cell* 9:547–557.

Chong, J., Baltz, R., Schmitt, C., Beffa, R., Fritig, B., and Saindrenan, P. 2002. Downregulation of a pathogen-responsive tobacco UDP-Glc: phenylpropanoid glucosyltransferase reduces scopoletin glucoside accumulation, enhances oxidative stress, and weakens virus resistance. *Plant Cell* 14:1093–1107.

Citovsky, V., Ghoshroy, S., Tsui, F., and Klessig, D.F. 1998. Non-toxic concentrations of cadmium inhibit systemic movement of turnip vein clearing virus by a salicylic acid-independent mechanism *Plant J.* 16:13–20.

Cole, A.B., Kiraly, L., Ross, K., and Schoelz, J.E. 2001. Uncoupling resistance from cell death in the hypersensitive response of *Nicotiana* species to *Cauliflower mosaic virus* infection. *Mol. Plant Microbe Interact.* 14:31–41.

Cooley, M.B., Pathirana, S., Wu, H.J., Kachroo, P., and Klessig, D.F. 2000. Members of the *Arabidopsis* HRT/RPP8 family of resistance genes confer resistance to both viral and oomycete pathogens. *Plant Cell* 12:663–676.

Covey, S.N., Al-Kaff, N.S., Langara, A., and Turner, D.S. 1997. Plants combat infection by gene silencing. *Nature* 385:781–782.

Culver J.N., and Dawson, W.O. 1989. Tobacco mosaic virus coat protein - an elicitor of the hypersensitive reaction but not required for the development of mosaic symptoms in *Nicotiana sylvestris*. *Virology* 173:755–758.

Culver, J.N., Stubbs, G., and Dawson W.O. 1994. Structure-function relationship between tobacco mosaic-virus coat protein and hypersensitivity in *Nicotiana sylvestris*. *J. Mol. Biol.* 242:130–138.

Cutt, J.R., Harpster, M.H., Dixon, D.C., Carr, J.P., Dunsmuir, P., and Klessig, D.F. 1989. Disease response to tobacco mosaic virus in transgenic tobacco plants that constitutively express the pathogenesis-related PR1b gene. *Virology* 173:89–97.

Dangl, J.L., and Jones, J.D.G. 2001. Plant pathogens and integrated defence responses to infection. *Nature* 411:826–833.

Dangl, J.L., Dietrich, R.A., and Richberg, M.H. 1996. Death don't have no mercy: cell death programs in plant-microbe interactions. *Plant Cell* 8:1793–1807.

Darby, R.M., Maddison, A., Mur, L.A.J., Bi, Y.M., and Draper, J. 2000. Cell-specific expression of salicylate hydroxylase in an attempt to separate localized HR and systemic signalling establishing SAR in tobacco. *Mol. Plant Pathol.* 1:115–123.

Delaney, T.P., Friedrich, L., and Ryals, J.A. 1995. *Arabidopsis* signal-transduction mutant defective in chemically and biologically induced disease resistance. *Proc. Natl Acad. Sci. USA* 92:6602–6606.

Delaney, T.P., Ukness, S., Vernoij, B., Friedrich, L., Weymann, K., Negrotto, D., Gaffney, T., Gut-Rella, M., Kessmann, H., and Ryals, J. 1994. A central role of salicylic acid in plant disease resistance. *Science* 266:1247–1250.

Dempsey, D.A., Shah, J., and Klessig, D.F. 1999. Salicylic acid and disease resistance in plants. *Crit. Rev. Plant Sci.* 18:547–575.

Dietrich, R.A. 2000. Emerging technologies and their application in the study of host-pathogen interactions. In (eds.), *Molecular Plant Pathology*, eds. M. Dickinson, and J. Beynon, Vol. 4. pp. 253–286. Sheffield, UK: CRC Press.

Dinesh-Kumar, S.P., Whitham, S., Choi, D., Hehl, R., Corr, C., and Baker, B. 1995. Transposon tagging of tobacco mosaic virus resistance gene N: its possible role in the TMV-N-mediated signal transduction pathway. *Proc. Natl Acad. Sci. USA* 92:4175–4180.

Di Serio, F., Schöb, H., Iglesias, A., Tarina, C., Bouldoires, E., and Meins, J., F. 2001. Sense- and antisense-mediated gene silencing in tobacco is inhibited by the same viral suppressors and is associated with accumulation of small RNAs. *Proc. Natl Acad. Sci. USA* 98:6506–6510.

Dixon, R.A. 2001. Natural products and plant disease resistance. *Nature* 411:843–847.

Erickson, F.L., Holzberg, S., Calderon-Urrea, A., Handley, V., Axtell, M., Corr, C., and Baker, B. 1999. The helicase domain of the TMV replicase proteins induces the N-mediated defence response in tobacco. *Plant J.* 18:67–75.

Fitchen J.H., and Beachy R.N. 1993. Genetically-engineered protection against viruses in transgenic plants. *Annu. Rev. Microbiol.* 47:739–763.

Flor, H.H. 1971. Current status of the gene-for-gene concept. *Annu. Rev. Phytopath.* 9:275–296.

Friedrich, L., Lawton, K., Ruess, W., Masner, P., Specker, N., Gut-Rella, M., Meier, B., Dincher, S., Staub, T., Uknes, S., Métraux, J.P., Kessmann, H., and Ryals, J. 1996. A benzothiadiazole derivative induces systemic acquired resistnce in tobacco. *Plant J.* 10:60–70.

Gaffney, T., Friedrich, L., Vernoij, B., Negrotto, D., Nye, G., Ukness, S., Ward, E., Kessmann, H., and Ryals, J. 1993. Requirement of salicylic acid for the induction of systemic acquired resistance. *Science* 261:754–756.

Genoud, T., and Métraux, J.P. 1999. Crosstalk in plant cell signalling: structure and function of the genetic network. *Trends Plant Sci.* 4:503–507.

Ghoshroy, S., Freedman, K., Lartey, R., and Citovsky, V. 1998. Inhibition of plant viral systemic infection by non-toxic concentrations of cadmium. *Plant J.* 13:591–602.

Gilchrist, D.G. 1998. Programmed cell death in plant disease: the purpose and promise of cellular suicide. *Annu. Rev. Phytopathol.* 36:393–414.

Gilliland, A., Singh, D.P., Hayward, J.M., Moore, C.A., Murphy, A.M., York, C.J., Slator, J., and Carr, J.P. 2003. Genetic modification of alternative respiration has differential effects on antimycin A-induced *versus* salicylic acid-induced resistance to *Tobacco mosaic virus*. *Plant Physiol.* 132:1518–1528.

Glazebrook, J., Rogers, E.E., and Ausubel, F.M. 1996. Isolation of *Arabidopsis* mutants with enhanced disease susceptibility by direct screening. *Genetics* 143:973–982.

Goodman, R.N., and Novacky, A.J. 1994. *The Hypersensitive Reaction in Plants to Pathogens. A Resistance Phenomenon.* St.Paul, MN: APS Press.

Görlach, J., Volrath, S., Knauf-Beiter, G., Hengy, G., Oostendorp, M., Staub, T., Ward, E., Kessman, H., and Ryals, J. 1996. Benzothiadiazole, a novel class of inducers of systemic acquire resistance, activates genes expression and disease resistance in wheat. *Plant Cell* 8:629–643.

Grant, S.R. 1999. Dissecting the mechanism of posttranscriptional gene silencing: divide and conquer. *Cell* 96:303–306.

Guo, H.S., and Ding, S.W. 2002. A viral protein inhibits the long range signalling activity of the gene silencing signal. *EMBO J.* 21:398–407.

Hamilton, A.J., and Baulcombe, D.C. 1999. A species of small antisense RNA in posttranscriptional gene silencing in plants.*Science* 286:950–952.

Hammerschmidt, R. 1999. Phytoalexins:What have we learned after 60 years? *Annu. Rev. Phytopathol.* 37:285–306.

Hammond-Kosack, K.E., and Jones, J.D.G. 1996. Resistance gene-dependent plant defense responses. *Plant Cell* 8:1773–1791.

Hanley-Bowdoin, L., Settlage, S.B., Orozco, B.M., Nagar, S., and Robertson, D. 1999. Geminiviruses: Models for plant DNA replication, transcription, and cell cycle regulation *Crit. Rev. Plant Sci.* 18:71–106.

Heath, M.C. 2000. Hypersensitive response-related death. *Plant Mol. Biol.* 44:321–334.

Heinlein, M. 2002. The spread of *Tobacco Mosaic Virus* infection: insights into the cellular mechanism of RNA transport. *Cell. Mol. Life Sci.* 59:58–82.

Hohn, T., and Futterer, J. 1997. The proteins and functions of plant pararetroviruses: knowns and unknowns. *Crit. Rev. Plant Sci.* 16:133–161.

Hooft van Huijsduijnen, R.A.M., Alblas, S.W., DeRijk, R. H., and Bol, J. F. 1986. Induction by salicylic acid of pathogenesis-related proteins and resistance to alfalfa mosaic virus in various plant species. *J. Gen. Virol.* 67:2135–2143.

Hull, R. 2002. *Matthews' Plant Virology.* Fourth Edition. London: Academic Press.

Jakab, G., Cottier, V., Toquin, V., Rigoli, G., Zimmerli, L., Métraux, J.P., and Mauch-Mani, B. 2001. β-Aminobutyric acid-induced resistance in plants. *Eur. J. Plant Pathol.* 107:29–37.

Ji, L.-H., and Ding, S.W. 2001. The suppressor of transgene RNA silencing encoded by *Cucumber mosaic virus* interferes with salicylic acid-mediated virus resistance. *Mol. Plant Microbe Interact.* 6:715–724.

Jones, D.A. 2000. Resistance genes and resistance protein function. In *Molecular Plant Pathology*, eds. M. Dickinson, and J. Beynon, Vol. 4, pp. 108–143. Sheffield, UK: CRC Press.

Kachroo, P. Yoshioka, K., Shah, J., Dooner, H.K., and Klessig, D.F. 2000. Resistance to turnip crinkle virus in *Arabidopsis* is regulated by two host genes and is salicylic acid dependent but *NPR1*, ethylene, and jasmonate independent. *Plant Cell* 12:677–690.

Kessmann, H., Staub, T., Hoffmann, C., Maetzke, T., Herzog, J., Ward, E., Uknes, S., and Ryals, J. 1994. Induction of systemic acquired resistance in plants by chemicals. *Ann Rev. Phytopathol.* 32:439–459.

Kim, C.H., and Palukaitis, P. 1997. The plant defense response to cucumber mosaic virus in cowpea is elicited by the viral polymerase gene and affects virus accumulation in single cells. *EMBO J.* 16:4060–4068.

Kombrink, E., and Schmelzer, E. 2001. The hypersensitive response and its role in local and systemic disease resistance. *Eur J. Plant Pathol.* 107:69–78.

Kuć, J. 1995. Phytoalexins, stress metabolism, and disease resistance in plants. *Ann. Rev. Phytopathol.* 33:275–297.

Kuć, J. 2001. Concepts and direction of induced systemic resistance in plants and its application. *Eur. J. Plant Pathol.* 107:7–12.

Lacomme, C., and Roby, D. 1999. Identification of new early markers of the hypersensitive response in *Arabidopsis thaliana.* *FEBS Lett.* 459:149–153.

Lacomme, C., and Santa Cruz, S. 1999. Bax-induced cell death in tobacco is similar to the hypersensitive response. *Proc. Natl. Acad. Sci. USA* 96:7956–7961.

Lam, E., and del Pozo, O. 2000. Caspase-like involvement in the control of plant cell death. *Plant Mol. Biol.* 44:417–428.

Laties, G.G. 1982. The cyanide-resistant, alternative path in plant mitochondria. *Annu. Rev. Plant Physiol.* 33:519–555.

Lawton, K.A., Friedrich, L., Hunt, M., Weymann, K., Delaney, T., Kessmann, H., Staub, T., and Ryals, J. 1996. Benzothiadiazole induces disease resistance in *Arabidopsis* by activation of the systemic acquired resistance signal transduction pathway. *Plant J* 10:71–82.

Lebel, E., Heifetz, P., Thorne, L., Uknes, S., Ryals, J., and Ward, E. 1998. Functional analysis of regulatory sequences controlling PR-1 gene expression in *Arabidopsis. Plant J.* 16:223–233.

Leisner, S.M., and Turgeon, R. 1993. Movement of virus and photoassimilate in the phloem—a comparative-analysis. *Bioessays* 15:741–748.

Lennon, A., Neuenschwander, U.H., Ribas-Carbo, M., Giles, L., Ryals, J.A., and Siedow, J.N. 1997. The effects of salicylic acid and tobacco mosaic virus infection on the alternative oxidase of tobacco. *Plant Physiol.* 115:783–791.

Li, W.X., and Ding, S.W. 2001. Viral suppressors of RNA silencing. *Curr. Opin. Biotech.* 12:150–154.

Li, H., Li, W.X., and Ding, S.W. 2002. Induction and suppression of RNA silencing by an animal virus. *Science* 296:1319–1321.

Linthorst, H.J.M., Meuwissen, R.L.J., Kauffmann, S., and Bol, J.F. 1989. Constitutive expression of pathogenesis-related proteins PR-1, GRP and PR-S in tobacco has no effect on virus infection. *Plant Cell* 1:285–291.

Lipardi, C., Wei, Q., and Paterson, B.M. 2001. RNAi as random degradative PCR: siRNA primers convert mRNA into dsRNAs that are degraded to generate new siRNAs. *Cell* 107:297–307.

Lucy, A.P., Guo, H.-S., Li, W.-X., and Ding, S.W. 2000. Suppression of post-transcriptional gene silencing by a plant viral protein localized in the nucleus. *EMBO J.* 19:1672–1680.

Malamy, J., Carr, J.P., Klessig, D.F., and Raskin, I. 1990. Salicylic acid: a likely endogenous signal in the resistance response of tobacco to viral infection. *Science* 250:1002–1004.

Maldonado, A.M., Doerner P., Dixon, R.A., Lamb, C.J., and Cameron, R.K. 2002. A putative lipid transfer protein involved in systemic resistance signalling in *Arabidopsis*. *Nature* 419:399–403.

Mallory, A.C., Ely, L., Smith, T.H., Marathe, R., Anandalakshmi, R., Fagard, M., Vaucheret, H., Pruss, G., Bowman, L., and Vance, V.B. 2001. HC-Pro suppression of transgene silencing eliminates the small RNAs but not transgene methylation or the mobile signal. *Plant Cell* 13:571–583.

Mauch, F., Hadwiger, L.A., and Boller, T. 1988. Antifungal hydrolases in pea tissue 1. Purification and characterization of two chitinases and two β-1, 3-glucanases differentially regulated during development and in response to fungal infection. *Plant Physiol.* 87:325–333.

Maxwell, D.P., Wang, Y., and McIntosh, L. 1999. The alternative oxidase lowers mitochondrial reactive oxygen production in plant cells. *Proc. Natl Acad. Sci. USA* 96:8271–8276.

Mayers, C.N., Palukaitis, P., and Carr, J.P. 2000. Sub-cellular distribution analysis of the cucumber mosaic virus 2b protein. *J. Gen. Virol.* 81:219–226.

Métraux, J.P., Signer, H., Ryals, J., Ward, E., Wyssbenz, M., Gaudin, J., Raschdorf, K., Schmid, E., Blum, W., and Inverardi, B. 1990. Increase in salicylic-acid at the onset of systemic acquired-resistance in cucumber. *Science* 250:1004–1006.

Mittler, R., Shulaev, V., Seskar, M., and Lam, E. 1996. Inhibition of programmed cell death in tobacco plants during pathogen-induced hypersensitive response at low oxygen pressure. *Plant Cell* 8:1991–2001.

Møller, S.G., and Chua, N.H. 1999. Interactions and intersections of plant signalling pathways. *J. Mol. Biol.* 293:219–234.

Moore, A.L., and Siedow, J.N. 1991. The regulation of the cyanide-resistant alternative oxidase of plant mitochondria. *Biochim. Biophys. Acta.* 1059:121–140.

Morrain, P., Béclin, C., Elmayan, T., Feuerbach, F., Gordon, C., Morel, J-B., Jouette, D., Lacombe, A-M., Nikic, S., Picault, N., Remoue., K., Sanial, M., Vo, T-A., and Vaucheret, H. 2000. *Arabidopsis SGS2* and *SGS3* genes are required for posttranscriptional gene silencing and natural virus resistance. *Cell* 101:533–542.

Mur, L.A.J., Bi, Y.M., Darby, R.M., Firek, S., and Draper, J. 1997. Compromising early salicylic acid accumulation delays the hypersensitive response and increases viral dispersal during lesion establishment in TMV-infected tobacco. *Plant J.* 12:1113–1126.

Murphy, A.M. and Carr, J.P. 2002. Salicylic acid has cell-specific effects on *Tobacco mosaic virus* replication and cell-to-cell movement. *Plant Physiol.* 128:543–554.

Murphy, A.M., Chivasa, S., Singh, D.P., and Carr, J.P. 1999. Salicylic acid-induced resistance to viruses and other pathogens: A parting of the ways? *Trends Plant Sci.* 4:155–160.

Nawrath, C., and Métraux, J.P. 1999. Salicylic acid induction-deficient mutants of *Arabidopsis* express PR-2 and PR-5 and accumulate high levels of camalexin after pathogen inoculation. *Plant Cell* 11:1393–404.

Naylor, M. 1999. The effects of salicylic acid on RNA plant viruses. PhD thesis, University of Cambridge.

Naylor, M., Murphy, A.M., Berry, J.O., and Carr, J.P. 1998. Salicylic acid can induce resistance to plant virus movement. *Mol. Plant Microbe Interact.* 11:860–868.

Nelson, R.S., and van Bel, A.J.E. 1998. The mystery of virus trafficking into, through and out of the vascular tissue. *Prog. Bot.* 59:476–533.

Niderman, T., Genetet, I., Bruyère, T., Gees, R., Stintzi, A., Legrand, M., Fritig, B., and Mösinger, E. 1995. Pathogenesis-related PR-1 proteins are antifungal. *Plant Physiol.* 108:17–27.

Nurnberger, T., and Scheel, D. 2001. Signal Transmission in the plant immune response. *Trends Plant Sci.* 6:372–379.

Oparka, K.J., and Santa Cruz, S. 2000. The great escape: Phloem transport and unloading of macromolecules. *Ann. Rev. Plant Physiol. Plant Mol. Biol.* 51:323–347.

Oostendorp, M. Kunz, W., Dietrich, B., and Staub, T. 2001. Induced resistance in plants by chemicals. *Eur. J. Plant Pathol.* 107:19–28.

Ordog, S.H., Higgins, V.J., and Vanlerberghe, G.C. 2002. Mitochondrial alternative oxidase is not a critical component of plant viral resistance but may play a role in the hypersensitive response. *Plant Physiol.* 129:1858–1865.

Osbourn, A. 1996. Preformed antimicrobial compunds and plant defense against fungal attack. *Plant Cell* 8:1821–1831.

Otsuki, Y., Shimomura, T., and Takebe, I. 1972. Tobacco mosaic virus multiplication and expression of the *N* gene in necrotic responding tobacco varieties. *Virology* 50:45–50.

Padgett, H.S., Watanabe, Y., and Beachy, R.N. 1997. Identification of the TMV replicase sequence that activates the *N* gene-mediated hypersensitive response. *Mol. Plant Microbe Interact.* 10:709–715.

Palauqui, J.C., Elmayan, T., Pollien, J.M., and Vaucheret, H. 1997. Systemic acquired silencing: transgene specific post-transcriptional silencing is transmitted by grafting from silenced stocks to non-silenced scions. *EMBO J.* 16:4738–4745.

Parker, J.E. 2000. Signalling in plant disease resistance. In *Molecular Plant Pathology*, eds. M. Dickinson, and J. Beynon, Vol. 4, pp. 144–174. Sheffield, UK: CRC Press.

Pennell, R.I., and Lamb, C. 1997. Programmed cell death in plants. *Plant Cell* 9:1157–1168.

Pieterse, C.M.J., Van Pelt, J.A., Van Wees, S.C.M., Ton, J., Leon-Kloosterziel, J.K.M., Keurentjes, J.J.B., Verhagen, B.W.M., Knoester, M., Van der Sluis, I., Bakker, P.A.H.M., and Van Loon, L.M. 2001. Rhizobacteria-mediated induced systemic resistance: triggering, signaling and expression. *Eur. J. Plant Path.* 107:51–61.

Ratcliff, F., Harrison, B.D., and Baulcombe, D.C. 1997. A similarity between viral defence and gene silencing in plants. *Science* 276:1558–1560.

Ratcliff, F.G., MacFarlane, S.A., and Baulcombe, D.C. 1999. Gene silencing without DNA: RNA-mediated cross-protection between viruses. *Plant Cell* 11:1207–1215.

Rauscher, M., Adam, A.L., Wirtz, S., Guggenheim, R., Mendgen, K., and Deising H.B. 1999. PR-1 protein inhibits the differentiation of rust infection hyphae in leaves of acquired resistant broad bean. *Plant J.* 19:625–633.

Ross, A.F. 1961a. Localized acquired resistance to plant virus infection in hypersensitive hosts. *Virology* 14:329–339.

Ross, A.F. 1961b. Systemic acquired resistance induced by localized virus infections in plants. *Virology* 14:340–358.

Ross, A.F. 1966. Systemic effects of local lesion formation. In *Viruses of Plants*, eds. A. B.R. Beemster, and J. Dijkstra, pp. 127–150. Amsterdam: North Holland Publishing.

Ryals, J., Lawton, K.A., Delaney, T.P., Friedrich, L, Kessmann, H., Neuenschwander, U.H., Uknes S., Vernooij, B., Weymann, K. 1995. Signal transduction in systemic acquired resistance. *Proc. Natl Acad. Sci. USA* 92:4202–4205.

Ryals, J., Weymann, K., Lawton, K., Friedrich, L., Ellis, D., Steiner, H. Y., Johnson, J., Delaney, T.P., Jesse, T., Vos, P., and Uknes, S. 1997. The *Arabidopsis* NIM1 protein shows homology to the mammalian transcription factor inhibitor I kappa B. *Plant Cell* 9:425–439.

Schenk, P.M., Kazan, K., Wilson, I., Anderson, J.P., Richmond, T., Somerville, S.C., and Manners, J.M. 2000. Coordinated plant defense responses in *Arabidopsis* revealed by microarray analysis. *Proc. Natl. Acad. Sci. USA* 97:11655–11660.

Schlumbaum, A., Mauch, F., Vogeli, U., and Boller, T. 1986. Plant chitinases are potent inhibitors of fungal growth. *Nature* 324:365–367.

Shah, J., Tsui, F., and Klessig, D.F. 1997. Characterization of a salicylic acid-insensitive mutant (*sai1*) of *Arabidopsis thaliana*, identified in a selective screen utilizing the SA-inducible expression of the *tms2* gene. *Mol. Plant Microbe Interact.* 10:69–78.

Shirasu, K., and Schulze-Lefert, P. 2000. Regulators of cell death in disease resistance. *Plant Mol. Biol.* 44:371–385.

Simons, B.H., Millenaar, F.F., Mulder, L., Van Loon, L.C., and Lambers, H. 1999. Enhanced expression and activation of the alternative oxidase during infection of *Arabidopsis* with *Pseudomonas syringae* pv tomato. *Plant Phys.* 120:529–538.

Soards, A.J., Murphy, A.M., Palukaitis, P., and Carr, J.P. 2002. Virulence and differential local and systemic spread of *Cucumber mosaic virus* in tobacco are affected by the CMV 2b protein. *Mol. Plant Microbe Interact.* 15:647–653.

Stuiver, M., and Custers, J.H.V. 2001. Engineering disease resistance in plants. *Nature* 411:865–868.

Ton, J. Van Pelt, J.A. Van Loon, L.C., and Pieterse, C.M.J. 2002. Differential effectiveness of salicylate-dependent and jasmonate/ethylene-dependent induced resistance in *Arabidopsis*. *Mol. Plant Microbe Interact.* 15:27–34.

Ueki, S., and Citovsky, V. 2001. Inhibition of systemic onset of post-transcriptional gene silencing by non-toxic concentrations of cadmium. *Plant J.* 28:283–291.

Ueki, S., and Citovsky, V. 2002. The systemic movement of a tobamovirus is inhibited by a cadmium ion-induced glycine-rich protein. *Nat. Cell Biol.* 4:478–485.

Vance, V., and Vaucheret, H. 2001. RNA silencing in plants-defense and counterdefense. *Science* 292:2277–2280.

van Loon, L.C., and van Strien, E.A. 1999. The families of pathogenesis-related proteins, their activities, and comparative analysis of PR-1 type proteins. *Physiol. Mol. Plant Pathol.* 55:85–97.

van Loon, L.C., Bakker, P.A.H.M., and Pieterse, C.M.J. 1998. Systemic resistance induced by rhizosphere bacteria. *Annu. Rev. Phytopathol.* 36:453–483.

Vernooij, B., Friedrich, L., Morse, A., Reist, R., Kolditzjawhar, R., Ward, E., Uknes, S., Kessmann, H., and Ryals, J. 1994. Salicylic-acid is not the translocated signal responsible for inducing systemic acquired-resistance but is required in signal transduction. *Plant Cell* 6:959–965.

Voinnet, O. 2001. RNA silencing as a plant immune system against viruses. *Trends Genet.* 17:449–459.

Voinnet, O., and Baulcombe, D.C. 1997. Systemic signalling in gene silencing. *Nature* 389:553.

Ward, E.R., Uknes, S.J., Williams, S.C., Dincher, S.S., Wiederhold, D.L., Alexander, D.C., Ahl-Goy, P., Metreaux, J.P., and Ryals, J.A. 1991. Coordinate gene activity in response to agents that induce systemic acquired resistance. *Plant Cell* 3:1085–1094.

Waterhouse, P.M., Smith, N.A., and Wang, M.B. 1999. Virus resistance and gene silencing: killing the messenger. *Trends Plant Sci.* 4:452–457.

Waterhouse, P.M., Wang, M.B., and Lough, T. 2001. Gene silencing as an adaptive defence against viruses. *Nature* 411:834–842.

Weber, H., and Pfitzner, A.J. 1998. Tm-2(2) resistance in tomato requires recognition of the carboxy terminus of the movement protein of tomato mosaic virus. *Mol. Plant Microbe Interact.* 11:498–503.

Weststeijn, E.A. 1981. Lesion growth and virus localization in leaves of *Nicotiana tabacum* cv. Xanthi nc. after inoculation with tobacco mosaic virus and incubation alternately at 22°C and 32°C. *Physiol. Plant Pathol.* 18:357–368.

White, R.F. 1979. Acetylsalicylic acid (aspirin) induces resistance to tobacco mosaic virus in tobacco. *Virology* 99:410–412.

White, R.F., Antoniw, J.F., Carr, J.P., and Woods, R.D. 1983. The effects of aspirin and polyacrylic acid on the multiplication and spread of TMV in different cultivars of tobacco with and without the *N*-gene. *Phytopathol. Z.* 107:224–232.

Wilson, T.M.A. 1993. Strategies to protect crop plants against viruses- pathogen-derived resistance blossoms. *Proc. Natl. Acad. Sci. USA* 90:3134–3141.

Wong, C.E., Carson, R.A.J., and Carr, J.P. 2002. Chemically induced virus resistance in *Arabidopsis thaliana* is independent of pathogenesis-related protein expression and the *NPR1* gene. *Mol. Plant Microbe Interact.* 15:75–81.

Wright, K.M., Duncan, G.H., Pradel, K.S., Carr, F., Wood, S., Oparka, K.J., and Santa Cruz, S. 2000. Analysis of the *N* gene hypersensitive response induced by a fluorescently tagged tobacco mosaic virus. *Plant Physiol.* 123:1375–1385.

Xie, Z., Fan, B., Chen, C., and Chen, Z. 2001. An important role of an inducible RNA-dependent RNA polymerase in plant antiviral defense. *Proc. Natl Acad. Sci. USA* 98:6516–6521.

Yoshioka, K., Nakashita, H., Klessig, D.F., and Yamaguchi, I. 2001. Probenozole induces systemic acquired resistance in *Arabidopsis* with a novel type of action. *Plant J.* 25:149–157.

Zehnder, G.W., Murphy, J.F., Sikora, E.J., and Kloepper, J.W. 2001. Application of rhizobacteria for induced resistance. *Eur. J. Plant Pathol.* 107:39–50.

Zhou, J.-M., Trifa, Y., Silva, H., Pontier, D.F., Lam, E., Shah, J., and Klessig, D.F. 2000. NPR1 differentially interacts with members of the TGA/OBF family of transcription factors that bind an element of the PR-1 gene required for induction by salicylic acid. *Mol. Plant Microbe Interact.* 13:191–202.

16

Mechanisms Underlying Plant Tolerance to Abiotic Stresses

MASARU OHTA, KAREN S. SCHUMAKER, AND JIAN-KANG ZHU

16.1 Introduction

During the course of their life cycle, land plants are exposed to numerous abiotic stresses such as drought, salinity, low and high temperatures, high light, and UV irradiation. It has been estimated that, due to abiotic stress, the yield of field-grown crops in the United States is only 22% of their genetic potential yield (Boyer, 1982). While all of these environmental factors can substantially reduce crop yield, drought, salinity, and low temperature have been especially problematic for agricultural productivity (Thomashow, 1999; Hasegawa et al., 2000; Shinozaki and Yamaguchi-Shinozaki, 2000; Zhu, 2002). As a result, much research has been performed to understand the physiological mechanisms underlying the ability of plants to tolerate these stresses.

Decades of research have been devoted to understand how various abiotic stresses affect the plant, and both stress-specific lesions and lesions that are common to more than one stress have emerged. For example, drought, salinity, and cold stress all lead to dehydration and osmotic imbalances in the plant (Levitt 1980). In addition to osmotic effects, exposure to high levels of salt also leads to ionic imbalances (Hasegawa et al., 2000; Zhu, 2002). In most cases, this ionic stress is caused by high concentrations of Na^+ and Cl^- in the soil. As Na^+ concentrations increase, the normal ratio of Na^+ to K^+ is altered and reduced cellular K^+ leads to decreased activities of numerous metabolic enzymes (Solomon et al., 1994; Papageorgiou and Murata, 1995; Schachtman and Liu, 1999). Stress-specific lesions from exposure to cold and freezing temperature occur as a result of reductions in membrane fluidity and the formation of ice in the intercellular spaces of plant tissues resulting in the physical disruption of cells and tissues, respectively.

A number of mechanisms have been implicated in the adaptation of plants to stress conditions (Thomashow, 1999; Hasegawa et al., 2000; Shinozaki and Yamaguchi-Shinozaki, 2000; Zhu, 2002). The physiological and metabolic changes that underlie these adaptations include production of osmoprotectants such as glycine betaine, proline, sucrose, and sugar alcohols that allow cellular osmotic adjustment for continued water uptake (drought, cold, salt stress; Hasegawa et al., 2000; Zhu, 2002), regulation of stomatal aperture (drought and salt stress; Schroeder et al., 1991) and alterations in the lipid composition (e.g., increased levels of unsaturated fatty acids and certain types of steroids and cerebrosides) of cellular membranes upon chilling and freezing stresses (Thomashow, 1999). In addition, regulation of ion homeostasis by transport proteins in various cellular membranes enables some plants to continue to grow during exposure to high levels of salt (Blumwald et al., 2000). In some cases, adaptation to abiotic stress has been shown to correlate with changes in the expression of subsets of genes that encode, for instance, polypeptides rich in hydrophilic amino acid residues (Thomashow, 1994; Ingram and Bartels, 1996; Bray, 1997). These osmotic stress and cold-responsive genes (OR and COR genes, respectively) may be induced by multiple stresses or a specific stress (Iwasaki et al., 1995; Yamaguchi-Shinozaki and Shinozaki, 1994). Evidence is also accumulating for the involvement of hormones in the response of plants to abiotic stresses. For example, the level of abscisic acid (ABA) increases in response to drought, salinity, and cold stresses, and this ABA plays a critical role in inducing the expression of OR/COR genes as well as in regulating stomatal aperture (Shinozaki and Yamaguchi-Shinozaki 1997, 2000; Schroeder et al., 2001). Recent evidence indicates that hormone-induced signal transduction under abiotic stresses involves changes in intracellular Ca^{2+} levels (Sanders et al., 1999; Knight and Knight, 2001).

In this chapter, we begin with an example that illustrates our current understanding of hormone-activated signal transduction pathways under abiotic stresses and the role of Ca^{2+} in this process. Analysis of the regulation of stomatal aperture in response to ABA provides insight into how cytosolic Ca^{2+} levels are controlled by ABA. Discussion of the Salt—Overly Sensitive pathway in plant responses to high levels of salt provides a model for our understanding of how increased cytosolic Ca^{2+} regulates downstream tolerance effectors. In the second part of the chapter, we describe recent progress using genetic and genomics approaches to dissect the pathways involved in plant responses to osmotic and cold stresses, and our current knowledge of the signaling components involved.

16.2 Molecules Involved in the Regulation of Cytosolic Ca^{2+} Levels in Response to ABA

16.2.1 Hydrogen Peroxide

Reactive oxygen species (ROS) can function as second messengers in cellular processes. For example, hydrogen peroxide (H$_2$O$_2$) plays an important role in plant defense responses triggered by pathogens and elicitors (Levine et al., 1994; Lamb and Dixon, 1997). Recently, H$_2$O$_2$ has been shown to act as a second messenger during ABA-induced stomatal closure in guard cells through its effect on Ca^{2+} channel activity (Pei et al., 2000; Zhang et al., 2001). This is supported by the following observations: (1) H$_2$O$_2$ production was induced by ABA treatment in guard cells of *Arabidopsis thaliana*, *Vicia faba* and in maize culture cells (Guan et al., 2000; Pei et al., 2000; Zhang et al., 2001), (2) inhibition of H_2O_2 production by treatment of cells with diphenylene iodonium (an inhibitor of NADPH oxidase which generates H_2O_2 from NADPH) in the presence of ABA partially abolished stomatal closure (Pei et al., 2000; Zhang et al., 2001), (3) ABA increased cytosolic Ca^{2+} levels through regulation of Ca^{2+} channels in guard cells (Schroeder and Hagiwara, 1990; McAinsh et al., 1990; Grabov and Blatt, 1998; Staxén et al., 1999; MacRobbie, 2000; Hamilton et al., 2000), (4) in the absence of ABA, plasma membrane Ca^{2+} channels were activated by H_2O_2 in *Arabidopsis* (Pei et al., 2000), (5) a precursor of H_2O_2 production, NADPH, was shown to be required for ABA activation of Ca^{2+} channels in protoplasts of *Arabidopsis* guard cells (Murata et al., 2001). Taken together, these studies demonstrate that H_2O_2 mediates ABA regulation of plasma membrane Ca^{2+} channels during guard cell signaling.

ABI1 and *ABI2* encode homologous type 2C protein phosphatases (PP2C; Koornneef et al., 1984; Leung et al., 1994; Meyer et al., 1994; Leung et al., 1997). Molecular and biochemical characterizations of the ABA-insensitive mutants *abi1* and *abi2* provided specific insights into the signaling pathways and components involved in ABA signaling. For example, guard cells of *abi1-1* and *abi2-1* do not exhibit ABA-activation of Ca^{2+} channel currents (Murata et al., 2001). While guard cells from the *abi1-1* mutants have impaired ABA-induced ROS production, H_2O_2-activation of Ca^{2+} channel currents is not affected. In contrast, guard cells from the *abi2-1* mutant produce ROS in response to ABA but have impaired responses of Ca^{2+} channels to H_2O_2. Thus, the *abi1-1* and *abi2-1* disrupt different steps of ABA signaling that leads to changes in intracellular Ca^{2+} levels (Murata et al., 2001).

16.2.2 IP₃, cADPR, and Heterotrimeric G Proteins

In mammalian cells, inositol (1,4,5)-trisphosphate (IP_3) and cyclic adenosine 5′-diphosphate ribose (cADPR) are known to act as second messengers that regulate cytosolic Ca^{2+} levels (Clapham, 1995; Cancela et al., 2000). IP_3 is produced from phosphatidylinositol 4,5-bisphosphate (PIP_2) by a phosphoinositide-specific phospholipase C (PI-PLC). cADPR is produced by ADP-ribosyl cyclase or by CD38, a lymphocyte protein, both of which use nicotinamide adenine dinucleotide (NAD^+) as a precursor in mammalian cells.

Information on the effects of IP_3 on Ca^{2+} levels in plant cells came originally from biochemical and cell biological studies. A number of years ago, *in vitro* experiments demonstrated that IP_3 can stimulate the release of Ca^{2+} from the vacuole, a major site of Ca^{2+} storage in plant cells (Schumaker and Sze, 1987). Subsequent studies identified specific processes that appear to involve IP_3 regulation. Microinjection of IP_3 into *Commelina communis* guard cells caused an increase in cytosolic Ca^{2+} levels leading to stomatal closure (Gilroy et al., 1990), and ABA was shown to cause a transient increase in IP_3 levels (Lee et al., 1996; Sanchez and Chua, 2001; Xiong et al., 2001a). An inhibitor of PI-PLC, U-73122, blocked ABA-induced stomatal closure and oscillations of cytosolic free Ca^{2+} in guard cells, providing additional evidence that PLC (and IP$_3$) may be involved in the regulation of cytosolic free Ca^{2+} during stomatal closure (Staxén et al., 1999).

Recently, genetic analyses have provided support for a role for IP$_3$ in ABA signaling. The *fiery1* (*fry1*) mutant (see, Section 15.5) shows enhanced expression of stress-responsive genes in response to cold, ABA and salt stress. *fry1* mutant plants exhibit ABA- and NaCl-sensitive phenotypes. *FRY1* encodes an inositol polyphosphate 1-phoshatase, which functions in the catabolism of IP$_3$ (Xiong et al., 2001a). The *fry1* mutant accumulates more IP$_3$ than wild-type plants after treatment with ABA. Thus, inositol polyphosphate 1-phosphatase negatively regulates ABA signaling, providing genetic evidence for the involvement of polyphosphoinositols (Xiong et al., 2001a). Additional evidence for the involvement of IP$_3$ in ABA signaling was obtained from studies where the activity of the *Arabidopsis* PI-PLC1 (*AtPLC1*) was inhibited by antisense suppression (Sanchez and Chua, 2001). Suppression of *AtPLC1* eliminated ABA-inhibition of seed germination and blocked the expression of stress-responsive genes induced by ABA (Sanchez and Chua, 2001), further establishing the role of PI-PLC and IP$_3$ in ABA signaling.

cADPR

cADPR has been shown to cause the release of Ca^{2+} from vacuoles in plant cells (Allen et al., 1995), and several reports provided evidence that cADPR also acts as a second messenger in ABA signaling (Wu et al., 1997; Leckie et al., 1998). Single-cell microinjection experiments into hypocotyls of the *aurea* mutant of tomato (deficient in phytochrome) have shown that cADPR mediates the transcriptional activation of *rd29A-GUS* and *kin2-GUS* reporter genes in response to ABA (Wu et al., 1997). Microinjection of a cADPR antagonist, 8-NH2-cADPR, or an antagonist of cADPR production, nicotinamide, reduced the rate of turgor loss in response to ABA and blocked ABA-induced stomatal closure in guard cells of *Commelina communis* (Leckie et al., 1998), thus providing additional evidence for the involvement of cADPR in ABA signal transduction.

Heterotrimeric G Proteins

Heterotrimeric G proteins are key regulators of ion channels in animal cells (Brown and Birnbaumer 1990). Upon activation, the G protein α subunit (Gα) binds GTP, resulting in the separation of the α subunit from the $\beta\gamma$ subunit pair (G$\beta\gamma$). Gα and G$\beta\gamma$ can then interact with downstream components of signaling pathways. Early pharmacological studies in plants suggested that G proteins may function in the response of guard cells to ABA (Lee et al., 1993; Ma, 1994; Kelly et al., 1995). Subsequent analysis of T-DNA knockout alleles of *gpa1-1* and *gpa1-2*, which encode the only prototypical Gα subunit in *Arabidopsis*, clearly demonstrated the role of G proteins in the regulation of ion channels and ABA signaling in guard cells (Wang et al., 2001). These studies showed that, in the *gpa1* mutant, stomata failed to close in response to ABA. Activities of K$^+$ and anion channels are important for the regulation of ion homeostasis and turgor changes underlying

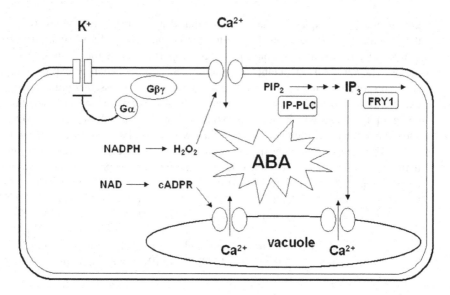

FIGURE 16.1. Abscisic acid regulation of cytosolic Ca^{2+} levels in guard cells. ABA triggers production of second messengers such as hydrogen peroxide (H_2O_2), cyclic ADP ribose (cADPR), and inositol 1,4,5-trisphosphate (IP_3). These second messengers activate Ca^{2+} transport systems leading to increases in cytosolic Ca^{2+} levels. Heterotrimeric G proteins are also shown to illustrate a mechanism underlying ABA-induced regulation of inward rectifying K^+ channels.

the regulation of stomatal aperture in plants (Schroeder et al., 2001). When the activities of these channels were measured in the *gpa1* mutants and compared to their activities in guard cells from wild-type plants, inwardly rectifying K^+ channels were not inhibited by ABA and anion channels were not activated by ABA (Wang et al., 2001), indicating heterotrimeric G protein involvement in ion homeostasis and stomatal regulation.

Information from the studies discussed above leads to a working model of the pathways and components underlying changes in cytosolic Ca^{2+} levels in response to ABA (Figure 16.1). Clearly, gaps in the pathway still exist. For example, genes encoding the ion channels that have been described electrophysiologically have not yet been identified. The physiological characterization of *Arabidopsis* T-DNA knockout mutants may provide one approach to identify these genes.

16.3 The *Salt Overly Sensitive* Signaling Pathway

Several *Arabidopsis* mutants that are more sensitive to salt than wild-type plants in their growth were isolated from a screen of 250,000 mutagenized seeds (Wu et al., 1996; Liu and Zhu, 1997; Zhu et al., 1998; Shi et al., 2002a). These *salt overly sensitive (sos1, 2, 3, 4)* mutants are hypersensitive specifically to Na^+ and Li^+, but not to Cs^+ and mannitol, suggesting that they are defective specifically in salt tolerance determinants (Liu and Zhu, 1997; Zhu et al., 1998; Shi et al., 2002a). Subsequent molecular and biochemical analyses have revealed that *SOS1* encodes a plasma membrane Na^+/H^+ exchanger (antiporter; Shi et al., 2000; Qiu et al., 2002), as supported by the following: (1) *SOS1* was found to complement the salt-sensitive phenotype of a yeast mutant lacking *nhx1* and *nha1*, genes that encode Na^+/H^+ antiporters on the vacuolar and plasma membranes, respectively (Shi et al., 2002b), (2) a SOS1-GFP fusion protein was detected on the surface of cells in hypocotyls and roots of *SOS1-GFP* transgenic *Arabidopsis* plants, (3) in contrast to Na^+/H^+ antiport activity in plasma membrane vesicles isolated from wild-type *Arabidopsis*, the activity was reduced in vesicles isolated from *sos1* plants (Qiu et al., 2002).

Subsequent cloning and characterization of the *SOS2* gene demonstrated that it encodes a serine/threonine protein kinase with an N-terminal domain similar to that of the yeast SNF1 kinase (Liu et al., 2000). SOS2 has an autophosphorylation activity and can phoshorylate synthetic oligopeptides with sequences similar to the recognition sequences of the SNF1/AMPK (*Arabidopsis*) kinases (Halfter et al., 2000; Liu et al., 2000; Guo et al., 2001). In addition to the N-terminal catalytic domain, the SOS2 kinase has a C-terminal regulatory domain. Yeast two-hybrid assays demonstrated that there is an interaction between the N-terminal catalytic and the C-terminal regulatory domains of the SOS2 protein (Guo et al., 2001). Deletion of the C-terminal domain enhanced the *in vitro* phosphorylation activity of the SOS2 protein indicating that, through interaction with the N-terminal catalytic domain, the C-terminal domain may be part of an auto-inhibitory mechanism regulating SOS2 kinase activity.

SOS3 encodes a calcium binding protein with sequence similarity to the β subunit of calcineurin (CNB) and to animal neuronal calcium sensors (Liu and Zhu, 1998; Ishitani et al., 2000). In yeast, *CNB* is necessary for salt tolerance (Nakamura et al., 1993; Mendoza et al., 1994). Subsequent *in vitro* studies demonstrated that SOS3 binds Ca^{2+} and that the protein can be N-myristoylated (Ishitani et al., 2000). Amino acid substitutions in the N-myristoylation consensus motif abolished the myristoylation of SOS3 *in vitro* and the modified protein failed to complement the salt-hypersensitive phenotype of the *sos3* mutant, indicating that the N-myristoylation of SOS3 is required for its function in plant salt tolerance.

The *SOS4* gene was identified based on its sensitivity to 100 mM NaCl, a salt concentration that was higher than that used to identify the *sos1, 2,* and *3* mutants (cf. 50 mM for the isolation of *sos1, 2, and 3*). Based on the gene sequence and its ability to complement an *E. coli* mutant defective in pyridoxal kinase activity, *SOS4* has been shown to encode a pyridoxal kinase (Shi et al., 2002a). Pyridoxal kinases are involved in the biosynthesis of pyridoxial-5-phosphate (PLP), an active form of vitamin B6 (Hanna et al., 1997). PLP and its derivatives are known to regulate ATP-gated P2X receptor ion channels in animals presumably by PLP antagonism of ATP (Ralevic and Burnstock, 1998). Since several plant K^+ channels also have putative cyclic nucleotide binding sites (Sentenac et al., 1992; Daram et al., 1997), it is possible that similar regulatory mechanisms operate in plant cells.

Interestingly, addition of pyridoxine to the growth medium can partially rescue the *sos4* salt sensitivity phenotype, suggesting that PLP may regulate Na^+ homeostasis by modulating the activity of ion transporters (Shi et al., 2002a).

Several lines of evidence indicate that *SOS1*, *SOS2*, and *SOS3* function in the same pathway: (1) the expression of *SOS1* is up-regulated by salt stress, (2) *SOS2* and *SOS3* are involved in the up-regulation of *SOS1* expression by salt stress (Shi et al., 2000), (3) the salt sensitivity of the *sos2 sos3* double mutant is virtually the same as that of the *sos2* mutant (no additive effects, Halfer et al., 2000), (4) SOS3 physically interacts with SOS2 in yeast two-hybrid and *in vitro* binding assays (Halfer et al., 2000), (5) SOS3 activates the activity of the SOS2 kinase in a Ca^{2+}-dependent manner *in vitro*. The interaction between SOS2 and SOS3 is mediated by a 21 amino acid region in the C-terminal regulatory domain of SOS2 (Guo et al., 2001). Recently, biochemical experiments have demonstrated that the transport activity of SOS1 is a direct target of the SOS2 kinase (Qiu et al., 2002). A constitutively active form of the SOS2 kinase (SOS2T/D308) enhances Na^+/H^+ antiport activity in plasma membrane vesicles isolated from the wild type, *sos2* and *sos3* plants. Similar results were obtained when the *Arabidopsis* SOS signaling pathway was reconstituted in yeast (Quintero et al., 2002); expression of SOS1 could complement the salt-sensitive phenotype of a yeast mutant lacking the *ENA1-4*, *NHA1*, and *NHX1* genes encoding Na^+-ATPases, and plasma membrane and vacuolar Na^+/H^+ antiporters, respectively (Quintero et al., 2002). When the *SOS2* and *SOS3* genes (Liu and Zhu, 1998; Liu et al., 2000), or when the constitutively active form of SOS2 kinase SOS2TD308 (Guo et al., 2001) was expressed in the yeast, there was a significant enhancement of the SOS1-mediated salt tolerance. Finally, SOS1 was shown to be phosphorylated by SOS3-activated SOS2 kinase *in vitro* (Quintero et al., 2002), suggesting that the Na^+/H^+ antiport activity of SOS1 is regulated by phosphorylation.

Based on these results, a current model for how *Arabidopsis* regulates the activity of a plasma membrane Na^+/H^+ antiporter in response to salt stress is shown in Figure 16.2. When cytosolic Ca^{2+} levels rise in response to cellular perception of salt stress (Lynch et al., 1989), increased Ca^{2+} activates SOS3, which, in turn, activates the SOS2 kinase. Activated SOS2 kinase phosphorylates SOS1 to stimulate its activity and to allow extrusion of Na^+ from the cytoplasm to protect sensitive metabolic activities. Since salt tolerance is a multigenic trait, the identification of additional substrates of the SOS2 kinase will help us understand how plants control molecules necessary for salt tolerance through Ca^{2+} signaling pathways.

16.4 Signaling Components Involved in Osmotic Stress Responses

16.4.1 AtHK1, a Possible Sensor of Osmotic Stress

Two-component histidine kinases have been shown to be functionally involved in osmosensing in bacteria, yeast and mammalian cells (Wurgler-Murphy and Saito,

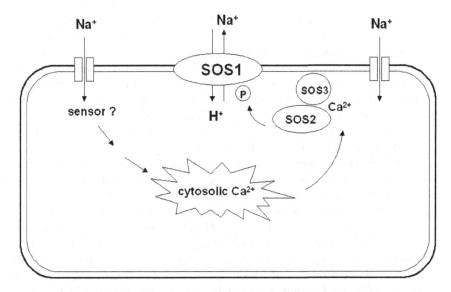

FIGURE 16.2. Components of the *Salt Overly Sensitive* signaling pathway. Ca^{2+} activates SOS3, which in turn activates the SOS2 kinase. Activated SOS2 kinase phosphorylates SOS1 to stimulate its activity and to allow extrusion of Na^+ from the cytoplasm to protect sensitive metabolic activities.

1997). *SNL1* in yeast encodes a two-component histidine kinase that serves as an osmosensor (Ota and Varshavsky 1993). *AtHK1* in *Arabidopsis* encodes a histidine kinase. Based on its ability to complement the osmotic stress response defect of the *snl1* mutant in yeast, AtHK1 in *Arabidopsis* may serve as an osmosensor in plants (Urao et al., 1999). Functional analysis of T-DNA knockout mutants of *AtHKT1* will be necessary to determine if indeed this protein functions as an osmosensor in plant cells.

16.4.2 Other Protein Kinases

Mitogen-activated protein (MAP) kinase cascades play crucial roles in signaling osmotic stress in yeast and mammalian cells (Widmann et al., 1999). MAP kinase cascades are usually composed of three protein kinases; a MAP kinase (MAPK) that is activated by a specific MAPK kinase (MAPKK), which, in turn, is activated by an upstream MAPKK kinase (MAPKKK). It has been shown that several MAPKs such as the tobacco salicylic acid-induced protein kinase (SIPK), the *Arabidopsis* ATMPK4/6, and the alfalfa salt stress-inducible MAP kinase (SIMK) are activated by osmotic stress (Munnik et al., 1999; Hoyos and Zhang, 2000; Ichimura et al., 2000; Mikołajczyk et al., 2000). SIMK and ATMPK4 are activated through specific interaction with alfalfa SIMK kinase (SIMKK) and *Arabidopsis* AtMEK1, respectively (Mizoguchi et al., 1998; Huang et al., 2000; Kiegerl et al., 2000; Matsuoka et al., 2002).

Calcium-dependent protein kinases (CDPKs) are a unique family of plant-specific kinases, distinguished by a C-terminal calmodulin-like regulatory domain with up to four calcium binding EF hand motifs. There are 40 different putative CDPKs in the *Arabidopsis* genome (Harmon et al., 2000). Several lines of evidence suggest that CDPKs are involved in osmotic stress signaling: (1) the transcript of *ATCDPK1* is up-regulated by drought and high-salinity stress (Urao et al., 1994), (2) constitutively active forms of AtCDPK1 and AtCDPK1a induce the expression of ABA-responsive, stress-induced genes in maize protoplasts (Sheen, 1996), (3) CDPK affects the activities of chloride and potassium channels that are important for the regulation of stomatal closure in guard cells (Pei et al., 1996; Li et al., 1998), (4) overexpression of the rice CDPK gene *OsCDPK7* enhanced the salt-induction of a number of stress-responsive genes in rice (Saijo et al., 2000).

Additional evidence for the involvement of kinases in the response of the plant to stress comes from studies on *AtGSK1*. AtGSK1 is a homolog of glycogen synthase kinase (GSK) 3/shaggy-like protein kinases. AtGSK1 can rescue the NaCl-stress sensitive phenotype of a yeast mutant in which the genes encoding the catalytic subunits of calcineurin have been deleted (Piao et al., 1999). In addition, overexpression of *AtGSK1* conferred enhanced tolerance to osmotic stress in *Arabidopsis* (Piao et al., 2001).

16.4.3 Transcription Factors

ABA plays a critical role in the signaling pathway leading to induction of *OR/COR* genes (Shinozaki and Yamaguchi-Shinozaki, 1997; 2000). ABA-induction of these genes is mediated through the ABA-responsive element/complex (ABRE; Yamaguchi–Shinozaki and Shinozaki, 1994; Shen and Ho, 1995; Vasil et al., 1995). Although ABA induces the expression of *OR/COR* genes, the expression of these genes is also observed in ABA-deficient and ABA-insensitive mutants in response to cold stress (Gilmour and Thomashow, 1991; Nordin et al., 1991). Analyses of *RD29A* and *COR15a* promoters have revealed that the dehydration-responsive element (DRE)/C-repeat mediates osmotic and cold stress signaling in an ABA-independent manner (Baker et al., 1994; Yamaguchi-Shinozaki and Shinozaki, 1994). Thus, ABA-dependent and ABA-independent pathways have been proposed to mediate gene expression in response to cold and osmotic stresses (Yamaguchi-Shinozaki and Shinozaki, 1994; Shinozaki and Yamaguchi-Shinozaki, 1997).

A group of basic leucine zipper (bZIP) proteins (ABF/AREBs) have been shown to bind to ABRE (Choi et al., 2000; Uno et al., 2000). The overexpression of *ABF3* or *ABF4* in *Arabidopsis* results in ABA hypersensitivity, reduced water loss from leaves, and enhanced drought tolerance (Kang et al., 2002). Conserved N-terminal regions of AREB1 and AREB2 have been shown to be phosphorylated by a 42 kDa ABA-activated protein kinase (Uno et al., 2000). These results illustrate that ABFs/ABREs mediate stress-responsive ABA signaling.

A group of transcription factors that bind to the DRE/C repeat element (CBFs/DREBs) has been cloned from *Arabidopsis* (Stockinger et al., 1997; Liu et al., 1998). Among these proteins, expression of *DREB2A* and *DREB2B* is specifically induced by drought and salinity but not by cold, suggesting *DREB2A* and *DREB2B* mediate osmotic stress signaling (Liu et al., 1998; Nakashima et al., 2000).

The expression of the *RD22* gene is induced by drought and ABA treatments (Yamaguchi-Shinozaki and Shinozaki, 1993). Analysis of the *RD22* promoter has revealed that a 67 bp region is responsive to dehydration and ABA in a protein synthesis-dependent manner (Iwasaki et al., 1995). A MYC-related bHLH DNA binding protein (RD22BP) has been isolated by screening an *Escherichia coli* expression library using the 67 bp *RD22* DNA fragment as a probe (Abe et al., 1997). RD22BP transactivates a *RD22* promoter-*GUS* reporter gene in *Arabidopsis* protoplasts, and its transactivation is enhanced by a MYB protein, AtMYB2. Activation by AtMYB2 is thought to take place cooperatively through a putative site for MYB-related DNA binding in the 67 bp region (Abe et al., 1997). Since the expression of *RD22BP* and *AtMYB2* is induced by drought and ABA treatments (Urao et al., 1993; Abe et al., 1997), these factors may mediate ABA or dehydration signaling in a protein synthesis-dependent pathway in which the expression of *RD22BP* and *AtMYB2* is activated by further upstream transcription factors.

Alfin1 is a Cys-4 and His/Cys-3 zinc-finger protein from alfalfa, which binds to promoter elements in the salt-inducible *MsPRP2* gene (Bastola et al., 1998). Overexpression of *Alfin1* enhances the expression of the endogeneous *MsPRP2* gene and improves tolerance to salinity in alfalfa, suggesting a role for the zinc finger in osmotic stress signaling (Winicov and Bastola, 1999).

16.5 Signaling Components Involved in Cold Stress Responses

16.5.1 Physical Changes in the Membrane as a Possible Sensor of Cold Stress

Although a two-component sensor has been implicated in cold sensing in cyanobacteria (Suzuki et al., 2000), it is not clear which molecules are involved in cold sensing in higher plants. It has been postulated that, in plants, cold is sensed via changes in membrane fluidity (Murata and Los, 1997) and cytoskeletal reorganization (Örvar et al., 2000). Since the level of cytosolic Ca^{2+} increases in plants in response to cold stress (Knight et al., 1991; 1996), it is possible that Ca^{2+} is also involved in the early stages of cold stress signaling and that physical changes in the plasma membrane may result in the activation of Ca^{2+} channels triggering downstream events.

16.5.2 Protein Kinases

In *Arabidopsis*, the MAP kinases, *AtMPK4* and *AtMPK6*, are also activated by low temperature in addition to their activation by osmotic stress (Ichimura et al.,

2000). Biochemical experiments have demonstrated that the protein kinase activity of AtMPK4 is stimulated by AtMEK1 in response to cold, drought, and salinity in *Arabidopsis* (Matuoka et al., 2002).

16.5.3 Transcription Factors

The transcription factors CBF1/DREB1B, CBF2/DREB1C, and CBF3/DREB1A bind to the DRE/C repeat element and are cold specific; the expression of these genes is induced by cold treatment but not by drought or by salinity (Liu et al., 1998; Medina et al., 1999). These three *CBF/DREB1* genes are arranged in tandem on chromosome 4 (Gilmour et al., 1998; Shinwari et al., 1998; Medina et al., 1999). Overexpression of *CBFs/DREB1s* has been shown to improve the tolerance of *Arabidopsis* to drought, salt, and freezing temperature (Liu et al., 1998; Jaglo-Ottosen et al., 1998; Kasuga et al., 1999; Gilmour et al., 2000). In *CBF/DREB1*-overexpressing plants, elevated levels of mRNA for downstream cold-inducible genes such as *RD29A* and *COR* were observed even without cold stress, indicating that these genes are targets for *CBF/DREB1s* (Jaglo-Ottosen et al., 1998; Kasuga et al., 1999; Gilmour et al., 2000). Overexpression of *CBF3/DREB1A* also led to elevated levels of proline and total soluble sugars including sucrose, raffinose, glucose, and fructose, mimicking multiple biochemical changes associated with cold acclimation (Gilmour et al., 2000). Since the expression of *CBFs/DREB1s* is induced by cold, it has been proposed that as yet unidentified transcription factors (inducer of *CBF* expression: ICE; Gilmour et al., 1998) activate the expression of the *CBF/DREB1s* after being modified by cold-induced signaling.

SCOF-1 is a C_2H_2-type zinc finger protein from soybean whose overexpression induces *COR* gene expression and enhances cold tolerance of nonacclimated transgenic *Arabidopsis* and tobacco plants (Kim et al., 2001). Although SCOF-1 fails to bind to the ABA-responsive element (ABRE) *in vitro*, SCOF-1 has been shown to enhance the DNA-binding activity of a soybean G-box binding bZIP transcription factor (SGBF-1) to the ABRE. This stimulation of activity is thought to take place through SCOF-1 interaction with SGBF-1, linking SCOF-1 to ABA-dependent signaling through SGBF-1 (Kim et al., 2001).

16.6 Genetic Dissection of Osmotic and Cold Stress Signaling Pathways

As has been found in yeast, molecular genetic approaches provide powerful tools for dissecting osmotic signaling pathways in *Arabidopsis* (O'Rourke et al., 2002). Using an *RD29A* promoter-luciferase reporter gene, we developed a system to screen for *Arabidopsis* mutants that are impaired in osmotic and cold-responsive gene expression (Ishitani et al., 1997). Screening the progeny from chemically mutagenized plants, 833 mutants were identified with over 100 of them exhibiting

TABLE 16.1. A list of mutants obtained from screening with *RD29A-LUC* imaging.

Locus	Phenotype	Gene/Protein	Stress tolerance[a]
LOS1[b]	los to cold	translation elongation factor2	sensitive to freezing
LOS2	los to cold	bifunctional enolase	sensitive to freezing
LOS5	los to cold, salt hos to ABA	ABA3/ molybdoprotein cofactor sulfurase	sensitive to freezing and salt
LOS6	los to salt hos to cold, ABA	ABA1/zeaxanthin epoxidase	N/D
HOS1[c]	hos to cold	RING-finger protein	sensitive to freezing
FRY1[d]	hos to all	Inositol polyphosphate 1-phosphatase	sensitive to cold, salt and drought
HOS2	hos to cold		sensitive to cold
SAD1[e]	hos to ABA, salt	similar to Sm protein	sensitive to salt and drought

ND: not determined.
[a] stress tolerance of respective mutants in these loci.
[b] LOS, low expression of osmotically responsive gene.
[c] HOS, high expression of osmotically responsive gene.
[d] FRY1, fiery1.
[e] SAD1, super sensitive to ABA and drought.

strong reporter gene expression phenotypes (Ishitani et al., 1997). These mutants were classified into three categories: (a) *cos* (constitutive expression of osmotically responsive genes), (b) *los* (low expression of osmotically responsive genes), (c) *hos* (high expression of osmotically responsive genes). After additional characterization of the mutants regarding their responses to cold, ABA and high levels of NaCl, the lines were grouped into 13 sub-categories. The mutants that have been characterized and those whose corresponding genes have been identified by positional cloning are shown in Table 16.1. In the following sections, we describe several of the mutants and discuss the roles of the corresponding genes in abiotic stress responses.

16.6.1 *LOS1*

The *los1-1* mutant exhibits a lower level of luminescence than what is observed in wild-type plants (wild-type refers to the unmutagenized *RD29A-LUC* parental line) when both are treated with low temperature (Guo et al., 2002). However, the response of the *los1-1* mutant to ABA or salinity is similar to what is seen in wild-type plants. Interestingly, *CBF/DREB1* transcripts are superinduced by cold in *los1-1*, although the expression of downstream target genes such as *RD29A* and *COR* (cold regulated) are significantly reduced in the mutant. *LOS1* encodes a translation elongation factor 2-like protein, and protein synthesis is significantly reduced when the mutant is exposed to low temperature (Guo et al., 2002). The *los1* studies revealed an interesting feedback repression of *CBF/DREB1* genes in the cold by their gene products or the products of their downstream target genes.

16.6.2 LOS2

The luminescence phenotype of the *los2* mutant is similar to what is seen in *los1-1* mutant plants: the *los2* mutant is defective in cold-responsive gene transcription of the *RD29A-LUC* and *COR* genes (Lee et al., 2002). A difference between the *los1-1* and *los2* mutants is that the *los2* mutation does not affect the cold-induction of *CBF* transcripts. *los2* plants are sensitive to chilling and freezing with the increased freezing sensitivity related to membrane damage; the plants show a dramatic increase in electrolyte leakage after chilling stress. In addition to alterations in membrane integrity, treatment with light and cold lead to features typical of programmed cell death (e.g., membrane blebbing, nuclear condensation, and DNA fragmentation). *LOS2* encodes an enolase that converts 2-phosphoglycerate to phoshpoenolpyruvate in the glycolytic pathway, and consistent with this fact, enolase activity is reduced in *los2* plants (Lee et al., 2002).

In human cells, a portion of the enolase (ENO1) enzyme (MBP-1, myc promoter-binding protein-1) has been shown to bind to a c-*myc* promoter and repress c-*myc* gene expression (Ray and Miller, 1991; Ghosh et al., 1999; Feo et al., 2000). A comparison of the LOS2 sequence with the human ENO1 sequence shows that the putative DNA binding and repression domains are highly conserved. Since LOS2 can bind to a target sequence of the human MBP1 *in vitro*, and LOS2-GFP is targeted to the nucleus as well as to the cytoplasm, LOS2 may act as a transcription factor (Lee et al., 2002). A computer search for possible target genes of *LOS2* revealed that the promoter region of a gene encoding a C_2H_2-type zinc finger transcription factor, *STZ/ZAT10* (Lippuner et al., 1996; Meissner and Michael 1997) has a sequence similar to the target sequence of the human MBP1. *In vitro* binding assays confirmed that *LOS2* can bind to this *STZ/ZAT10* promoter sequence (Lee et al., 2002). Cold stress transiently induces the accumulation of *STZ/ZAT10* mRNA in wild-type plants compared to higher levels and sustained induction in *los2* plants. These results suggest that LOS2 negatively regulates the expression of *STZ/ZAT10* by binding to its promoter.

STZ/ZAT10 has been shown to act as an active repressor of transcription in *Arabidopsis* leaves (Ohta et al., 2001), and recent transient expression assays have demonstrated that it can repress the expression of an *RD29A-LUC* reporter gene (Lee et al., 2002). These results indicate that LOS2 is a positive regulator of *RD29A* expression, and that this regulation may be achieved through the control of *STZ/ZAT10* expression, thus preventing STZ/ZAT10 repression of *RD29A* expression.

16.6.3 LOS5

Luminescence intensities in *los5* mutant seedlings are considerably lower than those in the wild-type plants when treated with cold or NaCl; however, luminescence intensities in response to ABA are unaffected in the mutant (Xiong et al., 2001b). The *los5* plants are sensitive to freezing and salt stress and RNA blot analysis demonstrated that NaCl-induction of *RD29A*, *COR* genes, *KIN1*, *RD22*, and *P5CS* is almost completely blocked in *los5* mutant plants. Furthermore, *los5* plants show typical features of ABA-deficient mutants and genetic analysis indicates that *los5* is allelic to *aba3*, which is an ABA-deficient mutant impaired in the incorporation of sulfur into molybdenum cofactor (MoCo) (Schwartz et al., 1997). Positional cloning revealed that *LOS5/ABA3* encodes a molybdopterin cofactor sulfurase (Xiong et al., 2001b). These studies on *los5* demonstrated a major role for ABA in the osmotic stress regulation of OR and *COR* genes, and suggested that the ABA-independent pathway(s) for the gene regulation may in fact require ABA for full function.

16.6.4 LOS6

Upon treatment with salt, *los6* plants show reduced luminescence compared to wild type (Xiong et al., 2001c). After drought treatment, *los6* plants show an ABA-deficient phenotype because ABA content in these plants does not increase. In contrast, the ABA content in wild-type plants can increase up to tenfold after an exposure to drought. Genetic analysis revealed that *LOS6* is allelic to *ABA1* and encodes a zeaxanthin epoxidase (ZEP).

16.6.5 HOS1

hos1-1 plants show cold-specific superinduction of luminescence (Ishitani et al., 1998). Consistent with this luminescence phenotype, the transcript levels for cold responsive genes, such as *RD29A*, *COR47*, *COR15A*, *KIN1*, and *ADH1* are higher than those in wild-type plants when both are treated with low temperature. Furthermore, the expression of *CBF/DREB1* genes, upstream regulators of the cold responsive genes, is also superinduced by cold and is more sustained in the *hos1-1* plants (Lee et al., 2001). These results suggest that the superinduction of *RD29A*, *COR47*, *COR15A*, *KIN1*, and *ADH1* is caused by the higher expression of the *CBF/DREB1* genes and that *HOS1* is a negative regulator of this cold-specific pathway.

hos1-1 plants flower earlier than wild type and are constitutively vernalized (Ishitani et al., 1998), suggesting that *HOS1* is also a negative regulator of vernalization. The early flowering phenotype of *hos1-1* plants can be explained by the reduced expression of *Flowering Locus C* (*FLC*), a negative regulator of flowering (Michaels and Amasino, 1999; Sheldon et al., 2000). The level of *FLC* transcript is significantly lower in *hos1-1* plants compared with the levels in wild-type plants (Lee et al., 2001).

HOS1 encodes a novel protein with a RING finger motif near the amino terminus (Lee et al., 2001). The subcellular localization of HOS1 appears to be regulated by cold stress. Green fluorescence from a HOS1-GFP fusion protein is observed in the nucleus only when the plants are treated with low temperature for one or two days, and this is in contrast to the cytoplasmic localization of the fluorescence when plants are grown in warm temperatures. A number of RING finger proteins have E3 ubiquitin ligase activities, so it is possible that *HOS1* regulates the level of

CBF/DREB1 expression by controlling the turnover of a regulator of *CBF/DREB1* (ICE; inducer of CBF expression) through ubiquitin-mediated protein degradation.

16.6.6 HOS5

hos5-1 mutant plants show enhanced expression of *RD29A-LUC* in response to ABA and osmotic stress, but not to cold stress (Xiong et al., 1999). The hyper-induction of *RD29A-LUC* by osmotic stress is likely to be ABA-independent, because hyper-induction is also observed in both the *hos5-1/aba1-1* and *hos5-1/abi1-1* double mutants. Although *hos5-1* shows enhanced expression of several *OR* genes and increased sensitivity of root growth to ABA, germination of *hos5-*1 seeds is more resistant to ABA compared to wild-type seeds. These results suggest that HOS5 is a negative regulator of both ABA-dependent and ABA-independent pathways; confirmation of this role for *HOS5* awaits cloning of the *HOS5* gene.

16.6.7 FIERY1/HOS2

The *fiery1* mutation enhances *RD29A-LUC* expression in response to cold, ABA, or hyperosmotic stress (Xiong et al., 2001a). In addition to enhanced expression of *OR* genes, *fiery1* plants show a more sustained expression of *CBF2* after cold treatment. These results suggest that the *fiery1* mutation affects an early step in the cold, osmotic stress, and ABA signal transduction pathways. Germination of *fiery1* seeds is more sensitive to ABA and NaCl, and *fiery1* plants are less tolerant to osmotic stress than wild type.

hos2 plants exhibit enhanced expression of *RD29A-LUC* specifically under cold stress (Lee et al., 1999b) and are less capable of developing freezing tolerance when treated with low nonfreezing temperatures. In contrast to *hos1* plants, the *hos2* mutation does not alter the vernalization response of *Arabidopsis*.

Positional cloning of the *FIERY1* (*FRY1*) gene revealed that *FRY1* encodes an inositol polyphosphate 1-phoshatase, which functions in the catabolism of IP_3 (Xiong et al., 2001a). Since the *fry1* mutant accumulates more IP_3 than wild-type plants upon ABA treatment, the inositol polyphosphate 1-phoshatase appears to negatively regulate ABA signaling. The *hos2* mutation has recently been mapped to the *FRY1* gene (Zhu, unpublished); however, *hos2* plants accumulate more IP_3 than wild-type plants only when treated with cold stress. Positional cloning has revealed that the *hos2* mutation is a cold-sensitive allele of the *FRY1* gene.

16.6.8 SAD1

In *sad1* (super sensitive to ABA and drought) plants, the expression of *RD29A-LUC* is dramatically higher when plants are treated with ABA or NaCl, but not with cold stress relative to levels in wild-type plants exposed to the same treatments (Xiong et al., 2001d). Consistent with the above results, the transcript levels of *RD29A*, *COR47*, and *KIN1* are also higher in *sad1* than in the wild-type plants in response to ABA or NaCl. In contrast, induction of the transcription factor

RD22BP is not affected in *sad1*, suggesting that *SAD1* regulates a subset of ABA- or drought-inducible genes. The *sad1* mutation enhances sensitivity to ABA or osmotic stress in both seeds and vegetative tissues as germination of *sad1* seeds and growth of *sad1* roots had increased sensitivity to ABA and higher concentrations of NaCl.

In the ABA-hypersensitive mutants *era1* and *adh1*, increased ABA sensitivity is accompanied by a reduction of transpirational water loss (Pei et al., 1998; Hugovieux et al., 2001). In contrast, the *sad1* plants show increased transpirational water loss, suggesting a defect in stomatal regulation (Xiong et al., 2001d). Since ABA content is not increased in the *sad1* plants in response to drought, the defect in stomatal regulation is likely to be due to a deficiency in ABA.

SAD1 encodes a 9.7 kDa polypeptide of 88 amino acids with sequence similarity to Sm domain proteins (Xiong et al., 2001d). Sm proteins are a family of small proteins that assemble the core components of the spliceosomal snRNP (Salgado-Garrido et al., 1999). Lsm (like-Sm) proteins are multifunctional molecules that modulate RNA metabolism including splicing, export, and degradation (He and Parker, 2000). Since *SAD1* shows a high degree of sequence similarity to the human Lsm5 protein (Achsel et al., 1999), SAD1 may be involved in mRNA metabolism in plants. The connection between mRNA metabolism and ABA and stress responses is very intriguing and the underlying mechanism remains a mystery.

16.7 Conclusions

Theoretically, the expression of genes is regulated both positively and negatively. Previous studies on abiotic stress signaling focused on positive regulation of gene expression. Characterization of mutants with aberrant expression of the *RD29A-LUC* reporter genes has provided genetic evidence that abiotic stress signaling is indeed regulated by both positive (*LOS* genes) and negative (*HOS* genes) factors. Among the negative regulators, identification of *FRY1* as an enzyme that functions in the catabolism of IP_3 gives important insight into the linkage between an upstream regulator of cytosolic Ca^{2+} increase and downstream events including the expression of ABA-, osmotic stress-, and cold-responsive genes (Xiong et al., 2001a).

CBFs/DREB1s are involved in the transcriptional induction of cold-responsive genes (Stockinger et al., 1997; Liu et al., 1998), and several of the mutations identified affect the expression of these *CBF/DREB1* genes. For example, a mutation in *los1* (encoding translation elongation factor 2) causes superinduction of *CBF/DREB1s* (Guo et al., 2002). Since *los1* plants fail to synthesize proteins under cold stress, the cold-induction of *CBFs/DREB1s* may be inhibited by feedback regulation by their gene products or products of downstream target genes. In *fry1* mutant plants, expression of the *CBF/DREB1s* transcripts is more sustained than that in the wild-type plants after six hours of cold treatments, suggesting that the expression of the *CBFs/DREB1s* is negatively regulated by a Ca^{2+} signaling pathway

(Xiong et al., 2001a). It has been established that the half-life for transcription factors is tightly regulated by ubiquitin-mediated protein degradation. The sustained induction of the *CBF/DREB1*s in the *hos1* mutants suggests that a similar mechanism might be involved in the regulation of *CBF/DREB1* expression (Lee et al., 2001). These studies demonstrate that the expression of the *CBF/DREB1*s is regulated by multiple mechanisms.

Previous research suggested that drought induces the expression of genes through both ABA-independent and ABA-dependent pathways (Shinozaki-Yamaguchi and Shinozaki, 1994; Shinozaki and Shinozaki-Yamaguchi, 1997), and that DRE/CTR and ABRE elements mediate ABA-independent and ABA-dependent gene expression, respectively. This model was supported by findings from cold stress signaling where cold treatments induced the expression of the cold-regulated genes in both ABA-deficient and ABA-insensitive mutants (Gilmour et al., 1991; Nordin et al., 1991). However, mutations in ABA biosynthetic genes (*los5/ABA3, los6/ABA*, and *abi1-1*) substantially block the induction of *RD29A, COR15A, COR47*, and *ADH* during osmotic stress (Xiong et al., 2001b, 20001c). These results demonstrate that ABA synthesis and ABA signaling are required for full expression of the osmotic stress-responsive genes and that interaction between DRE/CTR and ABRE elements plays a more important role in osmotic stress signaling than was previously thought.

The expression of *RD29A, COR15A,* and *COR47* is induced by cold and osmotic stresses and is thought to be important for the adaptation to abiotic stress, because overexpression of *COR15A* improves the freezing tolerance of *Arabidopsis* (Steponkus et al., 1998). However, *hos1* and *fry1* plants are impaired in freezing tolerance even though they show higher expression of these genes in response to cold stress (Ishitani et al., 1998; Xiong et al., 2001a). In addition, in *eskimo1*, a constitutively freezing-tolerant *Arabidopsis* mutant, expression of *COR15A* and *COR47* is virtually the same as in wild-type plants (Xin and Browse, 1998). These results suggest that several distinct signaling pathways regulate different aspects of cold acclimation. Microarray analysis of these mutants may identify clusters of genes that are regulated by the different pathways.

References

Abe, H., Yamaguchi-Shinozaki, K., Urao, T., Iwasaki, T., Hososkawa, D., and Shinozaki, K. 1997. Role of *Arabidopsis* MYC and MYB homologs in drought- and abscisic acid-regulated gene expression. *Plant Cell* 9:1859–1868.

Achsel, T., Brahms, H., Kastner, B., Bachi, A., Wilm, M., and Lührmann, R. 1999. A doughnut-shaped heteromer of human Sm-like proteins binds to the 3'-end of U6 snRNA, thereby facilitating U4/U6 duplex formation *in vitro*. *EMBO J.* 18:5789–5802.

Allen, G.J., Muir, S.R., and Sanders, D. 1995. Release of Ca^{2+} from individual plant vacuoles by both InsP3 and cyclic ADP-ribose. *Science* 268:735–737.

Baker, S.S., Wilhelm, K.S., and Thomashow, M.F. 1994. The 5'-region of *Arabidopsis thaliana cor15a* has *cis*-acting elements that confer cold-, drought- and ABA-regulated gene expression. *Plant Mol. Biol.* 24:701–713.

Bastola, D.R., Pethe, V.V., and Winicov, I. 1998. Alfin1, a novel zinc-finger protein in alfalfa roots that binds to promoter elements in the salt-inducible *MsPRP2* gene. *Plant Mol. Biol.* 38:1123–1135.

Blumwald, E., Aharon, G.S., and Apse, M.P. 2000. Sodium transport in plant cells. *Biochem. Biophys. Acta.* 1465:140–151.

Boyer, J.S. 1982 Plant productivity and environment. *Science* 218:443–448.

Bray, E. A. 1997. Plant responses to water deficit. *Trends Plant Sci.* 2:48–54.

Brown, A.M., and Birnbaumer, L. 1990. Ionic Channels And Their Regulation By G Protein Subunits. *Annu. Rev. Physiol.* 52:197–213.

Cancela, J.M., Gerasimenko, O.V., Gerasimenko, J.V., Tepikin, A.V., and Petersen, O.H. 2000. Two different but converging messenger pathways to intracellular Ca^{2+} release: the roles of nicotinic acid adenine dinucleotide phosphate, cyclic ADP-ribose and inositol trisphosphate. *EMBO J.* 19:2549–2557.

Choi, H., Hong, J., Kang, J., and Kim, S.Y. 2000. ABFs, a family of ABA-responsive element binding factors. *J. Biol. Chem.* 275:1723–1730.

Clapham, D.E. 1995. Calcium signaling. *Cell* 80:259–268.

Daram, P., Urbach, S., Gaymard, F., Sentenac, H., and Cherel, I. 1997. Tetramerization of the AKT1 plant potassium channel involves its C-terminal cytoplasmic domain. *EMBO J.* 16:3455–3463.

Feo, S., Arcuri, D., Piddini, E., Passantino, R. and Giallongo, A. 2000. *ENO1* gene product binds to the c-*myc* promoter and acts as a transcriptional repressor: relationship with *Myc* promoter-binding protein 1 (MBP-1. *FEBS Lett.* 473:47–52

Ghosh, A.K., Steele, R., and Ray, R.B. 1999. Functional domains of c-*myc* promoter binding protein 1 involved in transcriptional repression and cell growth regulation. *Mol. Cell. Biol.* 19:2880–2886.

Gilmour, S.J., and Thomashow, M.F. 1991. Cold acclimation and cold-regulated gene expression in ABA mutants of *Arabidopsis thaliana*. *Plant Mol. Biol.* 17:1233–1240.

Gilmour, S.J., Zarka, D.G., Stockinger, E.J., Salazar, M.P., Houghton, J.M., and Thomashow, M.F. 1998. Low temperature regulation of the *Arabidopsis* CBF family of AP2 transcriptional activators as an early step in cold-induced *COR* gene expression. *Plant J.* 16:433–442.

Gilmour, S.J., Sebolt, A.M., Salazar, M.P., Everard, J.D., and Thomashow, M.F. 2000. Overexpression of the *Arabidopsis CBF3* transcriptional activator mimics multiple biochemical changes associated with cold acclimation. *Plant Physiol.* 124:1854–1865.

Gilroy, S., Read, N.D., and Trewavas, A.J. 1990. Elevation of cytosolic calcium by caged inositol triphosphate initiates stomatal closure. *Nature* 346:769–771.

Grabov, A., and Blatt, M. 1998. Membrane voltage initiates Ca^{2+} waves and potentiates Ca^{2+} increases with abscisic acid in stomatal guard cells. *Proc. Natl. Acad. Sci. USA* 95:4778–4783.

Guan L.M., Zhao, J., and Scandalios, J.G. 2000. *Cis*-elements and *trans*-factors that regulate expression of the maize *Cat1* antioxidant gene in response to ABA and osmotic stress: H_2O_2 is the likely intermediary signaling molecule for the response. *Plant J.* 22:87–95.

Guo, Y., Halfter, U., Ishitani, M., and Zhu, J.K. 2001. Molecular characterization of functional domains in the protein kinase SOS2 that is required for plant salt tolerance. *Plant Cell* 13:1383–1399.

Guo, Y., Xiong, L., Ishitani, M., and Zhu, J.K. 2002. An *Arabidopsis* mutation in translation elongation factor 2 causes superinduction of *CBF/DREB1* transcription factor genes but blocks the induction of their downstream targets under low temperatures. *Proc. Natl. Acad. Sci. USA* 99:7786–7791.

Guy, C.L. 1990. Cold acclimation and freezing stress tolerance: Role of protein metabolism. *Ann. Rev. Plant Physiol. Plant Mol. Biol.* 41:187–223.

Halfter, U., Ishitani, M., and Zhu, J.K. 2000. The *Arabidopsis* SOS2 protein kinase physically interacts with and activated by the calcium-binding protein SOS3. *Proc. Natl. Acad. Sci. USA* 97:3735–3740.

Hamilton, D.W., Hills A., Köhler, B., and Blatt, M.R. 2000. Ca^{2+} channels at the plasma membrane of stomatal guard cells are activated by hyperpolarization and abscisic acid. *Proc. Natl. Acad. Sci. USA* 97:4967–4972.

Hanna, M.C., Turner, A.J. and Kirkness, E.F. 1997. Human pyridoxal kinase. cDNA cloning, expression, and modulation by ligands of the benzodiazepine receptor *J. Biol. Chem.* 272:10756–10760.

Harmon, A.C., Gribskov, M., and Harper, J.F. 2000. CDPKs-a kinase for every Ca^{2+} signal? *Trends Plant Sci.* 5:154–159.

Hasegawa P.M., Bressan, R.A., Zhu, J.K., and Bohnert, H.J. 2000. Plant cellular and molecular responses to high salinity. *Annu. Rev. Plant Physiol. Plant Mol. Biol.* 51:463–499.

He, W., and Parker, R. 2000. Functions of Lsm proteins in mRNA degradation and splicing. *Curr. Opin. Cell Biol.* 12:346–350.

Hoyos, M. E., and Zhang, S. 2000. Calcium-independent activation of salicylic acid-induced protein kinase and 40-kilodalton protein kinase by hyperosmotic stress. *Plant Physiol.* 122:1355–1363.

Huang, Y., Li, H., Gupta, R., Morris, P.C., Luan, S., and Kieber, J.J. 2000. ATMPK4, an *Arabidopsis* homolog of mitogen-activated protein kinase is activated *in vitro* by AtMEK1 through threonine phosphorylation. *Plant Physiol.* 122:1301–1310.

Hugouvieux, V., Kwak, J.M., and Schroeder, J.I. 2001. An mRNA cap binding protein, ABH1, modulates early abscisic acid signal transduction. *Cell* 106:477–487.

Ichimura, K., Mizoguchi, T., Yoshida, R., Yuasa, T., and Shinozaki, K. 2000. Various abiotic stresses rapidly activate *Arabidopsis* MAP kinases ATMPK4 and ATMPK6. *Plant J.* 24:655–665.

Ingram, J., and Bartels, D. 1996. The molecular basis of dehydration tolerance in plants. *Annu. Rev. Plant Physiol. Plant Mol. Biol.* 47:377–403.

Ishitani, M., Xiong, L., Stevenson, B., and Zhu, J.K. 1997. Genetic analysis of osmotic and cold stress signal transduction in *Arabidopsis*: interactions and convergence of abscisic acid-dependent and abscisic acid-independent pathways. *Plant Cell.* 11:1935–1949.

Ishitani, M., Xiong, L., Lee, H., Stevenson, B., and Zhu, J.K. 1998. *HOS1*, a genetic locus involved in cold-responsive gene expression in Arabidopsis. *Plant Cell.* 10:1151–1161.

Ishitani, M., Liu, J., Halfter, U., Kim, C.S., Shi, W., and Zhu, J.K. 2000. SOS3 function in plant salt tolerance requires N-myristoylation and calcium binding. *Plant Cell* 12:1667–1677.

Iwasaki, T., Yamaguchi-Shinozaki, K., and Shinozaki, K. 1995. Identification of a *cis*-regulatory region of a gene in *Arabidopsis thaliana* whose induction by dehydration is mediated by abscisic acid and requires protein synthesis. *Mol. Gen. Genet.* 247:391–408.

Jaglo-Ottosen, K.R., Gilmour, S.J., Zarka, D.G., Schabenberger, O., and Thomashow, M.F. 1998. *Arabidopsis CBF1* overexpression induces *COR* genes and enhances freezing tolerance. *Science* 280:104–106.

Kang, J., Choi, H., Im, M., and Kim, S.Y. 2002. *Arabidopsis* basic leucine zipper proteins that mediate stress-responsive abscisic acid signaling. *Plant Cell* 14:343–357.

Kasuga, M., Liu, Q., Miura, S., Yamaguchi-Shinozaki, K., and Shinozaki, K. 1999. Improving plant drought, salt, and freezing tolerance by gene transfer of a single stress-inducible transcription factor. *Nat. Biotech.* 17:287–291.

Kelly, W.K., Esser, J.E., and Schroeder, J.I. 1995. Effects of cytosolic calcium and limited, possible dual, effects of G protein modulators on guard cell inward potassium channels. *Plant J.* 8:479–489.

Kiegerl, S., Cardinale, F., Siligan, C., Gross, A., Baudouin, E., Liwosz, A., Eklöf, S., Till, S., Bögre, L., Hirt, H., and Meskiene, I. 2000. SIMKK, a mitogen-activated kinase MAPK kinase, is a specific activator of salt stress-induced MAPK, SIMK. *Plant Cell* 12:2247–2258.

Kim, J.C., Lee, S.H., Cheong, Y.H., Yoo, C.-M., Lee, S.I., Chun, H.J., Yun, D.-J., Hong, J.C., Lee, S.Y., Lim, C.O., and Cho, M.J. 2001. A novel cold-inducible zinc finger protein from soybean, SCOF-1, enhances cold tolerance in transgenic plants. *Plant J.* 25:247–259.

Knight, M.R., Campbell, A.K., Smith S.M., and Trewavas, A.J. 1991. Transgenic plant aequorin reports the effects of touch and cold-shock and elicitor on cytoplasmic calcium. *Nature* 352:524–526.

Knight, H., S.M., Trewavas, A.J., and Knight, M.R., 1996. Cold calcium signaling in Arabidopsis involves two cellular pools and a change in calcium signature after acclimation. *Plant Cell* 8:489–503.

Knight, H., and Knight, M.R. 2001. Abiotic stress signaling pathways: specificity and cross-talk. *Trends Plant Sci.* 6:262–267.

Koornneef, M., Reuling, G., and Karssen, C.M. 1984. The isolation and characterization of abscisic acid-insensitive mutants of *Arabidopsis thaliana*. *Plant Physiol.* 61:377–383.

Lamb, C.J., and Dixon, R.A. 1997. The oxidative burst in plant disease resistance. *Annu. Rev. Plant Physiol. Mol. Biol.* 48:251–275.

Leckie, C.P., McAinsh, M.R., Allen, G.J., Sanders, D., and Hetherington, A.M. 1998. Abscisic acid-induced stomatal closure mediated by cyclic ADP ribose. *Proc. Natl. Acad. Sci. USA* 95:15837–15842.

Lee, H.J., Tucker, E.B., Crain, R.C., and Lee, Y. 1993. Stomatal opening is induced in epidermal peels of *Commelina communis* L. by GTP analogs or pertussis toxin. *Plant Physiol.* 102:95–100.

Lee, Y., Choi, Y.B., Suh, J., Lee, J., Assmann, S.M., Joe, C.O., Keller, J.F., and Crain, R.C. 1996. Abscisic acid-induced phosphoinositide turnover in guard cell protoplasts of *Vicia faba*. *Plant Physiol.* 110:987–996.

Lee, H., Xiong, L., Ishitani, M., Stevenson, B., and Zhu, J.K. 1999b. Cold-regulated gene expression and freezing tolerance in an *Arabidopsis thaliana* mutant. *Plant J.* 17:301–308.

Lee, H., Xiong, L., Gong, Z., Ishitani, M., Stevenson, B., and Zhu, J.K. 2001. The *Arabidopsis HOS1* gene negatively regulates cold signal transduction and encodes a RING finger protein that displays cold-regulated nucleo-cytoplasmic partitioning. *Genes Dev.* 15:912–924.

Lee, H., Guo, Y., Ohta, M., Xiong, L., Stevenson, B., and Zhu, J.K. 2002. LOS2, a genetic locus required for cold-responsive gene transcription encodes a bi-functional enolase. *EMBO J.* 21:2692–2702.

Leung, J., Bouvier-Durand, M., Morris, P.C., Guerrier, D., Chefdor, F., and Giraudat, J. 1994. *Arabidopsis* ABA-responsive gene ABI1: Features of a calcium-modulated protein phosphatase. *Science* 264: 1448–1452.

Leung, J., Merlot, S., and Giraudat, J. 1997. The Arabidopsis ABSCISIC-ACID INSENSITIVE 2 ABI2 and ABI1 genes encode homologous protein phosphatase 2C involved in abscisic acid signal transduction. *Plant Cell* 9:759–771.

Levine, A., Tenhaken, R., Dixon, R., and Lamb, C. 1994. H_2O_2 from the oxidative burst orchestrates the plant hypersensitive disease resistance response. *Cell* 79:583–593.

Levitt, J. 1980 *Responses of Plants to Environmental Stresses.* Second Edition. Vol. 1., New York: Academic Press.

Li, J., Lee, Y.R.J., and Assmann, S.M. 1998. Guard cells possess a calcium-dependent protein kinase that phosphorylates the KAT1 potassium channel. *Plant Physiol.* 116:785–795.

Lippuner, V., Cyert, M.S., and Gasser, C.S. 1996. Two classes of plant cDNA clones differentially complement yeast calcineurin mutants and increase salt tolerance of wild-type yeast. *J. Biol. Chem.* 271:12859–12866.

Liu, J., and Zhu, J.K. 1997. An *Arabidopsis* mutant that requires increased calcium for potassium nutrition and salt tolerance. *Proc. Natl. Acad. Sci. USA* 97:14960–14964.

Liu, J., and Zhu, J.K. 1998. A calcium sensor homolog required for plant salt tolerance. *Science* 280:1943–1945.

Liu, J., Ishitani, M., Halfter, U., Kim, C., and Zhu, J.K. 2000. The *Arabidopsis thaliana* SOS2 gene encodes a protein kinase that is required for salt tolerance. *Proc. Natl. Acad. Sci. USA* 97:3730–3734.

Liu, Q., Kasuga, M., Sakuma, Y., Abe, H., Miura, S., Yamaguchi-Shinozaki, K., and Shinozaki, K. 1998. Two transcription factors, DREB1 and DREB2, with an EREBP/AP2 DNA binding domain separate two cellular signal transduction pathways in drought- and low-temperature-responsive gene expression, respectively, in *Arabidopsis. Plant Cell* 10:1392–1406.

Lynch, J., Polito, V.S., and Lauchili, A. 1989. Salinity stress increases cytoplasmic Ca activity in maize root protoplasts. *Plant Physiol.* 90:1271–1274.

Ma, H. 1994. GTP-binding proteins in plants: new members of an old family. *Plant Mol. Biol.* 26:1611–1636.

MacRobbie, E.A.C. 2000. ABA activates multiple Ca^{2+} fluxes in stomatal guard cells, triggering vacuolar K^+ (Rb^+) release. *Proc. Natl. Acad. Sci. USA* 97:12361–12368.

Matsuoka, D., Nanmori, T., Sato, K., Fukami, Y., Kikkawa, U., and Yasuda, T. 2002. Activation of *AtMEK1*, *Arabidopsis* mitogen-activated protein kinase kinase, *in vitro* and *in vivo*: analysis of active mutants expressed in *E. coli* and generation of the active form in stress response in seedlings. *Plant J.* 29:637–647.

McAinsh, M.R., Brownlee, C., and Hetherington, A.M. 1990. Abscisic acid-induced elevation of guard cell cytoplasmic Ca^{2+} precedes stomatal closure. *Nature* 343:186–188.

Medina, J., Bargues, M., Terol, J., Pérez-Alonso, M., and Salinas, J. 1999. The *Arabidopsis* CBF gene family is composed of three genes encoding AP2 domain-containing proteins whose expression is regulated by low temperature but not by abscisic acid or dehydration. *Plant Physiol.* 119:463–469.

Meissner, R., and Michael, A.J. 1997. Isolation and characterization of a diverse family of *Arabidopsis* two and three-fingered C_2H_2 zinc finger protein genes and cDNAs. *Plant Mol. Biol.* 33:615–624.

Mendoza, I., Rubio, F., Rodriguez-Navarro, A., and Pardo, J.M. 1994. The protein phosphatase calcineurin is essential for NaCl tolerance of Saccharomyces cerevisiae. *J. Biol. Chem.* 269:8792–8796.

Meyer, K., Leube, M.P., and Grill, E. 1994. A protein phosphatase 2C involved in ABA signal transduction in *Arabidopsis thaliana. Science* 264:1452–1455.

Michaels, S.D., and Amasino, R.M. 1999. *FLOWERING LOCUS C* encodes a novel MADS domain protein that acts as a repressor of flowering. *Plant Cell* 11:949–956.

Mikoajcyk, M., Awotunde, O., Muszyńska, G., Klessig, D. F., and Dobrowolska, G. 2000. Osmotic stress induces rapid activation of salicylic acid-induced protein kinase and a homolog of protein kinase ASK1 in tobacco cells. *Plant Cell* 12:165–178.

Mizoguchi, T., Ichimura, K., Irie, K., Morris, P., Giraudat, J., Matsumoto, K., and Shinozaki, K. 1998. Identification of a possible MAP kinase cascade in *Arabidopsis thaliana* based on pairwise yeast two-hybrid analysis and functional complementation tests of yeast mutants. *FEBS Lett.* 437:56–60.

Munnik, T., Ligterink, W., Meskiene, I., Calderini, O., Beyerly, J., Musgrave, A., and Hirt, H. 1999. Distinct osmo-sensing protein kinase pathways are involved in signaling modulate and serve hyper-osmotic stress. *Plant J.* 20:381–388.

Murata, N., and Los, D.A. 1997. Membrane fluidity and temperature perception. *Plant Physiol.* 115:875–879.

Murata, Y., Pei, Z.M., Mori, I.C., and Schroeder, J.I. 2001. Abscisic acid activation of plasma membrane Ca^{2+} channels in guard cells requires cytosolic NAD(P)H and differentially disrupted upstream and downstream of reactive oxygen species production in *abi1-1* and *abi2-1* protein phosphatase 2C mutants. *Plant Cell* 13:2513–2523.

Nakamura, T., Liu, Y., Hirata, D., Namba, H., Harada, S., Hirokawa, T., and Miyakawa, T. 1993. Protein phosphatase type 2B (calcineurin)-mediated, FK506-sensitive regulation of intracellular ions in yeast is an important determinant for adaptation to high salt stress conditions. *EMBO J.* 12:4063–4071.

Nakashima, K., Shinwari, Z., Sakuma, Y., Seki, M., Miura, S., Yamaguchi-Shinozaki, K., and Shinozaki, K. 2000. Organization and expression of two *Arabidopsis* DREB2 genes encoding DRE-binding proteins involved in dehydration- and high-salinity- responsive gene expression. *Plant Mol. Biol.* 42:657–665.

Nordin, K., Vahala, T., and Palva, E.T. 1991. Differential expression of two related, low-temperature-induced genes in *Arabidopsis thaliana* (L.) Heynh. *Plant Mol. Biol.* 21:641–653.

Ohta, M., Matsui, K., Hiratsu, K., Shinshi, H., and Ohme-Takagi, M. 2001. Repression domains of class II ERF transcriptional repressors share an essential motif for active repression. *Plant Cell* 13:1959–1968.

O'Rourke, S.M., Herskowitz, I., and O'Shea, E.K. 2002. Yeast go the whole HOG for the hyperosmotic response. *Trends Genet.* 18:405–412.

Ota, I.M., and Varshavsky, A 1993. A yeast protein similar to bacterial two-component regulators. *Science* 262:566–569.

Örvar, B.L., Sangwan, V., Omann, F., and Dhindsa, R.S. 2000. Early steps in cold sensing by plant cells: the role of actin cytoskeleton and membrane fluidity. *Plant J.* 23:785–794.

Papageorgiou, G., and Murata, N. 1995. The unusually strong stabilizing effects of glycine betaine on the structure and function of the oxygen-evolving photosystem II complex. *Photosynth. Res.* 44:243–252.

Pei, Z.M., Ward, .M., Harper, J.F., and Schroeder, J.I. 1996. A novel chloride channel in *Vicia faba* guard cell vacuoles activated by the serine/ threonine kinase, CDPK. *EMBO J.* 15:6564–6574.

Pei, Z.M., Ghassemian, M., Kwak, C.M., McCourt, P., Klüsener, B., and Schroeder, J.I. 1998. Role of farnesyltransferase in ABA regulation of guard cell anion channels and plant water loss. *Science* 282:287–290.

Pei, Z.M., Murata, Y., Benning, G., Thomine, S., Klüsener, B., Allen, G.J., Grill, E., and Schroeder, J.I. 2000. Calcium channels activated by hydrogen peroxide mediate abscisic acid signaling in guard cells. *Nature* 406:731–734.

Piao, H.L., Pih, K.T., Lim, J.H., Kang, S.G., Jin, J.B., Kim, S.H., and Hwang, I. 1999. An *Arabidopsis GSK3/shaggy-like* gene that complements yeast salt stress-sensitive mutants is induced by NaCl and abscisic acid. *Plant Physiol.* 119:1527–1534.

Piao, H.L., Lim, J.H., Kim, S.H., Cheong, G.W., and Hwang, I. 2001. Constitutive over-expression of *AtGSK1* induces NaCl stress responses in the absence of NaCl stress and results in enhanced NaCl tolerance in *Arabidopsis. Plant J.* 27:305–314.

Qiu, Q.S., Guo, Y., Dietrich, M.A., Schumaker, K.S., and Zhu, J.K. 2002. Regulation of SOS1, a plasma membrane Na^+/H^+ exchanger in *Arabidopsis thaliana*, by SOS2 and SOS3. *Proc. Natl. Acad. Sci. USA* 99:8436–8441.

Quintero, F.J., Ohta, M., Shi, H., Zhu, J.K., and Pardo, J.M. 2002. Reconstitution in yeast of the *Arabidopsis* SOS signaling pathway for Na^+ homeostasis. *Proc. Natl. Acad. Sci. USA* 99:9061–9066.

Ralevic, V., and Burnstock, G 1998. Receptors for purines and pyrimidines. *Phrmacol. Rev.* 50:413–492.

Ray, R., and Miller, D.M. 1991. Cloning and characterization of a human c-*myc* promoter-binding protein. *Mol. Cell. Biol.* 11:2154–2161.

Saijo, Y., Hata, S., Kyozuka, J., Shimamoto, K., and Izui, K., 2000. Over-expression of a single Ca^{2+}-dependent protein kinase confers both cold and salt/drought tolerance on rice plants. *Plant J.* 23:319–327.

Salgado-Garrido, J., Bragado-Nilsson, E., Kandels-Lewis, S., and Séraphin, B. 1999. Sm and Sm-like proteins assemble in two related complexes of deep evolutionary origin. *EMBO J.* 18:3451–3462.

Sanchez, J.P., and Chua, N.H. 2001. *Arabidopsis* PLC1 is required for secondary responses to abscisic acid signaling. *Plant Cell* 13:1143–1154.

Sanders, D., Brownlee, C., and Harper, J.F. 1999. Communicating with calcium. *Plant Cell* 11:691–706.

Schachtman, D., and Liu, W. 1999. Molecular pieces to the puzzle of the interaction between potassium and sodium uptake in plants. *Trends Plant Sci.* 4:281–287.

Schroeder, J.I., and Hagiwara, S. 1990. Repetitive increases in cytoplasmic Ca^{2+} of guard cells by abscisic acid activation of nonselective Ca^{2+} permeable channels. *Proc. Natl. Acad. Sci. USA* 87:9305–9309.

Schroeder, J.I., Allen, G.J., Hugouvieux, V., Kwak, J.M., and Waner, D. 2001. Guard cell signal transduction. *Ann Rev. Plant Physiol. Plant Mol. Biol.*

Schumaker, K.S., and Sze, H. 1987. Inositol 1,4,5-triphosphate release Ca^{2+} from vacuolar membrane vesicles of oat roots. *J. Biol. Chem.* 262:3944–3946.

Schwartz, S.H., Léon-Kloosterziel, K.M., Koornneef, M., and Zeevaart, J.A.D. 1997. Biochemical characterization of the *aba2* and *aba3* mutants in *Arabidopsis thaliana. Plant Physiol.* 114:161–166.

Sentenac, H., Bonneaud, N., Minet, M., Lacroute, F., Salmon, J.M., Gaymard, F., and Grignon, C. 1992. Cloning and expression in yeast of a plant potassium ion transport system. *Science* 256:663–665.

Sheen, J. 1996. Ca^{2+}-dependent protein kinases and stress signal transduction in plants. *Science* 274:1900–1902.

Sheldon, C.C., Rouse, D.T., Finnegan, E.J., Peacock, W.J., and Dennis, E.S. 2000. The molecular basis of vernalization: The central role of *FLOWERING LOCUS C (FLC). Proc. Natl. Acad. Sci. USA* 97:3753–3758.

Shen, Q., and Ho, T.H.D. 1995. Functional dissection of an abscisic acid (ABA)-inducible gene reveals two independent ABA-responsive complexes each containing a G-box and a novel *cis*-acting element. *Plant Cell* 7:295–307.

Shi, H., Ishitani, M., Kim, C., and Zhu, J.K. 2000. The *Arabidopsis thaliana* salt tolerance gene *SOS1* encodes a putative Na^+/H^+ antiporter. *Proc. Natl. Acad. Sci. USA* 97:6896–6901.

Shi, H., Xiong, L., Stevenson, B., Lu, T., and Zhu, J.K. 2002a. The *Arabidopsis salt overly sensitive 4* mutants uncover a critical role for vitamin B6 in plant salt tolerance. *Plant Cell* 14:575–588.

Shi, H., Quintero, F.J., Pardo, J.M., and Zhu, J.K. 2002b. The putative plasma membrane Na^+/H^+ antiporter SOS1 controls long-distance Na^+ transport in plants. *Plant Cell* 14:465–477.

Shinozaki, K., and Yamaguchi-Shinozaki, K. 1997. Gene expression and signal transduction in water-stress response. *Plant Physiol.* 115:327–334.

Shinozaki, K., and Yamaguchi-Shinozaki, K. 2000. Molecular responses to dehydration and low temperature: Differences and cross-talk between two stress signaling pathways. *Curr. Opin. Plant Biol.* 3:217–223.

Shinwari, Z.K., Nakashima, K., Miura, S., Kasuga, M., Seki, M., Yamaguchi-Shinozaki, K., and Shinozaki, K 1998. An *Arabidopsis* gene family encoding DRE/CRT binding protein involved in low-temperature-responsive gene expression. *Biochem. Biophys. Res. Commun.* 250:161–170.

Solomon, A., Beer, S., Waisel, Y., Jones, G.P., and Paleg, L.G. 1994. Effects of NaCl on the carboxylating activity of Rubisco from *Tamarix jordanis* in the presence and absence of proline-related compatible solutes. *Plant Physiol.* 90:198–204.

Staxén, I., Pical, C., Montgomery, L.T., Gray, J.E., Hetherington, A.M., and McAinsh, M.R. 1999. Abscisic acid induces oscillations in guard-cell cytosolic free calcium that involve phosphoinoside-specific phospholipase C. *Proc. Natl. Acad. Sci. USA* 96:1779–1784.

Steponkus, P.L., Uemura, M., Joseph, R.A., Gilmour, S.J., and Thomashow, M.F. 1998. Mode of action of the *COR15a* gene on the freezing tolerance of *Arabidopsis thaliana*. *Proc. Natl. Acad. Sci. USA* 95:14570–14575.

Stockinger, E.J., Gilmour, S.J., and Thomashow, M.F. 1997. *Arabidopsis thaliana CBF1* encodes an AP2 domain-containing transcriptional activator that binds to the C-repeat/DRE, a *cis*-element that stimulates transcription in response to low temperature and water deficit. *Proc. Natl. Acad. Sci. USA* 94:1035–1040.

Suzuki, I., Los, D.A., Kanehisa, Y., Mikami, K., and Murata, N. 2000. The pathway for perception and transduction of low-temperature signals in *Synechocystis*. *EMBO J.* 19:1327–1334.

Thomashow, M.F. 1994. *Arabidopsis thaliana* as a model for studying mechanisms of plant cold tolerance. In *Arabidopsis*, eds. E. Meyerowitz, and C. Somerville, pp. 807–834. New York: Cold Spring Harbor.

Thomashow, M.F. 1999. Plant cold acclimation: freezing tolerance genes and regulatory mechanisms. *Ann Rev. Plant Physiol. Plant Mol. Biol.* 50:571–599.

Uno, Y., Furihata, T., Abe, H., Yoshida, R., Shinozaki, K., and Yamaguchi-Shinozaki, K. 2000. *Arabidopsis* basic leucine zipper transcription factors involved in an abscisic acid-dependent signal transduction pathway under drought and high-salinity conditions. *Proc. Natl. Acad. Sci. USA* 97:11632–11637.

Urao, T., Yamaguchi-Shinozaki, K., Urao, S., and Shinozaki, K. 1993. An *Arabidopsis* myb homolog is induced by dehydration stress and its gene product binds to the conserved MYB recognition sequence. *Plant Cell* 5:1529–1539.

Urao, T., Katagiri, T., Mizoguchi, T., Yamaguchi-Shinozaki, K., Hayashida, N., and Shinozaki, K. 1994. Two genes that encode Ca^{2+}-dependent protein kinases are induced by drought and high-slat stresses in *Arabidopsis thaliana*. *Mol. Gen. Genet.* 244:331–340.

Urao, T., Yakubov, B., Satoh, R., Yamaguchi-Shinozaki, K., Seki, M., Hirayama, T., and Shinozaki, K. 1999. A transmembrane hybrid-type histidine kinase in *Arabidopsis* functions as an osmosensor. *Plant Cell* 11:1743–1754.

Vasil, V., Marcotte, W.R., Rosenkras, L., Cocciolone, S.M., Vasil, I.K., Quantrano, R.S., and McCarty, D.R. 1995. Overlap of Viviparous1 VP1) and abscisic acid response elements in the Em promoter: G-box elements are sufficient but not necessary for VP1 transactivation. *Plant Cell* 7:1511–1518.

Wang, X.Q., Ullah, H., Jones, A.M., and Assmann, S.M. 2001. G protein regulation of ion channels and abscisic acid signaling in *Arabidopsis* guard cells. *Science* 292:2070–2072.

Winicov, I., and Bastola, D.R. 1999. Transgenic overexpression of the transcription factor *Alfin1* enhances expression of endogenous *MsPRP2* gene in Alfalfa and improves salinity tolerance of the plants. *Plant Physiol.* 120:473–480.

Widmann, C., Gibson, S., Jarpe, M.B., and Johnson, G.L. 1999. Mitogen-activated protein kinase: conservation of a three-kinase module from yeast to human. *Physiol. Rev.* 79:143–180.

Wu, S.J., Ding, L., and Zhu, J.K. 1996. *SOS1*, a genetic locus essential for salt tolerance and potassium acquisition. *Plant Cell* 8:617–627.

Wu, Y., Kuzma, J., Maréchal, E., Graeff, R., Lee, H.C., Foster, R., and Chua, N.H. 1997. Abscisic acid signaling through cyclic ADP-ribose in plants. *Science* 278:2126–2130.

Wurgler-Murphy, S.M., and Saito, H. 1997. Two-component signal transduction and MAPK cascades. *Trends Biochem. Sci.* 22:172–176.

Yamaguchi-Shinozaki, K., and Shinozaki, K. 1993. The plant hormone abscisic acid mediates the drought-induced expression but not the seed-specific expression of *rd22*, a gene responsive to dehydration stress in *Arabidopsis thaliana*. *Mol. Gen. Genet.* 238:17–25.

Yamaguchi-Shinozaki, K., and Shinozaki, K. 1994. A novel *cis*-element in Arabidopsis gene is involved in responsiveness to drought, low-temperature, or high-salt stress. *Plant Cell* 6:251–264.

Xin, Z., and Browse, J. 1998. *eskimo1* mutants of *Arabidopsis* are constitutively freezing-tolerant. *Proc. Natl. Acad. Sci. USA* 95:7799–7804.

Xiong, L., Ishitani, M., Lee, H., and Zhu, J.K. 1999. *HOS5*—a negative regulator of osmotic stress-induced gene expression in *Arabidopsis thaliana*. *Plant J.* 19:569–578.

Xiong, L., Lee, B., Ishitani, M., Lee, H., Zhang, C., and Zhu, J.K. 2001a. *FIERY1* encoding an inositol polyphosphate 1-phosphatase is a negative regulator of abscisic acid and stress signaling in *Arabidopsis*. *Genes Dev.* 15:1971–1984.

Xiong, L., Ishitani, M., Lee, H., and Zhu, J.K. 2001b. The *Arabidopsis LOS5/ABA3* locus encodes a molybdenum cofactor sulfurase and modulates cold stress- and osmotic stress-responsive gene expression. *Plant Cell* 13:2063–2083.

Xiong, L., Lee, H., Ishitani, M., and Zhu, J.K. 2001c. Regulation of osmotic stress-responsive gene expression by the *LOS6/ABA1* locus in *Arabidopsis*. *J. Biol. Chem.* 277:8588–8596.

Xiong, L., Gong, Z., Rock, C.D., Subramanian, S., Guo, Y., Xu, W., Galbraith, D., and Zhu J.K. 2001d. Modulation of abscisic acid signal transduction and biosynthesis by an Sm-like protein in *Arabidopsis*. *Dev. Cell.* 1:771–781.

Zhang, X., Zhang, L., Dong, F., Gao, J., Galbraith, D.W., and Song, C.P. 2001. Hydrogen peroxide is involved in abscisic acid-induced stomatal closure in *Vicia faba. Plant Physiol.* 126:1438–1448.

Zhu, J.K., Liu, J., and Xiong, L. 1998. Genetic analysis of salt tolerance in *Arabidopsis*: evidence for a critical role of potassium nutrition. *Plant Cell* 10:1181–1191.

Zhu, J.K. 2002. Salt and drought stress signal transduction in plants. *Annu. Rev. Plant Physiol. Plant Mol. Biol.* 53:247–273.

17

Commercialization of Plant Systemic Defense Activation: Theory, Problems and Successes

ANNE J. ANDERSON, KRIS A. BLEE, AND KWANG-YEOL YANG

17.1 Introduction

Crop protection can reduce losses by 10% to 60% depending on the disease, the locality and the crop (Crop Protection Compendium, 2002). An array of different strategies to reduce the consequences of pathogen pressure is available. Of these methods, the use of commercial products that stimulate defense reactions in the plant host to reduce plant pathogen success is in its infancy. Although the activation of systemic resistance has been demonstrated reproducibly in the laboratory for many plant species, utilizing a wide range of activating materials, it is not yet a proven technology widely accepted in commerce. A general view is that field results are too variable, and therefore risky, for many farmers when the alternative strategies for protection are perceived as more reliable. Systemic defense activation, however, offers attractive features:

- Ecological compatibility, with some products fulfilling the requirements for the "organic" farming label.
- Protection for the whole plant, with effects extending post harvest.
- Protection against pathogens that are not controlled by available methods, which is especially valuable for those pathogens with resistance to a chemical pesticide.
- Function through plant-based mechanisms rather than a direct attack on the pathogenic organism, thus, avoiding direct but undesired effects on non-pathogenic organisms.
- Provision of protection to a broad range of challenges including microbes, insects and nematodes.
- Compatibility with short time reentry and short time preharvest applications.
- Applications may be teamed with other differently based strategies to provide better protective coverage.
- The array of genes activated in systemic resistance may be beneficial in thwarting other stresses in the field, such as heat, cold, drought, and damage from the blue to UV irradiances of sunlight.

Detriments to commercial use include:

- The protection requires time to become effective in the plant, especially when the stimulating treatment is not applied to the whole plant.
- Variability in performance, especially in instances where biologicals are used to activate the defenses.
- Activation of defense against one pathogen may promote greater susceptibility to other pathogens using different strategies to attack the plant.
- Fine-tuning of the activation mixture, the method of application and the timing between applications for maximum effectiveness.
- Overexpression of defenses may lead to stunting and reduced productivity.

In this review, we introduce the mechanisms leading to induced plant defenses and illustrate some peculiarities of systemic resistance compared with the hypersensitive response (HR). We discuss how molecular and biochemical knowledge has participated in the development and understanding of the mode of action of commercial products that stimulate plant systemic defense in the field. We describe the nature of products that are commercially available with their division into chemical and microbial categories. We close with summaries and speculations.

17.1.1 Molecular Understanding of the Pathways for Systemic Resistance

Two pathways for systemic resistance that have drawn the main attention of researchers involve salicylic acid (SA) or jasmonic acid (JA)/ethylene as key signaling compounds (Dong, 1998; Reymond and Farmer, 1998). As discussed in detail by Nawrath et al. and Pieterse et al. (Chapters 7 and 8, this volume) these pathways result in the accumulation of the products of different defense genes. Examples of these differences are illustrated in Figure 17.1.

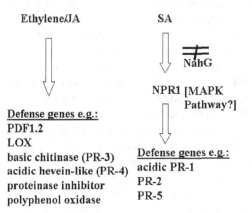

FIGURE 17.1. Differential defense gene activation by pathways involving ethylene/JA or SA.

The defense participants include the pathogenesis related (PR) proteins, discussed by Tuzun et al. (Chapter 6, this volume). The functions of this group are diverse and some are not as yet fully resolved, e.g., some members of the PR-1 group are antifungal by unknown mechanisms (Alexander et al., 1993). Other PR proteins have enzymatic activities that will degrade components in fungal cell walls (glucanases and chitinases) or help to generate phenolic radicals (the peroxidases) to produce barriers, such as cell wall lignification, or other antifungal materials in the plant. The marker protein most commonly ascribed to the SA pathway is the acidic PR-1, whereas PDF1.2 and the protease inhibitor genes are correlated with the JA/ethylene pathway. Expression of genes encoding the basic PR proteins, generally ascribed to a vacuolar location, is attributed more to the JA/ethylene-regulated defense pathway (van Loon, 1997). In contrast, the acidic PR proteins are generally thought to be apoplastic and associated with the SA-regulated pathway. However, global gene expression analysis reveals that several defense and metabolic genes are coregulated by both SA and JA/ethylene (e.g., Schenk et al., 2000). Crosstalk between metabolic pathways that involve genes encoding defense proteins controlled by such different plant growth regulators as JA, ethylene, SA, and abscisic acid is observed (e.g., Audenaert et al., 2002a; Gazzarrini and McCourt, 2003; Kunkel and Brooks, 2002). Thus, although these pathways can be viewed academically as being distinct, it is likely that effective resistance in the field will arise as a result of crosstalk between several pathways controlling defense gene expression.

The effectors that activate the SA and JA/ethylene regulated defense pathways may differ, as illustrated in Figure 17.2. Activation of the SA-regulated pathway is associated with events that cause necrosis. Thus, the pathway is aligned with the hypersensitive response (HR) where programmed plant cell death is part of the mechanism by which a pathogen is constrained to the initial invasion site. Cell death by HR initiates resistant events, termed local resistance, in the cells surrounding the containment site (Dangl et al., 1996).

With time, expression of defense genes occurs at greater distance to result in a systemic effect (Epple et al., 2003). Pathogens that cause necrosis as part of their symptomology also elicit the SA-regulated pathway (Ward et al., 1991). The classic findings of the significance of the SA pathway stemmed in part from studies with the lesion-causing virus, tobacco mosaic virus (Ross, 1961). Other pioneering work from Kuć (1982) showed that necrotizing bacterial and fungal pathogens would confer induced systemic resistance. An increase in the level of SA is associated with the induction of the systemic resistance phenomenon (Ward et al., 1991). Thus, this effect is not apparent in plants that are transformed to express the *nahG* gene encoding a bacterial salicylic hydrolase (Delaney et al., 1994). Metabolism of SA to catechol by the hydrolase in these plants is presumed to limit the accumulation of SA and prevent the expression of the SA-regulated genes (Figure 17.1; Neuenschwander et al., 1995).

In contrast, the JA/ethylene-regulated pathway of defense is associated with chewing insects where both wounding and specific components from insects participate in the stimulation (Figure 17.2; Kessler and Baldwin, 2002; Korth and

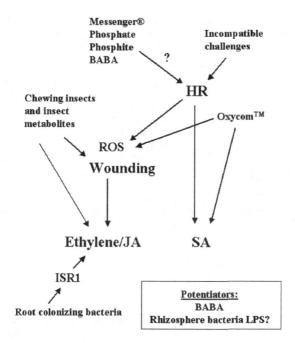

FIGURE 17.2. Differential effectors for pathways regulated by ethylene/JA or SA.

Thompson, Chapter 11 of this volume). Bacterial lipopolysaccharides also activate genes in this pathway (Dow et al., 2000). Elucidation of the JA/ethylene-regulated defense pathways was founded with the observation of systemic induction of proteinase inhibitors in the plants as a response to chewing (Ryan and Pearce, 1998; Kessler and Baldwin, 2002). Impaired insect digestion is correlated to the induced accumulation of proteinase inhibitors as well as to the effects of induced polyphenol oxidases in the plant tissues (Kessler and Baldwin, 2002). Ryan's studies in solanaceous plants revealed the crucial role of the synthesis of a novel peptide systemin in the signaling pathway which leads to oxylipin production and to altered gene expression. Although systemin appears to be restricted to certain solanaceous plants, the oxylipin pathway has been demonstrated for many plants (Turner et al., 2002). Interestingly, some of the volatile oxylipins are associated indirectly with plant defense because they act as attractants for predators of the insect pests (Kessler and Baldwin, 2002; van Poecke and Dicke, 2002). Although the JA/ethylene-regulated pathway is involved in insect resistance, other studies now reveal that it also is a major player in resistance to certain microbial pathogens (see Table 17.1). Likewise, the SA-regulated pathway is associated with resistance to an insect, the gall midge (Ollerstam and Larsson 2003).

For both the JA/ethylene- and SA-regulated pathways, signaling events include activation of members of the MAPK-cascade of protein kinases. Phosphorylation of the tobacco signal transduction MAPK member, salicylic acid-induced protein

TABLE 17.1. Spectrum of pest suppression associated with the salicylic acid (SA) and jasmonic acid (JA)/ethylene-regulated pathways.

	Pest	Reference
SA-regulated pathway	Downy mildew	Thomma et al. (2001); Ton et al. (2002)
	Powdery mildew	Thomma et al. (2001)
	Tobacco mosaic virus	Delaney et al. (1994)
	Turnip crinkle virus	Ton et al. (2002)
	Gall midge	Ollerstam and Larsson (2003)
JA/ethylene-regulated pathway	*Alternaria brassicicola*	Kunkel and Brooks (2002); Thomma et al. (2001); Ton et al. (2002)
	Botrytis cinerea	Diaz et al. (2002); Kunkel and Brooks (2002); Thomma et al. (2001)
	Erwinia carotovora	Kunkel and Brooks (2002); Thomma et al. (2001)
	Fusarium oxysporum	Garaats et al. (2002)
	Pythium spp.	Garaats et al. (2002); Kunkel and Brooks (2002); Thomma et al. (2001)
	Rhizopus stolonifer	Garaats et al. (2002)
	Thielaviopsis basicola	Garaats et al. (2002)
	Beet armyworm	Kessler and Baldwin (2002)
	Colorado potato beetle	Kessler and Baldwin (2002)
	Egyptian cotton worm	Stotz et al. (2002)
	Manduca sexta	Kessler and Baldwin (2002)
	Noctuid moth	Kessler and Baldwin (2002); Stout et al. (1999)
SA- and JA/ethylene-regulated pathways	Powdery mildew	Ellis et al. (2002)
	Pseudomonas syringae	Ellis et al. (2002)
	Xanthomonas campestris	Ton et al. (2002)
	Green peach aphid	Ellis et al., (2002); Kessler and Baldwin (2002)

kinase (SIPK), is rapid after SA treatment (Zhang and Klessig, 1997) and activation of wound-induced protein kinase (WIPK) initiates JA synthesis (Turner et al., 2002). Both SIPK and WIPK activation occurs as a result of the recognition event between the products of the *Cladosporium fulvum* avirulence gene, *avr9*, and its cognate resistance gene, *cf9*, responsible for HR (Romeis et al., 1999). The on–off-switch protein, CTR1, in ethylene signaling is believed to be a MAPKKK (Wang et al., 2002). In *Arabidopsis* a MAPK, MAPK4, acts as a repressor for the SA-regulated pathway, thus, promoting JA/ethylene effects (Turner et al., 2002; Wang et al., 2002). A plethora of transcriptional activators are implicated in altering defense gene expression (Eulgem et al., 1999; Chen and Chen, 2002; Turner et al., 2002; Wang et al., 2002). This complex situation means that defense genes are expressed and proteins are produced at different times in the response, e.g., phenylalanine ammonia-lyase (PAL) versus PR-1 [Guo et al., 2000]).

Other factors such as plant age also influence when defense genes are expressed. Certain defense genes are increased in expression by elevated sugar levels *in planta* (Ehness et al., 1997; Herbers et al., 1996). Studies by several groups find increased expression of certain defense genes in senescent tissues (e.g., Hanfrey et al., 1996; Quirino et al., 1999; Zhu et al., 2001). A recent paper (Yoshida et al., 2002) indicates that the *cpr5* gene, which causes constitutive expression of defense genes, is allelic with *hys1* that regulates senescent-induced defense gene expression. Further exploration is needed to clarify how the SA-independent expression of defense genes in these aging tissues relates to sugar sensing (Rolland et al., 2002). Another speculation is that gene regulation by the plant growth regulators ABA and ethylene may explain the sugar-linked expression of the defense genes (Gazzarrini and McCourt, 2003). Likewise, how plant aging affects systemic expression of the defense genes also has been little studied, although this factor is of vital importance for field efficacy.

17.1.2 Induced Plant Defense Responses and Field Protection

Induction of systemic resistance in crops is an attractive protective strategy because it can activate defenses throughout the plant. It complements existing plant-based strategies of preformed defenses and the localized induced response of HR. Cell death in HR is localized to the challenged cell and is initiated by recognition between the host and pathogen factors conditioned by resistance genes and avirulence genes, respectively. Because the response is dependent upon single genes for recognition, breeding for plant genes to confer HR has been a primary strategy to provide high-level protection against specific pests. However, frequently the pathogen population change to lose the effective avirulence gene. Thus, control based only on HR-based resistance may have limited time efficacy in the field. In contrast to the hypersensitive response, the plant cells expressing systemic resistance do not undergo programmed cell death *en masse*. Consequently, the systemically resistant plant maintains growth and production while offering pest protection. Because so many different types of stimuli may be involved in induction of the process, and its implementation may involve crosstalk between several defense pathways, pathogen resistance to plant systemic mechanisms may be less likely to develop.

Plants utilize some of the same chemical and physical ploys of the hypersensitive response to limit pathogen ingress in systemic resistance. Early inhibition of ingress and growth is a typical response observed upon challenge of a systemically protected plant (Hammerschmidt, 1999a). For systemic resistance to be effective, activation before the pathogen pressure reaches a crisis point is essential. Depending on the trigger used, such as the biologicals where signals have to be transduced from the root to leaf tissues, considerable pretreatment time is required for commercial applications to be successful.

Altered transcription and protein synthesis associated with defense gene activation may bring about a cost to the plant (Heil, 2002; Heil and Baldwin, 2002; Heil and Bostock, 2002). Choices must be made by the plant in how to allocate energy and metabolic resources. The view of Heil and Baldwin (2002) is that overexpression of defense traits in either of the pathways will result in poor growth and impaired reproduction. They cite the occurrence of stunted growth for 11 plant lines that were transformed to have increased expression of defense-related genes. However, they make the case for the need of more studies on the trade-off of protection versus metabolic cost under natural environmental conditions. Chemical overstimulation of defense also may result in poor plant performance. Although resistance to bacterial spot in bell pepper was induced by BTH, acibenzolar-S-methyl, weekly applications during the entire crop season reduced yield (Romero et al., 2001).

Responses in addition to protection against pathogen challenge may result from activation of the SA- and JA/ethylene-regulated pathways for gene expression. For example, tomatoes have an enhanced resistance to low temperatures when these pathways are stimulated (Ding et al., 2002). Protection against heat-induced oxidative damage in *Arabidopsis* involves responses orchestrated by ethylene, ABA, and SA (Larkindale and Knight, 2002). As revealed by gene microarray analyses (e.g., Chen and Chen, 2002; Cheong et al., 2002), these same growth regulators are key players in governing expression of defense genes.

17.1.3 Consequences of Multiple Defense Pathways: The Good and The Bad

The genes encoding defense functions associated with the SA- or the JA/ethylene-regulated pathways are differentially effective against different pests (Thomma et al., 2001; Ellis et al., 2002; Garaats et al., 2002; Kunkel and Brooks, 2002; Table 17.1). Which defense gene products are key in limiting each pathogen have not been resolved. For instance, although *Pseudomonas syringae* pv. *tabaci* incites the production of PR-1 in tobacco, neither this protein nor the PR proteins 3 and 5 appear to account for inhibition of growth of this pathogen (Thomma et al., 2001). The fact that different defense ploys are effective against different pathogens means that activation of only one pathway, (e.g., the SA-regulated pathway), may leave plants protected against some but not against all pests (i.e., chewing insects).

In the field there will be multiple interactive effectors and pathogenic challenges (Cui et al., 2002) and these may affect the responses of the plant. Systemic resistance effective against the cabbage looper in *Arabidopsis* was induced only by a pathogen-induced hypersensitive response and not by mutations that result in increased levels of SA, although this is one of the consequences of HR believed to be involved in establishing systemic resistance (Cui et al. 2002). These findings illustrate that there are complexities in these pathways that currently we do not understand. New studies continue to bring more questions of the accepted pathways, for example the evidence for an SA defense response independent of the transcription regulator, NPR1 (Figure 17.1; Wang et al., 2002). Such branching and crosstalk between pathways (Feys and Parker, 2000; Gazzarrini and McCourt, 2003) is of significance when considering protection against an array of pathogens.

Protection from one set of pathogens over another may also result from negative interactions between the two pathways. SA applications strongly impair the functioning of the JA-regulated pathway, in part by inhibiting key enzymes in oxylipin synthesis (Thaler et al., 2002). JA appears to be inhibitory, but to a lesser extent, to the SA-regulated pathway (e.g., Seo et al., 1997; Ellis et al., 2002). Abscisic acid (ABA) antagonizes SA-regulated responses (Audenaert et al., 2002a). Again the commercial impact of such antagonism would be that, although protected against one set of pathogens, the plants might be more susceptible to others. For SA-treated plants, an increase in feeding by insects that normally would be repressed by the JA/ethylene-regulated defense genes has been reported (Felton et al., 1999; Preston et al., 1999; Stout et al., 1999; Thaler et al., 1999).

Not all interactions between pathways are negative. Studies of the expression of distinct defense genes reveal synergism in effectors. Ethylene and SA act synergistically on the expression of several defense genes including *PR-2c*, *PR-3a*, *PR-3b*, *PR-4* and *PR-5* (van Loon, 1997). Such crosstalk between pathways may depend to some extent on potentiation. Certain activators of systemic resistance when present with very low levels of SA result in very effective expression of such genes as *PR-1* (Conrath et al., 2002). Because microbial challenge of plants can act to potentiate effects, perhaps through modification of SA or ethylene levels, the presence or absence of microbial challenges under field conditions within a time frame of application of a systemic resistance inducer may have dramatic effects.

Reactive oxygen species (ROS) are another important class of chemicals with field significance for defense. These arise during the early events in hypersensitivity, or they are produced naturally through plant metabolism or as a result of plant irradiation by the UVA/B spectrum of sunlight (Mittler, 2002; Neill et al., 2002). Recent gene-chip array studies to detect hydrogen peroxide-responsive plant genes confirm induction of a subset of defense-associated genes (Desikan et al., 2001). ROS signaling includes certain members of the MAPK families and transcriptional activators that are also involved in the SA pathway, so some crossover in defense products exists (Kovtun et al., 2000; Mittler, 2002). For instance, interaction between SA and hydrogen peroxide was suggested from studies of tobacco with a catalase deficit that under oxidative stress, imposed by high light, responded with elevated expression of *PR-1* (Chamnongpol et al., 1996, 1998). Although SA applications triggered localized increases in *PR-1*, the systemic response in the transformed plants was observed only under high light, suggesting that ROS was involved for long distance signaling. Also the increased expression of *PR-1* occurred without plant cell death, possibly because the ROS caused ethylene to be produced which enhanced the effect of SA on gene expression (Chamnongpol et al., 1998). However, negative as well as positive interactions between ROS and ethylene have been noted for other systems (Wang et al., 2002). These findings have relevance to certain of the commercial products discussed in the next section.

Another finding in research on systemic resistance is that in some cases tolerance rather than resistance is induced (Kloek et al., 2001). Although the treatment leads to loss in symptom formation, assessment of pathogen numbers reveals that colonization has not been impeded. Thus, the induced defense responses may act to reduce symptom formation rather than limiting pathogen growth. In the field, this could be a problem in that the method would not reduce inoculum input for another growing cycle.

At present the extent to which the SA- and JA/ethylene-regulated pathways are represented in each plant genus and the level to which there is cultivar specificity is unknown. Indeed, resistance in bean to the necrotrophic fungus *Botryis cinerea* requires the SA pathway whereas for *Arabidopsis*, the JA/ethylene pathway is more important (Díaz et al., 2002). The SA-regulated pathway is also required in tomato for defense against *B. cinerea* (Audenaert et al., 2002a). Moreover, it appears that the JA and ethylene pathways in tomato act independently whereas they are intertwined in *Arabidopsis* (Diaz et al., 2002). Another example of plant variability is that the application of the systemic inducer β-aminobutyric acid (BABA) is more effective against late blight in tomato than potato (Cohen, 2002). Crop variability in induced protection by BTH is also documented (Oostendorp et al., 2001). Thus, our knowledge is far from complete in understanding how a treatment inducing a systemic response in one plant under laboratory conditions will have an impact on the wide spectrum of crops in agriculture, horticulture, and forestry. Ease of genetic transformation and the information from genomic sequencing projects has favored acquiring knowledge in *Arabidopsis* and tobacco with other plants being less studied, especially the monocots. As we identify key genes involved in the pathways in these model plants, the variability of responses in other crop plants will be more easily predicted. The current 2002/3 NSF Initiative to understand the functioning of all of the *Arabidopsis* genes will spearhead this effort. Similarly, the completion of genomic sequencing for other plants (corn, tomato, rice) is hastening our ability to harness the power of the plant in defense strategies.

17.2 Current Commercial Products

Products in commerce that induce systemic resistance include chemicals and biologicals. Those products registered by EPA in the USA as biopesticides are listed at http://www.epa.gov/pesticides/biopesticides. This site covers all products with biocontrol activity irrespective of mechanism, including those that induce systemic resistance. A reoccurring statement for most of these products is their relative safety to the environment and to human health. A listing with references of chemicals inducing systemic resistance, updated to May 2003, is provided by the Scottish Research Institute in Dundee, Scotland (http://www.scri.sari.ac.uk/). Microbes and their metabolites that have biocontrol activity are also listed in the review article by McSpadden-Gardener and Fravel (2002). A similar list is compiled and updated, currently to April 2003, by the American Phytopathological Society Committee on Biological Control (available at http://www.oardc.ohio-state.edu/apsbcc). In these lists only four microbial products (three bacilli and a *Streptomyces* species) are cited with plant defense activation being a proven mechanism. However, as discussed by McSpadden-Gardener and Fravel (2002), not all products with plant defense-inducing potential are registered currently as a pesticide but rather, perhaps because of the expenses associated with registering a product as a pesticide, they only have the classification of fertilizers or plant growth promoters. As illustrated by the list of about 40 companies achieving EPA biopesticide registration between 1995 and 2000, most of these companies are relatively small with niche markets in comparison to the larger companies associated with production of the synthetic, chemically based, direct-impact pesticides. The politics of registration is posing problems. For instance, in California there has been a legal issue on whether a substance that is only registered as a fertilizer, phosphite, but which has proven resistance potential against the oomycete pathogens, can be used in attempts to control sudden oak death caused by a *Phytophthora*-like fungus.

17.2.1 The Chemical Inducers

Our review of the chemicals that induce resistance extends the review of Oostendorp et al. (2001). The chemical products with the potential to induced resistance fall under three classifications: inorganic, synthesized, and natural products.

Inorganic

Phosphates and Phosphites. Both phosphite and phosphate salts are demonstrated to induce systemic resistance. When applied as a foliar spray phosphate salts induce resistance under field conditions (Reuveni and Reuveni, 1998). Di- and tri-basic sodium and potassium salts at alkaline pH were proven effective (Gottstein and Kuć, 1989) as part of the pioneering studies of induced resistance from Kuć group. Systemic protection against fungi, bacteria, and viruses is reported (Mucharromah and Kuć, 1991).

Interpretation of the findings with phosphites is more complex because of debates on their mode of action. Salts are termed phosphites when in dry powder form. In water they are converted to phosphonates. Phosphonates are taken up and redistributed in the plant through the xylem and then the phloem (Rickard, 2000). They are used commercially as alternative phosphate ("P") fertilizers, and increase plant growth. Oxidation to phosphates is a presumed mechanism. A direct fungicidal effect of phosphonates is observed, especially for the fungal-like pathogens, *Pythium, Phytophthora* and downy mildews. This knowledge has, in part, stemmed from studies with the commercial registered fungicide, Aliette, that produces aluminum tris-ethyl phosphonate. However, the same antifungal potential is displayed by inorganic phosphonates. The phosphonates are believed to exert their effect by limiting polyphosphate formation in the fungi (Niere et al., 1994). Activation of plant defenses is another proposed mode of action of the phosphonates (Smillie et al., 1989). Product information from Bayer for the commercial fungicide phosphonate marketed as Chipco indicates that enhanced plant defenses including the production of antimicrobial phytoalexins are part of the modes of action of this chemical. Products formulated to produce inorganic phosphonates include Nutri-Phite® (Biagro Western, USA), Ele-Max® (Helena Chemical Co, USA), and Phytogard® (CATE, France).

Both commercial and technical grade phosphites were effective in controlling the root and crown rot caused by *Phytophthora capsici* (Förster et al., 1998). Studies in lettuce (Pajor et al., 2001) showed that Phytogard® protected against downy mildew in a dose and systemic manner. Current work with Nutri-Phite® on citrus by the team of Graham and McLean (personal communication) reveals increased resistance in fruit as it develops on the tree against *Phytophthora palmivora*, between 30 and 60 days after application.

How the phosphates and phosphites are perceived by the plant, or which pathways are involved in the induced resistance phenomena, are little resolved. In cucumber, dipotassium hydrogen phosphate treatments were associated with localized cell death at the sites of application (Orober et al., 2002). This treatment caused systemic protection against cucumber anthracnose in cucumber. The chemical applications mimicked the hypersensitive response further because both superoxide anion and hydrogen peroxide were detected. The response was likened to HR induced by the tobacco necrosis virus (Orober et al., 2002). However, in lettuce, treatment with Phytogard® did not increase the PR-1 protein anticipated from activation of a SA pathway (Bécot et al., 2000). None of the PR proteins (PR-1, PR-5 and PR-9) examined were elevated in level. Studies in our laboratory confirm activation of defense or growth related genes. Rapid, strong, and lasting increased expression of transcripts for genes encoding phenylalanine ammonia lyase, peroxidase, chalcone synthase, and the cell wall protein hydroxyproline rich glycoprotein were stimulated in bean (Kim et al., unpublished data) after sprays with Nutri-Phite®. Small lesions were seen on the bean foliage within two days following application.

Oxycom^{TM}. Oxycom^{TM} is produced by Redox Chemicals, USA and currently it is not registered as a biopesticide, although laboratory and field tests have demonstrated promotion of plant health and productivity of several crops under conditions of pathogen pressure (Kim et al., 2001; Yang et al., 2002). The active product is a mixture of reactive oxygen species, salicylic acid, and compounds with fertilizer activity. Application is by spray and by drench with repeat applications as needed for each crop. Oxycom^{TM} protects tobacco against infection by *Pseudomonas syringae* pv. *tabaci* (Yang et al., 2002). Abuse of the application system by repeated root saturation results in stunting of tomatoes in a greenhouse trial (Anwar et al., 2003). In contrast, a single application prior to inoculation of root knot nematodes conferred a tolerance response. Although nematode populations were not reduced there was no deleterious effect on foliar growth (Anwar and McKenry, 2002). Our studies with Oxycom™ further illustrate how application method may be important. We found that spraying on leaves induced confluent activation from the PR-1 promoter whereas application to roots induced a veinal pattern of activation of this promoter in the leaves (Blee et al., 2004). Thus, targeting application to the feeding strategies of the pathogen may be important for field control.

Systemic induction of defense genes associated with the SA- and the ethylene-regulated pathways has been observed in bean and tobacco after Oxycom™ applications (Kim et al., 2001; Yang et al., 2002). Gene chip array data analysis of the response of *Arabidopsis* to Oxycom™ treatments supports the concept of activation of the SA- and JA/ethylene-regulated pathways. We speculate that, like the findings of Chamnongpol et al. (1996, 1998), it is the simultaneous presence of ROS with SA that, in part, determines the defense activation potential of the product.

Synthesized Organic Chemicals

BABA β-aminobutyric acid. Induced resistance by the nonprotein amino acid (BABA) was reviewed recently by Cohen (2002). Registration is being pursued currently. BABA treatment results in different plant responses (induced physical barriers such as lignification, phytoalexin, and PR production) for each pathosystem studied (Cohen, 2002). Effective resistance is generated for a wide range of plant–pathogen systems (e.g., Shailasree et al., 2001). Curative effects of BABA treatment, a feature not observed with other chemical systemic resistance activators, are observed in some pathosystems. In tobacco, cell death accompanied by the generation of superoxide and hydrogen peroxide was induced by BABA treatment (Siegrist et al., 2000). Thus, association with the SA-regulated pathway would be expected. However, in cauliflower (Silue et al., 2002) induction of the typical barrage of PR proteins expected from this pathway (PR-1, PR-2, PR-5) were not detected. Rather only PR-2 accumulated significantly after challenge with downy mildew for which protection was apparent. In common with other activators, the involvement of the SA pathway in BABA-stimulated resistance is variable with the pathogen studied. It is possible that some of the variability in response relates to effective dose. As discussed by Conrath et al. (2002), activators of systemic resistance responses may cause plant cell death at high concentrations yet act at lower doses to potentiate the defense response in conjunction with other effectors. Studies with BABA in *Arabidopsis* suggest that potentiation of defense gene expression in response to another agonist is a likely mode of action (Zimmerli et al., 2000). Such potentiation was demonstrated with the observed resistance to the necrotrophic fungus *Botrytis cinerea* in BABA-treated plants (Zimmerli et al., 2001).

BTH Benzo(1,2,3)thiadiazole-7-carboxylic acid derivatives. One of the most academically studied chemicals with systemic inducing activity is BTH, marketed by Syngenta (www.syngenta.cropprotection-us.com) under the name of Bion® in Europe or Actigard® in the USA. The compound is formulated as a water-dispersible granule to be applied as a drench. It was approved in 2002 for use on tobacco, tomato, lettuce, and spinach in the USA. The longevity of the protection afforded by BTH is variable, with a longer efficacy in monocots than in dicots (Staub, 2001). Data compiled by Tally et al. (1999) illustrate that BTH has crop specificity. Although resistance is induced in tomato against late blight there was no activation of defenses for potato late blight. Thus, the spectrum for effectiveness must be determined for each plant–pathogen system (Tally et al., 1999).

Effectiveness on 12 crops with activity against bacteria, viruses, fungi, insects, and nematodes was summarized by Oostendorp et al. (2001). Efficacy of Bion® against rhizoctonia leaf spot and wild fire in tobacco has been reported in other field studies (Cole, 1999). Suggested use of BTH is not as a "stand-alone" product but in conjunction with other protection methods. For example, a mixture of Bion® and copper hydroxide was more effective than single treatments in controlling bacterial spot of pepper (Buonaurio et al., 2002).

The BTH compounds are believed to stimulate the SA pathway downstream of SA before the transcriptional activator NPR1 (Figure 17.1). Early laboratory studies showed applications of BTH to barley promoted a rapid and effective HR-like response in the treated plant when challenged by powdery mildew (Gorlach et al., 1996) although other defense mechanisms were also stimulated. As listed in the report of SAR activators from the Scottish Research Institute, BTH is associated with increased accumulations of acidic PR-1, PR-2, and PR-5, each of which is an accepted marker for the SA-regulated pathway. BTH has potentiator activity enhancing the production of PR-1 and PAL with treatments by SA (Conrath et al., 2002). Potentiation after infection with a pathogen also has been noted (e.g., Benhamou and Belanger, 1998; Latunde-Dada and Lucas, 2001), a response requiring the *NPR1* gene (Kohler et al., 2002). Such potentiation means that under field conditions where the SA pathway may already be activated by challenge with a necrotizing microbe, BTH may enhance the activation of defense pathways.

Probenazole (3-allyloxy-1,2-benzisothiazole-1,1-dioxide). Probenazole, formulated as Oryzemate, is used in rice to provide protection against the rice blast fungus *Magnaporthe grisea* and bacterial blight, *Xanthomonas oryzae* (Watanabe et al., 1977, 1979). Spray or paddy applications result in uptake and metabolism into benzoate and saccharin-based products. A leucine-rich repeat (LRR) nuclear binding protein, RPR1, changes in level upon application of probenazole to rice, suggesting the potential for an interaction that resembles the recognition between microbial avirulence effectors and the resistance gene products that trigger the hypersensitive response (Sakamoto et al., 1999). Additional studies show that RPR1 belongs to the Pib family of genes associated with rice blast resistance genes (Wang et al., 2001; Chauhan et al., 2002) and resides on the chromosome in regions that show extensive cultivar variability. This situation resembles the clustering of genes encoding LRR proteins genes that are part of the signaling pathways determining resistance against *Pseudomonas syringae* pv. *tomato* (Salmeron et al., 1996). Gene expression of the *Pib* family is regulated by several environmental factors, including SA. Indeed, Probenazole causes SA to accumulate in Arabidopsis and requires NPR1-dependent defense gene activation (Yoshioka et al., 2001). Failure of *NahG* plants to show defense indicates that the chemical acts apparently upstream of SA. These findings illustrate that, although debated, a defense pathway involving SA regulation is likely to operate in rice and that this pathway can be successfully activated under commercial conditions to boost plant productivity.

Natural Products

Chitin and Chitosan.   A chitosan product called Elexa® is sold by SafeScience, USA, and is EPA approved for use on cucumber, vines, potatoes, strawberry, and tomato as "an alternative for traditional fungicides" and a "plant defense booster". Although direct effects of growth inhibition of fungal pathogens are reported as a mode of action, chitosan also activates plant defenses (Hadwiger et al., 1994; Chang et al., 1995). The activity of chitosan in stimulating plant defenses was established when researchers were screening fungal cells wall components as elicitors of HR. Chitosan treatments of pea caused an array of defense genes to be expressed and phytoalexins to accumulate (Hadwiger et al., 1994). Responses in other plants include elicitation of both PAL and peroxidase activities in wheat leaves (Vander et al., 1998). Treatment of tomato with chitosan enhanced resistance to the crown and root rot pathogen *Fusarium oxysporum* f. sp. *radicis-lycopersici* and stimulated defense responses, such as reinforcement of the plant cell wall and the alteration of the plasma membrane (Benhamou and Theriault, 1992). Synergism between chitosan and a root-colonizing protectant *Bacillus* isolate against Fusarium wilt infection has been observed (Benhamou et al., 1998). Rapid formation of plant cell wall modifications (chitin-enriched and callose deposits) was cited as limiting penetration of the fungus into the bacterized root.

Perception of chitosan may be initiated by electrostatic disruptions in the plant plasmalemma (Benhamou and Theriault, 1992). The signaling pathway for chitosan involves rapid induction of a 48 kDa MAPK activity in tomato that is independent of JA signaling (Stratmann and Ryan, 1997) and hydrogen peroxide production through the oxylipin pathway (Orozco-Cardena and Ryan, 1999). A burst of oxylipin synthesis was detected after rice was treated with chitosan (Rakwal et al., 2002). Involvement of ROS and MAPK activation (the ROS responsive-AtMAPK3 in *Arabidopsis*) after chitin treatments was demonstrated (Link et al., 2002; Zhang et al., 2002). However, the activation of two chitin-stimulated genes in *Arabidopsis* was independent of functional ethylene, JA and SA pathways (Zhang et al., 2002), although another required JA or SA regulation. The ethylene/JA-regulated pathways also were implicated in the defense response induced by chitin in pepper, where a specific chitin-binding protein was detected (e.g., Lee et al., 2001).

Messenger®.   Messenger® is marketed by Eden Biosciences, USA, and is a preparation of a secreted peptide from the bacterium *Erwinia amylovora*. This peptide, termed a "harpin", triggers changes in plant tissues typical of HR (Yang et al., 1993; Desikan et al., 1998). Consistent with these findings is the observation that harpin elicits disease resistance in Arabidopsis in a SA-dependent manner (Dong et al., 1999). The water-soluble powdered product is applied as a foliar spray and is stated to exert the required changes in the plant within three to five days. Eden Biosciences indicates efficacy on 40 crops, including specialty crops of strawberries, citrus, and ornamentals. Data sheets for applications for several crops are available from their website http://www.edenbio.com. Studies described in a patent for Messenger® indicate that there is also a strong plant growth-promoting activity associated with the product.

Strobilurins.   Several products from wood-associated fungi are marketed as stro-bilurins, which have both indirect and direct effects on fungal pathogens (Ypema and Gold, 1999). Formulations include: Quadris and Abound, containing azoxys-trobin, Trifloxystrobin, formulated as Flint, Stratego, and Compass; and pyra-clostrobin, formulated as Cabrio EG and Headline, Amistar, Bankit, Priori, Ortiva, and Heritage. New products are being commercialized (e.g., Acanto, a picoxys-trobin from Syngenta targeted toward emergent wheat). They are approved for 85 crops ranging from cereals including rice, to vines, fruits, vegetables, turf, and ornamentals. Strobilurins are designated as "reduced risk" products by the EPA. The products have direct fungicidal activity, by inhibiting mitochondrial respiration in the fungus at the site of complex III, the ubiquinine oxidation center. For some of the products their mobility in the plant is a benefit. However, they also activate plant defenses. The formulation Pyraclostrobin F 500 from BASF Inc., demon-strated NahG-independent protection against *Pseudomonas syringae* pv. *tabaci* (Herms et al., 2002). Although the strobilurin did not cause PR-1 accumulation itself, it primed tobacco for greater production when subsequently challenged with the wild-fire pathogen. Resistance to TMV generated by the strobilurin treatment was variable and cultivar dependent.

   Although the rapid development of resistance in pathogens to strobilurins ap-pears to be a problem, causing restricted and intregrated use with chemicals of different modes of action, studies as yet do not reveal whether strobilurins' ability to induce resistance will still have commercial importance.

Summary for Chemical Activators

The chemicals that stimulate systemic resistance display a wide range of structures and activate a diversity of plant defense genes. At present there are no commer-cial products based on the stimulating components that are naturally produced by insects (Kessler and Baldwin, 2002). A common thread for many of the activators is that under some conditions they mimic events occurring in HR (Messenger®, BABA, phosphite, Oxycom™). The commercial use of SAR inducers that func-tion through causing "local lesions" was questioned by Oostendorp et al. (2001). However, field studies with these compounds demonstrate that any induced phytotoxicity is not adverse because beneficial effects against pathogen pressure have been shown. Several of the products also have potentiation activity. Although at low concentration these seem to be only weak activators of defense, in combination with other factors they promote more rapid and greater activation of resistance.

Enhanced plant growth is another observation associated with use of chemicals that induce systemic resistance (e.g., phosphate, BABA, Messenger®, Oxycom™). Improved growth under field conditions is also a common effect of colonization of plant roots with beneficial microbes, hence the descriptive term "plant growth-promoting-bacteria". Reasons for improved growth are not resolved. One debated theory is that the growth of minor pathogens is reduced and, thus, the plants have more energy to divert to plant growth. Additionally, the metabolites, such as SA, involved in plant defense may also participate in regulating cell size. Expression of an effector gene, AvrBs3, results in enlarged mesophyll cells and increased transcripts of auxin-related genes and expansin, genes associated with cell expansion (Marois et al., 2002). Plant cells surrounding isolated dead cells, generated by changes in SA accumulation or by infection with a necrotic pathogen, were observed to grow abnormally large (Vanacker et al., 2001). Thus, roles for SA and the regulatory protein, NPR1, in controlling the balance between plant cell death and cell growth are suggested (Vanacker et al., 2001). Understanding the value of this growth effect of chemicals associated with plant defense toward their field efficacy will be most interesting.

Probably the most neglected factor involved in the significance to field protection is the role of nutrition to the plant. Nutrition may not be a notable factor in controlled greenhouse/laboratory studies where long-term plant growth is not the norm. In the field, the plants must have adequate nutrition to permit the required changes in gene expression to be accomplished. Whether plants purposely treated with effectors of systemic resistance under commercial conditions require specialized nutrition awaits rigorous examination.

17.2.2 Microbial Stimulants of Plant Defense

The EPA-registered Biopesticides with stated ability to induce resistance include bacilli. YieldShield from Gustafson, Inc. (www.gustafson.com) is a powdered formulation of *Bacillus pumilus* GB34 and is used as a seed treatment to confer protection on soybean for root pathogens. The APS listing indicates that Yield-Shield is currently under registration as a biopesticide. Serenade from AgraQuest, Inc. (www.agraquest.com) is based on *Bacillus subtilis* QST716. The preparation is reported to control a variety of pathogens (powdery mildew, downy mildew, Cercospora leaf spot, early blight, late blight, brown rot, and the bacteria, *Erwinia amylovora*) on a range of crops (vegetables, cucurbits, grapes, hops, peanuts, pome fruits, stone fruits). The product description indicates that the mode of action of the bacterium includes activation of host defenses but no further information is available.

The marketing of organisms as biocontrol agents that stimulate plant defenses, as opposed or in addition to a direct effect on the pathogen, is strongly supported by laboratory studies. Indeed the YieldShield *Bacillus* species were initially discovered in screens of bacteria for plant growth promoting and protection activities

(Raupach and Kloepper, 1998). Colonization of the plant by these biological control agents activates genes associated with both the SA- and the ethylene/JA-regulated pathways. For instance, certain fluorescent pseudomonads and *Bacillus* isolates stimulate expression from the *PR-1* gene in colonized tobacco (Park and Kloepper, 2000). Accumulation of protein regulated by the *PR-1* gene promoter is time dependent, requiring about 10 days for sizable activation (Park and Kloepper, 2000). This finding stresses the need for treatments with microbial inducers well before the disease pressure exists so that the plant is preconditioned for resistance. By comparison, defense genes associated with systemic resistance pathways are activated generally less than 24 hours after chemical application.

SA-independent activation of systemic resistance is reported after colonization of plant roots with the fluorescent pseudomonad WCS417r (Pieterse et al., 1996). The term induced systemic resistance, ISR, has been used to determine such microbially induced resistance [the term "ISR" has also been used to indicate resistance that is induced and systemic, regardless of the eliciting agent]. Resistance was induced toward the fungal root rot pathogen, *Fusarium oxysporum* f. sp.*raphani*, and the leaf pathogens blue mold (*Peronospora tabacina*), *Xanthomonas campestris* and *Pseudomonas syringae* pv. *tomato*. The lack of induced resistance in the JA-response mutant, *jar1*, and the ethylene-response mutant, *etr1*, is consistent with involvement of the JA/ethylene-regulated pathway. However, although sensitivity to the JA/ethylene pathway is essential, activation of ACC synthase or defense gene expression associated with these pathways was not observed (Knoester et al., 1999). Rather a rapid increase in a specific JA-regulated gene was observed only after pathogen challenge was detected (van Wees et al., 1999), suggesting that potentiation is occurring. A locus conditioning this sensitivity, ISR1, has been identified in *Arabidopsis* (Ton et al., 2001). Cultivars that fail to develop induced resistance are altered in this locus and lack ethylene sensitivity in their roots (Ton et al., 2001). Similar genetic differences in commercial crops could result in differential effectiveness of the microbials in inducing ISR.

How do the bacteria induce the response? The activity of some bacteria may correspond to their production of SA (Mercado-Blanco et al., 2001). Other activators for systemic resistance are extracellular bacterial surface structures, flagellin and its major structural protein, lipopolysaccharides (LPS) or the secreted siderophores. In 1999, a conserved domain from the N-terminus of flagellins was shown to stimulate alkalization of the medium of cultured plant cells, K^+ efflux and elicit ROS production, thus mimicking HR (Felix et al., 1999; Gómez-Gómez et al., 1999). A flagellin from *Pseudomonas (Acidivorax) avenae* incompatible on rice was also shown to cause ROS production and HR in rice (Che et al., 2000; Tanaka et al., 2003). In contrast, flagellins from compatible isolates were inactive (Che et al., 2000).

Using a synthetic peptide corresponding to a conserved 15-amino acid sequence from the N terminus of flagellin, a receptor was identified in *Arabidopsis* as a leucine-rich repeat kinase encoded by a single locus (Gómez-Gómez et al., 2001). The signal transfer chain involved in flagellin perception in *Arabidopsis* was further probed and was shown to include specific members of the MAPK pathway

(Asai et al., 2002). These MAPKs (AtMAPK3/6) are also known participants in oxidative-stress signaling (Kovtun et al., 2000) and again these findings are consistent with flagellin stimulating a HR-like response. Commercialization of chemical inducers based on flagellin structure seems unlikely at present because their use stunts plant growth.

Early work on plant recognition of LPS structures demonstrated that infusions of LPS from a range of enteric bacteria created a localized effect that nullified growth of both incompatible and compatible challenges from *Ralstonia solanacearum* on a temporary basis. This work, from Sequiera's lab, is placed into context with current findings in the review of Dow et al. (2000). Whereas the core of the enteric LPS was needed for a localized protective response, in other systems the core, its conserved sugar residues or the variable O-antigen side chains was involved. Although LPS from xanthomonads alone has weak elicitor activity, exposure to the LPS potentiated defense processes upon subsequent microbial challenge (Newman et al., 2002). This finding shows that the LPS effects are similar to the chemical inducers, BABA and BTH or the strobilurins, for which potentiation has been demonstrated. Speculation is raised that the *hrp* system required for microbial pathogenesis suppresses the potential for the LPS to otherwise induce resistance Dow et al. (2000).

LPS from saprophytic root-colonizing pseudomonads also induces systemic resistance responses. LPS from *Pseudomonas fluorescens* accounted for the systemic resistance against Fusarium wilt induced in radish when roots were colonized by this bacterium (Leeman et al., 1995). LPS from *P. fluorescens* strain WCS417r also was an inducing factor in certain *Arabidopsis* ecotypes (van Wees et al., 1997).

Bacterial siderophores, iron-binding compounds that are secreted when iron is limited, are demonstrated to cause ISR. The activity has been demonstrated with siderophores from *P. putida* (Leeman et al., 1995); *P. fluorescens* (Mauhofer et al., 1994); *P. aeruginosa* (Audenaert et al., 2002b), and a *Serratia marcescens* strain (Press et al., 2001). An interaction between the antifungal phenazine, pyocyanin, and the siderophore, pyochelin, both produced by *P. aeruginosa* 7NSK2, is proposed to account for the ability of this strain to cause ISR (Audenaert et al., 2002b). SA-regulated genes are demonstrated to be important in this system (Audenaert et al., 2002b). Because of the dependence on iron availability to induce siderophore production, the use of such ISR-inducing bacteria as an inoculant to induce resistance may be effective only in iron-deficient soils, such as those that a have a basic pH.

These findings raise the possibility of whether synthetic chemicals based on LPS or siderophore structures could be commercially marketed. Such products could be used for specialty high-profit crops, such as ornamentals.

Summary and Comments on Microbials as Inducers of Systemic Resistance

The potential for commercialization of microbes with defense stimulating properties seems endless. Surveys of published findings suggest that many microbial isolates have the potential to activate defenses. For instance, although the hydrogen peroxide that is produced by *Talaromyces* species is assumed biocidal in its biological control potential (Stosz et al., 1996), this ROS could play a role in stimulating plant defense. The elicitor activity of the xylanase secreted from *Trichoderma viride* (Yano et al., 1998) suggests that induced resistance may also account for biocontrol activity of such *Trichoderma* isolates. Additionally, the chitosan and/or glucan oligomers released from fungal walls being degraded by *Trichoderma* could have elicitor activity. Such factors could explain why root colonization by a *Trichoderma* isolate was suggested to induce a systemic resistance response (Yedidia et al., 1999).

The limitation of commercial development of the microbials themselves, rather than the products they produce (e.g., harpin, chitosan), is in our weak ability to manipulate the field environment to provide the beneficial organisms at the right time, at the right place, and with expression of the needed set of genes. Basic studies on genes involved in colonization and survival may provide the understanding to better implement microbials in the field. For instance, identification of genes that underlie effective root colonization by pseudomonads may provide tools for better screening for isolates excelling in the field (Lugtenberg et al., 2001).

Work on formulations of the organism so that field applications have maximal effect is needed. Here the understanding of how microbials overcome adverse environmental conditions (e.g., Beattie and Lindow, 1995; Lindow and Leveau, 2002) will be useful. Generally microbials are raised under conditions where cells are produced at maximum growth rates to highest density. However, such rich-medium growth conditions may not generate cells that are optimum in expressing traits required for field survival. Expression of the genes for resistance to heat, dessication, and UV light may be stimulated by modified culture conditions and result in microbials that survive better when applied in the field. Genetic engineering of plants to excel as hosts for beneficial microbes may come into play. Recent findings (e.g., Fray et al., 1999) with plants engineered to produce the acyl homoserine lactones that are signals for altered bacterial expression of genes involved in quorum sensing, survival, and competition illustrate how we can manipulate the behavior of associated microbials to minimize pathogen and maximize biocontrol effects.

17.3 Summary

The recent years of laboratory studies are starting to explain at the molecular levels the complexities of pathogen-resistance mechanisms. More studies under commercial field condition are required to test the robustness of stimulation of the systemic defenses that lab studies demonstrate plants possess. With all of the natural modes of stimulation from microbial contacts, is it feasible to use these induced mechanisms in the field? In answering this question, Heil and Baldwin (2002) indicate that the mechanisms associated with chewing insect defense in field plants can still be elevated. Enhanced levels of control over what is available from nature may result from genetic modification of genes in the defense pathways or genes controlling the effector structures in the microbes. Teaming the inducers of systemic resistance with traditional methods may be beneficial (e.g., Friedrich et al., 2001).

We need to maximize protection against crop losses in yield and quality through studies of dose, application frequency, and application techniques to understand the load in altered metabolism that plants can endure under field conditions. These studies of the profitable side of pest control must be balanced with long-term studies to deduce possible environmental consequences on biodiversity incurred by purposeful manipulation of the plants systemic defense responses. Many questions arise in this area. How will insect visitation be altered if activation of the SA pathway changes the emission of volatiles that other insect use as cues for predation or for finding food? Will some pathogens evolve into superpathogens as they mutate to avoid the systemic resistance measures? Will changes in PR proteins, such as the proteinase inhibitors or polyphenol oxidases, alter the digestibility of the foods for desirable consumers (animals and humans)? Will the fact that several PR proteins are allergens in humans (Salcedo et al., 1999; Ebner et al., 2001) have an effect on workers when the plant materials are processed or the products are ingested? These questions illustrate the increased need for interaction between researchers with expertise in such different areas as plant pathology, entomology, microbiology, and immune responses to work together to help formulate successful products to stimulate systemic resistance in the field with optimal effectiveness.

References

Alexander, D., Goodman, R.M., Gut-Rella, M., Glascock, C., Weymann, K., Friedrich, L., Maddox, D., Ahl-Goy, P., Luntz, T., Ward, E., and Ryals, J. 1993. Increased tolerance to two oomycete pathogens in transgenic tobacco expressing pathogenesis-related protein la. *Proc. Natl. Acad. Sci. USA* 90:7327–7331.

Anwar, S.A., and McKenry, M.V. 2002. Effect of Oxycom™ on growth of tomato and reproduction of *Meloidogyne incognita. Nematology* 4:141.

Anwar, S.A., Mckenry, M.V., Yang, K.Y., and Anderson, A.J. 2003. Induction of tolerance to root knot nematodes by Oxycom. *Journal of Nematology.* 35:306–313.

Asai, T., Tena, G., Plotnlkova, J., Willmann, M.R., Chiu, W., Gómez-Gómez, L., Boller, T., Ausubel, F.M., and Sheen, J. 2002. MAP kinase signaling cascade in *Arabidopsis* innate immunity. *Nature* 415:977–983.

Audenaert, K., De Meyer, G.B., and Höfte, M. 2002a. Abscisic acid determines basal susceptibility of tomato *Botrytis cinerea* and suppresses salicylic acid-dependent signaling mechanisms. *Plant Physiol.* 128:491–501.

Audenaert, K., Pattery, T., Cornelis, P., and Höfte, M. 2002b. Induction of systemic resistance to *Botrytis cinerea* in tomato by *Pseudomonas aeruginosa* 7NSK2: role of salicylic acid, pyochelin, and pyocyanin. *Mol. Plant Microbe Interact.* 15:1147–1156.

Beattie, G.A., and Lindow, S.E. 1995. The secret life of foliar bacterial pathogens on leaves. *Annu. Rev. Phytopathol.* 33:145–172.

Bécot, S., Pajot, E., Le Corre, D., Monot, C., and Silué, D. 2000. Phytogard® (K₂HPO₃) induces localized resistance in cauliflower to downy mildew of crucifers. *Crop Prot.* 19:417–425.

Benhamou, N., and Thériault, G. 1992. Treatment with chitosan enhances resistance of tomato plants to the crown and root rot pathogen *Fusarium oxysporum* f. sp. *radicis-lycopersici*. *Physiol. Mol. Plant Pathol.* 45:33–52.

Benhamou, N., and Bélanger, R.R. 1998. Benzothiadiazole-mediated induced resistance to *Fusarium oxysporum* f. sp. *radicis-lycopersici* in tomato. *Plant Physiol.* 118:1203–1212.

Benhamou, N., Kloepper, J.W., and Tuzun, S. 1998. Induction of resistance against *Fusarium* wilt of tomato by combination of chitosan with an endophytic bacterial strain: ultrastructure and cytochemistry and the host response. *Planta* 204:153–168.

Blee, K.A., Yang, K.-Y., Anderson, A.J. 2004. Activation of defense pathways: synergism between reactive oxygen species and salicylic acid and consideration of field applicability. *European Journal of Plant Pathology* 110:203–212.

Buonaurio, R., Scarponi, L., Ferrara, M., Sidoti, P., and Bertona, A. 2002. Induction of systemic acquired resistance in pepper plants by acibenzolar-S-methyl against bacterial spot disease. *Eur. J. Plant Pathol.* 108:41–49.

Chamnongpol, S., Willekens, H., Langebartels, C., van Montagu, M., Inze, D., and van Camp, W. 1996. Transgenic tobacco with a reduced catalase activity develops necrotic lesions and induces pathogenesis-related expression under high light. *Plant J.* 10:491–503.

Chamnongpol, S., Willekens, H., Moeder, W., Langebartels, C., Sandermann, H., van Montagu, M., Inzé, D., and van Camp, W. 1998. Defense activation and enhanced pathogen tolerance induced by H₂O₂ in transgenic tobacco. *Proc. Natl. Acad. Sci. USA* 95:5818–5823.

Chang, M.M., Horovitz, D., Culley, D., and Hadwiger, L.A. 1995. Molecular cloning and characterization of a pea chitinase gene expressed in response to wounding, fungal infection and the elicitor chitosan. *Plant Mol. Biol.* 1:105–111.

Chauhan, R.S., Farman, M.L., Zhang, H.B., and Leong, S.A. 2002. Genetic and physical mapping of a rice blast resistance locus, *Pi-CO39(t)* that corresponds to the avirulence gene *AVR1-CO39* of *Magnaporthe grisea*. *Mol. Genet. Genom.* 267:603–612.

Che, F.S., Nakajima, Y., Tanaka, N., Iwano, M., Yoshida, T., Takayama, S., Kadota, I., and Isogai, A. 2000. Flagellin from an incompatible strain of *Pseudomonas avenae* induces a resistance response in cultured rice cells. *J. Biol. Chem.* 275:32347–32356.

Chen, C., and Chen, Z. 2002. Potentiation of developmentally regulated plant defense response by AtWRKY18, a pathogen-induced *Arabidopsis* transcription factor. *Plant Physiol.* 129:706–716.

Cheong, Y.H., Chang, H-S., Gupta, R., Wang, X., and Luan, S. 2002. Transcriptional profiling reveals novel interactions between wounding, pathogen, abiotic stress, and hormonal responses in *Arabidopsis*. *Plant Physiol.* 129:661–677.

Cohen, Y. 2002. β-aminobutyric acid-induced resistance against plant pathogens. *Plant Dis.* 86:448–457.

Cole, D.L. 1999. The efficacy of acibenzolar-S-methyl, an inducer of systemic acquired resistance, against bacterial and fungal diseases of tobacco. *Crop Prot.* 18:267–273.

Conrath, U., Pieterse C.M.J., and Mauch-Mani, B. 2002. Priming in plant–pathogen interactions. *Trends Plant Sci.* 7:210–216.

Crop Protection Compendium. 2002. CABI Publishing. Available at http://www.cabicompendium.org/cpc/ecomonic.asp.

Cui, J., Jander, G., Racki, L. R., Kim, P.D., Pierce, N.E., and Ausubel F.M. 2002. Signals involved in *Arabidopsis* resistance *to Trichoplusia ni* caterpillars induced by virulent and avirulent strains of the phytopathogen *Pseudomonas syringae*. *Plant Physiol.* 129:551–564.

Dangl, J.L., Dietrich, R.A., and Richberg, M.H. 1996. Death don't have no mercy: cell death programs in plant-microbe interactions. *Plant Cell* 8:1793–1807.

Delaney, T.P., Uknes, S., Vernooij, B., Friedrich, L., Weymann, K., Negrotto, D., Gaffney, T., Gut-Rella, M., Kessmann, H., Ward, E., and Ryals, J. 1994. A central role of salicylic acid in plant disease resistance. *Science* 266:1247–1250.

Desikan, R., A.-H.-Mackerness, S., Hancock, J.T., and Neill, S.J. 2001. Regulation of the *Arabidopsis* transcriptome by oxidative stress. *Plant Physiol.* 127:159–172.

Desikan, R., Reynolds, A., Hancock, J.T., and Neill, S.J. 1998. Harpin and hydrogen peroxide both initiate programmed cell death but have differential effects on defence gene expression in *Arabidopsis* suspension cultures. *Biochem. J.* 330:115–120.

Díaz, J., ten Have, A., and van Kan, J.A.L. 2002. The role of ethylene and wound signaling in resistance of tomato to *Botrytis cinerea*. *Plant Physiol.* 129:1341–1351.

Ding, C-K., Wang, C.Y., Gross, K.C., and Smith, D.L. 2002. Jasmonate and salicylate induce the expression of pathogenesis-related-protein genes and increase resistance to chilling injury in tomato fruit. *Planta* 214:895–901.

Dong, H., Delaney, T.P., Bauer, D. W., and Veer, S.V. 1999. Harpin induces disease resistance in *Arabidopsis* through the systemic acquired resistance pathway mediated by salicylic acid and the *NIM1* gene. *Plant J.* 20:207–215.

Dong, X. 1998. SA, JA, ethylene, and disease resistance in plants. *Curr. Opin. Plant Biol.* 1:316–323.

Dow, M., Newman, M.A., and von Roepenack, E. 2000. The induction and modulation of plant defense responses by bacterial lipopolysaccharides. *Annu. Rev. Phytopathol.* 38:241–261.

Ebner, C., Hoffmann-Sommergruber, K., and Breiteneder, H. 2001. Plant food allergens homologous to pathogenesis-related proteins. *Allergy* 56:S67:43–44.

Ehness, R., Ecker, M., Godt, D.E., and Roitsch, T. 1997. Glucose and stress independently regulate source and sink metabolism and defense mechanisms via signal transduction pathways involving protein phosphorylation. *Plant Cell* 9:1825–1841.

Ellis, C., Karafyllidis, I., and Turner, J.G. 2002. Constitutive activation of jasmonate signaling in an *Arabidopsis* mutant correlates with enhanced resistance to *Erysiphe cichoracearum*, *Pseudomonas syringae* and *Myzus persicae*. *Mol. Plant Microbe Interact.* 10:1025–1030.

Epple, P., Mack, A.A., Morris, V.R., and Dangl, J.L. 2003. Antagonistic control of oxidative stress-induced cell death in *Arabidopsis* by two related, plant-specific zinc finger proteins. *Proc. Natl. Acad. Sci. USA* 100:6831–6837.

Eulgem, T., Rushton, P.J., Schmelzer, E., Hahlbrock, K., and Somssich, I.E. 1999. Early nuclear events in plant defense signalling: rapid gene activation by WRKY transcription factors. *EMBO J.* 18:4689–4699.

Felix, G., Duran, J.D., Volko, S., and Boller, T. 1999. Plants have a sensitive perception system for the most conserved domain of bacterial flagellin. *Plant J.* 18:265–276.

Felton, G.W., Korth, K.L., Bi, J.L., Wesley, S.V., Huhman, D.V., Mathews, M.C., Murphy, J.B., Lamb, C., and Dixon, R.A. 1999. Inverse relationship between systemic resistance of plants to microorganisms and to insect herbivery. *Curr. Biol.* 9:317–320.

Feys, B.J., and Parker, J.E., 2000. Interplay of signaling pathways in plant disease resistance. *Trends Genet.* 16:449–455.

Förster, H., Adaskaveg, J.E., Kim, D.H., and Stanghellini, M.E. 1998. Effect of phosphite on tomato and pepper plants and on succeptibility of pepper to Phytophthora root and crown rot in hydroponic culture. *Plant Dis.* 82:1165–1170.

Fray, R.G., Throup, J.P., Daykin, M., Wallace, A., Williams, P., Stewart, G.S., and Grierson, D. 1999. Plants genetically modified to produce N-acylhomoserine lactones communicate with bacteria. *Nat. Biotechnol.* 17:958–959.

Friedrich, L., Lawton, K., Dietrich, R., Willits, M., Cade, R., and Ryals, J. 2001. *NIM1* overexpression in *Arabidopsis* potentiates plant disease resistance and results in enhanced effectiveness of fungicides. *Mol. Plant Microbe Interact.* 14:1114–1124.

Garaats, B.P.J., Bakker, P.A.H.M., and van Loon, L.C. 2002. Ethylene insensitivity impairs resistance to soilborne pathogens in tobacco and *Arabidopsis thaliana*. *Mol. Plant Microbe Interact.* 15:1078–1085.

Gazzarrini, S., and McCourt, P. 2003. Cross-talk in plant hormone signaling: what *Arabidopsis* mutants are telling us. *Ann. Bot.* 91:605–612.

Gómez-Gómez, L., Bauer, Z., and Boller, T. 2001. Both the extracellular leucine-rich repeat domain and the kinase activity of FLS2 are required for flagellin binding and signaling in *Arabidopsis*. *Plant Cell* 13:1155–1163.

Gómez-Gómez, L., Felix, G., and Boller, F. 1999. A single locus determines sensitivity to bacterial flagellin in *Arabidopsis thaliana*. *Plant J.* 18:277–284.

Gorlach, J., Volrath, S., Knauf-Beiter, G., Hengy, G., Beckhove, U., Kogel, K.H., Oostendorp, M., Staub, T., Ward, E., Kessmann, H., and Ryals, J. 1996. Benzothiadiazole, a novel class of inducers of systemic acquired resistance, activates gene expression and disease resistance in wheat. *Plant Cell* 8:629–643.

Gottstein, H.D., and Kuć, J. 1989. Induction of systemic resistance to anthracnose in cucumber by phosphates. *Phytopathology* 79:176–179.

Guo, A., Salih, G., and Klessig, D.F. 2000. Activation of a diverse set of genes during the tobacco resistance response to TMV is independent of salicylic acid; induction of a subset is also ethylene independent. *Plant J.* 21:409–418.

Hadwiger, L.A., Ogawa, T., and Kuyama, H. 1994. Chitosan polymer sizes effective in inducing phytoalexin accumulation and fungal suppression are verified with synthesized oligomers. *Mol. Plant Microbe Interact.* 4:531–533.

Hammerschmidt, R. 1999a. Induced disease resistance: how do induced plants stop pathogens? *Physiol. Mol. Plant Pathol.* 55:77–84.

Hammerschmidt, R. 1999b. Phytoalexins: what have we learned after 60 years? *Annu. Rev. Phytopathol.* 37:285–306.

Hanfrey, C., Fife, M., and Buchanan-Wollaston, V. 1996. Leaf senescence in *Brassica napus*: expression genes encoding pathogenesis-related proteins. *Plant Mol. Biol.* 30:597–609.

Heil, M. 2002. Ecological costs of induced resistance. *Curr. Opin. Plant Biol.* 5:1–6.

Heil, M., and Baldwin, I.T. 2002. Fitness costs of induced resistance: emerging experimental support for a slippery concept. *Trends Plant Sci.* 7:61–67.

Heil, M., and Bostock, R. 2002. Induced systemic resistance (ISR) against pathogens in the context of induced plant defences. *Ann. Bot.* 89:503–512.

Herbers, K., Meuwly, P., Metraux, J. P., and Sonnewald, U. 1996. Salicylic acid-independent induction of pathogenesis-related protein transcripts by sugars is dependent on leaf developmental stage. *FEBS Lett.* 397:239–244.

Herms, S., Seehaus, K., Koehle, H., and Conrath, U. 2002. A strobilurin fungicide enhances the resistance of tobacco against tobacco mosaic virus and *Pseudomonas syringae* pv. *tabaci*. *Plant Physiol.* 130:120–127.

Kessler, A., and Baldwin, I.T. 2002. Plant responses to insect herbivory: the emerging molecular analysis. *Annu. Rev. Plant Biol.* 53:299–328.

Kim, Y.C., Blee, K.A., Robins, J., and Anderson A.J. 2001. Oxycom™ under field and laboratory conditions increases resistance responses in plants. *Eur. J. Plant Pathol.* 107:129–136.

Kloek A.P., Verbsky M.L., Sharma S.B., Schoelz J.E., Vogel J., Klessig D.F., and Kunkel B.N. 2001. Resistance to *Pseudomonas syringae* conferred by an *Arabidopsis thaliana* coronatine-insensitive (*coi1*) mutation occurs through two distinct mechanisms. *Plant J.* 25:509–522.

Knoester, M., Pieterse, C.M., Bol, J.F., and van Loon, L.C. 1999. Systemic resistance in *Arabidopsis* induced by rhizobacteria requires ethylene-dependent signaling at the site of application. *Mol. Plant Microbe Interact.* 8:720–727.

Kohler, A., Schwindling, S., and Conrath, U. 2002. Benzothiadiazole-induced priming for potentiated responses to pathogen infection, wounding, and infiltration of water into leaves requires the *NPR1/NIM1* Gene in *Arabidopsis*. *Plant Physiol.* 128:1046–1056.

Kovtun, Y., Chiu, W.L., Tena, G., and Sheen, J. 2000. Functional analysis of oxidative stress-activated mitogen-activated protein kinase cascade in plants. *Proc. Natl Acad. Sci. USA* 97:2940–2945.

Kuć, J. 1982. Induced immunity to plant disease. *BioScience* 32:854–860.

Kunkel, B.N., and Brooks, D.M. 2002. Crosstalk between signaling pathways in pathogen defense. *Curr. Opin. Plant Biol.* 5:325–331.

Larkindale, J., and Knight, M.R. 2002. Protection against heat stress-induced oxidative damage in *Arabidopsis* involves calcium, abscisic acid, ethylene, and salicylic acid. *Plant Physiol.* 128:682–695.

Latunde-Dada, A.O., and Lucas, J.A. 2001. The plant defence activator acibenzolar-S-methyl primes cowpea [*Vigna unguiculata* (L.) Walp.] seedlings for rapid induction of resistance. *Physiol. Mol. Plant Pathol.* 58:199–208.

Lee, S.C., Kim Y.J., and Hwang, B.K. 2001. A pathogen-induced chitin-binding protein gene from pepper: its isolation and differential expression in pepper tissues treated with pathogens, ethephon, methyl jasmonate or wounding. *Plant Cell Physiol.* 12:1321–1330.

Leeman, M., van Pelt, J.A., Den Ouden, F.M., Heinsbroek, M., Bakker, P.A.H.M., and Schippers, B. 1995. Induction of systemic resistance against Fusarium wilt of radish by lipopolysaccharides of *Pseudomonas fluorescens*. *Phytopathology* 85:1021–1027.

Lindow, S.E., and Leveau, J.H. 2002. Phyllosphere microbiology. *Curr. Opin. Biotechnol.* 13:238–243.

Link, V.L., Hofmann, M.G., Sinha, A.K., Ehness, R., Strnad, M., and Roitsch, T. 2002. Biochemical evidence for the activation of distinct subsets of mitogen-activated protein kinases by voltage and defense-related stimuli. *Plant Physiol.* 128:271–281.

Lugtenberg, B.J.J., Dekkers, L., and Bloemberg, G.V. 2001. Molecular determinants of rhizosphere colonization by *Pseudomonas*. *Annu. Rev. Phytopathol.* 39:461–490.

Marois, E., van den Acherveken, G., and Bonas, U. 2002. The *Xanthomonas* type III effector protein AvrBs3 modulates plant gene expression and induces cell hypertrophy in the susceptible host. *Mol. Plant Microbe Interact.* 7:637–646.

Mauhofer, M., Hase, C., Meuwly, P., Metraux, J.P., and Defago, G. 1994. Induction of systemic resistance of tobacco to tobacco necrosis virus by the root-colonizing strain *Pseudomonas fluorescens* CHA0. *Phytopathology* 84:139–146.

McSpadden-Gardener, B., and Fravel, D.R. 2002. Biological control of plant pathogens: Research, commercialization, and application in the USA. Online. *Plant Health Progress* doi:10.1094/PHP-2002-0510-01-RV.

Mercado-Blanco, J., van der Drift, K.M., Olsson, P.E., Thomas-Oates, J.E., van Loon, L.C., and Bakker, P.A. 2001. Analysis of the *pmsCEAB* gene cluster involved in biosynthesis of salicylic acid and the siderophore pseudomonine in the biocontrol strain *Pseudomonas fluorescens* WCS374. *J. Bacteriol.* 6:1909–1920.

Mittler, R. 2002. Oxidative stress, antioxidants and stress tolerance. *Trends Plant Sci.* 9:405–410.

Mucharromah, E., and Kuć, J. 1991. Oxalate and phosphates induce systemic resistance against diseases caused by fungi, bacteria and viruses in cucumber. *Crop Prot.* 10:265–270.

Neill, S., Desikan, R., and Hancock. J. 2002. Hydrogen peroxide signaling. *Curr. Opin. Plant Biol.* 2:282–290.

Neuenschwander, U., Vernooij, B., Friedrich, L., Uknes, S., Kessmann, H., and Ryals, J. 1995. Is hydrogen peroxide a second messenger of salicylic acid in systemic acquired resistance? *Plant J.* 8:227–233.

Newman, M.A., von Roepenack-Lahaye, E., Parr, A., Daniels, M.J., and Dow, J.M. 2002. Prior exposure to lipopolysaccharide potentiates expression of plant defenses in response to bacteria. *Plant J.* 29:487–495.

Niere, J.O., DeAngelis, G., and Grant, B.R. 1994. The effect of phosphonate on the acid-soluble phosphorus components in the genus *Phytophthora. Microbiol.* 140:1661–1670.

Ollerstam, O., and Larsson, S. 2003. Salicylic acid mediates resistance in the willow *Salix viminalis* against the gall midge *Dasineura marginemtorquens. J. Chem. Ecol.* 29:163–174.

Oostendorp, M., Kunz, W., Dietrich, B., and Staub, T. 2001. Induced disease resistance in plants by chemicals. *Eur. J. Plant Pathol.* 107:19–28.

Orober, M., Siegrist, J., and Buchenhauer, H. 2002. Mechanisms of phosphate-induced disease resistance in cucumber. *Eur. J. Plant Pathol.* 108:345–353.

Orozco-Cardena, M., and Ryan, C.A. 1999. Hydrogen peroxide is generated systemically in plant leaves by wounding and systemin via the octadecanoid pathway. *Proc. Natl. Acad. Sci. USA* 96:6553–6557.

Pajor, E., Le Corre, D., and Silué, D. 2001. Phytogard® and DL-β-amino butyric (BABA) induce resistance to downy mildew (*Bremia lactucae*) in lettuce (*Lactuca sativa* L). *Eur. J. Plant Pathol.* 107:861–869.

Park, K.S., and Kloepper, J.W. 2000. Activation of PR-1a promoter by rhizobacteria that induce systemic resistance in tobacco against *Pseudomonas syringae* pv. *tabaci. Biol. Control* 18:2–9.

Pieterse, C.M.J., van Wees, S.C., Hoffland, E., van Pelt, J.A., and van Loon, L.C. 1996. Systemic resistance in *Arabidopsis* induced by biocontrol bacteria is independent of salicylic acid accumulation and pathogenesis-related gene expression. *Plant Cell* 8:1225–1237.

Press, C.M., Loper, J.E., and Kloepper, J.W. 2001. Role of iron in rhizobacteria-mediated induced systemic resistance of cucumber. *Phytopathology* 91:593–598.

Preston, C.A., Lewandowski, C., Enyedi, A.J., and Baldwin, I.T. 1999. Tobacco mosaic virus inoculation inhibits wound-induced jasmonic acid-mediated responses within but not between plants. *Planta* 209:87–95.

Quirino, B.F, Normanly, J., and Amasino, R.M. 1999. Diverse range of gene activity during *Arabidopsis thaliana* leaf senescence includes pathogen-independent induction of defense-related genes. *Plant Mol. Biol.* 40:267–278.

Rakwal, R., Tomogami, S., Agrawal, G.K., and Iwahashi, H. 2002. Octadecanoid signaling component "burst" in rice (*Oryza sativa* L.) seedling leaves upon wounding by cut and treatment with fungal elicitor chitosan. *Biochem. Biophys. Res. Commun.* 5:1041–1045.

Raupach, G.S., and Kloepper, J.W. 1998. Mixtures of plant growth-promoting rhizobacteria enhance biological control of multiple cucumber pathogens. *Phytopathology* 88:1158–1164.

Reuveni, R., and Reuveni, M. 1998. Foliar-fertilizer therapy-a concept in integrated pest management. *Crop Prot.* 17:111–118.

Reymond, P., and Farmer E.E. 1998. Jasmonate and salicylate as global signals for defense gene expression. *Curr. Opin. Plant Biol.* 1:404–411.

Rickard, D.A. 2000. Review of phosphorus acid and its salts as fertilizer materials. *J. Plant Nutr.* 2:161–180.

Rolland, F., Moore, B., and Sheen, J. 2002. Sugar sensing and signaling in plants. *Plant Cell.* S185–205.

Romeis, T., Piedras, P., Zhang, S., Klessig, D., and Hirt, H. 1999. Rapid Avr9- and Cf-9 Dependent activation of MAP kinases in tobacco cell cultures and leaves: convergence of resistance gene, elicitor, wound, and salicylate responses. *Plant Cell* 11:273–287.

Romero, A.M., Kousik, C.S., and Ritchi, D.F. 2001. Resistance to bacterial spot in bell pepper induced by acibenzolar-S-methyl. *Plant Dis.* 85:189–194.

Ross, A.F. 1961. Systemic acquired resistance induced by localized virus infections in plants. *Virology* 14:340–358.

Ryan, C.A., and Pearce G. 1998. Systemin: a polypeptide signal for plant defensive genes. *Annu. Rev. Cell Dev. Biol.* 14:1–17.

Sakamoto, K., Tada, Y., Yokozeki, Y., Akagi, H., Hayashi, N., Fujimura, T., and Ichikawa, N. 1999. Chemical induction of disease resistance in rice is correlated with the expression of a gene encoding a nucleotide binding site and leucine-rich repeats. *Plant Mol. Biol.* 40:847–855.

Salcedo, G., Díaz-Perales, A., and Sánchez-Monge, R. 1999. Fruit allergy: plant defense proteins as novel potential panallergens. *Clin. Exp. Allergy* 29:1158–1160.

Salmeron, J.M., Oldroyd G.E., Rommens, C.M., Scofield, S.R., Kim H.S., Lavelle, D.T., Dahlbeck, D., and Staskawicz, B.J. 1996. Tomato *Prf* is a member of the leucine-rich repeat class of plant disease resistance genes and lies embedded within the *Pto* kinase gene cluster. *Cell* 1:123–133.

Schenk, P.M., Kazan, K., Wilson, I., Anderson, J.P., Richmond, T., Somerville, S.C., and Manners, J.M. 2000. Coordinated plant defense responses in *Arabidopsis* revealed by microarray analysis. *Proc. Natl Acad. Sci. USA* 97:11655–11660.

Seo, S., Sano, H., and Ohasi, Y. 1997. Jasmonic acid in wound signal transduction pathways. *Plant Physiol.* 101:740–745.

Shailasree, S., Sarosh, B.R., Vasanthi, N.S., and Shetty H.S. 2001. Seed treatment with β-aminobutyric acid protects *Pennisetum glaucum* systemically from *Sclerospora graminicola*. *Pest Manage. Sci.* 57:721–728.

Siegrist, J., Orober M., and Buchenauer, H. 2000. β-Aminobutyric acid-mediated enhancement of resistance in tobacco to tobacco mosaic virus depends on the accumulation of salicylic acid. *Physiol. Mol. Plant Pathol.* 56:95–106.

Silue, D., Pajot, E., and Cohen, Y. 2002. Induction of resistance to downy mildew (*Peronospora parasitica*) in cauliflower by DL-β-Amino-n-butanoic acid (BABA). *Plant Pathol.* 51:97–102.

Smillie, R., Grant, B.R., and Guest, D. 1989. The mode of action of phosphite: Evidence for both direct and indirect modes of action on three *Phytophthora* spp. in plants. *Phytopathology* 59:924–926.

Staub, T. 2001. Induced disease resistance in crop health management. *Plant Health Prog.*

Stosz, S.K., Fravel, D.R., and Roberts, D.P. 1996. *In vitro* analysis of the role of glucose oxidase from *Talaromyces flavus* in biocontrol of the plant pathogen *Verticillium dahliae. Appl. Environ. Microbiol.* 62:3183–3186.

Stotz, H.U., Koch, T., Biedermann, A., Weniger, K., Boland, W., and Mitchell-Olds, T. 2002. Evidence for regulation of resistance in *Arabidopsis* to Egyptian cotton worm by salicylic and jasmonic acid signaling pathways. *Planta* 214:648–652.

Stout, M.J., Fidantsef, A.L., Duffey, S.S., and Bostock, R.M. 1999. Signal interactions in pathogen and insect attack: systemic plant-mediated interactions between pathogens and herbivores of the tomato, *Lycopersicon esculentum. Physiol. Mol. Plant Pathol.* 54:115–130.

Stratmann, J.W., and Ryan, C.A. 1997. Myelin basic protein kinase activity in tomato leaves is induced systemically by wounding and increases in response to systemin and oligosaccharide elicitors. *Proc. Natl Acad. Sci. USA* 94:11085–11089.

Tally, A., Oostendorp, M., Lawton, K., Staub, T., and Bassi, B. 1999. Commercial development of elicitors of induced resistance to pathogens. In *Induced Plant Defense Against Pathogens and Herbivores: Biochemistry, Ecology, and Agriculture*, eds. A.A. Agrawal, S. Tuzun, and E. Bent, pp. 357–369. APS Press, St. Paul MN USA.

Tanaka, N., Che, F.S., Watanbe, N., Gijiwara, S., Takayama, S., and Isogai, A. 2003. Flagellin from an incompatible strain of Acidovorax avenae mediates H_2O_2 generation accompanying hypersensitive cell death and expression of PAL, Chit-1 and PBZ1 but not of Lox in rice. *Mol. Plant Microbe Interact.* 16:422–428.

Thaler, J.S., Fidantsef, A.L., Duffey, S.S., and Bostock, R.M. 1999. Trade-offs in plant defense against pathogens and herbivores: a field demonstration of chemical elicitors of induced resistance. *J. Chem. Ecol.* 25:1597–1609.

Thaler, J.S., Fidantsef, A.L., and Bostock, R.M. 2002. Antagonism between jasmonate- and salicylate-mediated induced plant resistance: effects of concentration and timing of elicitation on defense-related proteins, herbivore, and pathogen performance in tomato. *J. Chem. Ecol.* 28:1131–1159.

Thomma, B., Penninckx, I., Broekaert, W.F., and Cammue, B. 2001. The complexity of disease signaling in *Arabidopsis. Curr. Opin. Immunol.* 13:63–68.

Ton, J., Davison, S., van Wees, S.C., van Loon, L.C., and Pieterse, C.M.J. 2001. The *Arabidopsis ISR1* locus controlling rhizobacteria-mediated induced systemic resistance is involved in ethylene signaling. *Plant Physiol.* 125:652–661.

Ton, J., van Pelt, J.A., van Loon, L.C., and Pieterse, C.M.J. 2002. Differential effectiveness of salicylate-dependent and jasmonate/ethylene-dependent induced resistance in *Arabidopsis. Mol. Plant Microbe Interact.* 15:27–34.

Turner, J.G., Ellis, C., and Devoto, A. 2002. The jasmonate signal pathway. *Plant Cell* S153–164.

Vanacker, H., Lu, H., Rate, D.N., and Greenberg, J.T. 2001. A role for salicylic acid and NPR1 in regulating cell growth in *Arabidopsis. Plant J.* 28:209–216.

Vander, P., Vårum, K.M., Domard, A., Gueddari, N.E.E., and Moerschbacher, B.M. 1998. Comparison of the ability of partially N-acetylated chitosans and chitooligosaccharides to elicit resistance reactions in wheat leaves. *Plant Physiol.* 118:1353–1359.

van Poecke, R.M., and Dicke, M. 2002. Induced parasitoid attraction by *Arabidopsis thaliana*: involvement of the octadecanoid and the salicylic acid pathway. *J. Exp. Bot.* 53:1793–1799.

van Loon, L.C. 1997. Induced resistance in plants and the role of pathogenesis-related proteins. *Eur. J. Plant Pathol.* 103:753–765.

van Wees, S.C., Luijendijk, M., Smoorenburg, I., van Loon, L.C., and Pieterse, C.M.J. 1999. Rhizobacteria-mediated induced systemic resistance (ISR) in *Arabidopsis* is not associated with a direct effect on expression of known defense-related genes but stimulates the expression of the jasmonate-inducible gene *Atvsp* upon challenge. *Plant Mol. Biol.* 41:537–549.

van Wees, S.C., Pieterse, C.M.J., Trissenaar, A., van't Westende, Y.A., Hartog, F., and van Loon, L.C. 1997. Differential induction of systemic resistance in *Arabidopsis* by biocontrol bacteria. *Mol. Plant-Microbe Interact.* 10:716–724.

Wang, K.L-C., Li, H., and Ecker, J.R. 2002. Ethylene biosynthesis and signaling networks. *Plant Cell* 14:S131–S151.

Wang, Z.C.,Yamanouchi, U., Katayose, Y., Saaskai, T., and Yano, M. 2001. Expression of the Pib rice-blast-resistance gene family is upregulated by environmental conditions favoring infection and by chemical signals that trigger secondary plant defences. *Plant Mol. Biol.* 47:653–661.

Ward, E.R., Uknes, S.J., Williams, S.C., Dincher, S.S., Wiederhold, D.L., Alexander, D.C., Ahl-Goy, P., Métraux, J.P., and Ryals, J.A. 1991. Coordinate gene activity in response to agents that induce systemic acquired resistance. *Plant Cell* 3:1085–1094.

Watanabe, T., Igarashi, H., Matsumoto, K., Seki, S., Mase, S., and Sekizawa, Y. 1977. The characteristics of probenazole (Oryzemate) for the control of rice blast. *J. Pesticide Sci.* 2:291–296.

Watanabe, T., Sekizawa, Y., Shimura, M., Suzuki, Y., Matsumoto, K., Iwata, M., and Mase, S. 1979. Effects of probenazole (Oryzemate) on rice plants with reference to controlling rice blast. *J. Pesticide Sci.* 4:53–59.

Yang, H.S., Hung, H.C., and Collmer, A. 1993. *Pseudomonas syringae* pv. *syringae* Harpin: a protein that is secreted via the Hrp pathway and elicits the hypersensitive response in plants. *Cell* 73:1255–1266.

Yang, K.Y., Blee K.A., Zhang, S., and Anderson, A.J. 2002. Oxycom™ treatment suppresses *Pseudomonas syringae* infection and activates a mitogen-activated protein kinase pathway in tobacco. *Physiol. Mol. Plant Pathol.* 61:249–256.

Yano, A., Suzuki, K., Uchimiya, H., and Shinshi, H. 1998. Induction of hypersensitive cell death by a fungal protein in cultures of tobacco cells. *Mol. Plant Microbe Interact.* 11:115–123.

Yedidia, I.I., Benhamou, N., and Chet, I.I. 1999. Induction of defense responses in cucumber plants (*Cucumis sativus* L.) by the biocontrol agent *Trichoderma harzianum*. *Appl. Environ. Microbiol.* 65:1061–1070.

Zhu, T., Budworth, P., Han, B., Brown, D., Change, H-S., Zou, G., and Wang, X. 2001. Toward elucidating the global gene expression patterns of developing *Arabidopsis*: parallel analysis of 8300 genes by a high density oligonucleotide probe array. *Plant Physiol Biochem.* 39:221–342.

Yoshida, S., Ito, M., Nishida, I., and Watanabe, A. 2002. Identification of a novel gene HYS1/CPR5 that has a repressive role in the induction of leaf senescence and pathogen-defence responses in *Arabidopsis thaliana*. *Plant J.* 29:427–437.

Yoshioka, K., Nakashita, H., Klessig, D.F., and Yamaguchi, I. 2001. Probenazole induces systemic acquired resistance in *Arabidopsis* with a novel type of action. *Plant J.* 25:149–157.

Ypema, H.L., and Gold, R.E. 1999. Kresoxim-methyl: modification of a naturally occurring compound to produce a new fungicide. *Plant Dis.* 83:4–16.

Zhang, B., Ramonell, K., Somerville, S., and Stacey, G. 2002. Characterization of early, chitin-induced gene expression in *Arabidopsis*. *Mol. Plant Microbe Interact.* 15:963–970.

Zhang, S., and Klessig D.F. 1997. Salicylic acid activates a 48-kD MAP kinase in tobacco. *Plant Cell* 9:809–824.

Zimmerli, L., Metrauz, J.P., and Mauch-Mani, B. 2001. β-aminobutyric acid-induced protection of *Arabidopsis* against the necrotrophic fungus *Botrytis cinerea*. *Plant Physiol.* 126:517–527.

Zimmerli, L., Jakab, G., Metrauz, J.P., and Mauch-Mani, B. 2000. Potentiation of pathogen-specific defense mechanisms in *Arabidopsis* by β-aminobutyric acid. *Proc. Natl Acad. Sci. USA* 97:12920–12925.

18

Engineering Plants for Durable Disease Resistance

J. Gilbert, M. Jordan, D.J. Somers, T. Xing, and Z.K. Punja

18.1 Introduction

As our knowledge of the cellular and genetic mechanisms of plant disease resistance increase, so does the potential for modifying these processes to achieve
broad-spectrum durable disease resistance. A number of approaches have been
taken by researchers to identify and understand the complex chain of events that is
set in motion when a plant is challenged by a pathogen (Table 18.1) (reviewed by
Broekaert et al., 2000; Cornelissen and Schram, 2000; Punja, 2001). Most effort
has been applied to studying the constitutive production in transgenic plants of antifungal compounds. These include production of naturally occurring pathogenesis-
related (PR) proteins that may inhibit or prevent pathogen growth in the plant, such
as hydrolytic enzymes, antifungal proteins, antimicrobial peptides, ribosome inactivating proteins, and phytoalexins. Others involve the expression of gene products
that are either antagonistic to pathogen virulence products, such as polygalacturonase, oxalic acid and lipase, or which enhance the structural defenses within the
plant, such as peroxidases and lignins. There has also been research into modifying
pathways such as those regulated by salicylic acid, jasmonic acid, ethylene, and
hydrogen peroxide that are important in plant defenses. Such resistance mechanisms occur naturally in the plant and the objective is to manipulate the system so
that gene products are expressed at levels that defend the plant against pathogen
attack, or render the pathogen incapable of attack. Alternative approaches concern
the interactions between R genes in plants and the corresponding dominant Avr
genes in the pathogen that culminate in the hypersensitive response (HR) in incompatible reactions, and the molecular genotyping of plant lines using DNA-based
techniques to facilitate the "pyramiding" of desirable disease resistance traits into
elite germplasm.

TABLE 18.1. Plant species genetically engineered to enhance resistance to fungal diseases (1991–2002).

Strategy used and plant species engineered	Expressed gene product	Effect on disease development	Reference
(a) Expression of hydrolytic enzymes			
Alfalfa (*Medicago sativa* L.)	Alfalfa glucanase	Reduced symptom development due to *Phytophthora megasperma*; no effect on *Stemphylium alfalfae*	Masoud et al. (1996)
American ginseng (*Panax quinquefolius* L.)	Rice chitinase	Not tested	Chen and Punja (2001)
Apple (*Malus* × *domestica*)	*Trichoderma harzianum* endochitinase	Reduced lesion number and lesion area due to *Venturia inaequalis*	Bolar et al. (2000); Wong et al. (1999)
Barley (*Hordeum vulgare* L.)	*Trichoderma* endo-1,4-ß-glucanase	Not tested	Nuutila et al. (1999)
Canola (*Brassica napus* L.)	Bean chitinase	Reduced rate and total seedling mortality due to *Rhizoctonia solani*	Broglie et al. (1991)
Broccoli (*Brassica oleracea* var. *italica*)	*Trichoderma harzianum* endochitinase	Reduced lesion size due to *Alternaria brassicicola*	Mora and Earle (2001)
Canola (*B. napus* L.)	Tomato chitinase	Lower percentage of diseased plants due to *Cylindrosporium concentricum* and *Sclerotinia sclerotiorum*	Grison et al. (1996)
Carrot (*Daucus carota* L.)	Tobacco chitinase	Reduced rate and final incidence of disease due to *Botrytis cinerea, Rhizoctonia solani* and *Sclerotium rolfsii*; no effect on *Thielaviopsis basicola* and *Alternaria radicina*	Punja and Raharjo (1996)
Chrysanthemum (*Dendranthema grandiflorum*) (Ramat.) Kitamura	Rice chitinase	Reduced lesion development due to *Botrytis cinerea*	Takatsu et al. (1999)
Cucumber (*Cucumis sativus* L.)	Petunia and tobacco chitinases	No effect on disease development due to *Colletotrichum lagenarium* and *Rhizoctonia solani*	Punja and Raharjo (1996)

TABLE 18.1. (Cont.)

Strategy used and plant species engineered	Expressed gene product	Effect on disease development	Reference
Cucumber (*C. sativus* L.)	Rice chitinase	Reduced lesion development due to *Botrytis cinerea*	Tabei et al. (1998)
Grape (*Vitis vinifera* L.)	Rice chitinase	Reduced development of *Uncinula necator* and fewer lesions due to *Elisinoe ampelina*	Yamamoto et al. (2000)
Grape (*V. vinifera* L.)	*Trichoderma harzianum* endochitinase	Reduction of *Botrytis cinerea* development in preliminary tests	Kikkert et al. (2000)
Potato (*Solanum tuberosum* L.)	*Trichoderma harzianum* endochitinase	Lower lesion numbers and size due to *Alternaria solani*; reduced mortality due to *Rhizoctonia solani*	Lorito et al. (1998)
Rice (*Oryza sativa* L.)	Rice chitinase	Fewer numbers of lesions and smaller size due to *Rhizoctonia solani*	Lin et al. (1995); Datta et al. (2000, 2001)
Rice (*O. sativa* L.)	Rice chitinase	Delayed onset and reduced severity of disease symptoms due to *Magnaporthe grisea*	Nishizawa et al. (1999)
Rose (*Rosa hybrida* L.)	Rice chitinase	Reduced lesion diameter due to black spot (*Diplocarpon rosae*)	Marchant et al. (1998)
Sorghum (*Sorghum bicolor* (L.) Moench)	Rice chitinase	Increased resistance to *Fusarium thapsinum*	Krishnaveni et al. (2001)
Strawberry (*Fragaria* × *ananassa* Duch.)	Rice chitinase	Reduced development of powdery mildew (*Sphaerotheca humuli*)	Asao et al. (1997)
Tobacco (*Nicotiana tabacum* L.)	Bean chitinase	Lower seedling mortality due to *Rhizoctonia solani*; no effect on *Pythium aphanidermatum*	Broglie et al. (1991); Broglie et al. (1993)
Tobacco (*N. tabacum* L.)	Peanut chitinase	Not tested	Kellmann et al. (1996)

(*cont.*)

TABLE 18.1. *(Cont.)*

Strategy used and plant species engineered	Expressed gene product	Effect on disease development	Reference
Tobacco (*N. tabacum* L.)	*Serratia marcescens* chitinase	Reduced disease incidence due to *Rhizoctonia solani* on seedlings; no effect on *Pythium ultimum*	Howie et al. (1994)
Tobacco (*N. tabacum* L.)	*Serratia marcescens* chitinase	Reduced development of *Rhizoctonia solani*	Jach et al. (1992)
Tobacco (*N. tabacum* L.)	*Streptomyces* chitosanase	Not tested	El Quakfaoui et al. (1995)
Tobacco (*N. tabacum* L.)	*Rhizopus oligosporus* chitinase	Reduced rate of development and size of lesions on leaves due to *Botrytis cinerea* and *Sclerotinia sclerotiorum*	Terakawa et al. (1997)
(b) Expression of PR-proteins			
Canola (*B. napus* L.)	Pea defense response gene, defensin	Reduced infection and development of *Leptosphaeria maculans*	Wang et al. (1999)
Carrot (*D. carota* L.)	Rice thaumatin-like protein	Reduced rate and final disease incidence due to *Botrytis cinerea* and *Sclerotinia sclerotiorum*	Chen and Punja (unpublished)
Potato (*S. tuberosum* L.)	Tobacco osmotin	Delayed onset and rate of disease due to *Phytophthora infestans*	Liu et al. (1994)
Potato (*S. commersonii* Dun.)	Potato osmotin-like protein	Enhanced tolerance to infection by *Phytophthora infestans*	Zhu et al. (1996)
Potato (*S. tuberosum* L.)	Pea PR10 gene	Reduced development of *Verticillium dahliae*	Chang et al. (1993)
Potato (*S. tuberosum* L.)	Potato defense response gene STH-2	No effect against *Phytophthora infestans*	Constabel et al. (1993)
Rice (*O. sativa* L.)	Rice thaumatin-like protein	Reduced lesion development due to *Rhizoctonia solani*	Datta et al. (1999)
Rice (*O. sativa* L.)	Rice Rir1b defense gene	Fewer lesions due to *Magnaporthe grisea*	Schaffrath et al. (2000)

TABLE 18.1. (*Cont.*)

Strategy used and plant species engineered	Expressed gene product	Effect on disease development	Reference
Tobacco (*N. tabacum* L.)	Tobacco PR1a	Reduced rate and final disease due to *Peronospora tabacina* and *Phytophthora parasitica*	Alexander et al., (1993)
Tobacco (*N. tabacum* L.)	Tobacco osmotin	No effect on *Phytophthora parasitica* var. *nicotianae*	Liu et al. (1994)
Wheat (*T. aestivum* L.)	Rice thaumatin-like protein	Delayed development of *Fusarium graminearum*	Chen et al. (1999)
(c) Expression of antimicrobial proteins/peptides/compounds			
Arabidopsis (*A. thaliana* L.)	Mistletoethionin viscotoxin	Reduced infection and development of *Plasmodiophora brassicae*	Holtorf et al. (1998)
Arabidopsis (*A. thaliana* L.)	Arabidopsis thionin	Reduced development and colonization by *Fusarium oxysporum*	Epple et al. (1997)
Carrot (*Daucus carota* L.)	Human lysozyme	Enhanced resistance to *Erysiphe heraclei* and *Alternaria dauci*	Takaichi and Oeda (2000)
Geranium (*Pelargonium* sp.)	Onion antimicrobial protein	Reduced development and sporulation of *Botrytis cinerea*	Bi et al. (1999)
Indian mustard (*Brassica juncea* L.)	Hevea chitin-binding lectin (hevein)	Smaller lesion size and reduced rate of development due to *Alternaria brassicae*	Kanrar et al. (2002)
Potato (*S. tuberosum* L.)	Alfalfa defensin	Enhanced resistance to *Verticillium dahliae*	Gao et al. (2000)
Potato (*S. tuberosum* L.)	*Bacillus amyloliquefaciens* barnase (RNase)	Delayed sporulation and reduced sporangia production by *Phytophthora infestans*	Strittmatter et al. (1995)
Potato (*S. tuberosum* L.)	Synthetic cationic peptide chimera	Reduced development of *Fusarium solani* and *Phytophthora cactorum*	Osusky et al. (2000)
Potato (*S. tuberosum* L.)	Human lactoferrin	Not tested	Chong and Langridge (2000)

(*cont.*)

TABLE 18.1. (*Cont.*)

Strategy used and plant species engineered	Expressed gene product	Effect on disease development	Reference
Rice (*O. sativa*)	Maize ribosome-inactivating protein	No effect on *Magnaporthe grisea* or *R. solani*	Kim et al. (1999)
Rice (*O. sativa* L.)	Trichosanthes ribosome-inactivating protein	Reduced lesion size due to *Pyricularia oryzae* and enhanced seedling survival	Yuan et al. (2002)
Rice (*O. sativa* L.)	Wheat puroindoline peptide	Reduced symptoms due to *Magnaporthe grisea* and *Rhizoctonia solani*	Krishnamurthy et al. (2001)
Tobacco (*N. tabacum* L.)	Amaranthus hevein-type peptide, Mirabilis knottin-type peptide	No effect on *Alternaria longipes* or *Botrytis cinerea*	De Bolle et al. (1996)
Tobacco (*N. tabacum* L.)	Radish defensin	Reduced infection and lesion size due to *Alternaria longipes*	Terras et al. (1995)
Tobacco (*N. tabacum* L.)	Stinging nettle (*Urtica dioica* L.) isolectin	Not tested	Does et al. (1999)
Tobacco (*N. tabacum* L.)	Pokeweed antiviral protein	Lower rate of infection and mortality due to *Rhizoctonia solani*	Wang et al. (1998); Zoubenko et al. (1997)
Tobacco (*N. tabacum* L.)	Synthetic magainin-type peptide	Reduced lesion development due to *Colletotrichum destructivum* and *Peronospora tabacina*	DeGray et al. (2001); Li et al. (2001)
Tobacco (*N. tabacum* L.)	Sarcotoxin peptide from *Sarcophaga peregrina*	Enhanced seedling survival following inoculation with *R. solani, Pythium aphanidermatum* and *Phytophthora nicotianae*	Mitsuhara et al. (2000)
Tobacco (*N. tabacum* L.)	Barley ribosome-inactivating protein	Reduced incidence and severity of *Rhizoctonia solani*	Logemann et al. (1992)
Tobacco (*N. tabacum* L.)	Maize ribosome-inactivating protein	Lower damage due to *Rhizoctonia solani*	Maddaloni et al. (1997)
Tobacco (*N. tabacum* L.)	Antifungal (killing) protein from *Ustilago maydis*-infecting virus (dsRNA)	Not tested	Park et al. (1996)

TABLE 18.1. *(Cont.)*

Strategy used and plant species engineered	Expressed gene product	Effect on disease development	Reference
Tobacco (*N. tabacum* L.)	Chloroperoxidase from *Pseudomonas pyrrocinia*	Reduced lesion development by *Colletotrichum destructivum*	Rajasekaran et al. (2000)
Tobacco (*N. tabacum* L.)	Synthetic antimicrobial peptide	Reduced lesion size due to *Colletotrichum destructivum*	Cary et al. (2000)
Tobacco (*N. tabacum* L.)	Human lysozyme	Reduced colony size and conidial production by *Erysiphe cichoracearum*	Nakajima et al. (1997)
Tomato (*L. esculentum* Mill.)	Radish defensin	Reduced number and size of lesions due to *Alternaria solani*	Parashina et al. (2000)
Wheat (*T. aestivum* L.)	Barley ribosome-inactivating protein	Slightly reduced development of *Blumeria graminis*	Bieri et al. (2000)
Wheat (*T. aestivum* L.)	Antifungal (killing) protein from *Ustilago maydis*-infecting virus (dsRNA)	Inhibition of *Ustilago maydis* and *Tilletia tritici* development on seeds	Clausen et al. (2000)
(d) Expression of phytoalexins			
Alfalfa (*M. sativa* L.)	Alfalfa isoflavone O-methyltransferase	Reduced lesion size due to *Phoma medicaginis*	He and Dixon (2000)
Alfalfa (*M. sativa* L.)	Peanut resveratrol synthase	Reduced lesion size and sporulation of *Phoma medicaginis*	Hipskind and Paiva (2000)
Barley (*H. vulgare*)	Grape stilbene (resveratrol) synthase	Reduced colonization by *Botrytis cinerea*	Leckband and Lörz (1998)
Grape (*V. vinifera* L.)	Grape stilbene (resveratrol) synthase	Reduced colonization by *Botrytis cinerea*	Coutos-Thevenot et al. (2001)
Tobacco (*N. tabacum* L.)	Synthetic magainin-type peptide	Reduced lesion size and sporulation due to *Peronospora tabacina*	Li et al. (2001)
Rice (*O. sativa* L.)	Grape stilbene (resveratrol) synthase	Reduced lesion development due to *Pyricularia oryzae*	Stark-Lorenzen et al. (1997)
Tobacco (*N. tabacum* L.)	*Fusarium trichodience* synthase	Not tested	Zook et al. (1996)

(cont.)

TABLE 18.1. *(Cont.)*

Strategy used and plant species engineered	Expressed gene product	Effect on disease development	Reference
Tobacco (*N. tabacum* L.)	Grape stilbene (resveratrol) synthase	Reduced colonization by *Botrytis cinerea*	Hain et al. (1993)
Tomato (*L. esculentum* Mill.)	Grape stilbene (resveratrol) synthase	Reduced lesion development by *Phytophthora infestans*; no effect on *Alternaria solani* or *Botrytis cinerea*	Thomzik et al. (1997)
Wheat (*T. aestivum* L.)	Grape stilbene (resveratrol) synthase	Not tested	Fettig and Hess (1999)
(e) Inhibition of pathogen virulence products			
Canola (*B. napus* L.)	Barley oxalate oxidase	Not tested	Thompson et al. (1995)
Poplar (*Populus* × *euramericana*)	Wheat oxalate oxidase	Delayed development of *Septoria musiva*	Liang et al. (2001)
Soybean (*Glycine max* L.) Merrill	Wheat oxalate oxidase (germin)	Reduced lesion length and disease progression due to *Sclerotina sclerotiorum*	Donaldson et al. (2001)
Tobacco (*N. tabacum* L.)	*Fusarium* trichothecene-degrading enzyme	Not tested	Muhitch et al. (2000)
Tobacco (*N. tabacum* L.)	Mutant RpL3 gene for mycotoxin insensitivity	Enhanced tolerance to *Fusarium graminearum* mycotoxin	Harris and Gleddie (2001)
Tobacco (*N. tabacum* L.)	Wheat oxalate oxidase (germin)	Not tested	Berna and Bernier (1997)
Tobacco (*N. umbratica* L.)	Tomato Asc-1 gene for insensitivity to fungal toxins	Enhanced resistance to *Alternaria alternata* f.sp. *lycopersici*	Brandwagt et al. (2002)
Tomato (*L. esculentum* Mill.)	Bean polygalacturonase inhibiting protein	No effect on disease due to *Fusarium oxysporum, Botrytis cinerea, Alternaria solani*	Desiderio et al. (1997)
Tomato (*L. esculentum* Mill.)	Pear polygalacturonase inhibiting protein	Reduced rate of development of *Botrytis cinerea*	Powell et al. (2000)
Tomato (*L. esculentum* Mill.)	Collybia velutipes oxalate decarboxylase	Enhanced resistance to *Sclerotinia sclerotiorum*	Kesarwani et al. (2000)

TABLE 18.1. (Cont.)

Strategy used and plant species engineered	Expressed gene product	Effect on disease development	Reference
(f) Alteration of structural components			
Potato (*S. tuberosum* L.)	Cucumber peroxidase	No effect on disease due to *Fusarium sambucinum* and *Phytophthora infestans*	Ray et al. (1998)
Tomato (*L. esculentum* Mill.)	Tobacco anionic peroxidase	No effect on disease due to *Fusarium oxysporum* and *Verticillium dahliae*	Lagrimini et al. (1993)
Wheat (*T. aestivum* L.)	Wheat germin (no oxalate oxidase activity)	Reduced penetration by *Erysiphe blumeria* into epidermal cells	Schweizer et al. (1999)
(g) Regulation of plant defense responses			
Arabidiopsis (*A. thaliana* L.)	Arabidopsis NPR1 protein	Reduced infection and growth of *Peronospora parasitica*	Cao et al. (1998)
Arabidiopsis (*A. thaliana* L)	Arbidopsis ethylene-response-factor 1(ERF1)	Enhanced tolerance to *Botrytis cinerea* and *Plectosphaerella cucumerina*	Berrocal-Lobo et al. (2002)
Canola (*Brassica napus* L.)	Tomato *C*19 gene	Delayed disease development due to *Leptosphaeria maculans*	Hennin et al. (2001)
Cotton (*G. hirsutum* L.), tobacco (*N. tabacum* L.)	*Talaromyces flavus* glucose oxidase	Enhanced protection against *Rhizoctonia solani* and *Verticillium dahliae*; no effect on *Fusarium oxysporum*	Murray et al. (1999)
Potato (*S. tuberosum* L.)	*Aspergillus niger* glucose oxidase	Delayed lesion development due to *Phytophthora infestans*; reduced disease development due to *Alternaria solani* and *Verticillium dahliae*	Wu et al. (1995; 1997)
Potato (*S. tuberosum* L.)	Tobacco catalase	Reduced lesion size due to *Phytophthora infestans*	Yu et al. (1999)
Potato (*S. tuberosum* L.)	Bacterial salicylate hydroxylase	No effect on *Phytophthora infestans*	Yu et al. (1997)

(cont.)

TABLE 18.1. (*Cont.*)

Strategy used and plant species engineered	Expressed gene product	Effect on disease development	Reference
Tobacco (*N. tabacum* L.)	*Aspergillus niger* glucose oxidase	Delayed disease development due to *Phytophthora nicotianae*	Lee et al. (2002)
Tobacco (*N. tabacum* L.), Arabidopsis (*A. thaliana* L.)	Bacterial salicylate hydroxylase	Enhanced susceptibility to *Phytophthora parasitica*, *Cercospora nicotianae*, *Peronospora parasitica*	Delaney et al. (1994); Donofrio and Delaney (2001)
Tobacco (*N. tabacum* L.)	Bacterial salicyclic acid-generating enzymes	Enhanced resistance to *Oidium lycopersicon*	Verberne et al. (2000)
Tobacco (*N. tabacum* L.)	*Arabidopsis* ethylene-insensitivity gene	Enhanced susceptibility to *Pythium sylvaticum*	Knoester et al. (1998)
Tobacco (*N. tabacum* L.)	*Phytophthora cryptogea* elicitor (ß-cryptogein)	Reduced infection by *Phytophthora parasitica*	Tepfer et al. (1998)
Tobacco (*N. tabacum* L.)	*Phytophthora cryptogea* elicitor (cryptogein)	Enhanced resistance to *Phytophthora parasitica*, *Thielaviopsis basicola*, *Botrytis cinerea* and *Erysiphe cichoracearum*	Keller et al. (1999)
Tomato (*L. esculentum* Mill.)	*Enterobacter* ACC deaminase	Reduced symptom development due to *Verticillium dahliae*	Robinson et al. (2001)
(h) Expression of combined gene products			
Apple (*Malus* × *domestica*)	*Trichoderma atroviride* endochitinase + exochitinase	Increased resistance to *Venturia inaequalis*	Bolar et al. (2001)
Carrot (*D. carota* L.)	Tobacco chitinase + β-1,3-glucanase, osmotin	Enhanced resistance to *Alternaria dauci*, *A. radicina*, *Cercospora carotae* and *Erysiphe heraclei*	Melchers and Stuiver (2000)
Tobacco (*N. tabacum* L.)	Barley chitinase + β-1,3-glucanase, or chitinase + ribosome-inactivating protein	Reduced disease severity due to *Rhizoctonia solani*	Jach et al. (1995)
Tobacco (*N. tabacum* L.)	Rice chitinase + alfalfa glucanase	Reduced rate of lesion development and fewer lesions due to *Cercospora nicotianae*	Zhu et al. (1994)
Tomato (*L. esculentum* Mill.)	Tobacco chitinase +β-1,3-glucanase	Reduced disease severity due to *Fusarium oxysporum* f.sp. *lycopersici*	Jongedijk et al. (1995); van den Elzen et al. (1993)

Adapted from: Punja (2001).

Reproduced by permission from the Canadian Phytopathological Society (2001).

18.2 Antimicrobial Compounds

18.2.1 Pathogenesis-Related and Antimicrobial Proteins

Pathogenesis-related (PR) proteins that show antibiotic activity against pathogens *in vitro* are termed antimicrobial, whereas those that are induced *in planta* after pathogen attack are referred to as PR genes (Broekaert et al., 2000). The hydrolytic enzymes, chitinase and glucanase, are among the most extensively researched PR proteins. They have been classified as PR-3 and PR-2 types, respectively (Broekaert et al., 2000). Chitins and glucans are components of fungal cell walls and the overexpression of the enzymes in the host tissues may lead to fungal cell lysis, retarding or preventing pathogen colonization (Mauch and Staehelin, 1989). Chitinases from many sources have been used to transform as many field, fruit, horticultural, and vegetable crops, although tobacco has been used most extensively as a model system. The reports of transformation with glucanases are fewer in number, but have similar results. No immunity has been expressed to pathogens after transformation, but in many cases, delayed onset of symptom development and fewer, or smaller sized lesions, have resulted in less disease (Punja, 2001). The strategy of transforming host plants with both a chitinase and a glucanase has been shown to be more effective in reducing disease than when deployed singly, as in carrot (Melchers and Stuiver, 2000), tobacco (Zhu et al., 1994), and tomato (Jongedijk et al., 1995).

Other PR or antimicrobial proteins (AMP), including osmotin and thaumatin-like proteins (TLP), expressed in various transgenic plants have also been shown to delay disease development or reduce final disease severity; e.g., tobacco PR-1a against *Peronospora tabacina* and *Phytophthora parasitica* in tobacco (Alexander et al., 1993) and rice TLP against *Fusarium graminearum* in wheat (Chen et al., 1999). Osmotin and TLP belong to subgroup PR-5, members of which exhibit antifungal activity and are produced in plants under conditions of abiotic or biotic stress (Zhu et al., 1995; Chen et al., 1999). The antimicrobial proteins, thionins and defensins, were characterized in the mid-1990s (Bohlman, 1994; Broekaert et al., 1995) and recently reviewed in depth by Broekaert et al. (2000). Of more than 100 studies listed, only 50% or so report that plants overexpressing an AMP (including chitinases and glucanases) show increased resistance to one or more pathogen. There are reasons for this result: the inoculation conditions may affect a plant's ability to resist attack. For example, inoculum concentration affected the reaction of transgenic potato plants, which were constitutively overexpressing a PR-5 protein. At an inoculum concentration of 50 zoospores/5 μl of *Phytophthora infestans*, transgenic potato plants were resistant to the pathogen, but at 100 zoospores/5 μl they were as susceptible as the nontransformed control plants (Zhu et al., 1996). The subcellular location of chitinases, whether in the extracellular spaces of plant tissue or within vacuoles, has been found to affect the response to attack by transgenic plants that are overexpressing an AMP. A basic vacuolar PR-3 type tobacco chitinase in *Nicotiana sylvestris* plants increased resistance to *Rhizoctonia solani*. However, *N. sylvestris* plants transformed with the same chitinase (but acidic and lacking the carboxy-terminal propeptide), targeted to the extracellular space were as susceptible to *R. solani* as the control plants (Vierheilig et al., 1993). The proteins also show specificity in their antimicrobial activity, in some cases showing increased resistance to one group of pathogens; e.g., PR-5 proteins have activity against oomycetes such as *Peronospora tabacina* and *Phytophthora parasitica*, but not to viruses (Lusso and Kuć, 1996). In fact, PR proteins with a specific antiviral activity have not been characterized (Verberne et al., 2000).

Ribosome inactivating proteins (RIPs) are N-glycosidases that remove a specific adenine from the sarcin/ricin (S/R) loop of the large rRNA, thus arresting protein synthesis at the translation step. They are divided into two groups: type 1, comprising single A-chain molecules, which exhibits the N-glycosidase activity to inactivate the ribosome, and type 2 in which a glycosylated lectin B-chain is connected to the active A-chain through a single disulphide bond (Stirpe et al., 1992). Several studies have shown type 1 RIPs to reduce fungal development both in the plant and *in vitro*. A type 1 RIP was obtained from pokeweed (*Phytolacca americana*) antiviral protein and shown to reduce development of the fungal pathogen, *Rhizoctonia solani*, when expressed in transgenic tobacco plants (Wang et al., 1998). Another type-1 RIP, PAP-R, was purified from *Agrobacterium rhizogenes*-transformed hairy roots of pokeweed and found to have *in vitro* antifungal activity against soilborne fungi (Park et al., 2001b). While the above studies examined activity of RIPs alone, activity of a type 1 RIP from barley seed was enhanced when expressed in combination with a barley seed glucanase or chitinase (Leah et al., 1991; Jach et al., 1995).

One of the longest-studied defense response mechanisms in plants is the production of phytoalexins (Hammerschmidt, 1999). These are defined as low-molecular weight antimicrobial compounds that are produced after infection and that have been demonstrated to delay disease development and reduce symptoms (Paxton, 1981). Despite the fact they have been studied since the 1940s, conclusive proof of their role in defense remains to be confirmed (Hammerschmidt, 1999). Some evidence for their role in plant defense was provided by studies with the phytoalexin resveratrol, which requires the enzyme stilbene synthase for biosynthesis. In grapevine, high levels of resveratrol are associated with resistance to *Botrytis cinerea*. While tobacco has the substrate for stilbene synthase it lacks the gene encoding for the enzyme. Introduction of stilbene synthase from grapevine into tobacco resulted in expression of the gene and production of resveratrol. Moreover, the transformed plants showed enhanced resistance to *B. cinerea*, albeit at levels too low for commercial interest (Hain et al., 1993).

18.2.2 Structural Defense Responses in the Host

Another strategy to engineer plants with durable plant disease resistance is to target the host cell wall, which presents a complex physical barrier to colonization by pathogens. An arsenal of enzymes and proteins involved in host defense is secreted from the cell wall. One group that has received some attention is the polygalacturonase-inhibiting proteins, PGIPs, that have a high affinity for fungal, but not plant, endopolygalacturonases. PGIPs are structurally related to a superfamily of leucine-rich repeat proteins that specialize in recognition of nonself molecules and rejection of pathogens. The action of PGIP in the plant *in vivo* may be to counteract fungal invasion by causing fungal polygalacturonases to increase their elicitation of plant defense responses (Cervone et al., 1987). In a study in which extracellular polygalacturonase from *Verticillium dahliae* was inhibited by a PGIP from cotton, the authors concluded that *in vitro* the enzyme-inhibitor complex resulted in an alteration of the balance between the release of elicitor-active oligosaccharides and the depolymerisation of the active oligogalacturonides into inactive molecules, favoring the accumulation of the elicitor active molecules at the site of infection and the activation of the defense responses of the host (James and Dubery, 2001). Similar to PR proteins and AMPs, expression of a PGIP may slow disease development and lesion expansion without conferring total host resistance. Leaves of transgenic tomato plants expressing a PGIP from pear were initially effectively colonized by the gray mould pathogen, *Botrytis cinerea*, whereas lesion expansion was later retarded in the transgenic tomato plants compared to the control plants (Powell et al., 2000). However, PGIPs with different specificities may be expressed in plants. Desiderio et al. (1997) found that high-level expression in transgenic plants of one member of the PGIP family, PGIP-1 from *Phaseolus vulgaris*, was not enough to confer general enhanced resistance to fungal pathogens.

As oxalic acid can act as a pathogenicity factor in plant disease, the degradation of oxalic acid may enhance resistance. Germin-like oxalate oxidases are glycoproteins, which are induced during seed germination and in response to fungal infection and abiotic stress (Dumas et al., 1995; Zhang et al., 1995; Berna and Bernier, 1997). The degradation of oxalic acid by oxalate oxidase also releases the reactive oxygen species H_2O_2 that may contribute to the activation of plant defense responses. The wheat germin gene, which has oxalate oxidase activity, has been introduced into soybean (Donaldson et al., 2001) and poplar (Liang et al., 2001) and increased resistance was observed to *Sclerotinia sclerotiorum* and *Septoria musiva*, respectively. The role of germin-like proteins in plant defense may not be related solely to oxalate oxidase activity as Schweizer et al. (1999) found that engineered germin genes lacking oxalate oxidase activity were capable of reducing the penetration ability of *Blumeria graminis* f.sp. *tritici* when transiently introduced into wheat leaves prior to infection. At sites of attempted fungal penetration, production of H_2O_2 was observed and the germin gene product became insolubilised, indicating a possible structural role in strengthening the cell wall during fungal attack.

18.2.3 Antibody-Based Resistance

Antibody-based resistance, or immunomodulation, is a novel genetic engineering strategy. The expressed antibodies that are involved in pathogenesis, either antibody fragments or recombinant antibodies (rAbs) against essential proteins, are expressed in plants to interfere in the pathogenesis process. The approach was first successfully applied to engineer resistance to viral disease and attempts to engineer resistance to other pathogens have been promising (De Jaeger et al., 2000; Schillberg et al., 2001; Zimmermann et al., Chapter 19 of this volume).

The constitutive expression of a cytosolic single-chain variable-fragment (scFv) against the coat protein of artichoke mottled crinkle virus in transgenic tobacco led to reduction in viral infection and a delay in symptom development (Tavladoraki et al., 1993). This result supported the hypothesis that transgenically expressed antibodies or antibody fragments recognizing critical epitopes on structural or nonstructural proteins of invading viruses may interfere with viral infection and confer viral resistance. Besides using the technique to obtain virus resistance, production of a scFv directed against the major membrane protein of the stolbur (big bud) phytoplasma proved to be effective in controlling mycoplasma (phytoplasma and spiroplasma) infection (Le Gall et al., 1998). Pathogenic phytoplasmas and spiroplasmas are located exclusively in the sieve tubes of the phloem tissue, into which they are inoculated by insect vectors. ScFv proteins were produced in tobacco and targeted to the apoplast with the bacterial *pelB* leader sequence. Transgenic tobacco shoots were grafted on a phytoplasma-infected tobacco root stock and grew free of symptoms, while the untransformed tobacco shoots showed severe stolbur symptoms and eventually died.

Antibody-based resistance against complex eukaryotic, multicellular pathogens, such as nematodes, fungi, and insects, remains a major challenge. Expression of full-size antibodies specific to root-knot nematode stylet secretions had no influence on parasitism of transgenic plants because antibodies were secreted to the apoplast (Baum et al., 1996). Most probably, the antibodies accumulated apoplastically whereas nematode stylet secretions were injected into the cytosol of the parasitized cells. When the antibody was fused to a C-terminal KDEL, the presence of which is necessary for ER retention and is sufficient to reduce the secretion of proteins from the ER, high cytosolic accumulation of the scFv was obtained (Rosso et al., 1996). Transgenic plants expressing this recombinant antibody seemed to show a significant reduction in the number of egg masses (Marie-Noëlle Rosso, personal communication in de Jaeger et al., 2000).

The benefits of application of antibodies to fungal pathogens are twofold: reduction of disease symptoms and neutralization of fungal toxins. Prevention of fungal infection using this approach was reported after conidia were mixed with the antibodies prior to inoculation (Wattad et al., 1997). However, symptom development for avocado, mango, and banana infected with *Colletotrichum gloeosporioides* was only inhibited using polyclonal antibodies against fungal pectate lyase. Direct evidence to show that expression of an antibody can protect plants from fungal infection is still lacking. A wide variety of fungal metabolites are toxic to plants, animals, and humans, having dramatic effects on animal and human health (Desjardins and Hohn, 1997; Gilbert and Tekauz, 2000). Several monoclonal and recombinant antibodies against mycotoxins produced by *Fusarium* have been isolated (Feuerstein et al., 1985; Hunter et al., 1985; Yuan et al., 1997). Some of these antibodies were shown to neutralize *in vitro* the inhibitory effect of a fungal toxin on protein synthesis in human B-lymphoblastoid cultures.

For insects, immunization of vertebrate hosts with antigens from the parasite was shown to protect against hematophagous arthropods (East et al., 1993). When the insects ingest antibodies from their host, a large fraction remains functional in the insect's gut and a small fraction passes into the hemolymph (Lehane, 1996), both enabling the antibodies to bind their antigen and to interfere with the insect's metabolism. Similar observations were made for herbivorous insects (Ben-Yakir and Shochat, 1996). The studies suggest that this approach would also work against insect attack of plants.

Remarkable progress has been made in antibody engineering and antibody isolation (e.g., by phage display). It will become more straightforward to isolate the genes encoding plant pathogen-specific antibodies. The more sophisticated design of the antibodies may reduce the use of pathogen-derived sequences and the use of agrochemicals.

18.3 Signal Perception and Transduction Pathways

Pathways that lead to plant disease resistance have recently been identified that are independent of those that culminate in the hypersensitive response (HR), the activation of pathogenesis related (PR) genes and concomitantly systemic acquired resistance (SAR, also known as induced systemic resistance, or ISR) (Pieterse and Van Loon, 1999). The pathways leading to HR and defense gene activation are separate (Heath, 2000), but there is considerable evidence that crosstalk between these pathways is common (Reymond and Farmer, 1998). Endogenous molecules such as salicylic acid (SA), jasmonic acid (JA), and ethylene are involved in these signaling pathways and play a role in the crosstalk (reviewed by Pieterse et al., Chapter 8 of this volume). A better understanding of the roles these molecules play and the identification of key elements in the signaling pathways could ultimately lead to an efficient, broad-spectrum durable disease resistance.

In gene-for-gene interactions the presence of a specific avirulence gene in the pathogen and a corresponding resistance gene in the host lead to localized cell death around the infection sites. The steps in this pathway include the recognition of the avirulence gene product, ion fluxes across cell membranes, the production of reactive oxygen intermediates (ROI, oxidative burst), the role of SA and JA, the need for protein synthesis, and, at least in some plant–microbe interactions, transcription (Heath, 2000). Using traditional molecular techniques, many components of defense mechanisms have been characterized and their underlying genes cloned. Now, genomics and proteomics approaches are bringing new tools, such as gene expression profiling and comparative proteome analysis, to the study of plant defense. We will review the progress made in the study of signal perception and signal transduction.

18.3.1 Signal Perception

At the frontier of plant–pathogen interaction, *R* genes represent a strategy in creating durable disease resistance. Many *R* genes have been cloned from dicot and monocot species and can be reintroduced via plant transformation. Compared with traditional breeding this not only reduces the time to incorporate new resistance into crops, but also allows *R* genes to be transferred into a sexually incompatible species. One way to increase the durability is to pyramid different *R* genes into the same plant. This strategy faces some challenges. The first is the availability of all the *R* genes that have a combined resistance spectrum to all existing pathogen races. The second is that the transfer of several *R* genes into the same plant is technically difficult, especially for cereal crops.

Another way to engineer disease resistance is to modify an *R* gene so that it can recognize an essential and common structure of various pathogens, pathogen races, or avirulence gene products. It is reasonable to believe that the biological cost to the pathogen of losing an essential component is high, and the chance of the pathogen overcoming a modified *R* gene that targets this essential component is relatively low. However, a thorough understanding of the products of both *R* and *avr* genes is required for such modification and only a few avirulence proteins have been identified from pathogens. A third strategy is to express the avirulence gene under a pathogen-inducible promoter in a plant containing the corresponding *R* gene. The expression of avirulence genes is triggered by infection, and subsequently the HR and SAR will occur, which will effectively control a broad range of pathogens. The recent success in expressing the *Phytophthora* elicitin for resistance to several virulent fungal pathogens is a good indication of the potential success of this strategy (Keller et al., 1999). Hennin et al. (2001) and Bertioli et al. (2001) expressed both the *Avr9* gene from *Cladosporium fulvum* and the corresponding tomato *R* gene *Cf-9* in *Brassica napus* and tobacco, respectively. In *Brassica* there was enhanced resistance to the blackleg pathogen, *Leptosphaeria maculans*, even though cell death was not observed. It was concluded that in these plants the level of Cf-9 was below threshold level for cell death, as *Cf-9* gene expression was only detectable by using reverse transcriptase-PCR even when a strong promoter, such as enhanced 35S, was used. Increased resistance was perhaps due to the induction of PR proteins for a longer period in the transgenics than in the wild type. In tobacco, spontaneous cell death was seen in the absence of the pathogen even though promoters that were pathogen responsive were used. Use of this avirulence gene strategy will depend on being able to tightly regulate the expression of the genes to prevent spontaneous necrosis to a degree that threatens crop productivity. It may be possible to achieve promising levels of resistance by expressing the *avr* genes constitutively but at a basal level. Such an expression of the *Pseudomonas syringae* pv. *syringae* avirulence gene *hrmA* was observed to confer enhanced resistance in tobacco to multiple pathogens (Shen et al., 2000).

One of the interesting cases in *R* gene transformation is tomato *Pto*. *Pto* confers a race-specific resistance to strains of *Pseudomonas syringae* pv. *tomato* containing the *avrPto* gene (Martin et al., 1993). The resistance is triggered by a direct interaction of the *Pto* protein and the *avrPto* protein that is secreted by the bacterium into the plant cell (Scolfield et al., 1996; Tang et al., 1996). Overexpression of *Pto* in tomato enhanced resistance to both bacterial and fungal pathogens in the absence of the corresponding avirulence gene *avrPto* (Tang et al., 1999). The study has suggested that the cloned *R* genes can be used to enhance resistance to unrelated pathogens or pathogen races. In such cases, the avirulence gene is not required for activation of the defense mechanisms and the resistance is likely to be durable.

18.3.2 Signal Transduction

Certain natural mutations known as lesion mimic mutants have decoupled the pathways leading to the HR from the pathogen recognition step. This leads to spontaneous lesions in the absence of the pathogen, which may often be accompanied by constitutive systemic resistance to a range of pathogens. The use of transgenes to create lesion mimic mutants is one method of enhancing resistance and several classes of transgenes can be used for this purpose (Mittler and Rizhsky, 2000). After signal perception, signals pass through various dynamic downstream pathways. One of the most significant approaches in studying the pathways is the development of microarrays. Microarrays are ordered high-density collections of genes, represented by DNA fragments or sets of oligoprimers, deposited onto glass slides. With the high density of arrays, all *Arabidopsis* genes (25,498) can be displayed on one or two standard slides. Slides are simultaneously hybridized with contrasting pairs of probes made of fluorescently labeled cDNA populations. These cDNA population pairs, prepared from RNA extracted from cells, tissues, or whole organisms, differ due to developmental stage or exposure to a specific biotic or abiotic stress.

By using microarray analysis, multiple defense-related genes can be identified. Microarray analysis has been used to simultaneously monitor changes in the expression patterns of 2,375 selected genes in *Arabidopsis* after inoculation with the incompatible fungal pathogen *Alternaria brassicicola* or treatment with the defense-related signaling molecules SA, methyl JA, or ethylene. Of the nine mitogen-activated protein kinases (MAPKs) analyzed, eight were significantly induced by the pathogen, *A. brassicola*, four by SA, seven by JA, and seven by ethylene (Schenk et al., 2000). Regardless of the complex interactions between the MAPK pathways and SA, JA, and ethylene (Genoud and Métraux, 1999; Fays and Parker, 2000), this analysis has clearly indicated the central role of MAPK pathways in plant defense.

In the MAPK signal transduction cascade, MAPK kinase is activated by an upstream MAPK kinase kinase, and in turn activates MAPK. The active MAPK may allow the activation of other protein kinases, catalyze the phosphorylation of cytoskeletal components or activate transcription factors once translocated to the nucleus (Hirt, 1997). The cascades involve parallel or redundant components, signal convergence and divergence, both positive and negative regulatory mechanisms, and scaffold proteins (Xing and Jordan, 2000; Xing et al., 2002). Overexpression in tomato of tMEK2^{MUT}, a constitutively activated tomato MAPK kinase, enhanced resistance to the virulent bacterial pathogen *Pseudomonas syringae* pv. *tomato* (Figure 18.1).

Control   Transgenic

FIGURE 18.1. Over-expression of $tMEK2^{MUT}$ and resistance to *Pseudomonas syringae* pv. *tomato*. Shown are a nontransgenic line (Control) and a representative transgenic line.

Modifications have been made on other phosphorylation pathways, including that of calmodulin domain-like protein kinase (CDPK) in an attempt to improve stress tolerance. A typical example is the manipulation of OsCDPK7, a rice CDPK, induced by cold and by high salinity in shoots and roots (Saijo et al., 2000). When the wild-type OsCDPK7 was constitutively overexpressed in rice, plants showed increased tolerance to cold and salt stress. In the tomato-*Cladosporium fulvum* pathosystem, a tomato CDPK appeared to mediate the activation of NADPH oxidase on the host plasma membrane (Xing et al., 1997). Another study suggested that tomato CDPK is independent of, or is located upstream from, a signaling pathway that is required for the production of ROI (Romeis et al., 2000). Genes encoding the CDPK(s) in the two studies have not been identified. Ectopic expression of AK1-6H, an *Arabidopsis* CDPK, in tomato protoplasts activated NADPH oxidase on the host plasma membrane and enhanced oxidative burst (Xing et al., 2001b). The complexity and sophistication of the defense mechanisms have made experimental studies challenging, and many attempts to identify a target gene for manipulation have failed owing to the complexity of the signaling process. Care has to be taken in developing manipulation strategies, for example, if the role of a protein or a gene has been evaluated for a gene belonging to a multigene family, upon subsequent development of manipulation strategies, it is not always possible to be sure that the intended member of that family is being manipulated. *NtMEK2*, a tobacco MAPK kinase gene, belongs to a multigene family (Yang et al., 2001). Overexpression of its constitutively-activated mutant, $NtMEK2^{DD}$, induced HR-like cell death in tobacco, which was preceded by the activation of endogenous salicylic acid-induced protein kinase (SIPK) and wound-induced protein kinase (WIPK) (Yang et al., 2001). However, the expression of constitutively activated mutants of other tobacco MAPKKs, including $NtMEK1^{DD}$, $NtMEK7^{DD}$, and $NtMEK8^{DD}$, did not activate SIPK or WIPK, even though the proteins were expressed at similar levels. Their expression did not cause HR-like cell death either. The work suggested that the activation of SIPK and WIPK is selective and that HR-like cell death is caused by a specific MAPK pathway. Identification of the gene(s) from a multigene family responsible for a particular pathway may become critical when we try to manipulate homologues from other species.

Signal transduction-inducing genes can originate from several different points in the signaling pathways. One of the first steps in the pathway leading to HR is a change in ion flux across cellular membranes due to enhanced proton pumping. Expression of a bacterial proton pump (bacterio-opsin) in transgenic tobacco and tomato plants resulted in enhanced disease resistance (Rizhsky and Mittler, 2001). Prevention of spontaneous lesions was achieved through the use of a wound inducible promoter.

Membrane associated heterotrimeric GTP-binding proteins (G-proteins) mediate signal transduction from cell surface receptors, and inactivation of G-proteins results in constitutive activation of downstream signaling components (Simon et al., 1994). Beffa et al. (1995) used the cholera toxin gene, the product of which blocks the GTPase activity of G-proteins resulting in inactivation, to produce transgenic tobacco plants. These plants exhibited SAR in the absence of infection and were much less susceptible to pathogen attack.

The oxidative burst results in the production of ROI that are thought to be mediators of HR cell death (Heath, 2000). GTPase proteins designated Rac in mammals act as regulators of ROI production. A rice homolog of Rac GTPase has been cloned and expressed constitutively in transgenic rice. The transgenic plants show HR-like responses and display increased resistance to virulent strains of two rice pathogens through enhanced phytoalexin production and altered expression of defense-related genes (Ono et al., 2001).

Enhanced synthesis of SA has been found to be necessary both as a signal in the SAR pathway and for the establishment of SAR. This is supported by the work of Verberne et al. (2000) who expressed two bacterial genes necessary for the conversion of chorismate into SA in tobacco resulting in the constitutive expression of PR genes and enhanced disease resistance. In plants the *NPR1* gene is required for SA-mediated SAR. The product of this gene is a 66-kD protein with ankyrin repeats that are involved in protein–protein interactions (Cao et al., 1997). Yeast two-hybrid experiments have shown that *NPR1* binds to and enhances the binding activity of members of a conserved family of leucine zipper proteins called TGA transcription factors (Zhou et al., 2000) that in turn bind to the promoters of certain PR genes. Pathogen infection or SA induces expression of the *NPR1* gene two- to threefold (Cao et al., 1997). It has been hypothesized that overexpression of *NPR1* at higher levels may provide increased resistance to pathogens. This has been borne out by several studies in which constitutive overexpression of *Arabidopsis NPR1* in *Arabidopsis* (Cao et al., 1998, Friedrich et al., 2001) and in rice (Chern et al., 2001) has provided increased levels of resistance to pathogens. Interestingly, Friedrich et al. (2001) also observed that plants overexpressing *NPR1* were more responsive to SA than wild type. This would provide the opportunity to combine *NPR1* overexpression with enhanced SA production to further increase disease resistance. Although SA induces *NPR1* expression it is expressed constitutively in an inactive state. SA treatment not only increases expression of *NPR1*, but also activates the protein. The mechanism of how SA or an SA-dependent signal activates *NPR1* may rely on a class of proteins designated *NIM1* interacting (NIMIN) (Weigel et al., 2001). These proteins are induced by SA treatment and via *NPR1* are able to interact with TGA transcription factors. It is suggested that the NIMIN proteins, through interaction with *NPR1* modify *NPR1* activity after pathogen challenge and may serve to activate *NPR1*. There is evidence to suggest that the induction of *NPR1* gene expression following SA treatment or pathogen challenge is mediated by a class of DNA binding proteins that contain the WRKY amino acid motif (Yu et al., 2001). The promoter of the *NPR1* gene contains W-box motifs that are recognized by particular SA and pathogen induced WRKY-type transcription factors. These W-boxes are necessary for the induction of *NPR1* transcription and *NPR1*-mediated activation of SAR. In addition to overexpression of *NPR1*, it appears that improved disease resistance could be obtained through modification of key elements of the entire *NPR1* pathway including WRKY and TGA transcription factors and the NIMIN proteins.

Jasmonic acid and ethylene play important roles in plant defense through distinct signaling pathways, which are independent of SA (Reymond and Farmer, 1998; Pieterse and Van Loon, 1999). As in the SA pathway, enhanced disease resistance can be envisioned via manipulation of either the synthesis of the signaling molecule and its perception or the endogenous genes induced by the signaling molecule. Seo et al. (2001) hypothesized that methyl jasmonate (MeJA) was a candidate for the intra- and intercellular signal-transducing molecule that mediates the JA-induced plant responses as it can diffuse through membranes. To test this hypothesis Seo et al. (2001) cloned the *Arabidopsis* gene for jasmonic acid carboxyl methyltransferase that catalyzes the methylation of JA to MeJA and overexpressed it in *Arabidopsis*. The resulting plants had enhanced levels of MeJA as well as enhanced expression of jasmonate-responsive genes including defense-related genes. The plants exhibited enhanced resistance to the pathogen *Botrytis cinerea*. In rice it has been found that JA is an inducer of SAR and that certain transcription factors of the *myb* family are involved (Lee et al., 2001). The transcription factor gene *Jamyb* is induced at much higher levels in susceptible than resistant varieties after inoculation with the rice blast fungus, and the gene is also highly induced in certain lesion mimic mutants. It appears that in rice the JA pathway is associated with the HR response and the identification of genes involved in this pathway may open the door to resistance strategies in monocots similar to those described above for the SA-mediated hypersensitive-response pathway in dicots.

Similarly, certain transcription factors have been identified that are involved in ethylene-mediated signaling pathways. These factors, dubbed EREBP for ethylene-responsive element binding proteins, bind to a particular motif in the promoter elements of PR genes induced by ethylene (Ohme-Takagi and Shinshi,

1995). Park et al. (2001a) isolated an EREBP transcription factor from tobacco that was induced by salt stress. Overexpression of this transcription factor in tobacco induced PR genes and resulted in enhanced disease resistance as well as salt tolerance, indicating that pathways responsive to biotic and abiotic stress share key components.

Ethylene sensitivity may play a role in limiting cell death either during attack from a virulent pathogen or during the hypersensitive response. Reduced ethylene sensitivity will limit cell death by inhibiting the induction of ethylene-responsive genes involved in HR. Ciardi et al. (2000) found that reducing ethylene sensitivity through the overexpression of an ethylene receptor resulted in greatly reduced necrosis after inoculation with virulent *Xanthomonas campestris* pv. *vesicatoria*.

The JA/ethylene mediated signaling pathway can also be activated by inhibitors of translation such as pokeweed antiviral protein (PAP). Zoubenko et al. (2000) reported that expression of a mutant form of PAP, which does not bind to ribosomes but inhibits translation by depurinating capped RNA in transgenic plants leads to increased resistance to viral and fungal pathogens. It is possible that certain negative regulators of the signaling pathways are inhibited in translation by such proteins as PAP, thereby activating a SA-independent pathway leading to disease resistance.

Defense pathways are complex and strategies have to be designed so that the manipulations will achieve the desired results without having detrimental effects on plant growth and development. For example, in the highly expressed $tMEK2^{MUT}$ transgenic tomato lines, yellowing or death of leaves was observed on the lower branches (Xing, unpublished). It is thus worth emphasizing that the strategies must build on an understanding of the defense mechanisms and their components in the global biological context. There are some advantages in manipulating signal transduction pathways for sustainable defense: (1) transferability; i.e., signaling elements isolated from one species may work in other species, (2) multiple barriers established in signaling cascade manipulations may strengthen defense capability and durability, (3) reduction of the possibility that pathogens will evolve new strategies to overcome disease resistance in transgenic plants, since specific recognition steps of plant–pathogen interactions are bypassed, (4) high potential for broad-spectrum disease resistance, and (5) new alternatives for systems in which disease resistance is polygenic, or where information on gene-for-gene interactions is still limited. In terms of achieving sustainable resistance, a few particular advances are to be emphasized. First, the signal transduction cascade involves a hierarchy of signal transduction and signal amplification. If a particular pathway is well understood, then engineering on multiple components of this particular pathway is possible and multiple barriers established in signaling cascade manipulations may strengthen defense capability and durability. Second, pathway elucidation will identify key components that merge different upstream signals that are perceived by the host with those that diversify the signals to different downstream components. These components may broadly enhance and amplify downstream defense mechanisms. A MAPK kinase was identified as such a key component. It acts as a convergence point for biotic and abiotic stress signals and activates a large array of defense genes (Xing et al., 2001a). A recent proteomic study confirmed the coenhancement of two PR genes with enhanced expression level through two downstream kinases (Xing et al., 2003). Third, the manipulation of downstream signaling components may reduce the possibility that pathogens will evolve new strategies to overcome disease resistance in transgenic plants, since specific recognition steps of plant–pathogen interaction are bypassed. To prove this advantage it will be essential to have field studies and to place transgenic plants under enhanced pathogen pressure. Even though there are many advantages in manipulating signal transduction pathways to achieve desirable outcomes, strategies must build on an understanding of the pathways and their components in the global biological context. The wealth of knowledge and recent technological advances will profoundly alter the ways in which we select and approach problems in the study of plant–pathogen interactions.

18.4 Molecular Breeding for Durable Disease Resistance

Molecular breeding is the science of using DNA fingerprinting or genotyping of plants to determine the allelic makeup of a plant on which we can predict its phenotype. If plant phenotypes can be predicted accurately, then selections can be made to retain plants with specific genotypes, which have the potential to express a desirable phenotype. Molecular breeding can be directly applied to developing durable disease resistance in plants. This is a form of engineering plants that relies on sophisticated laboratory technology, prior knowledge of disease resistance genes, and gene interactions. Molecular breeding can circumvent more cumbersome, established pedigree breeding strategies and even generate plant genotypes unattainable by conventional methods (Young, 1999). The implementation of molecular breeding techniques requires certain genetic resources, the most important being a high-density genetic map of the crop species as well as access to genetic markers. With these tools in hand, molecular breeding can: (1) accelerate elite line production; (2) establish novel combinations of disease resistance genes/alleles; (3) assemble complex genotypes across the genome; and (4) create gene pyramids for more durable disease resistance.

18.4.1 Basic Strategies

Molecular breeding begins with the generation of a genetic map of the crop species on to which the location of disease resistance genes are mapped. Once accomplished, markers that are closely linked or associated with the resistance genes can be identified. There are numerous examples of this in the literature, largely related to single, dominant genes. The molecular markers can then be used to track and select the disease resistance gene in backcross (BC) population or other segregating population. This strategy is particularly effective for diseases where plant phenotyping is expensive, highly variable, and time consuming, such as Fusarium head blight in wheat (Gilbert and Tekauz, 2000) and blast infection in rice (Hittalmani et al., 2000).

In addition to selection of single genes in backcross (BC) population, molecular breeding technology is now sufficiently advanced in most major crop species to permit the assembly of complex genotypes across the whole genome. Molecular breeding is not restricted to single, small intervals of the genome. The strategy includes the use of many markers, high throughput techniques, and analyses capable of dealing with large datasets and plant population. Whole genome genotyping permits one to design or engineer a plant's genetic makeup with reasonable precision and to take into account all mapped genetic traits segregating in the cross being examined. Computer modeling and software design have been important aspects of this field of research in recent years. Software for quantitative trait locus (QTL) analysis (Lincoln et al., 1992) indicates the position of candidate genes controlling traits of interest. Other software analysis can identify the progeny that should be intercrossed to maximize the recovery of elite alleles (Charmet et al., 1999).

18.4.2 Microsatellite Markers

Engineering plants using molecular breeding requires molecular makers that are amenable to high throughput techniques and that are easily interpreted. Microsatellite markers are the most popular today and have been developed in many major crop species (Akkaya et al., 1995; Röder et al., 1998; Pestova et al., 2000; Saal et al., 2001). They are PCR-based and can be detected by high-resolution, automated electrophoresis equipment. Typically, they are able to detect polymorphism at a single locus. Other advantages include the fact that, like other markers, microsatellites have good genome coverage, and thus have a high propensity to be linked to genes of interest. Molecular breeding using microsatellite markers can also be very cost effective when high throughput methods are implemented. We recently evaluated the cost to be approximately Can$0.40/data point, including labor and consumables.

18.4.3 Molecular Mapping and Fusarium Resistance in Wheat

Fusarium head blight (FHB) resistance in wheat provides an example of engineering disease resistance in plants. Resistance to this disease is inherited quantitatively and there are multiple genes acting in combination to provide the maximum level of resistance. Complicating this pathosystem further is the fact that FHB resistance is characterized by several mechanisms, including reduced initial infection, reduced spread of the disease, and reduced levels of mycotoxins, all of which may be controlled by separate gene loci (Mesterhazy, 1983; Bai and Shaner, 1994; Mesterhazy, 1995; Ban, 2000). The principal causal agent of this devastating disease in North

America is *Fusarium graminearum*. Current methods of screening for FHB are time consuming and labor intensive. Wheat plants are susceptible to infection by *F. graminearum* at flowering and therefore must be screened as adults. The disease reaction is dependent on environmental conditions and evaluations can thus vary from year to year, requiring repeated evaluation of breeding lines to identify resistant phenotypes. Finally, FHB resistance in wheat is controlled by several genes, each of which may independently affect disease progression and provide a variable contribution to the final level of resistance (Waldron et al., 1999; Buerstmayr et al., 2000; Anderson et al., 2001). The complexity of genetic resistance controlling FHB provides an excellent target for the development of molecular breeding tools and their implementation in plant breeding to create elite wheat varieties with durable FHB resistance.

Breeding for FHB resistance may not be attainable using conventional breeding and phenotypic selection. Extensive studies are now complete and many more planned to discover and characterize the genes controlling resistance to different aspects of FHB (Bai et al., 1999; Waldron et al., 1999; Anderson et al., 2001; Buerstmayr et al., 2002). This knowledge is expected to lead to accelerated development of elite breeding population, enriched for FHB resistance and pyramiding known FHB resistance genes/alleles in combinations not yet detected in wheat germplasm.

18.4.4 QTL Analysis

When phenotypic and genotypic data are merged, an analysis of variance can be performed that locates defined chromosome intervals, which explain the variation in phenotype and thus identifies QTLs (reviewed by Kelly and Vallejo, Chapter 3 of this volume). This approach forms the first phase of engineering plants using molecular breeding. There have been several analyses in recent years to locate FHB resistance QTLs in wheat. QTLs controlling variable aspects of FHB infection have been located on chromosome 3B (Bai et al., 1999; Waldron et al., 1999; Ban, 2000), 5A (Buerstmayr et al., 2002), and 6B (Anderson et al., 2001). Our laboratory has recently identified or confirmed QTLs for FHB resistance on chromosome 2D, 3B, 4B, and 5A (Somers et al., unpublished data).

Further, the resistance source used in these studies differs and thus there is potential for allelic variation. For example, the FHB resistant Chinese variety Sumai3 has been the most intensively studied line, and our lab has extended similar analyses to two additional FHB resistant lines, Wuhan-1 (Chinese) and Maringa (Brazilian). We note that there are indeed allelic differences between Sumai3 and Wuhan-1 or Maringa, at the major QTL located on 3BS, based on microsatellite polymorphisms (Figure 18.2). This raises the potential to study different alleles, and to determine which alleles and allele combinations are the most effective to resist infection. The contribution of each QTL toward resisting infection can be quite variable, ranging from <10% to >40%. These measurements may be influenced by the susceptible parent used in the cross, and thus, the same QTL in different populations may appear to contribute differently to FHB resistance.

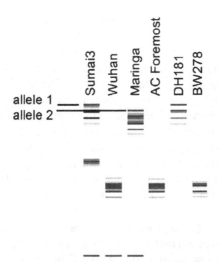

FIGURE 18.2. Amplification of GWM533 microsatellite in six wheat accessions. Sumai3, Maringa and DH181 (a Sumai3 derivative) all carry a Fusarium head blight (FHB) resistance QTL on chromosome 3BS, marked by GWM533. Wuhan, AC Foremost and BW278 all carry FHB susceptibility alleles at this locus on 3BS. Allele 1 is present in Sumai3 and DH181, allele 2 is present in Maringa, representing a potential alternate resistance allele at this locus.

Regardless, most plant breeders would agree that a combination of the correct QTLs assembled via molecular breeding has the greatest potential to substantially enrich breeding population for disease resistance.

18.4.5 Gene Pyramiding and Alternate Alleles

The greatest challenge in molecular breeding of complex traits is the number of genes required for pyramiding. Durable disease resistance can be achieved by combining different genes with similar alleles in a single line. This strategy protects against a single gene being defeated by an evolving pathogen, via the possession of additional, similar genes in combination. There are relatively few examples of marker-assisted gene pyramiding in crop species, but the ones reported suggest the theories and strategies are successful. These include durable resistance toward anthracnose in bean (Young and Kelley, 1997) and bacterial blight resistance in rice (Huang et al., 1997). In these cases, PCR-based or RFLP markers were discovered and linked to different disease resistance genes that were pyramided by recurrent backcrossing, enabling selection of the genes with markers. Gene pyramiding was also successful in rice to combine QTLs for improved root depth (Shen et al., 2001). This study differed from the previous examples in that selection was made on chromosome intervals containing QTLs rather than a single chromosome locus/marker near a disease resistance gene.

In the case of FHB, pyramiding genes is required because maximum protection against the pathogen is achieved by a combination of genes having an additive effect (Anderson et al., 2001; Buerstmayr et al., 2002). Molecular markers give a distinct advantage in both situations described above, where direct genotypic evidence is generated to show that gene pyramiding was achieved.

Pathology testing to combine genes that have similar effects is difficult. For example, if two resistance genes against the prevalent races of wheat leaf or stem rust are combined, pathology screening may not be able to distinguish between plants with one or two genes. In the case of FHB, the subtle additive effects of genes may be difficult to resolve. In addition, the expression of FHB resistance genes in different genetic backgrounds can be variable. Thus, pathology testing can be effective, but does not ensure that the desired combination of genes has been achieved.

Alternate alleles at a single locus can be specifically studied and combined with other specific alleles at other loci. The alleles at one locus can be distinguished by polymorphisms in microsatellite markers surrounding the locus. Figure 18.2 shows a microsatellite marker near the major FHB QTL on 3BS in wheat. The polymorphisms between Sumai3 and Maringa, both of which have FHB resistance QTLs detected in this region, suggest there may be different alleles of the same gene. Continuing efforts toward discovery of new sources of FHB resistance may lead to new genes/alleles to be studied and combined to enhance FHB resistance (Gilchrist et al., 2000).

18.4.6 High-Throughput Molecular Breeding

Engineering plants for durable disease resistance is not achieved by simply selecting for the correct combination of genes/alleles. Since disease resistance gene sources are often accessions that are unadapted to local environmental conditions, it is equally important to restore an elite genetic background to deliver the newly combined disease resistance genes with desirable agronomic and quality traits. In conventional breeding programs the elite genetic background is restored by repeated backcrossing to the recurrent parent that is time consuming and may result in linkage drag of undesirable chromosome intervals. Direct selection of alleles across the genome from the elite parent is one method to restore the remainder of the genome. Many spring crops flower two months after planting, and in order to accelerate plant breeding, it is desirable to make crosses to selected plants in each generation, rather than wait for data from progeny testing. Therefore, high-throughput genotyping techniques, facilitated by appropriate equipment, must be ready to select plants based on DNA fingerprints in time to perform a subsequent backcross. The combination of DNA extraction and PCR in 96 or 384 well plate formats, coupled with automated capillary electrophoresis of fluorescent tagged microsatellites, is very effective for this purpose.

In addition to the need for high-throughput technology, molecular breeders need to consider the number of disease resistance genes being introgressed for pyramiding, the size of the genome, level of polymorphism in the cross, and availability of microsatellites covering the genome. In our experience, a population of 100 BC1F1 wheat plants, all of which are heterozygous for the introgressed genes, is sufficient to select a few plants with 87% restored elite alleles. These results demonstrate the gains that can be made, since they mimic selection of plants from a typical BC2F1 population.

18.5 Summary and Conclusions

The advantages of successfully engineering plants for disease resistance are evident: increased yields and improved quality, avoidance of grain contamination by toxic secondary metabolites associated with certain fungal diseases, and reduction of fungicide use and chemical release into the environment, to name but a few. Despite the rapid increase in knowledge gained over the past decade, durable engineered plant disease resistance against fungal pathogens so far eludes us. The constitutive overproduction of antifungal proteins does not render plants resistant to fungal pathogens, but rather delays the onset or extent of symptom development. Used singly, the products of overexpressed antimicrobial compounds or PR proteins have not consistently provided target resistance levels in host plants; several elements of the plant's own defense system have to be introduced in order to enhance resistance (Honée, 1999). In all cases the acquired resistance is effective against a limited number of fungal pathogens only. For example, transgenic crops overexpressing a chitinase gene may be resistant to several pathogens with chitin in their cell walls, but not to fungi lacking chitin. It is possible that a fungus may adapt to a host and modify its cell wall composition by increasing biosynthesis of more chitosan or glucan instead of chitin and become insensitive to chitinases (Cornelissen and Schram, 2000). An understanding of how relatively quickly pathogens such as rusts can mutate to overcome host resistance has led to the perception in some parts of the world that widespread cultivation of crops genetically engineered for pest or disease resistance will impose intense selection pressure on pathogens to overcome novel resistance mechanisms, ultimately resulting in the development of new races or isolates of pests or diseases that would be difficult to control.

Strategies that may provide transgenic crops with potentially more durable resistance are based on general defense responses occurring in plants during incompatible plant–pathogen interactions involving local cell death induced by the attacking pathogen and directed toward a broader spectrum of pathogens. A tenet that was once held strongly was that only multigenic resistance was of a durable nature, but Johnson (1993) cautioned against discarding major genes and HR material based on the preconception that neither can provide durable resistance. The method of changing race-specific resistance into race-nonspecific resistance was pioneered in the early 90s using tomato-*Cladosporium fulvum* as a model pathosystem (De Wit and van Kan, 1993). The recent successful transformation of several species with both the tomato R gene and the avirulence gene, *Avr9*, from *C. fulvum,* provided researchers with the opportunity to examine this strategy for engineering broad-based durable resistance (Bertioli et al., 2001; Hennin et al., 2001). While a hypersensitive response was not observed in *Brassica napus* to the blackleg pathogen, enhanced resistance was observed. In tobacco, however, transgenic plants harboring both the resistance and avirulence gene constructs underwent spontaneous necrosis, emphasizing the need to examine these new strategies in the global context of the host biology.

The use of molecular markers to track genes, or chromosomal regions of interest, may facilitate the breeding of cultivars with resistance to several important diseases while maintaining essential agronomic and quality characteristics. As new sources of disease resistance genes and alternate alleles at these gene loci are characterized from genetic experiments, the prospects of engineering durable disease resistance in plants will increase. Today, the technical capability to pyramid multiple genes together is available through high-density molecular maps for some species and tightly linked markers. Crop production should begin to see the benefits of molecular breeding and engineering soon, not only for durable disease resistance, but also for improved quality and agronomic traits.

When the research is taken from the laboratory bench to the field, the true value and durability of these new strategies will become apparent. The definition of durable resistance is a resistance that remains effective in a crop species that is widely grown for many years in a region where environmental conditions are favorable for the disease to occur (Johnson, 1983). As far as we have come in a relatively short period to engineering durable resistance to plant diseases, the test will be to observe the action in the field.

References

Akkaya, M.S., Shoemaker, R.C., Specht, J.E., Bhagwat, A.A., and Cregan, P.B. 1995. Integration of simple sequence repeat DNA markers into a soybean linkage map. *Crop Sci.* 35(5):439–1445.

Alexander, D., Goodman, R.M., Gut-Rella, M., Glascock, C., Weymann, K., Friedrich, L., Maddox, D., Ahl-Goy, P., Luntz, T., Ward, E., and Ryals, J.A. 1993. Increased tolerance to two oomycete pathogens in transgenic tobacco expressing pathogenesis-related protein 1a. *Proc. Natl. Acad. Sci. USA* 90:7327–7331.

Anderson, J.A., Stack, R.W., Liu, S., Waldron, B.L., Field, A.D., Coyne, C., Moreno-Sevilla, B., Mitchell Fetch, J., Song, Q.J., Cregan, P.B., and Frohberg, R.C. 2001. DNA markers for Fusarium head blight resistance QTL in two wheat populations. *Theor. Appl. Genet.* 102:1164–1168.

Asao, H., Nishizawa, Y., Arai, S., Sato, T., Hirai, M., Yoshida, K., Shinmyo, A., and Hibi, T. 1997. Enhanced resistance against a fungal pathogen *Sphaerotheca humuli* in transgenic strawberry expressing a rice chitinase gene. *Plant Biotechnol.* 14:145–149.

Bai, G., Kolb, F.L., Shaner, G., and Domier, L.L. 1999. Amplified fragment length polymorphism markers linked to a major quantitative trait locus controlling scab resistance in wheat. *Phytopathology* 89:343–348.

Bai, G., and Shaner, G. 1994. Scab of wheat: prospects for control. *Plant Dis.* 78:760–766.

Ban, T. 2000. Review studies on the genetics of resistance to Fusarium head blight caused by *Fusarium graminearum* in wheat. *In Proceedings of the International Symposium on Wheat Improvement for Scab Resistance*, eds. J. Raupp, Z. Ma, P. Chen, and D. Liu, pp. 82–93, May 5–11, 2000. Suzhou and Nanjing, The Republic of China.

Baum, T.J., Hiatt, A., Parrott, W.A., Pratt, L.H., and Hussey, R.S. 1996. Expression in tobacco of a functional monoclonal antibody specific to stylet secretions of the root-knot nematode. *Mol. Plant Microbe Interact.* 9:82–387.

Beffa, R., Szell, M., Meuwly, P., Pay, A., Vögeli-Lange, R., Métraux, J.P., Neuhaus, G., Meins, F., Jr., and Nagy, F. 1995. Cholera toxin elevates pathogen resistance and induces pathogenesis-related gene expression in tobacco. *EMBO J.* 14:5753–5761.

Ben-Yakir, D., and Shochat, C. 1996. The fate of immunoglobulin G fed to larvae of *Ostrinia nubilalis*. *Entomol. Exp. Appl.* 81:1–5.

Berna, A., and Bernier, F. 1997. Regulated expression of a wheat germin gene in tobacco: oxalate oxidase activity and apoplastic localization of the heterologous protein. *Plant Mol. Biol.* 33:417–429.

Berrocal-Lobo, M., Molina, A., and Solano, R. 2002. Constitutive expression of ethylene-response-factor 1 in *Arabidopsis* confers resistance to several necrotrophic fungi. *Plant J.* 29:23–32.

Bertioli, D.J., Guimarces, P.M., Jones, J.D.G., Thomas, C.M., Burrows, P.R., Monte, D.C., and De M. Leal-Bertioli, S.C. 2001. Expression of Tomato Cf genes and their corresponding avirulence genes in transgenic tobacco plants using nematode responsive promoters. *Ann. Appl. Biol.* 138:333–342.

Bi, Y.-M., Cammue, B.P.A., Goodwin, P.H., Krishna Raj, S., and Saxena, P.K. 1999. Resistance of *Botrytis cinerea* in scented geranium transformed with a gene encoding the antimicrobial protein Ace-AmP1. *Plant Cell Rep.* 18:835–840.

Bieri, S., Potrykus, I., and Fütterer, J. 2000. Expression of active barley seed ribosome-inactivating protein in transgenic wheat. *Theor. Appl. Genet.* 100:755–763.

Bliffeld, M., Mundy, J., Potrykus, I., and Fütterer, J. 1999. Genetic engineering of wheat for increased resistance to powdery mildew disease. *Theor. Appl. Genet.* 98:1079–1086.

Bohlmann, H. 1994. The role of thionins in plant protection. *Crit. Rev. Plant Sci.* 13:1–16.

Bolar, J.P., Norelli, J.L., Harman, G.E., Brown, S.K., and Aldwinckle, H.S. 2001. Synergistic activity of endochitinase and exochitinase from *Trichoderma atroviride* (*T. harzianum*) against the pathogenic fungus (*Venturia inaequalis*) in transgenic apple plants. *Transgenic Res.* 10:533–543.

Bolar, J.P., Norelli, J.L., Wong, K.W., Hayes, C.K., Harman, G.E., and Aldwinckle, H.S. 2000. Expression of endochitinase from *Trichoderma harzianum* in transgenic apple increases resistance to apple scab and reduces vigor. *Phytopathology* 90:72–77.

Brandwagt, B.F., Kneppers, T.J.A., Nijkamp, H.J.J., and Hille, J. 2002. Overexpression of the tomato Asc-1 gene mediates high insensitivity to AAL toxin and fumonisin B_1 in tomato hairy roots and confers resistance to *Alternaria alternata* f.sp. *lycopersici* in *Nicotiana umbratica* plants. *Mol. Plant Microbe Interact.* 15:35–42.

Broekaert, W.F., Terras, F.R.G., Cammue, B.P.A., and Osborn, R.W. 1995. Plant defensins: novel antimicrobial peptides as components of the host defense system. *Plant Physiol.* 108:1353–1358.

Broekaert, W.F., Terras, F.R.G., and Cammue B.P.A. 2000. Induced and pre-formed antimicrobial proteins. In *Mechanisms of Rresistance to Plant Diseases*, eds. A.J. Slusarenko,

R.S.S. Fraser, and L.C. Van Loon, pp. 371–477. Dordrecht, The Netherlands: Kluwer Academic Publishers.

Broglie, R., Broglie, K., Roby, D., and Chet, I. 1993. Production of transgenic plants with enhanced resistance to microbial pathogens. In *Transgenic plants*, eds. S.D. Kung, and R. Wu, Vol. 1, pp. 265–276. New York: Academic Press.

Broglie, K., Chet, I., Holliday, M., Cressman, R., Biddle, P., Knowlton, S., Mauvais, C.J., and Broglie, R. 1991. Transgenic plants with enhanced resistance to the fungal pathogen *Rhizoctonia solani. Science* 254:1194–1197.

Buerstmayr, H., Lemmens, M., Hartl, L., Doldi, L., Steiner, B., Stierschneider, M., and Ruckenbauer, P. 2002. Molecular mapping of QTL for Fusarium head blight resistance in spring wheat. I. Resistance to fungal spread (type II resistance). *Theor. Appl. Genet.* 104:84–91.

Buerstmayr, H., Steiner, B., Lemmens, M., and Ruckenbauer, P. 2000. Resistance to Fusarium head blight in winter wheat: heritability and trait associations. *Crop Sci.* 40:1012–1018.

Cao, H., Glazebrook, J., Clarke, J.D., Volko, S., and Dong, X. 1997. The *Arabidopsis* NPR1 gene that controls systemic acquired resistance encodes a novel protein containing ankyrin repeats. *Cell* 88:57–63.

Cao, H., Li, X., and Dong, X. 1998. Generation of broad-spectrum disease resistance by overexpression of an essential regulatory gene in systemic acquired resistance. *Proc. Natl. Acad. Sci. USA* 95:6531–6536.

Cary, J.W., Rajasekaran, K., Jaynes, J.M., and Cleveland, T.E. 2000. Transgenic expression of a gene encoding a synthetic antimicrobial peptide results in inhibition of fungal growth *in vitro* and *in planta*. *Plant Sci.* 154:171–181.

Cervone, F., De Lorenzo, G., Degré, L., Salvi, G., and Bergami, M. 1987. Purification and characterization of a polugalacturonase-inhibiting protein from *Phaseolus vulgaris* L. *Plant Physiol.* 85:631–637.

Chang, M.M., Chiang, C.C., Martin, M.W., and Hadwiger, L.A. 1993. Expression of a pea disease resistance response gene in the potato cultivar Shepody. *Am. Potato J.* 70:635–647.

Charmet, G., Robert, N., Perretant, M.R., Gay, G., Sourdille, P., Groos, C., Bernard, S., and Bernard, M. 1999. Marker-assisted recurrent selection for cumulating additive and interactive QTLs in recombinant inbred lines. *Theor. Appl. Genet.* 99:1143–1148.

Chen, W.P., Chen, P.D., Liu, D.J., Kynast, R., Friebe, B., Velazhahan, R., Muthukrishnan, S., and Gill, B.S. 1999. Development of wheat scab symptoms is delayed in transgenic wheat plants that constitutively express a rice thaumatin-like protein gene. *Theor. Appl. Genet.* 99:755–760.

Chen, W.P., and Punja, Z.K. 2002. *Agrobacterium*-mediated transformation of American ginseng with a rice chitinase gene. *Plant Cell Rep.* 20:1039–1045.

Chern, M.S., Fitzgerald, H.A., Yadav, R.C., Canlas, P.E., Dong, X., and Ronald, P.C. 2001. Evidence for a disease-resistance pathway in rice similar to the NPR1-mediated signaling pathway in *Arabidopsis. Plant J.* 27:101–113.

Chong, D.K.X., and Langridge, W.H.R. 2000. Expression of full-length bioactive antimicrobial human lactoferrin in potato plants. *Transgenic Res.* 9:71–78.

Ciardi, J.A., Tieman, D.M., Lund, S.T., Jones, J.B., Stall, R.E., and Klee, H.J. 2000. Response to *Xanthomonas campestris* pv. *vesicatoria* in tomato involves regulation of ethylene receptor gene expression. *Plant Physiol.* 123:81–92.

Clausen, M., Kräuter, R., Schachermayr, G., Potrykus, I., and Sautter, C. 2000. Antifungal activity of a virally encoded gene in transgenic wheat. *Nat. Biotechnol.* 18:446–449.

Constabel, P.C., Bertrand, C., and Brisson, N. 1993. Transgenic potato plants overexpression the pathogenesis-related *STH-2* gene show unaltered susceptibility to *Phytophthora infestans* and potato virus X. *Plant Mol. Biol.* 22:775–782.

Cornelissen, B.J.C., and Schram, A. 2000. Transgenic approaches to control epidemic spread of diseases. In *Mechanisms of Resistance to Plant Diseases*, eds. A. Slusarenko, R.S.S. Fraser, and L.C. van Loon, pp. 575–599. Dordrecht, The Netherlands: Kluwer Academic Publishers.

Coutos-Thevenot, P., Poinssot, B., Boromelli, A., Yean, H., Breda, C., Buffard, D., Esnault, R., Hain, R., and Boulay, M. 2001. *In vitro* tolerance to *Botrytis cinerea* of grapevine 41B rootstock in transgenic plants expressing the stilbene synthase *Vst1* gene under the control of a pathogen-inducible PR10 promoter. *J. Expt. Bot.* 52:901–910.

Dai, Z., Hooker, B.S., Anderson, D.B., and Thomas, S.R. 2000. Expression of *Acidothermus cellulolyticus* endoglucanase E1 in transgenic tobacco: biochemical characteristics and physiological effects. *Transgenic Res.* 9:43–54.

Datta, K., Koukoliková-Nicola, Z., Baisakh, N., Oliva, N., and Datta, S.K. 2000. *Agrobacterium*-mediated engineering for sheath blight resistance of indica rice cultivars from different ecosystems. *Theor. Appl. Genet.* 100:832–839.

Datta, K., Tu, J., Oliva, N., Ona, I., Velazhahan, R., Mew, T.W., Muthukrishnan, S., and Datta, S.K. 2001. Enhanced resistance to sheath blight by constitutive expression of infection-related rice chitinase in transgenic elite indica rice cultivars. *Plant Sci.* 160:405–414.

Datta, K., Velazhahan, R., Oliva, N., Ona, I., Mew, T., Khush, G.S., Muthukrishnan, S., and Datta, S.K. 1999. Overexpression of cloned rice thaumatin-like protein (PR-5) in transgenic rice plants enhances environmental-friendly resistance to *Rhizoctonia solani* causing sheath blight disease. *Theor. Appl. Genet.* 98:1138–1145.

De Bolle, M.F.C., Osborn, R.W., Goderis, I.J., Noe, L., Acland, D., Hart, C.A., Torrekens, S., Van Leuven, F., and Broekaert, W.F. 1996. Antimicrobial peptides from *Mirabilis jalapa* and *Amaranthus caudatus*: expression processing, localization and biological activity in transgenic tobacco. *Plant Mol. Biol.* 31:993–1008.

de Gray, G., Rajasekaran, K., Smith, F., Sanford, J., and Daniell, H. 2001. Expression of an antimicrobial peptide via the chloroplast genome to control phytopathogenic bacteria and fungi. *Plant Physiol.* 127:852–862.

de Jaeger, G., De Wilde, C., Eeckhout, D., Fiers, E., and Depicker, A. 2000. The plantibody approach: expression of antibody genes in plants to modulate plant metabolism or to obtain pathogen resistance. *Plant Mol. Biol.* 43:419–428.

Desjardins, A.E., and Hohn, T.M. 1997. Mycotoxins in plant pathogenesis. *Mol. Plant Microbe Interact.* 10:147–152.

Delaney, T.P., Uknes, S., Vernooij, B., Friedrich, L., Weymann, K., Negrotto, D., Gaffney, T., Gut-Rella, M., Kessman, H., Ward, E., and Ryals, J. 1994. A central role of salicylic acid in plant disease resistance. *Science* 266:1247–1250.

Desiderio, A., Aracri, B., Leckie, F., Mattei, B., Salvi, G., Tigelaar, H., Van Roekel, J.S.C., Baulcombe, D.C., Melchers, L.S., Lorenzo, G., and Cervone, F. 1997. Polygalacturonase-inhibiting proteins (PGIPs with different specifities are expressed in *Phaseolus vulgaris*. *Mol. Plant Microbe Interact.* 10:852–860.

De Wit, P.J.G.M., and van Kan, J.A.L. 1993. Is durable resistance against fungi attainable through biotechnological procedures? In *Durability of Disease Resistance*, eds. Th. Jacobs, and J.E. Parlevliet, pp. 57–70. Dordrecht, The Netherlands: Kluwer Academic Publishers.

Does, M.P., Houterman, P.M., Dekker, H.L., and Cornelissen, B.J.C. 1999. Processing, targeting, and antifungal activity of stinging nettle agglutin in transgenic tobacco. *Plant Physiol.* 120:421–431.

Donaldson, P.A., Anderson, T., Lane, B.G., Davidson, A.L., and Simmonds, D.H. 2001. Soybean plants expressing an active oligomeric oxalate oxidase from wheat gf-2.8 (germin) gene are resistant to the oxalate-secreting pathogen *Sclerotinia sclerotiorum*. *Physiol. Mol. Plant Pathol.* 59:297–307.

Donofrio, N.M., and Delaney, T.P. 2001. Abnormal callose response phenotype and hypersusceptibility to *Peronospora parasitica* in defense-compromised *Arabidopsis nim 1-1* and salicylate hydroxylase-expressing plants. *Mol. Plant Microbe Interact.* 14:439–450.

Dumas, B., Freyssinet, G., and Pallett, K.E. 1995. Tissue-specific expression of germin-like oxalate oxidase during development and fungal infection of barley seedlings. *Plant Physiol.* 107:1091–1096.

East, I.J., Fitzgerald, C.J., Pearson, R.D., Donaldson, R.A., Vuocolo, T., Cadogen, L.C., Tellam, R.L., and Eisemann, C.H. 1993. *Lucilla cuprina*: inhibition of larval growth induced by immunization of host sheep with extracts of larval peritrophic membrane. *Int. J. Parasitol.* 23:221–229.

El Quakfaoui, S., Potvin, C., Brzezinksi, R., and Asselin, A. 1995. A *Streptomyces* chitosanase is active in transgenic tobacco. *Plant Cell Rep.* 15:222–226.

Epple, P., Apel, K., and Bohlmann, H. 1997. Overexpression of an endogenous thionin enhances resistance of *Arabidopsis* against *Fusarium oxysporum*. *Plant Cell* 9:509–520.

Fays, B., and Parker, J. 2000. Interplay of signaling pathways in plant disease resistance. *Trends Genet.* 16:449–455.

Fettig, S., and Hess, D. 1999. Expression of a chimeric stilbene synthase gene in transgenic wheat lines. *Transgenic Res.* 8:179–189.

Feuerstein, G., Powell, J.A., Knower, A.T., and Hunter, K.W., Jr. 1985. Monoclonal antibodies to T-2 toxin. *In vitro* neutralization of protein synthesis inhibition and protection of rats against lethal toxemia. *J. Clin. Invest.* 76:2134–2138.

Friedrich, L., Lawton, K., Dietrich, R., Willits, M., Cade, R., and Ryals, J. 2001. NIM1 Overexpression in *Arabidopsis* potentiates plant disease resistance and results in enhanced effectiveness of fungicides. *Mol. Plant Microbe Interact.* 14:1114–1124.

Gao, A.G., Hakimi, S.M., Mittanck, C.A., Wu, Y., Woerner, B.M., Stark, D.M., Shah, D.M., Liang, J., and Rommens, C.M.T. 2000. Fungal pathogen protection in potato by expression of a plant defensin peptide. *Nat. Biotechnol.* 18:1307–1310.

Genoud, T., and Métraux, J. 1999. Crosstalk in plant cell signaling: structure and function of the genetic network. *Trends Plant Sci.* 4:503–507.

Gilbert, J., and Tekauz, A. 2000. Review: recent developments in research on fusarium head blight in Canada. *Can. J. Plant Pathol.* 22:1–8.

Gilchrist, L., Rajaram, S., and Crossa, J. 2000. New sources of scab resistance and breeding progress at CIMMYT. In *Proceedings of the International Symposium. Wheat Improvement for Scab Resistance*, eds. J. Raupp, Z. Ma, Z., P. Chen, and D. Liu, pp. 194–199, May 5–11, 2000. Suzhoe and Nanjing, China.

Grison, R., Grezes-Besset, B., Scheider, M., Lucante, N., Olsen, L., Leguay, J.L., and Toppan, A. 1996. Field tolerance to fungal pathogens of *Brassica napus* constitutively expressing a chimeric chitinase gene. *Nat. Biotechnol.* 14:643–646.

Hain, R., Reif, H.-J., Krause, E., Langebartels, R., Kindl, H., Vornam, B., Wiese, W., Schmeltzer, E., Schreier, P.H., Stöker, R.H., and Stenzel, K. 1993. Disease resistance results from foreign phytoalexin expression in a novel plant. *Nature* 36:153–156.

Hammerschmidt, R. 1999. Phytoalexins: what have we learned after 60 years? *Annu. Rev. Phytopathol.* 37:285–306.

Harris, L.J., and Gleddie, S.C. 2001. A modified Rpl3 gene from rice confers tolerance of the *Fusarium graminearum* mycotoxin deoxynivalenol to transgenic tobacco. *Physiol. Mol. Plant Pathol.* 58:173–181.

He, X.Z., and Dixon, R.A. 2000. Genetic manipulation of isoflavone 7–O-methyltransferase enhances biosynthesis of 4'-O-methylated isoflavonoid phytoalexins and disease resistance in alfalfa. *Plant Cell* 12:1689–1702.

Heath, M.C. 2000. Hypersensitive response-related death. *Plant Mol. Biol.* 44:321–334.

Hennin, C., Höfte, M., and Diederichsen, E. 2001. Functional expression of *Cf9* and *Avr9* genes in *Brassica napus* induces enhanced resistance to *Leptosphaeria maculans*. *Mol. Plant Microbe Interact.* 14:1075–1085.

Hipskind, J.D., and Paiva, N.L. 2000. Constitutive accumulation of a resveratrol-glucoside in transgenic alfalfa increases resistance to *Phoma medicaginis*. *Mol. Plant Microbe Interact.* 13:551–562.

Hirt, H. 1997. Multiple roles of MAP kinases in plant signal transduction. *Trends Plant Sci.* 2:11–15.

Hittalmani, S., Parco, A., Mew, T.V., Zeigler, R.S., and Huang, N. 2000. Fine mapping and DNA marker-assisted pyramiding of three major genes for blast resistance in rice. *Theor. Appl. Genet.* 1007:1121–1128.

Holtorf, S., Ludwig-Müller, J., Apel, K., and Bohlmann, H. 1998. High-level expression of a viscotoxin in *Arabidopsis thaliana* gives enhanced resistance against *Plasmodiophora brassicae*. *Plant Mol. Biol.* 36:637–680.

Honée, G. 1999. Engineered resistance against fungal pathogens. *Eur. J. Plant Pathol.* 105:319–326.

Howie, W., Joe, L., Newbigin, E., Suslow, T., and Dunsmuir, P. 1994. Transgenic tobacco plants which express the *chiA* gene from *Serretia marcescens* have enhanced tolerance to *Rhizoctonia solani*. *Transgenic Res.* 3:90–98.

Huang, N., Angeles, E.R., Domingo, J., Magpantay, G., Singh, S., Zhang, G., Kumaravadivel, N., Bennett, J., and Khush, G.S. 1997. Pyramiding of bacterial blight resistance genes in rice: marker-assisted selection using RFLP and PCR. *Theor. Appl. Genet.* 95(3):313–320.

Hunter, K.W., Jr., Brimfield, A.A., Miller, M., Finkelman, F.D., and Chu, S.F. 1985. Preparation and characterization of monoclonal antibodies to the trichothecene mycotoxin T-2. *Appl. Environ. Microbiol.* 49:168–172.

Jach, G., Görnhardt, B., Mundy, J., Logemann, J., Pinsdorf, E., Leah, R., Schell, J., and Mass, C. 1995. Enhanced quantitative resistance against fungal disease by combinatorial expression of different barley antifungal proteins in transgenic tobacco. *Plant J.* 8:97–109.

Jach, G., Logemann, S., Wolf, G., Oppenheim, A., Chet, I., Schell, J., and Logemann, J. 1992. Expression of a bacterial chitinase leads to improved resistance of transgenic tobacco plants against fungal infection. *Biopractice* 1:33–40.

James, J.T., and Dubery, I.A. 2001. Inhibition of polygalacturonase from *Verticillium dahliae* by a polygalacturonase inhibiting protein from cotton. *Phytochemistry* 57:149–156.

Johnson, R. 1993. Durability of disease resistance in crops: some closing remarks about the topic and the symposium. In *Durability of Disease Resistance*, eds. Th. Jacobs, and J.E. Parlevliet, pp. 283–300. Dordrecht, The Netherlands: Kluwer Academic Publishers.

Johnson, R. 1983. Genetic background of durable resistance. In *Durable Resistance in Crops,* eds. F. Lamberti, J.M. Aller, and N.A.van der Graaff, pp. 5–26. New York: Plenum Press.

Jongedijk, E., Tigelaar, H., Van Roekel, J.S.C., Bres-Vloemans, S.A., Dekker, I., Van den Elzen, P.J.M., Cornelissen, B.J.C., and Melchers, L.S. 1995. Synergistic activity of chitinases and β-1,3-glucanases enhances fungal resistance in transgenic tomato plants. *Euphytica* 85:173–180.

Kanrar, S., Venkateswari, J.C., Kirti, P.B., and Chopra, V.L. 2002. Transgenic expression of hevein, the rubber tree lectin, in Indian mustard confers protection against *Alternaria brassicae. Plant Sci.* 162:441–448.

Keller, H., Pamboukdjian, N., Ponchet, M., Poupet, A., Delon, R., Verrier, J.L., Roby, D., and Ricci, P. 1999. Pathogen-induced elicitin production in transgenic tobacco generates a hypersensitive response and nonspecific disease resistance. *Plant Cell* 11:223–235.

Kellmann, J.-W., Kleinow, T., Engelhardt, K., Philipp, C., Wegener, D., Schell, J., and Schreier, P.H. 1996. Characterization of two class II chitinase genes from peanut, and expression studies in transgenic tobacco plants.*Plant Mol. Biol.* 30:351–358.

Kesarwani, M., Azam, M., Natarajan, K., Mehta, A., and Datta, A. 2000. Oxalate decarboxilase from *Collybia velutipes.* Molecular cloning and its overexpression to confer resistance to fungal infection in transgenic tobacco and tomato. *J. Biol. Chem.* 275:7230–7238.

Kikkert, J.R., Ali, G.S., Wallace, P.G., Reisch, B., and Reustle, G.M. 2000. Expression of a fungal chitinase in *Vitis vinifera* L., 'Merlot' and 'Chardonnay' plants produced by biolistic transformation. *Acta Horticult.* 528:297–303.

Kim, J.K., Duan, X., Wu, R., Seok, S.J., Boston, R.S., Jang, I.C., Eun, M.Y., and Nahm, B.H. 1999. Molecular, and genetic analysis of transgenic rice plants expressing the maize ribosome-inactivating protein b-32 gene and the herbicide resistance bar gene. *Mol. Breed.* 5:85–94.

Knoester, M., Van Loon, L.C., Van den Heuvel, J., Hennig, J., Bol, J.F., and Linthorst, H.J.M. 1998. Ethylene-insensitive tobacco lacks nonhost resistance against soil-borne fungi. *Proc. Natl. Acad. Sci. USA* 95:1933–1937.

Krishnamurthy, K. Balconi, C., Sherwood, J.E., and Giroux, M.J. 2001. Wheat puroindolines enhance fungal disease resistance in transgenic rice. *Mol. Plant Microbe Interact.* 14:1255–1260.

Krishnaveni, S., Jeoung, J.M., Muthukrishnan, S., and Liang, G.H. 2001. Transgenic sorghum plants constitutively expressing a rice chitinase gene show improved resistance to stalk rot. *J. Genet. Breed.* 55:151–158.

Lagrimini, L.M., Vaughn, J., Erb, W.A., and Miller, S.A. 1993. Peroxidase overproduction in tomato: wound-induced polyphenol deposition and disease resistance. *Hortcult. Sci.* 28:218–221.

Leah, R., Tommerup, H., Svendsen, I., and Mundy, J. 1991. Biochemical and molecular characterization of three barley seed proteins with antifungal properties. *J. Biol. Chem.* 266:1564–1573.

Leckband, G., and Lörz, H. 1998. Transformation and expression of a stilbene synthase gene of *Vitis vinifera* L. in barley and wheat for increased fungal resistance. *Theor. Appl. Genet.* 96:1004–1012.

Lee, M.-W., Qi, M., and Yang, Y. 2001. A novel jasmonic acid-inducible rice *myb* gene associates with fungal infection and host cell death. *Mol. Plant Microbe Interact.* 14:527–535.

Lee, Y.H., Yoon, I.S., Suh, S.C., and Kim, H.I. 2002. Enhanced disease resistance in transgenic cabbage and tobacco expressing a glucose oxidase gene from *Aspergillus niger*. *Plant Cell Rep.* 20:857–863.

Le Gall, F., Bové, J.M., and Garnier, M. 1998. Engineering of a single-chain variable-fragment scFv) antibody specific for the stolbur phytoplasma (mollicute) and its expression in *Escherichia coli* and tobacco plants. *Appl. Environ. Microbiol.* 64:4566–4572.

Lehane, M.J. 1996. Digestion and fate of the vertebrate blood-meal in insects. In *The Immunology of Host-Ectoparasitic Arthropod Relationships*, ed. S.K. Wikel, pp. 131–149. Wallingford, United Kingdom: CAB International.

Li, Q., Lawrence, C.B., Xing, H.-Y., Babbitt, R.A., Bass, W.T., Maiti, I.B., and Everett, N.P. 2001. Enhanced disease resistance conferred by expression of an antimicrobial magainin analog in transgenic tobacco. *Planta* 212:635–639.

Liang, H., Maynard, C.A., Allen, R.D., and Powell, W.A. 2001. Increased *Septoria musiva* resistance in transgenic hybrid poplar leaves expressing a wheat oxalate oxidase gene. *Plant Mol. Biol.* 45:619–629.

Lin, W., Anuratha, C.S., Datta, K., Potrykus, I., Muthukrishnan, S., and Datta, S.K. 1995. Genetic engineering of rice for resistance to sheath blight. *Bio/Technology* 13:686–691.

Lincoln, S., Daley, M., and Lander, E. 1992. Mapping genes controlling quantitative traits with MAPMAKER/QTL 1.1. Whitehead Institute Technical Report. Second Edition. Cambridge, MA: Whitehead Technical Institute.

Liu, D., Raghothama, K.G., Hasegawa, P.M., and Bressan, R. 1994. Osmotin overexpression in potato delays development of disease symptoms. *Proc. Natl. Acad. Sci. USA* 91:1888–1892.

Logemann, J., Jach, G., Tommerup, H., Mundy, J., and Schell, J. 1992. Expression of a barley ribosome-inactivating protein leads to increased fungal protection in transgenic tobacco plants. *Bio/Technology* 10:305–308.

Lorito, M., and Scala, F. 1999. Microbial genes expression in transgenic plants to improve disease resistance. *J. Plant Pathol.* 81:73–88.

Lorito, M., Woo, S.L., D'Ambrosio, M., Harman, G.E., Hayes, C.K., Kubicek, C.P., and Scala, F. 1996. Synergistic interaction between cell wall degrading enzymes and membrane affecting compounds. *Mol. Plant Microbe Interact.* 9:206–213.

Lusso, M., and Kuć, J. 1996. The effect of sense and antisense expression of the *PR-N* gene for β-1,3-glucanase on disease resistance of tobacco to fungi and viruses. *Physiol. Mol. Plant Pathol.* 49:267–283.

Maddaloni, M., Forlani, F., Balmas, V., Donini, G., Stasse, L., Corazza, L., and Motto, M. 1997. Tolerance to the fungal pathogen *Rhizoctonia solani* AG4 of transgenic tobacco expressing the maize ribosome-inactivating protein b-32. *Transgenic Res.* 6:393–402.

Marchant, R., Davey, M.R., Lucas, J.A., Lamb, C.J., Dixon, R.A., and Power, J.B. 1998. Expression of a chitinase transgene in rose (*Rosa hybrida* L.) reduces development of blackspot disease (*Diplocarpon rosae* Wolf). *Mol. Breed.* 4:187–194.

Martin, G.B., Brommonschenkel, S., Chunwongse, J., Frary, A., Ganal, M.W., Spivey, R., Wu, T., Earle, E.G., and Tanksley, S.D. 1993. Map-based cloning of a protein kinase gene conferring disease resistance in tomato. *Science* 262:1432–1436.

Masoud, S.A., Zhu, Q., Lamb, C., and Dixon, R.A. 1996. Constitutive expression of an inducible β-1,3-glucanase in alfalfa reduces disease severity caused by the oomycete pathogen *Phytophthora megasperma* f. sp. *medicaginis*, but does not reduce severity of chitin-containing fungi. *Transgenic Res.* 5:313–323.

Mauch, F., and Staehelin, L.A. 1989. Functional implications of the subcellular localization of ethylene-induced chitinase and β-1,3-glucanase in bean leaves. *Plant Cell* 1:447–457.

Melchers, L.S., and Stuiver, M.H. 2000. Novel genes for disease-resistance breeding. *Curr. Opin. Plant Biol.* 3:147–152.

Mesterhazy, A. 1983. Breeding wheat for resistance to *Fusarium graminearum* and *Fusarium culmorum*. *Z. Pflanzenzuchtg.* 91:295–311.

Mesterhazy, A. 1995. Types and components of resistance to Fusarium head blight. *Plant Breed.* 114:377–386.

Mitsuhara, I., Matsufuru, H., Ohshima, M., Kaku, H., Nakajima, Y., Murai, N., Natori, S., and Ohashi, Y. 2000. Induced expressed of sarcotxin IA enhanced host resistance against both bacterial and fungal pathogens in transgenic tobacco. *Mol. Plant Microbe Interact.* 13:860–868.

Mittler, R., and Rizhsky, L. 2000. Transgene-induced lesion mimic. *Plant Mol. Biol.* 44:335–344.

Mora, A.A., and Earle, E.D. 2001. Resistance to *Alternaria brassicicola* in transgenic broccoli expressing a *Trichoderma harzianum* endochitinase gene. *Mol. Breed.* 8:1–9.

Muhitch, M.J., McCormick, S.P., Alexander, N.J., and Hohn, T.M. 2000. Transgenic expression of the *TR1 101* or *PDR 5* gene increases resistance of tobacco to the phytotoxic effects of the trichothecene 4,15-diacetoxyscirpenol. *Plant Sci.* 157:201–207.

Murray, F., Llewellyn, D., McFadden, H., Last, D., Dennis, E.S., and Peacock, W.J. 1999. Expression of the *Talaromyces flavus* glucose oxidase gene in cotton and tobacco reduces fungal infection, but is also phytotoxic. *Mol. Breed.* 5:219–232.

Nakajima, H., Muranaka, T., Ishige, F., Akatsu, K., and Oeda, K. 1997. Fungal and bacterial disease resistance in transgenic plants expressing human lysozyme. *Plant Cell Rep.* 16:674–679.

Neuhaus, J.M., Ahl-Goy, P., Hinz, U., Flores, S., and Meins, F., Jr. 1991. High-level expression of a tobacco chitinase gene in *Nicotiana sylvestris*. Susceptibility of transgenic plants to *Cercospora nicotianae* infection. *Plant Mol. Biol.* 16:141–151.

Nielsen, K.K., Mikkelsen, J.D., Kragh, K.M., and Bojsen, K. 1993. An acidic class III chitinase in sugar beet: induction by *Cercospora beticola*, characterization, and expression in transgenic tobacco plants. *Mol. Plant Microbe Interact.* 6:495–506.

Nishizawa, Y., Nishio, Z., Nakazono, K., Soma, M., Nakajima, E., Ugaki, M., and Hibi, T. 1999. Enhanced resistance to blast (*Magnaporthe grisea* in transgenic Japonica rice by constitutive expression of rice chitinase. *Theor. Appl. Genet.* 99:383–390.

Nuutila, A.M., Ritala, A., Skadsen, R.W., Mannonen, L., and Kauppinen, V. 1999. Expression of fungal thermotolerant endo-1,4-β-glucanase in transgenic barley seeds during germination. *Plant Mol. Biol.* 41:777–783.

Ohme-Takagi, M., and Shinshi, H. 1995. Ethylene-inducible DNA binding proteins that interact with an ethylene-responsive element. *Plant Cell* 7:173–182.

Ono, E., Wong, H.L., Kawasaki, T., Hasegawa, M., Kodama, O., and Shimamoto, K. 2001. Essential role of the small GTPase rac in disease resistance of rice. *Proc. Natl. Acad. Sci. USA* 98:759–764.

Osusky, M., Zhou, G., Osuska, L., Hancock, R.E., Kay, W.W., and Misra, S. 2000. Transgenic plants expressing cationic peptide chimeras exhibit broad-spectrum resistance to phytopathogens. *Nat. Biotechnol.* 18:1162–1166.

Parashina, E.V., Serdobinksii, L.A., Kalle, E.G., Lavrova, N.V., Avetisov, V.A., Lunin, V.G., and Naroditskii, N.V. 2000. Genetic engineering of oilseed rape and tomato plants expressing a radish defensin gene. *Russ. J. Plant Physiol.* 47:417–423.

Park, C.-M., Berry, J.O., and Bruenn, J.A. 1996. High-level secretion of a virally encoded anti-fungal toxin in transgenic tobacco plants. *Plant Mol. Biol.* 30:359–366.

Park, J.M., Park, C.J., Lee, S.B., Ham, B.K., Shin, R., and Paek, K.H. 2001a. Overexpression of the tobacco Tsi1 gene encoding an EREBP/AP2-type transcription factor enhances resistance against pathogen attack and osmotic stress in tobacco. *Plant Cell* 13:1035–1046.

Park, S.W., J.C. Linden, Vivanceo, J.M. 2001b. Characterization of a novel ethylene-inducible ribosome-inactivating protein exuded from root cultures of *Phytolacca americana*. http://abstracts.aspb.org/aspp2001/public/P28/0063.html.

Paxton, J.D. 1981. Phytoalexins—a working redefinition. *Phytopathology Z*. 101:106–109.

Pestsova, E., Ganal, M.W., and Röder, M.S. 2000. Isolation and mapping of microsatellite markers specific for the D genome of bread wheat. *Genome* 43:689–697.

Pieterse, C.M.J., and Van Loon, L.C. 1999. Salicylic acid-independent plant defence pathways. *Trends Plant Sci*. 4:52–58.

Powell, A.L.T., van Kan, J., ten Have, A., Visser, J., Greve, L.C., Bennett, A.B., and Labavitch, J.M. 2000. Transgenic expression of pear PGIP in tomato limits fungal colonization. *Mol. Plant Microbe Interact*. 13:942–950.

Punja, Z.K. 2001. Genetic engineering of plants to enhance resistance to fungal pathogens—a review of progress and future prospects. *Can. J. Plant Pathol*. 23:216–235.

Punja, Z.K., and Raharjo, S.H.T. 1996. Response of transgenic cucumber and carrot plants expressing different chitinase enzymes to inoculation with fungal pathogens. *Plant Dis*. 80:999–1005.

Rajasekaran, K., Cary, J.W., Jacks, T.J., Stromberg, K.D., and Cleveland, T.E. 2000. Inhibition of fungal growth in planta and *in vitro* by transgenic tobacco expressing a bacterial nonheme chloroperoxidase gene. *Plant Cell Rep*. 19:333–338.

Ray, H., Douches, D.S., and Hammerschmidt, R. 1998. Transformation of potato with cucumber peroxidase: expression and disease response. *Physiol. Mol. Plant Pathol*. 53:93–103.

Reymond, P., and Farmer, E.E. 1998. Jasmonate and salicylate as global signals for defense gene expression. *Curr. Opin. Plant Biol*. 1:404–411.

Rizhsky, L., and Mittler, R. 2001. Inducible expression of bacterio-opsin in transgenic tobacco and tomato plants. *Plant Mol. Biol*. 46:313–323.

Robinson, M.M., Shah, S., Tamot, B., Pauls, K.P., Moffatt, B.A., and Glick, B.R. 2001. Reduced symptoms of Verticillium wilt in transgenic tomato expressing a bacterial Acc deaminase. *Mol. Plant Pathol*. 2:135–145.

Röder, M.S., Korzun, V., Wendehake, K., Plaschke, J., Tixier, M.H., Leroy, P., and Ganal, M.W. 1998. A microsatellite map of wheat. *Genetics* 149:2007–2023.

Romeis, T., Piedras, P., and Jones, J.D.G. 2000. Resistance gene-dependent activation of a calcium-dependent protein kinase in the plant defense response. *Plant Cell* 12:803–815.

Rosso, M.N., Schouten, A., Roosien, J., Borst-Vrenssen, T., Hussey, R.S., Gommers, F.J., Bakker, J., Schots, A., and Abad, P. 1996. Expression and functional characterization of a single chain Fv antibody directed against secretions involved in plant nematode infection process. *Biochem. Biophys. Res. Commun*. 220:255–263.

Saal, B., Plieske, J., Hu, J., Quiros, C.F., and Struss, D. 2001. Microsatellite markers for genome analysis in Brassica. II. Assignment of rapeseed microsatellites to the A and C genomes and genetic mapping in *Brassica oleracea* L. *Theor. Appl. Genet*. 102(5):695–699.

Saijo, Y. Hata, S., Kyozuka, J., Shimamoto, K., and Izui, K. 2000. Over-expression of a single Ca^{2+}-dependent protein kinase confers both cold and salt/drought toleranceon rice plants. *Plant J*. 23:319–328.

Schaffrath, U., Mauch, F., Freydl, E., Schweizer, P., and Dudler, R. 2000. Constitutive expression of the defense-related *Rir1b* gene in transgenic rice plants confers enhanced resistance to the rice blast fungus *Magnaporthe grisea*. *Plant Mol. Biol.* 43:59–66.

Schenk, P.M., Kazan, K., Wilson, I., Anderson, J.P., Richmond, T., and Somerville, S.C. 2000. Coordinated plant defense responses in *Arabidopsis* revealed by microarray analysis. *Proc. Natl. Acad. Sci. USA* 97:11655–11660.

Schillberg, S., Zimmerman, S., Zhang, M.Y., and Fischer, R. 2001. Antibody-based resistance to plant pathogens. *Transgenic Res.* 10:1–12.

Schweizer, P., Christoffel, A., and Dudler, R. 1999. Transient expression of members of the germin-like gene family in epidermal cells of wheat confers disease resistance. *Plant J.* 20:541–552.

Scolfield, S.R., Tobias, C.M., Rathjen, J.P., Chang, J.H., Lavelle, D.T., Michelmore, R.W., and Staskawicz, B.J. 1996. Molecular basis of gene-for-gene specificity in bacterial speck disease of tomato. *Science* 274:2063–2065.

Seo, H.S., Song, J.T., Cheong, J.J., Lee, Y.H., Lee, Y.W., Hwang, I., Lee, J.S., and Choi, Y.D. 2001. Jasmonic acid carboxyl methyltranferase: A key enzyme for jasmonate-regulated plant responses. *Proc. Natl. Acad. Sci. USA* 98:4788–4793.

Shen, L., Courtois, B., McNally, K.L., Robin, S., and Li, Z. 2001. Evaluation of near-isogenic lines of rice introgressed with QTLs for root depth through marker-aided selection. *Theor. Appl. Genet.* 103(1):75–83.

Shen, S., Li, Q., He, S.Y., Barker, K.R., Li, D., and Hunt, A.G. 2000. Conversion of compatible plant-pathogen interactions into incompatible interaction by expression of the seudomonas syringae pv. syringae 61 hrmA gene in transgenic tobacco plants. *Plant J.* 23:205–213.

Shi, J., Thomas, C.J., King, L.A., Hawes, C.R., Posee, R.D., Edwards, M.L., Pallett, D., and Cooper, J.I. 2000. The expression of a baculovirus-derived chitinase gene increased resistance of tobacco cultivars to brown spot (*Alternaria alternata*). *Ann. Appl. Biol.* 136:1–8.

Simon, M.I., Strathmann, M.P., and Gautam, N. 1994. Diversity of G proteins in signal transduction. *Science* 252:802–808.

Stark-Lorenzen, P., Nelke, B., Hänbler, G., Mühlbach, H.P., and Thomzik, J.E. 1997. Transfer of a grapevine stillbene synthase gene to rice (*Oryza sativa* L.). *Plant Cell Rep.* 16:668–673.

Stirpe, F., Barbieri, L., Battelli, L.G., Soria, M., and Lappi, D.A. 1992. Ribosome-inactivating proteins from plants: present status and future prospects. *Bio/Technology* 10:405–412.

Strittmatter, G., Janssens, J., Opsomer, C., and Botterman, J. 1995. Inhibition of fungal disease development in plants by engineering controlled cell death. *Bio/Technology* 13:1085–1089.

Tabaeizadeh, Z., Agharbaoui, Z., and Harrak, H. 1999. Transgenic tomato plants expressing a *Lycopersicon chilense* chitinase gene demonstrate improved resistance to *Verticillium dahliae* race 2. *Plant Cell Rep.* 19:197–202.

Tabei, Y., Kitade, S., Nishizawa, Y., Kikuchi, N., Kayano, T., Hibi, T., and Akutsu, K. 1998. Transgenic cucumber plants harboring a rice chitinase gene exhibit enhanced resistance to gray mold (*Botrytis cinerea*). *Plant Cell Rep.* 17:159–164.

Takaichi, M., and Oeda, K. 2000. Transgenic carrots with enhanced resistance against two major pathogens, *Erysiphe heraclei*, and *Alternaria dauci*. *Plant Sci.* 153:135–144.

Takatsu, Y., Nishizawa, Y., Hibi, T., and Akutsu, K. 1999. Transgenic chrysanthemum (*Dendranthema grandiflorum* (Ramat.) Kitamura) expressing a rice chitinase gene shows enhanced resistance to gray mold *Botrytis cinerea*). *Sci. Horticult.* 82:113–123.

Tang X., Frederick, R.D., Zhou, J., Halterman, D.A., Jia, Y., and Martin, G.B. 1996. Initiation of plant disease resistance by physical interaction of AvrPto and the Pto kinase. *Science* 274:2060–2063.

Tang, X., Xie, M., Kim, Y.J., Zhou, J., Klessig, D.F., and Martin, G.B. 1999. Overexpression of Pto activates defense responses and confers broad resistance. *Plant Cell* 11:15–29.

Tavladoraki, P., Benvenuto, E., Trinca, S., De Martinis, D., Cattaneo, A., and Galeffi, P. 1993. Transgenic plants expressing a functional single-chain Fv antibody are specifically protected from virus attack. *Nature* 366:469–472.

Tepfer, D., Boutteaux, C., Vigon, C., Aymes, S., Perez, V., O'Donohue, M.J., Huet, J.-C., and Pernollet, J.-C. 1998. *Phytophthora* resistance through production of a fungal protein elicitor (β-cryptogein) in tobacco. *Mol. Plant Microbe Interact.* 11:64–67.

Terakawa, T., Takaya, N., Horiuchi, H., Koike, M., and Takagi, M. 1997. A fungal chitinase gene from *Rhizopus oligosporus* confers antifungal activity to transgenic tobacco. *Plant Cell Rep.* 16:439–443.

Terras, F.R.G., Eggermont, K., Kovaleva, V., Raikhel, N.V., Osborn, R.W., Kester, A., Rees, S.B., Torrekens, S., van Leuven, F., and Vanderleyden, J. 1995. Small cysteine-rich antifungal proteins from radish: their role in host defense. *Plant Cell.* 7: 573–588.

Thompson, C., Dunwell, J.M., Johnstone, C.E., Lay, V., Ray, J., Schmitt, M., Watson, H., and Nisbet, G. 1995. Degradation of oxalic acid by transgenic oilseed rape plants expressing oxalate oxidase. *Euphytica* 85:169–172.

Thomzik, J.E., Stenzel, K., Stöcher, R., Schreier, P.H., Hain, R., and Stahl, D.J. 1997. Synthesis of a grapevine phytoalexin in transgenic tomatoes (*Lycopersicon esculentum* Mill.) conditions resistance against *Phytophthora infestans*. *Physiol. Mol. Plant Pathol.* 51:265–278.

Van den Elzen, P.J.M., Jongedijk, E., Melchers, L.S., and Cornelissen, B.J.C. 1993. Virus and fungal resistance: from laboratory to field. *Philos. Trans. R. Soc. Lond. B Biol. Sci.* 342:271–278.

Van Loon, L.C., and Van Strien, E.A. 1999. The families of pathogenesis-related proteins, their activities, and comparative analysis of PR-1 type proteins. *Physiol. Mol. Plant Pathol.* 55:85–97.

Verberne, M.C., Verpoorte, R., Bol, J.F., Mercado-Blanco, J., and Linthorst, H.J.M. 2000. Overproduction of salicylic acid in plants by bacterial transgenes enhances pathogen resistance. *Nat. Biotechnol.* 18:779–783.

Vierheilig, H., Alt, M., Neuhaus, J.M., Boller, T., and Wiemken, A. 1993. Colonization of transgenic *Nicotiana sylvestris* plants, expressing different forms of *Nicotiana tabacum* chitinase, by the root pathogen *Rhizoctonia solani* and by the mycorrhizal symbiont *Glomus mosseae*. *Mol. Plant Microbe Interact.* 6:261–264.

Waldron, B.L., Moreno-Sevilla, B., Anderson, J.A., Stack, R.W., and Frohberg, R.C. 1999. RFLP mapping of QTL for Fusarium head blight resistance in wheat. *Crop Sci.* 39:805–811.

Wang, Y., Nowak, G., Culley, D., Hadwiger, L.A., and Fristensky, B. 1999. Constitutive expression of a pea defense gene DRR206 confers resistance to blackleg (*Leptosphaeria maculans*) disease in transgenic canola (*Brassica napus*). *Mol. Plant Microbe Interact.* 12:410–418.

Wang, P., Zoubenko, O., and Tumer, N.E. 1998. Reduced toxicity and broad spectrum resistance to viral and fungal infection in transgenic plants expressing pokeweed antiviral protein H. *Plant Mol. Biol.* 38:957–964.

Wattad, C., Kobiler, D., Dinoor, A., and Prusky, D. 1997. Pectate lyase of *Colletotrichum gloeosporioides* attacking avocado fruits—cDNA cloning and involvement in pathogenicity. *Physiol. Plant Pathol.* 50:197–212.

Weigel, R.R., Bäuscher, C., Pfitzner, A.J.P., and Pfitzner, U.M. 2001. NIMIN-1, NIMIN-2 and NIMIN-3, Members of a novel family of proteins from *Arabidopis* that interact with NPR1/NIM1, a key regulator of Systemic Acquired Resistance in plants. *Plant Mol. Biol.* 46:143–160.

Wong, K.W., Harman, G.E., Norelli, J.L., Gustafson, H.L., and Aldwinckle, H.S. 1999. Ghitinase-transgenic lines of 'Royal Gala' apple showing enhanced resistance to apple scab. *Acta Horticult.* 484:595–599.

Wu, G., Shortt, B.J., Lawrence, E.B., Léon, J., Fitzsimmons, K.C., Levine, E.B., Raskin, I., and Shah, D.M. 1997. Activation of host defense mechanisms by elevated production of H_2O_2 in transgenic plants. *Physiol. Mol. Plant Pathol.* 55:85–97.

Wu, G., Shortt, B.J., Lawrence, E.B., Levine, E.B., Fitzsimmons, K.C., and Shah, D.M. 1995. Disease resistance conferred by expression of a gene encoding H_2O_2-generating glucose oxidase in transgenic potato plants. *Plant Cell* 7:1357–1368.

Xing, T., and Jordan, M. 2000. Genetic engineering of signal transduction mechanisms. *Plant Mol. Biol. Rep.* 18:309–318.

Xing, T., Higgins, V.J., and Blumwald, E. 1997. Race-specific elicitors of *Cladosporium fulvum* promote translocation of cytosolic components of NADPH oxidase to the plasma membrane of tomato cells. *Plant Cell* 9:249–259.

Xing, T., Rampitsch, C., Miki, B.L., Mauthe, W., Stebbing, J., Malik, K., and Jordan, M. 2003. MALDI-Qq-TOF-MS and transient gene expression analysis indicated co-enhancement of β-1,3-glucanase and endochitinase by tMEK2 and the involvement of divergent pathways. *Physiol. Mol. Plant Pathol.* In Press.

Xing, T., Malik, K., Martin, T., and Miki, B.L. 2001a. Activiation of tomato PR and wound-related genes by a mutagenized tomato MAP kinase kinase through divergent pathways. *Plant Mol. Biol.* 46:109–120.

Xing,T., Wang, X-J., Malik, K., and Miki, B.L. 2001b. Ectopic expression of PR and *Arabidopsis* CDPK enhanced NADPH oxidase activity and oxidative burst in tomato protoplasts. *Mol. Plant Microbe Interact.* 14:1261–1264.

Xing, T., Ouellet, T., and Miki, B.L. 2002. Towards genomics and proteomics studies of protein phosphorylation in plant-pathogen interactions. *Trends Plant Sci.* 7:224–230.

Yamamoto, T., Iketani, H., Ieki, H., Nishizawa, Y., Notsuka, K., Hibi, T., Hayashi, T., Matsuka, N., and Matsuka, M. 2000. Transgenic grapevine plants expressing a rice chitinase with enhanced resistance to fungal pathogens. *Plant Cell Rep.* 19:639–646.

Yang, K.Y., Liu, Y., and Zhang, S. 2001. Activation of a mitogen-activated protein kinase pathway is involved in disease resistance in tobacco. *Proc. Natl. Acad. Sci.USA* 98:741–746.

Yang, Y., Shah, J., and Klessig, D.F. 1997. Signal perception and transduction in plant defense responses. *Genes Dev.* 11:1621–1639.

Yoshikawa, M., Tsuda, M., and Takeuchi, Y. 1993. Resistance to fungal diseases in transgenic tobacco plants expressing the phytoalexin elicitor-releasing factor, β-1,3-endoglucanase from soybean. *Naturwissenschaften* 80:417–420.

Young, N.D. 1999. A cautiously optimisitc vision for marker-assisted selection. *Mol. Breed.* 5(6):505–510.

Young, R.A., and Kelley, J.D. 1997. RAPD markers linked to three major anthracnose resistance genes in common bean. *Crop Sci.* 37(3): 940–946.

Yuan, H., Ming, X., Wang, L., Hu, P., An, C., and Chen, Z. 2002. Expression of a gene encoding trichosanthin in transgenic rice plants enhances resistance to fungus blast disease. *Plant Cell Rep.* 20:992–998.

Yuan, Q., Clarke, J.R., Zhou, H.R., Linz, J.E., Pestka, J.J., and Hart, L.P. 1997. Molecular cloning, expression, and characterization of a functional single-chain Fv antibody to the mycotoxin zearalenone. *Appl. Environ. Microbiol.* 63:263–269.

Yu, D., Chen, C., and Chen, Z. 2001. Evidence for an important role of WRKY DNA binding proteins in the regulation of NPR1 gene expression. *Plant Cell* 13:1527–1539. http://www.plantcell.org

Yu, D., Liu, Y., Fan, B., Klessig, D.F., and Chen, Z. 1997. Is the high basal level of salicylic acid important for disease resistance in potato? *Plant Physiol.* 115:343–349.

Yu, D., Xie, Z., Chen, C., Fan, B., and Chen, Z. 1999. Expression of tobacco class II catalase gene activates the endogenous homologous gene and is associated with disease resistance in transgenic potato plants. *Plant Mol. Biol.* 39:477–488.

Zhang, Z., Collinge, D.B., and Thordal-Christensen, H. 1995. Germin-like oxalate oxidase, a H_2O_2-producing enzyme, accumulates in barley attacked by the powdery mildew fungus. *Plant J.* 8:139–145.

Zhou, J.M., Trifa, Y., Silva, H., Pontier, D., Lam, E., Shah, J., and Klessig, D.F. 2000. NPR1 differentially interacts with members of the TGA/OBF family of transcription factors that bind an element of the PR-1 gene required for induction by salicylic acid. *Mol. Plant-Microbe Interact.* 13:191–202.

Zhu, B., Chen, T.H.H., and Li, P.H. 1995. Activation of two osmotin-like protein genes by abiotic stimuli and fungal pathogen in transgenic potato plants. *Plant Physiol.* 108:929–937.

Zhu, B., Chen, T.H.H., and Li, P.H. 1996. Analysis of late-blight disease resistance and freezing tolerance in transgenic potato plants expressing sense and antisense genes for an osmotin-like protein. *Planta* 198:70–77.

Zhu, Q., Maher, E.A., Masoud, S., Dixon, R.A., and Lamb, C.J. 1994. Enhanced protection against fungal attack by constitutive co-expression of chitinase and glucanase genes in transgenic tobacco. *Bio/Technology* 12:807–812.

Zoubenko, O., Ucken, F., Hur, Y., Chet, I., and Turner, N. 1997. Plant resistance to fungal infection induced by non-toxic pokeweed antiviral protein mutants. *Nat. Biotechnol.* 15:992–996.

Zoubenko, O., Hudak, K., and Tumer, N.E. 2000. A non-toxic Pokeweed antiviral protein mutant inhibits pathogen infection via a novel salicylic acid-independent pathway. *Pl. Mol. Biol.* 44:219–229.

Zook, M., Hohn, T., Bonnen, A., Tsuji, J., and Hammerschmidt, R. 1996. Characterization of novel sesquiterpenoid biosynthesis in tobacco expressing a fungal sesquiterpene synthase. *Plant Physiol.* 112:311–318.

19

Plantibody-Based Disease Resistance in Plants

Sabine Zimmermann, Neil Emans, Rainer Fischer, and Stefan Schillberg

19.1 Introduction

Crop diseases represent a significant threat to our food supply and are therefore of great economic importance. These diseases have been with us for as long as plants have been cultivated. However, intensified land use, monoculture, and modern crop cultivation methods have favored disease outbreaks on a large scale. Such outbreaks can have devastating economical and sociopolitical effects, and may even result in famine (Agrios, 1997). It is estimated that 10–15% of worldwide crop production is lost to pathogens and the effects of pathogens (Rangaswami, 1983).

The traditional approach to pathogen control is a combination of techniques, such as vector management, crop rotation, the production of pathogen-free seeds, and chemical control measures. Until recently, the agricultural industry has relied on plant breeding to provide resistant varieties and on the use of chemicals to protect susceptible plants. While plant breeding has been successful in some cases—indeed most crop varieties used today incorporate some form of genetic resistance (Crute and Pink, 1996)—it is a time-consuming process and the resistance provided may not be durable. Meanwhile, the continued use of agrochemicals can have undesirable environmental consequences.

An alternative route to pathogen control is the use of molecular biotechnology and plant transformation to produce resistant crop varieties. The advantages of this strategy are speed (resistant plant lines can be produced in a matter of months), scope (genes can be imported from any species, so sexual compatibility is no longer a limitation), and adaptability (if resistance is broken by the pathogen, new resistant plant lines can be developed quickly). Resistance has been engineered against a number of different pathogens, including viruses, bacteria, fungi, and invertebrates. Various strategies have been used including the transfer of natural plant resistance genes to susceptible crops, the expression of pathogen genes to provide pathogen-derived resistance, and the expression of heterologous resistance proteins from bacteria, fungi, and animals. Included within this last category are recombinant mammalian antibodies and their derivatives, which can be expressed in plants to provide plantibody-based resistance (Schillberg et al., 2001; Noelke et al., 2004).

In this chapter, we discuss molecular approaches to disease resistance but focus on the production of disease-resistant plant lines using recombinant antibodies. All organisms, from the simplest bacteria to the most complex multicellular species, possess some form of defense mechanism against pathogen activity. It is acknowledged that vertebrates possess the most sophisticated defense system, involving the production of antibodies recognizing pathogen-specific antigens. Among the vertebrates, mammals have the most advanced ability to make antibodies specific for a particular antigen. However, even this ability is now superseded by *in vitro* systems based on phage display (Winter and Milstein, 1991; Winter et al., 1994), which can yield antibody libraries of incredible complexity (over 10^{11} sequences) (Griffiths and Duncan, 1998; Sidhu, 2000). These technological advances mean that it is relatively straightforward to produce and isolate antibodies that bind with high specificity to any particular plant pathogen and to express these antibodies in plants (Fischer et al., 1999a).

19.2 Molecular Approaches to Pathogen Control

The development of disease-resistant plant varieties requires some knowledge of the normal way in which pathogens interact with, infect and replicate within plants. Once these principles are understood, it is possible to devise strategies to interfere with these processes in genetically modified plants. Three major concepts have evolved from such studies:

- The expression of DNA, RNA, or protein from the pathogen or a close relative can interfere with normal pathogen replication (pathogen-derived resistance).
- Natural defense response genes of plants can be modified, enhanced, and moved between species.
- Heterologous proteins, i.e., proteins derived from neither plants nor their pathogens, can be expressed that interfere with various stages of the pathogen life cycle.

19.2.1 Pathogen-Derived Resistance

The concept of pathogen-derived resistance arose from a phenomenon called cross-protection, where a plant infected with a mild virus is made resistant against super-infection by a related strain that causes more severe symptoms (Fulton, 1986). In an attempt to isolate the useful properties of the protective virus, it was suggested that the deliberate expression of individual viral proteins in plants could yield virus-resistant varieties (protein-mediated resistance). This was first demonstrated by Powell-Abel et al. (1986) who expressed the coat protein of tobacco mosaic virus (TMV) in transgenic tobacco and obtained plants showing strong resistance to TMV infection. Many other virus-resistant plants have been produced by the expression of coat proteins, including some that have reached commercial status (e.g., see Tricoli et al., 1995). A related strategy is the expression of a dominant negative protein, i.e., a nonfunctional viral protein that sequesters active components into inactive complexes. For example, nonfunctional replicase genes have been expressed to produce plants resistant to potato virus X (Braun and Hemenway, 1992; Longstaff et al., 1993).

Interestingly, the early work on protein-mediated resistance showed that even control transgenes that were designed to be untranslatable provided some level of protection (Lindbo and Dougherty, 1992). The expression of untranslatable viral RNA results in posttranscriptional silencing of the corresponding viral gene, i.e., there is high-level transcription but the RNA fails to accumulate in the plant cell. The basis of this RNA-mediated resistance is the sequence-specific degradation of viral transcripts caused by the appearance of small amounts of double stranded RNA homologous to the corresponding gene (reviewed by Hammond et al., 2001). The same effect can be brought about by the expression of antisense RNA (Bourque, 1995) and satellite RNAs (Baulcombe et al., 1986; Harrison et al., 1987).

The discovery of RNA-mediated resistance raised the question as to whether RNA was also the trigger in cases of coat protein-mediated resistance. However, RNA-mediated resistance requires transcription of the viral genome whereas protein-specific protective effects have been identified that take place before the viral genome is expressed (reviewed by Bendahmane and Beachy, 1999). For example, overexpression of the TMV coat protein gene is thought to block the disassembly of invading virions. It is therefore likely that the protection observed in transgenic plants expressing coat-protein genes results from the overlap of protein- and RNA-mediated effects.

A final category of pathogen-derived resistance involves the integration of defective interfering (DI) viral genomes, small derivatives of the viral genome that out-compete the full length genomes for components of the replication machinery. For example, a tandem copy of an African cassava mosaic virus (ACMV) DI-DNA remains inert in the plant genome and does not interfere with normal plant development. However, upon infection with wild-type ACMV, viral replicase is produced and the DI-DNA accumulates, at the expense of the full-length genome (e.g., Stanley et al., 1990).

Although pathogen-derived resistance has been successful, there are two major drawbacks to the method. Firstly, this type of resistance has been achieved for a number of viruses (reviewed by Lomonossoff, 1995; Beachy, 1997; 1999) but has been of limited use for cellular pathogens. Secondly, there may be undesirable side effects resulting from recombination events between the pathogen-derived transgene (or DI-genome) and any infecting virus (Greene and Allison, 1994). For example, this could create new strains with increased virulence and modified host range (Borja et al., 1999; Rubio et al., 1999; Aaziz and Tepfer, 1999).

19.2.2 Natural and Engineered Resistance Genes

Plant breeding has been used to incorporate natural resistance genes into commercial crop varieties, but this method suffers from the limitations of sexual compatibility and the species gene pool. Gene transfer to plants abolishes both these limitations by overcoming species boundaries and therefore extending the gene pool to all life on earth.

In some cases, natural resistance genes have been transferred from one crop plant to another. For example, the tobacco *N* gene, which confers resistance to TMV, has been transferred to tomato (Whitham et al., 1996). In other cases, pathogenesis-related proteins or defense peptides have been overexpressed, or plant metabolism has been modified to increase the production of antimicrobial compounds (Rommens and Kishore, 2000). Plant-derived ribosomal inactivating proteins have also been expressed in transgenic plants. These proteins possess a limited antifungal activity but also provide strong resistance against a broad spectrum of plant viruses (Moon et al., 1997; Tumer et al., 1997). It is also possible to modify resistance genes for improved performance and reintroduce such genes to the original host species. For example, an engineered version of the rice cysteine proteinase inhibitor protein oryzacystatin has been expressed in transgenic rice to confer resistance to nematodes (Vain et al., 1998).

The disadvantage of using natural plant-derived resistance genes is that the conferred resistance is often short term. Alternative strategies for more durable resistance have been investigated. These include the expression of heterologous proteins with antiviral or antimicrobial activities. Such proteins include antiviral ribonucleases and mammalian 2',5'oligoadenylate synthetases (Watanabe et al., 1995; Ogawa et al., 1996), insect-derived lytic peptides for bacterial resistance (e.g., Jaynes et al., 1993; Huang et al., 1997) and human lysozyme for resistance against bacteria and fungi (Nakajima et al., 1997). The use of ribozymes has also been investigated as a strategy for viral resistance (de Feyter et al., 1996; Kwon et al., 1997).

Recombinant antibodies can also be included in the category of heterologous resistance proteins. Various types of pathogen-specific antibodies have been expressed in plants, including full-length immunoglobulins and single chain Fv fragments (scFvs). These are summarized in Tables 19.1 and 19.2, and we discuss individual case studies in more detail below. First, however, we consider the practicalities of producing and expressing pathogen-directed antibodies in plants.

19.3 Antibody Cloning

19.3.1 Recombinant Antibodies

Recombinant antibodies originated in the 1970s, with the advent of hybridoma technology for the production of monoclonal antibodies against any conceivable antigen (Koehler and Milstein, 1975). By cloning cDNA sequences encoding the

TABLE 19.1. Plantibody-based resistance to plant pathogens.

Plant	Pathogen/Antigen	Antibody	Localization	Biological effect	References
N. benthamiana	Artichoke mottled crinkle virus (coat protein)	scFv	cytosol	reduction of infection and delay in symptom development	Tavladoraki et al. (1993)
N. tabacum cv. Xanthi nc	Tobacco mosaic virus (coat protein)	full-size IgG$_{2b}$	apoplast	70% reduction of local lesion number	Voss et al. (1995)
N. tabacum cv. Xanthi	Meloidogyne incognita (nematode stylet secretions)	full-size antibody	apoplast	no biological effect, probably due to mistargeting of the antibody	Baum et al. (1996)
N. benthamiana	Beet necrotic yellow vein virus (coat protein)	scFv	ER	delay in symptom development	Fecker et al. (1997)
N. tabacum cv. Xanthi nc	Tobacco mosaic virus (coat protein)	scFv	cytosol	>90% reduction of local lesion number, 11% of transgenic plants were fully resistant in systemic infection assays	Zimmermann et al. (1998)
N. tabacum	Stolbur phytoplasma (major membrane protein)	scFv	apoplast	transgenic tobacco shoots grew free of symptoms	Le Gall et al. (1998)
Maize	Spiroplasma kunkelii (membrane protein)	scFv	cytosol	no resistance	Chen & Chen (1998)
N. tabacum cv. Petite Havana SR1	Tobacco mosaic virus (coat protein)	scFv	plasmalemma membrane surface	13% of transgenic plants were fully resistant in systemic infection assays	Schillberg et al. (2000)
N. tabacum cv. W38	Potato virus strain Y and D, Clover yellow vein virus strain 300 (coat protein)	scFv	cytosol, apoplast	suppression of infection	Xiao et al. (2000)
N. tabacum cv. Samsun NN	Tobacco mosaic virus (coat protein)	scFv	cytosol	100% reduction of virus infection	Bajrovic et al. (2001)
N. benthamiana	Tombusviridae (conserved motif E of viral RdRp)	scFv	cytosol, ER	full to partial resistance	Boonrod et al. (2004)
Arabidopsis thaliana	Fusarium (cell-wall bound proteins, mycelium surface proteins, germinated spores)	scFv fusion proteins	apoplast	high levels of protection	Peschen et al. (2004)
N. tabacum	Stolbur phytoplasma (major membrane protein)	scFv	apoplast	short delay in symptom appearance and phytoplasma multiplication	Malembic-Maher et al. (2005)

TABLE 19.2. Pathogen specific antibodies expressed in plants (no biological activity tested).

Plant	Pathogen/Antigen	Antibody	Localization	References
N. tabacum cv Samsun NN	*Botrytis cinerea*-produced cutinase	full-size	apoplast	Van Engelen et al. (1994)
Tobacco protoplasts	*Meloidogyne incognita* (nematode salivary secretions)	scFv	cytosol, apoplast, ER	Rosso et al. (1996)
N. benthamiana	TSWV (glycoprotein G1)	scFv	apoplast	Franconi et al. (1999)
N. tabacum cv. Petite Havana SR1, *N. tabacum* cv. Xanthi nc	Tobacco mosaic virus (coat protein monomer)	full-size, scFv	apoplast, cytosol	Schillberg et al. (1999)
N. tabacum cv. Petite Havana SR1	Tobacco mosaic virus (coat protein, coat protein monomer)	bispecific scFv	cytosol, apoplast, ER	Fischer et al. (1999b)
A. thaliana	*Fusarium*-produced mycotoxin (zearalenone)	scFv	apoplast	Yuan et al. (2000)
N. tabacum	*Citrus tristeza* virus (coat protein)	scFv	cytosol ?	Galeffi et al. (2002)

heavy and light immunoglobulin chains, it became possible to express antibodies in a variety of alternative systems and to modify the native immunoglobulins in a number of ways (Kipriyanov and Little, 1999). For example, site-directed mutagenesis and the polymerase chain reaction have been widely used to replace or delete specific codons. The introduction of appropriately placed stop codons allows the direct expression of Fab or F(ab')$_2$ fragments (Figure 19.1).

Intact antibodies are usually multivalent so they bind to their target with high affinity and high specificity. They comprise Fc domains, which are important in immunotherapy through their ability to recruit cytotoxic effector functions. Mouse monoclonal antibodies used as immunotherapeutic reagents in human had to be modified because they induced anti-mouse immune responses. To avoid, mask, or redirect this human immune surveillance, different strategies were developed. Effector functions can be eliminated from full size antibodies and new functions introduced by converting mouse immunoglobulins into chimeric mouse/human polypeptides (Matsushita et al., 1992; Hastings et al., 1992) (Figure 19.1). Furthermore, modification of the hinge region, glycosylation sites, and Fc-receptor binding sites can enhance performance, biodistribution and stability (Leong et al., 2001). Monovalent single-chain antibodies (scFvs) and bispecific single-chain fragments can also be constructed (Figure 19.1). These possess only the variable

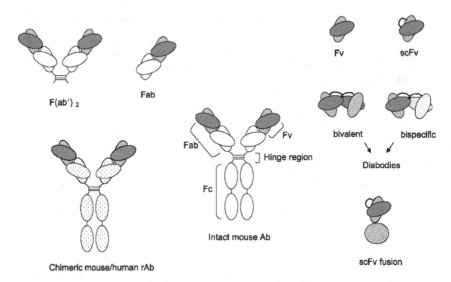

FIGURE 19.1. Schematic presentation of an intact mouse antibody with Fc, Fab and Fv fragments and single constant (white or dotted ovals) and variable regions (grey-coloured ovals). Recombinant antibodies (rAbs) are shown as monovalent as well as bivalent and bispecific scFv (different specificities for the bispecific scFv are presented as light grey ovals), combined with linkers (black line). Also shown are Fab and F(ab')$_2$ fragments, a chimeric mouse/human rAb (the human part is presented with dotted ovals), and, as an example for fusing a gene of an antibody fragment to another protein domain (toxin, enzyme, peptide), a scFv fragment with his fusion partner (presented as a hatched sphere).

domains of the parent immunoglobulin joined by a peptide linker (Fuchs et al., 1997; Kipriyanov, 2002). Because of the unaltered antigen-binding surface these small antibody fragments also provide full binding specificity in comparison with intact antibodies. They have been engineered also into multimeric conjugates for increasing the functional affinity through the use of either chemical or genetic cross-links (Tomlinson and Holliger, 2000). Multifunctional proteins can be generated by fusing the gene for an antibody fragment to genes encoding other protein domains such as toxins, biological response modifiers, or enzymes (Fan et al., 2002; Niv et al., 2003; Li et al., 1999; Rau et al., 2002).

19.3.2 Obtaining Antibody cDNA Sequences for Expression in Plants

Many monoclonal antibodies that are specific for plant pathogens already exist, e.g., as diagnostic tools used in research, and these can be a useful resource (Fischer, 1990). However, if a suitable monoclonal antibody is not available, it is necessary to produce a new antibody with the appropriate specificity. There are two major approaches to this (Kipriyanov and Little, 1999).

Traditionally, antibody cDNAs have been isolated from hybridoma cell lines. The normal process is to inject mice with the pathogen or plant-derived protein of interest, isolate B-lymphocytes and fuse these with immortalized myeloma cells under appropriate selective conditions to establish a hybridoma line. Hybridomas are then screened for their ability to produce antibodies that bind the pathogen-derived antigen. Hybridoma technology permits the production of highly specific monoclonal antibodies, but the process is labor intensive and requires the use of animals, animal cell culture, and expensive equipment. Many molecular biology laboratories have neither the facilities nor the experience to generate monoclonal antibodies in this manner. Importantly, hybridoma technology also does not allow the immediate and convenient isolation and cloning of immunoglobulin-encoding cDNAs.

In contrast, phage display technology is an *in vitro* method that allows the production and selection of specific antibodies linked to the cDNA sequences encoding them (McCafferty et al., 1990; Griffiths and Duncan, 1998). Antibody phage display involves the expression of a library of different antibody genes as fusion proteins on the surface of bacteriophage. This is a massively parallel technique in which the antibody and corresponding cDNA are coselected based on the antibody's affinity for a particular antigen. Since the invention of the technique in the early 1990s, the display of recombinant antibodies on the surface of filamentous bacteriophage has revolutionized antibody generation (Hoogenboom et al., 1991; Hoogenboom and Winter, 1992; Hoogenboom et al., 1992). The advantage of phage display is that artificial monoclonal antibodies can be isolated in virtually any molecular biology laboratory using basic recombinant DNA techniques. During library production, cloned heavy and light chain gene fragments are mixed, and novel combinatorial specificities occur that cannot be found in the original donor animal.

Several groups have shown the feasibility of phage display by isolating antibodies with nanomolar affinity from either immunized donor derived libraries or from naïve libraries, thereby completely bypassing the immune system (Hoogenboom and Winter, 1992). It is also possible to develop high-quality libraries in which the complementarity-determining regions of the antibodies are completely synthetic (Knappik et al., 2000). Low affinity antibodies can be improved by several *in vitro* approaches until suitable affinities, in the nanomolar or even picomolar range, are obtained. Such approaches include mutagenesis, chain or gene shuffling (Marks et al., 1992), and antibody complementarity determining region randomization followed by biopanning and screening to identify higher affinity clones (Hawkins et al., 1992; Schier and Marks, 1996; Hoogenboom, 1997). Phage display technology is a tool to generate antibodies with the desired specificity for any plant pathogen-derived antigen.

19.4 Antibody Expression in Plants

Functional full size recombinant antibodies were first expressed in transgenic plants in 1989 (Hiatt et al., 1989; Düring et al., 1990). Various immunoglobulin classes have been expressed in plants, including monotypic and chimeric IgG, IgM, and IgA (Ma et al., 1994; 1995; Voss et al., 1995; De Wilde et al., 1996; Baum et al., 1996). In addition to full size antibodies, various functional antibody derivatives have also been produced successfully in plants. These include Fab fragments (De Neve et al., 1993; De Wilde et al., 1996), scFvs (Owen et al., 1992; Firek et al., 1993; Tavladoraki et al., 1993; Artsaenko et al., 1995; Fecker et al., 1997; Fiedler et al., 1997; Schouten et al., 1997), bispecific scFvs (Fischer et al., 1999b) and membrane anchored scFvs (Schillberg et al., 2000; Vine et al., 2001).

Antibody expression is achieved by inserting the cloned cDNA into a plant expression cassette comprising a strong promoter, control elements that enhance protein synthesis and signals that direct the recombinant protein to the appropriate intracellular compartment (Fischer and Emans, 2000). Protein targeting is critical for efficient antibody production since this influences folding and assembly as well as posttranslational modification. Significant increases in recombinant antibody yield have been observed if the antibodies are targeted to the secretory pathway rather than the cytosol (Conrad and Fiedler, 1998; Schillberg et al., 1999). This is achieved by the inclusion of a signal peptide and generally results in the antibodies being secreted to the apoplast, the intracellular space beneath the cell wall (e.g., see Voss et al., 1995). If a transmembrane anchor sequence is included, however, the antibody will be inserted into the plasma membrane (Schillberg et al., 2000; Vine et al., 2001). The addition of a C-terminal KDEL sequence in addition to a signal peptide causes antibodies to be retrieved from the Golgi apparatus and returned to the lumen of the endoplasmic reticulum (ER). This can result in 10- to 100-fold higher yields compared to apoplast targeting (Conrad and Fiedler, 1998). However, as discussed below, the desire for high yields must be balanced against the need to block pathogen activity in the appropriate cell compartment and the cytosol may be a more-appropriate site for antibody accumulation. Cytosolic expression of recombinant antibodies has been difficult because the environment is not suitable for stable folding and assembly. For this reason, only scFv fragments can be reliably expressed in this manner (Owen et al., 1992; Tavladoraki et al., 1993; Schouten et al., 1996; Zimmermann et al., 1998).

19.5 Antibody Mediated Resistance—Case Studies

19.5.1 Antibody-Based Viral Resistance

The first antibody-mediated virus-resistant plants were reported by Tavladoraki et al. (1993). A scFv fragment specific for artichoke mottled crinkle virus (AMCV) was constructed from the parent monoclonal antibody and expressed in bacteria to show that it retained binding specificity. The scFv was then expressed in the plant cytosol and both transgenic protoplasts and plants were shown to be resistant to AMCV challenge.

Subsequently, full-length TMV-specific monoclonal antibodies were shown to protect tobacco plants against TMV infection when expressed in the apoplast

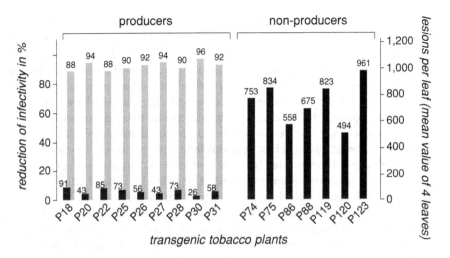

FIGURE 19.2. Improved TMV resistance by cytosolic scFv accumulation in transgenic plants. The amount of the TMV-specific scFv24 in transgenic *N. tabacum* cv. Xanthi nc plant lines was determined by ELISA. We applied 1 μg/ml TMV-*vulgare* to five leaves of each non-transgenic and transgenic plant. The average local lesion number per leaf was scored five days after inoculation (black columns) and the percentage reduction of infectivity compared to transgenic non-producers was calculated (gray columns).

(Voss et al., 1995). In this study, the degree of protection correlated with the expression level. When the antibody reached 0.23% of total soluble protein, TMV lesions were reduced by 70%; at 0.4% total soluble protein, the lesions were reduced by almost 95%. Interestingly, cytosolic expression of a scFv fragment derived from this antibody also conferred resistance, even though the level of accumulation was approximately 20,000-fold lower than in the plants expressing full-length immunoglobulins (Zimmermann et al., 1998). This study showed that even very small amounts of cytosolic antibody were able to neutralize invading virions. There was a greater than 90% reduction in local lesions (Figure 19.2) and at least 10% of transgenic plants were protected against systemic TMV infections.

Although the cytosol appears to be the most suitable site for anti-viral antibodies, the secretory pathway has been investigated as an alternative for scFv accumulation (Fecker et al., 1997; Xiao et al., 2000). For example, Fecker et al. (1997) expressed a scFv fragment against beet necrotic yellow vein virus (BNYVV) in *Nicotiana benthamiana* and used a KDEL signal to make the antibody accumulate in the ER lumen. They reported delayed onset of disease symptoms when plants were challenged with the virus. More recently, we have developed a novel expression system in which a scFv antibody is targeted to the plasma membrane of tobacco cells by fusion to a mammalian transmembrane domain (Schillberg et al., 2001). The membrane-anchored scFv, with the antigen-binding domain exposed to the apoplast, provided strong resistance against TMV.

Thus far, antibodies expressed in virus-resistant transgenic plants have been raised mainly against viral coat proteins, which are the most likely to mutate to overcome this new form of immunity. The effectiveness of antibody-based viral resistance may be increased if antibodies are targeted against viral proteins crucial for replication, movement, and transmission. These proteins are far more structurally constrained than coat proteins and offer the best hope for durable resistance against viral diseases. We have produced tobacco plants expressing a scFv fragment specific for the movement protein of tomato spotted wilt virus (TSWV). Ten anti-NS_M scFvs were isolated by phage display, characterized for binding activity by expression in *E. coli*, and expressed in the cytosol of transgenic plants. The antibodies were expressed at high levels (0.1–8% of total soluble protein) and significantly delayed the onset of disease symptoms when the plants were challenged with the pathogen (Zhang et al., unpublished data). Recently, Boonrod and colleagues (2004) have demonstrated that scFv antibody fragments binding to a conserved domain in a plant viral RNA-dependent RNA polymerase (RdRp), a key enzyme in virus replication, conferred varying degrees of resistance against four plant viruses. *N. benthamiana* plants producing selected scFvs showed high levels of resistance against infections with tomato bushy stunt virus (TBSV) and closely related viruses. Moreover, transgenic plants showed partial resistance against two more distantly related viruses.

19.5.2 Antibody-Based Resistance Against Bacteria and Phytoplasma

Different genetic strategies have been used to produce plants resistant to bacterial diseases, including the expression of lytic peptides and other antibacterial proteins, the inhibition of bacterial toxins and the enhancement of natural plant defenses (reviewed by Mourgues et al., 1998). Le Gall et al. (1998) showed that antibody-based resistance is also useful against bacterial pathogens. These investigators provided a potential strategy to control phytoplasma diseases by expressing a scFv specific for the stolbur phytoplasma major membrane protein. Stolbur phytoplasma are strictly limited to the sieve tubes within the phloem tissues, therefore phytoplasma specific scFvs were directed through the secretory pathway. Transgenic tobacco shoots expressing anti-phytoplasma scFvs were top-grafted onto tobacco plants heavily infected with phytoplasma. The shoots grew free of symptoms while nontransgenic grafted tobacco shoots showed severe disease symptoms. Recently, the group has demonstrated that also field trial experiments resulted in a short delay in symptom appearance and phytoplasma multiplication when the stolbur phytoplasma was graft-transmitted to tobacco plants producing the specific scFv. However, no significant resistance was observed when the phytoplasma was transmitted by its insect vector in greenhouse experiments.

In contrast, cytosolic expression of a scFv specific for corn stunt spiroplasma (CSS) in maize did not confer resistance to the pathogen (Chen and Chen, 1998). It would be interesting to repeat this study using antibodies directed to the secretory pathway in phloem cells since CSS are restricted to the sieve tubes.

19.5.3 Antibody-Based Fungal Resistance

Fungal pathogens are the most challenging target for antibody-mediated resistance because they affect crops in two ways: by destroying plants and seeds in the field and by contaminating the harvested crop with fungal toxins. During infection, invasive mycelia spread throughout the host plant, secreting enzymes and toxins that are essential for pathogenesis and parasitization. These proteins and toxins are suitable targets for recombinant antibodies, and if effective antibody-based strategies could be developed then environmental pollution caused by the extensive use of fungicides could be avoided.

Thus far, antibodies have been raised against a number of fungal antigens including conidiaspore proteins, secreted proteins and other compounds, cell wall fragments and cell surface antigens of mycelia (Pain et al., 1992; Robert et al., 1993; Goebel et al., 1995; Murdoch et al., 1998). *In vitro* studies showed that the development of disease symptoms in avocado, mango, and banana infected with *Colletotrichum gloeosporioides* was inhibited if the inoculum was first mixed with polyclonal antibodies specific for the fungal pectate lyase (Wattad et al., 1997).

The success of antibody-based resistance against plant viruses indicates that a similar approach may work against fungi. However, given the diversity of mechanisms that fungi use to attack plants, there are considerable technical challenges that need to be overcome. An attractive strategy is the use of recombinant antibody fusion proteins expressed in transgenic plants as "shuttles" to deliver anti-fungal polypeptides to the fungal cell surface, where they would attack and kill the invading hyphae. To prove this hypothesis we have generated scFvs against *Fusarium* surface proteins and fusions with anti-fungal polypeptides were produced in *Arabidopsis*. Bioassays demonstrated that recombinant antibody fusion proteins conferred resistance to the fungal pathogen (Peschen et al., 2004).

19.5.4 Engineering Nematode Resistance

The control of nematode disease is another important target for antibody-based resistance. However, attempts to inhibit nematode parasitism of tobacco roots through antibody expression have not been successful thus far. Full-length immunoglobulins raised against nematode stylet secretions had no effect on a root-knot nematode parasite when expressed in transgenic plants (Baum et al., 1996). The targeting of the antibody was probably an important factor in the above case as the antibody was targeted to the apoplast, where it would be unable to inactivate the stylet secretions injected into cells. A scFv fragment of corresponding specificity expressed in the cytosol would probably be more effective.

The identification of new target antigens could also improve the chances of antibody-based nematode resistance. One suitable candidate is the cellulase expressed by invading root-knot nematode larvae. This enzyme is thought to play an important role during intracellular migration of the larvae through the root cortex to the vascular cylinder. A number of cellulases have been cloned from the root-knot nematode *Meloidogyne incognita* (Smant et al., 1998; Rosso et al., 1999). Furthermore, expression of recombinant antibodies under the control of promoters induced by nematode invasion (Favery et al., 1998) will allow pathogen-specific antibody expression at the exact time and location when the pathogen is most vulnerable.

19.5.5 Optimization of Antibody-Based Resistance

Much research has been carried out to optimize antibody expression in plants, and high levels of production can now be achieved on a routine basis. However, this research has focussed on maximizing yields as a final objective. The antibodies, generally intended for diagnostic or therapeutic use, are then extracted from the plant and purified (Schillberg et al., 2002). Interestingly, Galeffi et al. (2002) have recently described the production of an scFv against *Citrus tristeza* virus in *E. coli* and transgenic tobacco plants, which could be applied topically to citrus plants in order to prevent infection.

Where antibodies are designed to function within the plant, high yields must be balanced against effectiveness in situ. In the case of antibodies for pathogen resistance, the intracellular compartment to which the antibody is targeted must intersect with the pathogen's life cycle, otherwise the benefits of high-level expression will be wasted.

It is now well established that the preferential site for high-level antibody accumulation in plants is the secretory pathway, because this provides a beneficial environment for protein folding and assembly. In the case of full size immunoglobulins, the secretory pathway is also essential for glycosylation, which is important for effector functions (not relevant in transgenic plants) and can also influence folding, assembly, and therefore stability. However, the secretory pathway may not always be the best choice when it comes to interfering with the pathogen life cycle. As discussed above, the low-level expression of scFv fragments in the cytosol provides good and in some cases better protection against viruses than high-level expression in the secretory pathway. This probably reflects the fact that the most important stages of the viral life cycle take place in the cytoplasm. Similarly, antibodies raised against nematode stylet secretions are ineffective when targeted to the apoplast, probably because the secretions are injected directly into the cytosol. Only scFvs have thus far been expressed in the cytosol and in the majority of cases antibody expression was only just detectable (Owen et al., 1992; Tavladoraki et al., 1993; Schouten et al., 1996; Zimmermann et al., 1998). In some reports, cytosolic scFvs have reached levels of up to 1.0 or 4% total soluble protein, suggesting that properties specific to each scFv may influence stability (De Jaeger et al., 1998; Noelke et al., 2005). Cytosolic antibody expression may be further enhanced in the future by the identification of intrinsic structural features or fusion protein partners that stabilize the antibodies (Spiegel et al., 1999; Worn et al., 2000).

Since the critical issue for antibody-based resistance is that the expressed antibody reaches the pathogen, the data on expression levels in subcellular compartments is useful for evaluating if a chosen strategy can be effective. With information on the life cycle of the pathogen at hand, a reasonable prediction can be made on the best way to approach engineering resistance. We have found that viral infection can be blocked in the cytoplasm, which is expected because most steps of viral multiplication, assembly, and cell-to-cell spread require the cytosol (Tavladoraki et al., 1993; Zimmermann et al., 1998). However, we have recently described a novel method for protecting plant cells from viral infection that does not require cytosolic expression. Anti-viral antibodies were fused to a mammalian transmembrane domain and then expressed at the plasma membrane, facing the apoplast. These membrane-displayed antibodies are at least as effective at creating resistance as those expressed as soluble proteins in the cytosol. We speculate that the surface displayed antibodies act to shield the plant from viral infection, but the actual mechanism of the resistance is still under investigation (Schillberg et al., 2000).

19.6 Perspective

The flexibility and specificity of antibodies is a well-known aspect of the vertebrate immune system and the use of antibodies to generate disease resistant plants is therefore an attractive approach. Antibodies specific for just about any target molecule, from small organic chemicals to native protein complexes, can be isolated and modified by antibody engineering. Modifications can be carried out to increase the binding specificity of the antibody or to increase its stability in a particular cellular compartment. This means that almost any pathogen structure could be targeted by recombinant antibodies expressed in plants. Furthermore, should the resistance be broken by the pathogen, new antibodies can be produced relatively quickly to reestablish resistance.

Another attractive property of antibody-based resistance is that multiple antibodies with different target specificities can be expressed in a single plant to engineer pyramidal resistance against an individual pathogen. This type of resistance will be more difficult to break because simultaneous adaptations will be required in different genes. The same strategy could be used to generate plant lines that are resistant to multiple pathogens. Alternatively, bispecific antibodies can be expressed (Fischer et al., 1999b), raising the interesting possibility of neutralizing two pathogen proteins simultaneously in the same cell, such as a viral coat protein and a replicase or movement protein. Finally, antibodies can be fused to other proteins, such as toxins, that have activity against the pathogen, providing a more targeted as well as more effective defense strategy.

It is therefore likely that antibody expression in plants will become the method of choice for producing disease-resistant varieties. The potential of this method is limited only by our understanding of plant–pathogen interactions. As our understanding grows and more targets are identified through genomic and proteomic approaches, antibodies can be designed and expression techniques tailored to suit each pathogen's profile. Creating transgenic crops that are resistant to pathogens, and need only minimal treatment with agrochemicals, will bring the goal of environmentally benign agriculture closer to reality.

References

Agrios, G.N. 1997. *Plant Pathology.* Fourth Edition. New York: Academic Press.

Aaziz, R., and Tepfer, M. 1999. Recombination in RNA viruses and in virus-resistant transgenic plants. *J. Gen. Virol.* 80:339–1346.

Artsaenko, O., Peisker, M., zur Nieden, U., Fiedler, U., Weiler, E.W., Müntz, K., and Conrad, U. 1995. Expression of a single-chain Fv antibody against abscisic acid creates a wilty phenotype in transgenic tobacco. *Plant J.* 8:745–750.

Bajrovic, K., Erdag, B., Atalay, E.O., and Cirakoclu, B. 2001. Full resistance to tobacco mosaic virus infection conferred by the transgenic expression of a recombinant antibody in tobacco. *Biotechnol. Equip.* 15:21–27.

Baulcombe, D.C., Saunders, G.R., Bevan, M.W., Mayo, M.A., and Harrison, B.D. 1986. Expression of biologically active satellite RNA from the nuclear genome of transformed plants. *Nature* 321:446–449.

Baum, T.J., Hiatt, A., Parrott, W.A., Pratt, L.H., and Hussey, R.S. 1996. Expression in tobacco of a functional monoclonal antibody specific to stylet secretions of the root-knot nematode. *Mol. Plant Microbe Interact.* 9:382–387.

Beachy, R.N. 1999. Coat-protein-mediated resistance to tobacco mosaic virus: discovery mechanisms and exploitation. *Philos. Trans. R. Soc. Lond. B Biol. Sci.* 354:659–664.

Beachy, R.N. 1997. Mechanisms and applications of pathogen-derived resistance in transgenic plants. *Curr. Opin. Biotechnol.* 8:215–220.

Bendahmane, M., and Beachy, R.N. 1999. Control of tobamovirus infections via pathogen-derived resistance. *Adv. Virus Res.* 53:369–86.

Borja, M., Rubio, T., Scholthof, H., and Jackson, A. 1999. Restoration of wild-type virus by double recombination of tombusvirus mutants with a host transgene. *Mol. Plant Microbe Interact.* 12:153–162.

Boonrod, K., Galetzka, D., Nagy, P. D., Conrad, U., Krczal, G. 2004. Single-chain antibodies against a plant viral RNA-dependent RNA polymerase confer virus resistance. *Nature Biotechnology* 22:856–862.

Bourque, J.E. 1995. Antisense strategies for genetic manipulations in plants. *Plant Sci.* 105:125–149.

Braun, C.J., and Hemenway, C.L. 1992. Expression of amino-terminal portions or full-length viral replicase genes in transgenic plants confers resistance to potato virus X infection. *Plant Cell* 4:735–744.

Chen, Y.D., and Chen, T.A. 1998. Expression of engineered antibodies in plants: a possible tool for spiroplasma and phytoplasma disease control. *Phytopathology* 88:1367–1371.

Conrad, U., and Fiedler, U. 1998. Compartment-specific accumulation of recombinant immunoglobulins in plant cells: an essential tool for antibody production and immunomodulation of physiological functions and pathogen activity. *Plant Mol. Biol.* 38:101–109.

Crute, I.R., and Pink, D.A.C. 1996. Genetics and utilization of pathogen resistance in plants. *Plant Cell* 8:1747–1755.

De Feyter, R., Young, M., Schroeder, K., Dennis, E., and Gerlach, W. 1996. A ribozyme gene and an antisense gene are equally effective in conferring resistance to tobacco mosaic virus on transgenic tobacco. *Mol. Gen. Genet.* 250:329–338.

De Jaeger, G., Buys, E., Eeckhout, D., De Wilde, C., Jacobs, A., Kapila, J., Angenon, G., Van Montagu, M., Gerats, T., and Depicker, A. 1998. High level accumulation of single-chain variable fragments in the cytosol of transgenic Petunia hybrida. *Eur. J. Biochem.* 259:1–10.

De Neve, M., De Loose, M., Jacobs, A., Van Houdt, H., Kaluza, B., Weidle, U., Van Montagu, M., and Depicker, A. 1993. Assembly of an antibody and its derived antibody fragment in*Nicotiana* and *Arabidopsis*. *Transgenic Res.* 2:227–237.

De Wilde, C., De Neve, M., De Rycke, R., Bruyns, A.M., De Jaeger, G., Van Montagu, M., Depicker, A., and Engler, G. 1996. Intact antigen-binding MAK33 antibody and Fab fragment accumulate in intercellular spaces of *Arabidopsis thaliana*. *Plant Sci.* 114:233–241.

Düring, K., Hippe, S., Kreuzaler, F., and Schell, J. 1990. Synthesis and selfassembly of a functional monoclonal antibody in transgenic *Nicotiana tabacum*. *Plant Mol. Biol.* 15:281–293.

Fan, D., Yano, S. Shinohara, H., Solorzano, C., Van Arsdall, M., Bucana, C.D., Pathak, S., Kruzel, E., Herbst R.S., Onn, A., Roach, J.S., Onda, M., Wang, Q.C., Pastan, I., and Fidler, I.J. 2002. Targeted therapy against human lung cancer in nude mice by high-affinity recombinant antimesothelin single-chain Fv immunotoxin. *Mol. Cancer Ther.* 1:595–600.

Favery, B., Lecomte, P., Gil, N., Bechtold, N., Bouchez, D., Dalmasso, A., and Abad, P. 1998. *RPE*, a plant gene involved in early developmental steps of nematode feeding cells. *EMBO J.* 17:6799–6811.

Fecker, L.F., Koenig, R., and Obermeier, C. 1997. *Nicotiana benthamiana* plants expressing beet necrotic yellow vein virus (BNYVV) coat protein-specific scFv are partially protected against the establishment of the virus in the early stages of infection and its pathogenic effects in the late stages of infection. *Arch. Virol.* 142:1857–1863.

Fiedler, U., Philips, J., Artsaenko, O., and Conrad, U. 1997. Optimisation of scFv antibody production in transgenic plants. *Immunotechnology* 3:205–216.

Firek, S., Draper, J., Owen, M.R., Gandecha, A., Cockburn, B., and Whitelam, G.C. 1993. Secretion of a functional single-chain Fv protein in transgenic tobacco plants and cell suspension cultures. *Plant Mol. Biol.* 23:861–870.

Fischer, R. 1990. Herstellung, Charakterisierung und Klonierung der Immunglobulingene von TMV-neutralisierenden Antikörpern. Eberhard-Karls-Universität Tübingen.

Fischer, R., Drossard, J., Schillberg, S., Artsaenko, O., Emans, N., and Naehring, J. 1999a. Modulation of plant function and plant pathogens by antibody expression. In *Metabolic Engineering of Plant Secondary Metabolism*, eds. R. Verpoorte, and A. Alfermann, pp. 87–109. Kluwer Academic Publishers, Dordrecht, The Netherland.

Fischer, R., Schumann, D., Zimmermann, S., Drossard, J., Sack, M., and Schillberg, S. 1999b. Expression and characterization of bispecific single chain Fv fragments produced in transgenic plants. *Eur. J. Biochem.* 262:810–816.

Fischer, R., and Emans, N. 2000. Molecular farming of pharmaceutical proteins. *Transgenic Res.* 9:279–299.

Franconi, R., Roggero, P., Pirazzi, P., Arias, F.J., Desiderio, A., Bitti, O., Pashkoulov, D., Mattei, B., Bracci, L., Masenga, V., Milne, R.G., and Benvenuto, E. 1999. Functional expression in bacteria and plants of an scFv antibody fragment against tospoviruses. *Immunotechnology* 4:189–201.

Fuchs, P., Breitling, F., Little, M., and Dübel, S. 1997. Primary structure and functional scFv antibody expression of an antibody against the human protooncogen c-myc. *Hybridoma* 16:227–233.

Fulton, R.W. 1986. Practices and precautions in the use of cross-protection for plant virus disease control. *Annu. Rev. Phytopathol.* 24:67–81.

Galeffi, P., Giunta, G., Guida, S., and Cantale, C. 2002. Engineering of a single chain variable fragment antibody specific for the *Citrus tristeza virus* and its expression in *Escherichia coli* and *Nicotiana tabacum*. *Eur. J. Plant Pathol.* 108:479–483.

Goebel, C., Hahn, A., and Hock, B. 1995. Production of polyclonal and monoclonal antibodies against hyphae from arbuscular mycorrhizal fungi. *Crit. Rev. Biotechnol.* 15:293–304.

Greene, A., and Allison, R. 1994. Recombination between viral RNA and transgenic plant transcripts. *Science* 263:1423–1425.

Griffiths, A., and Duncan, A. 1998 Strategies for selection of antibodies by phage display. *Curr. Opin. Biotechnol.* 9:102–108.

Hammond, S.M., Caudy, A.A., and Hannon, G.J. 2001. Post-transcriptional gene silencing by double-stranded RNA. *Nat. Rev. Genet.* 2:110–119.

Harrison, B.D., Mayo, M.A., and Baulcombe, D.C. 1987. Virus resistance in transgenic plants that express cucumber mosaic virus satellite RNA. *Nature* 328:799–805.

Hastings, A, Morrison S.L., Kanda, S., Saxton, R.E. and Irie, R.F. 1992. Production and characterization of a murine/human chimeric anti-idiotype antibody that mimics ganglioside. *Cancer Res.* 52:1681–1686.

Hawkins, R.E., Russell, S.J., and Winter, G. 1992. Selection of phage antibodies by binding affinity–mimicking affinity maturation. *J.Mol.Biol.* 226: 889–896.

Hiatt, A., Cafferkey, R., and Bowdish, K. 1989. Production of antibodies in transgenic plants. *Nature* 342:76–78.

Hoogenboom, H.R. 1997. Designing and optimising library selection strategies for generating high-affinity antibodies. *Trends Biotechnol.* 15:62–70.

Hoogenboom, H.R., Griffiths, A.D., Johnson, K.S., Chiswell, D.J., Hudson, P., and Winter, G. 1991. Multi-subunit proteins on the surface of filamentous phage: methodologies for displaying antibody (Fab) heavy and light chains. *Nucleic Acids Res.* 19:4133–4137.

Hoogenboom, H.R., Marks, J.D., Griffiths, A.D., and Winter, G. 1992. Building antibodies from their genes. *Immunol. Rev.* 130:41–68.

Hoogenboom, H.R., and Winter, G. 1992. By-passing immunisation. Human antibodies from synthetic repertoires of germline V_H gene segments rearranged *in vitro*. *J. Mol. Biol.* 20:381–388.

Huang, Y., Nordeen, R.O., Di, M., Owens, L.D., and McBeath, J.H. 1997. Expression of an engineered cecropin gene cassette in transgenic tobacco plants confers disease resistance to *Pseudomonas syringae* pv. *tabaci*. *Phytopathology* 87:494–499.

Jaynes, J.M., Nagpala, P., Destefanobeltran, L., Huang, J.H., Kim, J.H., Denny, T., and Cetiner, S. 1993. Expression of a cecropin-B lytic peptide analog in transgenic tobacco confers enhanced resistance to bacterial wilt caused by *Pseudomonas solanacearum*. *Plant Sci.* 89:43–53.

Kipriyanov, S.M. 2002. Generation of bispecific and tandem diabodies. *Methods Mol. Biol.* 178:317–331.

Kipriyanov, S.M., and Little, M. 1999. Generation of recombinant antibodies. *Mol. Biotechnol.* 12:173–201.

Knappik, A., Ge, L.M., Honegger, A., Pack, P., Fischer, M., Wellnhofer, G., Hoess, A., Wolle, J., Pluckthun, A., and Virnekas, B. 2000. Fully synthetic human combinatorial antibody libraries (HuCAL) based on modular consensus frameworks and CDRs randomized with trinucleotides. *J. Mol. Biol.* 296:57–86.

Koehler, G., and Milstein, C. 1975. Continuous cultures of fused cells secreting antibody of predefined specificity. *Nature* 256:495–497.

Kwon, C., Chung, W., and Paek, K. 1997. Ribozyme mediated targeting of cucumber mosaic virus RNA 1 and 2 in transgenic tobacco plants. *Mol. Cells* 7:326–334.

Le Gall, F., Bove, J.M., and Garnier, M. 1998. Engineering of a single-chain variable-fragment (scFv) antibody specific for the stolbur phytoplasma (Mollicute) and its expression in *Escherichia coli* and tobacco plants. *Appl. Environ. Microbiol.* 64:4566–4572.

Leong, S.R., DeForge, L., Presta, L., Gonzales, T., Fan, A., Reichert, M., Chuntharapai, A., Kim, K.J., Tumas, D.B., Lee, W.P., Gribling, P., Snedecor Chen, H., Hsei, V., Schoenhoff, M., Hale, V., Deveney, J., Koumenis, I., Shah, Z., McKay, P., Galan, W., Wagner, B., Narindray, D., Hebert, C., and Zapata, G. 2001. Adapting pharmacokinetic properties of a humanized anti-interleukin-8 antibody for therapeutic applications using site-specific pegylation. *Cytokine* 16:106–119.

Li, J.Y., Sugimura, K., Boado, R.J., Lee, H.J., Zhang, C., Duebel, S., and Pardridge, W.M. 1999. Genetically engineered brain drug delivery vectors: cloning, expression and *in vivo* application of an anti-transferrin receptor single-chain antibody-streptavidin fusion gene and protein. *Prot. Eng.* 12:787–796.

Lindbo, J.A., and Dougherty, W.G. 1992. Untranslatable transcripts of the tobacco etch virus coat protein gene sequence can interfere with tobacco etch virus replication in transgenic plants and protoplasts. *Virology* 189:725–733.

Lomonossoff, G.P. 1995. Pathogen-derived resistance to plant viruses. *Annu. Rev. Phytopathol.* 33:323–343.

Longstaff, M., Brigneti, G., Boccard, F., Chapman, S., and Baulcombe, D. 1993. Extreme resistance of potato virus X infection in plants expressing a modified component of the putative viral replicase. *EMBO J.* 12:379–386.

Ma, J.K.C., Lehner, T., Stabila, P., Fux, C.I., and Hiatt, A. 1994. Assembly of monoclonal antibodies with IgG1 and IgA heavy chain domains in transgenic tobacco plants. *Eur. J.Immunol.*24:131–138.

Ma, J.K., Hiatt, A., Hein, M., Vine, N.D., Wang, F., Stabila, P., van Dolleweerd, C., Mostov, K., and Lehner, T. 1995. Generation and assembly of secretory antibodies in plants. *Science* 268:716–719.

Malembic-Maher, S., Le Gall, F., Danet, J.L., de Borne, F.D., Bové, J.M., and Garnier-Semancik, M. 2005. Transformation of tobacco plants for single-chain antibody expression via apoplastic and symplasmic routes, and analysis of their susceptibility to stolbur phytoplasma infection. *Plant Science* 168:349–358.

Marks, J.D., Griffiths, A.D., Malmqvist, M., Clackson, T.P., Bye, J.M., and Winter, G. 1992. By-passing immunization: building high affinity human antibodies by chain shuffling. *Bio/Technol.* 10:779–83.

Matsushita, S., Maeda, H., Kimachi, K., Eda, Y., Maeda, Y., Murakami, T., Tokiyoshi, S., and Takatsuki, K. 1992. Characterization of a mouse/human chimeric monoclonal antibody (C beta 1 to a principal neutralizing domain of the human immunodeficiency virus type 1 envelope protein. *AIDS Res. Hum. Retroviruses* 8:1107–1115.

McCafferty, J., Griffiths, A.D., Winter, G., and Chiswell, D.J. 1990. Phage antibodies: filamentous phage displaying antibody variable domains. *Nature* 348:552–554.

Moon, Y., Song, S., Choi, K., and Lee, J. 1997. Expression of a cDNA encoding *Phytolacca insularis* antiviral protein confers virus resistance on transgenic potato plants. *Mol. Cells* 7:807–815.

Mourgues F., Brisset, M.N., and Chevreau, E. 1998. Strategies to improve plant resistance to bacterial diseases through genetic engineering. *Trends Biotechnol.* 16:203–210.

Murdoch, L., Kobayashi, I., and Hardham, A. 1998. Production and characterization of monoclonal antibodies to cell wall components of the flax rust fungus. *Eur. J. Plant Pathol.* 104:331–346.

Nakajima, H., Muranaka, T., Ishige, F., Akatsu, K., and Oeda, K. 1997. Fungal and bacterial disease resistance in transgenic plants expressing human lysozyme. *Plant Cell Reps.* 16:674–679.

Niv, R., Segal, D., and Reiter, Y. 2003. Recombinant single-chain and disulfide-stabilized Fv immunotoxins for cancer therapy. *Methods Mol. Biol.* 207:255–268.

Noelke, G., Fischer, R., and Schillberg, S. 2004. Antibody-based pathogen resistance in plants. *Journal of Plant Pathology* 86:5–17.

Noelke, G., Schneider, B., Fischer, R., and Schillberg, S. 2005. Immunomodulation of polyamine biosynthesis in tobacco plants has a significant impact on polyamine levels and generates a dwarf phenotype. *Plant Biotechnology Journal* doi: 10.1111/j.1467-7652.2005.00121.x.

Ogawa, T., Hori, T., and Ishida, I. 1996. Virus-induced cell death in plants expressing the mammalian 2'; 5' oligoadenylate system. *Nat. Biotechnol.* 14:1566–1569.

Owen, M., Gandecha, A., Cockburn, B., and Whitelam, G. 1992. Synthesis of a functional anti-phytochrome single-chain Fv protein in transgenic tobacco. *Bio/Technol.* 10: 790–4.

Pain, N., O'Connell, R., Bailey, J., and Green, J. 1992. Monoclonal antibodies which show restricted binding to four *Colletotrichum* species: *C. lindemuthianum, C. malvarum, C. orbiculare* and *C. trifolii. Physiol. Mol. Plant Pathol.* 41:111–126.

Peschen, D., Li, H.P., Fischer, R., Kreuzaler, F., and Liao, Y.C. 2004. Fusion proteins comprising a Fusarium-specific antibody linked to antifungal peptides protect plants against a fungal pathogen. *Nature Biotechnology* 22:732–738.

Powell-Abel, P., Nelson, R.S., De, B., Hoffmann, N., Rogers, S.G., Fraley, R.T., and Beachy, R.N. 1986. Delay of disease development in transgenic plants that express the tobacco mosaic virus coat protein gene. *Science* 232:738–743.

Rangaswami, G. 1983. Principal diseases of food crops. In *Chemistry and World Food Supplies*, ed. L.Shemilt, pp. 109–119. Pergamon Press, NY.

Rau, D., Kramer, K., and Hock, B. 2002. Single-chain Fv antibody-alkaline phophatase fusion proteins produced by one-step cloning as rapid detection tools for ELISA. *J. Immunoassay Immunochem.* 23:129–143.

Robert, A., Mackie, A., Hathaway, V., Callow, J., and Green, J. 1993. Molecular differentiation in the extrahaustorial membrane of pea powdery mildew haustoria at early and late stages of development. *Physiol. Mol. Plant Pathol.* 43:147–160.

Rommens, C.M., and Kishore, G.M. 2000. Exploiting the full potential of disease-resistance genes for agricultural use. *Curr. Opin. Biotechnol.* 11:120–125.

Rosso, M.N., Schouten, A., Roosien, J., Borst-Vrenssen, T., Hussey, R.S., Gommers, F.J., Bakker, J., Schots, A., and Abad, P. 1996. Expression and functional characterization of a single chain Fv antibody directed against secretions involved in plant nematode infection process. *Biochem. Biophys. Res. Commun.* 18: 255–263.

Rosso, M., Favery, B., Piotte, C., Arthaud, L., De Boer, J., Hussey, R., Bakker, J., Baum, T., and Abad, P. 1999. Isolation of a cDNA encoding a beta-1,4-endoglucanase in the root-knot nematode *Meloidogyne incognita* and expression analysis during plant parasitism. *Mol. Plant-Microbe Interact.* 12:585–591.

Rubio, T., Borja, M., Scholthof, H., and Jackson, A. 1999. Recombination with host transgenes and effects on virus evolution: an overview and opinion. *Mol. Plant Microbe Interact.* 12:87–92.

Schier, R., and Marks, J.D. 1996. Efficient *in vitro* affinity maturation of phage antibodies using BIAcore guided selections. *Hum. Antibodies Hybridomas* 7:97–105.

Schillberg, S., Zimmermann, S., Zhang, M.Y., and Fischer, R. 2001. Antibody-based resistance to plant pathogens. *Transgenic Res.* 10:1–12.

Schillberg, S., Emans, N., and Fischer, R. 2002. Antibody molecular farming in plants and plant cells. *Phytochem. Rev.* 1:45–54.

Schillberg, S., Zimmermann, S., Findlay, K., and Fischer, R. 2000. Plasma membrane display of anti-viral single chain Fv fragments confers resistance to tobacco mosaic virus. *Mol. Breed.* 6:317–326.

Schillberg, S., Zimmermann, S., Voss, A., and Fischer, R. 1999. Apoplastic and cytosolic expression of full-size antibodies and antibody fragments in *Nicotiana tabacum. Transgenic Res.* 8:255–263.

Schouten, A., Roosien, J., de Boer, J.M., Wilmink, A., Rosso, M.N., Bosch, D., Stiekema, W.J., Gommers, F.J., Bakker, J., and Schots, A. 1997. Improving scFv antibody expression levels in the plant cytosol. *FEBS Lett.* 415:235–241.

Schouten, A., Roosien, J., van Engelen, F.A., de Jong, G.A.M., Borst-Vrenssen, A.W.M., Zilverentant, J.F., Bosch, D., Stiekema, W.J., Gommers, F.J., Schots, A., and Bakker, J. 1996. The C-terminal KDEL sequence increases the expression level of a single-chain antibody designed to be targeted to both cytosol and the secretory pathway in transgenic tobacco. *Plant Mol. Biol.* 30:781–793.

Sidhu, S.S. 2000. Phage display in pharmaceutical biotechnology. *Curr. Opin. Biotechnol.* 11:610–616.

Smant, G., Stokkermans, J., Yan, Y., Deboer, J., Baum, T., Wang, X., Hussey, R., Gommers, F., Henrissat, B., Davis, E., Helder, J., Schots, A., and Bakker, J. 1998. Endogenous cellulases in animals—isolation of beta-1,4-endoglucanase genes from two species of plant-parasitic cyst nematodes. *Proc. Natl. Acad. Sci. USA* 95:4906–4911.

Spiegel, H., Schillberg, S., Sack, M., Holzem, A., Nähring, J., Monecke, M., Liao, Y.-C., and Fischer, R. 1999. Expression of antibody fusion proteins in the cytoplasm and ER of plant cells. *Plant Sci.* 149:63–71.

Stanley, J., Frischmuth, T., and Ellwood, S. 1990. Defective viral DNA ameliorates symptoms of geminivirus infection in transgenic plants. *Proc. Natl. Acad. Sci. USA* 87:6291–6295.

Tavladoraki, P., Benvenuto, E., Trinca, S., De Martinis, D., Cattaneo, A., and Galeffi, P. 1993. Transgenic plants expressing a functional single-chain Fv antibody are specifically protected from virus attack. *Nature* 366:469–472.

Tricoli, D.M., Carney, K.J., Russell, P.F., McMaster, J.R., Groff, D.W., Hadden, K.C., Himmel, P.T., Hubbard, J.P., Boeshore, M.L., and Quernada, H.D. 1995. Field evaluation of transgenic squash containing single or multiple virus coat protein gene constructs for resistance to cucumber mosaic virus, watermelon mosaic virus 2, and zucchini yellow mosaic virus. *Bio.Technol* 13:1458–1465.

Tomlinson, I., and Holliger, P. 2000. Methods for generating multivalent and bispecific antibody fragments. *Methods Enzymol.* 326:461–479.

Tumer, N., Hwang, D., and Bonness, M. 1997. C-terminal deletion mutant of pokeweed antiviral protein inhibits viral infection but does not depurinate host ribosomes. *Proc. Natl. Acad. Sci. USA* 94:3866–3871.

Vain, P., Worland, B., Clarke, M.C., Richard, G., Beavis, M., Liu, H., Kohli, A., Leech, M., Snape, J., Christou, P., and Atkinson, H. 1998. Expression of an engineered cysteine proteinase inhibitor Oryzacystatin-I Delta D86 for nematode resistance in transgenic rice plants. *Theor. Appl. Genet.* 96:266–271.

Van Engelen, F.A., Schouten, A., Molthoff, J.W., Roosien, J., salinas, J., Dirkse, W.G. et al. 1994. Coordinate expression of antibody subunit genes yields high-levels of functional antibodies in roots of transgenic tobacco. *Plant Mol. Biol.* 26:1701–1710.

Vine, N.D., Drake, P., Hiatt, A., and Ma, J.K. 2001. Assembly and plasma membrane targeting of recombinant immunoglobulin chains in plants with a murine immunoglobulin transmembrane sequence. *Plant Mol. Biol.* 45:159–167.

Voss, A., Niersbach, M., Hain, R., Hirsch, H., Liao, Y., Kreuzaler, F., and Fischer, R. 1995. Reduced virus infectivity in *N. tabacum* secreting a TMV-specific full size antibody. *Mol. Breed.* 1:39–50.

Watanabe, Y., Ogawa, T., Takahashi, H., Ishida, I., Takeuchi, Y., Yamamoto, M., and Okada, Y. 1995. Resistance against multiple plant viruses in plants mediated by a double stranded-RNA specific ribonuclease. *FEBS Lett.* 372:165–168.

Wattad, C., Kobiler, D., Dinoor, A., and Prusky, D. 1997. Pectase lyase of *Colletotrichum gloeospoprioides* attacking avocado fruits: cDNA cloning and involvement in pathogenicity. *Physiol. Mol. Plant Pathol.* 50:197–212.

Whitham, S., McCormick, S., and Baker, B. 1996. The *N* gene of tobacco confers resistance to tobacco mosaic virus in transgenic tomato. *Proc. Natl. Acad. Sci. USA* 93:8776–8781.

Winter, G., Griffiths, A.D., Hawkins, R.E., and Hoogenboom, H.R. 1994. Making antibodies by phage display technology. *Ann. Rev. Immunol.* 12:433–455.

Winter, G., and Milstein, C. 1991. Man-made antibodies. *Nature* 349: 293–299.

Worn, A., Auf Der Maur, A., Escher, D., Honegger, A., Barberis, A., and Plückthun, A. 2000. Correlation between *in vitro* stability and *in vivo* performance of anti-GCN4 intrabodies as cytoplasmic inhibitors. *J. Biol. Chem.* 275:2795–2803.

Xiao, X.W., Chu, P.W.G., Frenkel, M.J., Tabe, L.M., Shukla, D.D., Hanna, P.J., Higgins, T.J.V., Müller, W.J., and Ward, C.W. 2000. Antibody-mediated improved resistance to ClYVV and PVY infections in transgenic tobacco plants expressing a single-chain variable region antibody. *Mol. Breed.* 6:421–431.

Yuan, Q., Hu, W., Pestka., J.J., He, S.Y., and Hart, L.P. 2000. Expression of a functional antizearaleone single-chain Fv antibody in transgenic *Arabidopsis* plants. *Appl. Environ. Microbiol.* 66:3499–3505.

Zimmermann, S., Schillberg, S., Liao, Y.C., and Fischer, R. 1998. Intracellular expression of TMV-specific single-chain Fv fragments leads to improved virus resistance in *Nicotiana tabacum. Mol. Breed.* 4:369–379.

Index

O